Progress in Atomic Spectroscopy

Part A

PHYSICS OF ATOMS AND MOLECULES

Series Editors:

P.G. Burke, *Queen's University of Belfast, Northern Ireland*
and
H. Kleinpoppen, *Institute of Atomic Physics, University of Stirling, Scotland*

1976:
ELECTRON AND PHOTON INTERACTIONS WITH ATOMS
Edited by H. Kleinpoppen and M.R.C. McDowell

1978:
PROGRESS IN ATOMIC SPECTROSCOPY, Parts A and B
Edited by W. Hanle and H. Kleinpoppen

In preparation:
ATOM–MOLECULE COLLISION THEORY: A Guide for the Experimentalist
Edited by Richard B. Bernstein

THEORY OF ELECTRON–ATOM COLLISIONS
By P.G. Burke and C.J. Joachain

Progress in Atomic Spectroscopy

Part A

Edited by

W. Hanle

Ist Institute of Physics
Justus Liebig University
Giessen, Germany

and

H. Kleinpoppen

Institute of Atomic Physics
University of Stirling
Stirling, Scotland

Plenum Press · New York and London

Library of Congress Cataloging in Publication Data

Main entry under title:

Progress in atomic spectroscopy.

(Physics of atoms and molecules)
Includes bibliographical references and index.
1. Atomic spectra. I. Hanle, Wilhelm, 1901-   II. Kleinpoppen, Hans.
QC454.A8P76   539.7   78-18230
ISBN 0-306-31115-1 (Part A)

©1978 Plenum Press, New York
A Division of Plenum Publishing Corporation
227 West 17th Street, New York, N.Y. 10011

Printed in the United States of America

Contents of Part A

I. BASIC PROPERTIES OF ATOMS AND PERTURBATIONS

Chapter 1
Atomic Structure Theory
A. Hibbert

Chapter 2
Density Matrix Formalism and Applications in Spectroscopy
K. Blum

Chapter 3

Perturbation of Atoms
Stig Stenholm

Chapter 4

Quantum Electrodynamical Effects in Atomic Spectra
A. M. Ermolaev

II. METHODS AND APPLICATIONS OF ATOMIC SPECTROSCOPY

Chapter 7

New Developments of Classical Optical Spectroscopy
Klaus Heilig and Andreas Steudel

Chapter 8

Excitation of Atoms by Impact Processes
H. Kleinpoppen and A. Scharmann

Chapter 11

The Microwave–Optical Resonance Method
William H. Wing and Keith B. MacAdam

Chapter 12

Lamb-Shift and Fine-Structure Measurements on One-Electron Systems
H.-J. Beyer

Chapter 13

Anticrossing Spectroscopy
H.-J. Beyer and H. Kleinpoppen

Chapter 14

Time-Resolved Fluorescence Spectroscopy
J. N. Dodd and G. W. Series

Chapter 15
Laser High-Resolution Spectroscopy
W. Demtröder

Contents of Part B

Chapter 19

Multiphoton Spectroscopy
Peter Bräunlich

Chapter 20

Fast-Beam (Beam-Foil) Spectroscopy
H. J. Andrä

Chapter 21

Stark Effect
K. J. Kollath and M. C. Standage

Chapter 22

Stored Ion Spectroscopy
Hans A. Schuessler

Chapter 23

The Spectroscopy of Atomic Compound States
J. F. Williams

Chapter 24

Optical Oscillator Strengths by Electron Impact Spectroscopy
W. R. Newell

Chapter 25

Atomic Transition Probabilities and Lifetimes
W. L. Wiese

Chapter 26
Lifetime Measurement by Temporal Transients
Richard G. Fowler

Chapter 27
Line Shapes
W. Behmenburg

Chapter 28

Collisional Depolarization in the Excited State
W. E. Baylis

Chapter 29

Energy and Polarization Transfer
M. Elbel

Chapter 30
X-Ray-Spectroscopy
K.-H. Schartner

Chapter 31
Exotic Atoms
G. Backenstoss

Introduction

W. HANLE and H. KLEINPOPPEN

In 1919, in the first edition of *Atombau and Spektrallinien*, Sommerfeld referred to the immense amount of information which had been accumulated during the first period of 60 years of spectroscopic practice. Sommerfeld emphasized that the names of Planck and Bohr would be connected forever with the efforts that had been made to understand the physics and the theory of spectral lines. Another period of almost 60 years has elapsed since the first edition of Sommerfeld's famous monograph. As the editors of this monograph, *Progress in Atomic Spectroscopy*, we feel that the present period is best characterized by the large variety of new spectroscopic methods that have been invented in the last decades.

Spectroscopy has always been involved in the field of research on atomic structure and the interaction of light and atoms. The development of new spectroscopic methods (i.e., new as compared to the traditional optical methods) has led to many outstanding achievements, which, together with the increase of activity over the last decades, appear as a kind of renaissance of atomic spectroscopy.

The substantial improvement of the spectroscopic resolving power by many orders of magnitude has prepared the testing ground for the foundations of quantum mechanics and quantum electrodynamics. Traditional optical spectroscopy provided sets of measurements of wavelengths that were ordered by their representation in spectral series; modern atomic spectroscopy and its applications have links with many other branches of fundamental physics. High-precision spectroscopy, as a test for quantum electrodynamics, links atomic spectroscopy with the physics of leptonic interactions. The remarkable and continued success of quantum electrodynamics in predicting precisely the properties of atoms is a most impressive testament to the applicability of physical theory and mathematics to experimental physics. The fascination which results from precision measurements in atomic spectroscopy, and which matches the highly accurate quantum-electrodynamical predictions of atomic structure, still remains an important attraction to physicists working in that field. While links between atomic spectroscopy and the physics of leptonic interactions have been well established over the last two decades, present precision experiments with dye lasers on parity-nonconserving effects in atoms are about to become most sensitive test experiments of theoretical models of weak neutral currents in particle physics.

A wide range of fundamental processes of physics is applied in modern atomic spectroscopy. Quantum-mechanical coherence and perturbation effects in atomic states are involved in various methods such as level crossing and anticrossing spectroscopy, fast-beam spectroscopy, and other time-resolving methods. Laser applications have had, and certainly will have, of course, a most profound impact not only on the range of spectroscopic applications but also on the precision that can be obtained. Higher-order interactions between light and atoms have been applied not only as tools for precision measurements of atomic structures but also to studies of higher-order effects in intense electromagnetic fields. There are, however, other less spectacular but nevertheless important areas that have a significant impact on the overall field of applications in atomic spectroscopy; for instance, the field of interatomic interactions has successfully been studied with methods of atomic spectroscopy. The amount of data collected by atomic spectroscopists has vastly increased; new methods and improved old ones have provided a huge amount of information on parameters characterizing atomic states, e.g., lifetimes (or oscillator strengths), level splittings, multipole moments, Stark shifts, anisotropic magnetic susceptibilities, coupling coefficients, etc. Atomic compound state spectroscopy, which originated from atomic collision studies, requires both representation in spectral series and a detailed analysis of the configuration interactions, which links atomic collision physics with atomic spectroscopy. Perturbations have not only been used as spectroscopic tools but also play a decisive role in the interpretation of collisional depolarization and energy transfer from one excited state to another. Most impressive progress has been made in neighboring areas such as x-ray spectroscopy and the spectroscopy of exotic atoms and positronium. The traditional links between astronomy and atomic spectroscopy are as strong as ever, leading to the advancement of knowledge in both of these branches of science. Because of extreme conditions that are possible in space but not obtainable in laboratory investigations, data from astronomical observations often find their interpretation through knowledge based upon atomic structure.

Of course, this book can only attempt to summarize the most important methods and applications of atomic spectroscopy that are prevalent at present. We will be happy if the scientific community appreciates the efforts made by the distinguished authors in this book; these authors have enthusiastically taken up the laborious but important task of describing their fields of interest within the severe limitation of available space. In this connection we are most grateful to the publishing company for their willingness to extend the book to two volumes. We would also like to express our indebtedness to our secretaries Mrs. A. Dunlop (Stirling), Mrs. A. Füchtemann (Bielefeld), Mrs. H. Schaefferling (Bielefeld), and Ms. H. Wallbott (Giessen) for their assistance in communicating with the authors and for many very careful considerations. We acknowledge the benefit of discussions and advice from colleagues and students including in particular Professor G. W. Series, Dr. H.-J. Beyer, Dr. K. Blum, Dr. K. J. Kollath, Dr. M. Standage, Mr. W. Johnstone, Mr. H. Jakubowicz, Mr. N. Malik, and Mr. A. Zaidi. One of us (H.K.) is very grateful for the hospitality of the University of Bielefeld, where parts of the book were prepared.

I
Basic Properties of
Atoms and Perturbations

1

Atomic Structure Theory *

A. Hibbert

1. Introduction

During the 1960's, many theories were proposed for treating properly the correlation of the electrons of an atom. Up till then, most of the methods used in atomic structure calculations—notably the Hartree–Fock method—had treated the interaction between the electrons in an approximate way. Although such approximate methods frequently led to energy values that were reasonably accurate, the same could not be said of the calculation of energy splittings or of other atomic properties. The new, more accurate theories attempted to remedy this situation. Though they were conceptually straightforward, the details of these theories were often complex and the amount of computation required was considerable. The feasibility of calculations based on them was thus closely linked to the availability of good computing facilities. Some of the methods were never extended beyond test cases. But others have been and indeed still are being exploited in the accurate calculation of atomic properties.

The purpose of this chapter is to describe those methods that have proved themselves capable of producing such accurate results. The interrelationship between them and others that were proposed is discussed elsewhere.[1] But before discussing these methods, it is necessary to consider the theory of atomic properties to which they have most frequently been applied, and also to consider the older methods used for one- and two-electron atoms, since these form the basis for work on many-electron systems.

* Completed July 1976.

A. Hibbert • Département d'Astrophysique Fondamentale, Observatoire de Paris, Meudon, 92190 Meudon, France. Permanent address: Department of Applied Mathematics and Theoretical Physics, Queen's University of Belfast, Belfast BT7 1NN, Northern Ireland.

We therefore begin by discussing, first for one-electron, then for many-electron atoms, some general theory of atomic properties. We follow this by a discussion of accurate methods of calculation.

Most of these have been nonrelativistic theories, and we shall briefly compare them by examining a few of the many results that have been obtained. But for large atoms or ions, nonrelativistic theories based on Schrödinger's equation are inadequate and must be replaced by relativistic theories based on Dirac's equation. The sophistication of such theories has, for the most part, been on a much lower level than the nonrelativistic counterparts. But recently more accurate methods have been proposed, and the computer programs based on them have become available. We therefore conclude the chapter with a short account of this work.

2. General Theory

2.1 One-Electron Atoms

2.1.1. Nonrelativistic Theory

The attempt to formulate theories of atomic structure goes back much further than the beginning of "modern" quantum mechanics in 1926. Bohr[2] postulated discrete internal energy levels of an atom and a discrete set of values that the radius of the electron can assume. From this he deduced specific energy expressions for the case of hydrogen using the Balmer formula for the wavelength λ_{nm} of frequency ν_{nm} of a radiated photon:

$$hc/\lambda_{nm} = h\nu_{nm} = hcR_H Z^2 (m^{-2} - n^{-2}) \tag{1}$$

where the energy difference between the levels is given by Planck's relationship

$$E_n - E_m = h\nu_{nm} \tag{2}$$

h is Planck's constant, c is the speed of light, Z is the atomic number of the hydrogenic ion; R_H is the modified Rydberg, given in terms of the universal Rydberg constant R_∞ by

$$hcR_H = hcR_\infty \frac{M_H}{M_H + m_0} = \frac{m_0 e^4}{2\hbar^2} \frac{M_H}{M_H + m_0} \tag{3}$$

where M_H is the mass of the nucleus, m_0 is the mass of the electron, $(-e)$ is the electronic charge, and $\hbar = h/2\pi$. Then the energy E_n is

$$E_n = -hcR_\infty \frac{Z^2}{n^2} \frac{M_H}{M_H + m_0} \tag{4}$$

Bohr's semiclassical picture had a limited success. But because of the uncertainty principle, it is not possible to specify completely the orbit of an electron. Indeed, the uncertainty principle required the abandonment of a deterministic model in favor of a probabilistic one, in which we define a probability per unit volume $|\Psi_n|^2$ of finding the electron in a quantum state n. Then

$$\int |\Psi_n|^2 \, d\tau = 1 \tag{5}$$

The equation satisfied by Ψ_n was proposed by Schrödinger[3]:

$$H\Psi \equiv (T + V)\Psi = i\hbar \frac{\partial \Psi}{\partial t} \tag{6}$$

so that the Hamiltonian H is the quantum equivalent of the classical sum of kinetic and potential energies. For *stationary* states of atoms, the probability distribution is independent of time. Writing

$$\Psi(\mathbf{r}, t) = \Psi(\mathbf{r}) \exp(-iEt/\hbar) \tag{7}$$

we obtain the time-independent Schrödinger equation:

$$H\Psi = E\Psi \tag{8}$$

For hydrogenlike systems

$$H = T + V = -\frac{\hbar^2}{2m}\nabla^2 - \frac{Ze^2}{r} \tag{9}$$

where $m = m_0 M_A/(m_0 + M_A)$ is the reduced mass of the electron in an atom with nuclear mass M_A and we have assumed that the potential energy $V(\mathbf{r})$ consists simply of the electrostatic interaction between the electron and the nucleus. Solutions of (8) with Hamiltonian (9) are given in most basic texts on quantum mechanics. We merely quote the result:

$$\Psi_{nlm_l}(r, \theta, \phi) = R_{nl}(r) Y_l^{m_l}(\theta, \phi) \tag{10}$$

where R_{nl} is an associated Laguerre function and $Y_l^{m_l}$ is a spherical harmonic, while the corresponding energy is given by Bohr's formula (4). This energy is clearly degenerate, being independent of the orbital angular momentum (corresponding to the quantum number l) and its z component (m_l). But high-resolution observations note a splitting of at least part of this degeneracy into closely spaced lines. This *fine structure* is not accounted for by the nonrelativistic treatment above, and although it can be considered by modifying the arguments, a satisfactory quantitative treatment requires a theory based on the relativistic Dirac equation.

2.1.2. Fine Structure

Dirac's equation for an electron in an external field with vector and scalar potentials $\mathbf{A}$ and ϕ takes the form[4]

$$i\hbar \frac{\partial \Psi}{\partial t} = \left\{ c\boldsymbol{\alpha} \cdot \left(\mathbf{p} + \frac{e}{c}\mathbf{A} \right) + \beta mc^2 - e\phi \right\} \Psi \equiv H_D \Psi \qquad (11)$$

where $\boldsymbol{\alpha}, \beta$ are 4×4 matrices which may be partitioned into 2×2 matrices according to

$$\boldsymbol{\alpha} = \begin{pmatrix} 0 & \boldsymbol{\sigma} \\ \boldsymbol{\sigma} & 0 \end{pmatrix}, \qquad \beta = \begin{pmatrix} 1 & 0 \\ 0 & -1 \end{pmatrix} \qquad (12)$$

with $\boldsymbol{\sigma}$ being the set of Pauli spin matrices, and $0, 1, -1$ represent 2×2 diagonal matrices. The relativistic energy also includes the rest mass energy mc^2. If we wish to arrive at a nonrelativistic approximation, this must be separated out by writing [cf. (7)]

$$\Psi(\mathbf{r}, t) = \Phi(\mathbf{r}, t) \exp(-imc^2 t/\hbar) \qquad (13)$$

If we also partition Φ into two double components [cf. (12)]:

$$\Phi = \begin{pmatrix} \Phi_1 \\ \Phi_2 \end{pmatrix} \qquad (14)$$

then we obtain the equations

$$\left(i\hbar \frac{\partial}{\partial t} + e\phi \right) \Phi_1 = c\boldsymbol{\sigma} \cdot \left(\mathbf{p} + \frac{e}{c}\mathbf{A} \right) \Phi_2 \qquad (15)$$

$$\left(i\hbar \frac{\partial}{\partial t} + e\phi + 2mc^2 \right) \Phi_2 = c\boldsymbol{\sigma} \cdot \left(\mathbf{p} + \frac{e}{c}\mathbf{A} \right) \Phi_1 \qquad (16)$$

The components of Φ_2 are smaller than those of Φ_1 by $0(v/c)$. Thus in the nonrelativistic limit, for which $v \to 0$, those of Φ_1 dominate. Eliminating Φ_2 between (15) and (16), and replacing the time-dependent operator $i\hbar \, \partial/\partial t$ by E [cf. (6) and (8)], we obtain

$$\left[E + e\phi - \frac{1}{2m} \left\{ \boldsymbol{\sigma} \cdot \left(\mathbf{p} + \frac{e}{c}\mathbf{A} \right) \right\} k(\mathbf{r}) \left\{ \boldsymbol{\sigma} \cdot \left(\mathbf{p} + \frac{e}{c}\mathbf{A} \right) \right\} \right] \Phi_1 = 0 \qquad (17)$$

Here

$$k(\mathbf{r}) = \left[1 + \frac{E + e\phi}{2mc^2} \right]^{-1} \qquad (18)$$

has a radial dependence of the form $(1 + r_0/r)$, where r_0 has the order of the dimensions of the nucleus. It may therefore be taken as unity in purely electronic structure calculations, corresponding to an expansion of (15) and

(16) to order $1/c$. After some manipulation, (17) becomes

$$E\Phi_1 = i\hbar \frac{\partial \Phi_1}{\partial t} = \left[\frac{1}{2m}\left(\mathbf{p} + \frac{e}{c}\mathbf{A}\right)^2 - e\phi + \frac{e\hbar}{2mc}\boldsymbol{\sigma}\cdot\mathbf{B} \right]\Phi_1 \tag{19}$$

where $\mathbf{B} = \text{curl } \mathbf{A}$. The last term in (19), which does not appear in the nonrelativistic Schrödinger equation for an electron in an external field, corresponds to the potential energy of a magnetic dipole with moment

$$\boldsymbol{\mu}_s = -\frac{e\hbar}{2mc}\boldsymbol{\sigma} = -\frac{e\hbar}{mc}\mathbf{s} \tag{20}$$

in the external field. Equation (20) defines the spin $\mathbf{s}$ of the electron, and may be compared with the corresponding moment

$$\boldsymbol{\mu}_l = -\frac{e\hbar}{2mc}\mathbf{l} \tag{21}$$

due to the orbital motion of the electron. The density

$$\rho = |\Phi|^2 \equiv |\Phi_1|^2 + |\Phi_2|^2 = |\Phi_1|^2 \tag{22}$$

to order $1/c$. This therefore coincides with the nonrelativistic density $|\Phi^{\text{Sch}}|^2$. But if we expand the equations to $O(1/c^2)$, then (22) is replaced by

$$\rho = |\Phi_1|^2 + |\Phi_2|^2 = |\Phi_1|^2 + \frac{\hbar^2}{4m^2c^2}|\boldsymbol{\sigma}\cdot\nabla\Phi_1|^2 \tag{23}$$

Landau and Lifschitz[4] show that the relationship between the relativistic and nonrelativistic functions to order $1/c^2$ is

$$\Phi_1 = \left(1 - \frac{\mathbf{p}^2}{8m^2c^2}\right)\Phi^{\text{Sch}} \tag{24}$$

If we consider only an external electric field (so that $\mathbf{A} = \mathbf{0}$), then the Hamiltonian becomes

$$H = \frac{\mathbf{p}^2}{2m} - e\phi - \frac{\mathbf{p}^4}{8m^3c^2} + \frac{e\hbar}{4m^2c^2}\boldsymbol{\sigma}\cdot(\mathscr{E}\times\mathbf{p}) + \frac{e\hbar^2}{8m^2c^2}\text{div }\mathscr{E} \tag{25}$$

where $\mathscr{E} = -\nabla\phi$ is the electric field. For a central field, $\phi = \phi(r)$ and

$$\mathscr{E} = -\frac{\mathbf{r}}{r}\frac{d\phi}{dr} \tag{26}$$

so that the fourth term on the right in (25) becomes

$$-\frac{e\hbar^2}{2m^2c^2 r}\frac{d\phi}{dr}\mathbf{l}\cdot\mathbf{s} \tag{27}$$

hence its name—spin–orbit interaction. The last three terms in (25) represent to $O(1/c^2)$ the relativistic corrections to Schrödinger's equation (8). They can be estimated using first-order perturbation theory, since the zero-order solutions are given by (10). For hydrogenlike atoms, $\phi = Ze/r$, $\mathscr{E} = Ze\mathbf{r}/r^3$, div $\mathscr{E} = 4\pi Ze\delta^3(\mathbf{r})$ where $\delta^3(\mathbf{r})$ represents the three-dimensional Dirac delta function, and straightforward analysis yields a first-order correction

$$\Delta E = -m\frac{(\alpha Z)^4}{2n^3}\left\{\frac{1}{(j+\frac{1}{2})}-\frac{3}{4n}\right\} = -\frac{(\alpha Z)^2}{n^2}E_n\left\{\frac{3}{4}-\frac{n}{(j+\frac{1}{2})}\right\} \qquad (28)$$

where $\alpha = e^2/\hbar c$ is the (dimensionless) fine-structure constant. The degeneracy of the nonrelativistic energy levels is now partially lifted. For the nonrelativistic Hamiltonian (9), the orbital angular momentum $\mathbf{l}^2$ and the z component l_z are constants of the motion (i.e., they commute with the Hamiltonian). For (25), the total angular momentum $\mathbf{j}^2$ and j_z, where

$$\mathbf{j} = \mathbf{l} + \mathbf{s} \qquad (29)$$

together with $\mathbf{l}^2$ and $\mathbf{s}^2$ (but not l_z or s_2) are constants of the motion. The energy levels are characterized by n, l, and j. Following the spectroscopic notation, which makes the alphabetic assignment

$$
\begin{array}{ccccccccc}
l: & 0 & 1 & 2 & 3 & 4 & 5 & 6 & 7 \cdots \\
 & s & p & d & f & g & h & i & k \cdots
\end{array}
\qquad (30)
$$

the hydrogenic states are $1s_{1/2}$; $2s_{1/2}$, $2p_{1/2}$, $2p_{3/2}$; $3s_{1/2}$, $3p_{1/2}$, $3p_{3/2}$, $3d_{3/2}$, $3d_{5/2}$;..., where the subscript denoting the value of j is dropped in the purely nonrelativistic theory. Alternatively, we can denote the $j = l - \frac{1}{2}$ state by $n\bar{l}$, so that, for example, the pair $3d_{3/2}$, $3d_{5/2}$ may be written as $3\bar{d}$, $3d$. Not all the degeneracy is removed by the fine-structure splitting. For example, from (4) and (28), the $2s_{1/2}$ and $2p_{1/2}$ states are degenerate. The remaining degeneracy is removed by radiative corrections (Lamb shift) which are considered by other chapters in the book.

The relativistic counterparts of (10) are

$$\Psi_{nkm} = \begin{pmatrix} \Phi_1 \\ \Phi_2 \end{pmatrix} = \frac{1}{r}\begin{pmatrix} P_{nk}(r)\chi_{k,m}(\theta, \phi) \\ iQ_{nk}(r)\chi_{-k,m}(\theta, \phi) \end{pmatrix} \qquad (31)$$

where

$$\chi_{km} = \sum_{\sigma} Y_l^{m-\sigma}(\theta \cdot \phi)\Theta^\sigma C(l\tfrac{1}{2}j: m-\sigma, \sigma, m) \qquad (32)$$

with

$$k = -(j+\tfrac{1}{2})a, \qquad a = \pm 1 \quad \text{when} \quad l = j \mp \tfrac{1}{2} \qquad (33)$$

and where Θ is an eigenfunction of the spin operators s_z and s^2, and C denotes a Clebsch–Gordan coefficient. The equations satisfied by the radial functions P and Q are

$$\frac{dP_{nk}}{dr} + \left(\frac{k}{r}\right) P_{nk} - (\hbar c)^{-1}(E_{nk} + 2mc^2 + e\phi)Q_{nk} = 0$$

$$\frac{dQ_{nk}}{dr} - \left(\frac{k}{r}\right) Q_{nk} + (\hbar c)^{-1}(E_{nk} + e\phi)P_{nk} = 0 \tag{34}$$

where E_{nk} is the nonrelativistic energy.

2.1.3. Hyperfine Structure

Fine-structure levels are split further as a result of interactions between atomic electrons and nuclear magnetic moments,[5] which may be described in terms of the vector potential $\mathbf{A}$ and Eq. (17). For a nucleus of charge Ze and magnetic moment $\boldsymbol{\mu}$,

$$\phi(\mathbf{r}) = Ze/r, \qquad \mathbf{A} = (\boldsymbol{\mu} \times \mathbf{r})/r^3, \qquad \boldsymbol{\mu} = g_I \mu_n \mathbf{I} \tag{35}$$

where $\mu_n = e\hbar/2Mc$ is the nuclear magneton, $g_I = \mu_I/I$ is the nuclear g factor relative to μ_n, M is the proton mass, μ_I is the nuclear magnetic moment, and $\mathbf{I}$ is the nuclear spin. Then (17) may be written as

$$H\Phi_1 = (H^{(0)} + H^{(1)} + H^{(2)})\Phi_1 \tag{36}$$

where $H^{(0)}$ is the fine-structure Hamiltonian (25),

$$H^{(1)} = -(e/2mc)[(\boldsymbol{\sigma} \cdot \mathbf{p})k(\mathbf{r})(\boldsymbol{\sigma} \cdot \mathbf{A}) + (\boldsymbol{\sigma} \cdot \mathbf{A})k(\mathbf{r})(\boldsymbol{\sigma} \cdot \mathbf{p})] \tag{37}$$

is the hyperfine-structure correction, and

$$H^{(2)} = (e^2/2mc^2)(\boldsymbol{\sigma} \cdot \mathbf{A})k(\mathbf{r})(\boldsymbol{\sigma} \cdot \mathbf{A}) \tag{38}$$

corresponds to the self-energy of a point magnetic dipole, and is independent of orbital and spin angular momenta [since $(\boldsymbol{\sigma} \cdot \mathbf{A})(\boldsymbol{\sigma} \cdot \mathbf{A}) = \mathbf{A}^2$]—hence we shall not discuss this term any further. Using (35), the Hamiltonian (37) reduces to

$$H^{(1)} = gg_I \mu_B \mu_n \left[k(r)\{r^{-3}\mathbf{I} \cdot (\mathbf{l} - \mathbf{s}) + 3r^{-5}(\mathbf{I} \cdot \mathbf{r})(\mathbf{s} \cdot \mathbf{r})\} \right.$$

$$\left. + \frac{dk}{dr}\{r^{-2}(\mathbf{I} \cdot \mathbf{s}) - r^{-4}(\mathbf{I} \cdot \mathbf{r})(\mathbf{s} \cdot \mathbf{r})\} \right] \tag{39}$$

where g is the electronic g factor and $\mu_B = e\hbar/2mc$ is the Bohr magneton. Blinder[6] shows that $k(r) \approx 1$, $dk/dr \approx 4\pi r^2 \delta^3(\mathbf{r}) = 4\pi\delta(r)$ and that the angular average of the second term in curly brackets in (39) is $2r^{-2}(\mathbf{I} \cdot \mathbf{s})/3$. Thus the effective hyperfine-structure operator is

$$H^{(1)} = gg_I \mu_B \mu_n \mathbf{I} \cdot [r^{-3}\mathbf{l} + \{3r^{-5}\mathbf{r}(\mathbf{s} \cdot \mathbf{r}) - r^{-3}\mathbf{s}\} + (8\pi/3)\delta^3(\mathbf{r})\mathbf{s}]$$

$$\equiv H_l^{(1)} + H_d^{(1)} + H_c^{(1)} \tag{40}$$

showing the division of $H^{(1)}$ into the orbital, spin-dipolar, and Fermi contact terms, respectively. For the orbital term, the g factor of the electron is 2. For the spin-dependent $H_d^{(1)}$ and $H_c^{(1)}$, the g factor is modified according to quantum electrodynamics[7-9] giving the value[10]

$$g_s = 2.002\ 32 \tag{41}$$

The expression (36) may also be written in the form

$$H^{(1)} = \mathbf{I} \cdot \mathbf{T} \tag{42}$$

where $\mathbf{T}$ is an irreducible tensor of rank 1 with respect to $\mathbf{j}$.[11] If we treat $H^{(1)}$ as a perturbation to $H^{(0)}$ given by (25), then in zero order, $\mathbf{I}^2$, I_z, $\mathbf{j}^2$, j_z are separately quantized, so that first-order theory results in an energy correction

$$\langle Im_I jm_j | H^{(1)} | Im_I jm_j \rangle = \langle Im_I | \mathbf{I} | Im_I \rangle \cdot \langle jm_j | \mathbf{T} | jm_j \rangle$$
$$= \lambda \langle Im_I | \mathbf{I} | I_{mI} \rangle \cdot \langle jm_j | \mathbf{j} | jm_j \rangle$$
$$= \lambda \langle Im_I jm_j | \mathbf{I} \cdot \mathbf{j} | Im_I jm_j \rangle \tag{43}$$

on application of the Wigner–Eckart theorem,[11] where λ is independent of m values. When $H^{(1)}$ is added to $H^{(0)}$, the total Hamiltonian is no longer diagonal in the $Im_I jm_J$ representation. Instead, we must use a representation characterized by $Ijfm_f$, where the total angular momentum $\mathbf{f}^2$ is determined by

$$\mathbf{f} = \mathbf{I} + \mathbf{j} \tag{44}$$

[cf. (29)]. The two representations are connected by a unitary transformation involving Clebsch–Gordan coefficients, so that $H^{(1)}$ remains equivalent to $\lambda \mathbf{I} \cdot \mathbf{j}$. In the coupled representation, the hyperfine energy correction is then

$$\lambda \langle Ijfm_f | \mathbf{I} \cdot \mathbf{j} | Ijfm_f \rangle = \tfrac{1}{2} \lambda \langle Ijfm_f | \mathbf{f}^2 - \mathbf{I}^2 - \mathbf{j}^2 | Ijfm_f \rangle$$
$$= \tfrac{1}{2} \lambda [f(f+1) - I(I+1) - j(j+1)]$$
$$= \tfrac{1}{2} \lambda K \tag{45}$$

thus defining K. From (45), the splitting between consecutive hyperfine structure levels arising from a particular fine-structure level is given by

$$E(f) - E(f-1) = \lambda f \tag{46}$$

Since, from (43),

$$\langle jm_j | T_z | jm_j \rangle = \lambda \langle jm_j | j_z | jm_j \rangle = \lambda m_j \tag{47}$$

the magnitude of the hyperfine splitting may be obtained from the use of electronic wave functions. We shall postpone further discussion on the evaluation of λ until we consider many-electron atoms in subsection 2.2.

The Hamiltonian (37) describes magnetic interactions between the electrons and the nucleus. The corresponding electrostatic interaction may be written as

$$H^Q = -\sum_{\text{protons}} \frac{e^2}{|\mathbf{r}_p - \mathbf{r}_e|} = -e^2 \sum_{\text{protons}} \sum_k \frac{r_p^k}{r_e^{k+1}} C_e^{[k]} \cdot C_p^{[k]} \tag{48}$$

since $r_e \gg r_p$, where $\mathbf{r}_e, \mathbf{r}_p$ denote the electron and proton position vectors, and $C^{[k]}$ is a tensor of rank $k^{(11)}$ with components proportional to the spherical harmonics. The $k = 0$ term is independent of r_p, and so the sum over the protons reduces to $-Ze^2/r_e$; i.e., the electrostatic Coulomb interaction. The $k = 1$ term vanishes from parity considerations so that the leading hyperfine correction is the quadrupole ($k = 2$) term:

$$H_2^Q = -e^2 \left(\sum_{\text{protons}} r_p^2 C_p^{[2]} \right) \cdot (r_e^{-3} C_e^{[2]}) \equiv -e^2 q_p^{[2]} \cdot q_e^{[2]} \tag{49}$$

Then the corresponding energy correction is, in general,

$$\Delta E^Q = \langle Iifm_f | H_2^Q | Iifm_f \rangle$$
$$= -e^2 (-1)^{I+j'-f} W(IjIj; f2)[(2I+1)(2j+1)]^{1/2} (I \| q_p^{[2]} \| I)(j \| q_e^{[2]} \| j) \tag{50}$$

where W denotes a Racah coefficient and $(\| \ \|)$ represents a reduced matrix element.[11]

2.2. Many-Electron Atoms

2.2.1. Multiplet and Fine Structure

For atoms with more than one electron, the electrostatic parts of the Hamiltonian contain not only electron–nucleus interactions, but also electron–electron interactions. Thus for an atom with N electrons, (9) is replaced by

$$H_N = \sum_{i=1}^N \left(-\frac{\hbar^2}{2m} \nabla_i^2 - \frac{Ze^2}{r_i} \right) + \sum_{i<j} \sum \frac{e^2}{r_{ij}} \tag{51}$$

where the subscript i indicates the coordinates of electron i, $r_{ij} = |\mathbf{r}_i - \mathbf{r}_j|$ and the double summation is over all pairs of electrons. In this nonrelativistic approximation, the angular momenta of the individual electrons are not constants of the motion. Instead, the angular momentum operators which commute with H_N are $\mathbf{L}^2, \mathbf{S}^2$, where

$$\mathbf{L} = \sum_{i=1}^N \mathbf{l}_i, \qquad \mathbf{S} = \sum_{i=1}^N \mathbf{s}_i \tag{52}$$

together with a single component of **L** and **S** (usually taken as L_z and S_z). The associated wave functions Ψ satisfying Schrödinger's equation thus depend on four angular momentum quantum numbers LSM_LM_S associated with these operators. Since H_N is a tensor of rank zero, we may write the energy in the form

$$E = \langle \Psi(LSM_LM_S)|H_N|\Psi(LSM_LM_S)\rangle$$
$$= C(L0L; M_L0M_L)C(S0S; M_S0M_S)\langle\Psi(LS)\|H_N\|\Psi(LS)\rangle$$
$$= \langle\Psi(LS)\|H_N\|\Psi(LS)\rangle \tag{53}$$

using the Wigner–Eckart theorem[11] and the fact that both Clebsch–Gordan coefficients denoted by C in (53) are equal to unity. Hence the energy value depends only on L and S, and we obtain a *multiplet structure* with energy levels characterized by these quantum numbers, usually in the notation $^{(2S+1)}L$, with L replaced by its (capital) alphabetic equivalent given by (30).

As with one-electron atoms, this multiplet structure splits into fine-structure levels characterized by the quantum number J, where [cf. (29)]

$$\mathbf{J} = \mathbf{L} + \mathbf{S} \tag{54}$$

The corresponding correction to the Hamiltonian is again obtained from a relativistic treatment. Following Breit,[12] Grant[13] demonstrates that the relativistic Hamiltonian can be expanded in powers of the fine-structure constant (or alternatively in powers of $1/c$), with leading terms

$$H = H_1 + H_2 + \cdots$$

where

$$H_1 = \sum_{i=1}^{N} H_D(i)$$
$$H_2 = \tfrac{1}{2}\sum_{i\neq j} H(i,j) \tag{55}$$
$$H(i,j) = 1/r_{ij} + B(i,j)$$

and the Breit operator $B(i,j)$ is given by

$$B(i,j) = -\frac{1}{2r_{ij}}\left[\boldsymbol{\alpha}(i)\cdot\boldsymbol{\alpha}(j) + \frac{(\boldsymbol{\alpha}(i)\cdot\mathbf{r}_{ij})(\boldsymbol{\alpha}(j)\cdot\mathbf{r}_{ij})}{r_{ij}^2}\right] \tag{56}$$

with $\boldsymbol{\alpha}$ given by (12). The operator H_1 includes the conventional spin–orbit interaction (spin–own-orbit) having the form $\sum_{i=1}^{N}(Z/r_i^3)\mathbf{l}_i\cdot\mathbf{s}_i$. The Breit operator includes interactions between the orbital angular momenta of different electrons, between different spin angular momenta, and between the orbital angular momentum of one electron and the spin of another.

These last two interactions—spin–spin and spin–other-orbit—modify the fine-structure splitting. The spin–orbit operator is tensorially equivalent to $\zeta(\mathbf{L} \cdot \mathbf{S})$ for an appropriate parameter ζ, so that [cf. (45)]

$$\Delta E = \tfrac{1}{2}\zeta[J(J+1) - L(L+1) - S(S+1)] \tag{57}$$

and

$$E(J) - E(J-1) = \zeta J \tag{58}$$

is the Landé interval rule. The effect of the spin–other-orbit operator is to modify the parameter ζ while retaining the interval rule. Since the Breit operator contains a spin–own-orbit contribution, this is more properly grouped with the spin–own-orbit part of H_D.[14] The spin–spin operator, having a different J dependence, modifies the interval rule. Departures from the interval rule also occur as a result of interactions between different LS terms with the same value of J. Specific forms of the several parts of the Breit operator are given by Bethe and Salpeter.[15]

At this point, it is worth commenting on the various possible angular momentum couplings schemes. For atoms with small N and Z, the fine structure is a small perturbation and in many cases can in practice be neglected. For example, it is sufficient to consider transitions between multiplets rather than between levels with different J. In such cases, the LS (Russell–Saunders) coupling applies, in which the orbital and spin angular momenta of the electrons are coupled separately (though not entirely independently because the exclusion principle prevents certain combinations). For large atoms, for which a full relativistic treatment is necessary, the orbital and spin angular momenta of each electron must first be coupled, and $\mathbf{J}$ is the sum of the individual $\mathbf{j}_i$. This is the jj scheme. For medium-sized atoms, there are several possibilities lying between these two extremes. In many such cases, the intermediate coupling scheme is a good approximation, in which $\mathbf{L}$ and $\mathbf{S}$ are determined as in the LS scheme, but it is also necessary to consider J levels, with $\mathbf{J}$ defined by (54). In some cases (e.g., the $2p^5 4s$ states in Si V[16]), a better representation is given by the $j_c l$ scheme, in which the core $(2p^5)$ is coupled to give $\mathbf{j}_c$. This is then added to the $\mathbf{l}$ of the outer electron to give $\mathbf{K} = \mathbf{j}_c + \mathbf{l}$, and the final state is obtained by adding $\mathbf{K}$ to the $\mathbf{s}$ of the outer electron: $\mathbf{J} = \mathbf{K} + \mathbf{s}$. Thus for $nl^q n'l'$ states, the angular momentum coupling scheme would be represented by

$$nl^q(^{2S+1}L_{j_c})n'l'[K]_J, \qquad \mathbf{j}_c = \mathbf{L} + \mathbf{S}$$

2.2.2. Hyperfine Structure

The analysis of hyperfine structure of one-electron atoms goes over directly to the many-electron case simply by replacements of the form

$r^{-3}\mathbf{l} \rightarrow \Sigma_{i=1}^{N} r_i^{-3}\mathbf{l}_i$. The energy splitting associated with the leading magnetic plus electric interactions is[1]

$$E(F)-E(J)=\tfrac{1}{2}hA_J K + hB_J[K(K+1)-4I(I+1)J(J+1)/3] \quad (59)$$

in which [cf. (44)]

$$\mathbf{F}=\mathbf{I}+\mathbf{J} \quad (60)$$

and we have written

$$\lambda = hA_J \quad (61)$$

so that the magnetic hyperfine-structure constant A_J (and similarly the electric quadrupole constant B_J) are expressed in frequency units (usually MHz, in practice).

When the fine-structure splitting is a small correction to the multiplet structure, it is possible to treat both fine and hyperfine structure as independent perturbations (even if hyperfine structure is small compared with fine structure—essentially the percentage correction to hyperfine structure arising from the neglect of fine structure is approximately the same as to the multiplet structure; i.e., the perturbations are, to a good approximation, additive). In this case, the J dependence of A_J and B_J may be written down explicitly[1]:

$$A_J = G_{en}\left(\frac{\mu_I}{I}\right)\left[\frac{\langle \mathbf{L} \cdot \mathbf{J}\rangle}{LJ(J+1)}a_l + \frac{\{3\langle \mathbf{S} \cdot \mathbf{L}\rangle\langle \mathbf{L} \cdot \mathbf{J}\rangle - L(L+1)\langle \mathbf{S} \cdot \mathbf{J}\rangle\}}{SL(2L-1)J(J+1)}\frac{1}{2}g_s a_d\right.$$

$$\left. + \frac{\langle \mathbf{S} \cdot \mathbf{J}\rangle}{SJ(J+1)}\frac{1}{6}g_s a_c\right] \quad (62)$$

where

$$G_{en} = 2\mu_B\mu_n/ha_0^3 \quad (63)$$

$$a_l = \left\langle LSLS \left| \sum_{k=1}^{N} r_k^{-3}l_{z_k} \right| LSLS \right\rangle \quad (64)$$

$$a_d = \left\langle LSLS \left| \sum_{k=1}^{N} r_k^{-5}(3z_k^2 - r_k^2)s_{z_k} \right| LSLS \right\rangle \quad (65)$$

$$a_c = \left\langle LSLS \left| \sum_{k=1}^{N} 2r_k^{-2}\delta(r_k)s_{z_k} \right| LSLS \right\rangle \quad (66)$$

and [cf. (45) and (57)]

$$\langle \mathbf{L} \cdot \mathbf{J}\rangle = \tfrac{1}{2}[J(J+1)+L(L+1)-S(S+1)] \quad (67)$$

$$\langle \mathbf{S} \cdot \mathbf{J}\rangle = \tfrac{1}{2}[J(J+1)-L(L+1)+S(S+1)] \quad (68)$$

$$\langle \mathbf{S} \cdot \mathbf{L}\rangle = \tfrac{1}{2}[J(J+1)-L(L+1)-S(S+1)] \quad (69)$$

The parameters a_l, a_d, a_c (later referred to as the orbital, spin-dipolar and contact terms, respectively) are thus evaluated using LS-coupled wave functions with $M_L = L$, $M_S = S$, and are calculated in *atomic units* (a.u.) for which $e = \hbar = m = 1$. The conversion factor from the atomic units to the MHz units of A_J is given by $G_{en} = 95.4129$.[10]
    Similarly,

$$B_J = \frac{-3e^2 q_J Q}{(ha_0^3)8I(2I-1)J(2J-1)} \tag{70}$$

where Q is the nuclear quadrupole moment and

$$q_J = \frac{6\langle \mathbf{L} \cdot \mathbf{J} \rangle^2 - 3\langle \mathbf{L} \cdot \mathbf{J} \rangle - 2L(L+1)J(J+1)}{L(2L-1)(2J+3)(J+1)} b_q \tag{71}$$

$$b_q = \left\langle LSLS \left| \sum_{k=1}^{N} r_k^{-s}(3z_k^2 - r_k^2) \right| LSLS \right\rangle \tag{72}$$

again evaluating the quadrupole term b_q in atomic units with $M_L = L$, $M_S = S$.
    Since, in (42), $\mathbf{T}$ is a tensor of rank 1, it is possible to construct off-diagonal matrix elements $\langle J \| T \| J \pm 1 \rangle$, and hence off-diagonal hyperfine constants $A_{J,J\pm 1}$. Similarly, since the quadrupole operator is a tensor of rank 2, one may evaluate $B_{J,J\pm 1}$ and $B_{J,J\pm 2}$. Explicit formulas are given elsewhere.[1]

2.3. Oscillator Strengths

The probability that an atom will undergo a radiative transition from one state to another is a time-dependent quantity, and to evaluate it we must begin with the time-dependent Schrödinger equation, in which the radiation field is determined by the vector potential $\mathbf{A}$ in (17). In a nonrelativistic analysis, only the first two terms of (19) remain, and indeed the quadratic term in $\mathbf{A}$ may also be neglected in a first-order calculation. The wave function Ψ is then expanded in terms of the complete set of stationary-state wave functions $\{\psi_n\}$ with energies $\{E_n\}$

$$\Psi = \sum_n c_n(t) \exp(-iE_n t/\hbar)\psi_n \tag{73}$$

where the summation includes an integration over the continuum. If the atom is in a state $|i\rangle$ at time $t = 0$, then $c_i(0) = 1$, $c_n(0) = 0$ for $n \neq i$. If the velocity potential is expanded in a Fourier series of plane waves of the form $\mathbf{A}_0 \cos(\omega t - \mathbf{k} \cdot \mathbf{r})$, then the first-order perturbation theory gives,[17] for a one-electron atom,

$$t^{-1}|c_j(t)|^2 = (2\pi c^2/3h^2\nu_{ji}^2)|\langle j|(e/mc)\mathbf{p}\exp(i\mathbf{k}\cdot\mathbf{r})|i\rangle|^2\rho(\nu_{ji}) \tag{74}$$

$$= B_{ij}\rho(\nu_{ji}) \tag{75}$$

where B_{ij} is the Einstein coefficient for absorption, m is the reduced mass of the electron, and $\rho(\nu_{ji})$ is the energy density per unit frequency range. The *absorption* oscillator strength f_{ij} is related to B_{ij} by

$$f_{ij} = (m/\pi e^2)h\nu_{ji}B_{ij} \tag{76}$$

thus retaining the usual convention that the absorption oscillator strengths are positive.

The emission of radiation resulting in an atom deexciting can be spontaneous (for which the transition rate depends only on the population of the upper level) or induced (for which the rate depends also on $\rho(\nu_{ji})$). The respective Einstein coefficients A_{ji}, B_{ji} satisfy

$$A_{ji} = 8\pi\nu_{ji}^2 c^{-3}(E_j - E_i)B_{ji} \tag{77}$$

and

$$g_i B_{ij} = g_j B_{ji} \tag{78}$$

where, for example, g_i is the statistical weight (number of degenerate states) of the level $|i\rangle$. Then the absorption oscillator strength is related to the emission transition probability (per unit time) by

$$f_{ij} = (mc/8\pi^2 e^2)\lambda_{ji}^2(g_i/g_i)A_{ji} \tag{79}$$

where $\lambda_{ji} = c/\nu_{ji}$. By (76) and (78), we have

$$g_i f_{ij} = -g_j f_{ji} \tag{80}$$

since $\nu_{ij} = -\nu_{ji}$.

Dipole Approximation

The calculation of oscillator strengths (or transition probabilities) thus involves the evaluation of the matrix elements occurring in (74). The magnitude of the wave vector $\mathbf{k}$ is $k = 2\pi/\lambda_{ji}$. Hence for wavelengths large compared with the size of the atom, $|\mathbf{k}\cdot\mathbf{r}| \ll 1$ and the expansion

$$\exp(i\mathbf{k}\cdot\mathbf{r}) = 1 + i\mathbf{k}\cdot\mathbf{r} - \cdots \tag{81}$$

will converge rapidly. The use of the constant leading term of (81) in (74) constitutes the dipole approximation. Also

$$(i\hbar/m)\mathbf{p} = i\hbar\dot{\mathbf{r}} \equiv [\mathbf{r}, H_0] \tag{82}$$

whenever matrix elements of these operators are being taken with respect to exact eigenfunctions $\{\psi_n\}$. Hence in the dipole approximation, the matrix element may be reexpressed according to

$$\langle j|(e/mc)\mathbf{p}|i\rangle = (i/\hbar c)(E_j - E_i)\langle j|e\mathbf{r}|i\rangle \tag{83}$$

so that two equivalent forms for the absorption oscillator strength are obtained:

$$f_{ij}^l = (2m/3\hbar^2)(E_j - E_i)|\langle j|\mathbf{r}|i\rangle|^2 \tag{84}$$

$$f_{ij}^v = (2/3m)(E_j - E_i)^{-1}|\langle j|\mathbf{p}|i\rangle|^2 \tag{85}$$

These are the *length* and *velocity* forms, respectively. The identity (82) may be applied to $\dot{\mathbf{r}}$ instead of $\mathbf{r}$, and a similar analysis yields an *acceleration* form:

$$\begin{aligned} f_{ij}^a &= (2\hbar^2/3m)(E_j - E_i)^{-3}|\langle j|\boldsymbol{\nabla} V|i\rangle^2 \\ &= (2Z^2e^4\hbar^2/3m)(E_j - E_i)^{-3}|\langle j|r^{-3}\mathbf{r}|i\rangle|^2 \end{aligned} \tag{86}$$

where $V = -Ze^2/r$. Other forms may be derived,[18] but they have not been widely used.

If the level j is degenerate, with different states being distinguished by the $(2J_j + 1)$ parameters m_j, then several "channels" will be open for the transition. The total absorption oscillator strength is then

$$f_{ij}^l = (2m/3\hbar^2)(E_j - E_i)\sum_{m_j}|\langle jm_j|\mathbf{r}|i\rangle|^2 \tag{87}$$

It is convenient to define the line strength

$$\begin{aligned} S_{ij}^l = S_{ji}^l &= \sum_{m_i,m_j}|\langle jm_j|e\mathbf{r}|im_i\rangle|^2 \\ &= \sum_{m_i,m_j,\mu}|\langle jm_j|er(4\pi/3)^{1/2}Y_1^\mu|im_i\rangle|^2 \end{aligned} \tag{88}$$

allowing for the possible degeneracy of the level i. The m dependence of the matrix element in (88) is given by the Clebsch–Gordan coefficient $C(J_i 1 J_j; m_i\mu m_j)$, for which

$$\sum_{m_i,\mu}|C(J_i 1 J_j; m_i\mu m_j)|^2 = 1 \tag{89}$$

which is independent of m_j. Thus the sum over m_j in (88) gives $g_j = (2J_j + 1)$ equal contributions, and we may write

$$\begin{aligned} f_{ij}^l &= (2m/3\hbar^2 e^2)(E_j - E_i)S_{ij}^l/g_i \\ &= (2m/3\hbar^2 e^2)(E_j - E_i)|\langle J_i\|erC^{[1]}\|J_i\rangle|^2 g_j/g_i \end{aligned} \tag{90}$$

Because of the symmetry of S_{ij} expressed by (88), it is clear that (80) is satisfied.

For many-electron atoms, the analysis is very similar, if one makes the replacements

$$\mathbf{r} \to \sum_{k=1}^{N} \mathbf{r}_k, \qquad \mathbf{p} \to \sum_{k=1}^{N} \mathbf{p}_k, \qquad r^{-3}\mathbf{r} \to \sum_{k=1}^{N} r_k^{-3}\mathbf{r}_k \qquad (91)$$

although the third of these replacements in (86) is valid only since

$$\sum_{k=1}^{N} \nabla_k \left(\sum_{i<j} r_{ij}^{-1} \right) = \sum_{i<j} (\nabla_i + \nabla_j) r_{ij}^{-1} = \mathbf{0} \qquad (92)$$

Equation (90) expresses the oscillator strength in terms of reduced matrix elements for jj or intermediate coupling schemes. For the LS scheme, (88) becomes

$$S_{ij}^l = \sum_{\substack{M_{L_j}, M_{S_j} \\ M_{L_i}, M_{S_i}}} \left| \left\langle L_j S_j M_{L_j} M_{S_j} \left| \sum_{k=1}^{N} e\mathbf{r}_k \right| L_i S_i M_{L_i} M_{S_i} \right\rangle \right|^2$$

$$= (2S_j + 1)(2L_j + 1) \left| \left\langle L_j S_j \left\| \sum_{k=1}^{N} er_k C_k^{[1]} \right\| L_i S_i \right\rangle \right|^2 \delta_{S_i S_j} \qquad (93)$$

so that

$$f_{ij}^l = (2m/3\hbar^2 e^2)(E_j - E_i) \left| \left\langle L_j S_j \left\| \sum_{k=1}^{N} er_k C_k^{[1]} \right\| L_i S_i \right\rangle \right|^2 \frac{2L_j + 1}{2L_i + 1} \delta_{S_i S_j} \quad (94)$$

We should point out that the forms of the Wigner–Eckart theorem and hence the definition of reduced matrix elements vary from reference to reference. Here we have used that given by Rose.[11] However, an alternative expression of the theorem (see, for example, Fano and Racah[19]) includes an additional factor of $(2J+1)^{1/2}$ in the reduced matrix element. The effect on (90) and (94) is to require additional factors $(2J_j+1)^{-1}$ and $(2L_j+1)^{-1}$ when this alternative form of the theorem is used.

In the calculation of oscillator strengths, there is one rather important difference between the one-electron and many-electron cases. For a one-electron atom, the equivalence of the length and velocity forms follows since one knows the exact wave functions $\{\psi_n\}$ in (73). But for many-electron atoms, exact solutions of Schrödinger's equation are, from a practical point of view, unobtainable. Consequently, formulas (84) and (85) may give quite different results. Suppose $|i\rangle$ and $|j\rangle$ are exact eigenfunctions of the same *approximate* Hamiltonian

$$\mathcal{H} = \sum_{k=1}^{N} \mathbf{p}_k^2/2m + V \qquad (95)$$

Then

$$i\hbar \sum_{k=1}^{N} \dot{\mathbf{r}}_k = \left[\sum_{k=1}^{N} \mathbf{r}_k, \mathcal{H} \right] = (i\hbar/m) \sum_{k=1}^{N} \mathbf{p}_k + \sum_{k=1}^{N} [\mathbf{r}_k, V] \qquad (96)$$

so that

$$(E_j - E_i)\left\langle j \middle| \sum_{k=1}^{N} \mathbf{r}_k \middle| i \right\rangle = (i\hbar/m)\left\langle j \middle| \sum_{k=1}^{N} \mathbf{p}_k \middle| i \right\rangle + \left\langle j \middle| \sum_{k=1}^{N} [\mathbf{r}_k, V] \middle| i \right\rangle \qquad (97)$$

Thus (97) is equivalent to (83) only when $[\mathbf{r}_k, V] = 0$. This is true for the exact and for local potentials, but not for nonlocal potentials. Thus the Hartree approximation (see the next section) will give identical f^l and f^v, whereas in general the Hartree–Fock approximation will not. The equality of f^l and f^v is thus no guarantee of their accuracy.

Higher-Order Radiation

The reduced matrix elements in (90) and (94) will be zero unless $\Delta J = |J_j - J_i| = 0, 1, 0 \nleftrightarrow 0$; or $\Delta S = |S_j - S_i| = 0$ and $\Delta L = |L_j - L_i| = 0, 1, 0 \nleftrightarrow 0$, respectively, where $0 \nleftrightarrow 0$ implies that J_i (or $L_i) = 0$ and J_j (or $L_j) = 0$ also give zero matrix elements. When these *selection rules* are satisfied, the corresponding transition is optically *allowed*. Otherwise it is *forbidden*, although a nonzero value for the transition probability may be obtained by considering other terms in the expansion (81), resulting in higher-order electric- or magnetic-multipole effects. The corresponding formulas for 2^λ-pole radiation are [20,21]

$$A_{ji} = \frac{\lambda(2\lambda+1)}{\lambda(2\lambda+1)[(2\lambda-1)!!]^2} \frac{1}{g_j} \left(\frac{E_j - E_i}{\hbar c}\right)^{2\lambda+1} \frac{1}{\hbar} \sum_{m_i, m_j, \mu} |\langle jm_j | Q_{\lambda\mu} | im_i \rangle|^2 \qquad (98)$$

where

$$(2\lambda - 1)!! = 1 \times 3 \times 5 \times \cdots \times (2\lambda - 1) \qquad (99)$$

The electric and magnetic 2^λ-pole moments occurring in (98) are

$$Q_{\lambda\mu}^{(el)} = e \sum_{k=1}^{N} r_k^\lambda \left(\frac{4\pi}{2\lambda+1}\right)^{1/2} Y_\lambda^\mu \qquad (100)$$

$$Q_{\lambda\mu}^{(mag)} = \left(\frac{e}{mc}\right) \sum_{k=1}^{N} \mathbf{\nabla} \left[r_k^\lambda \left(\frac{4\pi}{2\lambda+1}\right)^{1/2} Y_\lambda^\mu \right] \cdot [(\lambda+1)^{-1} l_k + \tfrac{1}{2} g_s s_k] \qquad (101)$$

In addition to the angular momentum selection rules, which result from the moments $\{Q_{\lambda\mu}\}$ forming an irreducible tensorial set of rank λ, and which are listed up to $\lambda = 2$ in Table 1, there is a selection rule associated with

Table 1. Multipole Radiation Operators and Angular Momentum Selection Rules

Type of radiation	Selection rules
Electric dipole	$\Delta S = 0$; $\Delta L = 0, 1$; $0 \not\leftrightarrow 0$; $\Delta J = 0, 1$; $0 \not\leftrightarrow 0$
Electric quadrupole	$\Delta S = 0$; $\Delta L = 0, 1, 2$,; $0 \not\leftrightarrow 0, 1$;
	$\Delta J = 0, 1, 2$; $0 \not\leftrightarrow 0, 1$; $\frac{1}{2} \not\leftrightarrow \frac{1}{2}$
Magnetic dipole	$\Delta S = \Delta L = 0$; $\Delta J = 0, 1$; $0 \not\leftrightarrow 0$
Magnetic quadrupole	$\Delta J = 0, 1, 2$; $0 \not\leftrightarrow 0, 1$; $\frac{1}{2} \not\leftrightarrow \frac{1}{2}$;
	$\Delta S = 1$, $\Delta L = 0, 1$; $0 \not\leftrightarrow 0$ or $\Delta S = 0$;
	$\Delta L = 0, 1, 2$; $0 \not\leftrightarrow 0, 1$
Spin–orbit electric dipole	$\Delta S = 1$; $\Delta L = 0, 1, 2$; $\Delta J = 0, 1$; $0 \not\leftrightarrow 0$
Spin–orbit magnetic dipole	$\Delta S = 1$; $\Delta L = 0, 1, 2$; $\Delta J = 0, 1$; $0 \not\leftrightarrow 0$

parity. The electric-dipole and magnetic-quadrupole operators have odd parity. Hence $|i\rangle$ and $|j\rangle$ must have opposite parity. The electric-quadrupole and magnetic-dipole operators have even parity, so $|i\rangle$ and $|j\rangle$ must have the same parity.

The specification of "allowed" and "forbidden" may vary from coupling scheme to coupling scheme. For example, a transition of the form $^1S \rightarrow {}^3P$ is forbidden in the dipole approximation in LS coupling. But in intermediate coupling, the $J = 0 \rightarrow J = 1$ transition becomes allowed because of the different LS possibilities of each of the two states ($^1S_0$ with $^3P_0$), ($^3P_1$ with $^1P_1$). This is an example of an *intercombination* transition.

We have assumed in the above discussion that the expansion (81) converges so rapidly that one need consider in any transition probability matrix element only the leading term which allows the selection rules to be satisfied. For many transitions of interest, the wavelength is two or three orders of magnitude greater than the atomic dimensions, and this truncation of (81) is a very good approximation. Recently, some experiments and calculations of two-electron, one-photon transitions have been reported[22,23] for which $\lambda \lesssim 4a_0$ (a_0 is the first Bohr radius). For such transitions, one should not expect (81) to converge particularly rapidly, and more than just the first nonzero contribution may need to be calculated. In fact, because of the low angular momenta of the states involved, selection rules prevent higher-order contributions in most of the cases reported. But if transitions between states of higher angular momentum are being treated, higher-order terms in (81) should at least be considered.

3. Two-Electron Atoms

In this and the next two sections, we shall consider the methods that have been employed in the calculation of wave functions satisfying

Schrödinger's time-independent equation. Thus we are working with the *LS* angular momentum coupling scheme and a nonrelativistic formalism. It is useful at this point briefly to consider calculations that have been performed for helium and its isoelectronic sequence. Although it may seem a simple system, it presents one of the most fundamental problems in atomic structure—the inseparability of the Hamiltonian because of the electron–electron interaction e^2/r_{ij} and the consequent question of how best to treat the "correlation" of the motion of the electrons.

Two different schemes have been followed. The first is based on the variational principle: If the energy E and wave function Ψ correspond to the *lowest* energy of *any* given symmetry class $(L, S,$ parity), then the expression

$$E^t = \langle \Psi^t | H | \Psi^t \rangle / \langle \Psi^t | \Psi^t \rangle \tag{102}$$

forms an upper bound to E when Ψ^t belongs to the same symmetry class as Ψ, and where H is the two- (in general, N-) electron Hamiltonian given by (51). If the function Ψ^t is expressed in terms of variable parameters, then this variational principle enables us to choose the "best" values of these parameters as those that make E^t as low as possible. Depending on the flexibility of the form of Ψ^t, the resulting wave function should be a good representation of the true wave function, at least in those regions of space that are emphasized by the Hamiltonian operator. Nevertheless, a *pointwise* solution of Schrödinger's equation is not obtained.

The second method assumes that it is possible to decompose the Hamiltonian according to

$$H = H_0 + \lambda H_1 \tag{103}$$

so that H_0 is a "good" approximation to H, in the sense that the eigenfunctions and eigenvalues of H_0 form a good approximation to those of H, and that at least those in which one is interested can be calculated readily. Thus we assume we can solve the equation

$$H_0 \Psi_0 = E_0 \Psi_0 \tag{104}$$

and we shall assume, without loss of generality, that Ψ_0 is normalized to unity. The corresponding eigenfunction Ψ of H can formally be expanded in a power series in the perturbation parameter:

$$\Psi = \Psi_0 + \lambda \Psi_1 + \lambda^2 \Psi_2 + \cdots \tag{105}$$

$$E = E_0 + \lambda E_1 + \lambda^2 E_2 + \cdots \tag{106}$$

Substitution of these expressions into Schrödinger's equation leads to (104) and the set of equations

$$(H_0 - E_0)\Psi_n + H_1 \Psi_{n-1} = \sum_{i=1}^{n} E_i \Psi_{n-i} \qquad n \geq 1 \tag{107}$$

for higher-order terms. Then after some elementary manipulation

$$E_1 = \langle \Psi_0 | H_1 | \Psi_0 \rangle \tag{108}$$

$$E_2 = \langle \Psi_0 | H_1 - E_1 | \Psi_1 \rangle \tag{109}$$

$$E_3 = \langle \Psi_1 | H_1 - E_1 | \Psi_1 \rangle \tag{110}$$

Equations (108)–(110) show specific examples of the general result[24] that knowledge of the wave function to nth order determines the energy to order $(2n+1)$. We have assumed here that E_0 is nondegenerate. If E_0 is degenerate, Ψ_0 must be a linear combination of the independent eigenfunctions.[25]

In order to apply this perturbation analysis to atomic systems, it is convenient first to express H in atomic units and then to scale both length and energy so that Z^{-1} a.u. and Z^2 a.u. become the units of length and energy, respectively. Then (51) becomes

$$H = Z^2 \mathscr{H} = Z^2[H_0 + Z^{-1}H_1] \tag{111}$$

where H_0 is a sum of hydrogenlike Hamiltonians:

$$H_0 = \sum_{i=1}^{N} h_0^{(i)}, \qquad h_0^{(i)} = -\tfrac{1}{2}\nabla_i^2 - r_i^{-1} \tag{112}$$

and

$$H_1 = \sum_{i<j} r_{ij}^{-1} \tag{113}$$

so that Z^{-1} becomes the natural perturbation parameter λ. We may therefore separate the coordinates of H_0 to give

$$\Psi_0 = \prod_{i=1}^{N} u_0^{(i)}, \qquad E_0 = \sum_{i=1}^{N} \varepsilon_0^{(i)} \tag{114}$$

where

$$h_0^{(i)} u_0^{(i)} = \varepsilon_0^{(i)} u_0^{(i)} \tag{115}$$

from which E_1 may be calculated easily.

The solution of (107) is in practice no easier to obtain than that of Schrödinger's equation. One could expand Ψ_1 in terms of the complete set of eigenfunctions of H_0, but this involves an infinite summation and integration over the continuum. Instead, it is more convenient to use a variational principle. If E and E' in (102) are expanded in powers of λ, then (102) becomes

$$E_0 + \lambda E_1 + \lambda^2 E_2 + \lambda^3 E_3 + \cdots \leqslant E_0' + \lambda E_1' + \lambda^2 E_2' + \lambda^3 E_3' + \cdots \tag{116}$$

Since E_0^t and E_1^t are known exactly, and since (116) is valid for all and therefore small λ, we have

$$E_2 \leqslant E_2^t = \langle \Psi_1^t | H_0 - E_0 | \Psi_1^t \rangle + 2\langle \Psi_1^t | H_1 - E_1 | \Psi_0^t \rangle \tag{117}$$

If $\Psi_0^t, \Psi_1^t, \ldots, \Psi_{k-1}^t$ are known exactly (or sufficiently so), then

$$E_{2k} \leqslant E_{2k}^t \tag{118}$$

which provides a variational principle from which Ψ_k^t can be determined. This Hylleraas–Scherr–Knight[26,27] scheme has been carried to high order by Midtdal and co-workers.[28,29] For excited states, one must be careful to include in the basis for Ψ_k^t all lower-lying states, in order to maintain a valid variational principle.[30]

Pioneering calculations on the helium sequence using each of these methods were performed by Hylleraas.[26,31] For the variational scheme, his trial function took the form

$$\Psi^t = e^{-s/2} \sum_{n,l,m \geqslant 0} C_{nlm} s^n t^l u^m \tag{119}$$

where $s = k(r_1 + r_2)$, $t = k(r_2 - r_1)$, $u = r_{12}$, and we have omitted the spin part of Ψ^t since the spin integrals in (102) are equal to unity. The variable parameters are $\{C_{nlm}\}$ and k. With only six terms in (119), Hylleraas obtained the upper bound for the ground state energy $E^t = -2.903\,24$ a.u., in good agreement with the experimental value of $-2.903\,72$ a.u. The corresponding lower bound[32] is $-2.925\,56$ a.u., which is relatively poor. This suggests that, while (119) is capable of producing good upper bounds, it does not approximate the true wave function as well. In fact, the only formal solution of Schrödinger's equation in the form (119) is the trivial one: $\Psi^t \equiv 0$.[33]

To overcome this problem, Kinoshita[34,35] introduced the variables

$$s, \quad p = u/s, \quad q = t/u; \quad 0 \leqslant p \leqslant 1, \quad -1 \leqslant q \leqslant 1 \tag{120}$$

and showed that an expansion of the form

$$\Psi^t = e^{-s/2} \sum C_{nlm}' s^n p^l q^m \tag{121}$$

could formally satisfy Schrödinger's equation. By taking 80 terms in this expansion, he achieved agreement between upper and lower bounds of the energy to one part in 10^5. These variational calculations were taken even further by Pekeris,[36] using perimetric coordinates:

$$u = \varepsilon(r_2 + r_{12} - r_1), \quad v = \varepsilon(r_1 + r_{12} - r_2), \quad w = 2\varepsilon(r_1 + r_2 - r_{12}) \tag{122}$$

where $\varepsilon = (-E)^{1/2}$ and E is the experimental energy. These coordinates are independent and range from zero to infinity. Moreover, if we write

$$\Psi^t = \exp\left[-\tfrac{1}{2}(u+v+w)\right]F(u, v, w) \qquad (123)$$

we obtain the correct asymptotic form as $(r_1 + r_2) \to \infty$. By choosing 1078 terms in the expansion

$$F(u, v, w) = \sum A_{lmn}L_l(u)L_m(v)L_n(w) \qquad (124)$$

where L denotes a Laguerre polynomial, Pekeris achieved an upper bound to the energy with an accuracy better than one part in 10^9, when mass polarization, relativistic, and Lamb-shift corrections are included.

Results of high accuracy have also been obtained using the perturbation scheme, by Midtdal et al.[29] In effect they use the coordinates (120). In Table 2 we display a comparison between the two methods, which can be seen to be in excellent agreement to many decimal places.

The variational principle as written in (102) applies directly to excited states of atoms which are the lowest of their symmetry class (e.g., $1s2s\ ^3S$ of helium). If the excited state is not the lowest of its class (e.g., $1s2s\ ^1S$ of helium), then (102) holds only when Ψ^t is constrained to be orthogonal to the *exact* wave functions of all lower-lying states in the class. It is not possible to construct exact wave functions for many-electron systems, even if for helium the work discussed above comes very close to achieving this. However, for certain choices of Ψ^t it is possible to proceed rigorously. Suppose that we expand

$$\Psi^t = \sum_{i=1}^{m} a_i\Phi_i \qquad (125)$$

where $\{\Phi_i\}$ are *prescribed* functions and, for convenience, form an orthonormal set. Let $E_1^m < E_2^m < \cdots < E_m^m$ be the eigenvalues of the $m \times m$

Table 2. Ground-State Energies (in a.u.) of the Helium Isoelectronic Sequence

Z	Variational[a]	Perturbation[b]
1	−0.527751016	−0.527750041
2	−2.903724377	−2.903724362
3	−7.279913413	−7.279913387
4	−13.655566238	−13.655566206
5	−22.030971580	−22.030971543
6	−32.406246602	−32.406246562
7	−44.781445149	−44.781445106
8	−59.156595123	−59.156595077
9	−75.531712364	−75.531712317
10	−93.906806515	−93.906806467

[a] Frankowski and Pekeris.[29]
[b] Midtdal et al.[95]

Hamiltonian matrix with element $\langle \Phi_i | H | \Phi_j \rangle$. If the basis set $\{\Phi_i\}$ is now augmented by Φ_{m+1}, and if the eigenvalues of the resulting $(m+1) \times (m+1)$ Hamiltonian matrix are $E_1^{m+1} < \cdots < E_{m+1}^{m+1}$ then we have the Hylleraas–Undheim or MacDonald theorem[37,38]

$$E_{k-1}^m < E_k^{m+1} < E_k^m \tag{126}$$

If the set $\{\Phi_i\}$ is extended to form a complete set so that $\Psi' \to \Psi$, then[39]

$$E_k^m \geqslant E_k^{\text{exact}} \tag{127}$$

That is, the kth lowest eigenvalue of the $m \times m$ Hamiltonian matrix constructed using (125) is an upper bound to the exact energy of the kth lowest state of the given symmetry class. In addition, the coefficients $\{a_i\}$ in (125) are the components of the corresponding eigenvector. Thus we have a variational principle for excited states. In particular, the trial wave function given by (123) and (124) is of the form of (125). Accad et al.[40] have used this form (modified only to account for angular momentum symmetry requirements) in a very accurate study of excited states of helium. But the theorem is applicable to any trial wave function that may be expressed in the form (125). It is therefore very powerful.

The degree of accuracy achieved for two-electron atoms is not maintained for larger systems. Larsson[41] obtained the ground-state energy of lithium as $-7.478\,025$ a.u. when using 60 terms in an expansion similar to (119). Comparing this with the experimental value (or rather the value after relativistic and other small effects have been subtracted from the experimental value) of[42] $-7.478\,069$ a.u., we see that the difference is several orders of magnitude greater for lithium than for helium. Calculations using perturbation theory[43] have proved to be no more accurate, except perhaps for large Z, where the expansion (106) with $\lambda = Z^{-1}$ converges fairly rapidly. Perturbation calculations have been performed also for first row atoms.[44]

But it is clear that the introduction of interelectronic coordinates $\{r_{ij}\}$ directly into the wave function via expansions similar to (119), (121), or (123) and (124) soon becomes intractable as the number of electrons increases. The same problem arises for perturbation expansions, since variational forms [such as (117)] of the perturbation equations are in practice used to determine the $\{\Psi_n\}$. We must therefore investigate alternative methods of obtaining approximate but useful wave functions.

4. Hartree–Fock Theory

The approximations introduced in the last section were mathematical, associated with the truncation of a perturbation expansion or of a complete set expansion of the wave function. Although in principle the accuracy is capable of indefinite refinement, the practical computational difficulties encountered force us to consider a different approach, in which the loss in the level of accuracy is offset by the gain in simplicity both of calculation and of concept. Of course, one could use simply the hydrogenlike zero-order perturbation equations (104), (111)–(115). The computation is simple, but the loss of accuracy perhaps too great.

A very satisfactory compromise is the method originally proposed by Hartree[45,46] and subsequently modified by Fock.[47] The method has been frequently and extensively reviewed, notably by Hartree himself,[48] whose work and the extensions based upon it have recently been thoroughly discussed by Froese-Fischer.[49] Here we shall merely recall the essentials of the theory.

The scheme originally proposed by Hartree allowed each electron to move independently of the others. Hence the probability density function should be a product of the probability density functions of the individual electrons. This requires each electron to be described by a function of its own coordinates $U_k(k)$ and the Hartree wave function

$$\Psi^H = U_1(1)U_2(2)\cdots U_N(N) \tag{128}$$

Each *orbital* function $U_k(k)$ may be determined by setting up a Schrödinger-type (Hartree) equation for each electron:

$$[-\tfrac{1}{2}\nabla^2 - Z/r + V_k(\mathbf{r})]U_k = \varepsilon_k U_k \tag{129}$$

where V_k is the electrostatic potential experienced by the kth electron:

$$V_k(\mathbf{r}) = \sum_{i \neq k} \int \frac{|U_i(\mathbf{s})|^2}{|\mathbf{r}-\mathbf{s}|}\, d\mathbf{s} \tag{130}$$

A *simplification* may be made by taking a spherical average of $V_k(\mathbf{r})$ to produce a purely radial potential:

$$V_k(r) = (4\pi)^{-1} \int V_k(\mathbf{r})\, d\hat{\mathbf{r}} \tag{131}$$

With this radial potential in (129) in place of $V_k(\mathbf{r})$, the entire potential is radial and we obtain a *central field* model. Comparing (128) and (129) with (114) and (112) we see that the purely hydrogenic Ψ_0 is a special case of the Hartree wave function. Moreover, the angular dependence of U_k and $u_0^{(k)}$ is exactly the same—a spherical harmonic.

Fock's[47] modification of this method was to require that the wave function satisfy the exclusion principle. Then

$$\Psi^{HF} = (N!)^{-1/2} \det |U_1, U_2, \ldots, U_N| \tag{132}$$

is the Hartree–Fock (HF) wave function in which the (i, j) element of the determinant is $U_i(j)$. If $U_i = U_j$, the determinant is zero, so that the exclusion principle is satisfied. From other elementary properties of determinants, Ψ^{HF} is antisymmetric with respect to interchange of a pair of coordinates, and the functions $\{U_i\}$ can be chosen to form an orthonormal set. The factor $(N!)^{-1/2}$ in (132) ensures that Ψ^{HF} is normalized to unity.

The HF equations may be derived using the variational principle[50] by setting to zero the first-order variation with respect to $\{U_i\}$ of $\langle \Psi^{HF} | H | \Psi^{HF} \rangle$ subject to the orthonormality conditions

$$\langle U_i | U_j \rangle = \delta_{ij} \tag{133}$$

If $\{\varepsilon_{ij}\}$ are the Lagrange multipliers thus introduced, these equations take the form

$$F_i U_i = \sum_{j=1}^{N} \varepsilon_{ij} U_j \tag{134}$$

where F_i is an integrodifferential operator. Since (132) and (134) are invariant with respect to unitary transformations among the $\{U_i\}$, we may choose that transformation that diagonalizes the *symmetric* matrix ε, giving diagonal matrix elements $\{\varepsilon_i\}$. It is also customary to factorize $\{U_i\}$ into space and spin parts (hence the name *spin–orbital*):

$$U_i(\mathbf{r}, m_{s_i}) = \phi_i(\mathbf{r}) \chi(m_{s_i}) \tag{135}$$

where

$$\chi(\tfrac{1}{2}) \equiv \alpha = \begin{pmatrix} 1 \\ 0 \end{pmatrix}, \qquad \chi(-\tfrac{1}{2}) \equiv \beta = \begin{pmatrix} 0 \\ 1 \end{pmatrix} \tag{136}$$

are eigenfunctions of $\mathbf{s}^2$ and s_z. Then the HF equations become

$$F_i \phi_i = \varepsilon_i \phi_i \tag{137}$$

The Fock operator F_i is given by

$$F_i = h_i + V_i = h_i + J_i - K_i \tag{138}$$

where

$$h_i = -\tfrac{1}{2} \nabla_i^2 - z/r_i \tag{139}$$

and the *direct* operator is

$$J_i = \sum_{j=1}^{N} \int \frac{|\phi_j(\mathbf{r}_j)|^2}{r_{ij}} \, d\tau_j \tag{140}$$

while the nonlocal *exchange* operator K_i is defined by

$$K_i \phi_i(\mathbf{r}_i) = \sum_{j=1}^{N} \phi_j(\mathbf{r}_i) \int \frac{\phi_j^*(\mathbf{r}_j) \phi_i(\mathbf{r}_j)}{r_{ij}} \, d\tau_j \tag{141}$$

The HF equations thus differ from the Hartree equations by these exchange terms. In general, there will be an infinite number of solutions $\{\phi_i\}$. Those orbitals that for a given state occur in (132) are said to be *occupied*, the remainder being *unoccupied*. We may assign an approximate physical meaning to $\{\varepsilon_i\}$ according to Koopman's theorem.[51] If we assume that the orbitals of an atom A and its ion A^+ are identical, then $(-\varepsilon_i)$ is the removal energy (ionization potential) of electron i.

4.1. Restricted HF Method

Most practical attempts to solve (137) incorporate further restrictions. In the case of the Hartree equations, we saw that a spherical averaging (131) of the potential led to a central field model, with a spherical harmonic angular dependence for the orbitals. A similar dependence is normally assumed for the HF orbitals:

$$\phi_i(\mathbf{r}) = r^{-1} P_{n_i l_i}(r) Y_{l_i}^{m_{l_i}}(\theta, \phi) \tag{142}$$

The spectroscopic notation (30) is used, although the integers $\{n_i\}$ no longer represent good quantum numbers—rather they label different functions associated with the same l_i, with the convention $n_i > l_i$. The radial functions P_{nl} are independent both of m_l and m_s. The HF energy functional $\langle \Psi^{HF} | H | \Psi^{HF} \rangle$ may then be written in terms of *one-electron* radial integrals

$$I(n_i l_i) = \left\langle P_{n_i l_i} \left| -\frac{1}{2} \frac{d^2}{dr^2} - \frac{Z}{r} + \frac{l_i(l_i+1)}{2r^2} \right| P_{n_i l_i} \right\rangle \tag{143}$$

and two-electron *Slater integrals*

$$R^k(n_i l_i, n_j l_j; n_p l_p, n_q l_q) = \int \int \frac{r_<^k}{r_>^{k+1}} P_{n_i l_i}(r_i) P_{n_j l_j}(r_2) P_{n_p l_p}(r_1) P_{n_q l_q}(r_2) \, dr_1 \, dr_2 \tag{144}$$

In (144), $(r_<, r_>)$ are the (smaller, larger) of r_1 and r_2, and the possible values of k are defined by the inequalities

$$|l_i - l_p| \leq k \leq (l_i + l_p) \quad \text{and} \quad |l_j - l_q| \leq k \leq (l_j + l_q) \tag{145}$$

The following notation is frequently used:

$$F^k(nl, n'l') = R^k(nl, n'l'; nl, n'l') \tag{146}$$

$$G^k(nl, n'l') = R^k(nl, n'l'; n'l', nl)$$

$$= R^k(nl, nl; n'l', n'l') \tag{147}$$

The HF equations reduce to a set of coupled integrodifferential equations for the radial functions P_{nl}, with orthonormality conditions [cf. (133)]

$$\int_0^\infty P_{nl}(r)P_{n'l}(r)\, dr = \delta_{nn'} \tag{148}$$

Radial functions associated with different values of l need not be orthogonal, since orbital orthogonality (133) is maintained by the orthogonality of the spherical harmonics.

If both spin–orbitals $\phi_i(\mathbf{r})\alpha$ and $\phi_i(\mathbf{r})\beta$ occur in (132), the spatial orbital is said to be *doubly* occupied. Since P_{nl} is independent of m_l and m_s, it can occur in up to $2(2l+1)$ occupied orbitals. If it occurs in *exactly* $2(2l+1)$ occupied orbitals, the *subshell* (nl) is *closed*. If $n_{\max}$ is the largest value of n for which P_{nl} occurs in an occupied orbital, and if all subshells (nl) for $l < n$ and $n \leq n_{\max}$ are closed, then the atom or ion forms a *closed-shell system*. The HF equations for the radial functions take rather different forms depending on whether or not one is treating a closed-shell system—a shell consisting of all orbitals with a common value of n. For a closed-shell system, these equations have the form of (134) with F_i being a radial integrodifferential operator:

$$-\frac{d^2}{dr^2}(P_{n_il_i}) + \left[-\frac{2}{r}(Z - Y_i) + \frac{l_i(l_i+1)}{r^2} \right] P_{n_il_i} - \frac{2}{r}X_{n_il_i} = \sum_j \delta_{l_il_j}P_{n_jl_j}\frac{\varepsilon_{ij}}{q_j} \tag{149}$$

where q_j is the number of electrons in subshell (n_jl_j), and the direct and exchange operators Y_i and X_i are defined by

$$r^{-1}Y_i = \sum_{k,j} a_k(q_i, q_j) \int r_<^k r_>^{-k-1}[P_{n_jl_j}(r_2)]^2\, dr_2 \tag{150}$$

$$r^{-1}X_i = \sum_{k,j} b_k(q_i, q_j)P_{n_jl_j}(r) \int r_<^k r_>^{-k-1}P_{n_il_i}(r_2)P_{n_jl_j}(r_2)\, dr_2 \tag{151}$$

where the coefficients $\{a_k\}, \{b_k\}$ involve integrations over angles and spin. Then for a closed-shell system, $q_j = 2(2l_j+1)$ for all j occurring in (149) so that the ε matrix, being symmetric, can be chosen to be diagonal as in (134). But for non-closed-shell systems, the matrix $\boldsymbol{\lambda}$ of right-hand sides, with $\lambda_{ij} = \varepsilon_{ij}/q_j$, is not symmetric and may not be diagonalized.

From (149) we may deduce

$$\varepsilon_{ij} = q_iq_j(q_i - q_j)^{-1}\int_0^\infty 2r^{-1}[(Y_i - Y_j)P_{n_il_i}P_{n_jl_j} - X_iP_{n_jl_j} + X_jP_{n_il_i}]\, dr \tag{152}$$

If $q_i = q_j$ and the integral is zero, ε_{ij} is not determined and may be set to zero. However, if $q_i = q_j$ and the integral is not zero, the HF equations are inconsistent. This inconsistency is associated with the orthogonality of $P_{n_il_i}$

and $P_{n_i l_i}$. Such an orthogonality may be imposed as an additional constraint on the wave function, which may be determined directly from the variational principle rather than from the HF equations, but this may have the effect of raising the energy values substantially above the correct HF result.[51–53] If $q_i \neq q_j$, then (152) determines the value of ε_{ij}. For a single incomplete subshell outside closed subshells, ε_{ij} is normally small compared with the diagonal elements $\{\varepsilon_{ii}\}$. But if two or more incomplete subshells of the same l value are present, diagonal and off-diagonal elements may be of the same order of magnitude.

Two simplifications of the HF scheme as described above have had quite wide applications. In the *frozen-core* HF approximation for an N-electron atom, the orbitals of $(N-1)$ of the electrons are fixed as those of the $(N-1)$-electron ion, and the orbital of the Nth electron is obtained by assuming that this electron moves in the field of the ion. The equation for ϕ_N is then

$$\left[h_N + \sum_{j=1}^{N-1} \int \frac{|\phi_j(\mathbf{r}_j)|^2}{r_{jN}} d\tau_j \right] \phi_N$$

$$+ \sum_{j=1}^{N-1} \phi_j(\mathbf{r}_N) \int \frac{\phi_j^*(\mathbf{r}_j)\phi_N(\mathbf{r}_j)}{r_{jN}} d\tau_j \, \delta(m_{s_j}, m_{s_N}) = \varepsilon_N \phi_N \qquad (153)$$

An advantage of the scheme is that, since the outer orbitals of a series of excited states of the same symmetry (e.g., $1snl\,{}^3L$ in helium) are eigenfunctions of the same effective Hamiltonian (ϕ_i, $1 \leq i \leq N-1$ are the same in all cases), these states are automatically orthogonal. This approximation is particularly suited to excited states of atoms that approximate well to an ionic core plus an excited electron. It has been applied extensively to excited states in helium[54] and in photoionization calculations in a number of atoms.[55]

The second simplification is based on a treatment by Slater[56] of the exchange potential. In this Hartree–Fock–Slater (HFS) method, the exchange potential for different orbitals is replaced by a universal exchange potential found by considering the charge density of the atom as a free-electron gas. The HFS equations for the radial functions are then

$$\left[-\frac{d^2}{dr^2} + \frac{l(l+1)}{r^2} + V(r) \right] P_{nl} = \varepsilon_{nl} P_{nl} \qquad (154)$$

where $V(r)$ is the common potential for all electrons

$$\frac{1}{2} V(r) = -Z/r + \left(\frac{1}{r}\right) \sum_{nl} \int_0^r [P_{nl}(s)]^2 \, ds$$

$$+ \sum_{nl} \int_r^\infty \frac{[P_{nl}(s)]^2}{s} \, ds - \left[\left(\frac{81}{32\pi^2 r^2}\right) \sum_{nl} [P_{nl}(r)]^2 \right]^{1/3} \qquad (155)$$

The first term is the electron–nuclear electrostatic contribution, the next two form the direct term (150), while the last term replaces the exchange potential (151). Herman and Skillman[57] refined this technique by requiring $V(r)$ to be given by (155) for $r < r_0$, while for $r \geq r_0$

$$V(r) = -2(Z - N + 1)/r \tag{156}$$

Thus $V(r)$ has the correct asymptotic form, and r_0 is chosen as the radial distance for which (155) and (156) are equal. Because of the common potential in (154) for a fixed l, (148) is automatically satisfied. Because of the Coulombic nature of (156), $V(r)$ supports an unlimited number of discrete energy levels, and each Rydberg state of the atom may be associated with an excited-state determinant of the form (132). This is in contrast to the conventional HF theory, where unoccupied radial functions are either unbounded or become very diffuse. The method has been incorporated into computer programs by Herman and Skillman,[57] which have been updated by Desclaux[58] and by Beck and Zare,[59] who also allow for relativistic effects. The modification of Slater is similar in spirit to the Thomas–Fermi model, a version of which has been used successfully by Eissner and Nussbaumer[60] to generate radial functions. Other forms of (155) are discussed in Section 8.

4.2. Solutions of HF Equations

Two essentially different approaches have been proposed for the solution of the HF equations. The first method treats the equations (149) directly either numerically or by expressing the functions in terms of an analytic basis which in practice must be finite. In the numerical version, initial estimates for the $\{P_{nl}\}$ are chosen (frequently as screened hydrogenic functions) and the direct and exchange potentials Y_i and X_i are evaluated. The resulting *differential* equations are integrated to give new radial functions. The process is repeated until a satisfactory degree of convergence has been achieved. The final functions $\{P_{nl}\}$ satisfy (149) in a self-consistent manner. The HF potential is thus often termed the *self-consistent field* (SCF).

In many analytic treatments, it is customary to express the radial functions as

$$P_{nl}(r) = \sum_{j=1}^{k} c_{jnl} \chi_{jnl} \tag{157}$$

where

$$\chi_{jnl}(r) = \frac{(2\zeta_{jnl})^{I_{jnl}+1/2}}{[(2I_{jnl})!]^{1/2}} r^{I_{jnl}} \exp{(-\zeta_{jnl}r)} \tag{158}$$

is known as a Slater-type orbital (STO). For a given set of $\{I_{jnl}, \zeta_{jnl}\}$, the equations (149) reduce to a set of algebraic equations for the coefficients $\{c_{jnl}\}$, which may be solved self-consistently.[61,62] In practice, $\{I_{jnl}\}$ are usually treated as fixed parameters, but the exponents $\{\zeta_{jnl}\}$ are allowed to vary to minimize the HF energy. Usually, for a given l, $\{\chi_{jnl}\}$ is independent of n.

Numerical wave functions are rarely given in the literature. However, a widely used program for generating them is now available generally.[63,64] Tabulations of the optimum parameters in (157) and (158) are more compact. Perhaps the most extensive of these is that of Clementi and Roetti,[65] who have updated earlier work[66] to include wave functions for atoms and ions up to $Z = 54$. Other, more limited, sets include low-lying states of the lithium sequence,[67] the lowest ^1S, ^1P, ^3P states of the beryllium sequence,[68,69] ground and excited states of nitrogen and oxygen in several stages of ionization,[70] and ground and excited states of aluminium and copper.[71] As Z increases, the number of basis functions in (157) can become quite large, and several authors[65,69] include "double-zeta" functions, which assign $2(n - l)$ basis functions for a given l, where n is the maximum n value for occupied nl orbitals; there are two basis functions to emphasize the regions of space in which each radial function is large.

The second approach to the solution of the HF equations is based on a perturbation expansion similar to that used for the full Schrödinger equation. Again by changing the scale of length and energy, Z^{-1} becomes the perturbation parameter, with the zero-order equation being hydrogenic and the perturbation being the HF potential. Various authors have used this approach for the ground states of helium through neon.[72–75] It is interesting to compare the exact nonrelativistic and HF energy expansions. For the ground states of helium and lithium, the zero-order wave function Ψ_0 is the same in the two wave function expansions, so that the energy expansions differ only in second and higher orders. The difference between the exact nonrelativistic and HF energies (the *correlation energy*[76]) is

$$\Delta E = \Delta E_2 + Z^{-1} \Delta E_3 + \cdots \qquad (159)$$

As Z increases, the difference tends to a constant value. But for beryllium for instance, the exact zero-order equation has two $n = 2$ hydrogenic solutions: $2s$ and $2p$. The zero-order wave function is then a linear combination of two ^1S couplings, which may be written as $1s^2 2s^2$ and $1s^2 2p^2$, according to degenerate perturbation theory. On the other hand, the HF theory requires the association of each orbital with a *single* spherical harmonic (142) so that the $2s, 2p$ degeneracy is not accommodated. Although E_0 is the same in the two energy expressions, being a sum of hydrogenic energies, the E_1 terms are different. Hence the correlation energy is

$$\Delta E = Z \, \Delta E_1 + \Delta E_2 + Z^{-1} \, \Delta E_3 + \cdots \tag{160}$$

and the difference increases approximately linearly with Z. It is, of course, possible to modify the HF method to take into account such zero-order degeneracy ("near-degeneracy" in a nonperturbative approach), effectively by a configuration interaction scheme (see Section 5.2). The energy difference between this modified theory and the exact treatment is again given by (159).

For the most part, the HF perturbation equations have been solved only up to first order. For low Z, the expansions converge rather slowly; for neutral helium, one must take the energy expansion to eighth or ninth order to achieve accuracy comparable with the first approach.[77]

5. Correlation Methods

The HF method has been used widely in its own right, sometimes with considerable success. But its success is variable, and the method is often quite unreliable in predicting atomic properties. We must therefore consider it as a first approximation in the treatment of electronic inter-actions. We may collectively term as "correlation methods" those schemes that treat electron correlation in a way that is more accurate than the HF averaging of the electrostatic potential.

There are essentially five categories of method, although some schemes span two or more categories:

(a)   methods that retain the independent-particle philosophy of HF;
(b)   configuration interaction;
(c)   perturbative methods;
(d)   many-electron theories allowing the use of interelectronic coordinates;
(e)   model potentials.

In this section we discuss briefly the essentials of these schemes.

5.1. Independent-Particle Models

The HF method is rather more general than the form in which it is usually expressed and in which we described it in Section 4. We can summarize the customary restrictions as follows:

(i)   The spin–orbitals are factorized into a simple product of a space part and a spin part, (135).

(ii)   The spatial functions factorize according to (142).
(iii) The radial functions are independent of m_l.
(iv)   The radial functions are independent of m_s.

Independent-particle methods lift one or more of these restrictions while retaining the form of the wave function as a single determinant built from one-electron functions, or as a suitable projection of this. The development of these methods arose primarily from an interest in hyperfine-structure calculations, particularly of the ground state $1s^2 2s\ ^2S$ of lithium. The HF approximation gives rather poor agreement with experiment. But it was found that if restriction (iv) was lifted, giving an unnormalized wave function

$$\Phi_1 = \det |u\alpha, v\beta, w\alpha| \qquad (161)$$

where $u \neq v$ are the two spatial $1s$ functions and w corresponds to $2s$, then a much improved value is obtained. Two other determinants are possible:

$$\Phi_2 = \det |u\beta, v\alpha, w\alpha| \quad \text{and} \quad \Phi_3 = \det |u\alpha, v\alpha, w\beta| \qquad (162)$$

The *exact* wave function is an eigenfunction, not only of H, but also of $\mathbf{S}^2$ and $\mathbf{L}^2$ together with the z components S_z and L_z. Pratt[78] pointed out that the linear combinations

$$\Psi_1 = \Phi_1 - \Phi_2 \quad \text{and} \quad \Psi_2 = 2\Phi_3 - \Phi_1 - \Phi_2 \qquad (163)$$

are eigenfunctions of these angular momentum operators, although in general $\{\Phi_i\}$ are not. A more general form of the wave function is then

$$\Psi = A_1 \Psi_1 + A_2 \Psi_2 \qquad (164)$$

where the ratio A_2/A_1 may be treated as a variational parameter. The use of Φ_1 alone corresponds to the spin-polarized Hartree–Fock (SPHF) method.[79] It is a linear combination of doublet and quartet contributions. The latter may be removed by use of a spin projection operator[80] giving the spin-extended Hartree–Fock (SEHF)[81] or GF[82] scheme, assuming that the radial functions are determined *after* the projection has been performed. The SEHF method also gives a linear combination of $\{\Phi_i\}$:

$$\Psi^{\text{SEHF}} = 2\Phi_1 - \Phi_2 - \Phi_3 = \tfrac{1}{2}(3\Psi_1 - \Psi_2) \qquad (165)$$

thus fixing the ratio A_2/A_1 in (164), but at a value very different from the variational result of -0.00253.[79] The use of Ψ_1 alone is the G1[83] or maximally paired HF method,[84] while the more general (164) gives the spin-optimized SCF (SOSCF) method.[84] Lunell[79] shows that the use of SOSCF is equivalent to lifting restriction (i) by expressing U_i as a rather more general spin–orbital

$$U_i = \phi_i(\mathbf{r})(\alpha + a_i\beta) \qquad (166)$$

with $\{a_i\}$ determined variationally.

By analogy with spin, the removal of restriction (iii) may be termed the orbital-polarization HF (OPHF) scheme,[85] while the remaining restriction (ii) may be removed by writing

$$\phi_i(\mathbf{r}) = r^{-1} \left[P_{n_i l_i} Y_{l_i}^{m_{l_i}} + \sum_{l'} P_{n_i l'} Y_{l'}^{m_{l'}} \right] \tag{167}$$

where the summation includes those values of l' for which $|l_i - l'|$ is even, thus preserving parity, and $m_{l'} = m_{l_i}$, otherwise the energy expression separates into components corresponding to different m_l.[86] A lifting of all these restrictions may be termed the unrestricted HF (UHF) approximation,[87] although this term has been used in the past for a rather more limited restriction removal.

5.2. Configuration Interaction

We use the term "configuration" to denote the assignment of the N electrons of an atomic system to a set of N orbital functions whose angular momenta are coupled to form a specific LS state. For example, the HF configuration of the ground state of carbon may be written $1s^2 2s^2 2p^2$ and the coupling scheme may be included where necessary.

The assignment of an electron to a specific nl orbital is a restriction normally imposed in the HF method. We have already noted that, in the limit of infinite Z, the nonrelativistic wave function for the ground state of beryllium takes the form

$$\Psi = a_1 \Phi_1(1s^2 2s^2 \, {}^1S) + a_2 \Phi_2(1s^2 2p^2 \, {}^1S) \tag{168}$$

that is, a linear combination of two functions describing different configurations. Moreover, in (163), Ψ_1 and Ψ_2 correspond to configurations $(1s1s')\,{}^1S\, 2s\,{}^2S$ and $(1s1s')^3S\, 2s\,{}^2S$, respectively, where $1s$, $1s'$ are different $1s$ functions. In this case, the angular momentum coupling needs to be specified in order to distinguish between them. Wave functions (164) and (168) are examples of configuration interaction, which may be written more generally as

$$\Psi(LS) = \sum_{i=1}^{M} a_i \Phi_i(\alpha_i LS) \tag{169}$$

where $\{\alpha_i\}$ denote the coupling schemes, and M is, in practice, finite. For a given set $\{\Phi_i\}$, the coefficients $\{a_i\}$ may be determined by minimizing the energy subject to the normalization condition $\langle \Psi | \Psi \rangle = 1$. If we introduce the Lagrange multiplier E, then this process results in the equations

$$\sum_{j=1}^{M} (H_{ij} - ES_{ij})a_j = 0, \qquad i = 1, 2, \ldots, M \tag{170}$$

where $H_{ij} = \langle \Phi_i | H | \Phi_j \rangle$, $S_{ij} = \langle \Phi_i | \Phi_j \rangle$. The possible values of E are the eigenvalues $\{E^{(i)}\}$ of H with respect to the overlap matrix S, and the coefficients in (169) are the components of the corresponding eigenvectors $\{a_j^{(i)}\}$. Since (169) is of the form (125), MacDonald's theorem applies,[37-39] i.e., (127) holds. Thus $\{E^{(i)}\}$ form upper bounds to the exact nonrelativistic energies of the corresponding states of the atom. Therefore we have in principle a method of obtaining wave functions and energies of atomic states.

The practical calculation of these wave functions involves one of two rather different schemes as described below.

(i) The set of configurations should include the HF configuration. The superposition-of-configurations (SOC) scheme *augments* the HF radial functions by those radial functions needed to define the remaining configurations. These additional functions may be optimized on the basis of the variational principle (127). The functions occupied in the HF configuration are *not* reoptimized.

(ii) The reoptimization of the HF functions would allow greater flexibility, and this is a feature of the multiconfigurational HF (MCHF) method. This difference between SOC and MCHF may be seen in the beryllium wave function of the form (168)—in SOC, the $1s$ function would be identical to that of the single-configuration (HF) function; in MCHF, the $1s$ function would be reoptimized for (168) and in general would be different from the SOC function.

Unlike SOC, the MCHF scheme uses the variational principle, not directly, but in order to set up equations similar in character to the HF equations (149), but now containing the coefficients $\{a_i\}$ of (169). Again the solution is iterative, but this time in two parts: (a) For given estimates of $\{a_i\}$ and $\{P_{nl}\}$ the equations are solved self-consistently to give new radial functions; (b) for these radial functions, a new set of $\{a_i\}$ is determined from (170).

Although there are several privately circulating computer packages that perform these calculations, some in a restricted form, there are some packages generally available, both for SOC[88] and MCHF,[63,64] while a program using scaled Thomas–Fermi radial functions has also been described.[60,89] The radial functions for SOC are normally represented by analytic forms such as (157),[88,90] while the MCHF functions are normally numerical,[64] although analytic MCHF equations may also be constructed,[91] just as for the HF method.

5.2.1. Natural Orbitals

One of the main disadvantages of the configuration interaction (CI) scheme is its slow convergence, which we demonstrate for the ground state

Table 3. Convergence of the CI Energies of Helium

		$-E$ (a.u.)	
l_{max}	Weiss[92]	Froese-Fischer[93]	Bunge[94]
0	2.87896	2.87899	2.879025
1	2.90039	2.90040	2.900507
2	2.90258	2.90252	2.902749
3	2.90307	2.90291	2.903307
4	2.90320	2.90303	2.903470
10	—	—	2.903590

of helium in Table 3. The calculation of Weiss[92] used nonorthogonal orbitals with radial functions each consisting of a single STO. The calculation of Bunge[94] is extensive and probably represents for each l_{max} (the largest l value of the orbitals occurring in the configurations) a limit fairly well converged with respect to n. But even when $l_{max} = 10$, the energy is still some way from the $l = \infty$ limit of -2.90372 a.u. As l_{max} increases, the energy improvement is approximately proportional to l_{max}^{-4}. As one increases the number of orbitals to achieve some degree of convergence, the number of configurations in the expansion (169) builds up rapidly, particularly for atoms with several electrons in open subshells, where the number of different angular momentum coupling schemes is also large. It is important therefore to choose the radial functions carefully to minimize the number of configurations needed for a particular level of accuracy.

One means of doing this is to use *natural orbitals*. The helium ground state CI wave function takes the form

$$\Psi(^1S) = \sum_{m,n} a_{mn}(msns)^1S + \sum_{m,n} b_{mn}(mpnp)^1S + \sum_{m,n} c_{mn}(mdnd)^1S + \cdots$$
$$= \mathbf{s}^T\mathbf{A}\mathbf{s} + \mathbf{p}^T\mathbf{B}\mathbf{p} + \mathbf{d}^T\mathbf{C}\mathbf{d} + \cdots \tag{171}$$

where $\mathbf{A}, \mathbf{B}, \mathbf{C}, \ldots$ may be chosen to be symmetric matrices and $\mathbf{s} = (1s, 2s, \ldots)$, etc. It is possible to perform unitary transformations amongst the radial functions so that $\mathbf{A}, \mathbf{B}, \mathbf{C}, \ldots$ become diagonal. Then we may write (171) as

$$\Psi(^1S) = \sum_n a_n \overline{ns}^{2\ 1}S + \sum_n b_n \overline{np}^{2\ 1}S + \sum_n c_n \overline{nd}^{2\ 1}S + \cdots \tag{172}$$

The $\{\overline{ns}, \overline{np}, \overline{nd}, \ldots\}$ are the natural orbitals.[96] Thus we have considerably reduced the number of configurations without changing the corresponding energy value. Froese-Fischer[93] has performed an MCHF calculation using (172) (MCHF rather than SOC since $\overline{1s} \neq 1s^{HF}$). Although only a small number of n values were used ($n_{max} = 4$, plus $5g$), the results for different l_{max} compare favorably with the more extensive calculations in Table 3, yet with fewer configurations. The discrepancies for the higher l values are due to the use of only one or two orbitals for each l.

Similar techniques may be used for excited states. Thus the $^1P^0$ wave function

$$\Psi(^1P^0) = \sum_{m,n} a'_{mn}(msnp)^1P^0 + \sum_{m,n} b'_{mn}(mpnd)^1P^0 + \sum_{m,n} c'_{mn}(mdnf)^1P^0 + \cdots \tag{173}$$

may be rewritten[93]

$$\Psi(^1P^0) = \sum_n a'_n [\overline{ns}\,\overline{(n+1)p}]^1P^0 + \sum_n b'_n [\overline{np}\,\overline{(n+1)d}]^1P^0$$
$$+ \sum_n c'_n [\overline{nd}\,\overline{(n+1)f}]^1P^0 + \cdots \tag{174}$$

However, the transformation amongst the p orbitals required in the diagonalization of $\mathbf{A}'$ is different from that of $\mathbf{B}'$. Hence there are two sets of p-type natural orbitals, but these are not mutually orthogonal. The reduction in the number of configurations is achieved at the expense of the loss of the relative simplicity of orbitals that are orthogonal. But it may well be a price worth paying. Actually the CI expansion converges relatively well for excited states, partly because for them HF is already a more reasonable approximation than for ground states.

An extension of these ideas to many-electron atoms has been implemented by Weiss.[90] The scheme corresponds to considering the set of configurations in which a specific *pair* or orbitals is changed from the HF configuration, and performing the matrix diagonalization of the CI coefficient matrix for that effective two-electron situation. This results in a set of *pseudonatural* orbitals.

5.2.2. Brillouin's Theorem

If the wave function Ψ of an N-electron atom is approximated by a single determinant of spin–orbitals, (132), and if Ψ' is a single determinant describing a configuration differing from Ψ by just one orbital, then Brillouin's theorem

$$\langle \Psi | H | \Psi' \rangle = 0 \tag{175}$$

holds when a fully unrestricted HF formalism is used.[97] A modification of this result

$$\langle \mathbf{O}\Phi | H - E | \mathbf{O}\Phi_i^a \rangle = 0 \tag{176}$$

may be derived, where $\mathbf{O}$ is a spin projection operator making $\mathbf{O}\Phi$ the SEHF wave function and $\mathbf{O}\Phi_i^a$ obtained from it by replacing the occupied

ϕ_i by the unoccupied ϕ_a. This result has been used to generate SEHF (and also SOSCF) orbitals.[81,84] The scheme is further extended to multi-configurational wave functions by Grein and Chang.[98]

Brillouin's theorem does not automatically apply to the more restricted form of HF theory discussed in Section 4. Bauche and Klapisch[99] show that (175) does hold when Ψ is the restricted HF function and consists of one of the following:

(i) completely filled subshells;
(ii) completely filled subshells plus an open shell containing the electron to be excited.

It is tempting to use Brillouin's theorem as a means of reducing further the number of configurations needed in the wave function. One example will suffice to show the dangers of such a proposal. From (ii) above, the $1s^2 2p$ and $1s^2 3p$ configurations should satisfy (175), and one might therefore omit $1s^2 3p$ from an expansion of the $1s^2 2p \, ^2P$ state wave function. If the wave function has only the form $\Psi = a_1 \Phi_1(1s^2 2p \, ^2P) + a_2 \Phi_2(1s^2 3p \, ^2P)$ then indeed $a_2 = 0$. But when other configurations are added, differing from $1s^2 2p$ by two orbitals, a_2 is no longer zero. Indeed, the inclusion of $1s^2 3p$ at this stage is the principal cause of the improvement of the hyperfine-structure parameter a_l, (64), from its HF value. The existence of a zero in the first row of the Hamiltonian matrix does not necessarily imply a zero (or even small) coefficient a_i in (169).

5.3. Bethe–Goldstone Theory

This method, which has been discussed in detail by Nesbet,[100] is based on work by Bethe and Goldstone[101] in which it is assumed that the correlation problem may be solved independently for each pair of electrons in an atom. Essentially, a Schrödinger-type equation is set up, satisfied by a two-particle function, for each pair of electrons. This equation is the (second-order) Bethe–Goldstone equation. Nesbet[100] seeks to solve this equation variationally in terms of a CI expansion

$$\Psi_{ij} = \Phi_0 + \sum_a c_i^a \Phi_i^a + \sum_b c_j^b \Phi_j^b + \sum_{a,b} c_{ij}^{ab} \Phi_{ij}^{ab} \qquad (177)$$

for the two particles (electrons) i and j, where Φ_0 is the HF function built from orbitals $\phi_1, \ldots, \phi_N$, while $\Phi_{ij\ldots}^{ab\ldots}$ is obtained from Φ_0 by the replacements $\phi_i \rightarrow \phi_a$, $\phi_j \rightarrow \phi_b$, ..., where $1 \leq i < j \leq N < a < b$ and $\{\{\phi_i\}, \{\phi_a\}\}$ form a complete set. One may define *net increments* as follows. The one-particle net increment $e_i = \Delta E_i = E_i - E_0$, where E_i is the lowest eigenvalue of the Hamiltonian matrix obtained using the first two parts of (77). Then the two-particle net increment is $e_{ij} = \Delta E_{ij} - e_i - e_j$. In fact, two forms of this scheme have been used. *Orbital* excitation involves the use of (177) with simple orbital replacement in the HF determinant, without any attempt to construct eigenfunctions of $\mathbf{L}^2$ and $\mathbf{S}^2$. In *configurational* excitation, such eigenfunctions *are* constructed.

This procedure may be generalized to n particles. The nth order Bethe–Goldstone equation for n specified orbitals $i, j, k, \ldots$, occupied in HF, is equivalent to a variational calculation of a wave function $\Psi_{ijk\cdots}$ consisting of Φ_0 and all configurations constructed from Φ_0 by replacing $m(\leq n)$ of the specified orbitals. Higher-order net increments may be defined; for example, $e_{ijk} = \Delta E_{ijk} - e_i - e_j - e_k - e_{ij} - e_{jk} - e_{ik}$.

A sequence of Bethe–Goldstone equations of increasing order eventually terminates at order N with the exact solution of the N-electron problem. The increments of energy (or of any other physical quantity that may be expressed as an expectation value, and whose net increments may therefore be defined) form a sequence whose sum is the exact value. The method becomes practicable if the series of increments converges adequately at second or third order.

Thus the method is in essence perturbative, but its practical use has involved variational calculations in a CI formalism at each order, and as such it forms a bridge between Sections 5.2 and 5.4. It is worth pointing out that at order n the Bethe–Goldstone wave functions for each set of $n < N$ electrons are determined variationally, but that no wave function containing all effects up to and including nth order is obtained until order $(n + 1)$.

5.4. Perturbation Methods

Although the Bethe–Goldstone method is perturbative in character, equations of the form (107) are not set up, nor is the Hamiltonian partitioned according to (103). Nevertheless, it is possible to consider such an approach and to treat the HF function Φ_0 as the zero-order solution [set $\lambda = 1$ in (103) for simplicity of notation]:

$$H_0\Phi_0 = E_0\Phi_0 \tag{178}$$

$$H_0 = \sum_{i=1}^{N} F_i \tag{179}$$

where F_i is given by (138). From (137),

$$E_0 = \sum_{i=1}^{N} \varepsilon_i \tag{180}$$

$$E_1 = \langle \Phi_0 | H_1 | \Phi_0 \rangle$$

$$= \left\langle \Phi_0 \left| \sum_{i<j} r_{ij}^{-1} - \sum_{k=1}^{N} V_k \right| \Phi_0 \right\rangle$$

$$= -\sum_{i<j} [J_{ij} - K_{ij}\delta(m_{si}, m_{sj})] \tag{181}$$

where

$$J_{ij} = \int r_{ij}^{-1} |\phi_i(\mathbf{r}_i)|^2 |\phi_j(\mathbf{r}_j)|^2 \, d\tau_i \, d\tau_j \tag{182}$$

$$K_{ij} = \int r_{ij}^{-1} \phi_i^*(\mathbf{r}_i)\phi_j^*(\mathbf{r}_j)\phi_j(\mathbf{r}_i)\phi_i(\mathbf{r}_j) \, d\tau_i \, d\tau_j \tag{183}$$

There have been several applications of this scheme to the ground state of helium, by using the variational form (117) to determine the first-order function Ψ_1. Indeed, a variety of different zero-order functions have been tried.[1] One must conclude that the convergence of the energy series to third order is fairly independent of the choice of Ψ_0, and that a hydrogenic starting point seems just as adequate as starting from HF. Only an elaborate $\Psi_0^{(102)}$ involving interelectronic coordinates improves the convergence while retaining the bounds criterion (116). Even then, only moderate improvement is achieved, and the method is not readily applicable to larger atomic systems.

For many-electron atoms, Sinanoglu[103] proposed the grouping of electrons, with each group containing some correlation. The first-order wave function included determinants involving one- and two-orbital changes from HF; cf. (177). The second-order energy can be expressed as a sum of terms arising from the correlation of pairs of electrons and from interactions involving three or four electrons. But the formal theory includes infinite summations over the complete set of orbitals, occupied and unoccupied, which would need to be truncated. These infinite summations can be removed from the formalism (but not from the reality) by writing

$$\Psi_1 = 2^{-1/2}\mathscr{A} \sum_{i<j} (\phi_1, \phi_2, \ldots, \phi_{i-1}, \phi_{i+1}, \ldots, \phi_{j-1}, \phi_{j+1}, \ldots, \phi_N, u_{ij})$$

$$\tag{184}$$

where $\mathscr{A}$ is an antisymmetrizing operator,[103] assuming a HF function with a single determinant (e.g., closed-shell systems); more general open-shell systems are discussed by Silverstone and Yin.[104] Thus Ψ_1 is a sum of terms each obtained by replacing the occupied HF orbitals ϕ_i and ϕ_j by a function u_{ij} of two electrons. We thus have again the possibility of introducing interelectronic coordinates into the (first-order) wave function, a process that proved for helium to be more rapidly convergent than conventional CI. [Equation (177) amounts to using a CI expansion of the $\{u_{ij}\}$.] The scheme outlined here has not been followed beyond a few test cases. But similar schemes starting from a hydrogenic H_0 (and Ψ_0) have been used with effect for lithium[105] and for first-row atoms.[106] Expression

(184) is rewritten[107] exactly as

$$\Psi_1 = \sum_{\Gamma_1 S_1 L_1} \sum_{\Gamma_2 S_2 L_2} \langle \Gamma_1 S_1 L_1, \Gamma_2 S_2 L_2 |\} \Gamma S L\rangle [\Phi_0(\Gamma_1 S_1 L_1 | N - 2), \Phi_1(\Gamma_2 S_2 L_2 | 2)]$$

$$(185)$$

where $\langle \, |\} \, \rangle$ denotes a *two*-particle fractional parentage coefficient,[108] $[\Phi_0, \Phi_1]$ is a vector-coupled product of an $(N - 2)$-electron function Φ_0, constructed from hydrogenic (therefore known) zero-order wave functions, and a two-electron function satisfying

$$(-\tfrac{1}{2}\nabla_1^2 - \tfrac{1}{2}\nabla_2^2 - r_1^{-1} - r_2^{-1} - E_0)\Phi_1 + (r_{12}^{-1} - E_1)\Phi_0 = 0 \qquad (186)$$

in the scaled units used for Z^{-1} expansions: Z^{-1} a.u. for length, Z^2 a.u. for energy. Equation (186) is just the first-order equation for heliumlike systems, which have been solved to high accuracy for several states.[109] Thus the first-order function (185) may be immediately written down. The case of zero-order degeneracy has been discussed by Laughlin and Dalgarno.[106]

An alternative approach[110] is to expand the wave function to each order in CI form, with a common set of configurations. The perturbation equations then become purely algebraic, and may be solved to high order. For an isoelectronic sequence, this method has the advantage of requiring the Hamiltonian matrix to be computed only once, compared with a similar computation for each ion in a set of conventional CI calculations. Thus the wave function is obtained to high order, but at each order one introduces the approximations associated with a finite CI expansion. By contrast, the use of (185) is much more accurate to first order, but the method is not readily extended to higher orders. Comparing the two approaches, for lowly ionized systems, one might expect the high-order CI method to be better, since the corrections to the wave function from higher-order terms may be more important than those associated with a truncation of the representation of Ψ_1. For higher Z, the accurate representation of Ψ_1, as in (185), should be sufficient to produce results for energies and other properties more accurate than from a CI representation.

Many-Body Perturbation Theory

Probably the most widely used perturbation scheme has been the many-body perturbation theory (MBPT) originally developed by Brueckner[111, 112] and Goldstone,[113] and reviewed by Kelly,[114] who considers some of the earlier applications to electron correlation in atoms. It is based on a second quantization form of Rayleigh–Schrödinger perturbation theory, in which the zero-order Hamiltonian is a sum of one-electron terms:

$$H_0 = \sum_{i=1}^{N} \left[-\tfrac{1}{2}\nabla_i^2 + V(r_i) \right] \qquad (187)$$

In practice the HF potential is used for V, so that H_0 is given by (179). In second quantization notation[115]

$$H_0 = \sum_n \varepsilon_n \eta_n^+ \eta_n \qquad (188)$$

$$H_1 = \sum_{p,q,m,n} \langle pq | r_{12}^{-1} | mn \rangle \eta_p^+ \eta_q^+ \eta_n \eta_m - \sum_{p,m} \langle p | V | m \rangle \eta_p^+ \eta_m \qquad (189)$$

where ε_n is given by (137), η^+, η are creation and annihilation operators, and the sums are over all HF states (orbitals), occupied and unoccupied, discrete and continuous. Interactions such as those occurring in (189) are represented by Feynman diagrams. Diagrams that contain completely disconnected parts are termed *unlinked*; otherwise they are *linked*.

After a time-*dependent* perturbation analysis, the exact wave function may be expanded as

$$\Psi = \sum_{n=0}^{\infty} {}^{L} [(E_0 - H_0)^{-1} H_1]^n \psi_0 \qquad (190)$$

with E_0 given by (180) and the superscript L indicating that only linked diagrams are included in the summation. Then

$$E = E_0 + \sum_{n=0}^{\infty} {}^{L'} \langle \psi_0 | H_1 \{ (E_0 - H_0)^{-1} H_1 \}^n | \psi_0 \rangle \qquad (191)$$

where L' indicates that only linked diagrams arising in (190) are included. Again the interactions in (191) may be represented by Feynman diagrams, and the use of the HF operator in H_0 allows certain of the sums to be set to zero before calculations are performed, thus simplifying the problem somewhat. Further details may be found in the review article of Kelly.[114] The applications of the method have been extensive, and we shall discuss some of them in Section 6. Particularly impressive has been the number of HF states actually employed, and the ability to take the calculations to third order and often to estimate the higher-order contributions. An interesting comparison between MBPT and the Bethe–Goldstone treatment has been given by Nesbet,[116] who shows that the nth-order Bethe–Goldstone wave function involving the replacement of the HF occupied orbitals ϕ_i, ϕ_j, ϕ_k, ... is equivalent to a MBPT sum (190) to *infinite* order but containing only those linked diagrams in which the labels on the backward-directed lines are limited to the set $(i, j, k, \ldots)$.

5.5. Many-Electron Theory

We saw in an earlier section that for helium the convergence of a CI expansion is slow compared with an expansion involving interelectronic coordinates, but the direct use of the latter in computations for large systems is very time-consuming. The possibility of their use in the two-electron functions u_{ij} in (184) is therefore potentially very powerful. But it is not necessary to restrict their use to a perturbation formalism. The extension to a direct variational formalism was made by Sinanoglu.[117] Since the theory for closed-shell systems is somewhat different from that of open shells, we shall consider these cases separately.

5.5.1. Closed Shells

If we write the exact wave function as

$$\Psi = \Phi + \chi \tag{192}$$

where Φ is the HF single-determinant wave function and we choose the normalization according to

$$\langle \Phi|\Phi \rangle = \langle \Psi|\Phi \rangle = 1, \qquad \langle \Phi|\chi \rangle = 0 \tag{193}$$

then the correlation energy is given by

$$E - E^{HF} = [1 + \langle \chi|\chi \rangle]^{-1}[2\langle \Phi|H - E^{HF}|\chi \rangle + \langle \chi|H - E^{HF}|\chi \rangle] \tag{194}$$

where E^{HF} is the HF energy [the sum of expressions (180) and (181)]. If we introduce the operator $S_i(j)$, where

$$S_i(j)\phi_j = \phi_j(j)\left[\int r_{ij}^{-1}|\phi_i(i)|^2 d\tau_i\right] - \delta(m_{si}, m_{sj})\phi_i(j)\int r_{ij}^{-1}\phi_i^*(i)\phi_j(i) d\tau_i \tag{195}$$

then we may write

$$H - E^{HF} = \sum_{i=1}^{N} e_i + \sum_{i<j} m_{ij} \tag{196}$$

where

$$e_i = F_i - \varepsilon_i \tag{197}$$

$$m_{ij} = r_{ij}^{-1} - S_i(j) - S_j(i) + J_{ij} - K_{ij}\delta(m_{si}, m_{sj}) \tag{198}$$

and so the correlation energy is given by

$$E - E^{HF} = [1 + \langle \chi|\chi \rangle]^{-1}\left[2\left\langle \Phi\left|\sum_{i<j} m_{ij}\right|\chi\right\rangle + \left\langle \chi\left|\sum_{i=1}^{N} e_i + \sum_{i<j} m_{ij}\right|\chi\right\rangle\right] \tag{199}$$

Sinanoglu[117] makes two points:

(i) The conclusion principle tends to arrange electrons with like spin as spatially far apart as possible.

(ii) The m_{ij} are short range.

As a result, it is unlikely that three or more electrons could come under the influence of a single m_{ij}. The exact correlation function χ may be expanded as

$$\chi = \mathscr{A}\left[(12 \cdots N) \left\{ \sum_i \frac{f_i}{(i)} + \sum_{i,j} \frac{u_{ij}}{(ij)} + \sum_{i,j,k} \frac{U_{ijk}}{(ijk)} + \cdots + \frac{U_{12\cdots N}}{(12\cdots N)} \right\} \right] \tag{200}$$

where in each sum the function $U_{ijk\ldots}$ replaces the orbitals $i, j, k, \ldots$ in the HF determinant. The (184) corresponds to the second term in (200). The orthogonality condition (193) is equivalent to requiring the correlation functions to be *strongly* orthogonal to the HF orbitals:

$$\int U_{ijk\cdots}(\mathbf{r}_i, \mathbf{r}_j, \mathbf{r}_k, \ldots)\phi_l(\mathbf{r}_i)\, d\tau_i = 0 \tag{201}$$

The form of (200) is that of the Nth-order Bethe–Goldstone expansion [cf. (177)], and, as in that scheme, the present theory will be of use if the expansion (200) can be truncated rather rapidly.

For closed-shell systems, odd-order terms f_i, U_{ijk}, etc. are negligible.[118] Fourth-order terms arise in two ways:

(a) Two pairs of electrons interact in different parts of the atoms ("unlinked clusters") through m_{ij} and m_{kl}; the contribution to χ is $u_{ij}u_{kl}$.

(b) Four electrons interact together ("linked clusters").

Then we may write, for example,

$$U_{1234} = u_{12}u_{34} + u_{13}u_{24} + u_{14}u_{23} + U'_{1234} \tag{202}$$

where the linked cluster term U'_{1234}, could to a good approximation be neglected on account of the arguments (i) and (ii) above. Thus the functions $U_{ijk\ldots}$ of order $2n$ can be well represented by sums of products of n pair functions $\{u_{ij}\}$. To determine these pair functions, we could use (184) as a trial wave function in the variational form (199). If a single term of (184) is used alone, (199) becomes

$$\varepsilon_{ij} = 2\langle \mathscr{B}(\phi_i, \phi_j)|m_{ij}|u_{ij}\rangle + \langle u_{ij}|e_i + e_j + m_{ij}|u_{ij}\rangle \tag{203}$$

where $\mathscr{B}$ is a two-electron antisymmetrizing operator, and the second-order Bethe–Goldstone equation is obtained. If (199) is used in full, then

$$E - E^{HF} = \left[1 + \sum_{i<j}\langle u_{ij}|u_{ij}\rangle\right]^{-1}\left[\sum_{i<j}\varepsilon_{ij} + 8\sum_{i,j,k}\{\langle \phi_i(i)u_{jk}(j,k)|m_{ik}|u_{ij}(i_{\dagger}j)\phi_k(k)\rangle\right.$$
$$\left. - \langle \phi_i(k)u_{jk}(i,j)|m_{ik}|u_{ij}(i,j)\phi_k(k)\rangle\}\right] \tag{204}$$

We thus return to the time-consuming problem of computing three-electron integrals in (204). Sinanoglu[117] demonstrates that these integrals are small. Also, the inclusion of higher-order terms allows the replacement of (204) by

$$E - E^{\text{HF}} \simeq \sum_{i<j} [1 + \langle u_{ij}|u_{ij}\rangle]^{-1}\varepsilon_{ij} \qquad (205)$$

which may then be used as a variational form for determining the $\{u_{ij}\}$.

5.5.2. Open Shells

A non-closed-shell many-electron theory (NCMET) is given by Oksutz and Sinanoglu.[119] To describe it, let us first define the *HF sea*. This is a set of orbitals occupied in the HF configuration plus all other orbitals that have the same or smaller n values. Thus, for the ground state of beryllium, the HF sea is 1s, 2s, *and* 2p. Then three types of correlation may be defined:

(i)   *Internal.* Described by configurations built entirely from orbitals in the HF sea (this includes near-degeneracy).
(ii) *Semi-internal.* Described by configurations built from $(N-1)$ orbitals of the HF sea, plus one other.
(iii) *All-External.* Described by configurations in which two or more orbitals are outside the HF sea. This part is the entire contribution for closed-shell systems.

We note that (i) and (ii) can be obtained entirely from a CI calculation that is *finite*, in the sense that the number of possible angular momentum couplings is finite. By contrast, (iii) would be described by an infinite angular momentum expansion in CI, but as in the closed shell case it allows the possibility of an interelectronic coordinate representation instead. In fact, this has rarely been done, so that practical calculations are usually CI. But we shall see in Section 6 that this classification of correlation is extremely useful. A computer program based on NCMET has been described.[206]

5.6. Model Potentials

Some atomic structure effects (e.g., many transitions) directly concern only the outer electrons of the atom. A full correlation treatment of the inner electrons (and in particular, those lying in closed shells) is often unnecessary for accurate calculations. A possible treatment is to limit the wave function so that only correlation in outer (sub)shells is included. In CI calculations, one would include configurations in which the inner electrons were described by their HF assignment. For example for magnesium, all configurations might have a common $1s^2 2s^2 2p^6$ core, with correlation being introduced only for the two valence electrons. Thus as far as the correlation of the electrons is concerned, we have an effective two-electron problem.

We can look at this another way. Instead simply of making such approximations in the wave function, we can use an effective Hamiltonian for the valence electrons and obtain the solutions of the corresponding Schrödinger equation. The influence of the core on the outer electrons can be represented by a *model* potential, parametrically chosen so that at least the low-lying energy levels of the atom are obtained accurately. Thus for a single valence electron outside an N-electron core, Szasz and McGinn[120] have used the Hamiltonian $(h_{N+1} + V_H)$, where the outer electron is denoted by $(N+1)$, h_{N+1} is given by (139), and

$$V_H = (A/r) \exp(-2\kappa r) + (B/r) \exp(-2\lambda r) \qquad (206)$$

with A, B, κ, and λ being chosen to fit the low-lying experimental energy levels. V_H is then the model or pseudopotential. The equivalent Schrödinger equation for two valence electrons is[120]

$$(h_1 + h_2 + V_H(1) + V_H(2) + r_{12}^{-1})\Psi(1, 2) = E\Psi(1, 2) \qquad (207)$$

where we now denote the two electrons by the subscripts 1 and 2. Equation (207) may be solved either in terms of interelectronic coordinates or by a CI expansion. In the former case, only two-electron integrals are involved, thus circumventing the difficulty of three-electron integrals found, for example, in the many-electron theory discussed in Section 5.5. The potential V_H contains a high repulsive barrier close to the center of the atom, preventing the valence electrons penetrating the core. The strong orthogonality condition (201) produces the same physical effect, but is responsible for the many-electron integrals which are difficult to handle. Calculations using these pseudopotentials have been performed for large atoms by Szasz and McGinn.[120,121]

An alternative model potential that incorporates directly the polarization of the core by the valence electron has been used by Weisheit and Dalgarno.[122] The function ψ_{N+1} describing the single valence electron then satisfies

$$[h + v_0(\mathbf{r}) + v_1(\mathbf{r})]\psi_{N+1} = \varepsilon_{N+1}\psi_{N+1} \qquad (208)$$

where

$$v_0(\mathbf{r}_{N+1}) = \sum_i \int \frac{\Phi^*\Phi}{r_{i,N+1}} \, d\tau_i \qquad (209)$$

for the *core* HF wave function Φ, and

$$\nu_1(\mathbf{r}_{N+1}) = -\tfrac{1}{2}\alpha_d r_{N+1}^{-4} - \tfrac{1}{2}\alpha_q r_{N+1}^{-6} \tag{210}$$

represents the polarization. Here α_d and α_q are the static dipole and quadrupole core polarizabilities, respectively. The equivalent scheme for two valence electrons is discussed by Victor and Laughlin,[123] whose model potential also contains parameters that may be chosen to fit energy levels.

For large atoms, model potential methods appear as an attractive means of handling a large number of electrons which are essentially passive in the atomic process under consideration. One would like to see the formalism extended and applied to systems with more than two electrons in open shells.

6. Applications

It is not our intention to review all the applications of the methods described in the previous sections. This is done in part elsewhere by the author[1] and in part in other chapters of the present book. But since many calculations *have* been performed, it is interesting to make some general comments both on the calculations themselves and on the light they shed on the relative usefulness of the methods we have presented.

6.1. Energy Levels

6.1.1. Spectra

The first test of any theory of atomic structure must be its ability to predict energy values, and in particular the splitting between energy levels in a spectrum. A popular comparison between theory and experiment has used the HF approximation with the spin–orbit term $\sum_{i=1}^{N} \zeta_i(\mathbf{l}_i \cdot \mathbf{s}_i)$ representing the fine structure splitting. The calculations are performed in intermediate coupling with the electrostatic and spin–orbit parts of the Hamiltonian treated together, rather than the latter as a perturbation of the former. The multiplet energies of different LS terms of a configuration can be expressed in the HF approximation in terms of one- and two-electron integrals. The former are common to all LS terms of the configuration, as are all but a few of the latter. Then the multiplet splittings may be expressed as a sum of just a few F^k or G^k integrals. If one assumes the HF approximation, the level splittings are given in terms of F^k, G^k, and the spin–orbit integrals ζ_{nl}. These quantities, which are calculated in theoretical methods, may be treated as parameters chosen to give the best

Table 4. Energy Levels (in a.u.) of the $2p^5 3p$ levels in Si V—Single Configuration and Same $3p$ for All Levels[a]

Term	LS	J	ΔE (a.u.)	ΔE (cm^{-1})	ΔE^b (cm^{-1})
$^3S_1$	-281.07143	-281.07293			
$^3D_3$		-281.03336			
$^3D_2$	-281.02640	-281.02807	-0.00529	-1164	-1031
$^3D_1$		-281.01959	-0.00848	-1861	-1905
$^1D_2$	-281.00215	-281.00562			
$^1P_1$	-280.99639	-280.99718			
$^3P_2$		-280.99127			
$^3P_1$	-280.99639	-280.98694	-0.00433	-950	-1007
$^3P_0$		-280.98990	0.00296	650	600
$^1S_0$	-280.77902	-280.77856			

[a] Conversion factor: 1 a.u. $= 219\,474.62$ cm^{-1}.
[b] Experimental value, Brillet.[16]

least-squares fit to the energy spectrum. A still better fit can be obtained by including other parameters, which take into account interactions with higher configurations.[124]

From an *ab initio* point of view, one can represent this configuration interaction by means of correlation methods discussed in Section 5. Let us consider the specific example of the $2p^5 3p$ levels in Si V, recently given by Brillet.[16] In Table 4 we compare the experimental splittings with some simple calculations in which we assumed a Si VI $1s^2 2s^2 2p^5\ {}^2P$ (frozen) HF core with radial functions given by Clementi[66] and we optimized the exponents of an analytic $3p$ function [cf. (157)] using the program CIV3.[88] In constructing Table 4, we assumed CI between those terms of the $2p^5 3p$ configuration with the same J value, but no CI from outside the configuration. Thus the *multiplet* splittings can be expressed in terms of the

Table 5. Energy Separation of $2p^5 3p$ Levels in Si V

Levels	ΔE^a (cm^{-1})	ΔE^b (cm^{-1})	ΔE^c (cm^{-1})
$^3S_1 - {}^1D_2$	14808	15646	18039
$^1D_2 - {}^1P_1$	1857	1861	1656
$^1P_1 - {}^1S_0$	48096	41195	38658
$^3S_1 - {}^1S_0$	64761	58702	58353

[a] From Table 4; same $3p$ function for all levels, no interaction with other configurations.
[b] Different $3p$ functions for each LS term; configuration interaction with $2p^6\ {}^1S_0$.
[c] Experimental splitting, Brillet.[16]

integrals $F^0(2p, 3p)$, $G^0(2p, 3p)$, and $G^2(2p, 3p)$. On the whole, the splitting of the fine-structure levels is fairly good. The deviations from Landé's interval rule occur mainly because of CI amongst $2p^5 3p$ levels resulting in off-diagonal spin–orbit matrix elements. But the multiplet structure obtained from these assumptions is not at all in good agreement with experiment (compare columns 1 and 3 of Table 5, where we list the splitting between levels that show no fine structure). There are essentially three shortcomings involved in this calculation:

(i)   neglect of CI between $2p^5 3p$ ^1S and the lower-lying $2p^6$ ^1S;
(ii) the use of the same $3p$ function for all terms;
(iii) neglect of other CI, including core relaxation.

The first of these effectively breaks the rules of the usual variational principle, and to get an upper bound to the $2p^5 3p$ ^1S energy, one must take the *second*-lowest eigenvalue (MacDonald's theorem) in a CI calculation involving at least these two configurations. In the simplest case, in which just these two are taken, the effect is to *raise* the $2p^5 3p$ ^1S energy by about 10^4 cm^{-1}, thus setting the calculated splitting in even worse agreement with experiment.

It is the lifting of the second restriction that brings theory and experiment much closer into line. The second column of Table 5 corresponds to a calculation in which, although the core of each term is again described by the Si VI HF state, the $3p$ of each term is determined independently (again with a CI calculation involving $2p^6$ for the ^1S term). It is the ^3S and ^1S states that change the most, with an energy lowering of about 20,000 cm^{-1} for the latter. In the earlier calculation, the multiplet splitting depends only on the values of the three Slater integrals mentioned above. As a result, the agreement between calculated and experimentally fitted values of these integrals is not very good. But the improved splitting obtained from varying the $3p$ functions is not simply due to improved agreement for these three integrals. Rather, all integrals involving the $3p$ function change from state to state, and the change in the splitting is the accumulation of changes in the values of this larger number of integrals. The lifting of the third restriction above no doubt would result in further improvement in the comparison between theory and experiment. One might expect the convergence of this improvement to be relatively slow compared with the more dramatic effects associated with (i) and (ii), and we shall not pursue the discussion here. Consideration of the LS dependence of $p^n p'$ configurations for low-lying members of the neon and argon sequences has also been given by Hansen.[213,214] He used numerical HF functions to consider the variation of Slater integral values for different terms. The convergence of the SCF scheme when more than one subshell of a given l value is open is a major problem because of large off-diagonal elements of the right-hand side of (149), particularly for the ^1S term. By imposing orthogonality between the $2p^53p\ ^1S$ and $2p^6\ ^1S$ states, Hansen obtained convergence and the optimal orbitals were also orthogonal. Orthogonality of HF states, however, does not imply that the variational principle can be used directly. Configuration interaction between the two ^1S states with energies determined by MacDonald's theorem is still necessary, since Brillouin's theorem does not apply, the $2p$ function for the two states being quite different. For our own calculations, Brillouin's theorem did not apply either, since we used the $2p$ function of the core, not of the $2p^6$ ground state. On the other hand, the use of the $2p$ function of the ground state leads to a significant raising of the $2p^53p\ ^1S$ energy, making further core corrections essential.

Another important CI effect is discussed by Zare[125] with specific reference to the $^1D^e$–$^3D^e$ splittings in the magnesium sequence. There is strong interaction between $3s3d\ ^1D$ and $3p^2\ ^1D$ resulting in an inversion of the $3s3d$ levels for Mg I compared with Al II and Si III, although the order for P IV is the same as for Mg I. In fact, for Al II and Si III, the lowest ^1D state is $3p^2$ and again in a CI calculation one must use Mac-Donald's theorem to determine the bounded energy levels. Moreover, in such cases, the wave function is a strong mixture of several configurations. If we define the *spectral purity* of a state to be the square of the largest coefficient in the CI expansion (169), then such states are spectrally not very pure.

It seems, therefore, that many theoretical energy level calculations will have to involve CI, containing at least the close-lying states, and that the radial functions, at least those describing the outer electron of excited states, will have to be different for different terms of the same configuration, if there is to be any reasonable agreement between theory and experiment.

6.1.2. Pair Energies

The philosophy of some of the methods described in Section 5 was to treat the correlation of N electrons of an atom by looking at all possible pairs separately. It was hoped that the corrections to energies etc. due to each pair could simply be added to give an accurate estimate of the overall correction. This "sum-of-pairs" scheme could, of course, be augmented by including higher-order terms in the complete wave function expansion, but this adds greatly to the necessary computational effort.

Some idea of the reliability of the sum-of-pairs approximation is given by the MBPT results of Lee *et al.*[126] They perform three separate calculations: firstly using pair excitations with symmetry-unadapted orbitals[127] in which the wave function is not an eigenfunction of $\mathbf{L}^2$ or $\mathbf{S}^2$;

secondly with symmetry-adapted orbitals giving an eigenfunction of $\mathbf{L}^2$ and $\mathbf{S}^2$; and thirdly a full CI variational calculation including all pair effects together. The first two processes essentially correspond to "orbital" and "configurational" excitation in the Bethe–Goldstone scheme.[100] Unlike these two, the third calculation gives a rigorous bound on the correlation energy. The three values for the L-shell ($n = 2$) correlation energy obtained in this way are -0.413, -0.386, -0.389, in a.u. These may be compared with the estimated limit[128] of -0.390 a.u. The overestimate introduced by the first method is due to the neglect of pair–pair inter-actions [the unlinked clusters of (202)], which reduce the value of the correlation energy. The second method above implicitly includes these interactions between all $2s$–$2p$ pairs and between all $2p$–$2p$ pairs. The full CI method also includes interactions between $2s$–$2p$ and $2p$–$2p$ pairs. These latter are thus seen not to be very important in this case. But differences do show up in the calculation of electron affinities in first-row atoms. Thus Moser and Nesbet,[129] using second-order configurational excitations in the Bethe–Goldstone scheme, obtain results quite different from full CI,[130,131] whereas the third-order orbital excitation cal-culation[132] agrees well with CI and with experiment. These last two cal-culations contain all pair–pair interactions. They differ only in genuine three-particle contributions, which are seen to be small, as expected. In part, the difficulty arises because the electron affinity is a difference in the energies of two systems with different numbers of electrons. It is difficult at the "pairs" level to treat the two systems in an equally balanced way with the customary use of a common set of radial functions to describe each system's correlation. It would appear that the accurate evaluation of elec-tron affinities requires the Bethe–Goldstone scheme to be taken to third order, or the use of CI involving correlation in all pairs. Such calculations are time-consuming and expensive, and it may be preferable to use collisional methods to obtain good results.

One must conclude that the sum-of-pairs approximation is often an efficient and accurate way of calculating correlation energies if symmetry-adapted or configurational excitations are used. But it can break down either when those pair–pair interactions that are omitted are significant, or when the orbital set used is truncated at too small a value of l. For the first-row atoms, the $l > 3$ orbitals contribute up to about 10% of the correlation energy.[133]

6.2. Oscillator Strengths

The progress made in the calculation of oscillator strengths (or tran-sition probabilities) during the past decade has been stimulated by the use of beam-foil spectroscopy to determine the lifetimes of atomic states, and in turn the calculations have stimulated further experimental work.[134] The relationship between experimentally determined lifetimes τ and theoretical transition probabilities A is given by

$$\tau_j^{-1} = \sum_i A_{ji} \tag{211}$$

where the sum extends over all lower states i accessible from j.

After such a wealth of results, it is reasonable to ask a number of questions: How can one assess the accuracy of oscillator strengths? How accurately can they be calculated? And can one obtain reasonably accurate values with a relatively simple calculation? In this section, we shall try to provide at least partial answers to these questions.

For the most accurate results, one naturally turns to helium, and for low-lying transitions, the work of Schiff *et al.*[135] using perimetric coordinates is outstanding. We give a selection of their results in Table 6. For the lowest transition, we include the acceleration result to show the extent of the agreement between the different forms. Recalling that for the exact wave functions the different forms give the same result, we see that the agreement is very good. It would be nice to be able to use the level of agreement as a guide to accuracy. But generally one cannot assume that an oscillator strength is accurate to the agreement between, say, length and velocity values. As an illustration, we present in Table 7 the length and velocity values f^l and f^v for an increasing number of terms in the wave functions. The difference between f^l and f^v with 220 terms is an order of magnitude less than the change in either f^l or f^v in going from 220 to 364 terms. Agreement between f^l and f^v is indeed a *necessary* requirement for accuracy, but it must be supported by evidence of convergence of the separate values. Ideally, one would like to be able to put rigorous bounds

Table 6. Comparison of Oscillator Strengths for Helium Sequence Transitions

Transition	Type	SPA[a]	GJK[b]	CMC[c]	HF[d]	Bounds (AW)[e]
He $1^1S \to 2^1P$	l	0.2761	0.27537	0.281	0.258	0.2761 ± 0.0014
	v	0.2761	0.27586	0.255	0.239	
	a	0.2761	0.26908			
He $2^3S \to 3^3P$	l	0.06446	0.0644	0.0559	0.0569	0.0645 ± 0.0029
	v	0.06446	0.0668	0.0500	0.0503	
He $5^1S \to 5^1P$	l	1.07	1.0869	1.046		
	v	1.1	1.0901	1.043		

[a] Schiff *et al.*[135]
[b] Green *et al.*[142]
[c] Cameron *et al.*[54]
[d] Hartree–Fock.
[e] Anderson and Weinhold.[137]

Table 7. Oscillator Strengths of $2^1S \rightarrow 2^1P$ in Helium (Schiff *et al.*[135])

Number of terms in wave functions	56	120	220	364
f^l	0.375082	0.376124	0.376354	0.376413
f^v	0.375697	0.376152	0.376358	0.376414

on the oscillator strength value. Such a scheme has been devised by Weinhold,[136] which has been applied to transitions in He-like systems.[137] These results have been valuable in ruling out certain earlier values that were outside the limits. But one can see from Table 6 that the agreement between f^l and f^v obtained by Schiff *et al.* is much closer than that between the upper and lower bounds. Indeed, one feels on the whole that the error bounds are rather conservative. The same appears to be true for the applications that have been made for the lithium[138] and beryllium[139] sequences. The latter is shown in Table 8, and most of the recent calculations that include some electron correlation fall within the bounds. There is also good agreement for Be I between theory and the recent experiment of Martinson *et al.*[140] But some earlier experiments for Be I, and the current ones quoted in Table 8 for C III and O V, fall outside the rigorous bounds. To understand this, it must be remembered that the lifetimes from which these oscillator strengths were derived using (211) are not actually measured in beam-foil spectroscopy. Rather, they appear as parameters in a least-squares fit of the beam intensity measured as a function of distance from the foil. The errors quoted are usually the corresponding standard deviations. Probably the main source of the discrepancy between theory and experiment is cascade effects from higher

Table 8. The Resonance Transition in the Beryllium Sequence

Calculation	Type	Be I	C III	O V
HF[144]	l	1.68	1.12	0.743
	v	1.03	0.598	0.308
NCMET[144]	l	1.25	0.760	0.513
	v	1.14	0.780	0.525
CI[145]	l	1.37	0.776	0.518
	v	1.38	0.783	0.527
Bounds[139]	lower	1.248	0.706	0.464
	l	1.344	0.765	0.515
	upper	1.439	0.822	0.556
Experiment		1.34 ± 0.05^a	0.67^b	0.42 ± 0.05^c

^{a} Martinson *et al.*[140]
^{b} Poulizac and Buchet.[146]
^{c} Martinson *et al.*[147]

levels. This generally results in an apparent lengthening of the lifetime, and hence a smaller oscillator strength. The influence of cascades has recently been discussed by Muhlethaler and Nussbaumer.[141]

It would appear therefore that rigorous bounds can be useful in assessing the accuracy both of theory and of experiment. But the bounds obtained for the beryllium sequence and presented in Table 8 required very accurate wave functions for both states in the transition, involving a combination of CI and the use of interelectronic coordinates (in effect, an example of the many-electron theory discussed in the last section). When somewhat less accurate but nonetheless very respectable wave functions were used, the bounds were so wide as to be of little use, mainly because of the relatively poor estimate of the overlap of the approximate and exact wave functions. The extension of these bounds methods to larger systems in such a way that useful bounds are obtained does not at present seem a practical proposition. One must therefore resort to an attempt to obtain well-converged oscillator strengths by examining the improvement as the accuracy of the wave functions is increased.

The great danger in assessing such a process for convergence is the risk of omitting some important contribution from the wave function, and thus reaching a spurious limit. Such an omission may result in a disagreement between f^l and f^v, thus warning of the error. But it might not. Since nearly all calculations on systems with more than four electrons have used CI (even if under another name), one is concerned that no important configurations are omitted. As a guide to which configurations *are* important, Westhaus and Sinanoglu[148] argue that accurate oscillator strengths require an accurate distribution of charge, and that this is determined by internal and semi-internal correlation (see Section 5) in *both* states. It is this somewhat simplified form of NCMET that has been used in their calculations. It has the obvious advantage that, assuming sufficient flexibility in the possible form of the radial functions, only finite CI is necessary.

One begins then with HF, which in Table 6 results in oscillator strengths lying will outside the rigorous bounds, and in Table 8 in length and velocity calculations differing by a factor of 2. When internal and semi-internal effects are included (Table 8) the agreement is much better, as is the agreement between NCMET and the CI results which include all-external correlation as well. In fact the accuracy of NCMET is not as good for neutral beryllium as for the isoelectronic ions, but the improvement over HF is still considerable. The cause of the discrepancy in Be I is the configuration $1s^2 3p3d$ ^1P of the upper state, which has a large coefficient in the CI expansion but is omitted in the NCMET work.

Neutral and singly ionized atoms are, in fact, particularly difficult from a theoretical point of view, partly because of important all-external effects as seen above, and partly because of large cancellation effects either in the transition integral (e.g., the $3s \rightarrow 4p$ transition in Mg II) or as a result of strong configuration mixing (for example, the $2s2p \rightarrow {}^1P \rightarrow 1p^2 \ {}^1D$ transition of Be I, where the upper state is a strong mixture of $2p^2$ and $2s3d$; indeed, because of the degeneracy of these ${}^1D$ states, the $3d$ function might more properly be included in the HF sea [149]). Because of such cancellations, all-external effects that may normally be small become important when compared with the already small NCMET value. As a result, the latter can be somewhat inaccurate—even an order of magnitude, as in the $2s2p \ {}^1P \rightarrow 2p^2 \ {}^1D$ transition in Be I.[145] A factor of 2 is not unusual.[150] For such cases, all-external effects must be included. A practical extension of NCMET to do this has been proposed by Beck and Nicolaides.[149] They express the oscillator strengths in terms of perturbation expansions of both states, and include all configurations in the first-order functions which interact directly through the Hamiltonian or dipole operators with the zero-order states. Thus the NCMET *calculations* form a subset of those that use this first-order perturbation method (FOTOS), although FOTOS is itself a subset of the full NCMET *theory*. In FOTOS, certain all-external effects are included, in addition to internal and semi-internal correlation. Just one example must suffice to show the improvement obtained by FOTOS. For the $2s^2 2p^3 \ {}^4S \rightarrow 2s2p^4 \ {}^4P$ transition in neutral nitrogen, FOTOS gives a length value of 0.084,[149] in excellent agreement with experiment. Conventional NCMET gives 0.036,[205] although the velocity value is 0.080. The poor quality of the earlier NCMET result of 0.286[151] is attributed by Luken and Sinanoglu[143] to variational collapse of the upper state wave function, which is not the lowest of its symmetry class. In such cases, significant changes[151] from the correct form of the wave function[205] occur unless use is made of MacDonald's theorem (127) in choosing both the parameters defining the radial parts of the orbitals and the eigenvector specifying the CI expansion coefficients.

We see therefore that for transitions in many neutral atoms between low-lying levels, a fairly accurate account of electron correlation must be taken to ensure adequate convergence of the oscillator strengths. For transitions between more highly excited states, such large-scale calculations may not be necessary. In Table 6, we have included the $1s5s \ {}^1S \rightarrow 1s5p \ {}^1P$ transition in neutral helium. The very simple frozen-core HF calculation[143] gives f^l and f^v in good agreement with each other and with a more extensive CI calculation.[142] By contrast with lower transitions the perimetric coordinate calculation of Schiff et al.[135] does not seem to be quite so good. Perhaps this is not surprising, since the $n = 5$ electron is radially much further out than the $n = 1$ electron, so that the motion of the electrons is already well correlated in HF (and the frozen-core treatment is then a good representation of HF), and the use of one-electron functions in

CI is well suited to obtaining the remaining corrections. The use of inter-electronic coordinates is more suited to treating short-range interactions amongst electrons occupying the same regions of space. Thus where the electron undergoing the transition is well separated from the others (e.g., between highly excited states, or in alkalis), one might expect the HF approximation to be reasonably good.

To summarize this discussion of accuracy, the level of accuracy that can be achieved diminishes as the number of electrons increases. For two-electron atoms, accuracy of better than 1% is possible. For atoms with several electrons, perhaps 5% is a more realistic estimate of the best accuracy achievable, though isolated examples[139] have reduced this to 2%. In transitions where large cancellations occur, however, it ceases to be meaningful to talk about percentage error. (For example, in the $2s2p\ ^1P \to 2p^2\ ^1D$ transition in Be I, the inclusion of all-external effects reduces the oscillator strength by a factor of 10 to less than 10^{-3}; in the $3s\ ^2S \to 4p\ ^2P$ transition in Mg II, the MCHF value is three times that of HF.[152]) For transitions in atoms with ≥ 10 electrons, error estimates of 10%–25% are not uncommon. A further very useful test of accuracy is the variation of f against Z (or Z^{-1}), which, if no cancellations occur, should be smooth (see Chapter 25 of this work).

As a rough guide, HF calculations might be expected to be satisfactory within about 10% for transitions in which the electron involved is well separated from the others. Otherwise, some form of correlated wave function is needed. For systems with $(Z - N) \geq 2$, the NCMET scheme incorporating internal and semi-internal correlation should be adequate to a similar level. So also should the Z^{-1}-expansion treatment used by Laughlin and Dalgarno.[106] But for neutral or lowly ionized systems, or for transitions in other systems where large cancellations occur either in the transition integral or because of CI, then a calculation involving all-external correlation is essential. The choice of possible configurations is then very wide indeed, but a prescription for choosing them such as that implied by FOTOS appears very satisfactory.

Another guideline is, of course, the level of agreement between f^l and f^v. If they do not agree, at least one of them is wrong. (If they do agree, particularly in HF, they might both be wrong.[148]) There has been considerable discussion in the literature on which of the two is better, in such circumstances. Unfortunately, two slightly different questions have been asked. Starace[153] and Grant,[154] who come to opposite conclusions, are really discussing which form is *consistent* with the formalism used. The outcome depends on how that formalism is set up.[155] In a coordinate representation (cf. Section 2.3) the velocity form is deduced from the theory.[154] In a momentum representation, one obtains the length form.[153] Essentially it depends on which gauge is being used in a relativistic treatment. But the answer to this question is not necessarily the same as the answer to the question asked by *users* of oscillator strengths: Which form is closer to the correct result? Several rules of thumb have been suggested in the past.[156] For example, length values have been preferred for transitions in which the electron's principal quantum number (assuming a single configuration state designation) does not change. For others, the velocity form may be preferred if variational wave functions have been used, since the Hamiltonian operator emphasizes roughly the same region of space as the velocity operator. The author's work on transitions in the beryllium sequence[145] frequently bears out these rules. But there are exceptions. For example, for the beryllium sequence transition $2s3s\ ^3S \rightarrow 2s3p\ ^3P$, for which the above rules would prefer the length form, it is the *velocity* form in the HF approximation that agrees extremely well with the final CI values, whereas the HF length value is typically 15% too high. The author agrees with Beck and Nicolaides[157] that it is not possible, *a priori*, to prefer one form or the other, and it is preferable to proceed to a point where at least length and velocity forms agree to a satisfactory extent.

We have written this discussion with specific reference to allowed electric-dipole transitions. Forbidden transitions have also been considered using similar methods,[1] and although the number of these is not so extensive, one might expect similar conclusions to apply. Perhaps the main exception is the f^l versus f^v discussion, since the two operators must be treated rather differently, with only the velocity form sometimes requiring a specific relativistic correction term.[158]

6.3. Hyperfine Structure

The development of the independent particle extensions of HF, discussed in Section 5, was stimulated in part by a wish to obtain improved values for hyperfine-structure splittings, in particular that of the lithium ground state. Only the Fermi contact term (66) contributes to this splitting, the other operators having zero matrix elements between $L = 0$ states. The HF value for a_c is 2.095 a.u.[159] compared with the experimental value of 2.906 a.u. [160] The use of SPHF, although the wave function is not an eigenfunction of $\mathbf{S}^2$, leads to the much improved value of 2.825 a.u.[159] The SEHF method,[81] which *does* use a spin eigenfunction, gives the value 3.020 a.u.[161] even though the energy is scarcely better than HF. A different spin eigenfunction, corresponding to the G1 method,[83] does improve the energy but gives a relatively poor a_c value of 2.58 a.u.[79] The SOSCF method,[84] which allows the weight of the two independent spin eigenfunctions to be determined variationally, gives the value 2.85 a.u.[79] This is rather a spread of results, but one thing is clear: A good value for a_c is obtained only when the spin eigenfunction Ψ_2 of (163) is included in the wave function. In CI language, Ψ_2 corresponds to a triplet spin core: $(1s1s')^3S\,2s$. With this condition satisfied, the independent particle models do fairly well—the accuracy of about 3%–4% being almost as good as that of the more elaborate correlation methods.[1] The explanation of this success lies in the fact that the contact term involves a spin-dependent operator and a delta function, so that a scheme incorporating the removal of the spin restrictions in HF, coupled with a revariation of the s-type orbitals (the only ones that are nonzero at the origin) should perform fairly well.

A similar consideration should apply to the excited states of lithium, although a comparison between the independent particle methods[1] reveals that only the SOSCF scheme reproduces the correct a_c for the $1s^2 2p\,^2P$ state. Since this state has P symmetry, the other hyperfine parameters a_l, a_d, and b_q are also nonzero. But none of the models that lift only spin restrictions have any significant influence on these parameters. To obtain improvement here, one must consider at least an *orbital* polarized HF scheme, in which contributions from higher angular momenta are introduced. Again in CI language, one should include configurations such as $1s3d2p$. But such schemes only partially (50% for a_d and b_q, < 1% for a_l) improve these parameters.[87] Accurate results *have* been obtained in conventional CI,[162] Bethe–Goldstone[163] or MBPT[164] formalisms, but only after the inclusion of *two*-electron replacements of HF orbitals. This is interesting, for the hyperfine operators are one-electron operators, and since the HF configuration is very much the dominant one in a CI expansion, one would expect that the *one*-electron replacement configurations were the crucial ones. In fact, the two-electron replacements induce significant nonzero coefficients in the one-electron replacement configurations $1s^2 np$, and it is *their* interaction with $1s^2 2p$ that results in correct values. The calculation of hyperfine splittings is thus seen to involve an intricacy of small effects.

The problem becomes even more complicated as the number of electrons increases, mainly because of the increased difficulty of obtaining an accurate value for a_c. In Table 9 we compare various methods for the ground states of boron, nitrogen, and oxygen. We list the various magnetic parameters and the associated diagonal and off-diagonal coupling constants. One can see that the spin-dependent independent-particle models perform rather badly even for the contact term. Indeed, the only independent-particle treatment that approaches the correct values is a fully unrestricted HF scheme.[87] The major causes of the difficulty of obtaining a_c correctly are as follows: (a) The variational procedure is not too well adapted to an accurate evaluation of the wave function at a single point, especially at the origin; (b) the values of a_c shown in Table 9 are the differences between contributions from the $1s$ and $2s$ subshells; the

Table 9. Magnetic Hyperfine Structure for Certain First-Row Atoms[a]

	HF[b]	SEHF[c]	UHF[d]	NCMET[e]	CI[f]	BG[g]	MBPT[h]	Expt.
^{11}B a_c	0.0	0.455	0.274	0.052	0.098	0.048		
a_l	0.7755	0.7943	0.7796	0.7542	0.7752	0.7789		
a_d	−0.1551	−0.1585	−0.1710	−0.1633	−0.1670	−0.1674		
$A_{1/2}$	353.9	344.6	362.6	357.1	363.6	366.8		366.1[i]
$A_{3/2}$	70.7	89.7	79.8	69.7	73.0	71.5		73.35[j]
$A_{3/2,1/2}$	22.0	5.3	9.6	18.0	16.6	18.7		16.44[i]
^{15}N a_c	0.0	1.984		0.897	1.152	1.271	1.233	
$A_{3/2}$	0.0	−23.8		−10.8	−13.8	−15.3	−14.8	−14.65[k]
^{17}O a_c	0.0	2.685	2.438	0.789	1.314		1.510	
a_l	4.973	4.279	4.565	4.570	4.760		4.563	
a_d	0.995	0.929	1.001	1.020	1.038		1.034	
A_2	−215.6	−220.6	−230.6	−211.5	−225.3		−220.1	−218.6[l]
A_1	−0.2	−18.9	−13.2	9.64	−0.05		4.1	4.74[l]
$A_{2,1}$	−145.4	−109.4	−120.1	−132.5	−133.4		−127.8	−126.6[l]
$A_{1,0}$	−146.6	−62.4	−73.4	−103.5	−101.6		−86.7	−91.7[l]

[a] Units: a_c, a_l, a_d in atomic units; A_J and $A_{J,J'}$ in MHz.
[b] McLean and Yoshimine.[165]
[c] Goddard.[166]
[d] Hay and Goddard.[215]
[e] Schaefer et al.[168]
[f] Glass.[169]
[g] Bethe–Goldstone; B: Nesbet;[170], N: Nesbet.[171]
[h] N: Dutta et al.[172]; O: Kelly.[173]
[i] Lew and Title.[174]
[j] Harvey et al.[175]
[k] Holloway et al.[176]
[l] Harvey.[177]

absolute value of each is an order of magnitude greater than their difference. In an independent-particle model, one is taking the difference of two values that are already not too accurate, and one can hardly suppose that the resulting value will be accurate either. It is then necessary to include electron correlation more carefully. The use of internal and semi-internal correlation (NCMET)[168] improves the situation considerably, but with such cancellation effects occurring, all-external correlation is inevitably required. The difference between the CI[169] and Bethe–Goldstone[170,171] results probably reflects the different basis sets used for the radial functions: both calculations are of comparable order. The MBPT calculations[172,173] appear to have the edge as regards accuracy. This may be due to the ability of the MBPT formalism to eliminate certain cancellation effects by prior summing over Feynman diagrams, but it may simply be that a larger radial basis set was used, resulting in greater convergence, particularly with respect to l.

There is another reason for the unsuitability of the independent-particle models. In HF and its various spin-unrestricted modifications, it is possible to write a_l, a_d, and b_q in terms of the single radial integral $\langle P_{nl} | r^{-3} | P_{nl} \rangle$ with $nl = 2p$ for first-row atoms. Thus the hyperfine splitting is effectively expressed in terms of two parameters—this integral and a_c. The inadequacy of such a representation was shown by Harvey,[177] and indeed for the methods including more correlation (CI, MBPT), a_l, a_d, and b_q are no longer simply related to a single integral. Thus for the magnetic splitting, three parameters—a_c, a_l, a_d—are necessary.

Many-body perturbation theory has also been used with success in a small number of larger atoms. Thus, for the ground-state ^4S of phosphorus, Dutta et al.[178] obtain the value $A_{3/2} = 49.8$ MHz, in excellent agreement with the experimental value of 55.1 MHz.[179] The HF value is, of course, zero since the only open subshell has p-symmetry, and being an S state the hyperfine splitting is given entirely by the contact term. But it is significant that the SPHF results have the wrong sign, with values ranging from -71 MHz to -107 MHz.[180] For larger atoms still, a proper relativistic treatment is more appropriate. Frequently such treatments extend HF only as far as core polarization. There is no reason to suppose that a relativistic core polarization approach will be much better than the nonrelativistic counterpart.[181] The example of phosphorus does not therefore allow for great confidence in relativistic core polarization corrections, and it would seem that a more elaborate relativistic treatment is essential.

7. Autoionization

Autoionizing or resonance states which occur as intermediate "bound" states in a collision process are more usually treated by the methods of scattering theory, which are outside the scope of the present article. Since these states lie in the continuum, for many of them there is an infinite number of lower-lying states of the same symmetry, and the normal variational principle, which requires orthogonality to lower-lying states, is impossible to apply. However, some autoionizing states (e.g., $1s2s2p\ ^4P^0$ of lithium) *are* the lowest of their symmetry class and the variational principle applies directly, so that they may be treated like bound states. The purpose of this short section is to consider ways of treating, by bound state methods, those autoionizing states that are *not* the lowest of their symmetry class.

We shall summarize briefly the Feshbach[182,183] formalism. The Schrödinger equation for an N-electron atom plus a further electron may be written

$$(H_{N+1} - E)\Psi(\mathbf{X}_1, \mathbf{X}_2, \ldots, \mathbf{X}_{N+1}) = 0 \tag{212}$$

where $\{\mathbf{X}_i\}$ denote the space and spin coordinates of the electrons. The $(N+1)$-electron Hilbert space is then partitioned by means of the Feshbach projection operators P and Q, where

$$P + Q = 1, \qquad PQ = QP = 0 \tag{213}$$

Then (212) becomes

$$P(H_{N+1} - E)(P + Q)\Psi = 0$$
$$Q(H_{N+1} - E)(P + Q)\Psi = 0 \tag{214}$$

whence

$$P\left[H_{N+1} + PH_{N+1}Q\frac{1}{Q(E - H_{N+1})Q}QH_{N+1}P - E\right]P\Psi = 0 \tag{215}$$

The operators are chosen so that P projects Ψ onto open channels and Q projects Ψ onto closed channels. We now introduce the set of eigenfunctions $\{\Phi_n\}$ satisfying

$$QH_{N+1}Q\Phi_n = \mathscr{E}_n\Phi_n \tag{216}$$

The resonance energy E_n is given by[184]

$$E_n = \mathscr{E}_n + \Delta_n \tag{217}$$

where the shift Δ_n is caused by the coupling to the continuum. The eigenvalues of $QH_{N+1}Q$ below the appropriate excitation threshold are *discrete*. When the width of the resonance is small, Δ_n is small so that $\mathscr{E}_n$ is a good approximation to E_n. Moreover, since Q is a projection operator,

$Q^2 = Q$ so that we may write

$$\langle Q\Phi_n | H_{N+1} | Q\Phi_n \rangle = \langle \Phi_n | QH_{N+1}Q | \Phi_n \rangle$$
$$= \langle \Phi_n | Q^3 H_{N+1} Q | \Phi_n \rangle$$
$$= \mathscr{E}_n \langle \Phi_n | Q^2 | \Phi_n \rangle$$
$$= \mathscr{E}_n \langle Q\Phi_n | Q\Phi_n \rangle \qquad (218)$$

Thus since $\{\mathscr{E}_n\}$ are discrete, we may use the variational principle directly if the trial wave function $\Psi^t = Q\Phi^t$ is constructed to lie in Q space; that is, if Ψ^t is orthogonalized to bound states of the N-electron target lying below threshold. Of course, these are known exactly only for one-electron systems, but it has allowed an accurate calculation of resonance positions of He and H$^-$, in which Φ^t is expressed in a Hylleraas-type expansion (119) to which the projection operator Q is then applied.[185] An alternative treatment is based on a close coupling expansion which for two-electron systems may be put in the form[186]

$$\Psi(\mathbf{X}_1, \mathbf{X}_2) = \mathscr{A} \sum_{i,j=k+1}^{M} C_{ij} v_i(\mathbf{X}_i) v_j(\mathbf{X}_2) \qquad (219)$$

where $\{v_i\}$ are hydrogenic functions, and k is the number of energetically accessible target states. It is the exclusion of the i or $j = 1, \ldots, k$ terms that leaves Ψ in Q-space and allows the use of the variational principle to determine $\{\mathscr{E}_n\}$. But if these terms *are* included

$$\Psi(\mathbf{X}_1, \mathbf{X}_2) = \mathscr{A} \sum_{i,j=1}^{M} C_{ij} v_i(\mathbf{X}_1) v_j(\mathbf{X}_2) \qquad (220)$$

then the wave function is no longer in Q-space and the application of the variational principle directly leads to energy values that include part of the resonance shift Δ_n.[184] But the bounds are not automatically destroyed, since (220) has the form of (125) so that MacDonald's theorem can be applied. If K is the number of terms in (220) that are not in (219), then the $(K + I)$th lowest root of the secular equation associated with (220) lies above the Ith lowest root associated with (219). Therefore an upper bound theorem applies even in this case, giving a means of picking out resonance energies. Alternatively, Lipsky and Russek[187] chose the roots on physical grounds. As more terms were added to (220) they found that certain roots stabilized, suggesting (but not guaranteeing) that all the important components of the autoionizing states had been included. Other formulations of this stabilization-of-roots scheme use a more general close-coup-

Table 10. Autoionizing Levels (in eV) of Helium

State	QHQ[a]	MCEB[b]	CC[c]	Experiment
Below the $n = 2$ threshold				
$^1S(1)$	57.82	57.87	57.84	57.82^d
$^1S(2)$	62.06	62.13	62.13	62.15^d
$^1S(3)$	62.95	62.99	62.98	62.95^d
$^3S(1)$	62.61		62.62	
$^1P^0(1)$	60.15		60.15	60.14^e
$^3P^0(1)$	58.29		58.32	58.34^d
Below the $n = 3$ threshold				
$^1S(1)$	69.37	69.45	69.39	
$^1S(2)$	70.41	70.42	70.39	
$^1P^0(1)$	69.88		69.92	69.94^e

^{a} $n = 2$, Bhatia et al.[190]; $n = 3$, Oberoi.[191]
^{b} $n = 2$, Holøien and Midtdal[192]; $n = 3$, Holøien and Midtdal.[193]
^{c} Close coupling; $n = 2$, Burke and Taylor[194]; $n = 3$, Burke and Taylor.[195]
^{d} Rudd.[196]
^{e} Madden and Codling.[197]

ling form in which the product of two hydrogenic functions in (220) is replaced by the product $v_i(X_1)\phi_j(X_2)$, where ϕ_j is not restricted to being hydrogenic.[188]

It might be argued that the use of (219) is preferable to (220) since (219) is already in Q space and can be used directly in the variational principle. But one may make arbitrary linear transformations amongst the hydrogenic functions which leave the form of (220) unchanged:

$$\Psi(\mathbf{X}_1, \mathbf{X}_2) = \mathcal{A} \sum_{i,j=1}^{M} C'_{ij}\phi_i(\mathbf{X}_1)\phi_j(\mathbf{X}_2) \tag{221}$$

where $\{\phi_i\}$ are no longer restricted to being hydrogenic. Since (220) and (221) are equivalent, the energy-bound theorem[189] still holds: the $(K + I)$th lowest root of the secular equation associated with (221) is an upper bound to $\mathscr{E}_I$. The results which we show in Table 10 demonstrate this bound property. Indeed, the bound-state methods compare well with the close-coupling collisional method and with experiment.

When the target atom has two or more electrons, it is no longer possible to specify target wave functions exactly, so the energy-bound theorem does not immediately extend to larger systems. Modified forms are possible in certain restricted cases.[198] But if the autoionizing state of an atom can be represented by two outer electrons well separated from the remaining core electrons, then the energy-bound theorem should be satisfied approximately. Autoionizing states of various atoms and ions have been investigated in this way.[145,199]

8. Relativistic Methods

The methods and applications described in the last few sections have been based mainly on a nonrelativistic formalism. For small atoms, such formalisms are generally very satisfactory. But for systems with many electrons, a relativistic treatment is necessary. We noted in Section 2 that the relativistic Hamiltonian for an N-electron atom could be expressed approximately by equations (55) and (56), corresponding to an expansion in terms of the fine-structure constant α:

$$H = \sum_{i=1}^{N} H_D(i) + \frac{1}{2} \sum_{i \neq j} \left\{ \frac{1}{r_{ij}} + B(i,j) \right\} + \cdots \tag{222}$$

The Breit interaction $B(i,j)$ is of order $(\alpha Z)^2$ relative to the Coulomb term, and the approximations used in obtaining $B(i,j)$ effectively omit terms of higher order in (αZ). Consequently, in relativistic calculations, the wave functions are determined by omitting the Breit interactions from (222). Their effect is then calculated by first-order perturbation theory.

The most extensive set of relativistic calculations has employed the HF approximation. As for the nonrelativistic case, the HF equations are set up by treating the total energy as a functional to be varied with respect to the radial functions of the orbitals. The angular and spin integrals may be performed using Racah algebra techniques.[13,200] If we assume that the orbitals are expressed in the form (31), the HF equations become, in atomic units [13]

$$\frac{dP_i}{dr} + \frac{k_i}{r} P_i - \left[2c + \frac{1}{c} \left(\frac{Y_i}{r} - \varepsilon_{ii} \right) \right] Q_i + X_i^Q = \sum_{j \neq i} \frac{1}{c} \varepsilon_{ij} q_j \delta_{k_i k_j} Q_j$$

$$\frac{dQ_i}{dr} - \frac{k_i}{r} Q_i + \frac{1}{c} \left(\frac{Y_i}{r} - \varepsilon_{ii} \right) P_i - X_i^P = \sum_{j \neq i} \frac{1}{c} \varepsilon_{ij} q_j \delta_{k_i k_j} P_j \tag{223}$$

where Y_i and X_i are direct and exchange operators and we have assumed the rest energy of an electron is taken as the zero of energy. A clear comparison is possible between this relativistic HF scheme and the non relativistic HF equations (149), or the relativistic one-electron equations (34). Many of the nonrelativistic results, e.g., Koopmans' theorem and the diagonalizability of the right-hand side of (223) for closed shells, apply in this situation also.

Numerical techniques for solving (223) are given by Desclaux et al.[201] Kim[202] has set up the equations for the coefficients of the basis functions when the orbitals are expressed in terms of a finite analytic basis set—this work is parallel to the nonrelativistic procedure of Roothaan.[61] It is found[202] that, unlike the nonrelativistic case, the use of a rather limited basis set is generally inadequate for relativistic calculations; the relativistic orbitals must be allowed greater flexibility of form. In addition to these conventional HF calculations, Hartree–Fock–Slater calculations have also been performed. Indeed, various forms of Slater's universal exchange potential have been tried. The standard form [cf. (155)] is

$$V_{ex}(r) = -[81\rho(r)/32\pi^2 r^2]^{1/3} \tag{224}$$

with the radial electron density given by

$$\rho(r) = \sum_i q_i[\{P_i(r)\}^2 + \{Q_i(r)\}^2] \tag{225}$$

Cowan *et al.*[203] include a factor of 2/3, which arises if the statistical approximation for the exchange terms is made before rather than after the variational process. As a generalization of this, Rosen and Lindgren[204] proposed the form

$$V_{ex}(r) = -(C/r)[81r^n\{\rho(r)\}^m/32\pi^2]^{1/3} \tag{226}$$

with C, n, m as variational parameters, and (224) corresponds to $C = n = m = 1$. As Z increases, C and n converge towards the values 2/3 and 1 for the modified Slater potential, while m remains almost unity.

Quite a number of relativistic HF calculations of transition probabilities[207–209] and hyperfine-structure effects[210,211] have been performed. In the latter case, spin polarization is also included. While this undoubtedly has a significant effect upon hyperfine splittings, it is not at all clear from corresponding nonrelativistic calculations that spin polarization is the only important correction to the HF values. We saw in Section 6 that correlation effects could be of comparable importance and would sometimes wipe out the spin polarization effect. It seems reasonable to suppose that correlation effects would generally be important in a relativistic framework, too. One important program[212] has been published recently which would allow such calculations to be made. It is based on a relativistic MCHF scheme, and follows the pattern of the nonrelativistic equivalent.[63,64] Only a limited number of calculations have yet been performed. They include a recent evaluation of relativistic oscillator strengths of resonance transitions in the lithium and beryllium isoelectronic sequences.[167] One can expect this program to be used widely in the next few years.

Acknowledgments

The author would like to thank Dr. R. Glass, Dr. M. Cornille, and Dr. L. Brillet for useful discussion, and also Dr. Brillet for allowing him to use her Si V spectrum results prior to their publication. The author is also grateful to the Centre National de la Recherche Scientifique for a fellowship during the tenure of which this report was written. It is a pleasure to thank Dr. H. van Regemorter and members of his group for their friendship during the author's stay at Meudon.

References

1. A. Hibbert, *Rep. Prog. Phys.* **38**, 1217 (1975).
2. N. Bohr, *Phil. Mag.* **26**, 1 (1913).
3. E. Schrödinger, *Ann. Phys., Leipzig* **79**, 361 (1926).
4. L. Landau and E. M. Lifschitz, *Relativistic Quantum Theory*, Pergamon Press, Oxford (1971).
5. W. Pauli, *Naturwissenschaften* **12**, 741 (1924).
6. S. M. Blinder, *Adv. Quant. Chem.* **2**, 47 (1965).
7. G. Breit, *Phys. Rev.* **72**, 984 (1947).
8. J. Schwinger, *Phys. Rev.* **73**, 416 (1948).
9. J. Schwinger, *Phys. Rev.* **76**, 790 (1949).
10. B. N. Taylor, W. H. Parker, and D. N. Langenberg, *Rev. Mod. Phys.* **41**, 375 (1969).
11. M. E. Rose, *Elementary Theory of Angular Momentum*, Wiley, New York (1957).
12. G. Breit, *Phys. Rev.* **34**, 553 (1929).
13. I. P. Grant, *Adv. Phys.* **19**, 747 (1970).
14. M. Jones, *J. Phys. B: Atom. Molec. Phys.* **3**, 1571 (1970).
15. H. A. Bethe and E. E. Salpeter, *Handbuch der Physik*, Vol. XXXV, p. 88, Springer Verlag, Berlin (1957).
16. W.-U. L. Brillet, *Phys. Scr.* **13**, 289 (1976).
17. G. K. Woodgate, *Elementary Atomic Structure*, McGraw-Hill, New York (1970).
18. J. Y. C. Chen, *J. Chem. Phys.* **40**, 615 (1964).
19. U. Fano and G. Racah, *Irreducible Tensorial Sets*, Academic Press, New York (1959).
20. M. Mitzushima, *Phys. Rev.* **134**, A883 (1964).
21. B. W. Shore and D. H. Menzel, *Principles of Atomic Spectra*, Wiley, New York (1968).
22. W. Wölfli, C. Stoller, G. Bonani, M. Suter, and M. Stöckli, *Phys. Rev. Lett.* **35**, 656 (1975).
23. H. Nussbaumer, *J. Phys. B: Atom. Molec. Phys.* **9**, 1757 (1976).
24. A. Dalgarno and A. L. Stewart, *Proc. R. Soc. London A* **238**, 269 (1956).
25. H. J. Silverstone, *J. Chem. Phys.* **54**, 2325 (1971).
26. E. A. Hylleraas, *Z. Phys.* **65**, 209 (1930).
27. C. W. Scherr and R. E. Knight, *Rev. Mod. Phys.* **35**, 436 (1963).
28. J. Midtdal, *Phys. Rev.* **138**, A1010 (1965).
29. J. Midtdal, G. Lyslo, and K. Aashamar, *Phys. Norvegica* **3**, 163 (1969).
30. W. H. Miller, *J. Chem. Phys.* **44**, 2198 (1966).
31. E. A. Hylleraas, *Z. Phys.* **54**, 347 (1929).
32. L. Wilets and I. J. Cherry, *Phys. Rev.* **103**, 112 (1956).
33. J. H. Bartlett, J. J. Gibbons, and C. G. Dunn, *Phys. Rev.* **47**, 679 (1935).
34. T. Kinoshita, *Phys. Rev.* **105**, 1490 (1957).
35. T. Kinoshita, *Phys. Rev.* **115**, 366 (1959).
36. C. L. Pekeris, *Phys. Rev.* **112**, 1649 (1958).
37. E. A. Hylleraas and B. Undheim, *Z. Phys.* **65**, 759 (1930).

38. J. K. L. MacDonald, *Phys. Rev.* **43**, 830 (1933).
39. J. F. Perkins, *J. Chem. Phys.* **45**, 2156 (1965).
40. Y. Accad, C. L. Pekeris, and B. Schiff, *Phys. Rev. A* **4**, 516 (1971).
41. S. Larsson, *Phys. Rev.* **169**, 49 (1968).
42. C. W. Scherr, J. N. Silverman, and F. A. Matsen, *Phys. Rev.* **127**, 830 (1962).
43. S. Seung and E. B. Wilson, *J. Chem. Phys.* **47**, 5343 (1967).
44. R. E. Knight, *Phys. Rev.* **183**, 45 (1969).
45. D. R. Hartree, *Proc. Cambridge Phil. Soc.* **24**, 89 (1927).
46. D. R. Hartree, *Proc. Cambridge Phil. Soc.* **24**, 111 (1927).
47. V. Fock, *Z. Phys.* **61**, 126 (1930).
48. D. R. Hartree, *The Calculation of Atomic Structures*, Wiley, New York (1957).
49. C. Froese-Fischer, *The Hartree–Fock Method for Atoms—a Numerical Approach*, Wiley Interscience, New York (1977).
50. J. C. Slater, *Quantum Theory of Atomic Structure*, Vols. 1 and 2, McGraw-Hill, New York (1960).
51. M. Cohen and P. S. Kelly, *Can. J. Phys.* **44**, 3227 (1966).
52. C. Froese, *Can. J. Phys.* **45**, 7 (1967).
53. C. S. Sharma, *J. Phys. B: Atom. Molec. Phys.* **1**, 1023 (1968).
54. S. Cameron, R. P. McEachran, and M. Cohen, *Can. J. Phys.* **48**, 211 (1970).
55. G. N. Bates and P. L. Altick, *J. Phys. B: Atom. Molec. Phys.* **6**, 653 (1973).
56. J. C. Slater, *Phys. Rev.* **81**, 385 (1951).
57. F. Herman and S. Skillman, *Atomic Structure Calculations*, Prentice-Hall, Englewood Cliffs, New Jersey (1963).
58. J. P. Desclaux, *Comp. Phys. Commun.* **1**, 216 (1969).
59. D. R. Beck and R. N. Zare, *Comp. Phys. Commun.* **1**, 113 (1969).
60. W. Eissner and H. Nussbaumer, *J. Phys. B: Atom. Molec. Phys.* **2**, 1028 (1969).
61. C. C. J. Roothaan, *Rev. Mod. Phys.* **23**, 69 (1951).
62. C. C. J. Roothaan and P. S. Bagus, *Meth. Comp. Phys.* **2**, 47 (1963).
63. C. Froese Fischer, *Comp. Phys. Commun.* **1**, 151 (1970).
64. C. Froese Fischer, *Comp. Phys. Commun.* **4**, 107 (1972).
65. E. Clementi and C. Roetti, *At. Data* **14**, 177 (1974).
66. E. Clementi, *IBM J. Res. Dev. Suppl.* **9**, 2 (1965).
67. A. W. Weiss, *Astrophys. J.* **138**, 1262 (1963).
68. H. Pfennig, R. Steele, and E. Trefftz, *J. Quant. Spectrosc. Radiat. Transfer* **5**, 335 (1965).
69. H. Tatewaki, H. Tateka, and F. Sasaki, *Int. J. Quant. Chem.* **5**, 335 (1971).
70. C. C. J. Roothaan and P. S. Kelly, *Phys. Rev.* **131**, 1177 (1963).
71. M. Synek, *Phys. Rev.* **131**, 1572 (1963).
72. J. Linderberg, *Phys. Rev.* **121**, 816 (1961).
73. M. Cohen and A. Dalgarno, *Proc. Phys. Soc.* **77**, 165 (1961).
74. M. Cohen, *Rev. Mod. Phys.* **35**, 506 (1963).
75. R. G. Wilson, *J. Phys. B: Atom. Molec. Phys.* **4**, 311 (1971).
76. P.-O. Löwdin, *Adv. Chem. Phys.* **2**, 207 (1959).
77. A. Hibbert, *Proc. Phys. Soc.* **91**, 819 (1967).
78. G. W. Pratt, *Phys. Rev.* **102**, 1303 (1956).
79. S. Lunell, *Phys. Rev.* **173**, 85 (1968).
80. P.-O. Löwdin, *Phys. Rev.* **97**, 1509 (1955).
81. U. Kaldor, *J. Chem. Phys.* **48**, 835 (1968).
82. W. A. Goddard, *J. Chem. Phys.* **48**, 450 (1968).
83. W. A. Goddard, *Phys. Rev.* **169**, 120 (1968).
84. U. Kaldor and F. E. Harris, *Phys. Rev.* **183**, 1 (1969).
85. A. J. Freeman and R. E. Watson, *Phys. Rev.* **131**, 2566 (1963).

86. R. E. Watson and A. J. Freeman, *Phys. Rev.* **131**, 250 (1963).
87. S. Larsson, *Phys. Rev. A* **2**, 1248 (1970).
88. A. Hibbert, *Comp. Phys. Commun.* **9**, 141 (1975).
89. W. Eissner, M. Jones, and H. Nussbaumer, *Comp. Phys. Commun.* **8**, 270 (1974).
90. A. W. Weiss, *Phys. Rev.* **162**, 71 (1967).
91. N. Sabelli and J. Hinze, *J. Chem. Phys.* **50**, 684 (1969).
92. A. W. Weiss, *Phys. Rev.* **122**, 1826 (1961).
93. C. Froese Fischer, *J. Comp. Phys.* **13**, 502 (1973).
94. C. F. Bunge, *Theor. Chim. Acta* **16**, 126 (1970).
95. K. Frankowski and C. L. Pekeris, *Phys. Rev.* **146**, 46 (1966).
96. P.-O. Löwdin, *Phys. Rev.* **97** 1474 (1955).
97. L. Brillouin, *J. Phys. (Paris)* **3**, 373 (1932).
98. F. Grein and T. C. Chang, *Chem. Phys. Lett.* **12**, 44 (1971).
99. J. Bauche and M. Klapisch, *J. Phys. B: Atom. Molec. Phys.* **5**, 29 (1972).
100. R. K. Nesbet, *Adv. Chem. Phys.* **14**, 1 (1969).
101. H. Bethe and J. Goldstone, *Proc. R. Soc. London A* **238**, 551 (1957).
102. K. Jankowski, *Theor. Chim. Acta* **13**, 165 (1969).
103. O. Sinanoglu, *J. Chem. Phys.* **33**, 1216 (1960).
104. H. J. Silverstone and M.-L. Yin, *J. Chem. Phys.* **49**, 2026 (1968).
105. A. Dalgarno and E. M. Parkinson, *Phys. Rev.* **176**, 73 (1968).
106. C. Laughlin and A. Dalgarno, *Phys. Rev. A* **8** 39 (1973).
107. C. D. H. Chisholm and A. Dalgarno, *Proc. R. Soc. A* **292**, 264 (1966).
108. C. D. H. Chisholm, A. Dalgarno, and F. R. Innes, *Adv. Atom. Molec. Phys.* **5**, 297 (1969).
109. K. Aashamar, G. Lyslo, and J. Midtdal, *J. Chem. Phys.* **52**, 3324 (1970).
110. A. Dalgarno and G. W. F. Drake, *Chem. Phys. Lett.* **3**, 349 (1969).
111. K. A. Brueckner, *Phys. Rev.* **97**, 1353 (1955).
112. K. A. Brueckner, *Phys. Rev.* **100**, 36 (1955).
113. J. Goldstone, *Proc. R. Soc. London A* **239**, 267 (1957).
114. H. P. Kelly, *Adv. Chem. Phys.* **14**, 129 (1969).
115. N. H. March, W. H. Young, and S. Sampanthar, *The Many-Body Problem in Quantum Mechanics*, Cambridge University Press, Cambridge (1967).
116. R. K. Nesbet, *Phys. Rev.* **175**, 2 (1968).
117. O. Sinanoglu, *J. Chem. Phys.* **36**, 706 (1962).
118. O. Sinanoglu and D. F. Tuan, *J. Chem. Phys.* **38**, 1740 (1963).
119. I. Oksutz and O. Sinanoglu, *Phys. Rev.* **181**, 42 (1969).
120. L. Szasz and G. McGinn, *J. Chem. Phys.* **42**, 2363 (1965).
121. L. Szasz and G. McGinn, *J. Chem. Phys.* **56**, 1019 (1972).
122. J. C. Weisheit and A. Dalgarno, *Chem. Phys. Lett.* **9**, 517 (1971).
123. G. A. Victor and C. Laughlin, *Chem. Phys. Lett.* **14**, 74 (1972).
124. Z. B. Goldschmidt and J. Starkand, *J. Phys. B: Atom. Molec. Phys.* **3**, L141 (1970).
125. R. N. Zare, *J. Chem. Phys.* **45**, 1966 (1966).
126. T. Lee, N. C. Dutta, and T. P. Das, *Phys. Rev. A* **4**, 1410 (1971).
127. R. K. Nesbet, *Proc. R. Soc. A* **230**, 312 (1955).
128. E. Clementi and A. Veillard, *J. Chem. Phys.* **44**, 3050 (1966).
129. C. M. Moser and R. K. Nesbet, *Phys. Rev. A* **6**, 1710 (1972).
130. A. W. Weiss, *Phys. Rev. A* **3**, 126 (1971).
131. M. A. Marchetti, M. Krauss, and A. W. Weiss, *Phys. Rev. A* **5**, 2387 (1972).
132. C. M. Moser and R. K. Nesbet, *Phys. Rev. A* **4**, 1336 (1971).
133. F. Sasaki and M. Yoshimine, *Phys. Rev. A* **9**, 17 (1974).
134. I. A. Sellin and D. J. Pegg, eds., *Beam-Foil Spectroscopy*, Vols. 1 and 2, Plenum Press, New York (1976).

135. B. Schiff, C. L. Pekeris, and Y. Acad, *Phys. Rev. A* **4**, 885 (1971).
136. F. Weinhold, *J. Chem. Phys.* **54**, 1874 (1971).
137. M. T. Anderson and F. Weinhold, *Phys. Rev. A* **9**, 118 (1974).
138. J. S. Sims, S. A. Hagstrom, and J. J. Rumble, *Phys. Rev. A.* **13**, 242 (1976).
139. J. S. Sims and R. C. Whitten, *Phys. Rev. A* **8**, 2220 (1973).
140. I. Martinson, A. Gaupp, and L. J. Curtis, *J. Phys. B: Atom. Molec. Phys.* **7**, L463 (1974).
141. H. P. Muhlethaler and H. Nussbaumer, *Astron. Astrophys.* **48**, 109 (1976).
142. L. C. Green, N. C. Johnson, and E. K. Kolchin, *Astrophys. J.* **144**, 369 (1966).
143. W. L. Luken and O. Sinanoglu, *Phys. Rev. A* **13**, 1293 (1976).
144. C. A. Nicolaides, D. R. Beck, and O. Sinanoglu, *J. Phys. B: Atom. Molec. Phys.* **6** 62 (1973).
145. A. Hibbert, *J. Phys. B: Atom. Molec. Phys.* **7**, 1417 (1974).
146. M. C. Poulizac and J. P. Buchet, *Phys. Scr.* **4**, 191 (1971).
147. I. Martinson, H. G. Berry, W. S. Bickel, and H. Oona, *J. Opt. Soc. Am.* **61**, 519 (1971).
148. P. Westhaus and O. Sinanoglu, *Phys. Rev.* **183**, 56 (1969).
149. D. R. Beck and C. A. Nicolaides, *Chem. Phys. Lett.* **36**, 79 (1975).
150. A. Hibbert, in *Beam-Foil Spectroscopy* (I. A. Sellin and D. J. Pegg, eds.), Vol. 1, p. 29, Plenum Press, New York (1976).
151. C. A. Nicolaides, *Chem. Phys. Lett.* **21**, 246 (1973).
152. C. Froese Fischer, in *Beam-Foil Spectroscopy* (I. A. Sellin and D. J. Pegg, eds.), Vol. 1, p. 69, Plenum Press, New York (1976).
153. A. F. Starace, *Phys. Rev. A* **3**, 1242 (1971).
154. I. P. Grant, *J. Phys. B: Atom. Molec. Phys.* **7**, 1458 (1974).
155. I. P. Grant and A. F. Starace, *J. Phys. B: Atom. Molec. Phys.* **8**, 1999 (1975).
156. R. J. S. Crossley, *Adv. Atom. Molec. Phys.* **5**, 237 (1969).
157. C. A. Nicolaides and D. R. Beck, *Chem. Phys. Lett.* **35**, 202 (1975).
158. G. W. F. Drake, *J. Phys. B: Atom. Molec. Phys.* **9**, L169 (1976).
159. L. M. Sachs, *Phys. Rev.* **117**, 1504 (1960).
160. R. G. Schlecht and D. W. McColm, *Phys. Rev.* **142**, 11 (1966).
161. W. A. Goddard, *Phys. Rev.* **157**, 93 (1967).
162. R. Glass and A. Hibbert, *J. Phys. B: Atom. Molec. Phys.* **9**, 875 (1976).
163. R. K. Nesbet, *Phys. Rev. A* **2**, 661 (1970).
164. S. Garpman, I. Lindgren, J. Lindgren, and J. Morrison, *Phys. Rev. A* **11**, 758 (1975).
165. A. D. McLean and M. Yoshimine, *IBM J. Res. Dev. Suppl.* **12**, 206 (1968).
166. W. A Goddard, *Phys. Rev.* **182**, 48 (1969).
167. Y.-K. Kim and J. P. Desclaux, *Phys. Rev. Lett.* **36**, 139 (1976).
168. H. F. Schaefer, R. A. Klemm, and F. E. Harris, *Phys. Rev.* **181** 137 (1969).
169. R. Glass, Ph.D. thesis, Queen's University of Belfast (1974).
170. R. K. Nesbet, *Phys. Rev. A* **2**, 1208 (1970).
171. R. K. Nesbet, in *Quantum Theory of Atoms, Molecules and the Solid State* (P.-O. Löwdin, ed.), p. 157 Academic Press, New York (1966).
172. N. C. Dutta, C. Matsubara, R. T. Pu, and T. P. Das, *Phys. Rev.* **177**, 33 (1969).
173. H. P. Kelly, *Phys. Rev.* **180**, 55 (1969).
174. H. Lew and R. S. Title, *Can. J. Phys.* **38**, 868 (1960).
175. J. S. M. Harvey, L. Evans, and H. Lew, *Can. J. Phys.* **50**, 1719 (1972).
176. W. W. Holloway, E. Lüscher, and R. Novick, *Phys. Rev.* **126**, 2109 (1962).
177. J. S. M. Harvey, *Proc. R. Soc. London A* **285**, 581 (1965).
178. N. C. Dutta, C. Matsubara, R. T. Pu, and T. P. Das, *Phys. Rev. Lett.* **21**, 1139 (1968).
179. J. M. Pendlebury and K. F. Smith, *Proc. Phys. Soc.* **84**, 849 (1964).
180. N. Bessis, H. Lefebvre-Brion, C. M. Moser, A. J. Freeman, R. K. Nesbet, and R. E. Watson, *Phys. Rev.* **135**, A588 (1964).

181. J. P. Desclaux and N. Bessis, *Phys. Rev. A* **2**, 1623 (1970).
182. H. Feshbach, *Ann. Phys. N.Y.* **5**, 357 (1958).
183. H. Feshbach, *Ann. Phys. N.Y.* **19**, 287 (1962).
184. P. G. Burke, *Adv. Atom. Molec. Phys.* **4**, 173 (1968).
185. A. K. Bhatia, P. G. Burke, and A. Temkin, *Phys. Rev. A* **8**, 21 (1973).
186. P. L. Altick and E. N. Moore, *Phys. Rev. Lett.* **15**, 100 (1965).
187. L. Lipsky and A. Russek, *Phys. Rev.* **142**, 59 (1966).
188. M. F. Fels and A. U. Hazi, *Phys. Rev. A* **5**, 1236 (1972).
189. J. F. Perkins, *Phys. Rev.* **178**, 89 (1969).
190. A. K. Bhatia, A. Temkin, and J. F. Perkins, *Phys. Rev.* **153**, 177 (1967).
191. R. S. Oberoi, *J. Phys. B: Atom. Molec. Phys.* **5**, 1120 (1972).
192. E. Holøien and J. Midtdal, *J. Phys. B: Atom. Molec. Phys.* **3**, 592 (1970).
193. E. Holøien and J. Midtdal, *J. Phys. B: Atom. Molec. Phys.* **4**, 32 (1971).
194. P. G. Burke and A. J. Taylor, *Proc. Phys. Soc.* **88**, 549 (1966).
195. P. G. Burke and A. J. Taylor, *J. Phys. B: Atom. Molec. Phys.* **2**, 44 (1969).
196. M. E. Rudd, *Phys. Rev. Lett.* **15**, 580 (1965).
197. R. P. Madden and K. Codling, *Astrophys. J.* **141**, 364 (1965).
198. J. F. Perkins, *Phys. Rev. A* **4**, 489 (1971).
199. J. J. Matese, S. P. Rountree, and R. J. W. Henry, *Phys. Rev. A* **8**, 2965 (1973).
200. I. P. Grant, *Proc. R. Soc. London A* **262**, 555 (1961).
201. J. P. Desclaux, D. F. Mayers, and F. O'Brien, *J. Phys. B: Atom. Molec. Phys.* **4**, 631 (1971).
202. Y.-K. Kim, *Phys. Rev.* **154**, 17 (1967).
203. R. D. Cowan, A. C. Larson, D. Liberman, and J. T. Waber, *Phys. Rev.* **144**, 5 (1966).
204. A. Rosen and I. Lindgren, *Phys. Rev.* **176**, 114 (1968).
205. W. L. Luken and O. Sinanoglu, *J. Chem. Phys.* **64**, 3141 (1976).
206. O. Sinanoglu and W. Luken, in *Computers in Chemical Research and Education* (D. Hadzi, ed.), Elsevier, Dordrecht (1973).
207. L. Holmgren and S. Garpman, *Phys. Scr.* **10**, 215 (1974).
208. S. Garpman, L. Holmgren, and A. Rosen, *Phys. Scr.* **10**, 221 (1974).
209. L. Holmgren, *Phys. Scr.* **11**, 15 (1975).
210. L. Holmgren and A. Rosen, *Phys. Scr.* **10**, 171 (1974).
211. L. Holmgren, *Phys. Scr.* **12**, 119 (1975).
212. J. P. Desclaux, *Comp. Phys. Commun.* **9**, 31 (1975).
213. J. E. Hansen, *J. Phys. B: Atom Molec. Phys.* **6**, 1387 (1973).
214. J. E. Hansen, *J. Phys. B: Atom. Molec. Phys.* **6**, 1751 (1973).
215. P. J. Hay and W. A. Goddard, *Chem. Phys. Lett.* **9**, 356 (1971).

2

Density Matrix Formalism and Applications in Spectroscopy

K. BLUM

1. Introduction

Quantum mechanics has been mostly concerned with those states of sytems that are represented by state vectors. It often happens, however, that our knowledge of the system is incomplete. For example, we may be able to say no more than that the system has a certain probability of being in the dynamical state characterized by a state vector. Because of this incomplete knowledge, a need for statistical averaging arises in the same sense as in classical physics. The evaluation of averages and probabilities of the physical quantities characterizing the system might become extremely cumbersome. Representing these states by density matrices allows us to avoid the introduction of unnecessary variables, which is of particular importance in many-body theory. Furthermore, the use of density matrix methods has the advantage of providing a uniform treatment of all quantum mechanical systems, whether they are completely or incompletely known.

This paper is intended as an introduction to the increasingly important methods of the density matrix formulation. It is not an exhaustive treatment of the subject, rather, it represents an elementary but reasonably self-contained account of the underlying quantum mechanical principles and the basic mathematical techniques.

We start in Section 2 with a detailed characterization of pure and mixed quantum mechanical states. The density matrix is then introduced and its basic properties derived. The polarization density matrix of photons is discussed as an example. In Section 3 we examine briefly the changes of

---

K. BLUM • Institute of Atomic Physics, University of Stirling, Stirling, Scotland.

quantum mechanical states resulting from an interaction with another (observed or unobserved) system. Then we construct the density matrix for interacting systems and the density matrix representing the information on one subsystem. Finally we discuss an example. The arguments presented here may be helpful in an analysis of experiments.

More advanced topics are discussed in Sections 4 and 5. Quantum mechanical calculations for systems having symmetry can be divided into two parts. One part consists of deriving as much information as possible from the symmetry requirements. The other part consists of dynamical calculations for which no information can be obtained from symmetry considerations. Often these two parts are tangled. The irreducible tensor method is designed to separate dynamical and geometrical elements and to provide a well-developed and efficient way to make use of the symmetry. Sections 4 and 5 give an introduction to this technique with various examples. In Section 4 we derive the basic properties of irreducible tensor operators and we discuss how the information on a given system can be expressed conveniently in terms of mean values of these operators. Finally, we show how the number of independent multipole parameters is often restricted by symmetry requirements.

The time evolution of the density matrix is treated in Section 5. The quantum mechanical description of time evolution is briefly reviewed in Section 5.1 and the basic equation of motion of the density matrix is derived. In Section 5.2 we present the theory of radiation from "polarized" atomic ensembles. Expressions are derived for the density matrix elements characterizing the polarization state of the emitted photons. The formulation given here disentangles dynamical and geometrical elements. The connection between polarization and angular distribution of the emitted radiation and source parameters (orientation and alignment) is stressed. The formulas obtained are useful in the theory of lifetime measurements, quantum beat calculations, and angular correlation theory. In Section 5.3 we consider the time evolution of density matrix and state multipoles in the presence of external or internal fields. We discuss the angular distribution of radiation emitted from such "perturbed" atomic sources. Hyperfine interaction is discussed as an example. Finally, in Section 5.4, we give a brief introduction to the Liouville formalism, which allows us to treat perturbed angular distributions and correlations in an elegant way.

Theory and applications of the density matrix have been well summarized by various authors.[1-5] Some textbooks on quantum mechanics outline the formalism,[6-8] and for applications in statistical physics we cite Ref. 9. Because of the introductory nature of this paper, we refer in general to reviews of the subject. We concentrate in this paper on the formal aspects of the theory; various applications and discussions can be found in the other papers of this volume.

2. Basic Theory

2.1. Pure and Mixed Quantum Mechanical States

In classical mechanics the dynamical state of a system of particles is completely determined once we have given the values of all positions and momenta. For all subsequent times the state of the system can then be predicted with certainty. But often only averages of positions and momenta have been given. Because of this incomplete information, the methods of statistical mechanics must be applied. In this paper we are concerned with quantum mechanical systems about which we have not obtained the maximum possible information. The dynamical states of these systems are conveniently described in terms of density matrices. However, the phrase "maximum possible information" has in quantum mechanics a more restricted meaning than in classical physics. As is well known, not all physical observables can be measured simultaneously with precision, for example, not position and momentum. Our first task therefore is to discuss the meaning of "maximum information" in quantum mechanics.

As is well known, a simultaneous precise measurement of two phsyical variables is only possible if the two corresponding operators commute. If two operators $\hat{A}_1$, $\hat{A}_2$ commute we can find states in which $\hat{A}_1$ and $\hat{A}_2$ have definite values a_1, a_2. If we can find a third operator $\hat{A}_3$ commuting with both $\hat{A}_1$ and $\hat{A}_2$, we find states in which $\hat{A}_1, \hat{A}_2, \hat{A}_3$ have simultaneously definite values a_1, a_2, a_3, and so on. The simultaneous eigenvalues $a_1, a_2, a_3, \ldots$ may thus be used to give an increasingly precise classification of the state of the system. (The classification of states in terms of constants of the motion, that is, in terms of operators commuting with the Hamiltonian, is an example.) The largest set of mutually commuting independent operators $\hat{A}_1, \hat{A}_2, \ldots$ that can be found will give the most complete characterization possible. The measurement of another variable (corresponding to an operator not commuting with the sets $\hat{A}_1, \hat{A}_2, \ldots$) necessarily introduces uncertainty into at least one of those already measured. A sharper specification of the system is therefore not possible. We have obtained the "maximum possible knowledge" about the system.

Thus, in general, maximum information about a system (in the quantum mechanical sense) is obtained if the eigenvalues $a_1, \ldots, a_n$ of a complete, commuting set of operators have been measured ("complete measurement"). After the measurements have been performed, one can be sure that the state of the system is precisely the corresponding eigenstate of the set $\hat{A}_1, \ldots, \hat{A}_n$ associated with the measured eigenvalues $a_1, \ldots, a_n$. It is completely specified by assigning the eigenstate $|\psi_a\rangle = |a_1, \ldots, a_n\rangle$ to it (using Dirac's notation). If one *repeats* the measurement of the quantities $a_1, \ldots, a_n$ on the system $|a_1, \ldots, a_n\rangle$ one can be sure to find exactly the same values $a_1, \ldots, a_n$ again. The existence of such a set of measurements (for which the results can be predicted with certainty) gives a necessary and sufficient characterization of a state of "maximum knowledge."

States of "maximum knowledge" are called "pure states." They represent the ultimate limit of precise observation as permitted by the uncertainty principle and are the quantum mechanical analogs of those classical states with sharp values of all positions and momenta.

For example, the dynamical state of a beam of noninteracting electrons is completely specified if the eigenvalues $\mathbf{p}$ and m_s of the momentum operator $\hat{\mathbf{P}}$ and the z component $\hat{S}_z$ of the spin operator have been measured. After the measurements have been performed, the electrons are in the pure state characterized by $|\mathbf{p}, m_s\rangle$. The state of electrons moving in a central field is conveniently classified by the eigenvalues of the Hamiltonian, orbital angular momentum and its z component, and $\hat{S}_z$.

The choice of a complete, commuting operator set is, of course, not unique. Let us consider two such sets $\hat{A}_1, \ldots, \hat{A}_n$ with eigenstates $|\psi_a\rangle = |a_1, \ldots, a_n\rangle$ and $B_1, \ldots, B_n$ with eigenstates $|\psi_b\rangle = |b_1, \ldots, b_n\rangle$, where at least one of the operators $\hat{B}_i$ does not commute with the first set. Assuming that the system is in the pure state $|\psi_a\rangle$ it can always be written as a linear superposition of all states $|\psi_b\rangle$:

$$\left|\psi_a\right\rangle = \sum_b c_b^{(a)} \left|\psi_b\right\rangle \tag{1}$$

Equation (1) is the mathematical expression of the "principle of superposition." As is well known, magnitudes and relative phases of the coefficients $c_b^{(a)}$ can be calculated by using the relation

$$\langle\psi_b|\psi_a\rangle = c_b^{(a)} \tag{2}$$

This gives us both the probability $|c_b^{(a)}|^2$ of finding the pure state $|\psi_b\rangle$ and the phase relation between the different states of the expansion. We say, $|\psi_a\rangle$ is a "completely coherent" superposition of states $|\psi_b\rangle$.

From Eq. (1) it follows that a pure state can be identified in two ways. We can characterize it either by a single vector $|\psi_a\rangle$ (that is, by specifying the relevant complete measurement), or by the set of coefficients $c_b^{(a)}$ relating $|\psi_a\rangle$ to the eigenstates $|\psi_b\rangle$ of another conveniently chosen operator set.

In practice, a complete preparation of a system is rarely achieved. Most frequently, the dynamical variables measured during the preparation do not constitute a complete set. As a consequence, the state of the system is not a pure state and can therefore not be characterized by a single state vector $|\psi\rangle$ or by a linear superposition of states. Because of the incomplete information on the system a need for statistical averaging arises in the same sense as in classical mechanics. The state of such systems must be described by a statistical mixture of pure states each having a suitable statistical weight. These systems are called "mixed states" or "mixtures."

The following example might be helpful to clarify these concepts. A beam of electrons may be in an eigenstate $|\mathbf{p}, m_s\rangle$ of operators $\hat{\mathbf{P}}$ and $\hat{S}_z$. This can always be written as linear superposition of eigenstates of operators $\hat{\mathbf{P}}$ and $\hat{S}_{z'}$ (where $\hat{S}_z$ and $\hat{S}_{z'}$ are related to different z axes):

$$|\mathbf{p}, m_s\rangle = \sum_{m_s'} c_{m_s'}^{(m_s)} |\mathbf{p}, m_s'\rangle \tag{3}$$

We compare this with a system obtained in the following way. A beam of N_+ electrons may have been prepared in the pure state $|\mathbf{p}, m_s' = +\frac{1}{2}\rangle$ and a second beam of N_- electrons in the state $|\mathbf{p}, m_s' = -\frac{1}{2}\rangle$. Both beams may then be combined and considered to be noninteracting. It is not possible to represent the state of the joint beam by a single-state vector because no complete measurement has been performed on the *combined* system (only the momentum $\mathbf{P}$ is known with certainty but the spin state of the joint system has not been specified). Neither is it possible to write the state of the combined system as linear superposition of the two states $|\mathbf{p}, m_s'\rangle$. We know only the probabilities

$$|c_+|^2 = \frac{N_+}{N_+ + N_-} \qquad \left(|c_-|^2 = \frac{N_-}{N_+ + N_-}\right)$$

of finding the state with $m_s' = +\frac{1}{2}$ $(-\frac{1}{2})$. A knowledge of the absolute squares of the coefficients c_+ and c_- does not provide us with the phase relation between the two states, so that a state vector in the form (3) cannot be constructed. We can merely state that the joint system consists of the two pure states $|\mathbf{p}, m_s' = \pm\frac{1}{2}\rangle$ with statistical weights $|c_\pm|^2$ and this statistical mixture represents the system. Both the mixture and the pure state $|\mathbf{p}, m_s\rangle$ do not have definite values of m_s'. In the case of the pure state (3) this uncertainty is caused by the inherent nature of the quantum mechanical measuring process ($\hat{S}_z$ and $\hat{S}_{z'}$ do not commute and therefore the state cannot have simultaneously sharp eigenvalues m_s and m_s'). In the case of the mixture we have to use statistical methods because of the incomplete preparation of the combined system.

Each pure state $|\psi_a\rangle$ of a mixture may be expanded in terms of eigenstates $|\psi_b\rangle$ of a conveniently chosen operator set. We say then that the mixture is a "partially coherent" superposition of states $|\psi_b\rangle$.

In the following, where we are not interested in differentiating between "complete" and "partial coherence," the term "coherent super-position" will be used.

The result of a measurement of a dynamical variable corresponding to an operator Q is given by the expectation or mean value $\langle Q \rangle$ of this operator. To obtain $\langle Q \rangle$ for the mixed state one first has to calculate

$\langle Q_a \rangle = \langle \psi_a | Q | \psi_a \rangle$ for each pure state in the mixture and then take an average, attributing to each of the pure states the corresponding statistical weight W_a:

$$\langle Q \rangle = \sum_a W_a \langle \psi_a | Q | \psi_a \rangle \qquad (4a)$$

$$= \sum_{abb'} W_a c_b^{(a)} c_{b'}^{(a)*} \langle \psi_{b'} | Q | \psi_b \rangle \qquad (4b)$$

where we inserted Eq. (1) for each $|\psi_a\rangle$ in the expression (4a). We emphasize that in Eq. (4a) statistics enter in two ways: first in the usual quantum theoretical expectation value $\langle \psi_a | Q | \psi_a \rangle$ and second in the ensemble average over these numbers with the statistical weight W_a. While the first averaging is connected with the uncontrollable perturbation of the system during the measurement of Q and therefore inherent in the nature of quantization, the second averaging is introduced because of our lack of information on the system, and this closely resembles the case of statistical mechanics. It is this second averaging in particular that can be performed conveniently by using density matrix techniques.

Finally, we note that very often it is not necessary to have a complete preparation of a system. For example, if the spin of particles participating in a reaction remains unchanged, all states can be written as the product of an orbital part and a spin part and the latter can be neglected. The state of the particles with respect to the spin is then irrelevant. If the system has sharp values of all other variables it can be considered as a pure state as far as the particular interaction under consideration is concerned. In this restricted sense the phrase "state of maximum knowledge" will always be used in the following.

2.2 The Density Matrix and Its Basic Properties

Consider a mixture of pure states $|\psi_a\rangle$. Using Eq. (4), the expectation value of any operator of interest can be calculated. Any information about the behavior of a system can be expressed in terms of expectation values of suitably chosen operators (like the Hamiltonian, angular momentum and spin operators, etc.). Thus the basic problem is the calculation of expectation values. In general the use of Eq. (4) is very tedious. It is more convenient to use a method based on the density matrix which one allows to perform those averaging processes, connected with the lack of information on the system, in a convenient way. Moreover, this method can be applied to mixtures as well as to pure states, and both can be treated on the same footing.

The density operator ρ, which describes a mixture of m pure states $|\psi_a\rangle$ with statistical weights W_a, is defined by the expression

$$\rho = \sum_{a=1}^{m} W_a|\psi_a\rangle\langle\psi_a| \tag{5a}$$

Each state $|\psi_a\rangle$ may be expanded in terms of a conveniently chosen set of eigenstates $|\psi_b\rangle$. Inserting Eq. (1) into Eq. (5a) for $|\psi_a\rangle$ and the corresponding expression

$$\langle\psi_a| = \sum_{b'} c_{b'}^{(a)*} \langle\psi_{b'}|$$

for the Hermitian adjoint states, we obtain

$$\rho = \sum_{a=1}^{m} W_a \sum_{bb'} c_b^{(a)} c_{b'}^{(a)*} |\psi_b\rangle\langle\psi_{b'}| \tag{5b}$$

Let us assume for simplicity that N orthogonal eigenfunctions $|\psi_b\rangle$ exist. ρ can then be written as an N-dimensional matrix (density matrix), where the element in the bth row and the b'th column is obtained from Eq. (5b):

$$\rho_{bb'} \equiv \langle\psi_b|\rho|\psi_{b'}\rangle = \sum_a W_a c_b^{(a)} c_{b'}^{(a)*} \tag{5c}$$

where we used the orthogonality of the eigenstates $\langle\psi_b|\psi_b'\rangle = \delta_{bb'}$. The matrix (5c) is the density matrix in the $\{|\psi_b\rangle\}$ representation. From expression (5c) it follows immediately that ρ is Hermitian:

$$\rho_{b'b} = \rho_{bb'}^* \tag{6}$$

Since the probability of finding the system in the state $|\psi_a\rangle$ is W_a, and since the probability that $|\psi_a\rangle$ is in the state $|\psi_b\rangle$ is $|c_b^{(a)}|^2$, the probability of finding the system in state $|\psi_b\rangle$ is given by the diagonal element

$$\rho_{bb} = \sum_a W_a |c_b^{(a)}|^2$$

Because probabilities are positive numbers, we obtain the condition

$$\rho_{bb} \geq 0 \tag{7}$$

for all diagonal elements.

The trace of ρ is a constant independent of the representation. In this section we normalize to

$$\mathrm{Tr}\rho = \sum_a W_a = 1 \tag{8}$$

Let us assume that the states $|\psi_a\rangle$ of the mixture (5a) are orthogonal. ρ is then diagonal in the $\{|\psi_a >\}$ representation:

$$\rho_{aa'} = W_a \delta_{aa'} \tag{9}$$

We say that the mixture is an "incoherent superposition" of states $|\psi_a\rangle$. A system is a coherent superposition of states $|\psi_b\rangle$ if its density matrix has at least one nonvanishing off-diagonal element in the $\{|\psi_b\rangle\}$ representation.

In order to obtain a convenient criterion for "complete" coherence we consider a system in a pure state $|\psi_a\rangle$. From Eq. (5a) it follows that this state is represented by a density operator

$$\rho = |\psi_a\rangle\langle\psi_a| \tag{10}$$

In the $\{|\psi_a\rangle\}$ representation only one diagonal element of ρ is different from zero. Using the normalization (8) this element is equal to 1 and we have the relation

$$\mathrm{Tr}(\rho^2) = (\mathrm{Tr}\,\rho)^2 = 1 \tag{11}$$

Because the trace is invariant under unitarity transformations, Eq. (11) is valid in any representation of the density matrix (10), not only in the diagonal one. Equation (11) is thus a sufficient and necessary condition for identifying a pure state. If a density matrix, given in the $\{|\psi_b\rangle\}$ representation, fulfills Eq. (11) the state of the system can be represented by a linear superposition of states $|\psi_b\rangle$. In this sense Eq. (11) is a condition for "complete coherence."

In general, ρ has in its diagonal representation more than one non-vanishing element. It follows for the trace of ρ^2

$$\mathrm{Tr}(\rho^2) = \sum_n \rho_{nn}^2 \leqslant \left(\sum_n \rho_{nn}\right)^2 = (\mathrm{Tr}\rho)^2$$

and because of the invariance of the trace under unitary transformations we have the general result [with the normalization (8)]

$$\mathrm{Tr}(\rho^2) \leqslant 1 \tag{12}$$

where the equality sign holds if and only if ρ describes a pure state.

From Eqs. (5) and Eq. (4) it follows that the expectation value of any operator Q is given by the trace of the product of ρ and Q:

$$\langle Q \rangle = \sum_{bb'} \left[\sum_a W_a c_b^{(a)} c_{b'}^{(a)*} \right] \langle \psi_{b'}|Q|\psi_b \rangle$$

$$= \sum_{bb'} \langle \psi_b|\rho|\psi_{b'}\rangle\langle\psi_{b'}|Q|\psi_b\rangle$$

$$= \mathrm{Tr}\,\rho Q \tag{13}$$

[with the normalization (8)]. This expression is valid for both pure and mixed states. Thus, the expectation value of any operator can be calculated from a knowledge of the density matrix.

As noted above, the diagonal elements $\langle \psi_a | \rho | \psi_a \rangle$ of the density matrix are just the probabilities that the pure state $|\psi_a\rangle$ will be observed in an experimental observation. However, most experimental situations that can be devised will not respond to one particular pure state only. In general, the detector responds to several pure states $|\psi_a\rangle$ with relative probabilities ("efficiencies") ε_a. The total probability of response of the apparatus will then be given by

$$W = \sum_a \varepsilon_a \langle \psi_a | \rho | \psi_a \rangle \tag{14a}$$

($\{|\psi_a\rangle\}$ representation). We introduce the operator

$$\varepsilon = \sum_a |\psi_a\rangle\langle\psi_a| \tag{14b}$$

in analogy to the density operator (5a). Equation (14a) can then be written in the form

$$W = \mathrm{Tr}\,\rho\varepsilon \tag{14c}$$

ε is called the "efficiency matrix" of the measuring device which completely describes the response of the measuring apparatus. If the apparatus responds only to the pure state $|\psi_a\rangle$ with certainty (that is, it is a perfect filter) we have $\varepsilon = |\psi_a\rangle\langle\psi_a|$. In that case we can project out of the mixture the definite pure state $|\psi_a\rangle$ and Eq. (14a) reduces to $W = W_a = \langle\psi_a|\rho|\psi_a\rangle$ (ideal measurement). In the following we will always assume that the measuring apparatus under consideration is a perfect filter.

2.3. Example: Polarization Density Matrix of Photons

An electromagnetic plane wave with polarization vector **e** traveling in the direction of the unit vector **n** is represented by

$$A(\mathbf{r}, t) = \mathbf{e}\, e^{i(\mathbf{k}\cdot\mathbf{r} - \omega t)} \tag{15}$$

($\mathbf{k} = \omega\mathbf{n}$). We denote this plane wave by $|\omega, \mathbf{n}, \mathbf{e}\rangle$. In this section we assume that all photons under consideration have the same frequency ω and travel in the same direction **n** and we suppress the dependence on ω and **n**.

The direction of motion **n** may coincide with the z axis of our coordinate system. We denote photons linearly polarized along the x (y) axis by $|x\rangle$ ($|y\rangle$). A photon state with arbitrary but definite polarization vector **e** can be written as a linear superposition of these states:

$$|\mathbf{e}\rangle = a_x^{(e)}|x\rangle + a_y^{(e)}|y\rangle \tag{16a}$$

In general, it is more convenient to use as basis states the combinations

$$|\pm 1\rangle = \mp 2^{-1/2}[|x\rangle \pm i|y\rangle] \tag{16b}$$

denoting circularly polarized photons. We write

$$|\mathbf{e}\rangle = \sum_\lambda a_\lambda^{(e)}|\lambda\rangle \tag{17}$$

($\lambda = \pm 1$). The coefficients $a_\lambda^{(e)}$ are the amplitudes of finding the circularly polarized state $|\lambda\rangle$ if the system is in the pure state $|e\rangle$.

States with polarization $\mathbf{e} = e_{\pm 1}$ have definite angular momentum components ± 1 with respect to the direction of motion as z axis. This property allows us to account for angular momentum conservation in emission and absorption processes in an elegant manner. We therefore use the states (16b) as basis states throughout this paper. The component of the angular momentum with respect to the direction of motion is called "helicity" and we refer to the states (16b) as "helicity states."

A mixture of pure states $|e\rangle$ with statistical weights W_e is described by the density operator

$$\rho = \sum_e W_e |\mathbf{e}\rangle\langle\mathbf{e}| \tag{18}$$

In the $\{|\lambda\rangle\}$ representation ("helicity representation") (17) we have

$$\rho = \sum_{e\lambda\lambda'} W_e a_\lambda^{(e)} a_{\lambda'}^{(e)*} |\lambda\rangle\langle\lambda'| \tag{19}$$

($\lambda, \lambda' = \pm 1$). The density operator ρ can be written as a 2×2-dimensional matrix with the elements

$$\rho_{\lambda\lambda'} = \sum_e W_e a_\lambda^{(e)} a_{\lambda'}^{(e)*} \tag{20}$$

We assume the normalization $\mathrm{Tr}\,\rho = 1$. The diagonal element $\rho_{+1,+1}(\rho_{-1,-1})$ is then the probability for finding a photon with helicity $|+1\rangle$ ($|-1\rangle$) in the beam.

In general ρ has four complex elements corresponding to eight real parameters. From the Hermiticity condition (6) and the normalization $\mathrm{Tr}\,\rho = 1$ it follows that only three of these are independent. It is therefore convenient to parametrize in terms of three real parameters η_1, η_2, η_3, and we write for the general density matrix (in the "helicity" representation):

$$\rho = \frac{1}{2}\begin{pmatrix} 1+\eta_2 & -\eta_3 + i\eta_1 \\ -\eta_3 - i\eta_1 & 1-\eta_2 \end{pmatrix} \tag{21}$$

The physical meaning of these parameters can be obtained in the following way. First we note that from condition (12) it follows that

$$\eta_1^2 + \eta_2^2 + \eta_3^2 \leqslant 1 \tag{22}$$

so that each η_i can vary only inside the limits $-1 \leqslant \eta_i \leqslant 1$ [the equality sign in (22) holds for pure states]. Because the probability of finding a photon in the states $|\pm 1\rangle$ is given by the diagonal elements $\frac{1}{2}(1 \pm \eta_2)$ it follows that η_2 gives the degree of circular polarization. The probability of finding a linearly polarized photon with polarization vector in the x direction is given by $\langle x|\rho|x \rangle$. Expressing the state $|x\rangle$ in terms of the basis $|\lambda\rangle$ it can be shown that this probability is given by $\frac{1}{2}(1 + \eta_3)$. The probability that a photon is linearly polarized along the y axis is $\langle y|\rho|y \rangle = \frac{1}{2}(1 - \eta_3)$. The values $\eta_3 = \pm 1$ therefore correspond to complete polarization in these directions. Similarly it can be shown that $\frac{1}{2}(1 \pm \eta_1)$ is the probability for finding a photon linearly polarized at angles $\lambda = \pm 45°$ to the x axis. The parameters η_i, introduced formally by Eq. (21), are therefore the well-known Stokes parameters. They characterize the polarization state of a photon beam completely.

If the polarization density matrix (21) is known, we can find the probability $W_e = \langle e|\rho|e \rangle$ of detecting a photon with any given polarization vector $\mathbf{e}$. Let us parametrize the general polarization vector (16a) in the following way:

$$|e\rangle = \cos \beta |x\rangle + e^{i\delta} \sin \beta |y\rangle$$

where we have used the normalization $\langle e|e \rangle = 1$ and we have chosen the overall phase so that $a_x^{(e)}$ is real. In this expression δ is the phase necessary to describe elliptical polarization. Thus, for linear polarization we have $\delta = 0$ and β is the angle between $\mathbf{e}$ and the x axis. For circular polarization we have $\cos \beta = \sin \beta = 1/2^{1/2}$ and $\delta = \pi/2$, etc. Expressing $|x\rangle$ and $|y\rangle$ in the helicity basis, and applying Eq. (21), we obtain

$$W_e = \tfrac{1}{2}(1 + \eta_3 \cos 2\beta + \eta_1 \sin 2\beta \cos \delta + \eta_2 \sin 2\beta \sin \delta)$$

Formally one can define a "polarization" vector $\mathbf{P}$ with the "components" η_1, η_2, η_3. From condition (22) it follows that the magnitude of $\mathbf{P}$ is restricted:

$$|\mathbf{P}| = (\eta_1^2 + \eta_2^2 + \eta_3^2)^{1/2} \leqslant 1 \tag{23}$$

where the equality sign holds if and only if one is dealing with a pure polarization state. States of "maximum knowledge" are therefore necessarily completely polarized in the sense that $|\mathbf{P}| = 1$.

Finally, we note that it is often convenient to drop the normalization $\text{Tr}\, \rho = 1$. With arbitrary normalization constant given by $\text{Tr}\, \rho$ we have then instead of Eq. (21)

$$\rho = \frac{\mathrm{Tr}\,\rho}{2}\begin{pmatrix} 1+\eta_2 & -\eta_3+i\eta_1 \\ -\eta_3-i\eta_1 & 1-\eta_2 \end{pmatrix} \tag{24}$$

Often one normalizes so that $\mathrm{Tr}\,\rho$ gives the intensity of the beam or the total number of photons.

In Section 5.2 we discuss in detail how to calculate the polarization density matrix ρ for particular cases.

3. Interacting Systems

3.1. Basic Principles

In this section we discuss the quantum mechanical state of two inter-acting systems. In general an interaction between two systems gives rise to a loss of information unless both systems are detected and their state completely determined. Examples are the excitation of atoms by photons and the coherence properties of the excited atoms, depolarization of atoms by collisions with other atoms or with the walls of the container or by spin-relaxation processes, the influence of external fields on the state of atomic systems, and so on.

In this section we briefly review the underlying quantum mechanical principles. In Section 3.2 we show how to construct the density matrix describing interacting systems, and the density matrix describing one subsystem. Finally we discuss an example.

We start with a discussion of the following situation. The initial system may consist of two separate subsystems assumed to be in pure states $|\psi_i\rangle$ and $|\phi_j\rangle$ (for example, $|\psi_i\rangle$ may describe a pure state of electrons or photons and $|\phi_j\rangle$ an atomic system; note that completely polarized electrons are required or completely polarized photons in the sense $|P|=1$; the indices i and j characterize particular eigenstates). Both subsystems may interact for a certain time. We analyze the final state of the combined system after the interaction. (A more complete treatment including the full time depen-dence will be given in Section 5.)

Our discussion is based on the following general argument. We consider the composite space spanned by all possible direct products $|\psi_i\rangle$ $|\phi_j\rangle$ of eigenstates of the two subsystems. Before the interaction starts the state of the combined system is a particular state vector $|\psi_{in}\rangle$ in this space, given by the direct product $|\psi_{in}\rangle = |\psi_i\rangle\,|\phi_j\rangle$. The operators in quantum mechanics and quantum electrodynamics, describing the interaction between the two subsystems, are linear operators in the composite space. Linear operators transform a single initial-state vector into another single-state vector. Starting with an initial state $|\psi_{in}\rangle$ we can be sure that the final state of the combined system can be described by a single-state vector too.

We denote the final state by $|\psi_{out}^{(ij)}\rangle$. This state vector can always be expanded in terms of the basis set $|\psi_n\rangle\,|\phi_m\rangle$, and we write

$$|\psi_{out}^{(ij)}\rangle = \sum_{nm} a_{nm}^{(ij)}|\psi_n\rangle|\phi_m\rangle \qquad (25)$$

(The sum may include integrals over continuous variables.) $a_{nm}^{(ij)}$ is the amplitude describing the transition $|\psi_i\rangle|\phi_j\rangle \to |\psi_n\rangle|\phi_m\rangle$. The sum in Eq. (25) contains those eigenstates allowed by the conservation of energy, angular momentum, and so on. More precisely, a particular eigenstate $|\psi_n\rangle$ of one subsystem is correlated with eigenstates $|\phi_m\rangle$ of the other subsystem in such a way that all relevant conservation laws are fulfilled.

If one detects only one subsystem (the ψ system, for example), its state in general must be represented by a mixture of eigenstates $|\psi_n\rangle$ corresponding to different transitions. Only under particular circumstances is it possible to assign a single-state to the subensembles. For example, the state $|\psi_i\rangle$ may describe an ensemble of electrons and the state $|\phi_j\rangle$ an ensemble of heavy spinless particles at rest, contained in a small volume of space. The electrons may then be scattered by these particles. If the electron energy is low enough so that the recoil of the heavy particles can be neglected, the state of these particles will not change during the interaction. We can then write Eq. (26) in the form

$$|\psi_{out}^{(ij)}\rangle = |\phi_j\rangle\Big[\sum_{n} a_{nj}^{(ij)}|\psi_n\rangle\Big] \equiv |\phi_j\rangle|\psi\rangle$$

where now the state of the final electrons is the pure state $|\psi\rangle$ given by the linear superposition of states $|\psi_n\rangle$ (where different $|\psi_n\rangle$ may describe electrons scattered in different directions). This linear superposition is, of course, responsible for the diffraction of the electrons and the corresponding interference phenomena. If, however, the recoil of the heavy particles cannot be neglected, the electrons cannot be characterized by a "state of maximum knowledge." This "loss of information" corresponds to a loss of "coherency" and the interference phenomena will be less important.

3.2. The Density Matrix of Interacting Systems

We now discuss the more general case that the initial system is a mixture of states $|\psi_i\rangle|\phi_j\rangle \equiv |\psi_i\phi_j\rangle$ with statistical weight W_{ij} (with $W_{ij} = W_i W_j$ for uncorrelated systems). In this case the final state of the combined system cannot be characterized by a single-state vector $|\psi_{out}\rangle$. It must be represented by a mixture of states or, more conveniently, by its density matrix. The density matrix, describing the final state after the interaction, can be constructed easily with the help of the results of Section 3.1. The density operator characterizing the initial state is given by the expression

$$\rho_{\text{in}} = \sum_{ij} W_{ij} |\psi_i \phi_j\rangle\langle\psi_i \phi_j| \tag{26}$$

where the sum goes over all states present in the mixture. During the interaction each state vector $|\psi_i \phi_j\rangle$ evolves into a state vector $|\psi_{\text{out}}^{(ij)}\rangle$. The density matrix ρ_{out} of the final system can then be constructed from Eqs. (25) and (26):

$$\rho_{\text{out}} = \sum_{ij} W_{ij} \sum_{\substack{nm \\ n'm'}} a_{nm}^{(ij)} a_{n'm'}^{(ij)*} |\psi_n \phi_m\rangle\langle\psi'_n \phi'_m| \tag{27a}$$

Its matrix elements are given by

$$\langle\psi_n \phi_m|\rho_{\text{out}}|\psi'_n \phi'_m\rangle = \sum_{ij} W_{ij} a_{nm}^{(ij)} a_{n'm'}^{(ij)*} \tag{27b}$$

where each row and column is labeled by two indices (n, m) corresponding to the eigenstates $|\psi_n\rangle$ and $|\phi_m\rangle$ of the separate systems.

In particular for scattering problems it is convenient to write the coefficients $a_{nm}^{(ij)}$ in matrix form. We define a matrix M (scattering matrix) by its matrix elements

$$\langle\psi_n \phi_m|M|\psi_i \phi_j\rangle = a_{nm}^{(ij)} \tag{28}$$

We can write Eq. (27b) in the form

$$\langle\psi_n \phi_m|\rho_{\text{out}}|\psi_{n'}\phi_{m'}\rangle = \left\langle\psi_n \phi_m|M\left[\sum_{ij} |\psi_i \phi_j\rangle W_{ij}\langle\psi_j \phi_i|\right]M^+|\psi_{n'}\phi_{m'}\right\rangle$$

Using Eq. (26) we have the matrix equation

$$\rho_{\text{out}} = M\rho_{\text{in}}M^+ \tag{29}$$

The matrix M contains all information about the interaction, ρ_{in} the information about the initial state.

Often one is only interested in one of the two subsystems after an interaction (say, the $|\psi\rangle$ system), with the second one undetected. As discussed in Section 3.1, the state of the subsystem must in general be described by a density matrix, and we are going to construct the matrix $\rho(\psi)$ which contains the information on the ψ system alone. $Q(\psi)$ may be an operator, acting only on the $|\psi\rangle$ states, that is, we have the relation $\langle\psi_n \phi_m|Q(\psi)|\psi_{n'}\phi_{m'}\rangle = \langle\psi_n|Q(\psi)|\psi_{n'}\rangle\delta_{mm'}$ using the orthogonality of the eigenstates $|\phi_n\rangle$. The combined ensemble may be described by a density matrix with elements (27b). The expectation value follows from

Eq. (13):

$$\langle Q(\psi) \rangle = \mathrm{Tr}\ [\rho Q(\psi)]$$

$$= \sum_{\substack{nm \\ n'm'}} \langle \psi_n \phi_m | \rho | \psi_{n'} \phi_{m'} \rangle \langle \psi_{n'} \phi_{m'} | Q(\psi) | \psi_n \phi_m \rangle$$

$$= \sum_{nn'} \left[\sum_m \langle \psi_n \phi_m | \rho | \psi_{n'} \phi_m \rangle \right] \langle \psi_{n'} | Q(\psi) | \psi_n \rangle \qquad (30)$$

We define a density matrix $\rho(\psi)$ by its matrix elements:

$$\langle \psi_n | \rho(\psi) | \psi_{n'} \rangle = \sum_m \langle \psi_n \phi_m | \rho | \psi_{n'} \phi_m \rangle \qquad (31)$$

Equation (31) can then be written as

$$\langle Q(\psi) \rangle = \sum_{nm'} \langle \psi_n | \rho(\psi) | \psi_{n'} \rangle \langle \psi_{n'} | Q(\psi) | \psi_n \rangle$$

$$= \mathrm{Tr}\ [\rho(\psi) Q(\psi)] \qquad (32)$$

which resembles Eq. (13).

Because all information about the subsystem $|\psi\rangle$ can be expressed in terms of expectation values $\langle Q(\psi) \rangle$ of as many operators $Q(\psi)$ as necessary, it follows from Eq. (33) that all information about the subsystem $|\psi\rangle$ is given by the density matrix $\rho(\psi)$. This matrix is obtained from ρ' by putting $m = m'$ and summing over all m, as Eq. (31) shows.

The two systems are uncorrelated when

$$\langle Q(\psi) Q(\phi) \rangle = \langle Q(\psi) \rangle \langle Q(\phi) \rangle \qquad (33)$$

for all pairs of operators. In this case ρ can be factorized:

$$\rho = \rho(\psi) \times \rho(\phi) \qquad (34)$$

where the multiplication sign denotes the direct product.

3.3. Example: Analysis of Light Emitted by Atoms [1]

As an illustration we briefly discuss the polarization of light emitted by excited atoms. Initially the atoms may be in the pure state $|n_i j_i M_i\rangle$, where j_i, M_i denote angular momentum and its z component and n_i all other quantum numbers. We consider an optical transition to states $|n_f j_f M_f\rangle$, where we assume that all final atoms have the same values of n_f and j_f and that the magnetic sublevels are degenerate. The emitted photons observed in a fixed direction $\mathbf{n}$ have the same energy and differ only in the polarization state, which depends on M_f. The final state of the combined system can be written in the form (25):

$$|\psi_{\mathrm{out}}^{(M_i)}\rangle = \sum_{M_f} a_{M_f}^{(M_i)} |n_f j_f M_f\rangle |e(M_f)\rangle$$

$|e(M_f)\rangle$ describes a subensemble of photons with polarization $e(M_f)$, emitted in a transition to an atomic state with definite M_f.

A particular subensemble may be selected by filtering the atoms after light emission through a Stern–Gerlach device which is set to accept only atoms of a definite magnetic quantum number M_f. The filtered atoms may then be detected by a counter and the coincidences between this counter and the photon counter may be observed. The observation is then restricted to photons emitted by atoms in a transition between states of "maximum information" (namely, $|n_i j_i M_i\rangle$ and $|n_f j_f M_f\rangle$ with a specified M_f) and this light is in a pure polarization state represented by $|e(M_f)\rangle$. As discussed in Section 2.3, this light is completely polarized in the sense $|P| = 1$. If the total beam of photons is detected without analyzing the angular momentum state of the final atoms, the photon state must be regarded as incoherent superposition of the pure states $|e(M_f)\rangle$ emitted by atoms that land in states with different values of M_f. This incoherent superposition gives necessarily $|P| < 1$. In general, incomplete polarization is associated with incomplete determination of the atomic state after (or before) the emission.

4. Multipole Expansion of the Density Matrix

4.1. Irreducible Tensor Operators

In the previous sections we defined a mixed state as incoherent superposition of pure states. This description has several disadvantages. First it is not unique. For example, unpolarized light can equally well be considered as incoherent superposition of right and left circularly polarized photon beams (with equal statistical weights) or as an incoherent superposition of two linearly polarized beams. Secondly, instead of specifying the components of the mixture and their statistical weights, the initial information on the system is often more conveniently given in terms of expectation values $\langle Q_i \rangle$ of operators Q_i. For example, unpolarized particles can be characterized by the statement that the expectation values of each component of its spin vanishes. Another example is the description of photon beams in terms of Stokes parameters.

If we consider as the initial information the expectation values $\langle Q_i \rangle$ of as many independent operators as there are parameters in the density matrix ρ, then ρ can be obtained by solving the equations

$$\mathrm{Tr}\,(\rho Q_i) = \langle Q_i \rangle \tag{35}$$

An N-dimensional density matrix has N^2 complex elements, that is, $2N^2$ real parameters. From the Hermiticity condition (6) and the normalization condition it follows that ρ has $N^2 - 1$ independent parameters. That is, $N^2 - 1$ independent data are required to specify ρ in general. Once ρ is determined in this way the expectation value of any other operator can then be calculated from ρ. This "operational" point of view has been emphasized in Ref. 1. We give examples in Section 4.2.

The usefulness of this method of specifying ρ depends on the choice of the basis operator set Q_i. When the angular symmetries of the ensemble are of interest it is convenient to represent ρ in terms of spherical basis operators. The systematic use of these operators had been suggested in Ref. 10. They have been applied, for example, in optical pumping work[11] and angular correlation theory.[12] In this section we give the definition and basic properties of these operators. In the following we are only interested in the angular momentum variables. We assume that the system has definite values of all other variables and suppress the dependence on these. Consider an ensemble of particles (atoms, electrons, or photons) which may be a coherent or incoherent superposition of states with different angular momenta j and j'. From the angular momentum states we may construct spherical tensor operators by making use of the angular momentum addition rule

$$T(j'j)_{KQ} = \sum_{MM'} (-1)^{j-M} (j'M', j-M|KQ)|j'M'\rangle\langle jM| \qquad (36)$$

The Clebsch–Gordan coefficient $(j'M', j-M|KQ)$ "projects" irreducible components out of the product $(-1)^{j-M}|j'M'\rangle\langle jM|$ so that the operators $T(j'j)_{KQ}$ transform under rotation according to the representation $D^{(K)}$ of the rotation group. Angular momentum conservation gives the restriction

$$|j'-j| \le K \le j+j', \qquad -K \le Q \le K \qquad (37)$$

We obtain a matrix representation of the operators (36) by "sandwiching" the operators between states $|jM\rangle$ and $\langle j'M'|$ and making use of the orthogonality of the states $|jM\rangle$:

$$\langle j'M'|T(j'j)_{KQ}|jM\rangle = (-1)^{j-M}(j'M', j-M|K-Q) \qquad (38)$$

This can be written explicitly as a matrix with $(2j'+1)$ rows and $(2j+1)$ columns (for fixed values of j' and j):

$$
\begin{pmatrix}
\langle j'j'|T(j'j)_{KQ}|jj\rangle & \langle j'j'|T(j'j)_{KQ}|j, j-1\rangle & \cdots & \langle j'j'|T(j'j)_{KQ}|j, -j\rangle \\
\langle j', j'-1|T(j'j)_{KQ}|jj\rangle & & & \\
\vdots & \vdots & & \vdots \\
\langle j', -j|T(j'j)_{KQ}|jj\rangle & & &
\end{pmatrix}
$$

The matrix is a square matrix for $j = j'$.

We can write the elements (38) in the form

$$\langle j'M'|T(j'j)_{KQ}|jM\rangle^* = \langle jM|T(j'j)_{KQ}^+|j'M'\rangle \tag{39}$$

[where the asterisk is superfluous because the elements (38) are real]. This defines a matrix representation of the Hermitian adjoint operator $T(j'j)_{KQ}^+$. Using Eq. (38) and the symmetry of the Clebsch–Gordan coefficients we obtain

$$\langle j'M'|T(j'j)_{KQ}|jM\rangle = (-1)^{j-j'+Q}\langle j'M'|T(jj')_{KQ}^+|jM\rangle \tag{40a}$$

This gives us the operator relation

$$T(j'j)_{KQ} = (-1)^{j-j'+Q}T(jj')_{K,-Q}^+ \tag{40b}$$

[note that $T(jj')_{KQ}$ corresponds to a matrix with $(2j+1)$ rows and $(2j'+1)$ columns]. The trace of operator products can be obtained from Eqs. (38) and (40b):

$$\text{Tr } T(j'j)_{KQ}'T(j'j)_{K'Q'} = \delta_{KK'}\delta_{QQ'} \tag{41}$$

where we again used the symmetry properties of the Clebsch–Gordan coefficients. We say that $T(j'j)_{KQ}^+$ and $T(j'j)_{K'Q}$ are mutually orthogonal under the trace operation.

We define a reduced matrix element by applying the Wigner–Eckart theorem to the matrix elements. Using the conventions of Ref. 18 we have

$$\langle j'M'|T(j'j)_{KQ}|jM\rangle = (-1)^{j-M}(j'M', j-M|KQ)\frac{\langle j'\|T_K\|j\rangle}{(2K+1)^{1/2}}$$

Comparing with Eq. (38) gives

$$\langle j'\|T_K\|j\rangle = (2K+1)^{1/2} \tag{42}$$

The equations (38) and (42) show that the matrix elements of the tensor operators are functions of the angular momentum quantum numbers alone and therefore purely geometrical elements. This property may conveniently be used to separate dynamical and geometrical elements. For example, the various electric multipole operators Q_{KQ} of atoms can be related to the corresponding tensor operator T_{KQ}. Applying the Wigner–Eckart theorem to the matrix elements of both operators and eliminating the common Clebsch–Gordon coefficient we obtain

$$\langle j'M'|Q_{KQ}|jM\rangle = \langle j'\|Q_K\|j\rangle\frac{\langle j'M'|T_{KQ}|jM\rangle}{\langle j'\|T_K\|j\rangle}$$

where the dynamics is entirely contained in $\langle j'\|Q_K\|j\rangle$.

It is sometimes convenient to relate the tensor operators (42) to the angular momentum operators $J_q (q = 0, \pm 1)$:

$$J_0 = J_z, \qquad J_{\pm 1} = \mp \frac{1}{2^{1/2}} (J_x \pm iJ_y) \tag{43}$$

and their products. We give the relation for $K \leq 2$ and sharp angular momentum $j = j'$ $[T(jj)_{KQ} \equiv T(j)_{KQ}]$:

$$T(j)_{KQ} = N_K \sum_{q_1 q_2} (1q_1, 1q_2 | KQ) J_{q_1} J_{q_2} \tag{44}$$

The normalization factor N_K is given by

$$N_K = \left[(-1)^{K+2j} \begin{Bmatrix} 1 & 1 & K \\ j & j & j \end{Bmatrix} (2j+1)(j+1)j \right]^{-1} \tag{45}$$

Tensors which represent the polarization of light can be formed from the components of the polarization vector $\mathbf{e}$. In an arbitrary orthogonal xyz coordinate system, where the z axis does not coincide with the direction of propagation, the polarization vector is three-dimensional. Its spherical components are

$$e_0 = e_b, \qquad e_{\pm 1} = \mp (1/2)^{1/2} (e_x \pm ie_y)$$

Similar to Eq. (36) we construct polarization tensors

$$T_{KQ} = \sum_{qq'} (-1)^{1-q} (1q', 1-q | KQ) e_q \, e_q^*$$

$(q, q' = \pm 1, 0)$. These tensors are often used in the literature (see, for example, Ref. 21 and references therein). However, they have the disadvantage that they are constructed in a three-dimensional basis in which the components e_0, $e_{\pm 1}$ are not independent of each other. This dependence of the components must be taken into account explicitly when sums over q, q' must be carried out. This disadvantage can be avoided by using the helicity system as introduced in Section 2.3 and further discussed in Section 5.2.

Finally we list some other conventions and notations relating to tensor operators:

Omont and Meunier[22]:

$$^{j'j}T_Q^{(K)} = \sum_{MM'} (-1)^{J-M} (j'M', j-M | KQ) |j'M'\rangle\langle jM|$$

D'Yanokov and Perel[23]:

$$T_\alpha^{(\kappa)} = \frac{2\kappa + 1}{(2j+1)^{1/2}} \sum_{MM'} (-1)^{j-M} \begin{pmatrix} j & \kappa & j \\ -M' & \alpha & M \end{pmatrix} |j'M'\rangle\langle jM|$$

Happer and Mathur[24]:

$$T_M^{(L)}(j,j') = \sum_m (-1)^{m-M-j'}(jM, j'M-m|LM)|jM\rangle\langle j'M-m|$$

Blum, Kleinpoppen[25],[26]:

$$T(j)_{KQ} = (2j+1)^{1/2} \sum_{M'M} (-1)^{j-M}(jM', j-M|KQ)|jM'\rangle\langle jM|$$

4.2. Expansion of the Density Matrix in Terms of "State Multipoles"

Consider an ensemble of particles that is an incoherent superposition of states $|\psi_n\rangle$ described by the density operator

$$\rho = \sum_n W_n |\psi_n\rangle\langle\psi_n|$$

Each state $|\psi_n\rangle$ may be expanded in terms of angular momentum states $|j, M\rangle$, which is the natural choice in all cases where angular symmetry is of importance. We again suppress the dependence on all other quantum numbers, assuming that these have definite values. We write

$$|\psi_n\rangle = \sum_{jM} a_{jM}^{(n)}|jM\rangle$$

The density operator is then given by

$$\rho = \sum_{jj'}\left[\sum_{nMM'} W_n a_{jM}^{(n)} a_{j'M'}^{(n)*} |jM\rangle\langle j'M'|\right] \tag{46}$$

with matrix elements

$$\langle jM|\rho|j'M'\rangle = \sum_n W_n a_{jM}^{(n)} a_{j'M'}^{(n)*} \tag{47}$$

ρ is a $(2j+1)+(2j'+1)+\cdots$ dimensional square matrix. For each pair j, j' of angular momenta we define a submatrix $\rho(j, j')$ of ρ by the matrix elements (47) (j, j' fixed). $\rho(j, j')$ is a matrix with $(2j+1)$ rows and $(2j'+1)$ columns. The density matrix describing the system in all its possible states can be written in the form

$$\rho = \begin{vmatrix} \rho(j,j) & \rho(j,j') & \rho(j,j'') & \cdots \\ \rho(j',j) & \cdots \\ \vdots \end{vmatrix} \tag{48}$$

For example, let us consider an atomic system that is a coherent superposition of a ground state $(j = 0)$ and an excited state with $j = 1$

$$|\psi\rangle = \sum_{jM} a_{jM} |j, M\rangle \qquad (j = 0, 1)$$

The elements of the submatrices are given by

$$\rho(j, j')_{mm'} \equiv \langle jM | \rho(j, j') | j'M'\rangle = a_{jM} a^*_{j'M'} \qquad (j, j' = 0, 1)$$

The density matrix ρ describing the system can then be written in the explicit form

$$\rho = \begin{pmatrix} \rho(0, 0)_{0,0} & \rho(0, 1)_{0,+1} & \rho(0, 1)_{0,0} & \rho(0, 1)_{0,-1} \\ \rho(1, 0)_{+1,0} & \rho(1, 1)_{+1,+1} & \rho(1, 1)_{+1,0} & \rho(1, 1)_{+1,-1} \\ \rho(1, 0)_{0,0} & \rho(1, 1)_{0,+1} & \rho(1, 1)_{0,0} & \rho(1, 1)_{0,-1} \\ \rho(1, 0)_{-1,0} & \rho(1, 1)_{-1,+1} & \rho(1, 1)_{-1,0} & \rho(1, 1)_{-1,-1} \end{pmatrix}$$

Density matrices of this form occur, for example, in optical pumping theory corresponding to transitions $j = 0 \leftrightarrow j = 1$. $\rho(0, 0)$ and $\rho(1, 1)$ are the density matrices describing the atoms in states with $j = 0$ and $j = 1$. Their off-diagonal elements are generally called "Hertzian" coherences. The elements $\rho(0, 1)_{0m}[= \rho(1, 0)^*_{m0}]$ connect a magnetic sublevel to the ground state and are called "optical" coherences (because they evolve at optical frequencies).

As pointed out in the introduction to this section, it is preferable to define ρ by a set of conveniently chosen quantities $\langle Q_i \rangle$ rather than by expressions like (46). Dealing with the angular momentum properties of states we choose the operators (36) as a basis set. We make the ansatz

$$\rho(j, j') = \sum_{KQ} a(j, j')_{KQ} T(jj')_{KQ}$$

where K and Q are restricted by Eq. (37). The coefficients can be determined by multiplying both sides with $T(jj')^+_{KQ}$ and calculating the trace by using Eq. (41). This gives

$$\text{Tr} \left[\rho(j, j') T(jj')^+_{KQ} \right] = a(j, j')_{KQ}$$

[note that $\rho(j, j') \cdot T(jj')^+_{KQ}$ is a square matrix, so that the trace is defined]. We denote the coefficients $a(jj')_{KQ}$ by $\langle T(jj')^+_{KQ}\rangle$ and write

$$\rho(j, j') = \sum_{KQ} \langle T(jj')^+_{KQ}\rangle T(jj')_{KQ} \qquad (49)$$

This expression is of great importance in all problems where angular momentum properties play a role. It is the starting point for most discussions in optical pumping work, angular correlation theory, spin polarization phenomena, and so on.

The quantities $\langle T(jj')^+_{KQ}\rangle$ are referred to as "statistical tensors" or "state multipoles" owing to their relationship with multipole moments or interactions. For $K = 1$ the three quantities $\langle T(jj)^+_{1Q}\rangle$ transform as the spherical components of a vector ("orientation" vector). For $K = 2$ the five quantities $\langle T(jj)^+_{2Q}\rangle$ transform as the components of a second-rank tensor ("alignment" tensor), and so on. If at least one multipole with $K > 0$ is different from zero the system is said to be "polarized." Using Eq. (44) the state multipoles can be written as expectation values of angular momentum operators. $\langle T_{1,Q}\rangle$ is proportional to the expectation value $\langle J_Q\rangle$, $\langle T_{2,0}\rangle$ is proportional to $\langle 3J_z^2 - J^2\rangle$, and so on. The use of these quantities has been proposed recently.[17]

The density matrix is completely specified by the set $\langle T(jj')^+_{KQ}\rangle$. For example, for sharp angular momentum $j = j'$ the set of all $\langle T(jj)^+_{KQ}\rangle$ consists of $2(j+1)^2$ quantities. The "monopole" T_{00} is a constant fixed by the normalization of ρ, which leaves us with $(2j+1)^2 - 1$ independent parameters necessary to specify ρ completely.

Taking matrix elements of both sides of Eq. (49) and using Eq. (38) we obtain

$$\langle jM|\rho(j, j')|j'M'\rangle = \sum_{KQ} \langle T(jj')^+_{KQ}\rangle(-1)^{j'-M'}(jM, j' - M'|KQ) \qquad (50)$$

The inverse relation follows by multiplying both sides with $(jM, j' - M'|KQ)$, summing over M, M', and using the orthogonality relation of the Clebsch–Gordon coefficients. We obtain

$$\langle T(jj')^+_{KQ}\rangle = \sum_{MM'} (-1)^{-j'+M'}(jM, j' - M'|KQ)\langle jM|\rho(j, j')j'M'\rangle \qquad (51)$$

which shows explicitly the equivalence of the two descriptions in terms of density matrix elements and in terms of state multipoles.

The significance of expression (49) will become clear by the discussions in other articles in this volume. We give examples in Section 5.2 and give here only an illustration of the formalism. We consider an ensemble of states with sharp angular momentum $j = 1$ being an incoherent superposition of magnetic substates so that the density matrix is diagonal in the $\{|1, M\rangle\}$ representation:

$$\langle 1M'|\rho|1M\rangle = W_M\delta_{MM'} \qquad (52)$$

The statistical weights may be $W_{+1} = \frac{1}{6}$, $W_0 = \frac{1}{3}$, $W_{-1} = \frac{1}{2}$. The state multipoles $\langle T^+_{KQ}\rangle$ can be calculated using Eq. (51). Inserting the results into Eq. (49) we obtain

$$\rho = \frac{1}{3^{1/2}} T_{00} - \frac{1}{3(2)^{1/2}} T_{1,0} \qquad (j = j' = 1)$$

We briefly discuss the multipole expansion of a density matrix ρ describing two interacting systems. ρ is a matrix in the composite space spanned by the states $|j_1 M_1\rangle$ and $|j_2 M_2\rangle$ of the two systems. Any operator acting on these states can be expanded in terms of the direct products $T(j_1 j_1')_{K_1 Q_1} \times T(j_2 j_2')_{K_2 Q_2}$. Similar to Eq. (49) we obtain the expansion

$$\rho = \sum_{K_1 Q_1 K_2 Q_2} \langle T(j_1 j_1')_{K_1 Q_1}^+ \times T(j_2 j_2')_{K_2 Q_2}^+ \rangle [T(j_1 j_1')_{K_1 Q_1} \times T(j_2 j_2')_{K_2 Q_2}]$$

(53)

The parameters $\langle T(j_1 j_1')_{00}^+ \times T(j_2 j_2')_{K_2 Q_2}^+ \rangle$ give the various multipoles of the second subsystem, $\langle T(j_1 j_1')_{K_1 Q_1}^+ \times T(j_2 j_2')_{00}^+ \rangle$ of the first one. The parameters with $K_1 \neq 0$ and $K_2 \neq 0$ are correlation terms. For uncorrelated systems we have $\langle T_{K_1 Q_1}^+ \times T_{K_2 Q_2}^+ \rangle = \langle T_{K_1 Q_1}^+ \rangle \langle T_{K_2 Q_2}^+ \rangle$.

The two systems considered here may consist of two different ensembles of particles (for example, electrons and atoms) or of different characteristics of the same particle like electronic angular momentum J and nuclear spin I. In this latter case we can write the density matrix either in the form (49) (where j is then the total angular momentum $J + I$: coupled representation) or in the form (53) (uncoupled representation):

$$\rho = \sum_{\substack{L_1 Q_1 \\ K_2 Q_2}} \langle T(JJ)_{K_1 Q_1}^+ \times T(II)_{K_2 Q_2}^+ \rangle [T(JJ)_{K_1 Q_1} \times T(I\,I)_{K_2 Q_2}] \quad (54)$$

Both expressions are equivalent and can be transformed into each other. We give the transformation equation for the tensor operators

$$T(j'j)_{KQ} = \sum_{\substack{K_1 Q_1 \\ K_2 Q_2}} [(2K_1 + 1)(2K_2 + 1)(2j + 1)(2j' + 1)]^{1/2}$$

$$\times (K_1 Q_1, K_2 Q_2 | KQ) \begin{Bmatrix} K_1 & K_2 & K \\ J & I & j' \\ J & I & j \end{Bmatrix} [T(JJ)_{K_1 Q_1} \times T(II)_{K_2 Q_2}]$$

(55)

This relation can be derived from Eq. (36) and expressing the product $[j'm'\rangle\langle jm|$ in terms of the uncoupled basis.

4.3. Symmetry Properties

The excitation of an atomic ensemble can be achieved in several ways: by the interaction of external fields with the static atomic moments, by absorption of radiation, by electron impact, and so on. Let us assume that the excitation process is axially symmetric with respect to some preferred axis. This axis can be, for example, the direction of an external field. In excitation by electron impact, in which the final electrons are not detected, the symmetry axis is given by the direction of the initial electron beam. This preferred axis may be chosen as z axis of a coordinate system (axis of quantization). The choice of x and y axis perpendicular to z is then arbitrary and the physical properties of the system must be independent of this choice. In particular, the state multipoles, describing the system, must be invariant under a rotation about the z axis. This invariance condition requires (rotation around z with angle ζ)

$$\langle T(j)_{K,Q}\rangle_{\text{rot}} = \sum_{Q'} \langle T(j)_{KQ'}\rangle D_{QQ'}^{(K)*}(0, 0, \alpha) = \langle T(j)_{KQ}\rangle \qquad (56)$$

where we used the transformation properties of spherical tensors. In this subsection we discuss states with definite angular momentum $j = j'$ unless otherwise specified and we write $T(jj)_{KQ} = T(j)_{KQ}$. The element of the rotation matrix is, in the notation of Ref. 18,

$$D_{QQ'}^{(K)*}(0, 0, \alpha) = e^{-iQ'\alpha}\delta_{QQ'}$$

Since α is arbitrary the condition (56) for axial symmetry is only satisfied if $Q = 0$. The excited ensemble is therefore described by multipoles $\langle T(j)_{KQ}\rangle$ with $Q = 0$. From Eq. (51) it follows then that ρ is diagonal:

$$\langle jM|\rho|jM'\rangle = W_M\delta_{MM'} \qquad (57)$$

As discussed in Section 2, the ensemble can then be considered as being an incoherent superposition of states $|j, M\rangle$. On the other hand, if one knows that ρ is diagonal in a particular $\{|j, M\rangle\}$ representation, it follows from Eq. (50) that all multipoles with $Q \neq 0$ vanish.

Axially symmetric ensembles can be classified depending upon their transformation properties under reversal of the symmetry axis. This corresponds to a rotation about an arbitrarily chosen y axis by an angle π represented by the rotation matrix

$$D_{QQ'}^{(K)}(0, \pi, 0) = (-1)^{K+Q}\delta_{Q,-Q'}$$

The invariance condition is $\langle T(j)_{K,Q}\rangle_{\text{rot}} = \langle T(k)_{K,Q}\rangle$. We have

$$\langle T(j)_{K,0}\rangle_{\text{rot}} = \sum_{Q'}\langle T(j)_{KQ'}\rangle(-1)^{K+Q'}\delta_{0,Q'} = \langle T(j)_{K,0}\rangle(-1)^K$$

Hence, for a system invariant under this operation, only state multipoles with even rank K can contribute. From Eq. (51) it follows then that ρ satisfies the condition

$$\langle jM|\rho|jM\rangle = \langle j-M|\rho|j-M\rangle \qquad (58)$$

so that the states $|jM\rangle$ and $|j, -M\rangle$ are equally populated.

Another important case is systems that are invariant under arbitrary rotations, that is, where no preferred axis exists. Because the tensors transform under arbitrary rotations like

$$\langle T(j)_{KQ}\rangle_{\text{rot}} = \sum_{Q'} \langle T(j)_{KQ'}\rangle D^{(K)}_{QQ'}(\alpha\beta\gamma)$$

the invariance condition $\langle T(j)_{KQ}\rangle_{\text{rot}} = \langle T_{KQ}\rangle$ can only be satisfied if $D^{(K)}_{QQ'}$ is the unit matrix; that is, the only nonvanishing multipole is the monopole with $K = Q = 0$. ρ is then proportional to the unit operator:

$$\rho(j) = \frac{1}{2j+1} \cdot 1$$

This means that isotropic ensembles have necessarily equal population for all magnetic sublevels and vice versa.

Finally we discuss briefly the case that the geometry of the excitation process contains only a plane of symmetry, rather than a symmetry axis.[13],[14] Because of the lower symmetry, multipoles with $Q \neq 0$ are not necessarily zero. Thus, the density matrix has nonvanishing off-diagonal elements, indicating coherent excitation of the atomic sublevels. We choose the symmetry plane as the x–z of our coordinate system. Reflection symmetry with respect to this plane implies the condition

$$\langle T(j)_{KQ}\rangle = (-1)^{K+Q}\langle T(j)_{K-Q}\rangle \tag{59a}$$

This relation can be derived from Eq. (36) by using the transformation properties of the states $|j_iM_i\rangle$ under reflections (see, for example, Ref. 16). For $K = 1$ Eq. (59a) can easily be derived by remembering that the orientation vector transforms as a pseudovector; that is, under reflection at the x–z plane the x and z components change their sign but not the y component. Expanding the Cartesian components in terms of the spherical components analogous to Eq. (43) we obtain Eq. (59a).

From Eqs. (40b) and (59a) we obtain

$$\langle T(j)_{K,Q}\rangle = (-1)^K \langle T(j)_{K,Q}\rangle^* \tag{59b}$$

Thus, $\langle T(j)_{1,0}\rangle$ is zero, the quantities $\langle T(j)_{1,\pm 1}\rangle$ imaginary (and therefore only the y component of the orientation vector is nonvanishing), and the quantities $\langle T(j)_{2Q}\rangle$ are real.

5. Time Evolution of Statistical Mixtures

5.1. Equations of Motion

Let $|\psi(t_0)\rangle$ be the state vector representing the dynamical state of the system at time t_0 and $|\psi(t)\rangle$ the state at time t. As discussed briefly in

Section 3.1 the correspondence between $|\psi(t_0)\rangle$ and $|\psi(t)\rangle$ is linear. Consequently we can define a linear operator $U(t, t_0)$ by the relation

$$|\psi(t)\rangle = U(t, t_0)|\psi(t_0)\rangle \tag{60}$$

We recall briefly the main properties of the "time evolution operator" U. (For details see, for example, Ref. 6.)

Inserting Eq. (60) into the time-dependent Schrödinger equation we obtain the differential equation

$$i\hbar\frac{\partial U(t, t_0)}{\partial t} = H(t)U(t, t_0), \qquad U(t_0, t_0) = 1 \tag{61}$$

where H is the Hamiltonian. Since H is Hermitian, U is unitary. Equation (61) can be transformed into the integral equation

$$U(t,t_0) = 1 - \frac{i}{\hbar}\int_{t_0}^{t} H(\tau)U(\tau, t_0)\,d\tau \tag{62}$$

The time dependence of a density matrix

$$\rho(t) = \sum_{a} W_a|\psi_a(t)\rangle\langle\psi_a(t)|$$

follows from Eq. (60):

$$\rho(t) = U(t, t_0)\rho(t_0)U(t, t_0)^{+} \tag{63}$$

Differentiating Eq. (63) with respect to the time and using Eq. (61) we obtain ($\dot{\rho} \equiv d\rho/dt$):

$$\dot{\rho}(t) = -(i/\hbar)[H(t), \rho(t)] \tag{64}$$

Equation (64) is due to von Neumann and is frequently called the quantal Liouville equation, because it assumes the same form as the equation of motion for the phase space probability distribution in classical statistics (with the commutator replaced by the Poisson bracket).

Equations (13) and (64) are basic equations of the theory. It is the simultaneous solution of these equations that leads to equations of motion for the expectation value of any operator.

In the following we assume that $H(t)$ can be written in the form

$$H(t) = H_0(t) + V(t) \tag{65}$$

where $H_0(t)$ is the Hamiltonian of a Schrödinger equation whose solution is known and $V(t)$ describes an interaction inducing transitions between the eigenstates of H_0. $U_0(t, t_0)$ may be the time evolution operator corresponding to H_0. U_0 satisfies Eqs. (61) and (62) (with H replaced by H_0). If H_0 does not depend explicitly upon the time, Eq. (61) can be solved formally:

$$U_0(t, t_0) = e^{-(i/\hbar)H_0(t-t_0)} \tag{66}$$

As an example we discuss a system in the pure state $|\psi(t)\rangle$ satisfying the Schrödinger equation with the Hamiltonian (65). We expand $|\psi(t)\rangle$ in terms of eigenfunctions $|\psi_n(t)\rangle$ of H_0. The time dependence of these states follows from Eqs. (60) and (66).

$$|\psi_n(t)\rangle = U_0(t, t_0)|\psi_n(t_0)\rangle = e^{-(i/\hbar)E_n(t-t_0)}|\psi_n(t_0)\rangle \tag{67}$$

For $|\psi(t)\rangle$ we obtain

$$|\psi(t)\rangle = \sum a_n(t)\, e^{-(i/\hbar)E_n(t-t_0)}|\psi_n(t_0)\rangle \tag{68}$$

The coefficients $a_n(t)$ describe only the time dependence due to the interaction $V(t)$. They are time independent if $V(t) = 0$.

The density operator corresponding to the state $|\psi(t)\rangle$ is given by Eq. (10). Using Eq. (68) we obtain the density matrix with elements ($t_0 = 0$)

$$\langle \psi_n(0)|\rho(t)|\psi_m(0)\rangle = a_n(t)a_m(t)^* \, e^{-(i/\hbar)(E_n-E_m)t} \tag{69}$$

The expectation value of an operator Q at time t is given by Eq. (13):

$$\langle Q \rangle_t = \mathrm{Tr}\,\rho(t) \cdot Q = \sum_{nm} a_n(t)a_m(t)^* \, e^{-(i/\hbar)(E_n-E_m)t}\langle \psi_m(0)|Q|\psi_n(0)\rangle \tag{70}$$

where the off-diagonal elements of the density matrix (69) give the interference terms.

Equations (60) and (64) describe the time dependence of state vectors and density matrices in a particular "representation," the "Schrödinger picture." We can discuss the time dependence in another useful way. If we apply to the state vectors $|\psi(t)\rangle$ the unitary operator $U_0(t, t_0)^+$ we obtain a state vector

$$|\psi(t)_{\mathrm{int}}\rangle = U_0(t, t_0)^+|\psi(t)\rangle \tag{71}$$

From the expansion (68) and Eq. (66) we obtain an expansion of $|\psi(t)_{\mathrm{in}}\rangle$ in terms of eigenstates of H_0:

$$|\psi(t)_{\mathrm{int}}\rangle = \sum_n a_n(t)|\psi_n(t_0)\rangle \tag{72}$$

Equation (72) shows that only the time dependence created by the interaction $V(t)$ is assigned to the states $|\psi_{\mathrm{int}}\rangle$. The rapidly varying factors $\exp[-(i/\hbar)E_n(t-t_0)]$ do not occur in Eq. (72).

The transformation (71) (applied simultaneously to states and operators of the Schrödinger picture) defines the "interaction picture" with the state vectors $|\psi(t)_{\text{int}}\rangle$. At a time t_0 the states $|\psi(t)_{\text{int}}\rangle$ coincide with the states $|\psi(t_0)\rangle$ of the Schrödinger picture [because $U(t_0, t_0) = 1$]. From this, and Eqs. (60) and (71), it follows that the time evolution in the interaction picture is given by

$$|\psi(t)_{\text{int}}\rangle = U(t, t_0)_{\text{int}} |\psi(t_0)_{\text{int}}\rangle \tag{73}$$

where the time evolution operator $U(t, t_0)_{\text{int}}$ is defined by

$$U(t, t_0) = U_0(t, t_0) \cdot U(t, t_0)_{\text{int}} \tag{74}$$

$U(t, t_0)_{\text{int}}$ satisfies the equation

$$i\hbar \frac{\partial U(t, t_0)_{\text{int}}}{\partial t} = V(t)_{\text{int}} U(t, t_0)_{\text{int}} \tag{75}$$

or equivalently

$$U(t, t_0)_{\text{int}} = 1 - \frac{1}{\hbar} \int_{t_0}^{t} V(\tau)_{\text{int}} U(\tau, t_0) \, d\tau \tag{76}$$

$V(t)_{\text{int}}$ is defined by

$$V(t)_{\text{int}} = U_0(t, t_0)^{+} V(t) U_0(t, t_0) \tag{77}$$

If $V(t)$ can be considered as a small perturbation it can be shown that in first-order perturbation theory U_{int} is given by

$$U(t, t_0)_{\text{int}} = 1 - \frac{i}{\hbar} \int_{t_0}^{t} V(\tau)_{\text{int}} \, d\tau \tag{78}$$

The time evolution of the density matrix ρ_{int} in the interaction picture follows from Eq. (73):

$$\rho(t)_{\text{int}} = U(t, t_0)_{\text{int}} \rho(t_0)_{\text{int}} U(t, t_0)_{\text{int}}^{+} \tag{79}$$

From this and Eq. (75) the Liouville equation in the interaction picture can be derived:

$$\dot{\rho}(t)_{\text{int}} = -(i/\hbar)[V(t)_{\text{int}}, \rho(t)_{\text{int}}] \tag{80}$$

which has a simpler structure than Eq. (64).

5.2. Time Evolution of Atomic Ensembles Interacting with Radiation

As an illustration of the techniques discussed in the various sections of this paper we now discuss the radiative decay of an ensemble of excited atomic states. We assume that the atoms have been instantaneously excited at time $t = 0$ and their state is specified by a density matrix $\rho(0)$. We do not specify the excitation mechanism (it may have been electron impact, excitation by photons, foil interaction, etc.). All the details of the excitation process are given by $\rho(0)$, and in the following we assume that $\rho(0)$ is known.

The excited atomic ensemble may be a mixture of states $|j_iM_i\rangle$, where we assume that all atoms have definite values for all quantum numbers except angular momentum j_i and its third component M_i. The ensemble may decay to lower levels $|j_fM_f\rangle$. We assume that these levels have a definite value j_f and we neglect the finite lifetime of the final levels. The initial levels $|j_iM_i>$ are assumed to be degenerate in M_i but not in j_i. All information about the system of atoms and emitted photons at time t are given by the density matrix $\rho(t)$. We are going to derive an expression for the density matrix elements $\langle j_fM_f, \omega\mathbf{n}\lambda |\rho(t)|j_fM_f', \omega\mathbf{n}\lambda'\rangle$, where $|j_fM_f, \omega\mathbf{n}\lambda\rangle$ denotes a state with the atom in a state specified by j_f, M_f and a photon with frequency ω, helicity λ emitted in the direction of the unit vector $\mathbf{n}$.

We normalize $\rho(t)$ so that the diagonal matrix element $\langle j_fM_f, \omega\mathbf{n}\lambda |\rho(t)|j_fM_f, \omega\mathbf{n}\lambda\rangle$ is the probability for finding at time t a photon with frequency ω and helicity λ emitted in the direction $\mathbf{n}$ with the atom in the state $|j_fM_f\rangle$ (which is proportional to the number of photons emitted in the time interval $0, \ldots, t$). From Eqs. (63), (74), and (78) we obtain in first-order perturbation theory

$$
\langle j_fM_f, \omega\mathbf{n}\lambda |\rho(t)|j_fM_f', \omega\mathbf{n}\lambda'\rangle
$$

$$
= \frac{1}{\hbar^2}\left\langle j_fM_f, \omega\mathbf{n}\lambda \left| U_0(t)\left[\int_0^t U_0(\tau)^+ V(\tau)U_0(\tau)\,d\tau\right]\rho(0)\right.\right.
$$

$$
\left.\left.\times\left[\int_0^t U_0(\tau)^+ V(\tau)^+ U_0(\tau)\,d\tau\right]U_0(t)^+ \right|j_fM_f', \omega\mathbf{n}\lambda'\right\rangle \quad (81a)
$$

with the notation $U(t, 0) = U(t)$. The operator $V(t)$ describes the interaction between the atoms and the radiation field. We insert a complete set of initial states $|j_iM_i\rangle$. The evolution operator U_0 describing noninteracting atoms and photons is given by Eq. (66). From this we obtain

$$
U_0(t)|j_iM_i\rangle = e^{-i\omega_i t - \gamma_i t/2}|j_iM_i\rangle, \qquad U_0(t)|j_fM_f, \omega\mathbf{n}\lambda\rangle = e^{-i(\omega+\omega_f)t}|j_fM_f, \omega\mathbf{n}\lambda\rangle \quad (81b)
$$

where we included formally the decay constants γ_i. Inserting these relations into Eq. (81a) we obtain [with $\omega_{fi} = (1/\hbar)(E_f - E_i)$]

$$
\langle j_fM_f, \omega\mathbf{n}\lambda |\rho(t)|j_fM_f', \omega\mathbf{n}\lambda'\rangle
$$

$$
= \frac{1}{\hbar^2}\sum_{\substack{j_ij_i'\\M_iM_i'}} \langle j_fM_f, \omega\mathbf{n}\lambda |V|j_iM_i\rangle\langle j_iM_i|\rho(0)|j_i'M_i'\rangle\langle j_i'M_i'|V^+|j_fM_f', \omega\mathbf{n}\lambda'\rangle
$$

$$
\times\left[\int_0^t d\tau\, e^{i[\omega+\omega_{fi}+(i/2)\gamma_i]\tau}\right]\left[\int_0^t d\tau\, e^{i[-\omega-\omega_{fi}+(i/2)\gamma_i]\tau}\right]
$$

Taking the nonrelativistic limit of the transition matrix elements we obtain in dipole approximation

$$\langle j_f M_f, \omega \mathbf{n} \lambda | V | j_i M_i \rangle = -i \omega_{if} (2\pi \hbar / \omega)^{1/2} e \langle j_f M_f | e_\lambda^* \mathbf{r} | j_i M_i \rangle \qquad (82)$$

with $\mathbf{r}$ the dipole operator. Using a coordinate system spanned by the three orthogonal unit vectors $\mathbf{e}_{\pm 1}$, $\mathbf{n}$ we have $e_\lambda^* \mathbf{r} = r_{-\lambda}(-1)$. Carrying out the time integrations we obtain

$$\langle j_f M_f, \omega \mathbf{n} \lambda | \rho(t) | j_f M_f', \omega \mathbf{n} \lambda' \rangle$$

$$= \frac{2\pi \hbar \, e^2 \omega_{if}^2}{\omega \hbar^2} \sum_{\substack{j_i M_i \\ j_i' n_i'}} \langle j_f M_f | e_\lambda^* \mathbf{r} | j_i M_i \rangle \langle j_i M_i | \rho(0) | j_i' M_j' \rangle \langle j_i' M_i' | (e_{\lambda'}^* r)^+ | j_i M_i \rangle$$

$$\times \left(\frac{\exp\{i[\omega + \omega_{fi} + (i/2)\gamma_i]t\} - 1}{i[\omega + \omega_{fi} + (i/2)\gamma_i]} \right) \left(\frac{\exp\{i[\omega + \omega_{fi'} + (i/2)\gamma_{i'}]t\} - 1}{i[\omega + \omega_{fi'} + (i/2)\gamma_{i'}]} \right)^*$$

In the numerical factor multiplying the right-hand side we have set $\omega_{fi} \approx \omega_{fi}$ since the splitting of the upper levels is much smaller than the energy difference between upper and lower states. We multiply by the density of final states $\omega^2 d\Omega d\omega / (2\pi c)^3$ (where $d\Omega$ is the element of solid angle into which the photon is emitted). We then integrate over ω. Because the main contributions come from the region $\omega \approx \omega_{fi}$ the integral over ω can be extended to $(-\infty)$ with negligible error. The ω integration can then be carried out in the complex ω plane by using Cauchy's integral formula. We obtain a density matrix $\rho(t)'$ with elements

$$\langle j_f M_f, \mathbf{n} \lambda | \rho(t)' | j_f M_f', \mathbf{n} \lambda' \rangle$$

$$= C(\omega) \sum_{\substack{j_i M_i \\ j_i' n_i'}} \langle j_f M_f | e_\lambda^* r | j_i M_i \rangle \langle j_i M_i | \rho(0) | j_i' M_i' \rangle \langle j_f M_f' | e_\lambda^*, r | j_i' M_i' \rangle^*$$

$$\times \frac{1 - \exp\left[-(1/\hbar)(E_i - E_{i'})t - \frac{1}{2}(\gamma_i + \gamma_{i'}) \right]}{(1/\hbar)(E_{i'} - E_i) - (i/2)(\gamma_i + \gamma_{i'})} \qquad (83)$$

$C(\omega)$ is given by

$$C(\omega) = \frac{e^2 \omega^3 \, d\Omega}{2\pi c^3 \hbar^2} \qquad (84)$$

We derive an expression for the elements $\rho_{\lambda\lambda'} = \langle \lambda | \rho(t) | \lambda' \rangle$ of the photon polarization density matrix ($\mathbf{n}$ fixed). As shown in Section 3.2 we have to put $M_f = M_f'$ in Eq. (83) and to sum over M_f. In order to obtain a more useful expression for $\rho_{\lambda\lambda'}$ in which geometrical and dynamical elements are disentangled we proceed as follows. In Eq. (83) all angular momentum components are related to $\mathbf{n}$ as quantization axis. We expand $\rho(0)$ in terms of state multipoles using Eq. (50). The Wigner–Eckart theorem is then applied to the transition matrix elements. The sum over all third components of the angular momenta can be performed by using a standard formula in which a sum over three Clebsch–Gordan coefficients is related to a $6j$-symbol. Finally we transform to a coordinate system defined by the excitation process (for example, the z axis may be the direction of the initial electron beam in electron excitation or the foil axis in beam-foil excitation). This transformation can easily be carried out by using the simple transformation properties of the state multipoles and the invariance of the helicity λ under rotations. The final result is

$$\rho_{\lambda\lambda'}(t) = C(\omega) \sum_{j_i j_{i'}} \langle j_f\|\mathbf{r}\|j_i\rangle\langle j_f\|\mathbf{r}\|j_{i'}\rangle^*(-1)^{j_i + j_f + \lambda}$$

$$\times \sum_{KQQ'} [(2K+1)]^{1/2} \begin{pmatrix} 1 & 1 & K \\ -\lambda' & \lambda & Q \end{pmatrix} \begin{Bmatrix} 1 & 1 & K \\ j_i & j_{i'} & j_f \end{Bmatrix} \langle T(j_i j_{i'})_{KQ'}\rangle D_{QQ'}^{(K)*}(\mathbf{n})$$

$$\times \frac{1 - \exp\left[(i/\hbar)(E_{i'} - E_i)t - \frac{1}{2}(\gamma_i + \gamma_{i'})t\right]}{(1/\hbar)(E_i - E_{i'}) - (i/2)(\gamma_i + \gamma_{i'})} \tag{85a}$$

The diagonal element $\rho_{\lambda\lambda}(t)$ gives the probability for finding the atom at time t in the lower level j_f and the photon emitted in direction $\mathbf{n}$ with helicity λ. The probability $\rho_{\lambda\lambda}(t)$ is proportional to the number of photons emitted between $t = 0$ and t. By differentiating $\rho_{\lambda\lambda'}(t)$ with respect to t one obtains the derivative $\dot{\rho}_{\lambda\lambda'}(t)$ which is proportional to the number of photons emitted in a very short time interval $t \cdots t + dt$. We obtain

$$\dot{\rho}_{\lambda\lambda'}(t) = C(\omega) \sum_{j_i j_{i'}} \langle j_f\|\mathbf{r}\|j_i\rangle\langle j_f\|\mathbf{r}\|j_{i'}\rangle^*(-1)^{j_i + j_f + \lambda}$$

$$\times \sum_{KQQ'} [2K+1]^{1/2} \begin{pmatrix} 1 & 1 & K \\ -\lambda' & \lambda & Q \end{pmatrix} \begin{Bmatrix} 1 & 1 & K \\ j_i & j_{i'} & j_f \end{Bmatrix} \langle T(j_i j_{i'})_{KQ'}\rangle D_{QQ'}^{(K)}(\mathbf{n})$$

$$\times \exp\left[(i/\hbar)(E_{i'} - E_i)t - \frac{1}{2}(\gamma_i + \gamma_{i'})t\right] \tag{85b}$$

From the density matrix elements $\rho_{\lambda\lambda'}$ the Stokes parameters can be obtained by using Eq. (24). In our normalization Tr ρ is the probability of finding a photon in the direction $\mathbf{n}$ independent of its helicity λ.

Only the state multipoles $\langle T_{KQ'}\rangle$ depend on the excitation mechanism. Equations (85) apply therefore to a variety of processes, for example, to lifetime measurements,[15] beam-foil excitation,[16] and angular correlation theory.[13]

The theory given here is very similar to other formulations given in the literature (see Ref. 17 and references therein). However, the emphasis here is placed on calculating the polarization density matrix of the emitted photons, rather than the intensity of photons emitted in a given direction in a definite polarization state. Such states are characteristic not of the emission process itself but of the radiation detectors. In the density matrix formalism (with only the emission direction fixed) it is not necessary to specify the polarization of the final state from the beginning on.

In Eqs. (85) only the reduced matrix elements $\langle j_f \| \mathbf{r} \| j_i \rangle$ depend on the dynamics of the decay process. The time dependence is given explicitly. The angular dependence is given by the elements $D_{Q'Q}^{(K)*}$ of the rotation matrix. Each state multipole $\langle T_{KQ'} \rangle$ is related to the corresponding element $D_{Q'Q}^{(K)*}$, and, thus each multipole gives rise to a characteristic angular dependence of the emitted light. By measuring angular distribution and polarization as functions of $\mathbf{n}$ information about the atomic source can be obtained. This is in particular useful if the excitation process is not well understood (for example, beam-foil excitation).

The $3j$ and $6j$ symbols are universal functions of the angular momentum quantum numbers. These symbols show explicitly the influence of angular momentum conversation on the emission process. Note that the $6j$ symbol vanishes for $K > 2$. Thus, the emission of dipole radiation depends directly only on orientation vector and alignment tensor. In general, by observing dipole radiation, no information can be obtained about the higher multipole parameters characterizing the atomic source.[19]

As discussed in Section 4.3 the geometry of the excitation process often restricts the number of independent multipoles $\langle T_{KQ'} \rangle$. If all parameters $K \neq 0$ vanish (isotropic excitation) the photons are emitted isotropically. From the properties of the $3j$ symbol follows $\rho_{\lambda\lambda'} = 0$ for $\lambda \neq \lambda'$ and $\rho_{\lambda\lambda} = \rho_{-\lambda-\lambda}$. Thus, all Stokes parameters η_i are zero, that is, the photons are unpolarized. The $6j$ symbol is zero unless $j_i = j_{i'}$; therefore the time factor vanishes and no quantum beats can be observed.

Departures from spherically symmetric atomic sources can be represented in terms of orientation vector and alignment tensor. If the excitation mechanism has axial symmetry, the only nonvanishing multipole parameters are the monopole and $\langle T_{2,0} \rangle$. If the geometry of the excitation process contains only a plane of symmetry, Eqs. (59) apply. These relations show that the orientation and alignment can be described by four independent multipoles: one for the orientation and three for the alignment. From Eqs. (59) and the symmetry properties of the $3j$ symbol it follows that η_2 depends only on the orientation parameter, η_1 and η_3 only on the alignment parameters. Therefore measurement of the circular polarization is required in order to obtain information on the components $\langle T_{1,\pm1} \rangle$ of the orientation vector.

We discussed the main features of Eqs. (85) in some detail in order to show the importance and usefulness of the multipole expansion of the density matrix given in Section 4.

5.3. Perturbed Angular Distributions

In Section 5.2 we assumed that atomic states with different angular momenta $j_{i'}, j_i'$ had been coherently excited at the time $t = 0$. This coherency leads to a modulation of the exponential decay of the excited states ("quantum beats") as in Eqs. (85a) and (85b). No quantum beats will occur for incoherent excitation.

This conclusion may not be correct if the atomic lifetime is sufficiently long and if the excited states are perturbed by external or internal fields during the time between excitation and decay. (This assumes sharply defined excitation and emission times.) These perturbations will disturb the (coherently or incoherently) excited states and lead to a time modulation of the orientation and alignment parameters and, thus, of the angular distributions and correlations.

The theory of perturbed angular distributions has been developed in nuclear physics[12] and has also been reviewed recently.[17] We put forward here the basic principles of the theory and discuss the influence of hyperfine interaction as an example.

The unperturbed intensity $I(\mathbf{n}, t)$ of light, emitted at time t in the direction $\mathbf{n}$, can be written in the form

$$I(\mathbf{n}, t) = \overline{C(\omega)} \sum_{\substack{j_f M_f \lambda \\ j_i M_i M_i'}} \langle j_f M_f | r_{-\lambda} | j_i M_i \rangle \langle j_i M_i | U_0(t) \rho(0) U_0(t)^+ | j_i M_i' \rangle$$
$$\times \langle j_i M_i' | (r_{-\lambda})^+ | j_f M_f \rangle$$

$$= \overline{C(\omega)} \sum_\lambda \text{Tr} \left[r_{-\lambda} U_0(t) \rho(0) U_o(t)^+ (r_{-\lambda})^+ \right] \tag{86}$$

where

$$\overline{C(\omega)} = \hbar\omega C(\omega) = \frac{e^2 \omega^4 \, d\Omega}{2\pi c^3 \hbar}$$

Expression (86) can be derived from Eq. (85b) or Eq. (83) using the relation (81a). In this and the following section we discuss initial states with sharp angular momentum $j_i = j_i'$. The state of those atoms, which at time t are still in the state with initial angular momentum j_i, is characterized by the density matrix $\rho(t)$ which evolves from $\rho(0)$ according to

$$\overline{\rho(t)} = U_0(t)\rho(0)U_0(t)^+ \tag{87}$$

[Note that the density matrix $\rho(t)$, introduced in Section 5.2, describes the state of the whole ensemble at time t, that is, atoms in the initial state j_i, and final state j_f, and the emitted photons.] Using Eqs. (86) and (87) we obtain

$$I(\mathbf{n}, t) = \overline{C(\omega)} \sum_\lambda \text{Tr} \left[r_{-\lambda} \overline{\rho(t)} (r_{-\lambda})^+ \right] \tag{88}$$

We will now discuss the time evolution of the density matrices and state multipoles in the presence of a perturbation described by a Hamiltonian H'. We assume that the total Hamiltonian $H = H_0 + H'$ does not depend explicitly on the time. In particular we consider atomic excitations in which the eigenstates of both H_0 and H are fully identified by angular momentum quantum numbers j_i, M_i and j, M. The atomic states may be excited instantaneously at time $t = 0$. The perturbations H' considered are assumed to be weak and of little relevance to the initial excitation. The term H' of the Hamiltonian can then be neglected during the excitation and the atoms are then excited in the eigenstate representation $\{|j_iM_i\rangle\}$ of H_0, characterized by a density matrix $\rho(0)$. At times $t > 0$ the atoms are described by the total Hamiltonian H. The time evolution between excitation and decay is now determined by an operator $U(t)$:

$$U(t) = e^{-(i/\hbar)Ht - (1/2)\Gamma t} \tag{89}$$

where Γ is the decay operator. The states of those atoms which have not decayed at time t is characterized by the density matrix

$$\overline{\rho(t)} = U(t)\rho(0)U(t) \tag{90}$$

The probability of finding an undecayed atom at time t is given by the trace of $\overline{\rho(t)}$ (which decreases exponentially with time t).

We expand $\rho(0)$ and $\overline{\rho(t)}$ in terms of state multipoles

$$\rho(0) = \sum_{KQ} \langle T(j_i; 0)_{KQ}^+ \rangle T(j_i)_{KQ} \tag{91a}$$

$$\overline{\rho(t)} = \sum_{kq} \langle T(j, j'; t)_{kq}^+ \rangle T(j, j')_{kq} \tag{91b}$$

The tensor operators $T(j_i)_{KQ}$ are constructed from eigenstates $|j_iM_i\rangle$ of H_0 and the tensors $T(j, j')_{kq}$ from eigenstates $|j, M\rangle$ of H using Eq. (36). We now relate the state multipoles at time t to the multipole parameters characterizing the excited atomic ensemble at *time* $t = 0$. From Eqs. (90) and (91a) it follows that

$$\langle T(j, j'; t)_{kq}^+ = \mathrm{Tr}\, \overline{\rho(t)} T(j, j')_{kq}^+$$

$$= \mathrm{Tr}\, U(t)\rho(0)U(t)^+ T(j, j')_{kq}^+$$

$$= \sum_{KQ} \langle T(j_i; 0)_{KQ}^+ \rangle G(t)_{Kk}^{Qq} \tag{92}$$

where we have introduced the "perturbation coefficients"[12]

$$G(t)_{Kk}^{Qq} = \mathrm{Tr}\, U(t)T(j_i)_{KQ}U(t)^+ T(j, j')_{kq}^+ \tag{93}$$

These quantities contain all the information about the perturbation. They are the coefficients in the expansion of the state multipoles at time t in terms of multipoles excited at $t = 0$. Equations (92) and (93) are the fundamental equations for determining the time evolution of the state multipoles. For a particular interaction, the problem of evaluating $G(t)_{Kk}^{Qq}$ reduces to that of finding the connection between the eigenstates of H_0 and H. $U(t)$ is diagonal in the eigenstate representation $\{|j, M\rangle\}$ of H. Evaluating expression (93) in this representation and applying Eq. (89) we obtain

$$G(t)_{Kk}^{Qq} = \sum_{\substack{iM \\ j'M'}} \exp\left[-(i/\hbar)(E_j - E_{j'})t - \tfrac{1}{2}(\gamma_j + \gamma_{j'})t\right]$$

$$\times \langle jM|T(j_i)_{KQ}|j'M'\rangle\langle j'M'|T(j, j')_{kq}^+|jM\rangle \qquad (94)$$

which gives the time dependence explicitly. The calculation of the matrix elements of the operators $T(j_i)_{KQ}$ requires knowledge of the coefficients $\langle j_iM_i|jM\rangle$ occurring in the expansion of the eigenstates of H in terms of the eigenstates of H_0:

$$|jM\rangle = \sum_{M_i} |j_iM_i\rangle\langle j_iM_i|jM\rangle$$

These coefficients are obtained by diagonalizing the Hamiltonian matrix.

The angular distribution of the light emitted at time t is given by Eq. (88), where $\overline{\rho(t)}$ is now determined by Eqs. (90) and (91b). By inserting Eq. (91b) into Eq. (88) we obtain

$$I(\mathbf{n}, t) = \overline{C(\omega)} \sum_{\lambda kq} \mathrm{Tr}\left[r_{-\lambda}T(j, j')_{kq}r_{-\lambda}^+\right]\langle T(j, j'; t)_{kq}^+\rangle \qquad (95)$$

This expression is written in the helicity system, whereas the state multipoles at time t and $t = 0$ are most conveniently expressed in the collision system. We therefore transform Eq. (95) to the collision system as in Section 5.2, and inserting then Eq. (92) we obtain

$$I(\mathbf{n}, t) = \overline{C(\omega)} \sum_{\substack{\lambda kq \\ KQq'}} \mathrm{Tr}\left[r_{-\lambda}T(j, j')_{kq}(r_{-\lambda})^+\right]D_{qq'}^{(k)}(\mathbf{n})\, G(t)_{Kk}^{Qq'}\langle T(j_i;0)_{KQ}^+\rangle$$

$$(96)$$

We emphasize the point that this expression is separated into three parts describing the excitation at $t = 0$, the perturbation, and the decay at time t. This separation is made possible by application of the irreducible tensor method, that is, by using Eqs. (91) Equations (94) and (96) are the fundamental equations for a discussion of perturbed angular distributions and quantum beat calculations.

As an example of the above theory, we discuss in some detail the effect of hyperfine interaction upon the angular distribution. At time $t = 0$ electronic states $|j_iM_i\rangle$ have been excited leaving the nuclear spin states $|I, M_I\rangle$

unaffected. The uncoupled states $|j_iM_i\rangle|IM_I\rangle$ are the eigenstates of H_0. The density matrix of the excited electronic system

$$\rho(0)_{el} = \sum_{KQ} \langle T(j_{i;}\, 0)^+_{KQ}\rangle T(j_i)_{KQ}$$

must be generalized to a density matrix describing the state of the whole system of electronic and nuclear states. The joint matrix in the uncoupled representation is given by Eq. (54). Usually the nuclear spins are unpolarized. It follows from the discussion in Section 4.3 that only the monopole $T(I)_{00} = (2I+1)^{-1/2}1$ contributes to the nuclear spin matrix.

Because the electronic and nuclear states are uncorrelated at time $t = 0$ we obtain

$$\rho(0) = \frac{1}{2I+1} \sum_{KQ} \langle T(j_{i;}\, 0)^+_{KQ}\rangle[T(j_{i;}\, 0)^+_{KQ} \times 1] \tag{97}$$

At times $t > 0$ the joint system is described by a Hamiltonian $H = H_0 + H'$ with the hyperfine interaction operator $H' \sim j_iI$. The eigenstates of H are identified as $|(j_iI)FM_F\rangle$. The hyperfine interaction causes an exchange of angular momentum between electronic and nuclear states, which leads to a polarization of nuclear spins. The density matrix $\overline{\rho(t)}$ can either be written in the coupled representation with operators $T(F, F')_{kq}$ analogous to Eq. (91b) or, as will be more convenient for the following discussion, in the uncoupled representation (54):

$$\overline{\rho(t)} = \sum_{\substack{kq \\ k_Iq_I}} \langle T(j_{i;}\, t)^+_{kq} \times T(I;\, t)^+_{k_Iq_I}\rangle T(j_i)_{kq} \times T(I)_{k_Iq_I} \tag{98}$$

The terms $\langle T(j_{i;}\, t)^+_{kq} \times T(I;\, t)^+_{00}\rangle$ represent pure electronic multipoles, $\langle T(j_{i;}\, t)^+_{00} \times T(I;\, t)^+_{k_Iq_I}\rangle$ pure nuclear polarization, and the terms with $k \neq 0$ and simultaneously $k_I \neq 0$ represent correlation terms.

Inserting Eq. (98) into Eq. (88) we have that, since the light emission depends on the electronic variables alone, then only the term with $k_I = 0$ can contribute, i.e.,

$$\mathrm{Tr}\,\{r_{-\lambda}[T(j_i)_{kq} + T(I)_{k_Iq_I}](r_{-\lambda})^+\} = \mathrm{Tr}\,[r_{-\lambda}T(j_i)_{kq}(r_{-\lambda})^+] \cdot \mathrm{Tr}\,[T(I)_{k_Iq_I}]$$

$$= (2I+1)^{1/2}\delta_{k_I0}\delta_{q_I0}\,\mathrm{Tr}\,[r_{-\lambda}T(j_i)_{kq}(r_{-\lambda})^+]$$

From this it follows that

$$I(\mathbf{n}, t) = \overline{C(\omega)} \sum_{kq\lambda} \mathrm{Tr}\,[r_{-\lambda}T(j_i)_{kq}(r_{-\lambda})^+]\langle T(j_{i;}\, t)^+_{kq} \times 1\rangle$$

$$= \overline{C(\omega)} \sum_{\substack{\lambda kq \\ KQq'}} \mathrm{Tr}\,[r_{-\lambda}T(j_i)_{kq}(r_{-\lambda})^+]D^{(k)}_{qq'}(\mathbf{n})G(t)^{Qq}_{Kk}\langle T(j_{i;}\, 0)^+_{KQ}\rangle \tag{99}$$

The perturbation coefficient

$$G(t)_{Kk}^{Qq} = \frac{1}{2I+1} \operatorname{Tr} U(t)[T(j_i)_{KQ} \times 1]U(t)^+[T(j_i)_{kq}^+ \times 1] \qquad (100)$$

can be obtained easily: We calculate the trace in the coupled representation $|(j_iI)FM_F\rangle$ in which $U(t)$ is diagonal. The total Hamiltonian H is invariant under joint rotations of electronic and nuclear coordinates; thus

$$U(t)[T(j_i)_{KQ} \times 1]U(t)^+$$

is another tensor operator with the same rank K and component Q. It follows that the trace (100) vanishes unless $K = k$, $Q = q$. Using this result, we obtain immediately

$$G(t)_{Kk}^{Qq} = \frac{1}{2I+1} \sum_{\substack{FM_F \\ F'M_F}} \exp\left[-(i/\hbar)(E_F - E_{F'})t - \tfrac{1}{2}(\gamma_F + \gamma_{F'})t\right]$$
$$\times |\langle(j_iI)FM_F|T(j_i)_{KQ} \times 1|(j_iI)F'M_F'\rangle|^2 \delta_{Kk}\delta_{Qq}$$

The matrix element of the tensor operator in the coupled representation is given by a standard expression (see, for example, Chapter 7 of Ref. 18), which gives finally

$$G(t)_{Kk}^{Qq} = \sum_{FF'} \frac{(2F+1)(2F'+1)}{2I+1} \left\{ \begin{matrix} j_i & F & I \\ F' & j_i & K \end{matrix} \right\}^2 \exp\left[i(-/\hbar)(E_F - E_{F'})t \right.$$
$$\left. -\tfrac{1}{2}(\gamma_F + \gamma_{F'})t\right]\delta_{Kk}\delta_{Qq} \qquad (101)$$

Inserting Eq. (101) into Eq. (99) gives the final expression for the angular distribution. For a detailed discussion of expression (101) and further examples, we refer to the literature, particularly to Ref. 17. A detailed discussion of quantum beat experiments may be found in Ref. 16.

5.4. The Liouville Formalism

In this section we give a brief introduction into the theory of perturbed angular distribution and quantum beat measurements in the Liouville formalism. We follow the approach of Ref. 20.

The elements of the usual Hilbert space are state vectors $|\psi\rangle$. The set of all linear operators $A, B, \ldots$ acting on the Hilbert states $|\psi\rangle$ also span a linear space: Any linear combination of operators $A, B, \ldots$ is also a linear operator. The operators in Hilbert space can be interpreted as "elements" $|A\rangle, |B\rangle, \ldots$ in this space. This space is called the "Liouville space" if an inner product is defined by

$$\langle A|B\rangle = \operatorname{Tr}(A^+B) \qquad (102)$$

We say that two elements $|A\rangle, B\rangle$ of the Liouville space are orthogonal to each other if their inner product vanishes [see, for example, the remark following Eq. (41)].

Operators in Liouville space acting on the elements $|A\rangle$, $|B\rangle$, . . . can then be introduced. These "superoperators" will be denoted by $\hat{A}$, $\hat{B}$, ... Of particular importance is the "Liouvillian" $\hat{H}$ which is the equivalent of the Hamiltonian H, defined by

$$\hat{H}|A\rangle = |[H, A]\rangle \qquad (103)$$

where $[H, A]$ is the usual commutator in Hilbert space. The time evolution operator in Liouville space is

$$\hat{U}(t) = e^{-(i/\hbar)\hat{H}t} \qquad (104)$$

The time evolution of the density matrix $\overline{\rho(t)}$ interpreted as element $|\overline{\rho(t)}\rangle$ in Liouville space is given by

$$|\overline{\rho(t)}\rangle = e^{-(i/\hbar)\hat{H}t}|\rho(0)\rangle \qquad (105)$$

The set of tensor operators $T(jj')_{KQ}$ defined by Eq. (36) corresponds to the elements $|T(jj')_{KQ}\rangle$ in Liouville space. Applying $\hat{H}$ to these elements and using Eqs. (36) and (103) we obtain

$$\hat{H}|T(jj')_{KQ}\rangle = (E_j - E_{j'})|T(jj')_{KQ}\rangle \qquad (106)$$

(with $H|j, M\rangle = E_j|j, M\rangle$). It is useful to construct the unit operator

$$\hat{1} = \sum_{KQ} |T(jj')_{KQ}\rangle\langle T(jj')_{KQ}| \qquad (107)$$

[That this operator acting on the elements $|T(jj')_{KQ}\rangle$ is indeed a unit operator follows immediately from Eqs. (41) and (102).]

We use these concepts to derive an expression for the intensity $I(\mathbf{n}, t)$ of light, emitted at time t by an atomic ensemble characterized by the density matrix $\overline{\rho(t)}$. Interpreting the trace (88) as an inner product in Liouville space and applying Eq. (105) we obtain

$$I(\mathbf{n}, t) = \overline{C(\omega)} \sum_\lambda \text{Tr}\,[(r_{-\lambda})^+ r_{-\lambda}\overline{\rho(t)}]$$

$$= \overline{C(\omega)} \sum_\lambda \langle(r_{-\lambda})^+ r_{-\lambda}|e^{-(i/\hbar)\hat{H}t}|\rho(0)\rangle \qquad (108)$$

Inserting the unit operator (107) twice in the relation (108) we have

$$I(\mathbf{n}, t) = \overline{C(\omega)} \sum_{\substack{KQ\lambda \\ kq}} \langle(r_{-\lambda})^+ r_{-\lambda}|T(j, j')_{kq}\rangle\langle T(j, j')_{kq}|e^{-(i/\hbar)\hat{H}t}|T(j_i)_{KQ}\rangle$$
$$\times \langle T(j_i)_{KQ}|\rho(0)\rangle \qquad (109)$$

As in Eq. (95) the first factor describes the decay, the second the perturbation, and the third the excited state. Applying Eqs. (102) and (106) these factors can be transformed into the more familiar form introduced in Section 5.3.

Equation (109) may be used as the fundamental equation in a discussion of perturbed angular distributions and quantum beat measurements. For details we refer to Refs. 16 and 20 and to the article of H. J. Andrä in Chapter 20 of this volume.

Acknowledgments

I am grateful to H. Kleinpoppen, H. J. Andrä, and W. E. Baylis for helpful discussions and suggestions and to K. Kollath and H. Jakubowicz for reading the manuscript. I would also like to thank Ms. A. Dunlop for her help in preparing the manuscript.

References

1. U. Fano, *Rev. Mod. Phys.* **29**, 74 (1957).
2. R. McWeeny, *Rev. Mod. Phys.* **32**, 335 (1960).
3. D. Ter Haar, *Rep. Progr. Phys.* **24**, 304 (1961).
4. P. O. Löwdin, *Phys. Rev.* **97**, 1924 (1955).
5. C. Cohen-Tannouidji, *Ann. Phys.* (*Paris*) **7**, 423 (1962).
6. A. Messiah, *Quantum Mechanics*, North Holland, Amsterdam (1965).
7. P. Roman, *Advanced Quantum Mechanics*, Addison-Wesley, New York (1965).
8. K. Gottfried, *Quantum Mechanics*, Vol. 1, W. A. Benjamin, New York (1966).
9. R. C. Tolman, *The Principles of Statistical Mechanics*, Oxford University Press, London (1954).
10. U. Fano, *Phys. Rev.* **90**, 577 (1953).
11. W. Harper, *Rev. Mod. Phys.* **44**, 169 (1972).
12. H. Frauenfelder and R. M. Steffen, in *Alpha-, Beta-, Gamma-Ray Spectroscopy*, Ed. K. Siegbahn, North Holland, Amsterdam (1968).
13. J. Macek and O. H. Jaecks, *Phys. Rev.* **179**, 2288 (1971).
14. H. G. Berry, L. J. Curtiss, D. C. Ellis, and R. M. Schectman, *Phys. Rev. Lett.* **32**, 751 (1974).
15. J. Macek, *Phys. Rev. A* **4**, 618 (1970).
16. H. J. Andrä, *Phys. Scr.* **9**, 252 (1974).
17. U. Fano, and J. Macek, *Rev. Mod. Phys.* **45**, 553 (1973).
18. A. R. Edmonds, *Angular Momentum in Quantum Mechanics*, Princeton University Press, Princeton, New Jersey (1957).
19. J. Macek and I. V. Hertel, *J. Phys.* **B7**, 2173 (1974).
20. J. Bosse and H. Gabriel, *Z. Phys.* **266**, 283 (1974).
21. H. G. Berry *et al.*, *J. Phys.* (*Paris*) **33**, 947 (1972).
22. A. Omont and J. Meunier, *Phys. Rev.* **169**, 92 (1968).

23. M. J. D'Yanokov and V. F. Perel, *Sov. Phys. JETP* **20**, 997 (1965).
24. W. Happer and B. S. Mathur, *Phys. Rev.* **163**, 12 (1967).
25. K. Blum and H. Kleinpoppen, *J. Phys.* **B8**, 922 (1975).
26. K. Blum and H. Kleinpoppen, *Electron and Photon Interactions with Atoms*, Eds. H. Kleinpoppen and M. R. C. McDowell, Plenum Press, New York (1976).

3

Perturbation of Atoms

STIG STENHOLM

1. Introduction

The free atom exists only in physics textbooks. In practice literally everything in the world can perturb the atoms, and experimentalists have to take extreme care to approach the unreachable ideal atom they purport to observe. Both standard textbook effects like Zeeman splittings and vacuum fluctuations as well as exotic ones like gravitational red-shifts serve as examples of perturbations. Not only is it impossible to encompass all this richness in one article, but the task exceeds the power of any single author, as the field is bound to expand until it engulfs all of known physics. Thus we have to restrict ourselves to a few phenomena only, and the criteria used will be presented in this introduction.

Following the general trend of the present monograph, we will discuss only phenomena of interest to spectroscopy. The influence of static electric and magnetic fields, Zeeman and Stark effects, are regarded as part of the spectra to be determined, as well as hyperfine structure and relativistic effects. Also the effects of the electromagnetic vacuum will be omitted; the physics of Lamb shifts and spontaneous emission line widths will be discussed in other articles. Here we regard all these fundamental effects as forming one compound system, the eigenvalues of which provide the subject matter to be studied by spectroscopy. Instead we discuss phenomena that can, at least in principle, be imposed at will in the laboratory. These form a research area rich in physical effects that can be studied under systematically varied external conditions.

STIG STENHOLM • Department of Technical Physics, Helsinki University of Technology, and Research Institute for Theoretical Physics, University of Helsinki.

111

Finally we choose to restrict the scope of the article to some questions that have recently aroused the interests of physicists working in the field of spectroscopy. Here it becomes unavoidable that personal bias plays a role, and I cannot but apologize for important works neglected.

One major theme of this article will be the wealth of phenomena observed in strong radio frequency fields and the various theoretical methods used to explain the results. Similar phenomena occur in strong optical fields, but with the exception of a few phenomena of spectroscopic importance we will not attempt to cover the field of laser physics. Some aspects have been reviewed earlier,[1] and the rapidly growing amount of work certainly postpones the time for a full evaluation to the future.

An important perturbation, which spectroscopic data have to be extricated from, is collision effects. The tables can, of course, be turned, and the measurements can be used to learn about atomic collision phenomena. In either case one must possess an adequate theory in order to separate the various effects. There has been some progress in this area recently and we will briefly review the main formulation of the problem. A detailed discussion of the applications of the theory and the experimental situation falls outside the scope of this article.

2. Influence of Electromagnetic Fields

2.1. General Considerations

2.1.1. Survey of the Effects Encountered

When an atom is situated in a strong field, a variety of effects are encountered, and sometimes the terminologies used by various authors are inconsistent and confusing. In this section we will try to enumerate the various contributions and indicate their relationship.

We start by a two-level atom with the energy spacing $\hbar\omega$, which is situated in the perturbing potential

$$V = \hbar v \cos \Omega t \tag{1}$$

Denoting the upper state by $|+\rangle$ and the lower one by $|-\rangle$ we expand the general state as

$$|\Psi\rangle = a\, e^{-i\Omega t}|+\rangle + b|-\rangle \tag{2}$$

We then find that the coefficient vector $\{a, b\}$ develops with the Hamiltonian matrix

$$H = \hbar \begin{pmatrix} \omega - \Omega & \frac{1}{2}v \\ \frac{1}{2}v & 0 \end{pmatrix} + \frac{\hbar v}{2}\begin{pmatrix} 0 & e^{-2i\Omega t} \\ e^{2i\Omega t} & 0 \end{pmatrix} \tag{3}$$

The second term is rapidly oscillating and is neglected in the first approximation. Following the convention of nuclear magnetic resonance this is called "the rotating wave approximation" (RWA).

After the RWA the Hamiltonian can be diagonalized and we find the two eigenfrequencies

$$\lambda = \tfrac{1}{2}\{\omega - \Omega \pm [(\omega - \Omega)^2 + v^2]^{1/2}\} \tag{4}$$

which display a shift of the resonance line determined by the coupling strength v. This is the ac (or dynamic) Stark shift[2] in analogy with the dc case.

Each state $|\pm\rangle$ becomes a superposition of the field-free eigenstates, and radiation originating on one level will display both frequencies (4). Thus the ac Stark shift emerges as a field-induced splitting of the lines.

For small values of the coupling we expand

$$\varepsilon \equiv \lambda + \Omega = \omega + \frac{v^2}{4(\omega - \Omega)} \tag{5}$$

for the upper level. This is the result obtained directly from a second-order perturbation treatment in v. In this approximation the influences on one level i by the other levels j can be superimposed and we find

$$\varepsilon_{i0} = \omega_{i0} + \sum_{j(\neq i)} \frac{v_{ij}^2}{4(\omega_{ij} - \Omega)} \tag{6}$$

where we count the energies from the arbitrary level 0. This is the perturbation-theory shift of the level i by the field. If we have a continuum of radiation with the weight function $g(\Omega)$ the shift becomes

$$\varepsilon_{i0} = \omega_{i0} + \frac{1}{4} \int d\Omega \, g(\Omega) \sum_{j(\neq i)} \frac{[v_{ij}(\Omega)]^2}{\omega_{ij} - \Omega} \tag{7}$$

This is the level shift caused by a broad light source (e.g., thermal lamps) of high intensity. The result is called a light shift (or "lamp shift") and has been investigated experimentally by Cohen–Tannoudji et al.[3-7]

If we include the term oscillating at the frequency 2Ω from Eq. (3) and use second-order perturbation theory, we find that the levels are shifted by the amounts

$$\Delta\varepsilon = \pm \frac{1}{4} \frac{\hbar^2 v^2}{2\Omega} \tag{8}$$

The transition frequency is consequently shifted by the amount

$$\Delta\varepsilon = \frac{\hbar v^2}{4\Omega} \tag{9}$$

which was first calculated by Bloch and Siegert.[8] It is entirely due to the inclusion of the "counter-rotating term" neglected in the RWA.

These effects are well understood and experimentally confirmed in the rf range, and combined with the detection sensitivity of optical pumping they have been generalized in several directions.[5] This work will be the topic of Sections 2.2 and 2.3. The strong monochromatic light sources provided by laser technology have made similar effects within the reach of experiments. In Sections 2.4 and 2.5 we will consider some of the work in this field.

2.1.2. Historical Review

The main part of this article will be arranged according to the subject matter treated, but in this section I will give a historical account of the field. I will attempt to disentangle the various contributions that have merged to give the rather satisfactory description we have of effects caused by an intense field. We will find that many persons have contributed and that the important steps were taken simultaneously by several groups, and often again and again in different contexts.

The classic paper about a two-level system coupled by a perturbation was written by Rabi[9] in 1937. Here the periodic "flipping" was derived. The effect of the counter-rotating component was, in 1940, shown to give what is now called the Bloch–Siegert shift.[8] One important branch of rf resonance work was closely connected with the development of nuclear magnetic resonance in solids and liquids, which we will not discuss further here.

With strong fields available, one can observe transitions where the required energy is provided by several rf photons. These high-order transitions suffer a generalized Bloch–Siegert shift, which was calculated[10] and observed[11] in 1955.

The full problem of two discrete states coupled by an alternating field was attacked by Autler and Townes.[2] In 1955 they published a formally exact solution for the quantum state in terms of a continued fraction, and they also made extensive use of effective frequency diagrams arising from the eigenvalues of the coupled system of field and discrete levels. They verified the theory on the molecule OCS, where a clear Stark splitting due to the ac field is observed.

At the beginning of the 1960's the theory of optical pumping experiments was rapidly forthcoming. Coherence and modulation effects were described[12,13] and Barrat and Cohen-Tannoudji derived the basic equations[14,15] for pumping with incoherent light. This field rapidly expanded to an independent branch, which we will not try to follow here; see review articles.[5,16,17]

The next major step in the theory of oscillational perturbations of two-level atoms was taken by Shirley in his 1963 California Institute of Technology thesis. He applied Floquet's theory of differential equations with periodic coefficients and reduced the problem to solving the eigenvalues of an infinite matrix. He then applied a variety of techniques to obtain solutions: perturbation expansions, continued fractions, analytic approximations, and numerical computations. Unfortunately not all of the results found their way into the published papers.[18,19]

The two-level system with a periodically modulated level spacing can be solved exactly in terms of Bessel's functions.[20,21] The result is identical with the parametric resonances known from electronics. A quantum mechanical treatment was given by Polonsky and Cohen-Tannoudji.[22] It was shown that the atomic g factor is modified by the rf field to oscillate according to Bessel's function J_0, and this result was verified experimentally by Cohen-Tannoudji, Haroche, and co-workers.[23–26] The corresponding result for the oscillating components of the induced magnetization was obtained by Pegg and Series[27] and measured by Chapman.[28] Similar results were obtained by Yabuzaki et al.[29]

The work in the French group developed a point of view where the combined system of atom plus field was considered, and its energy levels were coupled by the perturbation. This "dressed atom" philosophy was applied to a large number of problems[23–26,30–32] and introduced in a unified way at the Cargése summer school, published in 1968.[5]

In the French group it was discovered that a generalized level crossing would arise when the atoms are subject to transverse pumping and an even number of rf photons match the energy difference between the levels. These "Haroche's resonances" were experimentally observed in Hg by Haroche and Cohen-Tannoudji,[24,26] and later by Tsukuda et al.,[33] who fitted their linearly oscillating field results by a semiclassical rotating wave description. This same theory was compared with the correct experimental arrangement by Aleksandrov and Sokolov.[34]

The semiclassical theory of the laser was formulated in terms of a density matrix description by Lamb in 1964.[35] Owing to the atomic motion in the standing wave the atoms see a modulated field with a velocity-dependent frequency. The continued fraction solution for the density matrix elements was rediscovered by Stenholm and Lamb[36] in 1968, and independently immediately after by Feldman and Feld.[37]

The relationship with rf resonances was early recognized, and in 1972 the present author could publish a semiclassical theory of the multiphoton resonances.[38–40] The generalized Bloch–Siegert shifts could be calculated to any desired accuracy even for very-high-order resonances. The major advantage over earlier approaches[2,19] was the ability to treat arbitrary relaxation rates phenomenologically.

The method used in the semiclassical approach is to expand in a Fourier series with multiples of the rf frequency. Already Shirley[19] pointed out that such a series makes the problem equivalent with the corresponding quantized field approach. The Fourier series method had also been introduced by Weber and Capeller[41–42] in 1971 in combination with an irreducible tensor expansion of the density matrix. They obtained the higher-order resonances numerically but did not find the continued fraction solution. The quantized field perturbation expansion for a strong field was discussed by Gush and Gush,[43] who utilized a continued fraction representation of the result.

At the time when these papers[38,39] were published there were no experimental results extending up to very strong rf fields in free atoms. The electron spin resonance measurements on metallic lithium by Morand and Theobald[44,45] followed both the one- and three-photon resonances until they disappeared down to zero field. The agreement with the semiclassical theory is excellent,[46] but it has been pointed out by Hannaford et al.[47] that the relaxation situation in a solid may not correspond to the simple one assumed by the theory. Arimondo and Moruzzi[48] attempted to verify the theory on Hg vapor. The results agreed in general but apparently owing to power broadening and alignment problems the measurement were inconclusive near zero dc fields.

In 1971 the interest in rf resonances increased thanks to two factors: First, Haroche published a monumental compilation[49,50] of the experimental work at the E.N.S. in Paris and its description within the dressed atom philosophy. Second, Chang and Stehle[51,52] appeared to find indications that the semiclassical description would fail at large intensities. If this were correct the rf experiments would constitute a crucial test of the validity of the semiclassical approach, and a great number of papers were devoted to this discussion.

Pegg and Series[53] criticized the conclusions and Pegg[54] reconsidered the problem. The Bloch–Siegert shift of the quantum theory was obtained as a perturbation series by Cohen-Tannoudji et al.,[55] which agreed with Shirley's semiclassical result.[19] The same series could be derived from the continued fraction result.[56] It thus seems that general agreement prevails. It has also been shown[57] how to derive the semiclassical equations from the quantum field Hamiltonian.

Independently of the previous work Swain[58,59] discovered the continued fraction solution of a two-level atom in one quantized field mode. A similar solution was presented by Stenholm[1] (page 53), but there it was not applied to any useful problem. In a series of articles[60–64] Swain could soon apply his approach to obtain the Bloch–Siegert shifts in agreement with the numerical results obtained by Hannaford et al.[47] and Ahmad and Bullough.[65,66] The general semiclassical theory is discussed by Pegg and Series.[67,68]

Several special treatments of the generalized Bloch–Siegert shifts deserve mentioning. Armstrong and Feneuille[69] relate photon statistics to classical phase questions. Hioe and Eberly[70] derive a nonperturbative resonance condition from a Bargmann analytic function representation of the field. A more mathematical approach is presented by Hioe and Montroll.[71] Wong et al.[72] derive low-order shifts from a theory introducing a hierarchy of time scales. Bialynicki-Birula and Bialynicka-Birula[73] have developed a method where only the quantum phase appears as a variable, and this way they can reformulate the problem of saturation effects in strong-field transitions.

It was early pointed out[5] that "Haroche's level crossings" were not affected by power broadening and hence could be measured to higher accuracy than the multiphoton resonances. The continued fraction theory was extended to the case of transverse pumping by Tsukada and Ogawa[74] and independently by Stenholm and Aminoff.[75] They showed that the theory agreed with the experiments of Tsukada et al.,[33] who published a more detailed comparison later.[76] Cohen-Tannoudji et al.[77] measured the coherence level-crossing resonance at twice the rf frequency and fitted the result well with perturbative expressions obtained from quantum theory. Their measurements were in excellent agreement with the result of the semiclassical continued fraction theory.[78] It thus appears that the experiments confirm both the semiclassical and the quantum results.

Even if the early continued fraction papers were devoted to numerical work, it was pointed out[56] that the truncated fractions provide analytic approximations much superior to power series expressions. Investigations of such truncated expressions have been carried out by Swain[79] and Aminoff.[80]

More general configurations of pumping beams and rf fields have been discussed by Dupont-Roc[32] and Pegg[81,82] and investigated both theoretically and experimentally by Tsukada and Yabuzaki et al.[83–86] Some results have also been derived by McClean and Swain.[87]

Arimondo and Moruzzi have discussed multiple photon transitions in a system described by irreducible tensors[88] and followed them through for an optical pumping cycle.[89] They find that an accurate theoretical description of the experiments would require good knowledge of the pumping and decay parameters involved.

The ac Stark shift also occurs when two atomic levels are coupled by an optical field. Owing to the inhomogeneous Doppler broadened line shape in a gaseous sample, the phenomenon appears in a rather different way. It can, however, be probed on another transition coupled to either one of the strongly coupled levels as described by Feld et al.[90,91] The occurrence of a splitting in the velocity-averaged response was discussed by Hänsch et al.[92,93] Investigations of three-level laser resonances form a very large field.[94] Recent observations of the complicated line shapes possible have been obtained by Schabert et al.[95,96]

The inhomogeneous broadening can be avoided when laser light is hitting an atomic beam in a perpendicular direction. The effects of the intense field on the spontaneous emission is seen as a broadening of the central peak and the appearance of two side peaks separated by the ac Stark shift frequency. The complete theoretical predictions have been derived by Mollow[97–99] and verified and extended by others.[100–107] The experimental work is still in progress.[109–111] A recent experiment[112] observes the Autler–Townes splitting directly on an atomic beam.

Vasilenko et al.[113] pointed out that Doppler-free spectra can be obtained by the absorption of two counterpropagating waves. This was soon verified experimentally,[114] and since then a large number of investigations have appeared. Liao and Bjorkholm[115] showed that the nonresonant nature of the intermediate state leads to a shift of the resonance condition, which is to be classified as a light shift. Salomaa and Stenholm[116,117] point out that the shift can be avoided when one achieves resonance in the intermediate state, with a considerable increase in signal intensity. This has been experimentally confirmed by Bjorkholm and Liao.[118] With a resonant intermediate level the situation corresponds to that treated earlier.[90–96] Some further theoretical treatments of the shift have also appeared.[119–121]

The field of optical pumping has undergone some further development because of the advances in laser source technology. This topic is, however, too large to be included in the present review, and we only refer the reader to the literature.[122]

2.2 Semiclassical Theory

2.2.1. Formulation of the Problem

In these sections we are going to consider atomic systems perturbed by the single frequency field

$$\hat{V} = +\hbar\hat{v} \cos \Omega t \qquad (10)$$

where $\hat{v}$ is an operator and Ω is the angular frequency. If the atomic Hamiltonian is written $\hat{H} = \hbar\hat{\omega}$ we find for the density matrix the equation of motion

$$i\frac{\partial}{\partial t}\hat{\rho} = [\hat{\omega}, \hat{\rho}] + [\hat{v}, \hat{\rho}] \cos \Omega t - iR(\hat{\rho}) + i\hat{L} \qquad (11)$$

where $R(\hat{\rho})$ is a relaxation term and $\hat{L}$ contains the pumping.

In accordance with the general method for differential equations with time-periodic coefficients we expand

$$\hat{\rho} = \sum_n \hat{\rho}(n) e^{i(n\Omega + Q)t} \tag{12}$$

where Q is a constant-frequency term. We obtain the equations

$$[\hat{\omega}, \hat{\rho}(n)] + (n\Omega + Q)\hat{\rho}(n) - iR(\hat{\rho}(n)) = -i\tfrac{1}{2}L(n) - \tfrac{1}{2}[\hat{v}, (\hat{\rho}(n+1) + \hat{\rho}(n-1))] \tag{13}$$

This equation is still quite general. In order to obtain coupled scaler equations, we have to choose a representation for the operators. In the general case one can use an expansion in terms of irreducible tensor components.[123] This has been discussed by Haroche[49,50] and Arimondo and Moruzzi.[88]

An important special case is the two-level system, and here we can use the three Pauli matrices $\hat{\sigma}_i$ ($i = 1, 2, 3$) with the commutation rules

$$[\hat{\sigma}_i, \hat{\sigma}_j] = 2i\varepsilon_{ijk}\hat{\sigma}_k \tag{14}$$

where ε_{ijk} is the antisymmetric unit tensor. For simplicity we restrict the following treatment to the two-level case, corresponding to a spin $\tfrac{1}{2}$. We write

$$\hat{\rho} = \tfrac{1}{2}(M + C\hat{\sigma}_1 + S\hat{\sigma}_2 + N\hat{\sigma}_3) \tag{15}$$

and (C, S, N) are equivalent to the three components of the Bloch vector.

The Hamiltonian is determined by

$$\hat{\omega} = \tfrac{1}{2}\omega\hat{\sigma}_3$$
$$\hat{v} = v\hat{\sigma}_1 + w\hat{\sigma}_3 \tag{16}$$

For generality we have assumed that the oscillating field is misaligned so that it has a longitudinal component w and a transverse component v.

Using the commutation rules (13) we obtain the coupled equations

$$(\gamma_1 + iQ + in\Omega)M(n) = L_0(n) \tag{17}$$

$$(\gamma_1 + iQ + in\Omega)N(n) = L_3(n) + v[S(n+1) + S(n-1)] \tag{18}$$

$$(\gamma_2 + iQ + in\Omega)C(n) + \omega S(n) = L_1(n) - w[S(n+1) + S(n-1)] \tag{19}$$

$$(\gamma_2 + iQ + in\Omega)S(n) - \omega C(n) = L_2(n) - v[N(n+1) + N(n-1)]$$
$$+ w[C(n+1) + C(n-1)] \tag{20}$$

From these equations, we can derive most of the results of the published papers.

In most cases of interest, the pumping term is time independent. In particular, there are no time variations in L which resonate with any harmonic of the field. It is, hence, appropriate to assume in the following that

$$L_i(n) = L_i \delta_{n0} \qquad (i = 1, 2, 3) \tag{21}$$

2.2.2. Purely Longitudinal Field

We first consider the situation of an oscillating field only along the static field, viz., $v = 0$, $w \neq 0$. In this case we choose to solve the initial-value problem and set $L_1 = L_2 = 0$.

Only the components C and S couple through w, and they enter in terms of the combination

$$R(n) = C(n) + iS(n) \tag{22}$$

which we assume to be time independent. For $n \neq 0$ we have

$$(\gamma_2 + iQ + in\Omega - i\omega)R(n) = iw[R(n+1) + R(n-1)] \tag{23}$$

If we choose $Q = \omega + i\gamma_2$ we find the recursion relation

$$R(n+1) + R(n-1) = (2n/z)R(n) \tag{24}$$

for $n > 0$. We have introduced

$$z = 2w/\Omega \tag{25}$$

The solution of Eq. (24), which goes to zero for large values of n is the Bessel function of order n

$$R(n) = AJ_n(z) \tag{26}$$

(see Watson,[124] pp. 45 and 66). The values for $n < 0$ are obtained in a similar way. For $n = 0$ we can take J_0 and determine A from the limit $w = 0$.

The off-diagonal density matrix element is given by

$$\rho_{-+} = \tfrac{1}{2} \sum_n (C + iS) = \tfrac{1}{2} \sum_n R(n) \exp\left[i(n\Omega + \omega + i\gamma_2)t\right]$$

$$= \tfrac{1}{2} A \sum_n J_n\left(\frac{2w}{\Omega}\right) \exp\left[i(n\Omega + \omega + i\gamma_2)t\right] \tag{27}$$

The same initial-value problems can be solved by the time evolution operator

$$U(t) = \exp\left[-i\int_0^t H(t')\,dt' - \frac{1}{2}\gamma_2 t\right]$$

$$= \exp\left[-i\hat{\sigma}_3\left(\frac{1}{2}\omega t - \frac{w}{\Omega}\sin\Omega t\right) - \frac{1}{2}\gamma_2 t\right] \tag{28}$$

which can be expanded in Bessel functions to give an expression exactly identical with Eq. (27) (see Series,[17] pp. 464, 465). The result (28) was first obtained by Aleksandrov *et al.*[20]

2.2.3. Transverse Oscillating Field

In Sections 2.2.3–2.2.5 we look for steady-state solutions, and then it proves possible to set $Q = 0$. When $w = 0$ we can eliminate $C(n)$ from the equations (17)–(20)

$$S(n) = -\frac{v}{2}\left[\frac{1}{\gamma_2 + i(n\Omega - \omega)} + \frac{1}{\gamma_2 + i(n\Omega + \omega)}\right][N(n+1) + N(n-1)]$$

$$+ \frac{\gamma_2}{\omega^2 + \gamma_2^2}\left(L_2 + \frac{\omega}{\gamma_2}L_1\right)\delta_{n0} \tag{29}$$

$$N(n) = \frac{v}{\gamma_1 + in\Omega}[S(n+1) + S(n-1)] + \frac{L_3}{\gamma_1}\delta_{n0} \tag{30}$$

If we let also $v = 0$, we can see that transverse pumping (L_1 or L_2 nonzero) gives the dispersive or absorptive part of the zero-field level crossing ($\omega = 0$) in $S(0)$ (the Hanle[125] effect).

When $L_3 \neq 0$, it induces even components into N and odd components into S; for nonzero L_1 or L_2 we get even components in S and odd ones in N. The general solution will be the sum of the two results because the equations are linear in S and N and the general solution is a superposition.

The equations are of the general form

$$z(n) = vD(n)[z(n+1) + z(n-1)] + A\delta_{n0} \tag{31}$$

where

$$z(n) = \begin{cases} S(n), & n \text{ odd} \\ N(n), & n \text{ even} \end{cases} \quad \text{if } L_3 \neq 0$$

$$z(n) = \begin{cases} N(n), & n \text{ odd} \\ S(n), & n \text{ even} \end{cases} \quad \text{if } L_1 \text{ or } L_2 \neq 0$$

From (31) it follows directly (see Stenholm[38]) that

$$\frac{z(1)}{z(0)} = \frac{vD(1)}{1 - \dfrac{v^2 D(1)D(2)}{1 - \dfrac{v^2 D(2)D(3)}{1 - \cdots}}} \tag{32}$$

$$\frac{z(-1)}{z(0)} = \frac{vD(-1)}{1 - \dfrac{v^2 D(-1)D(-2)}{1 - \cdots}} \tag{33}$$

which are in the form of continued fractions. From the equation with $n = 0$ we obtain

$$z(0) = \frac{A}{1 - v^2 D(0)M} \tag{34}$$

with

$$M = \frac{1}{v}\left[\frac{z(1)}{z(0)} + \frac{z(-1)}{z(0)}\right] \tag{35}$$

This solution is easy to evaluate numerically because the continued fraction converges very rapidly.

2.2.4. Longitudinal Pumping

When we populate the levels by incoherent processes, thermal light or collision events, we have only $L_3 \neq 0$ and only even components of $N(n)$ are nonvanishing.

The population on the upper level can be determined from the detection operator[126,127]

$$\begin{aligned}
W_- &= \frac{1}{2}\left(\frac{L_3}{\gamma_1} - \overline{\langle\hat{\sigma}_3\rangle}\right) \\
&= \frac{1}{2}\left[\frac{L_3}{\gamma_1} - \overline{\mathrm{Tr}\,(\hat{\rho}\hat{\sigma}_3)}\right] \\
&= \frac{1}{2}\left[\frac{L_3}{\gamma_1} - N(0)\right]
\end{aligned} \tag{36}$$

where the bar denotes a time average over many periods of the rf field.

From Eqs. (14) and (34) we find

$$W_- = \frac{L_3}{2\gamma_1}\left[1 - \frac{1}{1-(v^2/\gamma_1)M}\right] \tag{37}$$

$$W_- = \frac{L_3}{2\gamma_1}\left[1 - \cfrac{1}{1 - \cfrac{v^2}{\gamma_1}\cfrac{D(1)}{1 - \cfrac{v^2 D(1)D(2)}{1 - \cfrac{v^2 D(2)D(3)}{1 - \cdots}}} + \text{c.c.}}\right] \tag{38}$$

The odd D functions are of the form

$$D(m) = -\frac{1}{2}\left[\frac{1}{\gamma_2 + i(n\Omega - \omega)} + \frac{1}{\gamma_2 + i(n\Omega + \omega)}\right] \tag{39}$$

and have a resonant behavior for $n\Omega \approx \omega$. These are the multiphoton resonances, where the transition frequency ω is reached by n photons of frequency Ω. Near the resonance $n = 1$ we write

$$D(1) = -\frac{1}{2}\left[\frac{1}{\gamma_2 + i(\Omega - \omega)}\right]$$

$$D(2) = -\frac{i}{2\Omega} + O\left(\frac{\gamma}{\Omega}\right) \tag{40}$$

because $\Omega \gg \gamma$ in the usual cases. We find

$$1 - \cfrac{1}{1 + \cfrac{v^2}{2\gamma_1}\left[\cfrac{1}{\gamma_2 + i(\Omega - \omega) + iv^2/4\Omega} + \text{c.c.}\right]}$$

$$\approx \frac{\gamma_2}{\gamma_1}v^2\left[\left(\Omega - \omega - \frac{v^2}{4\Omega}\right)^2 + \gamma_2^2 + \frac{\gamma_2}{\gamma_1}v^2\right]^{-1} \tag{41}$$

where the line shape is seen to be a Lorentzian shifted to

$$\Omega = \omega + v^2/4\Omega \tag{42}$$

which is the well-known Bloch–Siegert shift.[8,5] The width is power broadened to

$$\Gamma = \gamma_2(1 + v^2/\gamma_1\gamma_2)^{1/2} \tag{43}$$

In general the $(2n+1)$th resonance is given by the approximations

$$D(2n) \approx -\frac{i}{2n\Omega}$$

$$D(2n+2) \approx -\frac{i}{2(n+1)\Omega} \tag{44}$$

$$D(2n+1) \approx -\frac{1}{2}\left\{\frac{1}{i[(2n+1)\Omega - \omega] + \gamma_2}\right\}$$

and we find the combination

$$\left[1 - \frac{v^2 D(2n)D(2n+1)}{1 - v^2 D(2n+1)D(2n+2)}\right] \approx 1 - \frac{v^2/4n}{[(2n+1)\Omega - \omega] - i\gamma_2 - v^2/4(n+1)\Omega} \tag{45}$$

which gives a resonance at the zero of (45) given by

$$\omega = (2n+1)\Omega - \frac{(2n+1)v^2}{4n(n+1)\Omega} \tag{46}$$

This is the generalized Bloch–Siegert shift.[5,1] A careful expansion of (38) shows that the strength of the resonance (46) is given by

$$A_n = 2\left(\frac{v^2}{4}\right)^{2n+1} \frac{1}{(4\Omega^2)^{2n}} \frac{1}{(n!)^4} \frac{\gamma_2}{\gamma_1} \tag{47}$$

(see Stenholm[57]). In the limit $\gamma_1 = \gamma_2$, this agrees with the result of quantum theory.[49,50]

When one wants to expand the Bloch–Siegert shift in a power series in v, one must express the resonance denominator of (38) in a consistent fashion in v (see Stenholm[56]) and then one obtains the series

$$\Omega = \omega + \frac{1}{\Omega}\left(\frac{v}{2}\right)^2 + \frac{5}{4\Omega^3}\left(\frac{v}{2}\right)^4 + \frac{61}{32\Omega^5}\left(\frac{v}{2}\right)^6 + \frac{407}{128\Omega^7}\left(\frac{v}{2}\right)^8 + \cdots \tag{48}$$

The second term of this series was obtained by Bloch and Siegert[8] in 1940, the third and fourth ones by Shirley[19] in 1965. The eighth- and tenth-order terms were obtained by Ahmad and Bullough[65] in 1974, whereas a recent work by I. and Z. Bialynicki-Birula[73] calculates the twelfth-order term.

Most semiclassical approaches and some quantum methods lead to results in nonperturbative form. Thus the resonance condition is obtained as an implicit function and can be solved without a power series expansion. This is the case with Shirley,[19] Hannaford et al.,[47] Ahmad and Bullough,[66] and Hioe and Eberly.[70] Truncating the continued fractions

(38) one can get a reliable analytic approximation to the exact expression (see Stenholm[56]), and Aminoff[80] shows that one easily reproduces all earlier approximations.

The quantum mechanical perturbation theory[55] derives a power series with the semiclassical coefficients in the limit $n \gg 1$. Thus all calculations agree on the theoretical results. Experimentally it is, however, harder to obtain reliable exact results because of the strong power broadening accompanying large shifts. Arimondo and Moruzzi[48] measured the one- and three-photon shifts in Hg vapor and were able to obtain an acceptable fit to the theory even if some alignment errors were present. A very good fit to theory is provided by the experiments of Morand and Theobald[44,45] even if their measurements are performed in solids and not on free atoms. The comparison between their measurements and theory is found in Stenholm.[46] Further discussion of experimental results will be presented in Section 2.2.5.

2.2.5. Transverse Pumping

The generalized level crossing phenomenon investigated by Haroche and collaborators[49,50] (coherence resonances) provides a high-order resonance without power broadening, as was pointed out by Cohen-Tannoundji.[5] The semiclassical theory for the phenomenon was developed independently by Tsukada and Ogawa[74] and Aminoff and Stenholm.[75]

To obtain the case of transverse pumping we set $L_2 = L_3 = 0$ in (29) and (30) and solve for the coherent dipole moment, which is observed,

$$W = \frac{1}{2}\left(\frac{L_1}{\gamma_2} - \overline{\langle \hat{\sigma}_1 \rangle}\right) = \frac{1}{2}\left[\frac{L_1}{\gamma_2} - C(0)\right] \tag{49}$$

which from (19) gives

$$W = \frac{1}{2}\left[\frac{L_1}{\gamma_2} + \frac{\omega}{\gamma_2} S(0)\right] \tag{50}$$

In the case of transverse pumping we have even terms in $S(n)$ only and $N(n)$ has got the odd ones. From Section 2.2.3 we obtain

$$S(0) = \frac{\omega}{\omega^2 + \gamma_2^2} L_1 \left[1 - v^2 D(0) \cfrac{D(1)}{1 - v^2 D(1)D(2) \cfrac{}{1 - v^2 D(2)D(3) \cfrac{}{1 - \cdots}}} + \text{c.c.}\right]^{-1} \tag{51}$$

Here we now have resonances in the even functions $D(2n)$ and not in the odd ones. The lowest-order resonance is given by

$$D(0) = -\frac{\gamma_2}{\omega^2 + \gamma_2^2} \tag{52}$$

$$D(1) = -\frac{i}{\Omega} + O\left(\frac{\gamma_1}{\Omega}\right) \tag{53}$$

$$D(2) = -\frac{1}{2}\left[\frac{1}{\gamma_2 + i(2\Omega - \omega)}\right] \tag{54}$$

$$D(3) = -\frac{i}{3\Omega} + O\left(\frac{\gamma_1}{\Omega}\right) \tag{55}$$

and hence we obtain from (51)

$$S(0) = \frac{\omega L_1}{\omega^2 + \gamma_2^2 - i\frac{v^2 \gamma_2}{\Omega}\left\{\left[1 - \frac{iv^2}{2\Omega}\frac{1}{\gamma_2 + i(2\Omega - \omega - 2v^2/3\Omega)}\right] - \text{c.c.}\right\}} \tag{56}$$

The resonance is shifted to

$$2\Omega = \omega + 2v^2/3\Omega \tag{57}$$

as given by Haroche.[49,50] In this case the general result is

$$2n\Omega = \omega + \frac{2v^2}{\Omega}\frac{n}{(4n^2 - 1)} \tag{58}$$

(see Stenholm and Aminoff[75]). The strength of the resonance is

$$A'_n = 8n^2 \left[\frac{n!}{(2n)!}\right]^4 \left(\frac{v^2}{\Omega^2}\right)^{2n-1} \tag{59}$$

and the width of the line is given by

$$\Gamma_2^2 = \gamma_2^2 \left\{1 + 4\left(\frac{v}{\Omega}\right)^{4n}\left[\frac{n!}{(2n)!}\right]^4\right\} \tag{60}$$

Because the multiphoton resonances of Section 2.2.4 have the width

$$\Gamma_1^2 = \gamma_2^2 + 2A_n \tag{61}$$

[see Eq. (47)] we find that the real transitions have a width of order Ω when $\Omega \approx v$, in which case the width of the coherence resonance (60) still is only of order γ_2 ($\ll \Omega$). Thus the position of these resonances can be measured with higher accuracy.

The coherence resonances had been measured by Tsukada *et al.*[33] and their result had been compared with the theory for a rotating wave.

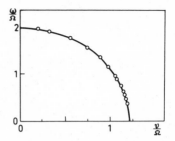

Figure 1. The shift of the two-photon coherence resonance in rf spectroscopy. The solid line is the continued fraction theory, the points are the measurements by Cohen-Tannoudji et al.[77]

These measurements could be successfully fitted[75] by the results of the full theory.

In the meantime Cohen-Tannoudji et al.[77] had measured the positions of the resonances accurately and their experiments showed an excellent fit to the theory by Stenholm and Aminoff; see Figure 1. Cohen-Tannoudji et al. used the quantum approach to obtain the first four terms in the power series

$$\omega = 2\Omega - \frac{2v^2}{3\Omega} - \frac{7}{54\Omega^3}v^4 - \frac{103}{2430\Omega^5}v^6 - \frac{6403}{3499200\Omega^7}v^8 - \cdots \qquad (62)$$

The eighth-order term was calculated by Aminoff.[80] For the coherence resonances no competition has occurred about the higher-order terms in a power series. It appears that the continued fraction method is the only nonperturbative method here, as a theory corresponding to Shirley's one[19] has not been worked out. Analytic approximations can, in any case, be obtained easily from a finite truncation of the continued fraction. To show the accuracy of the method we look[80] at the point where the 2Ω resonance is shifted enough to merge into the background at $\omega = 0$. This occurs at the zero of the equation $J_0(x) = 0$ or $2v = 2.4048\Omega$. We find

Power series to eighth order: 2.6605
Continued fraction: 1st approximation   2.4495
                      2nd   2.4084
                      3rd   2.4050
                      4th   2.4048

As can be seen already, the first CF approximation implies a considerable improvement over the power series. These results are only slightly affected by a finite line width,[75] but additional shifts of the order $(\gamma/\Omega)^4$ have been derived.[80]

An experimental verification of more complicated cases has been published by Yabuzaki et al.[76]

2.3. The Quantized Field Approach

2.3.1. The Fully Quantum Mechanical Hamiltonian

Corresponding to the semiclassical Hamiltonian introduced in Section 2.2, Eqs. (10)–(16), we now have the fully quantum mechanical one

$$H = \hbar[\tfrac{1}{2}\omega^2\hat{\sigma}_3 + \Omega b^+ b + (\Lambda\hat{\sigma}_1 + \Gamma\hat{\sigma}_3)(b + b^+)] \tag{63}$$

where b^+ is the boson operator creating one quantum of the rf field. Taking the matrix elements of (63) and Eq. (10) we can obtain the correspondence

$$\Gamma n^{1/2} = \tfrac{1}{2}w \tag{64}$$

$$\Lambda n^{1/2} = \tfrac{1}{2}v \tag{65}$$

where $n^{1/2}$ is the matrix element of the operators b, b^+.

The Hamiltonian (63) gives a convenient way to enumerate the basic processes of the interaction between radiation and matter in terms of photon creation and annihilation events. The number of photons involved in even moderate rf fields is very large ($\approx 10^{20}$), and hence the field can well be described by classical amplitudes (see Section 2.3.4). The identity of the few-photon processes is, however, retained in the semiclassical description (see Section 2.2) and they may be used to classify the physical processes induced by the field. Such an approach has been used extensively by the French research workers,[5,49,50] who regard the atom immersed in radiation as one system, the dressed atom. This approach has proved very fruitful in providing physical insight, even if most results can be derived just as easily from the semiclassical formulation of the problem.

2.3.2. The Atom Dressed by a Longitudinal Field

In Section 2.2.2 we saw how the case of a purely longitudinal field can be solved exactly. The same system can be obtained[22] from the quantized approach by setting $\Lambda = 0$.

We introduce the states

$$|\psi\rangle = |\eta\rangle|\phi_n\rangle \tag{66}$$

where

$$\hat{\sigma}_3|\eta\rangle = \eta|\eta\rangle \tag{67}$$

and hence $\eta = \pm 1$. With these states we obtain

$$h = \langle\eta|H/\hbar|\eta\rangle = \tfrac{1}{2}\omega\eta + \Gamma\eta(b + b^+) + \Omega b^+ b \tag{68}$$

Introducing the operator

$$\beta = b + \Gamma \eta / \Omega \qquad (69)$$

we can diagonalize h in terms of the eigenstates $|\phi_n\rangle$

$$h|\phi_n\rangle = \varepsilon(\eta, n)|\phi_n\rangle \qquad (70)$$

where

$$\varepsilon(\eta, n) = \tfrac{1}{2}\omega\eta + \Omega n - \Gamma^2/\Omega \qquad (71)$$

and n is the occupation number for the boson state created by β^+.

When we plot the eigenvalues $\varepsilon(\eta, n)$ as a function of ω we obtain the picture shown in Figure 2. A pair of states with slope $\pm\tfrac{1}{2}$ start at each value Ωn. These are the states of the dressed atom.

The eigenvalues become degenerate whenever

$$\varepsilon(+1, n) - \varepsilon(-1, n+k) = 0 \qquad (72)$$

for some value of k. This gives a level crossing at the points

$$k\Omega = \omega \qquad (73)$$

for all integer values of k. When we put in a perturbation proportional to $\hat{\sigma}_1$ we couple certain of these and the crossing is avoided, leading to an anticrossing signal. These cases will be discussed in Section 2.3.3.

If we calculate the time-dependent expectation value of the dipole moment we obtain

$$\mu(t) = \langle \gamma\hbar\hat{\sigma}_1 \rangle = \gamma\hbar\langle\psi(t)|\hat{\sigma}_1|\psi(t)\rangle$$

$$= \gamma\hbar\left[J_0(4\Gamma n^{1/2}/\Omega) + \sum_{k>0} J_{2k}(4\Gamma n^{1/2}/\Omega)\cos 2k\Omega t \right] \qquad (74)$$

The calculations are found in the references.[22,1] Together with Eq. (64) the result (74) is equivalent to (27). The dressed dipole moment has

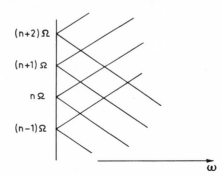

Figure 2. The unperturbed eigenvalues of the atomic levels in a static field ω, dressed by n rf photons.

been derived from semiclassical theory by Pegg and Series[27] and verified experimentally.[28,128,129]

2.3.3. The Dressed Atom in a Transverse Field

The Hamiltonian of the transverse problem is obtained from (63) by setting $\Gamma = 0$. The combination

$$H_0 = \hbar(\tfrac{1}{2}\omega\hat{\sigma}_3 + \Omega b^+ b) \tag{75}$$

is regarded as the unperturbed part with the eigenstates $|\eta, n\rangle$ and the eigenvalues $\varepsilon(\eta, n)$, which are the same as in Figure 2. The coupling term

$$H_1 = \hbar\Lambda\hat{\sigma}_1(b + b^+) \tag{76}$$

can couple only the sequence of states

$$|\eta, n\rangle \to |-\eta, n \pm 1\rangle \to |\eta, n \pm 2\rangle, \quad |\eta, n\rangle \to |-\eta, n \pm 3\rangle, \quad |-\eta, n \pm 1\rangle \to \cdots \tag{77}$$

As degeneracy occurs only between states of different eigenvalues η we find that the crossings at $\omega = (2k + 1)\Omega$ are removed, because the levels will be coupled in some order of the perturbation. The crossings at $\omega = 2k\Omega$, however, remain genuinely degenerate. The former ones correspond to the multiphoton transitions of Section 2.2.4, whereas the latter are coherence resonances of Haroche's type, discussed in 2.2.5.

When the intermediate levels participating in the coupling are eliminated, the crossing levels form a two-level subsystem with the effective Hamiltonian

$$\mathfrak{H} = \begin{pmatrix} \tfrac{1}{2}\omega + R_{++}(\omega) & R_{+-}(\omega) \\ R_{-+}(\omega) & n\Omega - \tfrac{1}{2}\omega + R_{--}(\omega) \end{pmatrix} \tag{78}$$

A particularly elegant way of achieving this is provided by the projection operator technique.[5,49] When $n = 2k + 1$, the coupling R_{+-} is nonzero in some order and an anticrossing signal is observed with the probability

$$W = \frac{2|R_{+-}(\omega)|^2}{\gamma^2 + 4|R_{+-}(\omega)|^2 + [(2k+1)\Omega - \omega - R_{++}(\omega) + R_{--}(\omega)]^2} \tag{79}$$

For a derivation see the references.[5,1]

Because the elements of the matrix R depend on ω we no longer have straight lines as the eigenvalues in Figure 2, but the behavior is deformed near the avoided crossing point as shown in Figure 3. The resulting picture shows that the resonance may be shifted from the uncoupled value; these are the generalized Bloch–Siegert shifts of Section 2.2.4. The factor $(R_{+-})^2$ in the denominator also provides the power broadening caused by the occurrence of real transitions between the levels.

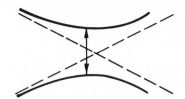

Figure 3. The shift of an avoided crossing when the
levels are coupled by the rf field.

When $n = 2k$ in (78) the coupling has to disappear: $R_{+-} = R_{-+} = 0$. Then a level-crossing signal is possible and we obtain the signal line shape

$$W \approx \frac{\gamma^2}{\gamma^2 + [2k\Omega - \omega - R_{++}(\omega) + R_{--}(\omega)]^2} \tag{80}$$

(See the references.) The crossing point is displaced by the ω dependence of $R_{++}(\omega) - R_{--}(\omega)$, but to lowest order no power broadening occurs because no real transitions take place (cf., however, Section 2.2.5); the situation is illustrated in Figure 4.

The experimental verification of the Bloch–Siegert shift and the theory of coherence resonances was discussed in Sections 2.2.4 and 2.2.5.

2.3.4. Relation to the Semiclassical Approach

For the quantized problem we expand the complete state of the coupled system as

$$|\psi_1\rangle = \sum_n (C_n^+ |+, n\rangle + C_n^- |-, n\rangle) \tag{81}$$

Using the Hamiltonian (75) and (76) we find the coupled equations

$$i\dot{C}_n^+ = (\Omega n + \tfrac{1}{2}\omega)C_n^+ + \Lambda(n+1)^{1/2}C_{n+1}^- + \Lambda n^{1/2}C_{n-1}^- \tag{82}$$

$$i\dot{C}_n^- = (\Omega n - \tfrac{1}{2}\omega)C_n^- + \Lambda(n+1)^{1/2}C_{n+1}^+ + \Lambda n^{1/2}C_{n-1}^+ \tag{83}$$

These equations are discussed in many papers; a review is given in Stenholm.[1]

Figure 4. The shift of a level crossing due to
perturbation by the rf field.

Here we want to compare the Eqs. (82) and (83) with those of the semiclassical problem in Section 2.2. The relevant Hamiltonian is

$$H = \hbar(\tfrac{1}{2}\omega\hat{\sigma}_3 + v\hat{\sigma}_1 \cos \Omega t) \tag{84}$$

Expanding the state in the form

$$|\psi_2\rangle = \sum_n e^{i\Omega n t}(C_n^+ |+\rangle + C_n^- |-\rangle) \tag{85}$$

we find the coupled equations

$$i\dot{C}_n^+ = (\Omega n + \tfrac{1}{2}\omega)C_n^+ + \tfrac{1}{2}v(C_{n+1}^- + C_{n-1}^-) \tag{86}$$

$$i\dot{C}_n^- = (\Omega n - \tfrac{1}{2}\omega)C_n^- + \tfrac{1}{2}v(C_{n+1}^+ + C_{n-1}^+) \tag{87}$$

We note these equations are the same as Eqs. (82) and (83) when we can neglect the n dependence of the square root $n^{1/2}$. When large photon numbers (nearly classical states) are involved this is approximately the case, and using the identification (65), we find the two formulations to be identical. This shows the close relationship between the quantized field approach and the Fourier series expansion in the frequency of the field. The present argument has been carried out for the density matrix formulation in Stenholm.[57]

Shirley[19] utilizes the relationship between the two approaches in a very sophisticated way. He writes the series (85) in terms of exact eigenstates of the periodically time-dependent Hamiltonian and uses Floquet theory to extract information about the eigenvalues and resonances.

Without coupling the eigenvalue plot of Figure 2 repeats itself exactly when the quantum number n is increased. In the semiclassical theory the constant coupling v preserves this symmetry even for the exact eigenvalues, but in quantum theory the factor $n^{1/2}$ makes the levels distort differently for different values of n. Only in the asymptotic region, where the semiclassical behavior is approached, does the quantized problem locally acquire the periodicity in n displayed by the semiclassical states. When this happens the two approaches become identical.

2.3.5. The Plots of Exact Eigenvalues

Following Shirley[19] we consider the exact eigenvalues $\lambda(\eta, n)$ of the full Hamiltonian. We do, however, choose to carry out this discussion in terms of the quantized Hamiltonian (75)–(76), which asymptotically has been seen to reproduce the semiclassical results. In this limit the eigenvalue diagram becomes locally periodic in n and we can write

$$H|\lambda(\eta, n)\rangle = \hbar\lambda(\eta, n)|\lambda(\eta, n)\rangle \tag{88}$$

$$\lambda(\eta, n) = \lambda_\eta^0 + \Omega n \tag{89}$$

Note that (89) can hold only in a limited range of large values, and we are not allowed to set $n = 0$. The matrix elements of H can be read off from Eqs. (82) and (83) and hence we obtain

$$\operatorname{Tr} H = \sum_n (\tfrac{1}{2}\omega + \Omega n - \tfrac{1}{2}\omega + \Omega n) = 2\Omega \sum_n n \tag{90}$$

To deal with the obvious divergence we assume a cutoff in the sum over n. The trace is, however, an invariant and hence also from (88) and (89)

$$\operatorname{Tr} H = \sum_{\eta n} \lambda(\eta, n) = \hbar\left(\lambda_+ + \lambda_- + 2\Omega \sum_n n\right) \tag{91}$$

If we look at the asymptotic behavior for large n we find that

$$\lambda_+ = q = -\lambda_- \tag{92}$$

If we start in the state $\eta = -$ and $n = 0$ at time $t = 0$ we have at a later time

$$|\psi(t)\rangle = e^{-iHt/\hbar}|-, 0\rangle \tag{93}$$

where $|-, 0\rangle$ is the unperturbed state. The probability of observing $\eta = +$ without regard for the field state is

$$P_{-+}(t) = \sum_n |\langle +n| e^{-iHt/\hbar}|-0\rangle|^2 \tag{94}$$

Introducing a complete set of eigenstates of H and taking a time average one obtains

$$P_{+-} = \overline{P_{-+}(t)} = \sum_n \sum_{\eta k} \sum_{\eta'k'} \overline{\exp\{i[\lambda(\eta, k) - \lambda(\eta', k')]t\}}$$

$$\times \langle +n|\lambda(\eta, k)\rangle\langle \lambda(\eta, k)|-0\rangle\langle -0|\lambda(\eta'k')\rangle\langle \lambda(\eta'k')|+n\rangle$$

$$= \sum_\eta \sum_{nk} \langle +n|\lambda(\eta k)\rangle\langle \lambda(\eta k)|-0\rangle\langle -0|\lambda(\eta, k)\rangle\langle \lambda(\eta, k)|+n\rangle \tag{95}$$

Introducing the notation ($\alpha, \beta = \pm 1$)

$$T_{\alpha\beta} = \sum_n \left|\langle \alpha n|\lambda(\beta, k)\rangle\right|^2 = \sum_k \left|\langle \alpha 0|\lambda(\beta, k)\rangle\right|^2 \tag{96}$$

we obtain

$$P_{-+} = \sum_n T_{+\eta} T_{-\eta} \tag{97}$$

(see Shirley[19]). Because of the normalization of the states we find

$$\sum_\beta T_{\alpha\beta} = \langle \alpha 0|\alpha 0\rangle = 1 = \langle \lambda(\beta, k)|\lambda(\beta, k)\rangle = \sum_\alpha T_{\alpha\beta} \tag{98}$$

The Hamiltonian depends parametrically on the field splitting ω and according to the Feynman–Hellmann theorem[108] we can write

$$\frac{\partial \lambda (+0)}{\partial \omega} = \frac{\partial q}{\partial \omega} = \left\langle \lambda (+, 0) \left| \frac{\partial H}{\partial \omega} \right| \lambda (+, 0) \right\rangle \tag{99}$$

We introduce a complete set of unperturbed eigenstates $|\eta, n\rangle$ and write

$$\frac{\partial}{\partial \omega} \left(H \sum_{\eta n} |\eta n\rangle\langle \eta n| \right) = \sum_{\eta n} \frac{1}{2} \eta |\eta n\rangle\langle \eta n| \tag{100}$$

because ω occurs only in the diagonal elements in the $\{|\eta, n\rangle\}$ representation; see Eqs. (82) and (83).

From (99) and (100) we obtain

$$\frac{\partial q}{\partial \omega} = \tfrac{1}{2} \sum_{\eta n} \eta \langle \lambda (+, 0)|\eta, n\rangle\langle \eta n|\lambda|+, 0\rangle$$

$$= \tfrac{1}{2} \left[\sum_n \left| \langle \lambda (+, 0|+, n\rangle \right|^2 - \sum_n \left| \langle \lambda (+, 0)|-, n\rangle \right|^2 \right]$$

$$= \tfrac{1}{2}(T_{++} - T_{-+}) \tag{101}$$

In a similar way we derive

$$\frac{\partial \lambda (-, 0)}{\partial \omega} = -\frac{\partial q}{\partial \omega} = \left\langle \lambda (-, 0) \left| \frac{\partial H}{\partial \omega} \right| \lambda (-, 0) \right\rangle = \frac{1}{2}(T_{+-} - T_{--}) \tag{102}$$

From the equations (98), (101), and (102) it follows that

$$T_{++} = T_{--} = \frac{1}{2} + \frac{\partial q}{\partial \omega}$$

$$T_{+-} = T_{-+} = \frac{1}{2} - \frac{\partial q}{\partial \omega} \tag{103}$$

which inserted into the transition probability (97) gives

$$P_{-+} = \frac{1}{2} \left[1 - 4 \left(\frac{\partial q}{\partial \omega} \right)^2 \right] \tag{104}$$

This is the equation derived by Shirley[19] from the Floquet Hamiltonian. Here we have chosen the derivation from a quantized field theory in order to be able to use the more familiar states $|\eta, n\rangle$. The two approaches are identical for large enough values of n.

Equation (104) is very useful when one wants to discuss multiphoton resonances. The exact eigenvalue plot of $\lambda (\eta, n)$ versus ω repeats itself for each value of n [see (89)], and hence all dependence on ω must reside in q.

From (104) we can see that there will be a resonance every time

$$\frac{\partial q}{\partial \omega} = 0 \tag{105}$$

which happens when the anticrossing distance achieves its smallest value (see Figure 3). The strength of the resonance is $\frac{1}{2}$ in this approach. Using the continued fraction theory of Section 2.2 it has been shown that only very strong relaxation processes cause additional shifts due to relaxations.[46,75] In addition to the work by Shirley the relations (104) and (105) have been used by Hannaford et al.,[47] Pegg,[68] and Hioe and Eberly.[70] Recently Swain[148] has derived generalizations of (104) when more than two atomic levels are involved.

2.4. Atoms in Strong Optical Fields

2.4.1. Strongly Coupled Two-Level System

The two-level system in Figure 5 has the equation of motion for the density matrix

$$i\dot{\rho}_{00} = -i\gamma_0\rho_{00} + i\kappa\rho_{11} + \alpha(\rho_{10} - \rho_{01}) \tag{106}$$

$$i\dot{\rho}_{10} = (\Delta - i\Gamma)\rho_{10} + \alpha(\rho_{00} - \rho_{11}) \tag{107}$$

$$i\dot{\rho}_{11} = -i\gamma_1\rho_{11} - \alpha(\rho_{10} - \rho_{01}) \tag{108}$$

where α is the coupling strength proportional to the field and Δ is the detuning between the field and the transition frequency. The levels decay with the rates γ_0 and γ_1, respectively, and κ denotes the fraction of γ_1 that is due to spontaneous emission to the ground state. The transverse relaxation rate is Γ.

The equations (106)–(108) form a system of four linear differential equations, because ρ_{10} is complex. The corresponding matrix is given

Figure 5. Two levels coupled by a laser field and allowed to return to equilibrium by spontaneous emission.

by

$$H = \begin{pmatrix} -i\gamma_0 & \alpha & -\alpha & i\kappa \\ \alpha & \Delta - i\Gamma & 0 & -\alpha \\ -\alpha & 0 & -(\Delta + i\Gamma) & \alpha \\ 0 & -\alpha & \alpha & -i\gamma_1 \end{pmatrix} \qquad (109)$$

If we for a moment neglect the damping terms we find the characteristic equation

$$\lambda^2(\lambda^2 - \Delta^2 - 4\alpha^2) = 0 \qquad (110)$$

with two roots equal to zero and two equal to the split pair

$$\lambda = \pm(\Delta^2 + 4\alpha^2)^{1/2} \qquad (111)$$

At resonance, $\Delta = 0$, this leads to the dynamically Stark split frequencies $\pm 2\alpha$, discussed by Autler and Townes.[2]

In order to see the influence of the relaxation rates we look at two special cases:

Firstly assume $\kappa = 0$ and $\gamma_0 = \gamma_1 = \Gamma$. Then all resonances acquire the imaginary part $i\gamma$ and all decay at the same rate. This result is, of course, quite obvious, but describes well a molecular system where the interaction time is determined by the apparatus.

In the case when the only longitudinal relaxation mechanism is spontaneous decay between the levels it follows that

$$\gamma_0 = 0, \qquad \gamma_1 = \kappa \qquad (112)$$

To see what the resonance condition is we write down the characteristic equation at resonance ($\Delta = 0$). We find

$$\lambda(\lambda - i\Gamma)[(\lambda - i\Gamma)(\lambda - i\kappa) - 4\alpha^2] = 0 \qquad (113)$$

Except for the unaffected root we find the following resonances:

$$\lambda = i\Gamma \qquad (114)$$

$$\lambda = (i/2)(\Gamma + \kappa) \pm [4\alpha^2 - \tfrac{1}{4}(\kappa - \Gamma)^2]^{1/2} \qquad (115)$$

As long as the splitting exceeds the difference in relaxation rates

$$4\alpha > |\kappa - \Gamma| \qquad (116)$$

we have two separated peaks of width $\tfrac{1}{2}(\Gamma + \kappa)$. These are the Stark-shifted levels (111). When only spontaneous emission contributes to the transverse decay rate we have[122]

$$\Gamma = \tfrac{1}{2}\kappa \qquad (117)$$

and with $\kappa \ll 8\alpha$ we have two Stark satellites of width

$$\tilde{\gamma} = \tfrac{3}{4}\kappa \tag{118}$$

as first derived by Mollow.[97] The spectrum of the scattered light is given by the approximate expression

$$R(\nu) = \mathrm{Im}\left[\frac{1}{\nu - \omega + i\tfrac{1}{2}\kappa} + \frac{1}{2}\frac{1}{\nu - \omega + 2\alpha + i\tfrac{3}{4}\kappa} + \frac{1}{2}\frac{1}{\nu - \omega - 2\alpha + i\tfrac{3}{4}\kappa}\right] \tag{119}$$

This result has been derived by many workers, and also generalized to other relaxation cases (see Refs. 98–107) and experimentally investigated in atomic beams.[109–111] Recent discussions have also included the photon correlations[107] and three-level schemes.[103] Some experiments are planned to measure correlations, and the Stark splitting of a three-level system has been observed.[112]

2.4.2. Stark Splitting in Doppler-Broadened Transitions

The dynamic Stark splitting of a two-level system interacting with a strong field is observable when one of the levels is coupled to a third one. This provides an opportunity to observe the effect in an atomic beam.[112] For an inhomogeneously broadened transition in a gas the situation becomes more complicated owing to the occurrence of the Doppler shift. Here we shall discuss the situation in terms of a simple model only.

Let the levels 0–1 be strongly coupled by the matrix element 2α (see Figure 5) and let all decay rates be equal to γ. Then the power broadened width of the transition becomes

$$\Gamma^2 = \gamma^2 + 16\alpha^2 \tag{120}$$

We assume the detuning to be Δ and couple the upper level to another one with the detuning $\delta - \Delta$. We assume only one wavelength and hence one value for the Doppler shift kv. The absorption on the coupled transition can be calculated and contains a term (see Shoemaker et al.,[130] Appendix A) proportional to the imaginary part of

$$A = \frac{1}{[(\Delta + kv)^2 + \Gamma^2]}\left[\frac{2(\delta + i\gamma) - (\Delta + kv - i\gamma)}{(\Delta - \delta + kv - i\gamma)(\delta + i\gamma) + 4\alpha^2}\right] \tag{121}$$

We first look at stationary atoms ($v = 0$) and find the resonance condition

$$\delta = i\gamma + \tfrac{1}{2}\Delta \pm (\tfrac{1}{4}\Delta^2 + 4\alpha^2)^{1/2} \tag{122}$$

which at resonance displays just the right Stark splitting $\pm 2\alpha$. The atoms, however, have a distribution of velocities, and when we integrate over all velocity groups only those nearly at resonance will contribute. When we assume a very broad velocity distribution ($\approx$constant) both roots (122) get an opportunity to contribute and a symmetric line shape arises.

The simplest way to perform the calculation is to close the velocity integration in the lower complex kv plane and then only the pole

$$kv = -\Delta - i\Gamma \tag{123}$$

will contribute. The result is

$$\langle A \rangle_{\mathrm{av}} \propto \frac{2(\delta + i\gamma) + i(\gamma + \Gamma)}{[\delta + i(\gamma + \Gamma)](\delta + i\gamma) - 4\alpha^2} = \frac{2}{\delta - i\frac{1}{2}(\gamma + \Gamma)} \tag{124}$$

where the roots

$$\delta = -i\gamma - \tfrac{1}{2}i\Gamma \pm \tfrac{1}{2}i\,(\Gamma^2 - 16\alpha^2)^{1/2} = -\tfrac{i}{2}(2\gamma + \Gamma \pm \gamma) \tag{125}$$

have been simplified by the use of (120). We can see how the Stark split result for $v = 0$ leads to a simple Lorentzian line shape after the velocity average.

The general theory of Doppler-broadened three-level resonances[90–94] shows a complicated line shape, which in special cases deviates considerably from a simple Lorentzian. There are doublet components[117,131] separated by a quantity proportional to α, but these have a more complicated origin than the simple Stark splitting (122) occurring before the velocity integration. A detailed experimental program has been devoted to their study by Toschek and collaborators[95,96] and they have verified the theoretical predictions.

2.5. Two-Photon Spectroscopy

2.5.1. Mathematical Reduction of a Three-Level System

We consider the situation in Figure 6, where three levels are coupled by two counterpropagating waves

$$\langle \phi_0 | V | \phi_1 \rangle = 2V_{01} \cos{(\Omega t - kz)}$$
$$\langle \phi_1 | V | \phi_2 \rangle = 2V_{12} \cos{(\Omega t + kz)} \tag{126}$$

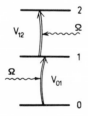

Figure 6.   Three levels coupled by two-photon transitions from counter-propagating beams.

We have included the Doppler shifts for this configuration because of its physical interest; other situations are as easily treated if needed. If the eigenstates are written $|\phi_i\rangle$ $(i = 0, 1, 2)$ we expand

$$|\psi(t)> = C_0|\phi_0\rangle + \exp[i(\Omega t - kz)]C_1|\phi_1\rangle + \exp(2i\Omega t)C_2|\phi_2\rangle \quad (127)$$

and obtain in the rotating-wave approximation from the Schrödinger equation

$$i\dot{C}_0 = V_{01}C_1 \quad (128)$$

$$i\dot{C}_1 = (\varepsilon_1 - \Omega + kv - i\tfrac{1}{2}\Gamma)C_1 + V_{10}C_0 + V_{12}C_2 \quad (129)$$

$$i\dot{C}_2 = (\varepsilon_2 - 2\Omega)C_2 + V_{21}C_1 \quad (130)$$

Here we have set the ground state energy equal to zero, $\varepsilon_0 = 0$, and introduced a relaxation rate Γ into the equation describing the intermediate level $|\phi_1\rangle$ only. The latter assumption is justified because in experimental situations the level $|\phi_2\rangle$ is often metastable and decays much more slowly. If the ground state $|\phi_0\rangle$ is not too heavily depleted we need not explicitly consider its repopulation due to spontaneous emission (for a general treatment, see the lecture notes[122]).

In order to see the implications of equations (128)–(130) we Laplace-transform them with the initial conditions at $t = 0$; $C_0 = 1$, $C_1 = C_2 = 0$. We obtain from (129) the result

$$\tilde{C}_1(s) = -D(s)(V_{10}\tilde{C}_0 + V_{12}\tilde{C}_2) \quad (131)$$

where

$$D(s) = (\varepsilon_1 - \Omega + kv - i\tfrac{1}{2}\Gamma - is)^{-1} \quad (132)$$

In cases of physical interest now the following situation arises: The amplitudes C_0 and C_2 develop with the rates determined by V_{01}, V_{21}, and $(\varepsilon_2 - 2\Omega)$, which are slow compared to either $(\varepsilon_1 - \Omega + kv)$ or Γ. It is then possible to investigate the time dependence of the slowly varying amplitudes in such a regime where we choose

$$|V|, |\varepsilon_2 - 2\Omega| \ll s \ll [(\varepsilon_1 - \Omega + kv)^2 + \tfrac{1}{4}\Gamma^2]^{1/2} \quad (133)$$

We can then neglect the s dependence of D and replace it by $D(0)$. After eliminating C_1 we obtain the equation for an effective two-level system in the form

$$i\frac{d}{dt}\begin{bmatrix} C_0 \\ C_2 \end{bmatrix} = \begin{pmatrix} -V_{01}D(0)V_{10} & -V_{01}D(0)V_{12} \\ -V_{21}D(0)V_{10} & \varepsilon_2 - 2\Omega - V_{21}D(0)V_{12} \end{pmatrix}\begin{pmatrix} C_0 \\ C_2 \end{pmatrix} \quad (134)$$

In this limit we have obtained the equations for an effective Hamiltonian as given by Grischkowsky et al.[120]

2.5.2. Physical Consequences

There are several physical consequences of Eq. (134). The first one was pointed out by Vasilenko et al.[113] and exploited by Cagnac et al.[114] and later by many others. This uses the fact that the level splitting resonance condition

$$\varepsilon_2 - 2\Omega - V_{21}D(0)V_{12} + V_{01}D(0)V_{10} = 0 \qquad (135)$$

is independent of the Doppler shift kv, when the intermediate level is far off resonance, viz.,

$$|\varepsilon_1 - \Omega| \gg kv \qquad (136)$$

over the whole Doppler profile. Then also the coupling matrix element becomes a constant $[V_{01}V_{12}/(\varepsilon_1 - \Omega)]$ and all atoms contribute equally to the absorbed signal. The drawback of this method is the appearance of the power shift given in (135). This has been investigated by Bjorkholm and Liao.[115] When the matrix elements are equal, $V_{12} = V_{10}$, the shift cancels exactly.[119]

Salomaa and Stenholm[116,117] have pointed out that going into resonance one can obtain the enhanced coupling strength $(2V_{01}V_{12}/\Gamma)$, which improves the signal even if only part of the Doppler profile contributes. The additional factor i makes the power shift turn into a power broadening, which puts less severe restrictions on the use of the method in spectroscopy. These predictions have been verified experimentally by Bjorkholm and Liao.[118]

When we consider time developments that take place over time scales satisfying

$$|V|, |\varepsilon_2 - 2\Omega| \ll t^{-1} \ll [(\varepsilon_1 - \Omega + kv)^2 + \tfrac{1}{4}\Gamma^2]^{1/2} \qquad (137)$$

we can use Eq. (134) to discuss coherent transients. We may thus obtain all the well-known phenomena of free-induction decay, nutations, photon echoes, and adiabatic following. Such work is presented in Refs. 119–121 and 132. Recently Baklanov et al.[133] have suggested a future frequency standard based on two-photon absorption in an atomic beam traversing two standing wave regions. The use of fast beams for two-photon spectroscopy is discussed by Salomaa and Stenholm.[134]

2.5.3. Validity of the Two-Level Approximation

We derived Eq. (134) under the assumption (133), which is equivalent to (137). This means that either the intermediate level decays fast (Γ large), or it is far off resonance ($\varepsilon_2 - \Omega$ large). In either case it is easy to understand the physics behind the disappearance of the intermediate level.

When Γ is large no population or coherence can be accumulated in level 1 and hence all information reaching it is rapidly transmitted to level 2. For a large detuning, the amplitude of level 1 oscillates rapidly with the frequency $\varepsilon_2 - \Omega$ and averages out over times satisfying the inequality (137). It is often said that the uncertainty principle allows the level to be populated only for a time

$$\tau \approx 1/|\varepsilon_2 - \Omega| \qquad (138)$$

but the exact meaning of this statement seems unclear. The physics is that the smearing of the intermediate state amplitude C_1 takes place at frequency $\varepsilon_2 - \Omega$, which greatly exceeds the exchange between levels 0–2, which remains slow.

When the inequality implied by (137) becomes invalid, one has to treat the full complication of the three-level system.

3. Collision Effects

3.1. Introduction

In spectroscopic applications one tries to work with samples that are as dilute as possible. Even then the atomic constituents are never entirely independent but affected by the interaction between the particles. Collision investigations have been carried out for a long time, and it is impossible to review the work in a short section. We choose to present a recent theoretical framework which is believed to provide the starting point for calculations of collision effects on laser spectroscopy. Some of its physical consequences are discussed but no detailed review of collisional work with lasers is attempted. Because of the rapid development in this field it may even be premature to try to summarize the experimental situation.

If the time during which the interparticle potential is appreciable remains shorter than the average free time, we can consider well-separated collision events and only two-particle encounters need be taken into account. This binary approximation describes well experiments performed on dilute gaseous samples.

3.2. Theoretical Considerations

3.2.1. The General Collision Equation

With well-separated collisions of short duration it is possible to neglect the influence of the electromagnetic field on the atomic system during encounters. It is possible to choose the energy eigenstates of the atom as a basis and the Hamiltonian is diagonal. When the collision time is long enough to displace the atomic energy levels adiabatically we obtain a phase perturbation. The requirements are thus that the collision time be short compared with the time of free flight but long compared with the optical period.

The basic conditions, to be satisfied by the sample, are assumed to be as follows:

(i) No resonant collisions included. This implies that the perturbers are either foreign atoms or the same atom in levels not affected by the optical field.

(ii) Impact approximation. The duration of the collision is assumed to be so brief that the atomic history can be divided into intervals dominated by the fields and separated by encounters with the perturbers.

(iii) Binary collisions. In the impact approximation limit the collisions are so infrequent that no more than two atoms will interact simultaneously.

Under these conditions the time evolution due to the fields and the perturbers can be added as independent phenomena. Berman[135,136] and Sobel'man and collaborators[137,138] have derived the general time dependence of the atomic density matrix due to collisions assuming (i)–(iii).

An instantaneous encounter at time t replaces the density matrix by a linear superposition of its elements before the collision, and the coefficients are given by the S matrix of the interaction

$$\rho_{\mu\mu'}(t+) = \sum_{\nu\nu'} S_{\mu\nu} S^+_{\nu'\mu'} \rho_{\nu\nu'}(t-) \tag{139}$$

where μ, ν label the states. For the collisional contribution to the time dependence we obtain in terms of the matrix

$$T = S - 1 \tag{140}$$

the equation

$$\frac{\partial}{\partial t}\rho_{\mu\mu'} = \sum_{\nu'} \overline{T^+_{\nu'\mu'}\rho_{\mu\nu'}} + \sum_{\nu} \overline{T_{\mu\nu}\rho_{\nu\mu'}} + \sum_{\nu\nu'} \overline{T_{\mu\nu}T^+_{\nu'\mu'}\rho_{\nu\nu'}} \tag{141}$$

where the bar denotes a complete average over the collisional event.

Berman and Sobel'man et al. have displayed the translational degree of freedom explicitly, and with $\mathbf{v} = \hbar\mathbf{k}/m$ we can write

$$\frac{\partial}{\partial t}\rho_{ab}(\mathbf{v}, t) = \sum_{cd} [t_{ac}(\mathbf{v})\delta_{bd} + t^*_{bd}(\mathbf{v})\delta_{ac}]\rho_{cd}(\mathbf{v}, t)$$

$$+ \sum_{cd} \int d^3\mathbf{v}'\, W^{cd}_{ab}(\mathbf{v}' \to \mathbf{v})\rho_{cd}(\mathbf{v}', t) \tag{142}$$

and it is possible to express the coefficients in terms of the scattering amplitude $f_{ac}(\mathbf{v}_r' \to \mathbf{v}_r)$ from the state $|c\rangle$ with relative velocity $\mathbf{v}_r'$ (in the collisional center-of-mass system) to the state $|a\rangle$ with relative velocity $\mathbf{v}_r$. We obtain[135,136]

$$t_{ac}(\mathbf{v}) = -N \int d^3 v_p \, W_p(\mathbf{v}_p) \frac{2\pi\hbar}{i\mu} f_{ac}(\mathbf{v}_r \to \mathbf{v}_r) \tag{143}$$

$$W_{ab}^{cd}(\mathbf{v}' \to \mathbf{v}) = N\left(\frac{m}{\mu}\right)^3 2 \int d^3 v_p' \int d^3 v_r W_p(\mathbf{v}_p')$$

$$\times \delta[\mathbf{v}_r + (m/m_p)\mathbf{v}' - (m/\mu)\mathbf{v} + \mathbf{v}_p'] \delta(v_r^2 - v_r'^2)$$

$$\times f_{ac}(\mathbf{v}_r' \to \mathbf{v}_r) f_{bd}^*(\mathbf{v}_r' \to \mathbf{v}_r) \tag{144}$$

Here N is the density of perturbers, $W_p(\mathbf{v}_p)$ their velocity distribution, and μ the collisional reduced mass. The delta functions express momentum and energy conservation. The functions $t_{\alpha\beta}$ are expressed in terms of the forward scattering amplitudes $f(\mathbf{v} \to \mathbf{v})$ only. An approach similar to the classical Boltzmann equation was earlier used by Snider.[139] Some aspects of the problem were discussed also by Omont;[140] see also the review in Ref. 123.

In addition to Eq. (142) time development due to the field and atomic Hamiltonian have to be included, and a quite general formulation is achieved.

In order to connect the present formulation with usual transport equations Berman introduces

$$\Gamma_{ab}^{cd}(\mathbf{v}) = \int d^3 v' \, W_{ab}^{cd}(\mathbf{v} \to \mathbf{v}') \tag{145}$$

$$-T_{ab}^{cd}(\mathbf{v}) = t_{ac}(\mathbf{v}) \, \delta_{bd} + t_{bd}^*(\mathbf{v}) \delta_{ac} + \Gamma_{ab}^{cd}(\mathbf{v}) \tag{146}$$

and he writes Eq. (142) in the form

$$\frac{\partial}{\partial t} \rho_{ab}(\mathbf{v}, t) = -\sum_{cd} T_{ab}^{cd}(\mathbf{v}) \rho_{cd}(\mathbf{v})$$

$$+ \sum_{cd} \int d^3 v' [W_{ab}^{cd}(\mathbf{v}' \to \mathbf{v}) \rho_{cd}(\mathbf{v}', t) - W_{ab}^{cd}(\mathbf{v} \to \mathbf{v}') \rho_{cd}(\mathbf{v}, t)] \tag{147}$$

where the second term contains the usual scattering-in and scattering-out terms of a Boltzmann equation.

3.2.2. Physical Implications

The physical meaning of Eq. (147) has been discussed by Berman,[141] who also gives an extensive list of references. Here we only pick out the simplest special cases and point out some of their spectroscopic consequences.

(α) No mixing between the levels. This implies

$$T^{cd}_{ab} \propto \delta_{ac}\delta_{bd}$$
$$W^{cd}_{ab} \propto \delta_{ac}\delta_{bd}$$

(148)

In this case we have elastic level perturbations only.

(i) If the velocity is changed but little in a collision

$$W(\mathbf{v} \to \mathbf{v}') \propto \delta(\mathbf{v} - \mathbf{v}')$$

(149)

from (147) we see that the transport part cancels exactly and we obtain the level perturbation

$$T^{ab}_{ab}(\mathbf{v}) = \gamma^{coll}_{ab}(\mathbf{v})$$

(150)

This is a complex quantity and provides a pressure broadening as well as a shift like the traditional theory (see Hindmarsh and Farr[142]). No other effects are present.

(ii) If the scattering amplitude f_a is completely independent of the state a, it can be shown that $T^{ab}_{ab} = 0$. This is a result of the unitarity of the S matrix as expressed in the optical theorem for the scattering amplitude. The velocity kernel $W^{ab}_{ab}(\mathbf{v} \to \mathbf{v}')$ is independent of the states a, b and (147) becomes a classical transport equation. It is then possible to assign a unique velocity to the whole atom, which scatters like a classical particle. It is noteworthy that no pressure shifts or broadenings are observable in this case. Whenever observed they imply unequal scattering amplitudes.

(iii) When the scattering amplitudes differ significantly ($f_a \neq f_b$) the two states of an off-diagonal element ρ_{ab} scatter along different trajectories and the outgoing part is eliminated owing to interference between the outgoing waves. This situation was first described by Berman and Lamb.[143] No velocity changes are associated with the off-diagonal elements but only a γ^{coll}_{ab} as in (150). However, the population probabilities ρ_{aa} and ρ_{bb} obey transport equations with the kernels W^{aa}_{aa} and W^{bb}_{bb}, respectively.

(β) Coupling between levels. If we take $c \neq a$ and $d \neq b$ in equation (147) we see from (146) that $-T = \Gamma$ and Γ cancels against the scattering-out term. Hence only the term

$$\frac{\partial}{\partial t}\rho_{ab}(\mathbf{v}, t) = \sum_{\substack{cd \\ (\neq ab)}} \int d^3\mathbf{v}' \, W^{cd}_{ab}(\mathbf{v}' \to \mathbf{v})\rho_{cd}(\mathbf{v}', t)$$

(151)

remains. This equation contains two possible processes[144]:

(i) The populations are coupled by W^{cc}_{aa}. This implies a transfer of population between atomic levels with a smearing of the velocity determined by the **v** dependence of W. Such transfer has been observed in laser spectroscopy.[144-147]

(ii) Transfer of coherence by W_{ab}^{cd} ($a \neq b$, $c \neq d$). This type of effect has not yet been observed but could give rise to coherent transients[144] in switched laser spectroscopy.

In the general case all consequences are present simultaneously. Then one needs very accurate measurements and a detailed analysis to disentangle the various collisional contributions. For more details we must refer the reader to the references.

In some cases the levels involved are magnetic sublevels and then one can show[123] that rotational symmetry allows $\langle m_a | \rho | m_b \rangle$ to couple only to $\langle m_c | \rho | m_d \rangle$ with

$$m_a - m_b = m_c - m_d \tag{152}$$

This restricts the range of possible experiments considerably. The physically interesting situation where one level is common, $a = c$, $m_a = m_c$ say, is not permitted by (153). It is unclear to what extent the anisotropy caused by a directed laser beam and external magnetic or electric fields will break the symmetry requirement. So far, only a general discussion is possible,[149] and further theoretical and experimental investigations are needed.

References

1. S. Stenholm, *Phys. Rep.* **6C**, 1 (1973).
2. S. H. Autler and C. H. Townes, *Phys. Rev.* **100**, 703 (1955).
3. C. Cohen-Tannoudji, *Ann. Phys. (Paris)* **7**, 423 (1962).
4. C. Cohen-Tannoudji, *Ann. Phys. (Paris)* **7**, 469 (1962).
5. C. Cohen-Tannoudji, in *Chargése Lectures in Physics*, Ed. M. Lévy, Vol. 2, p. 347, Gordon and Breach, New York (1968).
6. C. Cohen-Tannoudji and J. Dupont-Roc, *Opt. Comm.* **1**, 184 (1969).
7. C. Cohen-Tannoudji and J. Dupont-Roc, *Phys. Rev. A* **5**, 968 (1972).
8. F. Bloch and A. Siegert, *Phys. Rev.* **57**, 522 (1940).
9. I. I. Rabi, *Phys. Rev.* **51**, 652 (1937).
10. J. Winter, *C.R. Acad. Sci. Paris* **241**, 375 (1955).
11. J. Margerie and J. Brossel, *C.R. Acad. Sci. Paris* **241**, 373 (1955).
12. J. N. Dodd and G. W. Series, *Proc. R. Soc. London* **A263**, 353 (1961).
13. J. P. Barrat, *Proc. R. Soc. London* **A263**, 371 (1961).
14. J. P. Barrat and C. Cohen-Tannoudji, *J. Phys. Rad.* **22**, 329 (1961).
15. J. P. Barrat and C. Cohen-Tannoudji, *J. Phys. Rad.* **22**, 443 (1961).
16. G. W. Series, *Physica* **33**, 138 (1967).
17. G. W. Series, in *Quantum Optics*, the 1969 Scottish Universities Summer School, Eds. S. M. Kay and A. Maitland, p. 395, Academic Press, London (1970).
18. J. H. Shirley, *J. Appl. Phys.* **34**, 783 (1963).
19. J. H. Shirley, *Phys. Rev.* **138**, B979 (1965).
20. E. B. Aleksandrov, O. V. Konstantinov, V. I. Perel, and V. A. Khodovoi, *Sov. Phys. JETP*, **18**, 346 (1964).

21. V. A. Khodovoi, *Sov. Phys. JETP,* **19**, 227 (1964).
22. N. Polonsky and C. Cohen-Tannoudji, *J. Phys. Theor. Appl.* **26**, 409 (1965).
23. C. Cohen-Tannoudji and S. Haroche, *C.R. Acad. Sci. Paris* **261**, 5400 (1965).
24. C. Cohen-Tannoudji and S. Haroche, *C.R. Acad. Sci. Paris* **262**, 268 (1966).
25. C. Cohen-Tannoudji and S. Haroche, *J. Phys. (Paris)* **30**, 153 (1969).
26. C. Landré, C. Cohen-Tannoudji, J. Dupont-Roc, and S. Haroche, *J. Phys. (Paris)* **31**, 971 (1970).
27. D. T. Pegg and G. W. Series, *J. Phys. B: Atom. Molec. Phys.* **3**, L33 (1970).
28. G. D. Chapman, *J. Phys. B: Atom. Molec. Phys.* **3**, L36 (1970).
29. T. Yabuzaki, N. Tsukada, and T. Ogawa, *J. Phys. Soc. Japan,* **32**, 1069 (1972).
30. C. Cohen-Tannoudji and S. Haroche, *C.R. Acad. Sci. Paris* **264**, 626 (1967).
31. C. Cohen-Tannoudji and S. Haroche, *J. Phys. (Paris),* **30**, 125 (1969).
32. J. Dupont-Roc, *J. Phys. (Paris),* **32**, 135 (1971).
33. N. Tsukada, T. Yabuzaki, and T. Ogawa, *J. Phys. Soc. Japan,* **33**, 698 (1972).
34. E. B. Aleksandrov and A. P. Sokolov, *Sov. Phys. JETP,* **34**, 48 (1972).
35. W. E. Lamb, Jr., *Phys. Rev.* **134**, A1429 (1964).
36. S. Stenholm and W. E. Lamb, Jr., *Phys. Rev.* **181**, 618 (1969).
37. B. J. Feldman and M. S. Feld, *Phys. Rev. A* **1**, 1375 (1970).
38. S. Stenholm, *J. Phys. B: Atom. Molec. Phys.* **5**, 878 (1972).
39. S. Stenholm, *J. Phys. B: Atom. Molec. Phys.* **5**, 890 (1972).
40. S. Stenholm, *J. Phys. B: Atom. Molec. Phys.* **6**, 1097 (1973).
41. U. Cappeller and H. G. Weber, *Ann. Phys. Lepizig* **26**, 359 (1971).
42. H. G. Weber, *Phys.* **247**, 336 (1971).
43. R. Gush and H. P. Gush, *Phys. Rev. A* **6**, 129 (1972).
44. S. Morand and G. Theobald, *C.R. Acad. Sci. Paris* **269**, 503 (1969).
45. S. Morand and G. Theobald, *C.R. Acad. Sci. Paris* **272**, 569 (1971).
46. S. Stenholm, *Phys. Lett.* **44A**, 7 (1973).
47. P. Hannaford, D. T. Pegg, and G. W. Series, *J. Phys. B: Atom. Molec. Phys.* **6**, L222 (1973).
48. E. Arimondo and G. Moruzzi, *J. Phys. B: Atom. Molec. Phys.* **6**, 2382 (1973).
49. S. Haroche, *Ann. Phys. (Paris)* **6**, 189 (1971).
50. S. Haroche, *Ann. Phys. (Paris),* **6**, 327 (1971).
51. C. S. Chang and P. Stehle, *Phys. Rev. A* **4**, 641 (1971).
52. C. S. Cheng and P. Stehle, *Phys. Rev. A* **5**, 1087 (1972).
53. D. T. Pegg and G. W. Series, *Phys. Rev. A* **7**, 371 (1973).
54. D. T. Pegg, *Phys. Rev. A* **8**, 2214 (1973).
55. C. Cohen-Tannoudji, J. Dupont-Roc, and C. Fabre, *J. Phys. B: Atom. Molec. Phys.* **6**, L214 (1973).
56. S. Stenholm, *J. Phys. B: Atom. Molec Phys.* **6**, L240 (1973).
57. S. Stenholm, *J. Phys. B: Atom. Molec. Phys.* **6**, 1650 (1973).
58. S. Swain, *Phys. Lett.* **43A**, 229 (1973).
59. S. Swain, *J. Phys. A: Math. Nucl. Gen.* **6**, 192 (1973).
60. S. Swain, *J. Phys. A: Math. Nucl. Gen.* **6**, 1919 (1973).
61. S. Swain, *J. Phys. A: Math. Nucl. Gen.* **6**, L169 (1973).
62. S. Swain, *Phys. Lett.* **46A**, 435 (1974).
63. S. Swain, *J. Phys. A: Math. Nucl. Gen.* **8**, 1277 (1975).
64. J. Hermann and S. Swain, *Phys. Lett.* **55A**, 446 (1976).
65. F. Ahmad and R. K. Bullough, *J. Phys. B: Atom. Molec. Phys.* **7**, L147 (1974).
66. F. Ahmad and R. K. Bullough, *J. Phys. B: Atom. Molec. Phys.* **7**, L275 (1974).
67. D. T. Pegg and G. W. Series, *Proc. R. Soc. London* **332**, 281 (1973).
68. D. T. Pegg, *J. Phys. B: Atom. Molec. Phys.* **6**, 246 (1973).

69. L. Armstrong, Jr. and S. Feneuille, *J. Phys. B: Atom. Molec. Phys.* **7**, L182 (1973).
70. F. T. Hioe and J. H. Eberly, *Phys. Rev. A* **11**, 1358 (1975).
71. F. T. Hioe and E. W. Montroll, *J. Math. Phys.* **16**, 1259 (1975).
72. J. Wong, J. C. Garrison, and T. H. Einwohner, *Phys. Rev. A* **13**, 674 (1976).
73. I. Bialynicki-Birula and Z. Bialynicka-Birula, *Phys. Rev. A* **14**, 1101 (1976).
74. N. Tsukada and T. Ogawa, *J. Phys. B: Atom. Molec. Phys.* **6**, 1643 (1973).
75. S. Stenholm and C.-G. Aminoff, *J. Phys. B: Atom. Molec. Phys.* **6**, 2390 (1973).
76. T. Yabuzaki, S. Nakayama, Y. Murakami, and T. Ogawa, *Phys. Rev. A* **10**, 1955 (1974).
77. C. Cohen-Tannoudji, J. Dupont-Roc, and C. Fabre, *J. Phys. B: Atom. Molec. Phys.* **6**, L218 (1973).
78. S. Stenholm and C.-G. Aminoff, in *VIII International Quantum Electronics Conference*, San Francisco, Digest of Technical Papers, p. 23 (1974).
79. S. Swain, *J. Phys. B: Atom. Molec. Phys.* **7**, 2363 (1974).
80. C.-G. Aminoff, *Phys. Fennica*, **10**, 173 (1975).
81. D. T. Pegg, *J. Phys. B: Atom. Molec. Phys.* **6**, L356 (1973).
82. D. T. Pegg, *J. Phys. B: Atom. Molec. Phys.* **6**, 241 (1973).
83. N. Tsukada and T. Ogawa, *Phys. Lett.* **45A**, 159 (1973).
84. N. Tsukada, T. Koyama, and T. Ogawa, *Phys. Lett.* **44A**, 501 (1973).
85. N. Tsukada, T. Yabuzaki, and T. Ogawa, *J. Phys. Soc. Japan*, **35**, 230 (1973).
86. T. Yabuzaki, Y. Murakami, and T. Ogawa, *J. Phys. B: Atom. Molec. Phys.* **9**, 9 (1976).
87. W. A. McClean and S. Swain, *J. Phys. B: Atom. Molec. Phys.* **9**, 1673 (1976).
88. E. Arimondo and G. Moruzzi, *J. Phys. B: Atom. Molec. Phys.* **9**, 727 (1976).
89. E. Arimondo and G. Moruzzi, *J. Phys. B: Atom. Molec. Phys.* **9**, 709 (1976).
90. M. S. Feld and A. Javan, *Phys. Rev.* **177**, 540 (1969).
91. B. J. Feldman and M. S. Feld, *Phys. Rev. A* **5**, 899 (1972).
92. Th. Hänsch, R. Keil, A. Schabert, Ch. Schmelzer, and P. Toschek, *Z. Phys.* **226**, 293 (1969).
93. Th. Hänsch and P. Toschek, *Z. Phys.* **236**, 213 (1970).
94. I. M. Beterov and V. P. Chebotaev, in *Progress in Quantum Electronics*, Eds. J. H. Sanders and S. Stenholm, Vol. 3, p. 1, Pergamon Press, Oxford (1975).
95. A. Schabert, R. Keil, and P. Toschek, *Opt. Comm.* **13**, 265 (1975).
96. A. Schabert, R. Keil, and P. Toschek, *Appl. Phys.* **6**, 181 (1975).
97. B. R. Mollow, *Phys. Rev.* **188**, 1969 (1969).
98. B. R. Mollow, *Phys. Rev. A* **2**, 76 (1970).
99. B. R. Mollow, *Phys. Rev. A* **13**, 758 (1976).
100. E. V. Baklanov, *Sov. Phys. JETP*, **38**, 1100 (1974).
101. M. E. Smithers and H. S. Freedhoff, *J. Phys. B: Atom. Molec. Phys.* **7**, L432 (1974).
102. H. S. Freedhoff and M. E. Smithers, *J. Phys. B: Atom. Molec. Phys.* **8**, L209 (1975).
103. R. B. Higgins, *J. Phys. B: Atom. Molec. Phys.* **8**, L321 (1975).
104. S. S. Hassan and R. K. Bullough, *J. Phys. B: Atom. Molec. Phys.* **8**, L147 (1975).
105. H. J. Carmichael and D. F. Walls, *J. Phys. B: Atom. Molec. Phys.* **8**, L77 (1975).
106. C. Cohen-Tannoudji, in *Laser Spectroscopy*, Eds. S. Haroche, J. C. Pebay-Peyroula, Th. Hänsch, and S. E. Harris p. 324, Springer-Verlag, Heidelberg (1975).
107. H. J. Carmichael and D. F. Walls, *J. Phys. B: Atom. Molec. Phys.* **9**, L43 (1976).
108. R. P. Feynman, *Phys. Rev.* **56**, 340 (1939).
109. F. Schuda, C. R. Stroud, Jr., and M. Hercher *J. Phys. B: Atom. Molec. Phys.* **7**, L198 (1974).
110. H. Walther, in *Laser Spectroscopy*, Eds. S. Haroche, J. C. Pebay-Peyroula, Th. Hänsch, and S. E. Harris, p. 358, Springer-Verlag, Heidelberg (1975).
111. F. Y. Wu, R. E. Grove, and S. Ezekiel, *Phys. Rev. Lett.* **35**, 1426 (1975).

112. J.-L. Picqué and J. Pinard, *J. Phys. B: Atom. Molec. Phys.* **9**, L77 (1976).
113. L. S. Vasilenko, V. P. Chebotaev, and A. V. Shishaev, *JETP Lett.* **12**, 113 (1970).
114. B. Cagnac, in *Laser Spectroscopy*, Eds. S. Haroche, J. C. Pebay-Peyroula, Th. Hänsch, and S. E. Harris, p. 165, Springer-Verlag, Heidelberg (1975).
115. P. F. Liao and J. E. Bjorkholm, *Phys. Rev. Lett.* **34**, 1 (1975).
116. R. Salomaa and S. Stenholm, *J. Phys. B: Atom. Molec. Phys.* **8**, 1795 (1975).
117. R. Salomaa and S. Stenholm, *J. Phys. B: Atom. Molec. Phys.* **9**, 1221 (1976).
118. J. E. Bjorkholm and P. F. Liao, *Phys. Rev. A* **14**, 751 (1976).
119. R. G. Brewer and E. L. Hahn, *Phys. Rev. A* **11**, 1641 (1975).
120. D. Grischkowsky, M. M. T. Loy, and P. F. Liao, *Phys. Rev. A* **12**, 2514 (1975).
121. M. Sargent III and P. Horowitz, *Phys. Rev. A* **13**, 1962 (1976).
122. C. Cohen-Tannoudji, in *Frontiers in Laser Spectroscopy*, Eds. R. Balian, S. Haroche, and S. Liberman, p. 4, North-Holland, Amsterdam (1977).
123. A. Omont, in *Progress in Quantum Electronics*, Eds. J. H. Sanders and S. Stenholm, Vol. 5, part 2, p. 69, Pergamon Press, Oxford (1976).
124. G. N. Watson, *Theory of Bessel Functions*, 2nd ed., Cambridge University Press, Cambridge (1962).
125. W. Hanle, *Z. Phys.* **30**, 93 (1924).
126. H. G. Dehmelt, *Phys. Rev.* **105**, 1924 (1957).
127. T. R. Carver and R. B. Partridge, *Am. J. Phys.* **34**, 339 (1966).
128. N. Polonsky and C. Cohen-Tannoudji, *C.R. Acad. Sci. Paris* **260**, 5231 (1965).
129. N. Polonsky and C. Cohen-Tannoudji, *C.R. Acad. Sci. Paris* **261**, 369 (1965).
130. R. L. Shoemaker, S. Stenholm, and R. G. Brewer, *Phys. Rev. A* **10**, 2037 (1974).
131. N. Skribanovitz, M. J. Kelley, and M. S. Feld, *Phys. Rev. A* **6**, 2302 (1972).
132. J. C. Garrison, T. H. Einwohner, and J. Wong, *Phys. Rev. A* **14**, 731 (1976).
133. Ye. V. Baklanov, V. P. Chebotaev, and B. T. Dubetsky, *Appl. Phys.* **11**, 201 (1976).
134. R. Salomaa and S. Stenholm, *Opt. Comm.* **16**, 292 (1976).
135. P. R. Berman, *Phys. Rev. A* **5**, 927 (1972).
136. P. R. Berman, *Phys. Rev. A* **6**, 2157 (1972).
137. V. A. Alekseév, T. L. Andreéva, and I. I. Sobel'man, *Sov. Phys. JETP* **35**, 325 (1972).
138. V. A. Alekseév, T. L. Andreéva, and I. I. Sobel'man, *Sov. Phys. JETP* **37**, 413 (1973).
139. R. F. Snider, *J. Chem. Phys.* **32**, 1051 (1960).
140. P. A. Omont, *J. Phys. (Paris)*, **26**, 26 (1965).
141. P. R. Berman, *Appl. Phys.* **6**, 283 (1975).
142. W. R. Hindmarsh and J. M. Farr, in *Progress in Quantum Electronics*, Eds. J. H. Sanders and S. Stenholm, Vol. 2, p. 141, Pergamon Press, Oxford (1973).
143. P. R. Berman and W. E. Lamb, Jr., *Phys. Rev. A* **2**, 2435 (1970).
144. S. Stenholm, in *Laser Spectroscopy*, Eds. S. Haroche, J. C. Pebay-Peyroula, Th. Hänsch, and S. E. Harris, p. 429, Springer-Verlag, Heidelberg (1975).
145. J. W. C. Johns, A. R. W. McKellar, Y. Oka, and M. Römheld, *J. Chem. Phys.* **62**, 1488 (1975).
146. M. Ouhayoun and C. Bordé, *C.R. Acad. Sci. Paris* **274**, 411 (1972).
147. W. K. Bischel and C. K. Rhodes, *Phys. Rev. A* **14**, 176 (1976).
148. J. Herman and S. Swain, *J. Phys. A: Math. Gen.* **9**, 1947 (1976).
149. S. Stenholm, *J. Phys. B: Atom. Molec. Phys.* **10**, 761 (1977).

4

Quantum Electrodynamical Effects in Atomic Spectra

A. M. Ermolaev

1. Introduction

1.1. Some Recent Developments in Applications of QED to Atomic Spectroscopy

Atomic spectroscopy has recently provided several topics that must be dealt with by the application of quantum electrodynamics. In the field of atomic structure theories, quantum electrodynamics (QED) considers spin-$\frac{1}{2}$ particles (electrons, muons) and photons (which are assumed to be free with the exception of their interaction with one another and with external electromagnetic fields). This simplified model, which does not take into consideration hadron interactions and interactions with weak currents, is often referred to as "restricted QED."[1]

The current interest of QED effects in atomic structure theory can be regarded as a continuation of the traditional direction as laid out in the fundamental works by Lamb and Retherford[2] and by Bethe[3] on the fine structure of the hydrogen spectrum; these works laid down the basis of modern quantum electrodynamics.

I shall start this review by giving a summary of some recent theoretical QED contributions to the theory of atomic spectra.

The hydrogen atom is still at the focus of the QED applications. The fourth-order self-energy of the electron has been recomputed by Appelquist and Brodski[4,5] and Peterman.[6] These authors have corrected

A. M. Ermolaev • Department of Applied Mathematics and Theoretical Physics, The Queen's University of Belfast, Belfast BT7, N.I.

an old numerical result computed earlier by Soto.[7,8] The new theoretical value for the fourth-order electron self-energy, together with calculations by Erickson[9-11] of some higher-order binding terms, have brought the theoretical and experimental values of the Lamb shift in hydrogen and deuterium into better agreement.

Erickson,[12] Mohr,[13-15] and Cheng and Johnson[16] have developed methods for the numerical evaluation of the exact matrix element which gives the second-order self-energy correction (the main term in the Lamb shift), for a one-electron atom with high nuclear charge Z. The results of these authors are important for an accurate description of K-shell atomic electrons in heavy and superheavy atoms.

The computational efforts of several theoretical groups[17-27] have been directed towards the determination of the electron and muon magnetic anomalies in the sixth order of perturbation theory. This work, which requires extensive use of computers, has now been completed and new calculations aimed at applying eighth-order perturbation theory have already begun.[28,29]

In the theory of two-electron atoms the evaluation of the radiative level shifts requires a knowledge of some atomic properties.[30] Consequently, the studies of the Lamb shifts in helium and in some two-electron ions are the main driving force behind many accurate nonrelativistic calculations of two-electron atomic states[31-37] and photoionization cross sections.[38] As a result of these calculations, many of which were performed by Pekeris and co-workers, the existence of radiative shifts in low-lying S states of two-electron ions with low Z has been confirmed.[31-33]

A strong dependence of the triplet P level shift in Li II and Be III on the electron density at the nucleus was theoretically predicted by this author[39,40] and measured by Berry and co-workers[41-43] in Li II.

Several authors[44-47] have carried out extensive theoretical work on the determination of higher-order relativistic and QED corrections to the fine-structure intervals in neutral helium. This project, which was suggested some time ago by Schwartz,[48] was aimed at providing the necessary theoretical basis for an experimental determination, from the fine-structure intervals in helium, of an improved QED value for the fine-structure constant α.

1.2. Existing Reviews

Many of the developments mentioned above have already been reviewed and discussed. For an introduction, I can refer to a monograph by Taylor et al.[49] The conceptual basis of QED, renormalization procedures, and covariant formulation of the theory have recently been discussed in Refs. 1 and 50–53.

Calculations of the anomalous magnetic moments of the electron and muon have been reviewed by Calmet *et al.*[54]

The Lamb shift in muonic atoms has recently been reviewed by Borie.[55]

Computer application in QED has been discussed by Calmet[56] along three main lines: renormalization, algebraic manipulations, and numerical evaluation of multidimensional Feynman integrals.[59–64]

An account of theoretical work done on higher-order corrections to the fine-structure intervals in helium has recently been published by Douglas and Kroll.[57]

The QED corrections to the transition rates in helium have been discussed by Kelsey.[58]

1.3. The Layout of the Present Work

The limited number of questions that will be discussed in the present paper are related to the Lamb shift of energy levels in one- and two-electron atoms. In Section 2, the theoretical results for the shifts in a one-electron atom will be reviewed and the case of high Z discussed; comparisons with experimental results will be made. In Section 3, I shall discuss the Lamb shift investigations in two-electron atoms. The discussion will include $L \neq 0$ states. Section 4 will conclude the paper.

Throughout the paper, the atomic units ($e = m_e = h = 1$) will be used and the energy will be expressed in Rydberg units (unless stated otherwise).

2. The Lamb Shift in One-Electron Atoms

2.1. Summary of Main Theoretical Results

For the forthcoming discussion, it will be convenient to write the QED expansion of the total energy E of a bound state of a one-electron atom in the following form:

$$E = Z^2[\varepsilon_D + (\alpha/\pi)\varepsilon_1 + (\alpha/\pi)^2\varepsilon_2 + \cdots] \, \text{Ry} \qquad (1)$$

where the scaled Dirac energy, ε_D, and QED corrections, ε_i, $i = 1, 2, \ldots$, are assumed to be functions of two parameters: the binding parameter, αZ, and the mass ratio, m_e/M, where m_e and M are, respectively, the electron and nuclear masses. We shall further assume that, for small values of these parameters, $\alpha Z \ll 1$ and $m_e/M \ll 1$, ε_1 can be expanded in terms of the two parameters.

The QED shift $E_{L,1}$ (or the Lamb shift) of a one-electron atomic level is then defined as the difference

$$E_{L,1} = (E - Z^2 \varepsilon_D)\,\text{Ry} \tag{2}$$

2.1.1. The Dirac Energy $Z^2 \varepsilon_D$

The first term in (1) has a closed analytical form.[30] It depends upon the principal quantum number, n, total angular momentum, j, and on the binding parameter, αZ. One can expand ε_D as follows:

$$\varepsilon_D = -1/n^2 + (\alpha Z)^2 (3/4n^4 - 1/kn^3)$$
$$- (\alpha Z)^4 (5/8n^6 - 3/2kn^5 + 3/4k^2 n^4 + 1/4k^3 n^3) + \cdots \tag{3}$$

where $k = j + \frac{1}{2}$. The rest energy of the electron has been subtracted from (3). The first term in (3) gives the nonrelativistic energy and the second term is the relativistic correction of order $Z^2 (\alpha Z)^2$ Ry, which corresponds exactly to the Pauli approximation made in the Dirac equation for an electron in the Coulomb field due to the point nucleus of charge Z.

2.1.2. QED Corrections for Ions with $\alpha Z \ll 1$

a. Second-Order Radiative Shift. In the expansion (1), this shift is given by $Z^2 (\alpha / \pi) \varepsilon_1$. It corresponds to Feynman diagrams 1a (the electron self-energy) and 1b (the vacuum polarization) shown in Figure 1. For $Z \ll 137$, the matrix elements can be expanded in terms of the binding parameter αZ. Thus we have

$$\varepsilon_1 = C_{20}^1 (\alpha Z)^2 + C_{21}^1 (\alpha Z)^2 \log (\alpha Z)^{-2} + C_{30}^1 (\alpha Z)^3 + C_{40}^1 (\alpha Z)^4$$
$$+ C_{41}^1 (\alpha Z)^4 \log (\alpha Z)^{-2} + C_{42}^1 (\alpha Z)^4 \log^2 (\alpha Z)^{-2} + C_{50}^1 (\alpha Z)^5 + \cdots \tag{4}$$

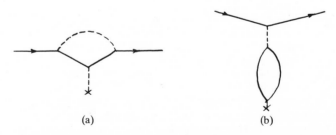

(a)   (b)

Figure 1. Feynman diagrams for the lowest-order Lamb shift in hydrogen. (a) Electron vertex and (b) vacuum polarization contributions to the shift.

If another expansion, in terms of m_e/M, is applied, then the first two coefficients in (4) become

$$C_{20}^1 = (8/3n^3)(1 - 3m_e/M)\{(11/24 + m_e/M)\delta_{l0} + \log[Z^2\,\mathrm{Ry}/K_0(n, l)]$$
$$+ (3/8)[c_{lj}/(2l+1)][1 + (1 - \delta_{l0})m_e/M] - (1/5)\delta_{l0}\} \qquad (5)$$

and

$$C_{21}^1 = (8/3n^3)(1 - 3m_e/M)\delta_{l0} \qquad (6)$$

where n, l, and j have the usual meaning and

$$c_{lj} = (l+1)^{-1}\delta_{j,1/2} - l^{-1}\delta_{j,-1/2} \qquad (7)$$

In (5), the last term comes from the vacuum polarization and the preceding terms in the same line are due to the anomalous magnetic moment of the electron. These expressions were obtained by several authors but without the mass corrections in (5) and (6),[65-69] and the mass corrections were derived in Ref. 70.

The dependence of the shift upon the quantity $K_0(n, l)$ comes from low-frequency terms of Figure 1a. The Bethe logarithm introduced in Ref. 2 is given by

$$\log\frac{K_0(Q)}{Z^2\,\mathrm{Ry}} = \frac{\sum_{Q'} F_{Q'Q}(E_{Q'} - E_Q)^2 \log(|E_{Q'} - E_Q|/Z^2\,\mathrm{Ry})}{\sum_{Q'} F_{Q'Q}(E_{Q'} - E_Q)^2} \qquad (8)$$

where the summation is extended to all states $\langle Q'|$ which are accessible via electric dipole transitions from the initial state $\langle Q|$; the oscillator strength for transitions is thus given by

$$F_{Q'Q} = \tfrac{2}{3}(\langle Q'|\bar{p}|Q\rangle)^2/(E_{Q'} - E_Q) \qquad (9)$$

where $E_{Q'}$ denotes the nonrelativistic energy of the atom in a state $\langle Q'|$ [the first term in the expansion (3)].

Numerical calculations of $K_0(n, l)$ were first reported in Ref. 71 and then by many authors (e.g., Ref. 72). Recent calculations by Klarsfeld and Marquet[73] cover hydrogenic states with $0 \leq l \leq n - 1$, $n \leq 8$.

Higher terms in the expansion (4) have been considered by several authors:

For the binding correction C_{30}^1 [74,75] we have

$$C_{30}^1 = (8/n^3)(1 - 3m_e/M)\delta_{l0}(1 + 11/128 - \tfrac{1}{2}\log 2 + 5/192) \qquad (10)$$

where the last term is due to the vacuum polarization.

For the higher-order term C_{40}^1, Erickson obtained[9,10,12]

$$C_{40}^1 = -(8/3n^3)\delta_{l0}(4\pi^2/3 + 4 + 4\log^2 2 + 0.28 \pm 0.5) + b(n, l) \qquad (11)$$

where b is a small state-dependent term; he also made an estimate for the next-order coefficient C_{50}^1 as

$$C_{50}^1 = (8/3n^3)\delta_{l0} \cdot 9.56\pi \tag{12}$$

Mass corrections have been neglected in (11) and (12) because of their small magnitude.

The coefficients C_{41}^1 and C_{42}^1 have been obtained exactly by several authors (Layzer,[76,79] Fried and Yennie,[77,78] Ericson and Yennie[9,10]):

$$C_{41}^1 = (8/3n^3)(1 - 3m_e/M)(4\log 2 - 1/10)\delta_{l0} + A_{nl}[1 + (m_e/M)(1 - \delta_{l0})] \tag{13}$$

where

$$A_n = \begin{cases} 3(\log(2/n) + S_n - 601/240 - 77/60n^2, & l = 0 \\[2mm] \dfrac{6 - 2l(l+1)/n^2}{(2l+3)l(l+1)(4l^2-1)} + (1 - 1/n^2)(1/10 + \delta_{j,1/2}/4)\delta_{l1}, & l \neq 0 \end{cases} \tag{14}$$

and S_n is the sum of the first n terms of the harmonic series,

$$S_n = 1 + 1/2 + 1/3 + \cdots + 1/n \tag{15}$$

Also,

$$C_{42}^1 = -(2/n^3)\delta_{l0} \tag{16}$$

In (13), the $1/10$ term is due to the vacuum polarization.

b. Fourth-Order Radiative Corrections. In the expansion (1), these corrections are given by the term $Z^2(\alpha/\pi)^2\varepsilon_2$. The vacuum polarization contribution at this order was obtained by Baranger *et al.*[80] and by Källen and Sabry[81]:

$$\varepsilon_2(\text{V.P.}) = -(2/n^3)(\alpha Z)^2(1 - 3m_e/M)\delta_{l0}(82/81) \tag{17}$$

An electron correction due to the anomalous magnetic moment of the electron can be written as follows

$$\varepsilon_2(\text{A.M.M.}) = (1/n^3)(\alpha Z)^2[1 - (2 + \delta_{l0})m_e/M](-a_0)c_{lj}/(2l+1) \tag{18}$$

For the numerical constant a_0 in (18), Sommerfield[82] and Peterman[83] both found that

$$a_0 = 197/72 + \pi^2/6 + 3\zeta(3)/2 - \pi^2 \log 2 = 0.656958 \tag{19}$$

First calculations of the self-energy contributions of a free electron were reported by Weneser *et al.*[84] and later by Soto.[7,8] These contributions can be written in the form

$$\varepsilon_2(\text{S.E.}) = (4/n^3)(\alpha Z)^2(1 - 3m_e/M)b_0\delta_{l0} \tag{20}$$

where, according to Soto's analytical calculation, $b_0 = 0.2153$. Appelquist and Brodski[4,5] pointed out that this numerical value for b_0 led to a noticeable disagreement between theory and experiment in the case of the $2S_{1/2} - 2P_{1/2}$ transition in hydrogen. Reevaluation of this constant using computer techniques gave $b_0 = 0.96 \pm 0.14$, which was in better agreement with experiment. In view of the importance of this correction, several other authors[85–88] repeated calculations of $\varepsilon_2(\text{S.E.})$ either analytically or numerically. At the present time, one can use the following trustworthy analytic expression:

$$b_0 = -4819/2592 - 49\pi^2/216 + \pi^2 \log 2/2 - 3\zeta(3)/2 = 0.93988 \qquad (21)$$

No estimates of the binding effects in ε_2 have been reported as yet.

c. *Higher-Order Corrections.* No estimates of the next term in (1) (ε_3) are available at present, with the exception of the contribution due to the magnetic anomaly of the electron. For the latter, one can write

$$\varepsilon_3(\text{A.M.M.}) = (1/n^3)(\alpha Z)^2 c_0 c_{lj}/(2l+1) \qquad (22)$$

where mass corrections have been omitted.

Accordingly to Kinoshita and Cvitanovic,[24,56] the sixth-order magnetic moment anomaly of the electron is $a_e^{(6)} = 1.29 \pm 0.06$. The constant c_0 in (22) will thus be taken to be

$$c_0 = 2.58 \pm 0.12 \qquad (23)$$

d. *Other Corrections to the Shift.* Among the other corrections which have a magnitude high enough to warrant inclusion in the Lamb shift (2), the recoil corrections[70,89,90] are the most important:

$$E_{\text{recoil}} = (8/3n^3)(\alpha Z)^2 Z^3 (\alpha/\pi)(m_e/M)\{B_n(l) + 2 \log [Z^2 \text{Ry}/K_0(n, l)]\} \text{ Ry}$$
$$(24)$$

where

$$B_n(0) = \log (\alpha Z)^{-2}/4 - 1/12 + (7/2)[\log (2/n) + S_n + 1 - 1/2n] \qquad (25)$$

and

$$B_n(l \neq 0) = -(7/4)[1/l(l+1)(2l+1)] \qquad (26)$$

The nuclear structure (the finite size of the nucleus) is taken into account[9,90] by means of a correction E_N:

$$E_N = (4/3n^3)Z^4(\langle r_N^2 \rangle/r_B^2)\delta_{l0} \text{ Ry} \qquad (27)$$

where $\langle r_N^2 \rangle$ is the mean-square radius of the nucleus, and r_B is the Bohr radius.

Despite a relatively small magnitude, E_N may produce substantial uncertainty in the theoretical value of the total shift (2).[91] Erickson[12] discussed this question in relation to the $2S_{1/2}$–$2P_{1/2}$ interval in deuterium. He pointed out that the theoretical value of the interval was six times standard deviation off the experimental one when the experimental value $r_d = 1.95 \pm 0.02$ F[92] was used for the effective radius of the deuteron wave function. However, with a modified definition of r_d which involved the deuteron form-factor slope according to

$$-6[dG_E(q^2)/d(q^2)] = \langle r_d^2 \rangle - \langle r_n^2 \rangle - \langle r_p^2 \rangle \qquad (28)$$

at $q^2 = 0$, he was able to obtain complete agreement with experiment by using $\langle r_n^2 \rangle = 0.116$ F^2 from the neutron–electron slope and $\langle r_p^2 \rangle$ from[93]; these values lead to 2.08 ± 0.02 F for the rms radius of the deuteron.

Finally, a small correction E_b[30] due to the nuclear motion will be included in (2). This correction,

$$E_b = -(1/4n^4)Z^2(\alpha Z)^2(m_e/M)\, \mathrm{Ry} \qquad (29)$$

appears after the reduction of the Breit equation for two particles of unequal masses and charges[30] to the center-of-mass frame of reference. E_b contributes equally to all levels with the same n.

2.1.3. Radiative Corrections for One-Electron Ions with High Z

Let us consider the matrix element for diagram (a) of Figure 2, which gives the second-order electron self-energy[75]:

$$\Delta E_n = -i\pi\alpha \int d(t_2 - t_1)d^3\bar{x}_2 d^3\bar{x}_1 \psi_n^*(\bar{x}_2)\gamma_\mu S_F^e(x_2, x_1)\gamma^\mu \psi_n(\bar{x}_1)D_F(x_2 - x_1)$$

$$- \delta_m \int d^3\bar{x}\psi_n^*(\bar{x})\psi_n(\bar{x}) \qquad (30)$$

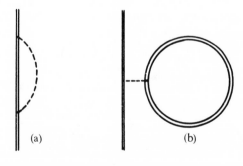

(a)   (b)

Figure 2. Feynman diagrams for the lowest-order radiative corrections in one-electron atom with high nuclear charge Z. The double line represents the electron propagation.

where units $c = \hbar = m_e = 1$ have been employed. In formula (30), ψ_n is a bound Dirac–Coulomb state of the electron, D_F is the photon propagator, and S_F^e is the propagator kernel related to the Dirac–Coulomb equation. The second term in (30) is the mass shift for a free electron. The observable shift of the bound electron energy is then given by the real part of ΔE_n.

For a Dirac–Coulomb equation with the point nucleus, $\alpha Z = 1$ is a singular value of the parameter αZ. An expansion of ΔE_n in terms of αZ can effectively be carried out only for $\alpha Z \ll 1$. For larger values, $Z \lesssim 1$, the evaluation of the matrix element (30) requires numerical procedures. For superheavy atoms with Z in the range of 137–170 the Dirac–Coulomb equation itself has to be modified to take into account the finite size of the nucleus.[128–130]

The first numerical evaluation of (30) was reported by Brown et al.,[94,95] who applied a method valid for large Z only to obtain an approximation to (30) for a $1s$ electron in the nuclear field of $Z = 80$. Since then the self-energy (30) has been computed numerically in the range of $Z = 70$–90 ($1S$-state calculations by Desiderio and Johnson[96]), $Z = 10$–110 (Mohr[14] for $1S$ state, and Mohr[15] for $2S$ and $2P$ states), and $Z = 70$–170 ($1S$-state calculations by Cheng and Johnson[16]).

Figure 3 compares numerical values of ΔE_n obtained with the help of the αZ expansion and those obtained in the course of the exact treatment of (30). As the graph indicates, for low Z the higher-order binding terms in

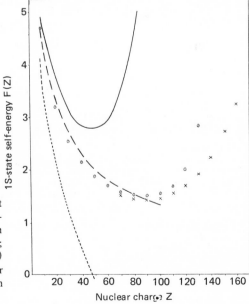

Figure 3. Comparison of different methods of computing the electron self-energy in one-electron atom. $(---)$ Formulas $(5)-(6)$; (———) formulas $(5)-(16)$; (— —) Erickson (1971); $(\bigcirc)$ Mohr (1974); $(\times)$ Cheng and Johnson (1976).

the expansion (4) are not negligible even for relatively light ions. Another conclusion that can be drawn from the graph is that even for superheavy elements with $Z > 137$ the self-energy (30) does not increase with Z as rapidly as some authors had earlier suggested.[97]

Another diagram that is to be discussed in relation to the ε_1 term of expansion (1) for strong nuclear field, is the vacuum polarization [(b) of Figure 2]. This term was considered by Wichmann and Kroll[98] and more recently by Rinker and Wilets[99] (who took into account the finite size of the nucleus) and by some other authors. For atomic systems, the vacuum polarization (electron–positron) contributes an amount small in comparison to diagram (a) (self-energy). The situation is completely reversed, however, in the muonic atoms where this correction dominates over all other contributions. For numerical examples the reader is referred to a recent review by Borie.[55] Even in light muonic atoms, this can be used in experimental determination of the nuclear sizes from observed muonic $2S_{1/2}$–$2P_{1/2}$ transitions.[100-103]

For a detailed discussion of the consequences of these theories for K-shell electrons in heavy and superheavy atoms, I refer the reader to the article by Professor Fricke.[104]

2.2. Comparison between the Theory and Experiment

2.2.1. The $2S_{1/2}$–$2P_{1/2}$ Interval in ^1H and ^2D

For hydrogen, there are three accurate *direct* measurements of the $n = 2$ radiative shift:

Triebwasser et al.[126]	1057.86 +0.06 MHz	(31)
Robiscoe and Shyn[108]	1057.90 ±0.06 MHz	(32)
Lundeen and Pipkin[109,125]	1057.893 ± 0.020 MHz	(33)

All three results are consistent with each other and with the theoretical values of the shift quoted by Lautrup et al.[52] and recomputed in the present work with updated values of some constants listed in Table 1.

Nevertheless, it now appears that new effort is required to produce a more accurate theoretical value of the shift. Revision of the numerical values of recoil corrections for some atoms has been discussed in Section 2.1.2.d. We may also note that Eq. (22) leads to ε_3(A.M.M.) ≈ 0.001 MHz, for $Z = 1$, which probably indicates that the total correction ε_3 would not contribute any appreciable amount to the shift in hydrogen. This leaves binding terms in the expansion for ε_1 (and possibly for ε_2) as the most likely main source of the uncertainty in the present theoretical value of the hydrogenic Lamb shift.

Table 1. Numerical Values of Some Physical Constants

Quantity	Numerical value	Reference
c	$2.99792458\,(1)\,10^8\,\mathrm{m\,sec}^{-1}$	Evenson[105]
$\alpha^{-1}_{\mathrm{(WQED)}}$	$1.37035987\,(29)\,10^2$	Olsen and Williams[106]
r_B	$5.2917715\,(81)\,10^{-9}\,\mathrm{cm}$	Taylor et al.[49]
m_e	$5.485930\,(34)\,10^{-4}$ a.m.u.	Taylor et al.[49]
m_e/M_p	$5.44630\,(10)\,10^{-3}$	Taylor et al.[49]
m_e/M_d	$2.72450\,(10)\,10^{-3}$	Taylor et al.[49]
m_e/M_{alpha}	$1.37097\,(10)\,10^{-3}$	Taylor et al.[49]
Ry	$1.097373143\,(10)\,10^5\,\mathrm{cm}^{-1}$	Hänsch et al.[107]
$\bar{r}_p$	$0.805\,(10)\,\mathrm{F}$	Erickson[9]
$\bar{r}_d$	$2.08\,(2)\,\mathrm{F}$	Erickson[12]
$\bar{r}_{\mathrm{alpha}}$	$1.68\,(3)\,\mathrm{F}$	Borie[55]

Mohr's Correction. Mohr[15] suggested a method of estimating higher-order terms in ε_1 based on the numerical computations of the self-energy (30) which he had performed for the $1S_{1/2}$, $2S_{1/2}$, and $2P_{1/2}$ states of one-electron atoms.

Let us write the self-energy correction as follows:

$$E_n(Z) = (8/3n^3)(\alpha/\pi)Z^2(\alpha Z)^2 F(Z)\,\mathrm{Ry} \qquad (34)$$

where $F(Z)$ has been tabulated by Mohr for Z in the range 10–110.* One may assume that the following expansion is valid for $F(Z)$:

$$F(Z) = \tilde{C}^1_{20} + \tilde{C}^1_{21}\log(\alpha Z)^{-2} + \tilde{C}^1_{30}(\alpha Z) + \tilde{C}^1_{41}(\alpha Z)^2\log(\alpha Z)^{-2}$$

$$+ \tilde{C}^1_{42}(\alpha Z)^2\log^2(\alpha Z)^{-2}$$

$$+ (\alpha Z)^2 \bar{C}^1_{40} - \bar{C}^1_{50}(\alpha Z)^3 + \bar{C}^1_{51}(\alpha Z)^3\log(\alpha Z)^{-2} \qquad (35)$$

where $\tilde{C}^1_{mk}$ are coefficients C^1_{mk} defined by (11)–(16) with the vacuum polarization terms omitted. The unknown $\bar{C}^1_{mk}$ are to be determined from a fitting procedure which matches the right-hand side of (35) with the tabulated function $F(Z)$. This method effectively takes into account the existence of higher-order terms in the expansion (35).

In order to complete the treatment of new coefficients in the expansion, the vacuum polarization terms are to be added to $\bar{C}^1_{mk}$. For this, the Uehling potential[127]

$$V(r) = (\alpha/\pi)(Z/r)\int_0^1 \theta(r, u)\,du \qquad (36)$$

---

* Because of a different choice of the numerical coefficient in (34), $F = \frac{3}{4}F_M$, where F_M is the actual function evaluated by Mohr.

where

$$\theta(r, u) = (1 - u^2) \int_{\infty}^{1} \exp\left[-(2r/r_B)(y/\varkappa)(1 - u^2)^{-1/2}\right](dy/y) \qquad (37)$$

$$\varkappa = (\alpha Z)[M/(m_e + M)] \qquad (38)$$

can be used since, as was found by Wichmann and Kroll,[98] it gives, for small Z, the dominant contribution to the vacuum polarization correction according to

$$E_{V.P.} = \int V(r)|\psi(n, l, j)|^2 \, d^3\bar{r} \qquad (39)$$

where $\psi(n, l, j)$ is the nonrelativistic wave function of the state in question. For the $2S_{1/2}$ state Mohr finds

$$C_{40}^1 = \bar{C}_{40}^1 - 1199/2100 \qquad (40)$$

$$C_{50}^1 = \bar{C}_{50}^1 + 1/2 \qquad (41)$$

$$C_{51}^1 = \bar{C}_{51}^1 + 5\pi/128 \qquad (42)$$

whereas for the $2P_{1/2}$ state the vacuum polarization contributions are zero.

Once the expansion (35) has been constructed, it can be used to estimate higher-order binding effects in the QED term ε_1 for *small* values of Z.

Mohr used[15] this method to study the Lamb shift in hydrogen by applying it to the interval $2S_{1/2}$–$2P_{1/2}$ and obtained $E_{L,1} = 1057.864 \pm 0.014$ MHz. This result is not consistent with the value of 1057.911 ± 0.011 MHz quoted by Lautrup *et al.*[52]

Table 2 presents coefficients $\bar{C}_{mk}^1$ that this author has obtained when the fitting procedure is applied to individual energy levels. The present calculations support Mohr's value of the shift and indicate that a better estimate would require both more terms in (35) and more accurate values

Table 2. Higher-Order Binding Terms in the Self-Energy Shift of Order (α/π); Coefficients $\bar{C}_{mk}^1$ in Formula (35) for $1S_{1/2}$ and $2S_{1/2}$ States

Coefficient $\bar{C}_{mk}^1$	Present estimates		Theoretical estimates of Erickson[10,12]
	$1S_{1/2}$	$2S_{1/2}$	
$\bar{C}_{40}^1$	−23.3	−24.8	−19.36 ± 0.5
$\bar{C}_{50}^1$	18.6	17.2	+30 ± ?
$\bar{C}_{51}^1$	5.1	7.3	unknown

of $F(Z)$ for low Z. For the present the existence of two different theoretical values for the shift in hydrogen has to be accepted.

For deuterium, the $n = 2$ Lamb shift was measured by Cosens,[110] who obtained 1059.28 ± 0.06 MHz. As in the case of hydrogen, the experimental accuracy is not sufficient to make a choice between the two theoretical values of the shift presented in Table 3.

2.2.2. The Lamb Shift in He II and Li III

The fine-structure and radiative shifts in ^{4}He II have been recently measured by Eibofner.[115] The theoretical intervals computed by the author without Mohr's correction are compared with the experimental intervals in Table 4.

Table 3. Theoretical Contributions to the $2S_{1/2}-2P_{1/2}$ Lamb Shift in Hydrogen and Deuterium

Order, relative to $\alpha Z^2 (\alpha Z)^2$ Ry	Contribution and relevant equations	Numerical value (in MHz) Hydrogen	Deuterium[b]
1	Self-energy, Eqs. (5)+(6)	1009.920^a 1009.922^b	1010.711
1	Magnetic moment, Eq. (5)	$67.720^{a,b}$	67.771
1	Vacuum polarization, Eq. (5)	$-27.084^{a,b}$	-27.107
(αZ)	Binding, Eq. (10)	$7.140^{a,b}$	7.140
$(\alpha Z)^m \log (\alpha Z)^{-k}$	Binding, Eqs. (11)–(16)[e]	-0.372^a -0.370^b	-0.370
$(\alpha Z)^m \log (\alpha Z)^{-k}$	Same, Eq. (35)	$-0.419^{b,d}$	-0.419^d
α	Self-energy, Eq. (20)	$0.444^{a,b}$	0.444
α	Magnetic moment, Eq. (18)	$-0.103^{a,b}$	-0.103
α	Vacuum polarization, Eq. (17)	$-0.239^{a,b}$	-0.239
(m_e/M)	Recoil corrections, Eq. (24)	$0.359^{a,b}$	0.180
$(\bar{r}_N/r_B)^2$	Nuclear size corrections, Eq. (27)	0.125^a 0.127^b	0.847
	Total shift:	$1057.911 (11)^a$ 1057.916^b $1057.857^{b,d}$	$1059.272 (25)^c$ 1059.274^b $1059.225^{b,d}$

[a] Table 1.1 of Lautrup et al.[52]
[b] Present calculation.
[c] From Erickson.[12]
[d] Present calculation with higher-order binding terms obtained by extrapolation of data from Mohr.[14,15]

[e] Term C_{40}^1 includes the state-dependent term $b(n, l)$, which contributes $+0.014$ MHz to the interval.[52]

Table 4. The $nS_{1/2}-2P_{1/2}$ and $nS_{1/2}-2P_{3/2}$ Intervals in ^{4}He II (in MHz)^a

n	Interval	Present work^b	Experiment	Reference
2	A	161548.0		
	B	14044.50	14045.4 ± 1.2	c
			14040.2 ± 1.8	d
3	A	47843.54	47843.8 ± 0.5	e
	B	4184.24	4183.17 ± 0.54	f
4	A	20180.12	20180.6 ± 0.8	g
	B	1769.05	1768.5 ± 0.8	g
5	A	10331.21	10332.9 ± 1.4	g
	B	906.70	905.0 ± 1.4	g
6	A	5978.37	5979.1 ± 1.2	g
	B	525.02	524.3 ± 1.2	g
7	A	3764.67		
	B	330.74		
8	A	2522.01		
	B	221.62		

^{a} nA: $nS_{1/2}-nP_{1/2}$; nB: $nS_{1/2}-nP_{3/2}$.
^{b} Equations (11)–(16) have been used for higher-order binding corrections.
^{c} Narasimham.[111]
^{d} Lipworth and Novick.[112]
^{e} Mader and Leventhal.[113]
^{f} Mader et al.[114]
^{g} Eibofner.[115]

For the $n = 2$ Lamb shift, a calculation using Mohr's correction gives 14042.59 MHz so that the present accuracy of the experiment does not allow a choice to be made between the two theoretical values. For $n = 3$, Mohr's correction appears to improve agreement with the experimental value of Mader et al.[114] since it gives $E_{L,1} = 4183.75$ MHz.

For the $n = 2$ Lamb shift in ^{6}Li III recently measured by Leventhal and Havey,[116] $E_{L,1\,\text{expt}} = 62765 \pm 21$ MHz; this compares well with the theoretical value of 62762 ± 9 MHz communicated to the authors by Erickson.

2.2.3. The Lamb Shift in the Ground $1S_{1/2}$ State of ^{1}H and ^{2}D

The isolated position of the ground $1S_{1/2}$ level from other levels of the one-electron atom makes it difficult to determine it accurately enough to observe the radiative shift of the level. For more than two decades, until the recent progress in multiphoton spectroscopy,[117–121] the work of Herzberg[123] was the only study of the ground-level shift (in ^{2}D).

In the first investigation of its kind the group of Hänsch[122] employed a Doppler-free two-photon spectroscopy, with both photons coming from a tunable laser, to excite $2S_{1/2}$ level from the ground $1S_{1/2}$ level. The authors also simultaneously measured the interval $2P_{3/2}-4D_{5/2}$, which they used as the reference line in their investigation of the ground level.

Let us write, for the $1S_{1/2}-2S_{1/2}$ and $2P_{3/2}-4D_{5/2}$ intervals,

$$I_1 = \tfrac{1}{4}[E(2S_{1/2}) - E(1S_{1/2})] = 3\,\mathrm{Ry}/16 + S(1S_{1/2})/4 - S(2S_{1/2})/4 \tag{43}$$

$$I_2 = E(4D_{5/2}) - E(2P_{3/2}) = 3\,\mathrm{Ry}/16 + S(2P_{3/2}) - S(4D_{5/2}) \tag{44}$$

where $S(n, l, j)$ is the sum of relativistic, QED, recoil, and proton size corrections given in Section 2.1.

For the interval ΔI,

$$\Delta I = I_1 - I_2 \tag{45}$$

the difference is mainly due to the displacement of the $1S_{1/2}$ level from its nonrelativistic position. In the experiment described, Hänsch and co-workers were able to measure the interval ΔI and deduce from it the $1S_{1/2}$ Lamb shift. Table 5 presents the experimental results[123] together with the theoretical values of ΔI and $E_{L,1}$, computed by the author, for ^{1}H and ^{2}D studied in Ref. 123 and for ^{4}He II (which has not yet been measured).

Multiphoton Doppler-free spectroscopy circumvents the difficulties of uv spectroscopy, where a similar investigation would require the very difficult determination of the series limit in order to establish the absolute wavelength of the Lyman α line and accurate value of the Rydberg constant for the atom in question. The present accuracy of multiphoton spectroscopy is less than 2% and an order of magnitude better than that of uv spectroscopy. Hänsch *et al.* pointed out that it would be possible to improve the resolution of the I_1 transition by narrowing the laser linewidth and bypassing a short-living P state by applying the Doppler-free two-photon spectroscopy to the $2S_{1/2}-4S_{1/2}$ D transition. This method may eventually exceed the accuracy of the present experimental determination of the $n = 2$ Lamb shift.

Table 5. The Lamb Shift $E_{L,1}$ of the Ground $1S_{1/2}$ Level in Some One-Electron Atoms (All Intervals Are Given in GHz)

Interval[a]	^{1}H	^{2}D	^{4}He II	Ref.
I_{th}, Eq. (45)	3.42098	3.41810	59.773	b
I_{expt}	3.408 ± 0.026	3.398 ± 0.027		c
$E_{L,1\,th}$, Eq. (2)	8.14942	8.17224	107.621	b
$E_{L,1\,expt}$		7.9 ± 1.1		d
	8.20 ± 0.10	8.25 ± 0.11		c

[a] In theoretical data, Eqs. (11)–(16) have been used for calculation of higher-order binding corrections.
[b] Present work.
[c] Lee *et al.*[122]
[d] Herzberg.[123]

2.2.4. One-Electron Ions with High Z

The progress in experimental studies of hot plasmas and in astrophysics may require the old spectroscopic tables of Garcia and Mack[124] to be updated in the near future.

3. The Lamb Shift in Two-Electron Atoms

3.1. General Form of the Total Energy Expansions

The total energy E of a bound state nLS of a two-electron atom of nuclear mass M and charge Z will be written in the form of the following general expansion:

$$E = Z^2\{\varepsilon_{\text{nrel}} + (\alpha Z)^2[\varepsilon_{\text{rel}} + (\alpha/\pi)\varepsilon_1 + (\alpha/\pi)^2\varepsilon_2 + \cdots]\}\,\text{Ry} \qquad (46)$$

In (46), the first term $\varepsilon_{\text{nrel}}$ is the scaled nonrelativistic energy of the atom. The second term, which is of order $(\alpha Z)^2$ with respect to the nonrelativistic level, is usually referred to as the relativistic correction to the level. This quantity is obtained[131,132,30] by the reduction of the Bethe–Salpeter equation[133] to the Breit–Pauli approximation. The energy corrections $\varepsilon_k, k = 1, 2, \ldots$, give higher-order QED and relativistic contributions to the energy E.

Apart from the similar dependence of these terms upon αZ and m_e/M as for the hydrogen atom, the expansion (46) depends upon one more parameter, say λ, which describes the correlation effects in E. The usual choice for this parameter is $\lambda = 1/Z$. It is known that an attempt to introduce the $1/Z$ expansion into relativistic equations such as the Hartree–Fock–Dirac equations for an atom, meets certain difficulties.[134] Nevertheless, this parameter will be used here in qualitative discussions since low-Z atoms will be considered, and $1/Z$ will thus refer to the nonrelativistic theory.

In terms of (46), the QED shift, E_L, is defined as the difference

$$E_{L,2\,\text{tot}} = E - Z^2\varepsilon_{\text{nrel}} - Z^2(\alpha Z)^2\varepsilon_{\text{rel}} \equiv E_L \qquad (47)$$

3.2. Calculations of the Nonrelativistic States

3.2.1. An Estimate of the Accuracy Required

Let us introduce the interval

$$\Delta E(n, L+1, S) = E(n, L+1, S) - E(n, L, S) \qquad (48)$$

which corresponds formally to the Lamb shift $E_{L,1}$ in a one-electron atom. Because of the electrostatic repulsion term $1/r_{12}$ in the Hamiltonian of a two-electron atom, the degeneracy of the energy levels with respect to the angular momentum quantum number L is already removed in the nonrelativistic approximation.

Let us consider, as an example, the $2^1S_0-2^1P_1$ interval in neutral helium. According to calculations of Accad et al.[36] and Suh and Zaidi[135]

$$\Delta E_{nrel}(210) = 4857.617 \text{ cm}^{-1} \tag{49}$$

$$\Delta E_{rel}(210) = -0.068 \text{ cm}^{-1} \tag{50}$$

$$\Delta E_L(210) = +0.104 \pm 0.014 \text{ cm}^{-1} \tag{51}$$

In terms of the wavelength of the transition, one obtains

$$\lambda_{nrel+rel} = 591.4072 \text{ Å} \qquad \lambda_{nrel+rel+L} = 591.4119 \text{ Å}$$

which can be compared with an experimental result of Herzberg[136] which was obtained by means of optical spectroscopy:

$$\lambda_{expt} = 591.4121 \pm 0.0005 \text{ Å} \tag{52}$$

In this particular example, $\Delta E_L(210)/\Delta E_{nrel}(210) = 2 \times 10^{-5}$, $\Delta E(210)/E(200) = 10^{-2}$, and the relative error in $\Delta E_L(210)$ is 10^{-1}. These lead us to the conditions

$$\eta(200) \leqslant 2.10^{-8}, \qquad \eta(210) \leqslant 2 \times 10^{-8} \tag{53}$$

where $\eta(n, L, S)$ denotes the relative error allowed in the determination of $E_{nrel}(n, L, S)$.

3.2.2. Solution of the Schrödinger Equation

For a two-electron atom with the nucleus of infinite mass, the Schrödinger equation is

$$H\psi_n = E_n\psi_n \tag{54}$$

$$H = -\tfrac{1}{2}(\nabla_1^2 + \nabla_2^2) - Z/r - Z/r_2 + 1/r_{12} \tag{55}$$

The nonrelativistic energy E_{nrel} is then given by

$$E_{nrel} = E_n - \varepsilon_M(n) \tag{56}$$

where the last term in (56) is the mass polarization correction which accounts for nonseparability of the equation (55), and E_{nrel} is expressed in terms of the Rydberg constant for the nucleus, R_M.

The standard method of solving (54) and (55) is based on the variational principle for the total energy, E_n, of the atom:

$$E_n[U] = \frac{(U, HU)}{(U, U)} = \text{minimum} \tag{57}$$

where the trial function $U(\vec{r}_1, \vec{r}_2)$ is expanded in terms of a given set $u_i(\vec{r}_1, \vec{r}_2)$, $i = 1, 2, \ldots, N$, such that U and the exact solution $\psi_n(\vec{r}_1, \vec{r}_2)$ are eigenfunctions of the operators $\vec{L}^2$ and $\vec{S}^2$. A set of functions u_i depending explicitly upon the electron separation r_{12} is usually referred to as the "Hylleraas set" or the "set of Hylleraas functions."

The low-lying S and P states of two-electron ions with Z in the range 1–10 have been computed by Pekeris and co-workers,[36] and low-lying D states of the same ions have been computed by Blanchard and Drake.[139]

A look back at the past three decades shows that the numerical difficulties associated with obtaining an approximate solution of (54) and (55) that satisfies conditions (53) or likewise, had not been realized for a long time. The pioneering work by Chandrasekhar et al.[32] discovered an unexpected gap of some 25 cm^{-1} between the theoretical position of the ground helium level, which they obtained on the basis of a ten-parametric variational solution of (54) and (55), and its experimental position. This result directly influenced subsequent theoretical and many numerical studies of the Schrödinger equation for a two-electron atom.[140,33] The level positions determined in Ref. 36 are accurate to within $0.01–0.001 \text{ cm}^{-1}$, and a solution of (54) and (55) to this accuracy has required trial functions, U, with up to 2300 parameters.

Another example of a strong interrelation between the variational problem (57) and the Lamb-shift studies is the long persistence of concentration on the ground state. This was entirely due to the slow convergence of the variational methods for excited states[137] and was quite contrary to the situation in one-electron atoms, where interest to the ground-state Lamb shift has just begun.

3.3. Relativistic Corrections ε_{rel}

All relativistic corrections of order $(\alpha Z)^2$ are expressed as the expectation values of some operators computed in the nonrelativistic state $\langle nLS|$.[30] These corrections for S and P states with Z in the range 2–10 have been tabulated in Ref. 36.

3.4. Radiative Corrections

3.4.1. General Expressions for the Shift of Second Order

In two-electron atoms, only radiative corrections of the lowest-order in α have been obtained up to now. These corrections form the term $Z^2(\alpha/\pi)(\alpha Z)^2 \varepsilon_1$ Ry in formula (46).

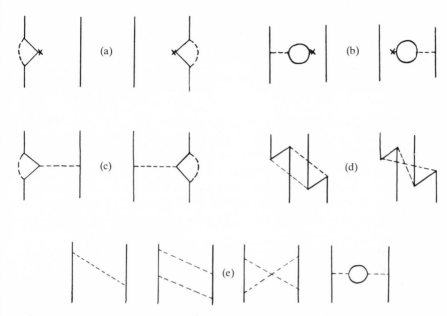

Figure 4. Feynman diagrams contributing to the lowest-order Lamb shift in two-electron atom (From Kabir and Salpeter[138]). (a) electron vertex (self-energy) and (b) vacuum polarization correction to the nuclear Coulomb field, (c) vertex correction to one-photon exchange between the atomic electrons, (d) two-photon exchanges between the electrons involving intermediate negative energy states of the electrons, (e) one- and two-photon exchanges between the electrons; first diagram represents the retarded Briet interaction; the last diagram represents vacuum polarization correction to one-photon exchange between the electrons.

Feynman diagrams contributing to this order are shown in Figure 4. Apart from diagrams for an electron in the external field there are also diagrams corresponding to photon exchanges between two electrons and radiative corrections to these processes.

Analytic expressions for various contributions to this order of the QED theory were obtained by Kabir and Salpeter,[138] Sucher,[141] and Araki.[142] Among these authors, Araki's published results are most complete and, according to Sucher,[141] are in agreement with his unpublished results.

For a state $\langle nLS| \equiv \langle q|$ the total radiative correction to the relativistic energy level of a two-electron atom will be written in the following form:

$$E_L = E_{L,2} + E'_{L,2} + E''_{L,2} + E'''_{L,2} + E_{L,2,\text{binding}} \qquad (58)$$

where the last term, $E_{L,2,\text{binding}}$, will be discussed later, and all other contributions in (58) are identified below.

3.4.2. Shifts Independent of the Total Momentum J

The main contribution comes from $E_{L,2}$, which has a form similar to that of the hydrogenic shifts (5) and (6):

$$E_{L,2} = (16Z/3)\alpha^3 \langle \delta(\bar{r}_1) \rangle \{ \log (\alpha Z)^{-2} + \log [Z^2 \, \text{Ry}/k_0(n, L, S)] + 19/30 \} \, \text{Ry} \tag{59}$$

In (59), the quantity $\langle \delta(\bar{r}_1) \rangle$ is defined as follows:

$$\langle \delta(\bar{r}_1) \rangle = \int \psi_n^2(0, \bar{r}_2) d^3 \bar{r}_2, \qquad \int \psi_n^2(\bar{r}_1, \bar{r}_2) d^3 \bar{r}_1 d^3 \bar{r}_2 = 1 \tag{60}$$

that is, $D(0) = 8\pi \langle \delta(\bar{r}_1) \rangle$, where $D(0)$ is the electron density at the nucleus.

The Bethe logarithm in (59) is defined, for a state $\langle q|$ of a two-electron atom, in a way similar to that for a hydrogenic ion:

$$\log \frac{k_0(q)}{Z^2 \, \text{Ry}} = \frac{\sum_{q'} f_{q'q} (E_{q'} - E_q)^2 \log (|E_{q'} - E_q|/Z^2 \, \text{Ry})}{\sum_{q'} f_{q'q} (E_{q'} - E_q)^2} \tag{61}$$

where the summation includes states $\langle q'|$ to which electric dipole transitions are possible from the initial state $\langle q|$, and the oscillator strength for transition is given by

$$f_{q'q} = \tfrac{2}{3} (\langle q'| \bar{p}_1 + \bar{p}_2 | q \rangle)^2 / (E_{q'} - E_q) \tag{62}$$

For $\langle \delta(\bar{r}_1) \rangle$, the first nonvanishing term of the $1/Z$ expansion is Z^3, in the case of S states, and Z^2 in the case of $L \neq 0$ states. The main contribution to the two-electron Lamb shift, that is, correction (59), is therefore of order Z^4 in S states and of order Z^3 in $L \neq 0$ states of the atom.

Let us now consider the two remaining terms, that is, $E'_{L,2}$ and $E''_{L,2}$:

$$E'_{L,2} = (28/3)\alpha^3 \langle \delta(\bar{r}_{12}) \rangle \log \alpha \, \text{Ry} \tag{63}$$

$$E''_{L,2} = \alpha^3 \langle \delta(\bar{r}_{12}) \rangle (178/15 - (40/3)\bar{r}_1 \cdot \bar{r}_2) - (28/3)\alpha^3 Q \, \text{Ry} \tag{64}$$

In (63) and (64),

$$\langle \delta(\bar{r}_{12}) \rangle = \int \psi_n^2(\bar{r}, \bar{r}) \, d^3 \bar{r} \tag{65}$$

so that this quantity is proportional to the average electron density at the location of one electron due to the presence of the second electron in the atom.

The quantity Q in (64) is the principal part of the logarithmically divergent quantity $\langle r_{12}^{-3} \rangle$, that is,[142]

$$Q = \lim_{a \to 0} [r_{12}^{-3}(a)/(4\pi) + \langle \delta(\bar{r}_{12}) \rangle \log a + \gamma] \tag{66}$$

where γ is Euler's constant, and the quantity $r_{12}^{-3}(a)$ has been defined as follows:

$$r_{12}^{-3}(a) = \begin{cases} r_{12}^{-3} & \text{if } r_{12} > a \\ 0 & \text{if } r_{12} < a \end{cases} \tag{67}$$

The corrections $E'_{L,2}$ and $E''_{L,2}$ are relatively insignificant for $Z \gg 1$ since they are smaller by a factor of $1/Z$ than the main term $E_{L,2}$. However, even for small Z, the numerical value of $\langle \delta(\bar{r}_{12}) \rangle$ constitutes only a few per cent of the factor $\langle \delta(\bar{r}_1) \rangle$ appearing in $E_{L,2}$.

Among these small corrections, $E'_{L,2}$ has an additional factor of $\log \alpha$ and is therefore numerically more important than others. The former comes from Feynman diagram (d) of Figure 4, which involves intermediate negative energy states of the atomic electrons.

The terms $E_{L,2}$ and $E'_{L,2}$ were first obtained by Kabir and Salpeter,[138] who also evaluated these corrections numerically for the ground state of He I and Li II. An analytic expression for the remaining term, $E''_{L,2}$, which entirely relates to the interactions between electrons represented by diagrams e of Figure 4, was independently obtained by Araki[142] and Sucher.[141] Because of the numerical evaluation of this correction for the $1^1 S_0$ state of helium in Ref. 141, $E''_{L,2}$ is usually referred to as the Sucher correction.

3.4.3. Shifts Dependent on the Total Momentum J

General expressions for this part of the total Lamb shift were derived by Fulton and Martin[89] for positronium, and by Araki[142] for a two-electron atom.

Let us introduce first the self-spin-orbit and mutual spin-orbit operators, H_{so}^s and H_{so}^m, respectively:

$$H_{so}^s = \tfrac{1}{2}\alpha^2 Z(\bar{S}_1 \cdot \bar{L}_1/r_1^3 + \bar{S}_2 \cdot \bar{L}_2/r_2^3) - \tfrac{1}{2}(\alpha^2/r_{12}^3)(\bar{S}_1 \cdot \bar{L}_{12} + \bar{S}_2 \cdot \bar{L}_{21}) \tag{68}$$

$$H_{so}^m = -(\alpha^2/r_{12}^3)(\bar{S}_1 \cdot \bar{L}_{21} + \bar{S}_2 \cdot \bar{L}_{12}) \tag{69}$$

The spin–spin operator H_{ss} will then be written as

$$H_{ss} = (\alpha^2/r_{12}^3)(\bar{S}_1 \cdot \bar{S}_2 - (3/r_{12}^2)(\bar{r}_{12} \cdot \bar{S}_1)(\bar{r}_{12} \cdot \bar{S}_2) \tag{70}$$

In (68) and (69), $\bar{L}_i = \bar{r}_i \times \bar{p}_i$, and $\bar{L}_{ik} = \bar{r}_{ik} \times \bar{p}_i$, where $\bar{r}_{ik} = \bar{r}_i - \bar{r}_k$, $i, k = 1, 2$.

The sum of operators (68)–(70) is identical to the fine-structure operator $H_3 + H_5$ of Bethe and Salpeter[30] with the exception of the contact term in H_5, which has now been moved to $E''_{L,2}$, formula (64).

I shall define[39] the energy correction $E'''_{L,2}$ as the diagonal element of the operator[142]

$$\bar{H}_{L,2} = (\alpha/\pi)(H_{so}^s + \tfrac{1}{2}H_{so}^m + H_{ss}) \tag{71}$$

that is,

$$E_{L,2}''' = 2\langle nLSJ|\bar{H}_{L,2}|nLSJ\rangle \, \text{Ry} \tag{72}$$

For large Z, the coupling between the spin and orbital momentum of the same electron due to the nuclear Coulomb field dominates over other J-dependent effects. As a result, the transformation properties of $\bar{H}_{L,2}$ and the fine-structure operator $H_3 + H_5$ become identical, and simple replacement of $\bar{S}_i$ by $(1 + \alpha/\pi)\bar{S}_i$ in the latter accounts for the leading term in $\bar{H}_{L,2}$. This term is of the same order as the main correction, $E_{L,2}$, that is, $Z^2(\alpha Z)^2(\alpha/\pi)\,\text{Ry}$. For low Z, $\bar{H}_{L,2}$ is no longer proportional to the fine-structure operator, an effect which does not arise in the corresponding order α in one-electron atoms.

The analogy with the fine-structure operator suggests that a reasonably accurate theory of the J-dependent Lamb shift in two-electron atoms can be developed in a simple approximation of hydrogenic functions, instead of employing more accurate solutions of the Schrödinger equation. (For atoms with $Z \leqslant 4$, more accurate estimates may require the use of Hylleraas wave functions.)

It should be noted that, for S states, the correction $E_{L,2}'''$ vanishes identically so that we need to consider $L \neq 0$ states only. For those, the exchange integrals can be neglected and the function $\langle 1s|$ of one of the electrons can be replaced by $\delta(\bar{r}_1)$. Assuming the effective nuclear charge to be $Z - 1$ for the excited electron, one then obtains

$$E_{L,2}''' = \frac{(A_{LJ}Z + B_{LJ})(Z - 1)^3 \alpha^3}{n^3 L(L+1)(2L+1)} \, \text{Ry}, \qquad S = 1 \tag{73}$$

and

$$E_{L,2}''' = 0, \qquad S = 0 \tag{74}$$

In (73) and (74),

$$A_{LJ} = \begin{cases} L, & J = L+1 \\ -1, & J = L \\ -(L+1), & J = L-1 \end{cases} \tag{75}$$

$$B_{LJ} = \begin{cases} -4L(L+1)/(2L+3), & J = L+1 \\ 0, & J = L \\ 4L(L+1)/(2L-1), & J = L-1 \end{cases} \tag{76}$$

3.4.4. The Lamb Shift of the Relativistic Potential of Ionization

In many applications it is convenient to deal with the Lamb shift of the potential of ionization rather than with the shift of an isolated level. The formulas required have already been given in previous sections, and I shall write

$$\Delta E_{L,2} = E_{L,1} - E_{L,2} \tag{77}$$

where both shifts relate to the same nuclear charge Z. Normally $E_{L,1}$ corresponds to the ground $1s$ state of the one-electron ion. As we have seen, many terms in both $E_{L,1}$ and $E_{L,2}$ depend on the electron density $D(0)$ at the nucleus. For this, (77) will lead to terms of the following typical form:

$$\Delta(Z)[\log (\alpha Z)^{-2} + 19/30 + \cdots] \tag{78}$$

where the function $\Delta(Z)$ is given by[39,40]

$$\Delta(Z) = \frac{2\pi\langle\delta(\bar{r}_1)\rangle}{Z^3} - 1 \tag{79}$$

The first term in (79) gives the ratio of the electron density at the nucleus for the two-electron ion to that for the one-electron ion. For S states $\Delta(Z) > 0$, and $n^3\Delta(Z) \to 1$ as Z increases. For $L \neq 0$ states, the excited electron does not contribute if its state is hydrogenic, and $\Delta(Z) = 0$. The departure of $\Delta(Z)$ from 0 is therefore a measure of correlation effects in $L \neq 0$ states of a two-electron atom, and the Lamb shift $\Delta E_{L,2}$ is proportional to these correlation effects.

3.4.5. Binding Corrections to the Lamb Shift in Two-Electron Atoms

Binding effects give an example of a phenomenon depending upon the electron density at the nucleus. For hydrogen this is shown explicitly by formula (10) and this suggests that the effect can be estimated in a many-electron atom by assuming that the electron density at the nucleus is determined by the self-consistent field of the atom. For a two-electron atom, where even more accurate treatment is available, the numerical values of $\langle\delta(\bar{r}_1)\rangle$ obtained by Pekeris and co-workers[31,36] will be used. In terms of the Lamb shift $\Delta E_{L,2}$, the binding correction is given as follows:

$$E_{L,2,\,\text{binding}} = -8\pi(\alpha/\pi)Z^2(\alpha Z)^3\Delta(Z)(1 + 11/128 - \tfrac{1}{2}\log 2 + 5/192)\,\text{Ry} \tag{80}$$

3.4.6. The Volume-Isotope Correction V_I

For the relativistic ionization potential of a two-electron atom, this correction is given by

$$\Delta V_I = -(4/3)Z^4\Delta(Z)(\langle r_N^2\rangle/r_B^2)\,\text{Ry} \tag{81}$$

which is similar to formula (27) for a hydrogenic ion.

3.5. Comparison between the Theory and Experiment (S States)

3.5.1. The Ground State of Two-Electron Atoms

The ground state of helium was studied by many authors, since it is a test case for many methods of calculating atomic properties (the average excitation energy and the electron density at the nucleus) involved in the theoretical determination of the shift.

Calculations of the Bethe logarithm, formulas (61) and (62), require a knowledge of all two-electron states and demand considerable computational effort. For He, the first computation of $\log k_0$ was performed by Kabir and Salpeter,[138] who also considered a similar problem for Li II. Apart from the method of direct summation in (61) applied also in,[163,164] the same problem was considered by Schwartz[165,166] by transforming it to a stationary problem for a first-order perturbation equation. Among other methods used for determining $\log k_0$, a method of Pekeris for He based on the oscillator sum rules[167] and its modification with improved asymptotic behavior for large Z discussed in[168] can be mentioned. Much more accurate results for $\log k_0$ can be obtained in a combined method employing both direct summation and fitting to the sum rules.[163,169] For a more complete reference list on calculations of photoionization cross sections, see, for instance, a recent review by Burke.[170]

For higher members of the heliumlike sequence, the excitation energy, k_0, is very near to the hydrogenic formula

$$K_0(1, 0) = 19.77 Z^2 \, \text{Ry} \tag{82}$$

and $\log k_0$ is usually replaced by $\log K_0$, leading to a probable error in the shift of few per cent only.

For helium, the computations included not only the main terms, that is, $E_{L,2}$ and $E'_{L,2}$,[138] but also the term $E''_{L,2}$ and an estimate of the binding correction.[163]

For heliumlike ions, the correction $E''_{L,2}$ was first obtained by Ermolaev,[39] who used the 20-parametric wave functions of Hart and Herzberg[171] to evaluate Q of formula (66). His results are accurate to within 10% and were later improved by the calculations of Aashamar,[145] who used 70-parametric Hylleraas functions. Nevertheless, the theoretical Lamb shift showed considerable deviation from recent experimental results.[147,149,151]

In my present work, I have investigated the source of this discrepancy and found that inclusion of the binding correction (80) brings theoretical and experimental shifts together to within the experimental error for all ions in which a comparison is possible. Table 6 presents these results.

Table 6. The Ground $1s^2\ {}^1S_0$ State of Two-Electron Atoms. The Radiative and Volume-Isotope Corrections to the Relativistic Potentials (in cm^{-1})

Z	Atom	$\Delta E_{L,2} + E'_{L,2}$ [a]	$E''_{L,2}$ [b]	$\Delta E_{L,2,\text{bind.}}$ [c]	ΔV_I [d]	Theory	Experiment	$E - T$	Reference
							Total shift		
2	^{4}He I	-1.267	-0.068	-0.026		-1.361	-1.28±0.15	0.08±0.15	e
3	^{7}Li II	-7.833	-0.492	-0.275		-8.600	-8.0±3	0.60±3	f
4	^{9}Be III	-27.11	-1.66	-1.34	-0.04	-30.15	-45±15	-14.5±15	g
5	^{11}B IV	-65.70	-4.01	-4.45	-0.12	-74.28	-93±20	-18.7±20	h
6	^{12}C V	-132.2	-7.99	-11.67	-0.28	-152.1	-178±30	-26.0±30	i
7	^{14}N VI	-235.2	-14.10	-26.15	-0.70	-276.2	-193±500	+83±500	j
8	^{16}O VII	-383.0	-22.85	-52.36	-1.14	-459.4	-518±600	-58±600	j
9	^{19}F VIII	-583.7	-34.74	-96.31	-2.10	-716.9	-993±800	-276±800	j
10	^{20}Ne IX	-844.5	-50.31	-165.77	-3.36	-1064			

[a] The Kabir and Salpeter terms[138] only. Calculations by Pekeris[31] and Aashamar.[143,144]
[b] Ermolaev[39] and Aashamar.[145]
[c] Present work.
[d] Ermolaev.[39]
[e] Herzberg,[136] Seaton,[153] Martin.[161]

[f] Herzberg and Moor.[146]
[g] Löfstrand.[147]
[h] Eidelsberg.[149]
[i] Edlen and Löfstrand.[151]
[j] Moor.[152]

Table 7. Lamb Shifts in $n = 2$ States of Be III, B IV, and C V (in cm^{-1})[a]

Term		Be III	B IV	C V
2^1S_0	Theory[b]	-3.28	-8.31	-17.22
	Experiment	-3.0 ± 0.1^c	-8.0 ± 0.5^d	-18 ± 4^e
	$E - T$	$+0.28 \pm 0.1$	$+0.31 \pm 0.5$	-0.78 ± 4
2^3S_1	Theory[b]	-4.06	-9.93	-20.05
	Experiment	-3.2 ± 0.2^c	-8.6 ± 0.5^d	-17 ± 1^e
	$E - T$	$+0.86 \pm 0.2$	$+1.33 \pm 0.5$	$+3 \pm 1$
2^1P_1	Theory[b]	-0.17	-0.41	-0.79
2^3P_1	Theory[b]	$+0.98$	$+2.04$	$+3.61$

[a] From Eidelsberg.[148,149]
[b] Ermolaev,[39] with the 10% error estimated there.
[c] Löfstrand.[147]
[d] Eidelsberg.[149]
[e] Edlen and Löfstrand.[151]

3.5.2. Excited S States

For He, the excitation energy k_0 was computed by Suh and Zaidi for $n = 2$,[135] and it was used in the computation of the Lamb shift of 2^1S_0 and 2^3S_1 levels in helium by Pekeris.[31] For heliumlike ions, estimates of K_0 for $n = 2$ states were obtained by Ermolaev.[39] Table 7 compares theoretical shifts with results of recent experimental studies of spectra in Be III, B IV, and C V. Similar results for higher members of the sequence are in Table 8.

3.6. Lamb Shift in P States of Two-Electron Atoms

For a considerable time, very little attention was paid to the Lamb shift of the $n = 2P$ levels in two-electron atoms, with the exception of

Table 8. The Lamb Shifts in Triplet States of He I and Li II[a] (All Shifts Are Given in cm^{-1})

	He I		Li II	
Term	2^3S_1	2^3P_1	2^3S_1	2^3P_1
$E_{L,2}$	-0.133	0.038	-0.995	-0.289
$-E'_{L,2} - E''_{L,2} - E'''_{L,2}$	0.000	-0.001	0.000	-0.006
E_{bind}	-0.002	0.001	-0.030	0.008
$E_{L,2\,\text{tot}}$	-0.135	$+0.038$	-1.025	$+0.291$
$E_{L,2\,\text{tot}}(S - P)$	-0.173		-1.316	
$E_{\text{expt}}(S - P)$	-0.175 ± 0.01^b		-1.255 ± 0.004^c	

[a] Ermolaev.[40]
[b] Martin.[161,162]
[c] Bacis and Berry.[43]

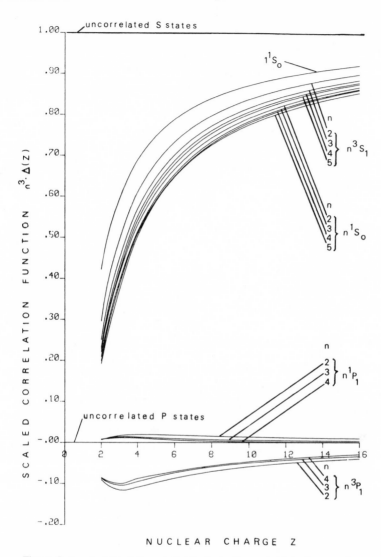

Figure 5. Lamb shift of the relativistic ionization potential of a two-electron atom. $\Delta(Z)$ defined by formula (79), determines the magnitude of the shift.

He.[57] The usual argument was that the low experimental accuracy did not warrant such investigations since the QED shift in P states was expected to be small. Estimates by Kastner[160] seemed to support this opinion.

A more accurate theoretical treatment has been given by Ermolaev,[39,40] who showed that correlation effects in the atom are the main factor determining the magnitude of the shift in a $L \neq O$ state (see Section 3.3.4 of the present work). Figure 5 shows the behavior of $\Delta(Z)$,

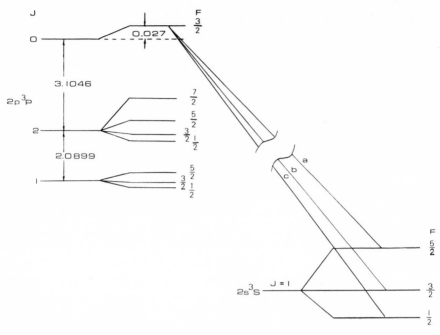

Figure 6. Energy diagram for the $2s^3S$–$2p^3P$ transition complex in ^{7}Li II (not to scale). The fine-structure separations and the upward shift of the $J = 0$, $F = \frac{3}{2}$ level are indicated in cm^{-1}. (From Berry and Bacis.[41])

defined by (79), for some S and P states. In the case of triplet P states, the function $\Delta(Z) < 0$, so the electron density at the nucleus is *lower* in a two-electron atom than that in the one-electron ion with the same Z. For light atoms with $Z = 2$–4 this anomaly is especially pronounced, making, for instance, Li II a suitable ion for the studies of the enhanced Lamb shift. In the latter case, the theoretical value of $E_L(2^3P_1)$, $Z = 3$, is[40] (see also Table 8)

$$E_L(2^3P_1) = +0.291 \pm 0.041 \text{ cm}^{-1} \qquad (83)$$

Beam-foil experimental studies of the Lamb shift in 2^3P_1 state of ^{7}Li II[41,43] and ^{6}Li II[43] have recently been reported by Berry and co-workers. Figure 6 shows the energy levels involved in the determination of the Lamb shift. The hyperfine structure a, b, c from $^3P_0$ ($F = \frac{3}{2}$) has been completely resolved in the experiments. The upwards shift of the common upper level ($J = 0$, $F = \frac{3}{2}$) is due to the hyperfine mixing of other levels with $F = \frac{3}{2}$, and equal to 0.027 cm^{-1} according to calculations of Berry and Bacis,[41] who used the off-diagonal elements of nuclear magnetic dipole interaction[172] and the theoretical fine structure of Accad et al.,[36] the singlet-triplet mixing of Ermolaev and Jones,[154] and the magnetic hyperfine coupling constant $A = 0.240 \pm 0.002$ measured by Berry *et al.*[173] A shift of the $^3P_0$ due to the nuclear quadrupole moment in ^{7}Li II was too small[174] to be taken into account. According to Bacis and Berry,[43] the experimental wave number difference for the $2s\ ^3S_1 - 2p\ ^3P_1$ interval obtained from the *a, b,* and *c* transitions of Figure 6 is

$$(2s\ ^3S_1 - 2p\ ^3P_1)_{\text{expt}} = 18226.1123 \pm 0.0040 \text{ cm}^{-1} \tag{84}$$

This can be compared with the relativistic value of the interval due to Accad *et al.*[36]:

$$(2s\ ^3S_1 - 2p\ ^3P_1)_{\text{rel}} = 18227.367 \pm 0.001 \text{ cm}^{-1} \tag{85}$$

The difference between these two intervals, $D_L = -1.2547 \pm 0.0040 \text{ cm}^{-1}$, is expected to be the difference between the Lamb shifts of the two levels. Assuming that $E_{L,2}(2^3S_1) = -1.025 \pm 0.056 \text{ cm}^{-1}$,[40] one finds that the experimental value of the Lamb shift of the 2^3P_1 level is

$$E_{L,2}(2^3P_1)_{\text{expt}} = +0.230 \pm 0.060 \text{ cm}^{-1} \tag{86}$$

which is in good agreement with the theoretical value of the shift given by formula (83).

The recent experimental studies of light two-electron atoms[42,149,150,156-159] confirm another prediction concerning P states,[40] namely, that the Lamb shifts in singlet P states are of small magnitude. Relevant theoretical and experimental data are presented in Table 9.

Table 9. Lamb Shifts in N VI–Ne IX (in cm^{-1})[a]

Ion	Transition	Wave-number difference[b] between relativistic theory and experiment	Theoretical Lamb shift[c] S state	P state	Total (P−S)
N VI	$2^3S_1 - 2^3P_1$	37.3 ± 1[d]	−35.7	5.79	41.5
	$3^3S_1 - 3^3P_1$	12.4 ± 0.8[e]	−10.6	1.72	12.3
	$2^1S_0 - 2^1P_1$	30 ± 1[f]	−31.3	−1.42	29.9
O VII	$2^3S_1 - 2^3P_1$	54.6 ± 5[g]	−58.3	8.65	67.0
	$2^1S_0 - 2^1P_1$	50 ± 2[f]	−51.8	−2.21	49.6
F VIII	$2^3S_1 - 2^3P_1$	92 ± 3[h]	−89.1	12.3	101.4
Ne IX	$2^3S_1 - 2^3P_1$	156 ± 3[h]	−129.0	16.7	146.0

[a] From Berry and Schectman.[42]
[b] Relativistic potentials from Accad *et al.*[36] corrected for the singlet–triplet mixing (Ermolaev and Jones[154]) according to Schiff *et al.*[155]
[c] Ermolaev.[39] For $n = 3$, the Lamb shifts obtained from $n = 2$ by means of n^{-3} scaling.
[d] Bockasten *et al.*[156]   [e] Berry and Schectman.[42]   [f] Baker.[157]
[g] An average of the experimental results of Baker,[157] Elton,[159] and Engelgardt and Sommer.[158]
[h] Engelhardt and Sommer.[158]

4. Concluding Remarks

Owing to a steady improvement of experimental techniques during the last few years and the introduction of new methods, the standard experimental accuracy in atomic spectroscopy has become so high that the quantum electrodynamical effects have to be included in the interpretation of the experimental data in many cases.

Although nothing indicates probability of any major breakdown of the theory in the near future, the trend of improving accuracy of the experiments may lead in the next few years to the necessity of including such difficult questions as the nuclear structure and the higher-order binding effects in the interpretation of the experimental data.

For many-electron atoms, one may expect that, along the lines of Section 3.2.1, the difficulties of the theoretical determination of the nonrelativistic positions of the levels will be the major factor in slowing down the Lamb-shift studies in atoms with several electrons.

As we have seen, in the case of two-electron atoms, the present experimental accuracy is already sufficiently high to require the QED corrections to be included. One may expect that this will stimulate further atomic structure calculations in two-electron atoms.

Other spectroscopic fields in which the quantum electrodynamical effects may be of importance are hot plasmas and astrophysics.

References

1. N. M. Kroll, in *Physics of One- and Two-Electron Atoms*, Eds. F. Bopp and H. Kleinpoppen, p. 179, North-Holland, Amsterdam (1969).
2. W. E. Lamb, Jr., and R. C. Retherford, *Phys. Rev.* **72**, 241 (1947).
3. H. A. Bethe, *Phys. Rev.* **72**, 339 (1947).
4. T. Appelquist and S. Brodski, *Phys. Rev. Lett.* **24**, 562 (1970).
5. T. Appelquist and S. Brodski, *Phys. Rev. A* **2**, 2293 (1970).
6. A. Peterman, *Phys. Lett.* **38B**, 330 (1972).
7. M. F. Soto, Jr., *Phys. Rev. Lett.* **17**, 1153 (1966).
8. M. F. Soto, Jr., *Phys. Rev. A* **2**, 734 (1970).
9. G. W. Erickson and D. R. Yennie, *Ann. Phys.* (*N.Y.*) **35**, 271 (1965).
10. G. W. Erickson and D. R. Yennie, *Ann. Phys.* (*N.Y.*) **35**, 447 (1965).
11. S. J. Brodski and G. W. Erickson, *Phys. Rev.* **148**, 148 (1966).
12. G. W. Erickson, *Phys. Rev. Lett.* **27**, 780 (1971).
13. P. J. Mohr, *Ann. Phys.* (*N.Y.*) **88**, 26 (1974).
14. P. J. Mohr, *Ann. Phys.* (*N.Y.*) **88**, 52 (1974).
15. P. J. Mohr, *Phys. Rev. Lett.* **34**, 1050 (1975).
16. K. T. Cheng and W. R. Johnson, *Phys. Rev. A* **14**, 1943 (1976).
17. J. Mignaco and E. Remiddi, *Nuovo cim.* **60A**, 519 (1969).
18. J. Aldins, S. Brodski, A. Dufner, and T. Kinoshita, *Phys. Rev. Lett.* **23**, 441 (1969).
19. J. Aldins, S. Brodski, A. Dufner, and T. Kinoshita, *Phys. Rev. D* **1**, 2378 (1970).

20. S. Brodski and T. Kinoshita, *Phys. Rev. D* **3**, 356 (1971).
21. J. Calmet and M. Perrottet, *Phys. Rev. D* **3**, 3101 (1971).
22. A. de Rujula, B. Lautrup, and A. Peterman, *Phys. Lett.* **33B**, 605 (1970).
23. M. Levine and J. Wright, *Phys. Rev. Lett.* **26**, 1351 (1971).
24. T. Kinoshita and P. Cvitanovic, *Phys. Rev. Lett.* **29**, 1534 (1972).
25. T. Kinoshita, *in* "Proceedings of the Third Colloquium on Advanced Computing Methods in Theoretical Physics," Marseille (1973).
26. R. Barbieri and E. Remiddi, *Phys. Lett.* **49B**, 468 (1974).
27. J. Calmet and A. Peterman, *Phys. Lett.* **58B**, 449 (1975).
28. J. Calmet and E. de Rafael, *Phys. Lett.* **56B**, 181 (1975).
29. J. Calmet and A. Peterman, *Phys. Lett.* **56B**, 383 (1975).
30. H. A. Bethe and E. E. Salpeter, *Quantum Mechanics of One- and Two-Electron Atoms*, Springer-Verlag, Berlin (1957).
31. C. L. Pekeris, *Phys. Rev.* **112**, 1649 (1958); **115**, 1216 (1959); **126**, 143 (1962); **126**, 1470 (1962); **127**, 509 (1962).
32. S. Chandrasekhar, G. Herzberg, and D. Elbert, *Phys. Rev.* **91**, 1172 (1953).
33. T. Kinoshita, *Phys. Rev.* **105**, 1490 (1957).
34. A. M. Ermolaev and G. B. Sochilin, *Sov. Phys.—Dokl.* **9**, 292 (1964); *Int. J. Quant. Chem.* **5**, 333 (1968).
35. K. Frankowski and C. L. Pekeris, *Phys. Rev.* **146**, 46 (1966); **160**, 1 (1967).
36. Y. Accad, C. L. Pekeris, and B. Schiff, *Phys. Rev. A* **4**, 516 (1971).
37. Y. Accad and C. L. Pekeris, *Phys. Rev. A* **11**, 1479 (1975).
38. B. Schiff, C. L. Pekeris, and Y. Accad, *Phys. Rev. A* **4**, 885 (1971).
39. A. M. Ermolaev, *Phys. Rev. A* **8**, 1651 (1973).
40. A. M. Ermolaev, *Phys. Rev. Lett.* **34**, 380 (1975).
41. H. G. Berry and R. Bacis, *Phys. Rev. A* **8** 36 (1973).
42. H. G. Berry and R. M. Schectman, *Phys. Rev, A* **9**, 2345 (1974).
43. R. Bacis and H. G. Berry, *Phys. Rev. A* **10**, 466 (1974).
44. M. Douglas, *Phys. Rev. A* **6**, 1929 (1972).
45. L. Hambro, *Phys. Rev. A* **5**, 2027 (1972).
46. L. Hambro, *Phys. Rev. A* **6**, 865 (1972).
47. J. Daley, M. Douglas, L. Hambro, and N. M. Kroll, *Phys. Rev. Lett.* **29**, 12 (1972).
48. C. Schwartz, *Phys. Rev.* **134**, A 1181 (1964).
49. B. N. Taylor, W. H. Parker, and D. N. Langenberg, *The fundamental Constants and Quantum Electrodynamics*, Academic Press, New York (1969).
50. S. J. Brodski and S. D. Drell, *Ann. Rev. Nucl. Sci.* **20**, 147 (1970).
51. S. J. Brodski, in *Proceedings of the 1971 International Symposium on Electron and Proton Interactions at Higher Energies*, Ed. N. B. Mistry, Laboratory of Nucl. Studies, Cornell University, Ithaca, New York (1972).
52. B. E. Lautrup, A. Peterman, and E. de Rafael, *Phys. Rep.* **3C**, 193 (1972).
53. N. M. Kroll, in *Atomic Physics 3. Proceedings of the Third International Conference on Atomic Physics*, Eds. S. J. Smith and G. K. Walters, Plenum Press, New York (1973).
54. J. Calmet, S. Narison, M. Perrottet, and E. de Rafael, *Rev. Mod. Phys.* **49**, 21 (1977).
55. E. Borie, *Z. Phys.* **275A**, 347 (1975).
56. J. Calmet, in "Proceedings of the Third Colloquium on Advanced Computing Methods in Theoretical Physics," Marseille (1973).
57. M. Douglas and N. M. Kroll, *Ann. Phys. (N.Y.)* **82**, 89 (1974).
58. E. J. Kelsey, *Ann. Phys. (N.Y.)* **98**, 462 (1976).
59. M. Levine, *J. Comput. Phys.* **1**, 454 (1967).
60. A. C. Hearn, "REDUCE 2 User's Manual," preprint No. UCP-19, University of Utah (1973).

61. M. Veltman, CERN preprint (1967).
62. W. Czyz, G. C. Sheppey, and J. D. Walecka, *Nuovo Cim.* **34**, 420 (1964).
63. A. J. Dufner, in "Proceedings of the First Colloquium on Advanced Computing Methods in Theoretical Physics," Marseille (1970).
64. B. E. Lautrup, in "Proceedings of the Second Colloquium on Advanced Computing Methods in Theoretical Physics," Marseille (1971).
65. R. P. Feynman, *Phys. Rev.* **74**, 1430 (1948); **76**, 769 (1949).
66. H. Fukuda, Y. Miyamoto, and S. Tomonaga, *Progr. Theoret. Phys.* (*Kyoto*) **4**, 47, 121 (1949).
67.. N. M. Kroll and W. E. Lamb, Jr., *Phys. Rev.* **75**, 388 (1949).
68. J. Schwinger, *Phys. Rev.* **75**, 898 (1949).
69. J. B. French and V. F. Weisskopf, *Phys. Rev.* **75**, 1240 (1949).
70. E. E. Salpeter, *Phys. Rev.* **87**, 328 (1952).
71. H. A. Bethe, L. M. Brown, and J. R. Stehn, *Phys. Rev.* **77**, 370 (1950).
72. C. L. Schwartz and J. J. Tiemann, *Ann. Phys.* (*N.Y.*) **6**, 178 (1959).
73. S. Klarsfeld and A. Marquet, *Phys. Lett.* **43B**, 201 (1973).
74. R. Karplus, A. Klein, and J. Schwinger, *Phys. Rev.* **86**, 288 (1952).
75. M. Baranger, H. A. Bethe, and R. P. Feynman, *Phys. Rev.* **92**, 482 (1953).
76. A. J. Layzer, *Phys. Rev. Lett.* **4**, 580 (1960).
77. H. M. Fried and D. R. Yennie, *Phys. Rev. Lett.* **4**, 583 (1960).
78. H. M. Fried and D. R. Yennie, *Phys. Rev.* **112**, 1391 (1958).
79. A. J. Layzer, *J. Math. Phys.* **2**, 292, 308 (1961).
80. M. Baranger, F. J. Dayson, and E. E. Salpeter, *Phys. Rev.* **88**, 680 (1952).
81. G. Källen and A. Sabry, *Kgl. Danske Vidensk, Selsk. Mat. Fys. Medd.*, No. 17 (1955).
82. C. M. Sommerfield, *Phys. Rev.* **107**, 328 (1957).
83. A. Peterman, *Helv. Phys. Acta* **30**, 407 (1957); *Nucl. Phys.* **3**, 689 (1957).
84. J. Weneser, R. Bersohn, and N. M. Kroll, *Phys. Rev.* **91**, 1257 (1953).
85. B. E. Lautrup, A. Peterman, and E. de Rafael, *Phys. Lett.* **31B**, 577 (1970).
86. R. Barbieri, J. A. Mignaco, and E. de Rafael, *Nuovo Cim. Lett.* **3**, 588 (1970).
87. R. Barbieri, J. A. Mignaco, and E. de Rafael, *Nuovo Cim.* **6A**, 81 (1971).
88. A. Peterman, *Phys. Lett.* **35B**, 325 (1971).
89. T. Fulton and P. C. Martin, *Phys. Rev.* **95**, 811 (1954).
90. H. Grotch and D. R. Yennie, *Rev. Mod. Phys.* **41**, 350 (1969).
91. G. W. Erickson, in *Physics of One- and Two-Electron Atoms*, Eds. F. Bopp and H. Kleinpoppen, p. 193, North-Holland, Amsterdam (1969).
92. F. A. Bumiller, F. R. Buskirk, J. W. Stewart, and E. B. Dally, *Phys. Rev. Lett.* **25**, 1774 (1970).
93. L. N. Hand, D. G. Miller, and R. Wilson, *Rev. Mod. Phys.* **35**, 335 (1963).
94. G. E. Brown, J. S. Langer, and G. W. Schaefer, *Proc. R. Soc.* (*London*) A **251**, 92 (1959).
95. G. E. Brown and D. F. Mayers, *Proc. R. Soc.* (*London*) A **251**, 105 (1959).
96. A. M. Desiderio and W. R. Johnson, *Phys. Rev.* A **3**, 1267 (1971).
97. L. N. Labzovski, *Sov. Phys.—JETP* **32**, 1171 (1971).
98. E. H. Wichmann and N. M. Kroll, *Phys. Rev.* **101**, 843 (1956).
99. G. A. Rinker, Jr., and L. Wilets, *Phys. Rev. Lett.* **26**, 1559 (1973).
100. A. G. Kovalevski and L. N. Labzovski, *Vest. Leningrad. Univ.* (*Ser. Fiz.*) No. 17, 16 (1974).
101. I. Sick, J. S. McCarthy, and R. R. Whitney, *Phys. Lett.* **64B**, 33 (1976).
102. J. Bernabeu and C. Jarlskog, *Phys. Lett.* **60B**, 197 (1976).
103. A. Bertin, G. Garboni, J. Ducles, U. Gastadli, G. Gorini, G. Neri, J. Picard, O. Pitzurra, A. Placci, E. Polacco, G. Torelli, A. Vitale, and E. Zavattini, *Phys. Lett.* **55B**, 411 (1975).

104. B. Fricke, in the present monograph, Chapter 5.
105. K. M. Evenson, *Bull. Am. Phys. Soc.* **20**, 10 (1975).
106. P. Olsen and E. R. Williams, in "Proceedings of the Fifth International Conference on Atomic and Fundamental Constants," Ed. J. H. Sanders, Paris (1975).
107. T. W. Hänsch, M. H. Nayfeh, S. A. Lee, S. M. Curry, and I. S. Shahin, *Phys. Rev. Lett.* **32**, 1336 (1974).
108. R. Robiscoe and T. Shyn, *Phys. Rev. Lett.* **24**, 559 (1970).
109. S. R. Lundeen, *Bull. Am. Phys. Soc.* **20**, 11 (1975).
110. B. Cosens, *Phys. Rev.* **173**, 49 (1968).
111. M. A. Narasimham, Ph. D. thesis, University of Colorado (1969).
112. E. Lipworth and R. Novick, *Phys. Rev.* **108**, 1434 (1958).
113. O. Mader and M. Leventhal, in "Proceedings of International Conference on Atomic Physics," New York University (1968).
114. O. Mader, M. Leventhal, and W. E. Lamb, Jr., *Phys. Rev. A* **3**, 1832 (1971).
115. A. Eibofner, *Z. Phys.* **A227**, 225 (1976).
116. M. Leventhal and P. E. Havey, *Phys. Rev. Lett.* **32**, 88 (1974).
117. B. Cagnac, G. Grinberg, and F. Biraben, *J. Phys. (Paris)* **34**, 845 (1973).
118. D. E. Roberts and E. N. Forston, *Phys. Rev. Lett.* **31**, 1539 (1973).
119. E. V. Baklanov and V. P. Chebotaev, *Opt. Commun.* **12**, 312 (1974).
120. C. Wieman and T. W. Hänsch, *Phys. Rev. Lett.* **36**, 1170 (1976).
121. F. M. Pipkin, *Comments Atom. Molec. Phys.* **5**, 45 (1975).
122. S. A. Lee, R. Wallenstein, and T. W. Hänsch, *Phys. Rev. Lett.* **35**, 1262 (1975).
123. G. Herzberg, *Proc. R. Soc. (London) A* **234**, 516 (1956).
124. J. D. Garcia and J. E. Mack, *J. Opt. Soc. Am.* **55**, 654 (1965).
125. S. R. Lundeen and F. M. Pipkin, *Phys. Rev. Lett.* **34**, 1368 (1975).
126. S. Triebwasser, E. S. Dayhoff, and W. E. Lamb, Jr., *Phys. Rev.* **89**, 98 (1953).
127. E. A. Uehling, *Phys. Rev.* **48**, 55 (1935).
128. I. Pomeranchuk and J. Smorodinski, *J. Phys. USSR* **9**, 97 (1945).
129. W. Pieper and W. Greiner, *Z. Phys.* **218**, 327 (1969).
130. V. S. Popov, *Yad. Fiz.* **12**, 429 (1970); **14**, 458 (1971); *Zh. Eksp. Teor. Fiz.* **60**, 1228 (1971).
131. J. Sucher and H. M. Foley, *Phys. Rev.* **95**, 966 (1954).
132. J. Sucher, *Phys. Rev.* **107**, 1448 (1957).
133. E. E. Salpeter and H. A. Bethe, *Phys. Rev.* **84**, 1232 (1951).
134. A. M. Ermolaev and M. Jones, *J. Phys. B: Atom. Molec. Phys.* **6**, 1 (1973).
135. K. S. Suh and M. H. Zaidi, *Proc. R. Soc. (London) A* **296**, 94 (1965).
136. G. Herzberg, *Proc. R. Soc. (London) A* **248**, 309 (1958).
137. A. M. Ermolaev, *J. Phys. B: Atom. Molec. Phys.* **7**, 1611 (1974).
138. P. K. Kabir and E. E. Salpeter, *Phys. Rev.* **108**, 1257 (1957).
139. P. Blanchard and G. W. F. Drake, *J. Phys. B: Atom. Molec. Phys.* **6**, 2495 (1973).
140. V. Fock, *Izv. Acad. Nauk SSSR, Ser. Fiz.* **18**, 161 (1954).
141. J. Sucher, *Phys. Rev.* **109**, 1010 (1958); Ph.D. thesis, Columbia University (1957).
142. H. Araki, *Progr. Theor. Phys. (Japan)* **17**, 619 (1957).
143. K. Aashamar, *Phys. Math. Univ. Osloensis*, 35 (1969).
144. K. Aashamar, *Nucl. Instr. Math.* **90**, 263 (1970).
145. K. Aashamar, *Phys. Norvegica* **7**, 169 (1974).
146. G. Herzberg and H. R. Moor, *Can. J. Phys.* **37**, 1293 (1959).
147. B. Löfstrand, *Phys. Scripta* **8**, 57 (1973).
148. M. Eidelsberg, *J. Phys. B: Atom. Molec. Phys.* **5**, 1031 (1972).
149. M. Eidelsberg, *J. Phys. B: Atom. Molec. Phys.* **7**, 1476 (1974).
150. M. Eidelsberg, Ph.D. thesis, University of Paris (1975).
151. B. Edlen and B. Löfstrand, *J. Phys. B: Atom. Molec. Phys.* **3**, 1380 (1970).

152. C. E. Moor, "Atomic Energy Levels," Natl. Bur. Stand. (U.S.) Circular No. 467 (1949).

153. M. J. Seaton, *Proc. Phys. Soc. (London)* **87**, 337 (1966).

154. A. M. Ermolaev and M. Jones, *J. Phys. B: Atom. Molec. Phys.* **5**, L225 (1972).

155. B. Schiff, Y. Accad, and C. L. Pekeris, *Phys. Rev. A* **8**, 2272 (1973).

156. K. Bockasten, R. Hallin, K. B. Johannson, and P. Tsui, *Phys. Rev. Lett.* **8**, 181 (1964).

157. S. C. Baker, *J. Phys. B: Atom. Molec. Phys.* **6**, 709 (1973).

158. W. Engelhardt and J. Sommer, *Astrophys. J.* **167**, 201 (1971).

159. R. C. Elton, *Astrophys. J.* **148**, 573 (1967).

160. S. O. Kastner, *Phys. Rev. A* **6**, 570 (1972).

161. W. C. Martin, *J. Phys. Ref. Data* **2**, 257 (1972).

162. W. C. Martin, *J. Res. Natl. Bur. Stand., Sect. A* **64**, 19 (1960).

163. E. E. Salpeter and M. H. Zaidi, *Phys. Rev.* **112**, 298 (1962).

164. A. Dalgarno amd A. L. Stewart, *Proc. Phys. Soc. (London)* **76**, 49 (1960).

165. C. Schwartz, *Phys. Rev.* **123**, 1700 (1961).

166. C. Schwartz and J. J. Tiemann, *Ann. Phys. (N.Y.)* **2**, 178 (1959).

167. C. L. Pekeris, *Phys. Rev.* **115**, 1216 (1959).

168. A. M. Ermolaev, Ph.D. thesis, Leningrad University (1965).

169. V. L. Jacobs, *Phys. Rev. A* **3**, 289 (1971).

170. P. G. Burke, in *Atomic processes and applications*, Eds. P. G. Burke and B. L. Moiseiwitsch, p. 200, North-Holland, Amsterdam (1976).

171. J. Hart and G. Herzberg, *Phys. Rev.* **106**, 79 (1957).

172. P. Güttinger and W. Pauli, *Z. Phys.* **67**, 743 (1931).

173. H. G. Berry, J. L. Subtil, E. H. Pinnington, H. J. Andrä, W. Wittman, and A. Gaupp, *Phys. Rev. A* **7**, 1609 (1973).

174. J. C. Browne and F. A. Matsen, *Phys. Rev.* **135**, 1227 (1964).

5

Inner Shells

B. FRICKE

1. Introduction

Interest in the physics of atomic inner shells is currently undergoing a renaissance after lying nearly dormant for some forty years. In the early days of quantum mechanics many basic problems were formulated in principle but had to be set aside as intractable in practice. Many of these can now be solved with modern experimental and theoretical techniques. From the experimental point of view the rapidly increasing interest in inner shells since the late 60s is the result of the availability of highly ionized heavy atoms with energies sufficient to reach the domain of inner-shells at a great number of small- and medium-sized accelerators. The experimental results together with the rapid development of faster and larger computers has stimulated this field from the theoretical point of view. The accuracy of the best Hartree–Fock calculations available nowadays already allows a quasiexperimental determination of the quantum electrodynamical contributions for the inner shells of high-Z atoms. This method as well as the results for the binding energies of inner shells for low Z, large Z, and superheavy Z (larger than 137) will be discussed in the first part of this chapter. In the second part the inner-shell vacancy production mechanisms such as Coulomb ionization and molecular excitation will be discussed very briefly. During the collision, combined atom or quasimolecular phenomena appear, which will be considered in the last part of this chapter.

B. FRICKE • Gesamthochschule Kassel, D 35 Kassel, Germany.

2. Binding Energies of Inner Electrons

2.1. Method of Calculation

How well do the experimental binding energies of atomic electrons compare with theoretical calculations? This is one of the basic questions in atomic physics. The method of calculation of binding energies from a theoretical point of view was given by Hartree and Fock[1] in the late 1920s.

2.1.1. Hartree–Fock Calculations

The so-called Hartree–Fock method is the basis of all good atomic calculations. This method with the various assumptions and approximations used nowadays is discussed in this book in Chapter 1. Only the nonrelativistic methods are reviewed there. Because we are interested here in the inner electrons, we must include the relativistic methods. As we shall see, even for neon ($Z = 10$) relativistic effects cannot be neglected and, for larger-Z atoms, nonrelativistic attempts to describe the atom are absolutely wrong. The difference between the nonrelativistic and the relativistic description first given by Dirac[2] is in the different kinetic energy operator. The relativistic Hamiltonian for the many-electron system may be expressed as

$$\mathbf{H} = \sum_{i}^{\substack{\text{all } N \\ \text{electrons}}} \left[-ic\boldsymbol{\alpha}_i \boldsymbol{\nabla}_i + \boldsymbol{\beta}_i c^2 - \frac{Z}{r_i} \right] + \sum_{i<j} \frac{1}{r_{ij}} \tag{1}$$

where $\boldsymbol{\alpha}$ and $\boldsymbol{\beta}$ are the Dirac 4×4 matrices

$$\boldsymbol{\alpha} = \begin{pmatrix} 0 & \boldsymbol{\sigma} \\ -\boldsymbol{\sigma} & 0 \end{pmatrix}, \qquad \boldsymbol{\beta} = \begin{pmatrix} \mathbf{1} & 0 \\ 0 & -\mathbf{1} \end{pmatrix}$$

with $\boldsymbol{\sigma}$ the Pauli matrices and $\mathbf{1}$ the unit matrix.

Using this relativistic Hamiltonian the resulting Hartree–Fock equations for the radial functions, where μ and ν are indices running over all occupied levels, take the following form:

$$\frac{dP_\mu}{dr} + \frac{k_\mu}{r} P_\mu = \left[2c + \frac{1}{c}(\varepsilon_\mu - V_\mu) \right] Q_\mu + W_Q + \frac{1}{c} \sum_{\mu \neq \nu}^{N} \varepsilon_{\mu\nu} Q_\nu \tag{2}$$

$$-\frac{dQ_\mu}{dr} + \frac{k_\mu}{r} Q_\mu = \frac{1}{c}(\varepsilon_\mu - V_\mu) P_\mu + W_P + \frac{1}{c} \sum_{\mu \neq \nu}^{N} \varepsilon_{\mu\nu} P_\nu \tag{3}$$

with

$$V_\mu = \frac{1}{r}\left[-Z + \sum_\nu \sum_k q_\nu A_k^E(\mu, \nu) Y_k(\nu, \nu)\right] \qquad (4)$$

$$W_{P\,\text{or}\,Q} = \frac{1}{rc}\sum_\nu \sum_k q_\nu B_k^E(\mu, \nu) Y_k(\mu, \nu) P_\nu \text{ or } Q_\nu \qquad (5)$$

$$Y_k(\mu, \nu) = r\int_0^\infty [P_\mu(s)P_\nu(s) + Q_\mu(s)Q_\nu(s)] U_k(r, s)\,ds \qquad (6)$$

$$U_k(r, s) = \begin{cases} r^k/s^{k+1} & \text{if } r \leqslant s \\ s^k/r^{k+1} & \text{if } r > s \end{cases}$$

P and Q are the large and small components of the wave functions, respectively; A^E and B^E are angular coefficients. For details of the relativistic Hartree–Fock method see Grant.[3] The relativistic Hartree–Fock equations (2)–(6) can easily be compared with the nonrelativistic expression in Chapter 1 of this book.

The main difference is that we now have a coupled system of $2N$ differential equations of first order for the large and the small components P and Q of the wave functions compared to the N coupled equations of second order in the nonrelativistic case.

These coupled differential equations have to be solved by iteration to self-consistency. As a result we get a value for the total energy. The binding energies E_{bind} of the electrons are given by the difference of the total energies E_T of the whole atom before and after the ionization:

$$E_{\text{bind}} = E_T(\text{atom}) - E_T(\text{ion})$$

If we use the results from such relativistic Hartree–Fock (or Dirac–Fock) calculations and compare them with experimental binding energies of inner electrons in very heavy atoms, we still find a discrepancy of the order of 1%. Approximations to this Dirac–Fock method such as Dirac–Slater or Dirac–Fock–Slater calculations (see Ref. 3) may yield better agreement with experimental results but may not be taken as better methods because they have an adjustable parameter and are, therefore, not fully *ab initio* methods.

2.1.2. Additional Effects (Magnetic, Retardation, QED)

At this level of sophistication all additional effects not taken into account so far have to be included. These are the magnetic effects, the influence of retardation in the interaction of the electrons, the quantum electrodynamic effects—vacuum polarization and self-energy (or vacuum fluctuation), and the correlation energy.

Although the total effect of the correlation energy may be large, the net effect can be neglected for the inner electrons in very heavy atoms because we have to deal here with differences of the same magnitude.[4]

The Hamiltonians of all four remaining additional effects are known only as perturbations, so their contribution can be computed consistently in a perturbation calculation only. The magnetic energy and the retardation[5] are given by the well-known Breit Hamiltonian

$$H_B = H_{\text{mag}} + H_{\text{retard}} = -\frac{1}{2}\left[\frac{\boldsymbol{\alpha}_i \cdot \boldsymbol{\alpha}_j}{r_{ij}} + \frac{(\boldsymbol{\alpha}_i r_{ij}) \cdot (\boldsymbol{\alpha}_j r_{ij})}{(r_{ij})^3}\right] \tag{7}$$

The vacuum polarization to first order in $(Z\alpha)$ can be expressed[6] by a local potential

$$V_{\text{V.P.}} = \frac{2\alpha e^2}{3\pi} \int \frac{\rho(\mathbf{r}')}{|\mathbf{r} - \mathbf{r}'|} Z_0(|\mathbf{r} - \mathbf{r}'|)\, dJ' \tag{8}$$

with

$$Z_n(|\mathbf{r}|) = \int_1^\infty \exp\left(-\frac{2}{\lambda_e}|\mathbf{r}|\xi\right)\left(1 + \frac{1}{2\xi^2}\right)\frac{(\xi^2 - 1)^{1/2}}{\xi^2} \cdot \frac{1}{\xi^n}\, d\xi \tag{9}$$

and $\lambda_e = 386$ fm and $\rho(\mathbf{r})$ is the nuclear charge density; the self-energy part, on the other hand, cannot be given by a simple expression. For these two QED corrections we also refer to Chapter 4 in this book on the Lamb shift. For large Z the vacuum fluctuation can be calculated according to the theory of Brown et al.[7] and the new calculations of Desiderio et al.[8] and Mohr.[9]

2.2. Results and Comparison with Experiments

2.2.1. Low-Z, Highly Ionized Atoms

A large number of excellent measurements of the spectra of highly ionized and excited atoms have been performed during recent years. For details see Chapter 17 of this work and Refs. 10 and 11. Because of the great complexity of the spectra many groups have focused their interest on highly ionized few-electron systems,[12] where it is still possible to handle the problem from the theoretical point of view. For example in Figure 1 part of the beam-foil excited Auger electron spectrum[13] of Ne on C is shown measured in coincidence with the outgoing projectile having the well-defined charge state 8^+. At the top of this figure the calculated transition energies are given; they fit the experimental data very well. In these calculations the even better and much more complicated multi-configuration Dirac–Fock method was used instead of the Dirac–Fock treatment discussed above.

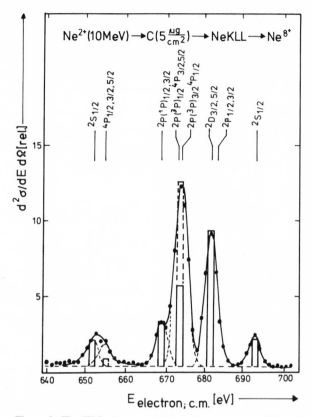

Figure 1. The KLL Auger spectrum of neon measured in coincidence with Ne^{8+} projectiles after bombarding a carbon foil. In the upper half comparison with levels from multiconfiguration Dirac–Fock calculations are given.

If we compare these results with the analogous nonrelativistic calculations we find a large shift of the two lines on the extreme right and extreme left of the spectrum in Figure 1. These two lines have large contributions from the $2s$ electron level, which is strongly influenced by the relativistic effects. This one example shows that only a full relativistic analysis of the spectra of highly ionized atoms with low Z leads to good agreement and thus relativistic effects have to be introduced even in this part of the periodic system.

2.2.2. Heavy Atoms with Inner Shell Vacancies

A comparison of theoretical and experimental binding energies of inner-shell electrons in very heavy atoms is very important for several

Table 1. Comparison of Experimental Binding Energies (in keV) of Inner Electrons in Fermium ($Z = 100$) with Theoretical Dirac–Fock Calculations Plus All Known Additional Corrections

Source	1s	2s	$2p_{1/2}$	$2p_{3/2}$	3s	$3p_{1/2}$
				Level		
Electric	−142.929	−27.734	−26.791	−20.947	−7.250	−6.815
Magnetic	+0.715	+0.091	+0.153	+0.092	+0.019	+0.033
Retardation	−0.041	−0.008	−0.013	−0.011	−0.001	−0.003
Vacuum fluctuation	+0.457	+0.096	+0.009	−0.003	+0.025	+0.003
Vacuum polarization	−0.155	−0.026	−0.004	+0.000	−0.006	−0.001
Theoretical total	−141.953(26)	−27.581(20)	−26.646(10)	−20.869(10)	−7.213(15)	−6.783(4)
Experimental value[a]	−141.963(13)	−27.573(8)	−26.664(7)	−20.868(7)	−7.200(9)	−6.779(7)

[a] Reference 16.

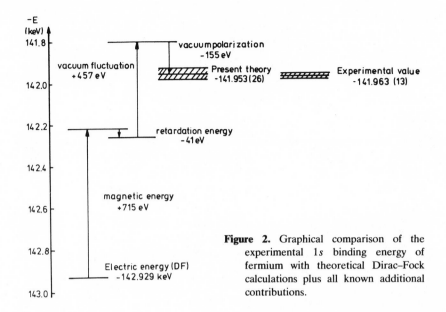

Figure 2. Graphical comparison of the experimental 1s binding energy of fermium with theoretical Dirac–Fock calculations plus all known additional contributions.

reasons. The first question is: "How good are the Dirac–Fock calculations?" This question is nontrivial because the Dirac equation itself is, in principle, a single-partial equation and the inclusion of more electrons cannot be justified from first principles. The second question is: "How large are the additional effects discussed in Section 2.1.2?" The third question is: "Do we need to take into account further effects such as nonlinear electrodynamic contributions, which were discussed by Greiner *et al.*[14]?"

Because all these effects, and thus the possible discrepancies, strongly increase with Z, the best way to answer these questions is to examine the highest element yet measured. In Table 1 a comparison[15] is given for the innermost electrons of fermium ($Z = 100$) measured by Porter *et al.*[16] In the calculated electric energies as well as the magnetic energies the relaxation effect due to the rearrangement from the atom to the ion is included.The error in the theoretical calculation still is of the order of 20 eV because of the uncertainty in the calculation of the self-energy. Table 1 definitely shows that the agreement for all the inner levels is within the error bars. This excellent agreement is also shown for the 1s level in Figure 2. A similar comparison of x-ray lines from plutonium and thorium[17] also leads to astonishingly good results.

This shows that the answers to our questions are as follows. The Dirac–Fock approximation using the solution of the Dirac equation is an excellent method even as far as the element fermium ($Z = 100$). The additional contributions are contained within the error limit of ±20 eV and we do not need to include, at this degree of accuracy, any additional effects. Since the experimental uncertainty is now within 1 eV,[17] however, much more work has to be done in the future from the theoretical point of view.

2.2.3. Level Behavior for Atoms with Z > 137

Bearing in mind the results in Table 1, which show an excellent agreement between theory and experiment, we can at least be relatively sure that theoretical calculations performed for even higher Z will be realistic approximations.

A magic limit is $Z = 137$ because at this point the solution of the Dirac equation for the $1s$ level of a hydrogenic atom, with a point nucleus, becomes imaginary. This limit can be circumvented if one uses an extended nucleus so that the divergency at $r = 0$ disappears. The resulting level behavior calculated from self-consistent Dirac–Slater calculations[18] is given in Figure 3.

Two interesting effects occur. The large spin–orbit splitting of levels such as the $2p_{1/2}$ and $2p_{3/2}$ or the $3p_{1/2}$ and $3p_{3/2}$ levels increases so

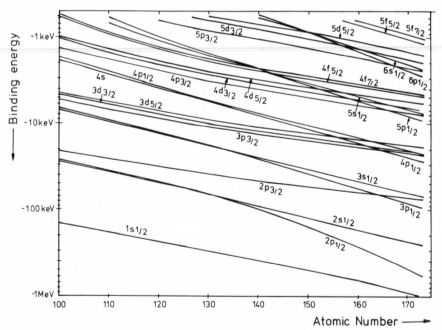

Figure 3. Inner atomic level for $100 < Z < 173$ from Dirac–Fock–Slater calculations.[18]

rapidly that their binding energy at $Z = 170$ differs already by factors of 6 and 4, respectively. The wave functions also change so much that the shielding for the $s_{1/2}$ electrons becomes much larger than the $p_{1/2}$ electrons having the same principal quantum number. As a result the $2p_{1/2}$ level becomes lower than the $2s_{1/2}$ level for Z above 130.

For even larger Z the ordering of the levels becomes very strange (see Figure 3). This will show up in the behavior of the outer levels and even in the predictions related to the chemistry of these elements.[19] The second, even more interesting result, is that at, or near, $Z = 173$ the $1s$ level dives into the negative continuum. For a long time it was not clear what would happen if such a bound level reached the negative continuum, until Müller et al.[20] found that this could well be interpreted in a way similar to a quasibound level in the positive continuum. They also predicted the creation of positrons when a hole is brought down into the negative continuum.

As we shall see in Section 4, this discussion of the level behavior is by no means academic since in the near future it will be possible to create quasi atoms with very large Z for a time of the order of 10^{-18} sec, by bombarding uranium with uranium, giving a united $Z = 184$. We refer to Section 4, where the united atom phenomena will be discussed briefly.

3. Inner-Shell Vacancy Production Mechanisms

Inner-shell excitation in atomic collisions has reached a refined stage within the past decade. Evidently, the relatively simplest collision system is that of a swift structureless point charge Z_1 of velocity v_1 penetrating a hydrogenlike atom (nuclear charge Z_2) while the electrons (orbital velocity u) are only slightly perturbed by the projectile. This condition is fulfilled,

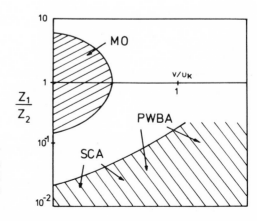

Figure 4. Regions of validity for various approximation schemes as a function of Z_1/Z_2 and collision velocity v (in units of the K-shell electron velocity u_K).

e.g., for inner shells in the case of $Z_1 \ll Z_2$ or $v_1 \gg u$. Such systems have been the subject of intense experimental and theoretical study,[21-23] which has resulted in a large body of data elucidating even the finer details of the interaction.[10,24] Generally, this region is termed "Coulomb excitation," indicating that excitation occurs mainly by direct Coulomb interaction between the projectile and the atomic electron. If the internal structure of the projectile and its influence on the target atom electrons cannot be neglected, particularly at low ion velocities $v_1 \ll u$ and comparable nuclear charges $(Z_1/Z_2 \approx 1)$, the interaction tends to become more complex. This is the region of "molecular excitation," where excitation occurs by way of coupling between promoted levels in the quasimolecule formed during the collision.[25] These different regions are shown in Figure 4. The distinction between both regimes is helpful though somewhat arbitrary as it separates in a qualitative way regions in which different methods are applied.

3.1. Coulomb Ionization Processes

3.1.1. Plane-Wave Born Approximation (PWBA)

The first treatment by quantum mechanics of the ionization process was based on the Bethe–Born approximation.[26,21] In this approximation the incident charged particles are treated as plane waves, whereas the target electrons are described by hydrogenic wave functions. The interaction between the projectile and the electron is treated to first order. The PWBA formula for the differential cross section for Coulomb ejection of a target electron with final energy E_f is given by

$$\frac{d\sigma}{dE_f} = \frac{4\pi}{\hbar^2} Z_1^2 e^4 \frac{M_1}{E_1} \int_{q_0}^{\infty} \frac{dq}{q^3} I \tag{10}$$

with

$$I = \sum_f \left| \int e^{i\mathbf{q}\cdot\mathbf{r}} \psi_f^*(r)\psi_i(r) \, d\tau \right|^2 \tag{11}$$

Here $Z_1 e$, M_1, and E_1 are the charge, mass, and energy of the projectile. Further, $\hbar q$ denotes the momentum transfer with $\hbar q_0$ its minimum value. The quantities $\psi_{i,f}$ are the electron wavefunctions in the initial and final states, respectively. The summation in Eq. (11) is extended over all final electron states.

The condition for the Born approximation to be valid for the description of a collision between two particles with charges Z_1e and Z_2e, projectile and target, respectively, is given by the inequality

$$Z_1Z_2e^2 \ll \hbar v_1 \qquad (12)$$

with v_1 being the relative velocity of the particles.[27] Nonrelativistic Coulomb wave functions have been used in Eq. (11) for the calculation of total ionization cross sections for various electron shells.[28,21] It appears that bombarding-energy dependences of the total Coulomb ionization cross sections are qualitatively described by PWBA calculations. This is well supported by experiments on K-shell Coulomb ionizations.[29] However, for K-shell ionizations, in particular, the agreement between experiment and nonrelativistic PWBA calculations is poor in the following two cases: (i) for very heavy target atoms, and (ii) at low projectile energies. For the heavy target atoms the PWBA Coulomb ionization cross sections are too small. Jamnik and Zupancic[30] have repeated the PWBA K-shell calculations with relativistic wave functions for the electrons. The relativistic increase of the electron density near the origin gives rise to an enlargement of the ionization cross sections, improving the agreement with experiment considerably. In the low-energy region, the inequality in Eq. (12) is no longer fulfilled. This manifests itself in PWBA cross sections that are too large compared to the experimental values. This fact was the impetus for the development of the semiclassical approximation model for atomic Coulomb excitation.

3.1.2. Binary Encounter Approximation (BEA)

Fairly recently Garcia[31] introduced the binary encounter approximation model (BEA) for the treatment of inner-shell ionizations by heavy charged particles based on the work of Gryzinski.[32] In this model the ionization process is considered as a classical impact between the projectile and a free target electron. The role ascribed to the rest of the atom is simply to provide the electron under consideration with a velocity distribution in its initial state. It should be noted that the BEA model may be looked upon as an example of the impulse approximation[33]; hence an agreement with the PWBA model at higher projectile energies is not surprising. From this rather simple but highly applicable model a scaling law is obtainable.[31] This law permits direct scaling of the total ionization cross sections for the respective inner electron shells for all target charges according to the binding energy of the electron shell in question. For the K-shell ionization the scaling law may be written in the following form:

$$E_K^2 \sigma_K = g(E_1/E_K) \qquad (13)$$

with g a general function to a good approximation, and the quantity E_K denoting the K-shell binding energy. For protons inducing K-shell ionization, in particular, the BEA model has yielded cross sections in very good agreement with experiment. At lower bombarding energies the agreement with measurement is not as good. However, approximate corrections for the nuclear repulsion of the projectile improves the agreement.[31]

Attempts have been made to improve the BEA model. In the original version of the model, relativistic effects on the velocity distribution and the mass of the target electron were not included. However, they have recently been considered by Hansen,[34] using rather complicated computations. Furthermore, McGuire[35] and McGuire and Omidvar[36] have established an impact parameter description of atomic K-shell ionization cross sections within the frame of the BEA picture, presumably inspired by the semiclassical approximation picture of Bang and Hansteen.[22] The method appears to result in relatively simple computations and rather satisfactory agreement with the few available experimental data. However, as pointed out by McGuire,[35] conceptually this extension of the BEA model does not seem to be entirely indisputable. The uncertainty principle does not appear to be satisfied, and also the applied identification procedure for introducing the projectile path, and thus the impact parameter, might seem somewhat artificial. Further work will show whether this most recent extension of the BEA model is well founded and proves to be fruitful.

3.1.3. Semiclassical Approximation (SCA)

The simplicity of the approach of the semiclassical approximation model for atomic Coulomb excitation should be stressed. The motion of the impinging particle in the field of the target nucleus is treated classically, whereas the transition of the inner-shell electron to the continuum is studied quantum mechanically.

The necessary and sufficient condition for a classical treatment of the incoming, ionizing particle is[27b]

$$2Z_1Z_2e^2 \gg \hbar v_1 \tag{14}$$

where, as above, the indices 1 and 2 refer to the projectile and target nucleus, respectively. Provided the condition in Eq. (14) is satisfied, the SCA theory permits calculations of differential as well as total Coulomb ionization cross sections.[22,23]

The treatment of the atomic Coulomb excitaton in the SCA formulation has hitherto been based on first-order time-dependent perturbation theory in impact parameter form and using unperturbed single-electron wave functions. The Coulomb interaction between the bound inner-shell electron and the bare projectile nucleus is used as the perturbing potential

$$V = \frac{Z_1 e^2}{|\mathbf{r} - \mathbf{R}(t)|} \tag{15}$$

with $\mathbf{R}(t)$ denoting the time-dependent position vector of the electron. By the introduction of hyperbolic paths in Eq. (15) it may be shown that the deflection of the bombarding particle in the Coulomb field of the target nucleus plays an important part in determining the magnitude of the ionization cross sections for low projectile energies.

One further advantage of the SCA model is that for high projectile energies, i.e., projectile paths degenerated into straight lines, the SCA model yields differential Coulomb ionization cross sections $d\sigma/dE_f$ exactly equivalent to the corresponding PWBA expression.[22,37]

The computational difficulties connected with the application of the complete SCA model are very large. The straight-line SCA modification is considerably more manageable and is, at present, in frequent use.[38]

3.1.4. Perturbed Stationary State Approximation

Very recently a formal framework for the theory of atomic inner-shell Coulomb ionizations by heavy charged particles has been given.[39–41] From standard perturbed stationary state theory[42] formulas containing perturbed atomic wave functions have been developed. Atomic wave functions at the distance of closest approach of the projectile are exploited. More specifically, the eigenfunctions used are those of the target atom perturbed by the projectile point charge at rest at the distance of closest approach from the target nucleus. In slow collisions, the approximation of Brandt and co-workers includes two effects not contained in the PWBA. They were originally incorporated into the theory a decade ago by Brandt et al.[43] These effects are: (1) an increase in the binding energy felt by the electron to be ejected ("binding effect") in the presence of the moving projectile inside the electron shell in question; (2) the Coulomb deflection of the projectile in the field of the target nucleus, this effect being incorporated in an approximate manner as inspired by the semiclassical model. This approximation has led to close agreement between theory and experiment.[39,41]

For a more detailed discussion of the theories of Coulomb ionization we refer to Hansteen[44] and Madison et al.[24] For a comparison with experimental results we refer to Chapter 17 of this book and references therein.

3.2. Molecular Excitation

In the course of the discussion of the inner-shell vacancy production mechanisms, we always have to bear in mind that the various methods discussed here are "valid" only for certain regions of Z_1/Z_2 and v_1/u as shown in Figure 4. The Coulomb ionization processes mainly occur in the lower part of this figure. When it became apparent that, for $Z_1 \approx Z_2$ and $v_1 \ll u$, the observed cross sections for inner-shell ionization (at least for small Z) were many orders of magnitude larger than predicted by any approximation discussed so far, Fano and Lichten[45] and Lichten[46] proposed that electron promotion via crossing molecular orbitals (MO) was the reason. This region is shown in Figure 4 (upper left). In this molecular orbital model, vacancies in the higher orbitals of the diatomic quasi-molecule can be filled by electrons from lower orbitals (electron promotion) giving rise to a finite probability that, after collision, the separated atoms have inner-shell vacancies. Lichten[46] and also later Barat and Lichten[47] discuss this process in terms of diabatic correlation diagrams. In Figure 5 we give the best known example of such a diabatic one-electron correlation diagram: the symmetric Ar–Ar collision[46] where, at large internuclear distances, the energy levels are those of Ar, whereas at zero internuclear distance they are those of the united Kr atom. The large decrease in the binding energy of the two $4f_\sigma$ electrons originating from the 2p shell of the separated atoms is the most dominant feature. Saris[48] and Saris and Onderdelinden[49] measured the L-shell, x-ray production cross

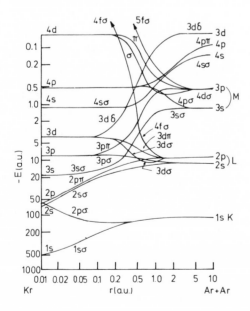

Figure 5. Nonrelativistic correlation diagram of the symmetrical collision Ar plus Ar.

sections in Ne and Ar gas targets for various projectiles from H^+ to Ti^+ at bombarding energies below 130 keV. These data clearly show that heavy ions produce x-rays at a rate that is orders of magnitude larger than that predicted by direct Coulomb ionization at these low energies. Similar results were found for K–vacancy production in many low-Z solid targets.[50-51] The enhanced cross sections and the existence of an apparent threshold (connected with the onset of the $4f_\sigma$ promotion) for x-ray production lend support to the MO model of inner-shell ionization.

Because of the large amount of experimental data and their (at least partial) description within this MO model we refer for details to Chapter 17 of this book as well as to several reviews.[52-55]

Because of the complexity of the colliding atoms or ions with their large numbers of electronic levels and transitions involved, an *ab initio* theoretical treatment that is tractable for large systems has not been given so far. Most calculations are done in a two- (few-) level approximation using either an adiabatic or diabatic representation.[55-57] For systems with a few electrons only, and thus very small Z, a full molecular treatment is possible but is out of practical reach for all other collisions. Often the exact number of electrons in the real system is virtually unknown over the full collision path. Evidently, K excitation is the most manageable problem for a quantitative analysis within a one-electron MO diagram model, since the $2p_\sigma$–$2p_\pi$ level crossing is isolated and the states that couple, by rotational interaction, are as near to hydrogenic as one can get. This process has been treated first by Briggs and Macek[58] and Astner *et al.*[59] using a scaled D_2^+ model. Not only do the experimental total cross sections agree well with the theoretical ones near excitation threshold, e.g., up to 300 keV Ne^+–Ne, but so also does the characteristic doubly peaked shape of the cross section as function of the impact parameter,[60] providing a clear distinction from the Coulomb excitation mechanism. But even this K-shell ionization, which is the relatively simplest problem in this connection, is not fully understood. In the schematic correlation Figure 6, we see all possible electronic transitions (a)–(e) that can leave K vacancies in one of the collision partners after the collision.[61] As just discussed, the $2p_\sigma$–$2p_\pi$ electron promotion process (a) via rotational coupling is the most powerful process as long as there are vacancies in the $2p(H)$ state. Generally, this is no longer true for $Z_1, Z_2 > 10$ although in solids such vacancies can be produced in multiple collision processes (see Refs. 10, 24, 26). Also, two-step processes, with long-range radial coupling of the $2p_\pi$ state to suitable vacant atomic orbitals at the beginning of the collision, can produce the vacancies needed to bring process (a) into play. Processes (b) and (c) are the direct excitation processes of the $2p_\sigma$ and the $1s_\sigma$ electrons, respectively, to the continuum and to vacant bound MO's. Also process (e) followed by (a) leads to a $2p_\sigma$ electron hole.

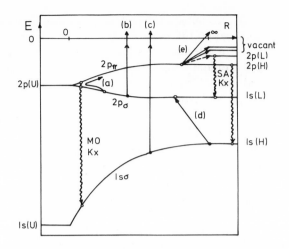

Figure 6. Schematic molecular orbital diagram of the
innermost levels for an assymetric collision. H, L, and
U refer to the higher-Z, lower-Z, and united atom,
respectively; (a) shows the $2p_\pi$–$2p_\sigma$ level coupling,
(b), (c), and (e) are direct ionization processes into
vacant bound levels or the continuum, and (d) stands
for the vacancy-sharing process of the two lowest
levels.

Once formed, a $2p_\sigma$ vacancy can be shared between the $1s(L)$ and
$1s(H)$ states. If w is the branching ratio, the cross sections for K-vacancy
production in the lower-Z and higher-Z collision partners are given
according to Meyerhof[62] by

$$\sigma_K(L) = (1 - w)\sigma(2p_\sigma)$$

$$\sigma_K(H) = w\sigma(2p_\sigma) + \sigma(1s_\sigma)$$

where in the absence of direct $2p_\pi$ vacancies as well as multiple collision
effects

$$\sigma(2p_\sigma) = \sigma_{(e)-(a)} + \sigma_{(b)}$$

$$\sigma(1s_\sigma) = \sigma_{(c)}$$

The $2p_\sigma \to 1s(H)$ vacancy transfer probability w has been computed[62]
using a schematic radial coupling process first proposed by Nikitin and also
examined by Demkov.* Because of the absence of good quantitative
single-mechanism cross sections, Meyerhof et al.[62] used scaling laws and
thus were able to fit the cross sections for the K vacancy productions quite

---

* For a review of this subject see Ref. 63.

well. This model explains the steep increase in the cross section for the projectile in the vicinity of the near-symmetric collision.

These semitheoretical considerations are far from the point where they could be generalized for all parts of the periodic system where different effects may be more important. It is worth mentioning here that, in very heavy systems, strong relativistic effects come into play and even change the structure of the correlation diagrams so that all scaling procedures fitted to low-Z results become very questionable.

In summary it appears that in special regions inner-shell ionization phenomena are well understood. For $Z_1 \approx Z_2$ and $v_1 \ll u$, the molecular promotion model gives a satisfactory and often a quantitative account of the excitation process. Particularly in medium-Z collision systems, the one-electron $2p_\sigma$–$2p_\pi$ coupling model in combination with the vacancy-sharing concept describes K excitation quite well. For $Z_1 \ll Z_2$ and $v_1 \approx u$, close agreement between theory (e.g., the SCA) and experiment can be obtained, and binding energy effects seem to be fairly well under control. Excitation of higher than K shells still gives some problems, probably due to uncertainties in the interatomic potential and the wave functions. For $v_1 > u$, little experimental data exist since the ion energies involved in inner-shell excitation are quite high in this relative velocity scheme. For increasing Z_1, a simple Z_1^2 scaling soon becomes insufficient, and other effects, such as charge exchange, grow in (as yet not quantified) importance.

The region between Coulomb excitation and molecular excitation is still a no-man's land. Experiments are scarce and a satisfactory theory does not exist. It is safe to predict that interesting things are going to happen here.

4. United Atom Phenomena and Related Processes

4.1. Molecular Orbital (MO)X-Rays; REC

A most interesting phenomenon is observed during heavy ion collisions. If a vacancy in an inner molecular level decays during the collision a quasimolecular noncharacteristic x-ray or MO x-ray can be observed. This process is schematically introduced in Figure 6, where a $2p_\sigma$–$1s_\sigma$ (usually called MO K x-ray) transition is shown. A very rough estimate of the probability of such a process is given by t_1/τ_K, where t_1 is the collision time and τ_K the K-vacancy lifetime. For 30-MeV Br–Br this ratio is of the order of $10^{-16}\,\text{sec}/10^{-13}\,\text{sec} \approx 10^{-3}$. For L and M vacancy decay this number can be of the order of 10^{-1}.

The pioneering experiment in this field was made in 1934 by Coates,[64] who measured the x-ray production cross section for several targets after the bombardment with 2.4-MeV mercury ions. To explain his observed large cross sections he proposed that the ionization mechanism is one of the time-varying molecular interactions rather than a direct Coulomb interaction. He also observed that some x-rays were not characteristic of either the beam or the target. These two important observations were, unfortunately, ignored for some 30 years.

The first experimental evidence and interpretation for MO L x-rays was found by Saris $et\ al.$[65] in Ar–Ar collisions in 1971 without knowledge of the experiments of Coates. Molecular orbital M x-rays were found by Mokler $et\ al.$[66] in I collisions with Au, Th, and U, and MO K x-rays were first detected by McDonald $et\ al.$[67] in C–C collisions. Meanwhile many other groups have reported the detection of several MO x-rays. A summary may be found in Refs. 68 and 69.

The observed spectra may be divided into two groups. The K MO x-rays show a continuum spectrum, whereas the M (and some L) MO spectra have a definite peak structure or at least shoulder behavior.

If a vacancy is present in a molecular level the transition can take place at any value of the internuclear distance R with the probability proportional to the time spent around a distance R and proportional to the multipole transition probability integrated over all possible Rutherford trajectories. For the K x-ray electronic transition the transition energy is a continuous function of the internuclear distance R. Hence the MO K x-ray spectrum is a continuum extending above the united atom limit because the collision is a dynamic process connected to the lifetime of the vacancy, the finite collision time, and the time dependence of the MO energy levels. Much effort has been put into the interpretation of these spectra, especially by the group of Greiner $et\ al.$,[70] as well as Meyerhof.[71]

A quantity that is discussed and measured nowadays by many groups is the anisotropy of the emitted radiation. This is a quantity where details of the sublevel occupation probabilities as well as the behavior of the wave functions and their variations with time at small internuclear distances come into effect. Details of this interesting field may be found in Refs. 69, 72, and 73.

A very simple quasistatic interpretation for MO M x-rays for the colliding I–Au system is given by Fricke $et\ al.$[74] They have calculated in a self-consistent way the correlation diagram of this system making use of the relativistic Hartree–Fock–Slater method so that all relativistic effects, spin–orbit splitting as well as the screening, are fully taken into account. From this correlation diagram they have extracted all possible transitions within the energy range of 5–11 keV, where the experimental MO M peak occurs. After exact integration over all possible Rutherford trajectories using dipole transition probabilities and a collision broadening of 0.5 keV, they obtain the spectrum (b) in Figure 7, where (a) is the observed spec-

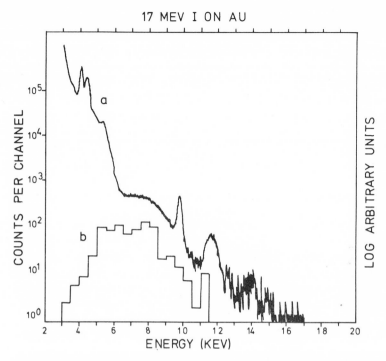

Figure 7. (a) Experimental spectrum of 17-MeV I on Au corrected for absorber; (b) calculated spectrum using a realistic Dirac–Fock–Slater correlation diagram[74] which reproduces the MO peak at 8 keV rather well.

trum. Using this method the correct structure and behavior is fairly well reproduced. A similar interpretation[75] for the L MO x-rays of the system I–Ag showed equally good results.

Since coincidence spectra as functions of the scattering angle of the colliding atom have begun to be available,[76] the experimental observations and theoretical interpretations of the molecular orbital x-rays have come into a second, even more interesting phase. The impression after the first five years of investigation of these noncharacteristic x-rays is that a basic interpretation can be given but that most details need further investigation.

During the course of these investigations another effect, long expected from astrophysicists, has been found by Schnopper *et al.*[77]: radiative electron capture (REC). This is the radiative capture into vacant projectile states of loosely bound target electrons. In such a transition, the normal transition energy of the bound atomic state is added to the energy of motion of the electron relative to the projectile. This makes the peak shift with varying projectile energy and also with the scalar product of the momentum of the projectile and the electron wave function, thus making it possible to extract important information on the momentum distribution of the electron from the detailed investigation of the shape of the REC peak.[78] More details, as well as information, on effects like radiative Auger effect (RAE), radiative electron rearrangement (RER) or radiative ionization (RI) may be found by K.-H. Schartner in Chapter 17 of this work.

For the discussion of the electron bremsstrahlung and nuclear bremsstrahlung, which may also become important in the analysis of the spectra, we can only refer to Refs. 71, 79, and 80.

4.2. Positron Emission in U–U Collision

If we go back to Figure 3, where we have considered the atomic behavior of atoms with $Z > 100$, we can see that at or about $Z \approx 173$ the $1s$ level reaches the negative continuum of electrons. At that point this discussion seemed to become purely academic.

Now in view of the MO x-rays and possible positron emission in $U–U$ collisions these values become most important even for the experimentalists. The values in Figure 3 are then united atom collision limits of very heavy atoms and are usually drawn on the left-hand side of the correlation diagrams (see Figure 5).

In the MO M x-rays observed by Mokler et al.,[66] information from quasiatoms with $Z_1 + Z_2$ up to 145 can already be obtained. Proceeding to even heavier colliding systems, at some point a limit is reached where the lowest level dives into the negative continuum. Müller et al.[70] have examined the physics of this process and found that nothing observable happens as long as a full $1s$ shell reaches the continuum. But if the $1s$ shell carries a vacancy down into the negative continuum it is possible that an electron from the negative continuum occupies this level thus leaving a hole in the continuum. This hole is a positron, which will be emitted immediately and may be observed. Greiner and co-workers[81] calculated the spectrum of the emitted positrons as a function of the projectile and target Z as well as their incident energy. Because a positron is easy to detect this experiment using the U−U, or even better the U−Cf, collision seems to be the easiest way to get information from a region where QED effects are most important.

In various respects the dynamics of the collision process not discussed so far play a most dominant role in the discussion of the positron experiment.* First, the dynamical broadening of the energy levels may already

---

* The experiment is well underway; first results were already presented at GSI Seminars 1977; see Ref. 82.

create a electron–positron pair before the adiabatic level reaches the negative continuum. Second, it is shown by first simplied coupled-channel calculations[83,84] that for small impact parameters there is a chance on the order of a few percent of direct vacancy formation in the $1s$ level during the collision. Third, there is a large probability of direct (Coulomb) formation[85] of electron–positron pairs in the time-varying field of the quasisuperheavy molecule as well as a secondary product following Coulomb ionization[86] of the nucleus.

In conclusion one may predict that the whole field of inner shells briefly discussed in this chapter will be one of the most interesting fields in the future of modern atomic physics.

Acknowledgments

To Professor J. M. Hansteen at Bergen, I am highly indebted for his contribution of the Coulomb ionization processes. I gratefully acknowledge very stimulating comments and discussions with many colleagues engaged in the physics of inner shells as well as partial financial support from Gesellschaft für Schwerionenforschung (GSI).

References

1. D. R. Hartree, *Proc. Cambridge Phil. Soc.* **24**, 89 (1927); V. Fock, *Z. Phys.* **61**, 126 (1930); **62**, 795 (1930).

2. P. A. M. Dirac, *Proc. R. Soc. London* **117**, 610 (1928); **118**, 351 (1928).

3. I. P. Grant, *Adv. Phys.* **19**, 747 (1970); F. P. Larkins, in *Atomic Inner Shell Processes*, Ed. B. Craseman, p. 377, Academic Press, New York (1975); I. Lindgren and A. Rosén, *Case Stud. At. Phys.* **4**, 93 (1974).

4. R. D. Cowan, *Phys. Rev.* **163**, 54 (1967).

5. G. Breit, *Phys. Rev.* **35**, 1447 (1930); **37**, 51 (1931); Y. K. Kim, *Phys. Rev.* **154**, 17 (1967); J.-P. Desclaux, Thèse, Université de Paris (1971).

6. J. Schwinger, *Phys. Rev.* **75**, 651 (1949); B. Fricke, *Z. Phys.* **218**, 495 (1969).

7. G. E. Brown, J. S. Langer, and G. W. Schaefer, *Proc. R. Soc.* **A251**, 92 (1959); G. E. Brown and D. F. Mayers, *Proc. R. Soc.* **A251**, 105 (1959).

8. A. M. Desiderio and W. R. Johnson, *Phys. Rev. A* **3**, 1267 (1971).

9. P. J. Mohr, *Ann. Phys. N.Y.* **88**, 26 (1974); **88**, 52 (1974).

10. J. D. Garcia, R. J. Fortner, and T. M. Kavanagh, *Rev. Mod. Phys.* **45**, 111 (1973).

11. N. Stolterfoht, in *Proceedings of the Eighth International Conference on the Physics of Electronic and Atomic Collisions*, Belgrade, Invited papers, Eds. B. C. Cobic and M. V. Kurepa, p. 117 (1974); papers in *Electronic and Atomic Collisions*, Eds. J. S. Risley and R. Geballe, University of Wash. Press, Seattle (1975).

12. P. Ziem, R. Bruch, and N. Stolterfoht, *J. Phys.* **B8**, L480 (1975); N. Stolterfoht and U. Leithäuser, *Phys. Rev. Lett.* **36**, 186 (1976); K. O. Groeneveld, G. Nolte, S. Schumann, and K. D. Sevier, *Phys. Lett.* **56A**, 29 (1976); R. Bruch, G. Paul, J. Andrä, and B. Fricke, *Phys. Lett.* **53A**, 293 (1975).
13. S. Schumann, K.-O. Groeneveld, and K. D. Sevier, *Abstracts of Contributed Papers of the Fifth International Conference on Atomic Physics,* Eds. R. Marrus and M. Prior, p. 191, Berkeley, July 1976.
14. J. Raefelski, B. Fulcher, and W. Greiner, *Nuovo. Comento* **13B**, 135 (1973).
15. B. Fricke, J.-P. Desclaux, and J. T. Waber, *Phys. Rev. Lett.* **28**, 714 (1972).
16. F. T. Porter and M. S. Freedman, *Phys. Rev. Lett.* **27**, 293 (1971).
17. G. L. Borchert, *Z. Naturforsch.* **31a**, 102 (1974); G. L. Borchert and B. Fricke, *Abstracts of Contributed Papers to the Fifth International Conference on Atomic Physics,* Ed. R. Marrus and M. Prior, p. 321, Berkeley, July 1976.
18. B. Fricke and G. Soff, Gesellschaft für Schwerionenforschung report No. GSI-T1-74 (1974).
19. B. Fricke and J. T. Webaber, *Actin. Rev.* **1**, 433 (1971); B. Fricke, *Struct. Bonding* (Berlin) **21**, 89 (1975); B. Fricke, *Naturwissenschaft* **63**, 162 (1976).
20. B. Müller, J. Rafelski, and W. Greiner, *Z. Phys.* **257**, 82 (1972); **257**, 163 (1972).
21. E. Merzbacher and H. W. Lewis, in *Handbuch der Physik*, Vol. **34**, p. 166, Ed. S. Flügge, Springer-Verlag, Heidelberg (1958).
22. J. Bang and J. M. Hansteen, *Mat. Fys. Medd. Dan. Vid. Selsk.* **31**, No. 13 (1959).
23. J. M. Hansteen and O. P. Mosebekk, *Nucl. Phys.* **A201**, 541 (1973).
24. D. H. Madison and E. Merzbacher, in *Atomic Inner Shell Processes*, Ed. B. Craseman, p. 1, Academic Press, New York (1975).
25. Q. C. Kessel and B. Fastrup, *Case Stud. At. Phys.* **3**, 137 (1973).
26. H. A. Bethe, *Ann. Phys. (Leipzig)*, **5**, 325 (1930).
27. (a) E. J. Williams, *Rev. Mod. Phys.* **17**, 217 (1945); (b) N. Bohr, *Kl. Dan. Vidensk. Selsk. Mat.-Fys. Medd.* **18**, No. 8 (1948).
28. W. Henneberg, *Z. Phys.* **86**, 592 (1933); G. S. Khandelwal, B. H. Choi, and E. Merzbacher, *At. Data* **1**, 103 (1969); B. H. Choi, E. Merzbacher, and G. S. Khandelwal, *At. Data* **5**, 291 (1973).
29. C. H. Rutledge and R. L. Watson, *At. Data Nucl. Data Tables* **12**, 195 (1973).
30. D. Jamnik and C. Zupancic, *Kl. Dan. Vidensk. Selsk. Mat.-Fys. Medd.* **31**, No. 2 (1957).
31. J. D. Garcia, *Phys. Rev.* **A 1**, 280 1402 (1970).
32. M. Gryzinski, *Phys. Rev.* **A138**, 336 (1965).
33. L. Vriens, *Case Stud. At. Phys.* **1**, 335 (1969).
34. J. S. Hansen, *Phys. Rev. A* **8**, 822 (1973).
35. J. H. McGuire, *Phys. Rev. A* **9**, 286 (1974).
36. J. H. McGuire and K. Omidvar, *Phys. Rev. A* **10**, 182 (1974).
37. H. A. Bethe and R. W. Jackiw, in *Intermediate Quantum Mechanics*, 2nd ed., Benjamin, New York (1968).
38. J. M. Hansteen, O. M. Johnson, and L. Kocbach, *J. Phys.* **B7**, L271 (1974), and references quoted therein.
39. G. Basbas, W. Brandt, and R. Laubert, *Phys. Rev. A* **7**, 983 (1973).
40. G. Basbas, W. Brandt, and R. H. Ritchi, *Phys. Rev. A* **7**, 1971 (1973).
41. W. Brandt and G. Lapicki, *Phys. Rev. A* **10**, 474 (1974).
42. N. F. Mott and H. S. W. Massey, *The Theory of Atomic Collisions*, 3rd ed., Oxford University Press, London (1965).
43. W. Brandt, R. Laubert, and I. Sellin, *Phys. Rev.* **151**, 56 (1966).
44. J. M. Hansteen, *Adv. At. Mol. Phys.* **11**, 299 (1975).
45. U. Fano and W. Lichten, *Phys. Rev. Lett.* **14**, 627 (1965).
46. W. Lichten, *Phys. Rev.* **164**, 131 (1967).

47. M. Barat and W. Lichten, *Phys. Rev. A* **6**, 211 (1972).
48. F. W. Saris, *Physica* **52**, 290 (1971).
49. F. W. Saris and D. Onderdelinden, *Physica* **49**, 441 (1970).
50. W. Brandt and R. Laubert, *Phys. Rev. Lett.* **24**, 1037 (1970).
51. R. C. Der, R. J. Fortner, T. M. Kavanagh, and J. M. Khan, *Phys. Rev. A* **4**, 556 (1971); M. Terasawa, T. Tamura, and H. Kamada *Abstracts of the Proceedings of the Seventh International Conference on the Physics of Electronic and Atomic Collisions*, Amsterdam, p. 410, North-Holland, Amsterdam (1970).
52. *Atomic Physics* 4, Eds. G. zu Putlitz, E. W. Weber, and A. Winnacker, Plenum Press, New York (1975).
53. W. Lichten, p. 249 of Ref. 52.
54. P. Richard, in *Atomic Inner Shell Processes*, Ed. B. Craseman, p. 73, Academic Press, New York (1975).
55. J. S. Briggs, Invited Talk, *Ninth International Conference on the Physics of Electronic and Atomic Collisions*, University of Washington, Seattle, July 1975.
56. F. T. Smith, *Phys. Rev.* **179**, 111 (1969); F. T. O'Malley, *Adv. At. Mol. Phys.* **7**, 223 (1971).
57. V. Sidis and H. Lefebvre-Brion, *J. Phys.* **B4**, 1040 (1971).
58. J. S. Briggs and J. Macek, *J. Phys.* **B5**, 579 (1972).
59. G. Astner, J. D. Garcia, and L. Liljeby, *J. Phys.* **B8**, L314 (1975).
60. G. Sackmann, H.-O. Lutz, and J. S. Briggs, *Phys. Rev. Lett.* **32**, 805 (1974); KFA report No. Jül-1154-KP (1975).
61. W. E. Meyerhof, "International Seminar on Ion-Atomic Collisions," Gif-sur-Yvette, France, July 1973; W. E. Meyerhof, *Comm. Atom. Mol. Phys.* **5**, 33 (1975).
62. W. E. Meyerhof, *Phys. Rev. Lett.* **31**, 1341 (1973); J. S. Briggs and K. Tjaulberg, *J. Phys.* **B8**, 1909 (1975).
63. E. E. Nikitin, *Adv. Quantum Chem.* **5**, 135 (1970).
64. W. M. Coates, *Phys. Rev.* **46**, 542 (1934).
65. F. W. Saris, W. F. van der Weg, H. Tavara, and R. Laubert, *Phys. Rev. Lett.* **28**, 717 (1972).
66. P. H. Mokler, H.-J. Stein, and P. Armbruster, *Phys. Rev. Lett.* **29**, 827 (1972).
67. J. R. McDonald, M. D. Brown, and T. Chiao, *Phys. Rev. Lett.* **30**, 471 (1973).
68. F. W. Saris and F. J. de Heer, p. 287 of Ref. 52; F. W. Saris and Th. Hoogkamer, in *Atomic Physics* 5, Eds. R. Marrus, M. Prior, and H. Shugort, Plenum Press, New York (1977).
69. P. H. Mokler, S. Hagmann, P. Armbruster, G. Kraft, H.-J. Stein, K. Rashid, and B. Fricke, p. 301 of Ref. 52.
70. B. Müller, R. K. Smith, and W. Greiner, p. 209 of Ref. 52, and references therein; B. Müller, in *Invited Lectures Review Papers and Progress Reports of the Ninth International Conference on the Physics of Electronic and Atomic Collisions*, Eds. J. S. Risley and R. Geballe, p. 481, University of Washington Press, Seattle (1976).
71. W. E. Meyerhof, T. K. Saylor, S. M. Lazarus, A. Little, B. B. Triplett, L. F. Chase, and R. Anholt, *Phys. Rev. Lett.* **32**, 1279 (1974).
72. *Abstracts of Contributed Papers, Second International Conference Inner Shell Ionization Phenomena*, Eds. W. Melhorn and R. Brenn, Freiburg University Press, Freiburg, March 1976.
73. Ch. Stoller, p. 1 of Ref. 72, and references therein.
74. B. Fricke, T. Morović, W.-D. Sepp, A. Rosén, and D. E. Ellis, *Phys. Lett.* **59A**, 375 (1976).
75. T. Morović, B. Fricke, W.-D. Sepp, A. Rosén, and D. E. Ellis, *Phys. Lett.*, **63A**, 12 (1977).
76. I. Tseruja, H. Schmidt-Böcking, R. Schulé, R. Schuch, H. Bethge, and H.-J. Specht, p. 9 of Ref. 72.

77. H. W. Schnopper, H.-D. Betz, J.-P. Delvaille, K. Kalata, A. R. Sohval, K. W. Jones, and H. E. Wegner, *Phys. Rev. Lett.* **29**, 898 (1972); P. Kienle, M. Kleber, B. Povh, R. M. Diamond, F. S. Stephens, E. Grosse, M. R. Maier, and D. Proetel, *Phys. Rev. Lett.* **31**, 1099 (1973).

78. M. Kleber and D. H. Jakubassa, *Nucl. Phys.* **A252**, 152 (1975), and references therein.

79. J. S. Greenberg, C. K. Davis, and P. Vincent, *Phys. Rev. Lett.* **32**, 1215 (1974).

80. D. H. Jakubassa and M. Kleber, *Z. Phys.* **A273**, 29 (1975).

81. H. Peitz, B. Müller, J. Rafelski, and W. Greiner, *Lett. Nuovo* Cimento **8**, 37 (1973); K. Smith, H. Peitz, B. Müller, and W. Greiner, *Phys. Rev. Lett.* **32**, 554 (1974); and Ref. 70.

82. R. Backe *et al.*, GSI Seminars 1977 (to be published).

83. D.-H. Jakubassa, *Phys. Lett.* **58A**, 163 (1976).

84. W. Betz, G. Soff, B. Müller, and W. Greiner, *Phys. Rev. Lett.* **37**, 1046 (1976).

85. G. Soff, J. Reinhardt, B. Müller, and W. Greiner, *Phys. Rev. Lett.* **38**, 592 (1977).

86. V. Oberacker, G. Soff, and W. Greiner, *Phys. Rev. Lett.* **36**, 1024 (1976).

6

Interatomic Potentials for Collisions of Excited Atoms

W. E. BAYLIS

1. Background

1.1. Key Role of Interatomic Potentials

Interatomic potentials serve an important function in the organization of data about collisions of excited atoms. They determine how atoms interact and thus represent the ultimate physical quantity responsible for the sizes of elastic and inelastic cross sections, reaction rates, and the profiles of pressure-broadened spectral lines.

Most measurements and calculations of interatomic potentials have concentrated on the interaction of ground-state atoms. However, for collisions of excited-state atoms, either both ground- and excited-state potentials are of roughly equal importance (e.g., in line broadening or quenching) or the excited-state potentials are dominant. Consequently, in this chapter *excited-state* interatomic potentials are emphasized. I try briefly to describe how they are (or can be) measured and calculated, and what their relation is to collision processes of excited-state atoms.

When an excited-state atom collides with a ground-state one, a large variety of processes can occur. Let the colliding pair be $A^*(m)+B$, where m labels degenerate or nearly degenerate sublevels of the excited atom

W. E. BAYLIS • Department of Physics, University of Windsor, Windsor, Ontario, Canada 3P4.

A^*. Most of the possible results, excluding those that involve ionization or charge exchange, can be summarized by the following:

  (i) $A^*(m) + B$ (elastic scattering)
 (ii) $A^*(m') + B$ (depolarization)
(iii) $A^{**}(m') + B$ (intra-atomic excitation transfer
                             or sensitized fluorescence)
                            (interatomic excitation transfer
 (iv) $A + B^*$ or sensitized fluorescence)
  (v) $A + B +$ kinetic energy (quenching)
 (vi) $AB + h\nu$ (associative deexcitation)

where A^{**} is another excited state of A, nondegenerate with A^*. Variants of these processes are possible. As discussed in Chapter 28, for example, depolarization (ii) can take place either as the collisional mixing of Zeeman sublevels or as the relaxation of excited-state coherence as evidenced, say, by the broadening of a Hanle signal; and (as discussed in Chapter 29) excitation transfer (iii) and (iv) can occur with the transfer of coherence or polarization. Furthermore, intra-atomic transfer (iii) can occur between fine-structure levels of a single multiplet [fine-structure transitions, referred to—together with depolarization (ii)—as intramultiplet mixing] or between different multiplets (intermultiplet mixing), and interatomic transfer (iv) can take place with various final states of both A and B^*.

Considering the rich variety of collisional data that can be accumulated about a given system, it is indeed fortunate that a relatively simple concept, namely, interatomic potentials, exists to relate and unify the data. The term "interatomic potentials" is used here in a broad sense: It means not only the Born–Oppenheimer energies of the system but also the corresponding adiabatic electronic states. With such interatomic potentials one can, given suitable numerical algorithms for the scattering process, compute all cross sections, rates, and line profiles. Conversely, even though the scattering data are in most cases difficult to "invert" in order to determine interatomic potentials, they do provide useful constraints for the potentials involved.

1.2. Some Practical Uses

Interatomic potentials and related collision data are important for understanding and predicting processes involving energy transfer through gases. Such processes include the cooling of dilute interstellar gas before its condensation and the possible formation of stars, as well as the energy balance in the upper atmosphere of, say, the earth, Jupiter, or the sun. In fact over a wide range of densities, whenever binary collisions play a significant role in the nonconvective transport of energy, we need collision rates in order to predict or—in some cases—even understand what happens. Frequently seemingly minor processes can exert a major influence on the properties of a system. For example, the fine-structure transition in O may strongly affect the temperature profile of the atmosphere of the earth,[1] and a small change in stellar opacity may substantially alter the density at the core of a star and hence its entire evolution.[2]

The search for powerful new lasers, largely for fusion research and military applications, can also make good use of interatomic potentials and collision rates. In one type of promising laser, the "excimer" or "exciplex" laser,[3,4] the upper level corresponds to the attractive well of an excited state, and the lower level to the repulsive wall of the ground state. Such a system, in which it is relatively simple to achieve population inversion, should also be tunable. A further advantage can be gained if the upper level can be made metastable at large internuclear separations R and radiatively unstable at small R. Interatomic potentials, in addition to their use in determining the possible laser transitions, can also be used to determine competing loss mechanisms arising from collisions or the reabsorption of resonance radiation.

Another potentially important application of excited-state collisions depends on the high selectivity possible through excitation by light plus the fact that at thermal velocities some reactions occur only if one of the reactants is excited. As long as the product has sufficiently different properties from the reactants, it is possible to perform efficient isotope separation. Tam et al.[5] have discussed the method in terms of collisions of alkalis and H_2. In the ground state, a very small fraction of the alkali atoms form weakly bound van der Waals molecules with H_2, but when the alkali is excited, the alkali hydride forms and precipitates out. Tam et al. have observed the reaction with Rb and Cs and propose reactions of the form

$$Cs^*(7p) + H_2 \rightarrow CsH + H$$

as dominant. [Collisions of $Cs(6s)$ with vibrationally excited H_2 might also contribute.] Isotope separation is achieved by isotopically specific excitation of the alkali. Where appropriate reactions can be found, separation by such reactions should be quite competitive with other optical methods employing two- or more-step ionization and electric fields to sweep out the ions.[6]

1.3. Standard Potential Forms

At sufficiently long range, two neutral atoms always attract one another, whereas at sufficiently small separations the force between them is repulsive. The potential energy $V(R)$ of the atomic pair must therefore contain at least one minimum $-\varepsilon$ at an intermediate internuclear separation R_m:

$$V(R_m) = -\varepsilon, \qquad V'(R_m) = 0 \tag{1}$$

Table 1. Well Parameters of a Few Diatomic Systems in
Their Ground States

Molecule	ε (eV)	R_m (Å)	Reference
CO	11.24	1.128	7
N_2	9.90	1.098	7
H_2	4.476	0.7416	7
Kr_2	1.74×10^{-2}	4.01	8
He_2	9.2×10^{-4}	2.96	9

where we take $V(R) \to 0$ as $R \to \infty$ and $V'(R) \equiv dV(R)/dR$. In the simplest cases, there is a single minimum (this seems usually true of ground states) and the values ε and R_m allow a fair characterization of the interatomic potential.

For ground-state atomic pairs, the values of ε range from 11.24 eV‡ (for CO) to $\sim 10^{-3}$ eV (for He_2) and the values of R_m, from 0.74 Å§ (for H_2) to 4 Å (for Kr_2), as shown in Table 1. Other well parameters frequently encountered are the "collision radius" σ, defined by $V(\sigma) = 0$, and the reduced curvature at the well minimum,

$$\varkappa = \frac{R_m^2}{\varepsilon} V''(R_m)$$

Many applications of interatomic potentials continue to use standard forms. Some of the most common forms for potentials with wells are given in Table 2. The use of any of these forms is of course an approximation, and one should not expect a two- or three-parameter potential determined by one experiment to provide an accurate basis for the description of another. This is especially true of excited states where the actual potentials frequently have more than one minimum and cannot be well described by standard forms.

Of the forms given in Table 2, all except the Morse potential have the long-range behavior $V(R) \sim R^{-m}$ with usually $m = 6$, characteristic of the asymptotic limit of van der Waals potentials (see Section 3.4). The Lennard–Jones LJ (12, 6) potential is the simplest and most often used form and is identical to the more flexible Kihara form with $\gamma = 0$. Note that the Buckingham potential has a maximum, typically near $z \simeq 1/4$, and the "collision radius" σ is given only through a transcendental equation. The

‡ An indication of the wide use of interatomic potentials in different fields is given by the many units in use. For convenience we note the following conversions:
$$1 \text{ eV} = 23.045 \text{ kcal/mole} = 8065.465 \text{ cm}^{-1} = 241.7966 \text{ THz}$$
$$= 1.60219 \times 10^{-12} \text{ erg} = 11604.85 \text{ °K} = 1/27.2107 \text{ a.u.}$$
§ 1 Å = 0.1 nm = $1.8897a_0$ ($1a_0$ = 1 Bohr radius = 1 a.u. of length).

Table 2. Common Potential Forms $V(R)^a$

Name and form	Typical parameter values
1. Lennard-Jones LJ (n, m) $$\frac{V(R)}{\varepsilon} = \frac{(mz^{-n} - nz^{-m})}{(n-m)}$$	$8 \leqslant n \leqslant 20$; $\dfrac{\sigma}{R_m} = \left(\dfrac{m}{n}\right)^{1/(n-m)}$ $m = 6$; $\varkappa = nm$
2. Kihara (n, m, γ) $$\frac{V(R)}{\varepsilon} = \frac{(ms^{-n} - ns^{-m})}{(n-m)}$$	$8 \leqslant n \leqslant 20$; $\dfrac{\sigma}{R_m} = \gamma + (1+\gamma)\left(\dfrac{m}{n}\right)^{1/(n-m)}$ $m = 6$; $\varkappa = \dfrac{nm}{(1-\gamma)^2}$ $-0.2 \leqslant \gamma \leqslant 0.4$
3. Buckingham exp (α, m) $$\frac{V(R)}{\varepsilon} = \frac{m\,e^{-\alpha(z-1)} - \alpha z^{-m}}{(\alpha - m)}$$	$8 \leqslant \alpha \leqslant 20$; $\dfrac{\sigma}{R_m}\left(\dfrac{\alpha}{m}\right)^{1/m} \exp\left[-\dfrac{\alpha}{m}\left(\dfrac{\sigma}{R_m} - 1\right)\right]$ $m = 6$; $\varkappa = \alpha m[1 - 1/(\alpha - m)]$
4. Morse (a) $$\frac{V(R)}{\varepsilon} = e^{-a(z-1)}[e^{-a(z-1)} - 2]$$	$4 \leqslant a \leqslant 8$; $\dfrac{\sigma}{R_m} = 1 - \dfrac{\ln 2}{a}$ $\varkappa = 2a^2$

^{a} The abbreviations $z = R/R_m$ and $s = (z-\gamma)/(1-\gamma)$ are used.

Morse potential is useful in bound-state spectroscopy because the vibrational energies are simply given:

$$E_\nu = -\varepsilon[1 - (\nu + \tfrac{1}{2})/\eta]^2, \qquad \nu = 0, 1, 2, \cdots$$

where $\eta \equiv 2\varepsilon/(\hbar\omega_0)$ and $\omega_0 \equiv [2\varepsilon a^2/(\mu R_m^2)]^{1/2}$, and the corresponding eigenfunctions can be readily expressed in terms of Laguerre polynomials.[10] However, because Morse potentials asymptotically approach zero too rapidly, they tend to contain fewer states, especially near the dissociation limit, than actual potentials of the same ε and R_m.

Many other analytic forms have been used, frequently with more parameters added for greater flexibility or splined onto a van der Waals expansion, $-C_6 R^{-6} - C_8 R^{-8} - \cdots$, for a more accurate asymptotic description. More ambitious constructions and calculations of potentials and the physical interactions underlying them are discussed in Section 3, as are monotonic potential forms for the repulsive part of the potential.

1.4. Scope of This Chapter and a Few References

Let me immediately apologize to authors whose works I have unintentionally slighted. The primary purpose of this chapter is not to present a literature survey, but rather to enable readers to enrich their understanding of the experimental and theoretical determination of interatomic potentials, the physics behind the interactions, and their relation to excited-state collision processes. Where experiments are discussed, apparatus and gadgetry are largely ignored in order to concentrate on the connection between the experimental observables and the theoretical parameters. In discussions of the theory, simpler cases are emphasized in order to convey the essential concepts involved. Literature citations here and elsewhere in the chapter are designed mainly to direct the reader to more detailed treatments of the theories and experiments than permitted here.

Interatomic potentials have been the subject of a recent book by Torrens[11] and two volumes by Goodisman.[12] Both authors restrict their discussions to ground-state potentials, but of course much of this is also applicable to excited states. Torrens emphasizes applications to metallic solids, whereas Goodisman's treatment is more oriented to physical chemistry applications involving binary atomic collisions. Particularly the Goodisman volumes contain extensive references of relevant papers up to 1972.

Bardsley[13] has reviewed pseudopotential calculations of atomic interactions with references up to 1974. He includes some discussion of excited states with emphasis on applications to electron scattering by atoms.

Although somewhat older, the excellent collection edited by Hirschfelder[14] is still a highly useful reference. Both theoretical calculations, particularly of long-range forces, and methods of experimental determination are reviewed.

Scattering theory in the classical-path formulation, especially as applied to excited-state collisions, and its relation to interatomic potentials has recently been the topic of a review by Nikitin.[15] The recent book by Child[16] contains an excellent chapter on the general theory of inelastic collisions as well as useful sections on semiclassical approximations.

2. Measurements

2.1. Beam Measurements in Ground States

Elastic-scattering cross sections, measured as a function of scattering angle and incident energy, have provided detailed information about a large number of ground-state potentials.[17-24] *Classically*, the cross section $d\sigma(\theta)/d\Omega$ is given in terms of a deflection function $\theta(b)$, which gives the scattering angle θ as a function of the impact parameter b:

$$\frac{d\sigma}{d\Omega} = \sum_i b_i \left| \sin \theta \frac{d\theta(b)}{db} \right|^{-1}_{b=b_i} \qquad (2)$$

where the sum is over all values of b such that $\theta = \theta(b_i)$. The classical deflection function can easily be calculated in the center-of-mass ("barycentric") system in terms of the potential $V(r)$ when the conservation law for angular momentum $L = -\mu R^2 \dot{\theta} = \mu v b$ is integrated (μ is the reduced mass and v the relative velocity):

$$\theta = \pi - 2b \int_{R_0}^{\infty} dR \, R^{-2} \left[1 - \frac{b^2}{R^2} - \frac{V(R)}{E} \right]^{-1/2} \qquad (3)$$

For a few simple monotonic potentials analytic cross sections exist[25,26] or numerical[27,28] tables have been given. It is also possible for monotonic potentials to derive "inversion" formulas[18,29] which give the interatomic potentials in terms of quantities such as the deflection function [which, according to Eq. (2), is obtained simply by integrating the measured cross section over angles].

However, the applicability of *classical* scattering theory is restricted to situations in which (1) there is no interference (monotonic deflection function with $0 < \theta \leqslant \pi$) or the interference oscillations are averaged over, (2) the de Broglie wavelength $\lambda = \hbar/\mu v$ is small compared with the distance scale of changes in the potential: $\lambda \ll |V(R)/V'(R)|$, and (3) the angular momentum L involved is large compared to $\hbar/\theta$: $L\theta \gg \hbar$. The classical theory is mainly useful as an approximation at high ($>$keV) energies.

In the *quantum* treatment, the elastic-scattering cross section is given in terms of the scattering amplitude $f(\theta)$ by

$$\frac{d\sigma}{d\Omega} = |f(\theta)|^2 \qquad (4)$$

where $f(\theta)$ is found from the asymptotic form of the scattered wave

$$\psi \sim e^{i\mathbf{k}\cdot\mathbf{R}} + f(\theta)R^{-1} e^{ikR}$$

Here $k = (2\mu E)^{1/2}/\hbar$ is the relative wave vector. In practice, one makes a partial wave expansion

$$\psi = \sum_L (2L+1)R^{-1}\mathcal{R}_L(R)P_L(\cos \theta)$$

and finds the radial wave function $\mathcal{R}_L(R)$ by integrating the Schrödinger equation

$$\mathcal{R}_L''(R)+\left[k^2+\frac{L(L+1)}{R^2}-\frac{2\mu}{\hbar^2}V(R)\right]\mathcal{R}_L(R)=0 \qquad (5)$$

outward from small R. Starting with the boundary condition at small R, namely,

$$\mathcal{R}_L(R)\xrightarrow{R\to 0}R^{L+1}/(2L+1)!$$

one finds the scattering phase shift η_L by matching $\mathcal{R}_L$ at large R to the asymptotic form

$$\mathcal{R}_L(R)\sim A_L k^{1/2}R[\cos\eta_L j_L(kR)-\sin\eta_L n_L(kR)]$$
$$\sim A_L k^{1/2}\sin(kR-\tfrac{1}{2}\pi L+\eta_L) \qquad (6)$$

where j_L and n_L are the spherical Bessel and Neumann functions[30] and A_L is a constant. A comparison of the asymptotic forms of the scattered wave and the partial-wave expansion yields

$$f(\theta)=(2ik)^{-1}\sum_L(2L+1)(e^{2i\eta_L}-1)P_L(\cos\theta) \qquad (7)$$

Although considerable effort has been expended on the formal inversion of scattering data in quantum mechanics,[31–33] practical inversion schemes[18,34–39] at present mainly use classical and semiclassical or JWKB[40] approximations. The classical approximation was discussed above. In the JWKB approximation, the radial wave function at large R has the form

$$\mathcal{R}_L(R)\sim k^{-1/2}\sin\left[\frac{\pi}{4}+\int_{R_0}^{R}dr\,k(r)\right] \qquad (8)$$

where

$$k(R)=\left[k^2-\frac{2\mu}{\hbar^2}V(R)-\frac{L(L+1)}{R^2}\right]^{1/2} \qquad (9)$$

and R_0 is the outermost zero of $k(R)$. When compared to Eq. (6), Eq. (8) gives the JWKB phase shift η_L^{JWKB}:

$$\eta_L^{\text{JWKB}}=\int_{R_0}^{\infty}dr[k(r)-k]-kR_0+\frac{\pi}{2}\left(L+\frac{1}{2}\right) \qquad (10)$$

With the substitution $(L+\tfrac{1}{2})=kb$, the Langer modification[40] [namely, replacing $L(L+1)$ by $(L+\tfrac{1}{2})^2$ in Eq. (9)], and the identity (easily proved by parts integration)

$$\int_b^\infty dr \left[\left(\frac{1-b^2}{r^2} \right)^{1/2} - 1 \right] = b \left(1 - \frac{\pi}{2} \right)$$

we find the alternate form

$$\eta_L^{\text{JWKB}} = \int_{R_0}^\infty dr\, k(r) - k \int_b^\infty dr \left(1 - \frac{b^2}{r^2} \right)^{1/2} \tag{11}$$

Differentiation of Eq. (10) with respect to L results in an important relation between the semiclassical phase shift and the scattering angle:

$$2 \frac{d\eta_L^{\text{JWKB}}}{dL} = \frac{2}{K} \frac{d\eta(b)}{db} = \theta \tag{12}$$

Classically, small deflections θ can be found from the impulse approximation

$$\theta = \frac{1}{\hbar k} \int_{-\infty}^\infty dt \frac{b}{R} \left[-\frac{d}{dR} V(R) \right] = \frac{2}{\hbar k} \frac{d}{db} \int_0^\infty dt\, V(t) \tag{13}$$

where $V(t) \equiv V[R(t)]$, which combined with Eq. (12) yields the "classical" phase shift

$$\eta \simeq \eta^{cl}(b) = -\frac{1}{\hbar} \int_0^\infty dt\, V(t) \tag{14}$$

For example, an interatomic potential of asymptotic form $V(R) \sim C_m R^{-m}$ gives a small angle phase shift

$$\eta \simeq \frac{C_m K_m}{\hbar v b^{m-1}}, \qquad K_m = \frac{\pi^{1/2}}{2} \frac{\Gamma(\frac{1}{2}(m-1))}{\Gamma(\frac{1}{2}m)} \tag{15}$$

and corresponding deflection angle

$$\theta = -\frac{(m-1)C_m K_m}{E b^m} \tag{16}$$

The classical cross section at small angles is thus [see Eq. (2)]

$$\frac{d\sigma}{d\Omega} = (m\theta \sin \theta)^{-1} \left[\frac{(m-1)C_m K_m}{E\theta} \right]^{2/m} \tag{17}$$

and is seen to increase as $\theta^{-2-2/m}$ as $\theta \to 0$. In the classical calculation, the total scattering cross section $\sigma = \int d\Omega\, d\sigma/d\Omega$ is thus infinite for any m.

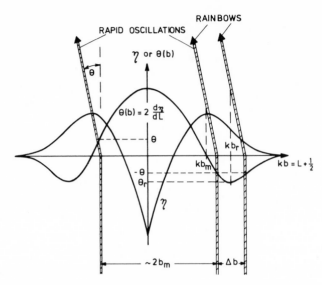

Figure 1. Origins of "rainbows" and rapid oscillations in elastic
scattering are shown by analogy to two-slit diffraction. The "slit
spacing" for rainbow oscillations observed at scattering angle θ
is Δb, and that for rapid oscillations is $\sim 2b_m$, where b_m is the
impact parameter for which the phase shift η is maximum and
for which the deflection function $\theta(b)$ is zero.

However, the quantum mechanical result [Eqs. (4) and (7)] is finite,

$$\sigma = 8\pi \int_0^\infty b\,db\,\sin^2 \eta(b) \tag{18}$$

$$= 2\pi \left(\frac{2C_m K_m}{\hbar v}\right)^{2/(m-1)} \cos\left(\frac{\pi}{m-1}\right)\Gamma\left(\frac{m-3}{m-1}\right)$$

$$= 8.0828\,(C_6/\hbar v)^{2/5} \qquad \text{for } m = 6 \tag{19}$$

and can be used with low-velocity measurements of σ to verify that the
interaction at long range has the form‡ $\sim C_6 R^{-6}$ and to determine the size
of C_6.

Since, as discussed in Section 1, all atom–atom interactions are attractive at long range and repulsive at short range, there is usually an intermediate impact parameter at which the deflection, θ, has a minimum value
$-\theta_r$. Near this "rainbow" angle, the scattering "piles up" and produces a

‡ At very long range ($R \gtrsim 100$ Å), retardation effects due to the finite velocity of light become
  important and $V(R)$ goes over to an R^{-7} dependence.[41] The effect is too small to be
  observed in normal scattering experiments.

maximum in the cross section in analogy to an optical rainbow.[23] Classically, $d\sigma/d\Omega$ becomes infinite at θ_r [put $d\theta/db = 0$ in Eq. (2)]. Quantum mechanically, the particle wave functions corresponding to trajectories on either side of b_r [defined by $\theta(b_r) \equiv \theta_r$] interfere as in two-slit diffraction, giving rise to a primary rainbow maximum in $d\sigma/d\Omega$ at an angle somewhat smaller in magnitude than $|\theta_r|$, plus smaller maxima ("supernumerary rainbows") at still smaller $|\theta|$ (see Figure 1). For scattering energies E rather larger than the well depth ε, the impulse approximation [Eq. (13)] shows that $E\theta_r$ depends only on the shape of the potential well. In fact

$$E\theta_r \simeq 2\varepsilon \qquad (20)$$

gives a good estimate (within about 10%) of ε for quite a range of potential shapes.[42] In addition, from the spacings $\Delta\theta$ of the supernumeraries, the shape of the deflection function can be mapped out in the neighborhood of θ_r. As in two-slit diffraction, $\Delta\theta$ gives roughly the ratio of the wavelength $2\pi/k$ to the "slit" separation Δb. Interference can also occur between scatterings at $+\theta$ with those at $-\theta$. Then, however, the "slit spacing" corresponds to $\sim 2b_m$ (see Figure 1) so that the spacings $\Delta\theta$ are much smaller. The resultant "rapid oscillations" give a measure of the minimum position $R_m \simeq b_m$. Figure 2 shows an experimental cross section displaying well-resolved rainbows and rapid oscillations.[43]

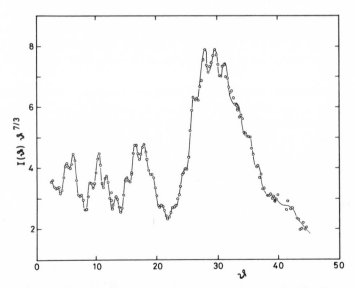

Figure 2. Rapid oscillations superimposed upon the primary and higher-order "rainbows," as observed in elastic scattering of Li on Hg at a relative barycentric energy of 0.305 eV (from Buck *et al.*[43]).

Another interference effect can be seen in the forward or total cross section at low velocities. Oscillations in σ as a function of v arise from the stationary-phase contribution, where η has its maximum value η_{max}. When η_{max} is $\frac{1}{2}\pi, \frac{3}{2}\pi, \ldots$, maxima occur in σ [see Eq. (18)], whose spacing can be shown to give a good determination of $\varepsilon^2 R_m$.[44,45]

2.2. Beam Measurements in Excited States

The entire discussion of Section 2.1 about beam measurements in the ground state is applicable here, but there are additional complications as well. One of the most obvious is that it is not generally possible to prepare a beam of purely excited atoms. With metastable excited states[46-54] most of the beam can be excited, but with states of radiative lifetimes (typically 10 ns) short compared to the time spent in the collision region (usually about 1 μs), not more than half the beam can be excited on the average and 25%–30% may be a practical upper limit. As a result the excited-state data must be sorted out of a background of ground-state scattering.

Excited-state scattering measurements on Na*–Ne[55] and Na*–Hg[56] have recently been reported. The narrow line of a stable, single–mode, continuous laser is generally used to excite the primary beam directly in the collision region. Those atoms that are illuminated by the laser and have their Doppler-shifted resonant frequencies lying within the laser linewidth are cycled up and down between the excited and ground states by absorption and spontaneous or induced emission. The cycle period depends on the laser intensity, but it is generally the same order of magnitude as the natural lifetime, and hence about 10^4 times larger than the duration time of a collision.

During collision, when the energy levels are perturbed, the atoms are removed from resonance with the laser and can thus not absorb or be induced to emit radiation. That portion of the radiation emitted by spontaneous decay during collision leads to a very small amount of line broadening and shift. In addition, some momentum is imparted on the average to the beam by the laser light.[57] These effects can generally be neglected in scattering experiments.

If there is more than one isolated ground state to which the excited atoms can decay, as for example the two hyperfine levels of the ground state of any naturally occurring alkali isotope, the laser will rapidly pump the atoms into that level from which no excitation occurs, effectively removing such atoms from the absorption–emission cycle. To prevent optical pumping by the laser, either (i) the ground-state levels are mixed, (ii) excitation is provided from all ground-state levels (Carter *et al.*[58] used the Doppler-shifted frequency of the reflected laser light to provide a second transition frequency), or (iii) the exciting transition is chosen so that the states excited can decay only to the original ground state.

The measurement is usually made by recording the *change* in scattered intensity when the laser is switched on. Since the laser serves not only to increase the excited-state density but also to decrease the ground-state one, the cross section measured corresponds to the difference $d(\sigma^* - \sigma_x)/d\Omega$ between that for the excited state $(d\sigma^*/d\Omega)$ and that for the ground state $(d\sigma_x/d\Omega)$. In comparing the results with pure ground-state measurements of $d\sigma_x/d\Omega$, one must allow for the effect of the laser in reducing the scattering volume and the width of the velocity distribution.

A further problem arises from the degeneracy or near degeneracy usually present in excited states although usually absent from ground states (we ignore possible nuclear- or electron-spin degeneracy). A complete analysis requires the integration of a coupled-channel form of the Schrödinger equation [Eq. (5)] to obtain the scattering matrix (Section 4.3). The amount of work required is formidable, especially if the potentials are to be adjusted to fit experimental results. The analysis is greatly simplified by making the "elastic approximation,"[59–62] in which the cross section for any excited atomic state is taken to be the statistically weighted sum of cross sections for elastic scattering from the different adiabatic potentials. Thus, for example, Carter *et al.*, in their analysis of Na*($3P$)–Ne scattering took the differential cross section $d\sigma^*/d\Omega$ to be[55]

$$\frac{d\sigma^*}{d\Omega} = \frac{1}{3}\frac{d\sigma^{B\Sigma}}{d\Omega} + \frac{2}{3}\frac{d\sigma^{A\Pi}}{d\Omega} \qquad (21)$$

where $d\sigma^{B\Sigma}/d\Omega$ and $d\sigma^{A\Pi}/d\Omega$ are the differential cross sections for elastic scattering from the $B^2\Sigma$ and $A^2\Pi$ states of Na*–Ne, respectively. Of course the elastic approximation can only be valid if the atomic states are well mixed by the collisions observed and, hence, if the atomic level splittings, $\Delta\varepsilon$, in the excited state are small compared to the nonadiabatic collision energies available, namely, $\hbar v/a$, where v is the relative velocity and a the atomic dimension over which the coupling between the levels changes appreciably. Recent calculations of Reid[61] and of Bottcher[59] show, however, that the elastic approximation [Eq. (21)] can drastically alter the position and spacing of "rapid" oscillations in $d\sigma^*/d\Omega$ even when the excited-state splitting $\Delta\varepsilon \ll \hbar v/a$.

Although scattering experiments with excited-state atomic beams promise to provide detailed and accurate information about excited-state interatomic potentials, severe difficulties in both the experimental measurement and the theoretical analysis will certainly restrict the number of applications. In addition to the measurements described above, in which the scattering of atoms by collisions with another beam[56] or with a gas[55] is measured directly, effects of the collisions can also be observed in the scattered fluorescence. Measurements of fine-structure transitions in a beam of K(4P) crossed with an Ar beam[63] as well as resonant depolarization measurements in Ba and Sr arising from collisions within a single beam[64-66] have been made.

2.3. Spectroscopic Measurements of Bound Molecules

If electronic transitions between bound states of a diatomic molecule can be observed, the frequencies of the vibration–rotation (vr) lines can be used to determine the initial and final potential-well shapes to high accuracy. A straightforward procedure is to adjust the potential wells to reproduce the observed energy levels.[67] One might start with the easily determined Morse potentials (see Section 1.3) and then make modifications. After each modification, the eigenvalues of bound states in the well must be recalculated. The procedure can be made more efficient by using first-order perturbation theory to relate potential modifications to changes in the eigenvalues.

The Rydberg–Klein–Rees (RKR) method[23,68] avoids the trial-and-error search for a potential by use of the (first-order) JWKB method.[40] Essentially the method builds the potential up from the well bottom by noting that every additional state (not counting spin degeneracy) requires an additional h^3 of phase space. According to the JWKB theory, the bound-state phase

$$\eta_L = (2\mu/\hbar^2) \int_{R_1}^{R_2} dr \left[E - V(r) - \frac{\hbar^2}{2\mu} \frac{L(L+1)}{r^2} \right]^{1/2} \tag{22}$$

where R_1 and R_2 are the classical turning points [zeros of the integrand in Eq. (22)], is simply

$$\eta_L = (\nu_L + \tfrac{1}{2})\pi \tag{23}$$

where ν_L is the vibrational quantum number. One constructs a continuous function $A(E, K)$, where $K = \hbar^2 L(L+1)/2\mu$ from the discrete data $E = E(K, \nu)$:

$$A(E, K) = \int_{R_1}^{R_2} dr (E - V - Kr^{-2}) = \hbar(2\mu)^{-1/2} \int_0^{\eta_L} d\eta (E - V - Kr^{-2})^{1/2}$$

$$\approx \pi\hbar(2\mu)^{-1/2} \sum_{\nu'=0}^{\nu} [E - E(K, \nu')]^{1/2} \tag{24}$$

and then notes that

$$\left(\frac{\partial A}{\partial E}\right)_K = R_2 - R_1, \qquad \left(\frac{\partial A}{\partial K}\right)_E = R_2^{-1} - R_1^{-1} \tag{25}$$

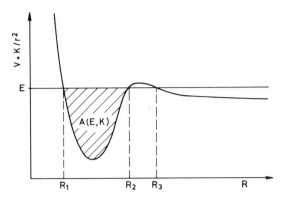

Figure 3. Bound-state "area" $A(E, K)$, where $K = (h^2/2m)L(L+1)$, used for Rydberg–Klein–Rees (RKR) analysis of potential curves from rotational and vibrational energies. When the total energy is E, the effective potential $V + K/r^2$ has two (or three) classical turning points for a bound (or "quasibound") state (see text).

[$A(E, K)$ is the area on the potential-energy diagram (Figure 3) bounded by E and $V + K/r^2$.] The potential $V(R)$ is mapped out by plotting the classical turning points R_1, R_2 as a function of energy. Higher-order JWKB corrections have been derived[69–71] but are rarely applied because the RKR method itself is usually sufficiently accurate.

When $L > 0$, wells can contain quasibound levels,[72,73] which can be observed in scattering as "orbiting collisions"[74–76] and in spectroscopy as rotational predissociation.[7] By measuring the level width Γ or equivalently the lifetime $\tau = \Gamma^{-1} = 2t_{tr}/T$ {where $t_{tr} \equiv \int_{R_1}^{R_2} dr \, \dot{r}^{-1} = \hbar \, \partial\eta/\partial E$ is the one-way transit time in the well and $T \simeq (1 + \theta^2)^{-1}$ with $\theta = \exp(\int_{R_2}^{R_3} dr \, [2\mu\hbar^{-2}(Kr^{-2} + V - E)]^{1/2})$ is the transmission coefficient through the barrier[73]} the long-range part of the potential can be accurately determined.[77,78]

Analyses such as described have been used for 30 years to obtain accurate potential curves for diatomic systems in which both ground and excited states are strongly bound (i.e., $\varepsilon \geqslant 1$ eV). By performing bound-state spectroscopy on beams that are super-cooled by supersonic expansion, it now appears possible to perform such analyses on systems with only weak van der Waals wells in the ground state.[79] Since all atom pairs have at least a van der Waals well in the ground state, many new systems are now amenable to the accurate spectroscopic determination of ground- and excited-state potential wells. In addition to the normal RKR analysis on the bound–bound transitions in such systems, the bound–free fluorescence emitted from specific laser-excited upper levels permits accurate determination of the repulsive wall of the lower state.

2.4. Spectroscopic Measurements of Unbound Molecules

Free–free and free–bound molecular transitions also contain valuable information about initial and excited potentials. They are observed as pressure-broadened atomic lines.[80,81] The width and shift of the line observed in the impact region near line center have been used to determine two parameters of the potential difference between ground and excited states,[81,82] which is then usually assumed to have a LJ (12, 6) form (see Section 1.3). Such determinations are of little value since difference potentials (usually more than one excited-state potential is involved) are rarely described adequately by a single LJ (n, 6) form.

Much more useful are measurements of the binary and tertiary (molecule formation and destruction in the excited state) contributions to the far wings of the lines, including satellite bands. By measuring profiles at different temperatures, both initial- and final-state potentials can be largely ascertained, as shown by Gallagher and co-workers[83–87] for many alkali–noble-gas systems. Details are given by Behmenberg in Chapter 27 of this work.

2.5. Transport Coefficients and Other Properties

Bulk properties of dilute gases usually depend largely on binary interactions. Where the property can be accurately measured and its relation to the interatomic potential derived, its value and possible dependence on the temperature T may give useful information about the potential. For example, in the virial expansion of the equation of state in reciprocal powers of the volume $\mathcal{V}$ per mole

$$\frac{P\mathcal{V}}{RT} = 1 + \frac{B(T)}{\mathcal{V}} + \frac{C(T)}{\mathcal{V}^2} + \cdots \qquad (26)$$

the second virial coefficient $B(T)$ represents the effects of binary collisions and is given in terms of the ground-state interaction potential $V(R)$ by

$$B(T) = 2\pi N_0 \int_0^\infty R^2 \, dR \, \{1 - \exp[-V(R)/kT]\} \qquad (27)$$

where N_0 is Avogadro's number. Measurements of $B(T)$ have been used to determine parameters in LJ (12, 6) potentials, but such potentials are usually not suitable for calculating other properties with much accuracy. Although in principle $B(T)$ can be used to determine a repulsive potential $V(R)$ uniquely, in practice this is difficult and one can best view the information contained in measurements of $B(T)$ as placing *constraints* on $V(R)$: The "true" potential should give $B(T)$ when substituted into Eq.

(27). Transport coefficients of viscosity and diffusion[88] as well as relaxation rates can provide similar constraints.

Properties of liquids and solids also contain information about the interactions,[88] and, indeed, lattice spacings and heats of formation have been used to estimate equilibrium separations and well depths. Because of many-body effects, however, the information thereby obtained about binary interactions is rather crude.

The same type of constraints placed on ground-state potentials by bulk properties of dilute gases is available for excited-state potentials from total cross sections or rates of various excited-state collision processes (see Section 1.1). Thus, for example, although fine-structure transition rates (sensitized fluorescent rates, see Chapter 29 of this work)[63,89–91] do not allow much of a determination of excited-state potentials, the "true" potential should reproduce the rates found. The recent formulation of fine-structure and m_j-mixing collisions as a quantum mechanical Boltzmann equation[92] makes the analogy between such measurements and those of transport coefficients even closer.

3. Calculations

3.1. Born–Oppenheimer Approximation and Adiabatic Potentials; Hellmann–Feynman Theorem

By "interatomic potential" is practically always meant the adiabatic potential energy, i.e., the potential energy of the system as a function of internuclear separation in the limit of fixed nuclei. That such a quantity is useful follows from the Born–Oppenheimer approximation for the total interaction. It is convenient because of its velocity and angular-momentum independence. In this section, the validity of the Born–Oppenheimer equation is investigated, and the use of adiabatic potentials as potential energies for nuclear motion (Hellmann–Feynman theorem) is discussed.

The system to be considered consists of nuclei of masses $M_1, M_2, M_3, \ldots$ and charges $Z_1e, Z_2e, Z_3e, \ldots$ and of electrons each of mass $m \simeq 10^{-4} M_J$. The Schrödinger equation to be solved is

$$H\psi(\mathbf{r}, \mathbf{R}) = E\psi(\mathbf{r}, \mathbf{R}) \tag{28}$$

where $\mathbf{r}$ stands for all the electronic coordinates $(\mathbf{r}_1, \mathbf{r}_2, \ldots)$, $\mathbf{R}$ stands for all the nuclear coordinates $(\mathbf{R}_1, \mathbf{R}_2, \ldots)$, and H is the nonrelativistic Hamiltonian (ignore spin–orbit and spin–spin couplings)

$$H = -\frac{\hbar^2}{2}\left(\sum_i \frac{\nabla_i^2}{m} + \sum_J \frac{\nabla_J^2}{M_J}\right) + V(\mathbf{r}, \mathbf{R}) \tag{29}$$

Here $V(\mathbf{r}, \mathbf{R})$ is the total Coulomb interaction which may be split into electron–electron, nucleus–electron, and nucleus–nucleus parts:

$$V(\mathbf{r}, \mathbf{R}) = V_{ee} + V_{eN} + V_{NN} \qquad (30)$$

with

$$V_{ee} = e^2 \sum_{i<k} |\mathbf{r}_i - \mathbf{r}_k|^{-1} \qquad (31a)$$

$$V_{eN} = -e^2 \sum_{iJ} Z_J |\mathbf{r}_i - \mathbf{R}_J|^{-1} \qquad (31b)$$

$$V_{NN} = e^2 \sum_{J<K} Z_J Z_K |\mathbf{R}_J - \mathbf{R}_K|^{-1} \qquad (31c)$$

Let $\phi_n(\mathbf{r}, \mathbf{R})$ be an electronic eigenstate of $(-\hbar^2/2m)\sum_i \nabla_i^2 + V(\mathbf{r}, \mathbf{R})$ with eigenvalue $\varepsilon_n(R)$:

$$\left[-\frac{\hbar^2}{m} \sum_i \nabla_i^2 + V(\mathbf{r}, \mathbf{R}) \right] \phi_n(\mathbf{r}, \mathbf{R}) = \varepsilon_n(\mathbf{R})\phi_n(\mathbf{r}, \mathbf{R}) \qquad (32)$$

Note that the differential operator does not contain $\mathbf{R}$. Indeed, one can view $\mathbf{R}$ here as a parameter. The eigenstates $\phi_n(\mathbf{r}, \mathbf{R})$ are taken to be orthonormal at every $\mathbf{R}$:

$$\int d\mathbf{r} \, \phi_m^*(\mathbf{r}, \mathbf{R})\phi_n(\mathbf{r}, \mathbf{R}) = \delta_{mn} \qquad (33)$$

Now if $\psi(\mathbf{r}, \mathbf{R})$ is expanded in the $\phi_n(\mathbf{r}, \mathbf{R})$,

$$\psi(\mathbf{r}, \mathbf{R}) = \sum_n \phi_n(\mathbf{r}, \mathbf{R})\chi_n(\mathbf{R}) \qquad (34)$$

the R-dependent expansion coefficients $\chi_n(\mathbf{R})$ describe the behavior of the nuclei. Indeed, the Schrödinger equation [Eq. (28)] becomes

$$E\chi_m(\mathbf{R}) = \sum_n \langle \phi_m | H | \phi_n \rangle \chi_n(\mathbf{R})$$

$$= \left[-\frac{\hbar^2}{2} \sum_J \frac{\nabla_J^2}{M_J} + \varepsilon_m(\mathbf{R}) \right] \chi_m(\mathbf{R}) + \sum_n U_{mn}\chi_n(\mathbf{R}) \qquad (35)$$

where the last term, with

$$U_{mn} = -\frac{\hbar^2}{2} \sum_J M_J^{-1} [\langle \phi_m | \nabla_J^2 \phi_n \rangle + \langle \phi_m | \nabla_J \phi_n \rangle \cdot \nabla_J] \qquad (36)$$

is the Born–Oppenheimer breaking term. In the Born–Oppenheimer approximation, the coupling terms $U_{mn}\phi_n$ in Eq. (35) are ignored, and the rest of Eq. (35) is simply the Schrödinger equation for nuclei moving in the adiabatic interatomic potential $\varepsilon_m(R)$. The approximation is usually justified[93] by noting that since the electronic coordinates are relative to the nuclei positions, $\langle \phi_m | \nabla_J \phi_n \rangle$ is about the same size as $\langle \phi_m | \nabla_i \phi_n \rangle$. As a result, the terms U_{mn} are about $(m_i/M_J)^{1/2}$ smaller than the kinetic energy terms $-\hbar^2 \nabla_i^2/2m_i$.

The Hellmann–Feynman theorem shows that when the system is in electronic state n, the adiabatic potential $\varepsilon_n(\mathbf{R})$ is effectively a potential energy for the motion of the nuclei, in that the negative gradient of $\varepsilon_n(\mathbf{R})$ with respect to $\mathbf{R}_J$ is the average value of the force on the Jth nucleus:

$$\nabla_J \varepsilon_n(\mathbf{R}) = \nabla_J \int d\mathbf{r} \, \phi_n^* \left[-\frac{\hbar^2}{2m} \sum_i \nabla_i^2 + V(\mathbf{r}, \mathbf{R}) \right] \phi_n$$

$$= \int d\mathbf{r} \, \phi_n^* \left[\nabla_J V(\mathbf{r}, \mathbf{R}) \right] \phi_n + \{ \quad \} \tag{37}$$

where

$$\{ \quad \} = \int d\mathbf{r} \, (\nabla_J \phi_n)^* \left[-\frac{\hbar^2}{2m} \sum_i \nabla_i^2 + V(\mathbf{r} \cdot \mathbf{R}) \right] \phi_n$$

$$+ \int d\mathbf{r} \, \phi_n^* \left[-\frac{\hbar^2}{2m} \sum_i \nabla_i^2 + V(\mathbf{r}, \mathbf{R}) \right] (\nabla_J \phi_n)$$

$$= \int d\mathbf{r} [(\nabla_J \phi_n)^* \varepsilon_n \phi_n + \phi_n^* \varepsilon_n (\nabla_J \phi_n)]$$

$$= \varepsilon_n(\mathbf{R}) \nabla_J \int d\mathbf{r} \, \phi_n^* \phi_n = 0$$

Thus

$$-\nabla_J \varepsilon_n(\mathbf{R}) = \langle \phi_n | \mathbf{F}_J | \phi_n \rangle \tag{38}$$

where $\mathbf{F}_J$ is the force on the Jth nucleus:

$$\mathbf{F}_J \equiv -\nabla_J V(\mathbf{r}, \mathbf{R}) \tag{39}$$

The Hellmann–Feynman theorem depends on the Born–Oppenheimer approximation, *not* for its derivation, but for its usefulness. In the Born–Oppenheimer approximation, the different adiabatic electronic states n are uncoupled, and two atoms approaching each other in the state n will—within the approximation—remain in that state and the motion of the nuclei is governed by the potential energy $\varepsilon_n(R)$ throughout the collision.

Only asymptotically $(R \to \infty)$ are the adiabatic states ϕ_n identical with the atomic ones. At smaller R they are linear combinations of the atomic states, and the coefficients of the linear combination generally change smoothly with R.

The Born–Oppenheimer approximation fails when the coupling terms U_{mn} [Eq. (36)] become important, owing to rapid changes in the adiabatic states with changing $\mathbf{R}$ or to large velocities of the nuclei, compared with the adiabatic energy separations[94] $\varepsilon_n(\mathbf{R}) - \varepsilon_m(\mathbf{R})$. A convenient measure of the applicability of the Born–Oppenheimer approximation is the "Massey parameter" $a\Delta\varepsilon/(\hbar v)$, where $\Delta\varepsilon$ is the energy difference, a is the length scale for $\mathbf{R}$ over which the states $\phi(\mathbf{r}, \mathbf{R})$ change appreciably, and v is the relative velocity of the nuclei. If and only if $a\Delta\varepsilon/(\hbar v) \gg 1$, transitions between the adiabatic states are negligible and the Born–Oppenheimer approximation is valid (see Section 4 for further discussion of the Massey parameter).

3.2. Hund's Cases and Correlation Diagrams

In describing the adiabatic electronic states of a diatomic system, the internuclear axis $\hat{\mathbf{R}}$, which is taken to be stationary, forms a natural symmetry and quantization axis. Let $\mathbf{l}$ and $\mathbf{s}$ be the total orbital and spin angular momenta of the electrons, respectively, and put $\mathbf{j} = \mathbf{l} + \mathbf{s}$. Quantum numbers Λ and Σ represent projections $\mathbf{l} \cdot \hat{\mathbf{R}}$ and $\mathbf{s} \cdot \hat{\mathbf{R}}$. Axial ($C_{\infty v}$) symmetry ensures that $[\mathbf{j} \cdot \hat{\mathbf{R}}, H_{\mathrm{el}}] = 0$, where $H_{\mathrm{el}} = -(\hbar^2/2m)\sum_i \nabla_i^2 + V(\mathbf{r}, \mathbf{R})$ is the electronic Hamiltonian. It follows that $\Omega = \Lambda + \Sigma$ is a good quantum number of the adiabatic state even when, owing to $\mathbf{l} \cdot \mathbf{s}$ coupling, Λ and Σ are not. In actual diatomic systems, radial and rotational motion mixes the adiabatic states together and may prevent Ω from being a good quantum number.

The total angular momentum of a rotating system in which the orbital motion of the nuclei is $\mathbf{L}$ (nuclear spins are ignored), is $\mathbf{J} = \mathbf{l} + \mathbf{s} + \mathbf{L}$. Generally $\mathbf{l}$, $\mathbf{s}$, and $\mathbf{L}$ are all coupled to each other, and in addition $\mathbf{l}$ is usually coupled to the electric field along the internuclear axis. Possible couplings are described by the various Hund's cases,[7] depending on which couplings are strongest (see Table 3). It should be recognized that the Hund's cases are idealizations which are often only partially realized. During a collision, a diatomic system usually "passes through" several Hund's cases, the approximate boundaries of which depend on both R and L. In the construction of adiabatic potentials, L is taken equal to zero and only the Hund's cases a, b, and c are possible.

At infinite R, adiabatic energies are given in terms of the atomic ones $E_i^{(A)}$ and $E_j^{(B)}$ simply by $E_i^{(A)} + E_j^{(B)}$. At small R, the adiabatic energies approach $E_k^{(AB)} + Z_A Z_B e^2/R$, where $E_k^{(AB)}$ are the energies of the united atom of atomic number $Z_A + Z_B$. A useful qualitative picture of the adiabatic potentials can be obtained if one knows how to correlate the states at large R with those at small R. The correlation is determined entirely by which quantum numbers are "good" at all R since this, in turn, determines which states are allowed to cross. A quantum number is "good" when the corresponding observable is conserved, i.e., commutes with H_{el}. States

Table 3. Common Hund's Coupling Cases

Case	Coupling hierarchy	Construction of $\mathbf{J}$	Remarks
a	$(\mathbf{l}\cdot\mathbf{s})<(\mathbf{l}\cdot\hat{\mathbf{R}})$ $(\mathbf{L}\cdot\mathbf{s})<(\Lambda\hat{\mathbf{R}}\cdot\mathbf{s})$	$\mathbf{l}\to\Lambda\,\hat{\mathbf{R}},\,\mathbf{s}\to\Sigma\hat{\mathbf{R}}\quad\Lambda\neq0$ $\Omega=\Lambda+\Sigma$ $\mathbf{J}=\mathbf{L}+\Omega\hat{\mathbf{R}}$	
b	$(\mathbf{l}\cdot\mathbf{s})<(\mathbf{l}\cdot\hat{\mathbf{R}})$ $(\Lambda\hat{\mathbf{R}}\cdot\mathbf{s})<(\mathbf{L}\cdot\mathbf{s})$	$\mathbf{l}\to\Lambda\hat{\mathbf{R}}$ $\mathbf{K}=\Lambda\hat{\mathbf{R}}+\mathbf{L}$ $\mathbf{J}=\mathbf{K}+\mathbf{S}$	Λ may be $=0$, Σ is not a good quantum number
c	$(\mathbf{l}\cdot\hat{\mathbf{R}})<(\mathbf{l}\cdot\mathbf{s})$ $(\mathbf{j}\cdot\mathbf{L})<(\mathbf{j}\cdot\hat{\mathbf{R}})$	$\mathbf{j}=\mathbf{l}+\mathbf{s}\to\Omega\hat{\mathbf{R}}$ $\mathbf{J}=\mathbf{L}+\Omega\hat{\mathbf{R}}$	Λ and Σ are not good quantum numbers; valid asymptotically when $L=0$
d	$(\mathbf{l}\cdot\hat{\mathbf{R}})<(\mathbf{l}\cdot\mathbf{L})$ $(\mathbf{l}\cdot\mathbf{s})<(\mathbf{l}\cdot\mathbf{L})$	$\mathbf{K}=\mathbf{L}+\mathbf{l}$ $\mathbf{J}=\mathbf{K}+\mathbf{s}$	$L\neq0$, Λ, Σ, and Ω are not good quantum numbers
e	$(\mathbf{l}\cdot\hat{\mathbf{R}})<(\mathbf{l}\cdot\mathbf{L})<(\mathbf{l}\cdot\mathbf{s})$	$\mathbf{j}=\mathbf{l}+\mathbf{s}$ $\mathbf{J}=\mathbf{j}+\mathbf{L}$	$L\neq0$, Λ, Σ, and Ω are not good quantum numbers; valid asymptotically when $L\neq0$

with different sets of good quantum numbers cannot be coupled by H_{el}; they can be degenerate and hence can cross. States with the same good quantum numbers are generally coupled by H_{el} and cannot cross. In the traditional correlation diagrams for diatomic systems, as given for example by Herzberg,[7] spin is ignored, and for heteronuclear systems the only good quantum number is Λ. Since states of different Λ can cross whereas those of the same Λ cannot, the correlation diagram is constructed simply by connecting the nth separated-atom state of given Λ to the nth united-atom state of the same Λ. For homonuclear diatomic systems, there is an additional good quantum number, namely, the parity (u for *ungerade* or odd and g for *gerade* or even) of the total electronic wave function. If spin–orbit $(\mathbf{l}\cdot\mathbf{s})$ coupling is taken into account, the procedure is similar except that Ω replaces Λ as a good quantum number.

3.3. Electron-Gas (Gombás) Models

The calculation of the adiabatic potentials $\varepsilon_n(\mathbf{R})$ is nontrivial, but there are some approximations that not only simplify the computations but also aid in understanding the physical bases for much of the interaction. In the electron-gas model, long used by Gombás for atomic calculations,[95–99] electron interactions are calculated statistically for fixed densities. The model is especially useful for spherically symmetric closed-shell atoms or atomic cores for which the relative change in electron density is small.[100-103]

Consider the interaction of two spherically symmetric electron distributions. By treating the densities as fixed, the interaction is computed only to first order. The two strongest contributions arise from Coulomb forces and the influence of the Pauli exclusion principle ("Pauli pressure"). Exchange and correlation contributions to the overlap effects are generally smaller but may nevertheless be significant near the potential minimum because of the near cancellation there of the Coulomb and Pauli components.

3.3.1. Coulomb Component

Let the ions A and B have electron number densities $\rho_A(r) = (4\pi r^2)^{-1}D_A(r)$ and $\rho_B(r') = (4\pi r'^2)^{-1}D_B(r')$, where $\mathbf{r}' = \mathbf{r} - \mathbf{R}$. If the nuclear charges are $Z_A e$ and $Z_B e$, then the Coulomb interaction is (see Figure 4)

$$V_{coul}(R) = e^2 \int d^3\mathbf{r}'' \int d^3\mathbf{r} \frac{[Z_A\delta(\mathbf{r}) - \rho_A(\mathbf{r})][Z_B\delta(\mathbf{r}'') - \rho_B(\mathbf{r}'')]}{|\mathbf{r}'' - \mathbf{r}'|} \quad (40)$$

Let $\mathring{z}_A$ indicate the charge on the ion A:

$$\mathring{z}_A = Z_A - \int d^3\mathbf{r}\,\rho_A(r) = Z_A - \int dr\, D_A(r) \quad (41)$$

and analogously for ion B. To simplify Eq. (40), the electrostatic potential $eW(R)$ at R, due to a positive charge $(Z - \mathring{z})e$ at the origin and the electron distribution $\rho(r)$, is introduced:

$$W(R) = (Z - \mathring{z})R^{-1} - \int d^3\mathbf{r}\,\rho(r)|\mathbf{r} - \mathbf{R}|^{-1}$$
$$= \int_R^\infty dr\, D(r)(R^{-1} - r^{-1}) \quad (42)$$

where use was made both of Eq. (41) and of the relation

$$\int d^3r\,\rho(r)|\mathbf{r} - \mathbf{R}|^{-1} = \int_0^R dr\, D(r)R^{-1} + \int_R^\infty dr\, D(r)r^{-1} \quad (43)$$

Substitution into Eq. (40) yields a result easily adaptable to numerical

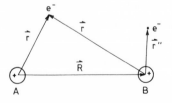

Figure 4. Position vectors for the interaction of atoms A and B. R is the internuclear separation and $\mathbf{r}' = \mathbf{r} - \mathbf{R}$ and $\mathbf{r}''$ are the positions of two electrons relative to the nucleus of atom B.

evaluation:

$$e^{-2}V_{\text{Coul}}(R) = \mathfrak{z}_A\mathfrak{z}_B R^{-1} + \mathfrak{z}_B W_A(R) + \mathfrak{z}_A W_B(R) - \int d^3r W_B(r')\rho_A(r) \quad (44)$$

As an example, for densities of the simple exponential form $D(r) = 4\beta^3(Z-\mathfrak{z})r^2 e^{-2\beta r}$, $e^{-2}V_{\text{Coul}}$ [Eq. (44)] may be found analytically:

$$e^{-2}V_{\text{Coul}}(R) = \mathfrak{z}_A\mathfrak{z}_B R^{-1} + \mathfrak{z}_A(Z_B-\mathfrak{z}_B)R^{-1}(1+\beta_B R)\,e^{-2\beta_B R}$$

$$+\,\mathfrak{z}_B(Z_A-\mathfrak{z}_A)R^{-1}(1+\beta_A R)\,e^{-2\beta_A R}$$

$$-\,(Z_A-\mathfrak{z}_A)(Z_B-\mathfrak{z}_B)R^{-1}[I(\alpha_A,\beta_A R)+I(\alpha_\beta,\beta_B R)] \quad (45)$$

where

$$I(\alpha,\beta R) = e^{-2\beta R}\alpha[\alpha(2\alpha-3)+(\alpha-2)\beta R] \quad (46)$$

and

$$\alpha_A = 1-\alpha_B = \beta_A^2/(\beta_A^2-\beta_\beta^2) \quad (47)$$

The result can be extended to electron densities consisting of several exponential shells. For the interaction of two identical neutral atoms ($\beta_A = \beta_B \equiv \beta$, $Z_A = Z_B \equiv Z$, $\mathfrak{z}_A = \mathfrak{z}_B = 0$) the simple result

$$V_{\text{Coul}}(R) = \frac{Z^2 e^2}{R} e^{-2x}\left(1+\frac{5x}{8}-\frac{3}{4}x^2-\frac{x^3}{6}\right), \qquad x = \beta R \quad (48)$$

is obtained. It is easily verified that $V_{\text{Coul}}(R)$ has a well depth of $\varepsilon = 0.01961\,(Z^2e^2\beta)$ at $R = 1.87\,\beta^{-1}$ (see Figure 5). The Coulomb interaction of the charge distributions is thus *attractive* at long range, where the nuclei are still well shielded from each other although considerable interaction exists between one nucleus and the electrons of the other atom. At *small R* ($R \lesssim \beta^{-1}$) $V_{\text{Coul}}(R)$ becomes repulsive and gradually approaches $Z^2e^2R^{-1}$ as $R \to 0$. It is instructive to compare the $V_{\text{Coul}}(R)$ results with the experimental ground-state well of H_2. In the $X\Sigma_g$ ground state the spins are paired so that there is no Pauli pressure and no exchange interaction (see below). Since $\beta = 1a_0^{-1}$, one finds a well depth $\varepsilon \simeq 0.54$ eV at $R_m \simeq 1$ Å. Disagreement with experimental values (~ 4.5 eV at 0.74 Å) arises because distortion in the electron density, which is very important in the bond formation, was ignored in the model. Of course the relative distortion in the total electron density is far greater for H_2 than for molecules with more electrons. In the formation of strong bonds, the electron density increases between the two nuclei and decreases elsewhere. In the complete electron-gas model, some of the decrease in energy due to the distortion is approximated by the correlation pseudopotential.

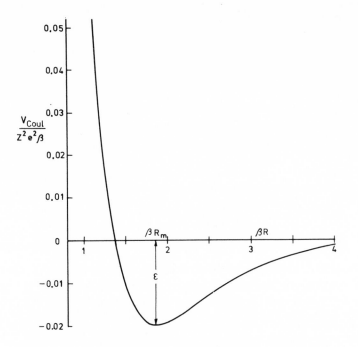

Figure 5. The Coulomb energy V_{Coul} for hypothetical atoms of fixed spherical electron density $\rho \propto e^{-2\beta r}$.

3.3.2. Pauli Pressure

The simple observation that each state per spin orientation requires a phase-space volume of h^3 leads directly to a relation between the Fermi energy $p_F^2/2m$ and the number density ρ of an electron gas (net spin assumed $= 0$):

$$\frac{p_F^2}{2m} = \frac{\hbar^2}{2m} (3\pi^2 \rho)^{2/3} \tag{49}$$

The total electronic kinetic energy of the system is therefore

$$\mathscr{E}_P[\rho(r)] = \int d^3r \int_0^{\rho(r)} d\rho \frac{\hbar^2}{2m} (3\pi^2 \rho) = \frac{3}{10} \frac{\hbar^2}{m} (3\pi^2)^{2/3} \int d^3r\, \rho^{5/3} \tag{50}$$

The minimization of $\mathscr{E}_P + V_{\text{Coul}}$ to determine electron densities constitutes the Thomas–Fermi statistical method.[96,98,99,104–106] Here, on the other hand, we start with more accurate densities and use $\mathscr{E}_P$ to calculate from them the Pauli *interaction* energy, i.e., the difference between $\mathscr{E}_P$ for the combined and for the separated systems[103]:

$$V_{\text{Pauli}}(R) = \frac{3\hbar^2}{10m}(3\pi^2)^{2/3}\int d^3r\,[(\rho_A + \rho_B)^{5/3} - \rho_A^{5/3} - \rho_B^{5/3}]$$

$$= \mathscr{E}_p[\rho_A(r) + \rho_B(r')] - \mathscr{E}_p[\rho_A(r)] - \mathscr{E}_p[\rho_B(r')] \qquad (51)$$

Note that the Pauli interaction as calculated in Eq. (51) is always repulsive.

As an illustration, V_{Pauli} can be found analytically for two overlapping spheres of constant electron density ρ. One finds

$$V_{\text{Pauli}}(R) = 3.373\rho^{5/3}(\hbar^2/m)\mathscr{V} \qquad (52)$$

where $\mathscr{V}$ is the intersecting volume, given for spheres of diameter d by

$$\mathscr{V} = \frac{\pi}{6}d^3\left[1 - \frac{3}{2}\frac{R}{d} + \frac{1}{2}\left(\frac{R}{d}\right)^3\right] \qquad (53)$$

(see Figure 6).

Note that, in contrast to the Coulomb interaction, the Pauli interaction is not linear in the densities. Consequently, the interaction due to several shells is not simply the sum of interactions for the individual shells.

The expression for $V_{\text{Pauli}}(R)$ simplifies when $\rho_A \ll \rho_B$:

$$V_{\text{Pauli}} \approx \frac{\hbar^2}{2m}(3\pi^2)^{2/3}\int d^3r\,\rho_A(r)\left[\rho_B(r')\right]^{2/3} \qquad (54)$$

The Coulomb and Pauli components give the largest contributions to the interaction potential of overlapping electron densities, and if the model densities are allowed to distort in response, they are the only electron-gas contributions that should be included. If distortion is *not* permitted, additional attractive pseudopotentials must be added to approximate the decrease in energy that the distortion would have achieved.

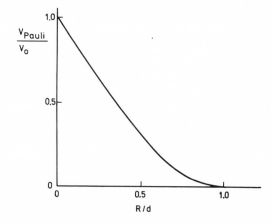

Figure 6. The Pauli pressure for two overlapping spheres of constant electron density ρ and diameter d. The constant V_0 is $1.766\ \hbar^2\rho^{5/3}\ d^3/m$, where m is the electron mass.

3.3.3. Exchange Pseudopotential‡

Pauli pressure causes electrons of the same spin state to avoid one another, and the resulting distortion decreases the Coulomb repulsion between them. The decrease in repulsion is represented by the attractive exchange potential. The calculation for an electron gas of constant density gives an exchange energy[96,107,108]

$$\mathcal{E}_x[\rho(r)] = -\frac{3}{4} e^2 \left(\frac{3}{\pi}\right)^{1/3} \int d^3r \, \rho^{4/3}(r) \tag{55}$$

The pseudopotential for the exchange interaction is

$$V_x(R) = \mathcal{E}_x[\rho_A(r) + \rho_B(r')] - \mathcal{E}_x[\rho_A(r)] - \mathcal{E}_x[\rho_B(r')] \tag{56}$$

3.3.4. Correlation Pseudopotential

Transient fluctuations in density can also reduce the Coulomb repulsion beyond what would be calculated from the average densities. The difference is called the correlation energy, and it encompasses both the polarization of an atomic core by a valence electron and long-range van der Waals forces (see below). The correlation energy for an electron gas has been evaluated at both low[109] and high[110] densities. A smooth function giving both limits is

$$\mathcal{E}_{\text{corr}}[\rho(r)] = -\frac{\alpha_0 + \alpha_1 r_s^{1/2} + \alpha_2 r_s}{1 + \beta_1 r_s^{1/2} + \beta_2 r_s} \ln\left(1 + \gamma_1 r_s^{-1} + \gamma_2 r_s^{-2}\right) \cdot \frac{me^4}{\hbar^2} \tag{57}$$

where $4\pi r_s^3/3 \equiv a_0^3 \rho^{-1}(r)$ and the constants α_i, β_i, and γ_i are given in Table 4. The corresponding interaction energy is

$$V_{\text{corr}}(R) = \mathcal{E}_{\text{corr}}[\rho_A(r) + \rho_B(r')] - \mathcal{E}_{\text{corr}}[\rho_A(r)] - \mathcal{E}_{\text{corr}}[\rho_B(r')] \tag{58}$$

The total interaction $V_{\text{Coul}}(R) + V_{\text{Pauli}}(R) + V_x(R) + V_{\text{corr}}(R)$ is readily found for two overlapping spherical systems. By transforming to confocal ellipsoidal coordinates[12,30] and summing integrands, one need evaluate only one double integral to obtain surprisingly accurate results.[100,111] Densities from single-configuration Hartree–Fock calculations are usually adequate, although neglect of relativistic[112] and correlation effects causes the outer electrons to be too weakly bound. Relativistic programs[113] can be used with polarization pseudopotentials (see Section 3.6) to avoid such problems.

‡ Some authors use "exchange" to include all the effects of electron antisymmetry. The "exchange interaction" is then dominated by Pauli pressure and is thus repulsive. Our use of the term conforms better to the traditional meaning in atomic-structure work (see for example Chapter 1 by Hibbert).

Table 4. Constants for Approximation to Correlation
Energy of an Electron Gas, Eq. (57)

i	α_i	β_i	γ_i
0	0.01555	—	—
1	0.005486	0.3528	6.64
2	0.00588	0.0891	21.76

For interactions at small enough R, the interaction approaches the Coulomb energy of the partially shielded nuclei, namely, $Z_A Z_B e^2 R^{-1} \Phi(R/a)$, where $\Phi(x)$ is the shielding function, sometimes represented by an exponential $[\Phi(x) = e^{-x}$ (Yukawa or screened-Coulomb potential) or xe^{-x} (Born–Mayer potential)][101,102] or the Thomas–Fermi screening function[28,114,115] with a given by[116] $a = 0.8853a_0(Z_A^{1/2} + Z_B^{1/2})^{-2/3}$ or by[117] $a = 0.8853a_0(Z_A^{2/3} + Z_B^{2/3})^{-1/2}$.

3.4. Induction and Dispersion (van der Waals) Forces

Interactions calculated with electron-gas models vanish when there is no overlap. However, the interaction of atoms at long range where overlap is negligible is usually an important factor for thermal-energy cross sections. The interaction of an ion with an atom is dominated at long range by *induction* forces: Electric moments induced in the atom by the field of the ion interact with that field. The long-range interaction of two neutral atoms is dominated by *dispersion*, or *van der Waals*, forces: Fluctuating electric moments in the two atoms are *correlated* so as to lower the total energy.[118,119]

Both induction and dispersion effects are simply illustrated with three-dimensional harmonic oscillators. Consider a negative charge $-e$ bound harmonically to a charge $+e$ fixed at the origin. The Hamiltonian for the system in an electric field $\mathbf{E}$ is

$$H = \frac{p^2}{2m} + \frac{1}{2}m\omega_0^2 r^2 - \mathbf{d} \cdot \mathbf{E} \qquad (59)$$

where $\mathbf{p}$ is the momentum, $\mathbf{r}$ the position, and m the mass of the negative charge, $\mathbf{d} = -e\mathbf{r}$ is the dipole moment operator of the system, and ω_0 is the resonant frequency of the oscillator. With $\mathbf{r}_0 = -e\mathbf{E}/m\omega_0^2$, Eq. (59) can be written

$$H = \frac{p^2}{2m} + \frac{1}{2}m\omega_0^2(\mathbf{r} - \mathbf{r}_0)^2 - \frac{1}{2}m\omega_0^2 r_0^2 \qquad (60)$$

i.e., as an oscillator of the same frequency ω_0 centered at $\langle \mathbf{r} \rangle = \mathbf{r}_0$, less the constant energy $\frac{1}{2}m\omega_0^2 r_0^2$. The effect of $\mathbf{E}$ is thus to induce a net electric dipole

$$\langle \mathbf{d} \rangle = -e\langle \mathbf{r} \rangle = -e\mathbf{r}_0 = \alpha \mathbf{E} \qquad (61)$$

where

$$\alpha = e^2/m\omega_0^2 \qquad (62)$$

is the polarizability of the system, and to lower the total energy by

$$\tfrac{1}{2}m\omega_0^2 r_0^2 = \tfrac{1}{2}\alpha E^2 \qquad (63)$$

For spherical systems in general, the interaction with a uniform electric field $\mathbf{E}$ is given by $-\frac{1}{2}\alpha E^2$, where the polarizability is readily found from second-order perturbation theory[120]:

$$\alpha = 2 \sum_n {}' \frac{\langle 0|\mathbf{d}|n\rangle \cdot \langle n|\mathbf{d}|0\rangle}{E_{0n}} \qquad (64)$$

where $|0\rangle$ is the initial state, the sum extends over all other states $|n\rangle$, and $E_{0n} = E_0 - E_n$ is the energy difference between states $|0\rangle$ and $|n\rangle$.

Higher-order expressions from perturbation theory give hyper-polarizability terms, with interactions varying as E^4, E^6, etc. In addition, field gradients can induce quadrupole moments. For the oscillator model, an axially symmetric field gradient with a potential $\phi(\mathbf{r})$ obeying

$$\frac{\partial^2}{\partial z^2} \phi \equiv \phi_{zz} = -\frac{1}{2}\phi_{xx} = -\frac{1}{2}\phi_{yy} \qquad (65)$$

and $|\phi_{zz}| \ll m\omega_0^2/e$ induces a quadrupole moment

$$Q_{zz} = -e\langle 2z^2 - x^2 - y^2 \rangle = C\phi_{zz} \qquad (66)$$

where $C = e^2\hbar/(4m^2\omega_0^3) = 0.25 \, (\alpha^3/a_0)^{1/2}$ is the quadrupole polarizability of the oscillator. The corresponding decrease in energy is $-\frac{1}{4}C\phi_{zz}^2$. Higher-order field derivatives induce octupole, hexadecipole, etc. moments, which also interact with the field to reduce the total energy.

The asymptotic interaction between a point charge $\mathfrak{z}e$ and an atom a distance R away is thus

$$-\tfrac{1}{2}\alpha E^2 - \tfrac{1}{4}C\phi_{zz}^2 = -\tfrac{1}{2}\alpha \mathfrak{z}^2 e^2 R^{-4} - c\mathfrak{z}^2 e^2 R^{-6} \qquad (67)$$

If the field E changes rapidly, however, as for example when its source is an electron, there are important *dynamic* corrections varying as R^{-6} which are positive and may well overshadow the quadrupole terms.[121–124]

The *dispersion* ("London" or "van der Waals") interaction is a quantum correlation phenomenon which can be illustrated by the dipole–dipole interaction

$$e^2 R^{-3}(x_A x_B + y_A y_B - 2z_A z_B) \tag{68}$$

of two oscillators where $\mathbf{r}_A = (x_A, y_A, z_A)$ and $\mathbf{r}_B = (x_B, y_B, z_B)$ are the positions of the negative charges relative to the oscillator centers at the origin and at $\mathbf{R} = R\hat{\mathbf{z}}$, respectively. The Hamiltonian for the interacting oscillators, known as the "Drude model,"[88,125,126] can be written

$$H = H_x + H_y + H_z \tag{69}$$

where

$$H_x = -\frac{\hbar^2}{2m}\left(\frac{\partial^2}{\partial x_A^2} + \frac{\partial^2}{\partial x_B^2}\right) + \frac{1}{2}m(\omega_A^2 x_A^2 + \omega_B^2 x_B^2) + \beta_x x_A x_B \tag{70}$$

and similarly for H_y and H_z. Here ω_A and ω_B are the natural frequencies of the oscillators and

$$\beta_x = \beta_y = -\tfrac{1}{2}\beta_z = e^2 R^{-3} \tag{71}$$

By the orthogonal transformation

$$\begin{pmatrix} x_1 \\ x_2 \end{pmatrix} = \begin{pmatrix} \cos\theta_x & \sin\theta_x \\ -\sin\theta_x & \cos\theta_x \end{pmatrix}\begin{pmatrix} x_A \\ x_B \end{pmatrix} \tag{72}$$

with

$$\tan 2\theta_x = \frac{2\beta_x}{m(\omega_A^2 - \omega_B^2)} \tag{73}$$

the Hamiltonian H_x becomes the sum of Hamiltonians for two uncoupled oscillators, and the eigenvalues of H_x can be immediately given:

$$E_x(N_x, n_x) = \hbar\Omega_x(N_x + \tfrac{1}{2}) + \hbar\omega_x(n_x + \tfrac{1}{2}) \tag{74}$$

where $N_x, n_x = 0, 1, 2, \ldots$ and

$$\Omega_x^2 = \tfrac{1}{2}(\omega_A^2 + \omega_B^2) + \{(\beta_x/m)^2 + [\tfrac{1}{2}(\omega_A^2 - \omega_B^2)]^2\}^{1/2} \tag{75a}$$

$$\omega_x^2 = \tfrac{1}{2}(\omega_A^2 + \omega_B^2) - \{(\beta_x/m)^2 + [\tfrac{1}{2}(\omega_A^2 - \omega_B^2)]^2\}^{1/2} \tag{75b}$$

The expressions for eigenvalues E_y and E_z are fully analogous. The total energy of the ground state ($\mathbf{N} = \mathbf{n} = 0$) thus decreases by

$$\Delta E = \tfrac{1}{2}\hbar(\Omega_x + \Omega_y + \Omega_z + \omega_x + \omega_y + \omega_z - 3\omega_A - 3\omega_B) \tag{76}$$

which at large R (small β) has the asymptotic form

$$\Delta E \simeq -C_6 R^{-6} \tag{77}$$

where the "van der Waals constant" C_6 is

$$C_6 = \tfrac{3}{2}\hbar e^4 m^{-2}[\omega_A \omega_B(\omega_A + \omega_B)]^{-1} \tag{78a}$$

$$= \tfrac{3}{2}\hbar \alpha_A \alpha_B \omega_A \omega_B/(\omega_A + \omega_B) \tag{78b}$$

The general expression for the $2l$-pole–$2l'$-pole interaction between atoms A and B in states $|A\rangle$ and $|B\rangle$, respectively, is given by second-order perturbation theory to be[127]

$$\epsilon(l, l') = e^4 R^{-2(L+1)}\binom{2L}{2l} \sum_{n,n'} \frac{|\langle A|\sum_i r_i^{(l,0)}|n\rangle|^2 |\langle n'|\sum_j r_j''^{(l,0)}|B\rangle|^2}{E_{nA} + E_{n'B}} \tag{79}$$

where $\mathbf{r}_i$ and $\mathbf{r}_j''$ designate positions of electrons of atoms A and B, respectively, relative to the nuclei (see Figure 4), $r^{(l,0)} \equiv r^l P_l(\hat{\mathbf{r}} \cdot \hat{\mathbf{R}})$, $L \equiv l + l'$, and $\binom{2L}{2l} = (2L)![(2l)!(2l')!]^{-1}$. Asymptotically, the leading dispersion term has $l = l' = 1$ and gives the R^{-6} van der Waals interaction [Eq. (77)] with

$$C_6 = \frac{3}{2}\hbar^4 e^2 m^{-2} \sum_{n,n'} \frac{f_{nA} f_{n'B}}{E_{nA} E_{n'B}(E_{nA} + E_{n'B})} \tag{80}$$

where the f's are oscillator strengths

$$f_{nA} = \tfrac{2}{3} m \hbar^{-2} E_{nA} |\langle A| \sum_i \mathbf{r}_i |n\rangle|^2 \tag{81}$$

Since for the harmonic oscillator, $f_{1A} = 1$ and $f_{nA} = 0$ for $n > 1$, the perturbation-theory result Eqs. (80) and (81) is seen to agree with that from the Drude model Eq. (78) for large R. Numerical values of C_6 and higher-order coefficients are available for many atom pairs.[128–130]

Long-range correlation effects are not included in the electron-gas model (Section 3.3). Several schemes have been proposed for adding van der Waals attraction to the electron-gas potentials[131,132] or to simple analytic forms of the repulsive potential[133,134] (Section 3.3).

3.5. SCF and Variational Methods

Self-consistent-field (SCF) and variational calculations of adiabatic potentials of diatomic systems are essentially extensions of atomic Hartree–Fock and variational treatments to two-center problems.[12,135–137] The total electronic wave function $\phi(\mathbf{r}, \mathbf{R})$ at a fixed internuclear separation R is written as a linear combination of "Slater determinants" $\Phi_k(\mathbf{r})$:

$$\phi(\mathbf{r}, \mathbf{R}) = \sum_k C_k \Phi_k(\mathbf{r}) \tag{82}$$

where each $\Phi_k(\mathbf{r})$ is an antisymmetric linear combination of products of one-electron molecular spin–orbitals φ_s:

$$\Phi_k(\mathbf{r}) = \mathscr{A}\pi\varphi_s \tag{83}$$

and $\mathscr{A}$ is the usual antisymmetrization operator.[138]

The spin orbitals φ_s are usually written as the product of a spin part $[\alpha \equiv \binom{1}{0}$ or $\beta \equiv \binom{0}{1}]$ together with a spatial part containing numerous variational parameters. The parameters commonly appear as expansion coefficients of φ_s in a basis of Slater or Gaussian atomic orbitals $r^{n-1}e^{-ar}Y_{lm}$ or $r^{n-1}e^{-ar^2}Y_{lm}$, respectively, or in terms of two-center functions expressed in terms of confocal ellipsoidal coordinates.[139,140] The form of the orbitals φ_s is often restricted ("restricted Hartree–Fock method") by symmetry requirements.[12]

If there is only one determinant $\Phi_k(\mathbf{r})$ in the expansion [Eq. (82)] of the electronic wave function, the parameters are found by iterative solution of the Hartree–Fock equations which result when $\langle\Phi_k|H_{el}|\Phi_k\rangle$, $H_{el} = -\hbar^2/(2m)\Sigma_i\nabla_i^2 + V(\mathbf{r}, \mathbf{R})$ is minimized with respect to the spin–orbitals φ_s, subject to orthonormalization, $\langle\varphi_s|\varphi_t\rangle = \delta_{st}$:

$$h_1\varphi_s(1) + \sum_t [\langle\varphi_t(2)|g_{12}|\varphi_t(2)\rangle\varphi_s(1) - \langle\varphi_t(2)|g_{12}|\varphi_s(2)\rangle\varphi_t(1)] = \varepsilon_2\varphi_s(1) \tag{84}$$

where [see Eqs. (30) and (31)]

$$h_1 = -\frac{\hbar^2}{2m}\nabla_1^2 - \frac{Z_A e^2}{|\mathbf{R}_A - \mathbf{r}_1|} - \frac{Z_B e^2}{|\mathbf{R}_B - \mathbf{r}|} + V_{NN} \tag{85}$$

and

$$g_{12} = e^2|\mathbf{r}_1 - \mathbf{r}_2|^{-1} \tag{86}$$

In most calculations with open shells, several determinants are necessary to give the required spin symmetry. This is even true of single-configuration calculations, where occupation of the spatial orbitals is fixed. Then off-diagonal energy terms ε_{st} exist between open and closed shells, which complicate the optimization of the orbitals. Various extended Hartree–Fock schemes to handle open-shell calculations have been discussed in the literature.[141–144]

The Hartree–Fock method of course involves approximations, the two most important of which are the neglect of some correlation and of relativistic effects. In a calculation with single configurations, each electron "sees" only the average field of all the others: No correlation is included. Yet for thermal atom–atom interactions the correlation is by no means negligible. Van der Waals forces arise entirely from interatomic correlation, and at large internuclear separations such forces easily dominate the interaction (see Section 3.4). Even the intra-atomic correlation can be important at separations where the atomic wave functions on the two centers begin to overlap. As pointed out in Section 3.3, neglect of intra-atomic correlation generally results in too small a binding energy and hence too large an extent for the valence electrons.

One way to include correlation in the SCF calculations is to employ several configurations in the sum [Eq. (82)] and to vary the coefficients C_K so as to minimize the total average energy $\langle \phi(\mathbf{r}, \mathbf{R}) | H_{el} | \phi(\mathbf{r}, \mathbf{R}) \rangle$. The orbitals used, φ_s can be reoptimized by the SCF procedure as the C_K are varied ("multiconfiguration Hartree–Fock" method) or the determinants Φ_k can be fixed after the initial optimization and *only* the coefficients C_K varied ("configuration interaction" or CI method). Only a small number of excess one-electron molecular spin orbitals permit the construction of many configurations, and computations involving thousands of configurations are no longer rare.[145]

Figure 7 shows how the inclusion of six configurations can permit much of the van der Waals interaction to be calculated between two neutral atoms at large internuclear separations. Individual configurations correspond to correlated dipole moments in the two atoms. Only when the configurations are summed in pairs is the expected spatial symmetry of the total wave function obtained.

1 CONFIGURATION   6 CONFIGURATIONS

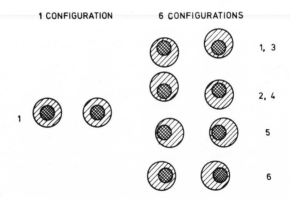

Figure 7. A single-configuration Hartree–Fock calculation predicts no van der Waals interaction (left-hand side). Relative distributions calculated for valence electrons and ion cores are shown schematically. Addition of six polarized configurations (numbered on the right-hand side) allows much of the van der Waals energy to be computed. (Configurations 1 and 2 represent polarization along the x axis, 3 and 4 represent polarization along the y axis, and 5 and 6 represent polarization along the z axis.)

Unfortunately convergence to the correct correlation energy is generally slow and unpredictable. Nevertheless, with a careful and experienced choice of configurations, it seems possible to calculate van der Waals forces with reasonable accuracy, at least in the ground state of relatively simple systems.[146–148] For heavier atoms the time-consuming correlation calculation can be reduced by the frozen-core[149–153] or polarized-core[154,155] approximations, in which core electrons esssentially retain their atomic form. Especially the polarized-core approximation seems well suited for calculations of long-range interactions (see Section 3.6).

Another approach to calculating the correlation contribution is that which Hylleraas[156,157] applied years ago to computations of atomic helium. Instead of using products of one-electron orbitals, the Hylleraas-type wave function includes terms of the form $r_{12}^n e^{-a(r_1+r_2)}$, which can reproduce much of the actual electron correlation.[93,158] Although highly successful in He[159,160] and in H_2,[161–164] the use of Hylleraas-type or pair functions[165] ("geminals") in molecules with more than two electrons has been limited,[166] largely owing to complications of calculating matrix elements of r_{ij}^{-1} between pairs of geminals.

One usually argues that relativistic effects are not important in atom–atom interactions, because it is mainly the *inner* electrons that are affected, and these are essentially unchanged during the interaction. The argument is flawed in that changes in the wave functions of the innermost electrons affect all the electrons, and the *relative* change is largest for the valence electrons, which play the most important role in the interaction. Use of relativistic Hartree–Fock ("Dirac–Fock") methods is now fairly common in atomic calculations[112,167] but still rare in molecular work. They have been employed in connection with pseudopotential calculations, however,[168] and relativistic corrections have been applied to some other results.[163]

Most SCF molecular calculations have been made on ground states. Their application to those excited states that are the lowest state of a given symmetry is, however, no more difficult. One simply ensures that the total wave function $\phi(\mathbf{r}, \mathbf{R})$ has the correct symmetry. The problem does become more difficult when the state desired is not the lowest one of given symmetry, since it is then necessary to solve for the lower states as well. An additional problem is caused by the correlation calculation, which tends to converge more slowly for excited than for ground states.

3.6. Pseudopotential and Semiempirical Methods

In order to make meaningful comparisons between theory and excited-state collision or line-broadening measurements, reliable potentials of excited states are frequently desired for heavy, many-electron systems. Accuracies measured in wave numbers $(1\ cm^{-1} \simeq 4.556 \times 10^{-6}$ a.u.) are sought for systems such as CsXe and Hg_2, and the long-range behavior is often especially important. Such accuracies are probably not attainable from *ab initio* calculations except for the simplest diatomics, with only two or three electrons, and even calculations of considerably less accuracy can be extremely difficult and time consuming. Pseudopotentials with semiempirical parameters offer a possible means of simplifying the computations while increasing their accuracy. Here some applications of semiempirical pseudopotentials to the problem of diatomic interactions are discussed.

It is useful to distinguish three types of pseudopotential. (i) The nonlocal pseudopotential such as that proposed by Phillips and Kleinman[169] and others,[170-172] in which that part of the valence orbitals not orthogonal to core orbitals is projected out. The resulting equations are useful for formal studies, but in the nonlocal form they are not significantly faster to use than a "frozen-core" or "polarized-core" approximation.[149-155] Although they have been applied directly in molecular calculations,[168,173-178] the nonlocal‡ pseudopotentials are more often approximated by (ii) effective or model potentials of simple analytic form.[182-187] The original form suggested by Hellmann[188] was designed to represent the net effect of Pauli pressure and Coulomb interaction with the core. More recent forms[189-193] have included core polarization terms. Parameters in the model potential are usually chosen empirically. (iii) Gombás[95-103] or electron-gas pseudopotentials, on the other hand, contain no adjustable parameters. (Consequently calculations with them are sometimes described—correctly if misleadingly—as *ab initio*.[100]) Applications of Gombás pseudopotentials to the interaction of closed-shell systems has been discussed in Section 3.3. In our formulation of diatomic-potential calculations with pseudopotentials, we concentrate on model and Gombás-type potentials.

The basic aim is to represent a system of two many-electron atoms by closed-shell polarizable cores plus valence electrons. One starts with model potentials $V_A(r)$ and $V_B(r)$ for the isolated atoms. These are constrained to have the correct limiting forms for one-electron interactions at small and large r. For example $V_A(r)$ obeys

$$V_A(r) \xrightarrow{r \to 0} -Z_A e^2/r \qquad (87a)$$

$$V_A(r) \xrightarrow{r \to \infty} -(\mathcal{Z}_A+1)e^2/r - \tfrac{1}{2}\alpha_A e^2/r^4 \qquad (87b)$$

‡ Goddard[179] and co-workers[177,178] use nodeless wave functions from the G1 method[180,181] to derive a unique local form for Phillips–Kleinman-type pseudopotentials.

where $Z_A e$ is the nuclear charge, $\jmath_A e$ the ionic charge, and α_A the polarizability of the core of atom A. The potential at intermediate r is chosen so as to reproduce spectroscopic energies or scattering phase shifts. The potential $V_A(r)$ may be a function of angular momentum,[13] but, in contrast to Phillips–Kleinman potentials, it is usually taken to be independent of energy, ensuring that the one-electron eigenfunctions of the Hamiltonian are orthogonal. Arbitrariness in the form of $V_A(r)$ for intermediate r can be reduced by starting with the Hartree–Fock Coulomb potential of the core.[124]

Usually the parameters of $V_A(r)$ are chosen so as to represent both Coulomb and Pauli-pressure interactions. All eigenfunctions of the Hamiltonian then correspond to physical states and the number of nodes in the model-potential one-electron wave functions of angular momentum l is n_l less than in the physical wave function, where n_l is the number of l-shells occupied by core electrons. If the inner nodal structure of the valence wave functions is important it may be preferable to omit the Pauli pressure and include the (attractive) exchange interaction in $V_A(r)$. The average kinetic energy of the electrons of given energy E is then much higher and the approximate inner nodal structure can be reproduced. The lowest-energy eigenfunctions of the Hamiltonian are, however, unphysical and correspond to core states. In calculations of atomic properties, the low-lying unphysical states are easily avoided, but in variational molecular computations their avoidance must be ensured by imposing orthogonality constraints, a significant complication.

Note that, when two or more valence electrons are present, the polarization term $\frac{1}{2}\alpha E^2$ resulting from the dipole induced in the core by the electric field $\mathbf{E}$ of all valence electrons (see Section 3.4) gives a coupling between electrons of the form

$$-\alpha e^2 \mathbf{r}_1 \cdot \mathbf{r}_2 / r_1^2 r_2^2 \qquad (88)$$

when the radial positions r_1 and r_2 of the electrons are large compared to the core size. The two-electron "direct" and "exchange" integrals over the coupling [Eq. (88)] are, however, less difficult to perform than those for the interelectronic repulsion, $e^2|\mathbf{r}_1 - \mathbf{r}_2|^{-1}$, and frequently vanish by symmetry considerations, as for example when one of the orbitals represents an atomic s state.

Recent calculations have demonstrated the power of atomic model potentials for one- and two-electron atoms. The comparisons made and reviewed by Bardsley[13] show that when correlation effects such as core polarization and interaction among valence electrons are adequately treated, atomic properties involving primarily the valence electron can be calculated to high accuracy with efforts that are small compared to Hartree–Fock calculations of much lower accuracy.

Adiabatic potentials for the diatomic system are constructed from (1) the total energy of the valence electrons of both atoms moving in $V_A + V_B$ plus (2) the interaction of the cores as calculated, say, by the electron-gas model (Section 3.3) plus (3) long-range van der Waals interaction of the *bare* cores. The larger van der Waals interaction arising at large R from the polarization of one core by valence electrons on the other is automatically included in the Hamiltonian of the valence electrons.[103,192,193]

There are questions about the formal justification of the procedure.[12] Once the cores begin to overlap, the potentials V_A and V_B are not strictly additive, partially because of distortion in the cores at small R and partially because the Pauli-pressure effects themselves are nonadditive (see Section 3.3). The nonadditive Pauli-pressure contribution to the valence-electron contribution is readily found from the overlapping core densities, and some of the distortion effects, for example the interaction between the dipoles induced in the two cores by the valence electrons, are also easy to include. Even without such corrections, however, recent calculations of one- and two-valence electron systems[103,189-197] demonstrate the ability of pseudopotential calculations to provide quantitatively useful information about both ground and excited states, and it appears possible to blame most inaccuracies, especially prevalent in earlier calculations (see Goodisman's comparisons[12]), on neglect of correlation, inadequate core–core interactions, and insufficiently flexible valence-electron wave functions.

4. Relation to Collision Processes; Scattering Theory

4.1. Time-Dependent Classical-Path Formulations

Books and numerous review articles have been written about scattering theory, and many are relevant here.[15,16,198-201] Although many variants of scattering theory have been employed, treatments applied to atom-atom interactions can be organized under the three subheadings of this section: time-dependent classical-path formulations, semiclassical and JWKB approximations, and full quantum treatments. While we touch on some concepts and approximations to scattering theory needed later in this book, we will emphasize the practical computation of cross sections or relaxation rates and relations to interatomic potentials. The classical-path formulation, although less rigorous than full quantum treatments, provides a more intuitive picture of the collision dynamics and has therefore been preferred for the introduction of basic concepts. For details about the scattering process and discussion of the measurements, the reader is referred to Chapters 20 and 27–29 of this work.

4.1.1. Impact-Parameter S Matrix

In the time-dependent classical-path treatment, one solves the time-dependent Schrödinger equation in the interaction picture (see Chapter 2 of this book)

$$i\hbar\dot{\boldsymbol{\Psi}}(t) = \mathbf{V}_I(t) \cdot \boldsymbol{\Psi}(t), \qquad \mathbf{V}_I(t) = e^{i\mathbf{H}_0 t/\hbar} \mathbf{V}(t) e^{-i\mathbf{H}_0 t/\hbar} \qquad (89)$$

where $\boldsymbol{\Psi}$ is the electronic state of the system, written as a time-dependent vector in the space of product atomic states, $\mathbf{H}_0$ is the sum of the two atomic Hamiltonians, and $\mathbf{V}(t) \equiv \mathbf{V}[R(t)]$ is the collisional perturbation whose time dependence is determined by the time dependence of the classical trajectory $R(t)$, not necessarily a straight line. Integration of Eq. (89) once from $t \to -\infty$ to $t \to +\infty$ for each dimension of the basis permits one to relate $\boldsymbol{\Psi}(+\infty)$ to $\boldsymbol{\Psi}(-\infty)$:

$$\boldsymbol{\Psi}(+\infty) = \mathbf{S}(b, v) \cdot \boldsymbol{\Psi}(-\infty) \qquad (90)$$

The relation is given by the S matrix, which is a function of impact parameter b and initial relative velocity v.

The determination of $\mathbf{S}$ is usually simpler and its symmetry better displayed by use of the time-development operator $\mathbf{U}(t, -t)$, defined by

$$\boldsymbol{\Psi}(t) = \mathbf{U}(t, -t)\boldsymbol{\Psi}(-t) \qquad (91)$$

Substitution of Eq. (91) into the Schrödinger equation (89) gives

$$i\hbar\dot{\mathbf{U}}(t, -t) = \mathbf{V}_I(t) \cdot \mathbf{U}(t, -t) + \mathbf{U}(t, -t) \cdot \mathbf{V}_I(-t) \qquad (92)$$

Since $\mathbf{U}(0, 0) = \mathbf{1}$, the unit matrix, a single integration from $t = 0$ to ∞ gives

$$\mathbf{S}(b, v) \equiv \mathbf{U}(\infty, -\infty) \qquad (93)$$

The integration can be checked at any t by testing for unitarity $\mathbf{U}^+(t, -t) = \mathbf{U}^{-1}(t, -t)$, which follows from the hermiticity of H_0 and $V(t)$. If in addition time and coordinate axes are chosen so that

$$\mathbf{V}^*(t) = \mathbf{V}(-t) \qquad (94)$$

then $\mathbf{U}(t, -t)$ is symmetric,

$$\mathbf{U}(t, -t) = \tilde{\mathbf{U}}(t, -t) \qquad (95)$$

which allows a considerable reduction in the integration of $\dot{\mathbf{U}}$ [Eq. (92)].

4.1.2. Cross Sections and Thermal Averages

Off-diagonal elements of $\mathbf{S}(b, v)$ give net transition amplitudes for collisions characterized by (b, v). By combining $|S(b, v)|^2$ with the deflection function (Section 2.1) for the classical-path potential, differential inelastic cross sections can be found, but these are less likely to be accurate than the total inelastic cross sections, given by

$$\sigma_{nm}(E) = 2\pi \int_0^\infty b\,db\,|S_{nm}(b, v)|^2 \tag{96}$$

where n and m label final and initial states and E is the initial relative kinetic energy $\frac{1}{2}\mu v^2$.

Measurements usually give a thermally averaged isotropic cross section Q_{fi} for transitions from a set of g_i initial states to g_f final states. For a Maxwellian distribution of velocities Q_{fi} is simply

$$Q_{fi} = \frac{1}{g_i} \sum_{\substack{n \in f \\ m \in i}} \int_0^\infty dx\, x\, e^{-x} \langle \sigma_{nm}(E) \rangle_{\hat{\mathbf{b}},\hat{\mathbf{v}}}, \qquad x = \frac{E}{kT} \tag{97}$$

where $\langle\ \rangle_{\hat{\mathbf{b}},\hat{\mathbf{v}}}$ indicates an average over all orientations $(\hat{\mathbf{b}}, \hat{\mathbf{v}})$ of the collision frame. The average over $(\mathbf{b}, \hat{\mathbf{v}})$ can be performed—without new S-matrix calculations—by applying rotation matrices directly to the atomic basis (see, for example, Chapter 28 of this work).

4.1.3. Detailed Balance

Let E_{fi} be the internal energy gained during collision and assume for simplicity that there is no net internal angular momentum change. The cross section for the reverse reaction $|n\rangle \to |m\rangle$ is

$$\sigma_{mn}(E) = 2\pi \int_0^\infty b'\,db'\,|S_{mn}^{-1}(b', v')|^2 \tag{98}$$

where by conservation of energy and angular momentum

$$E' = \tfrac{1}{2}\mu v'^2 = E - E_{fi} = \tfrac{1}{2}\mu v^2 - E_{fi} \tag{99}$$

and

$$Eb^2 = E'b'^2 \tag{100}$$

Because of the unitarity of $\mathbf{S}$, $|S_{mn}^{-1}|^2 = |S_{nm}|^2$ and Eqs. (98)–(100) give directly the detailed-balance relation

$$E\sigma_{nm}(E) = E'\sigma_{mn}(E'), \qquad E > E_{fi} \tag{101}$$

E_{fi} is a threshold energy for the reaction $|m\rangle \to |n\rangle$. When substituted into Eq. (97), detailed balance yields

$$Q_{fi} = \frac{g_f}{g_i} e^{-E_{fi}/kT} Q_{if} \tag{102}$$

a useful relation between forward and reverse reaction thermal cross sections.

4.1.4. Density-Matrix Formulation; Relaxation Rate

For collisions in a gaseous medium, the atomic density matrix ρ (see Chapter 2 of this book) is often more convenient than the wave function Ψ. In the interaction picture, a single collision changes ρ to $S \cdot \rho \cdot S^+$, and the net result of collisions in a thermal, isotropic medium of collision partners of density n is to cause a time-rate of change of density-matrix elements

$$\dot{\rho}_{nn'} = -(\gamma \cdot \rho)_{nn'} = -\sum_{mm'} \gamma_{nn'}^{mm'} \rho_{mm'} \qquad (103)$$

where the collisional relaxation rate is a "Liouville" operator γ defined by

$$\gamma \cdot \rho = n\bar{v} \int_0^\infty dx\, x\, e^{-x} 2\pi \int_0^\infty b\,db \langle \rho - S \cdot \rho \cdot S^+ \rangle_{\hat{v},\hat{b}} \qquad (104)$$

with, again, $x = E/kT$. Comparison to Eqs. (96) and (97) leads to the identification

$$\gamma_{nn}^{mm} = n\bar{v} Q_{nm} \qquad (105)$$

4.1.5. Perturbation Expansion

The dynamics of the collision, and hence the connection to the interatomic potentials, is contained in the S matrix. The integration of Eq. (89) or (92) to obtain S can be written formally

$$S = \mathcal{O}_T \exp\left[-i \int_{-\infty}^\infty \frac{V_I(t)}{\hbar}\right] \qquad (106)$$

where $\mathcal{O}_T$ is the time-ordering operator whose application in Eq. (106) gives the Dyson expansion:

$$\mathcal{O}_T \exp\left[-i \int_{-\infty}^\infty dt\, \frac{V_I(t)}{\hbar}\right] = 1 - \frac{i}{\hbar} \int_{-\infty}^\infty dt_1\, V_I(t_1)$$
$$- \frac{1}{\hbar^2} \int_{-\infty}^\infty dt_1 \int_{-\infty}^\infty dt_2\, V_I(t_1)V_I(t_2) + \cdots \qquad (107)$$

The "physics" of the collision is clearer in the approximations to S. In nth-order perturbation theory, only terms in the Dyson expansion containing up to n factors of V_I are retained and the result can be represented by a sum of all distinct diagrams containing no more than n interaction lines. First-order perturbation theory, corresponding to the first Born approximation when calculated with straight paths, is valid only for very weak interactions, for which the largest matrix element of $\int_{-\infty}^\infty dt\, V_I(t)$ is small compared to $\hbar$. If $V_I(t)$ is too large, the approximated S is seriously nonunitary.

4.1.6. Magnus Approximation

A unitary version of the first-order theory is the "exponential" or Magnus[202–204] approximation in which the time-ordering in Eq. (106) is omitted:

$$\mathbf{S} = e^{2i\eta} \tag{108}$$

where from the definition of $\mathbf{V}_I$ [Eq. (89)], the phase-shift matrix η has matrix elements

$$\hbar \eta_{nm} = -\frac{1}{2} \int_{-\infty}^{\infty} dt\, [\mathbf{V}_I(t)]_{nm} = -\frac{1}{2} \int_{-\infty}^{\infty} dt\, e^{i\omega_{nm}t} V_{nm}(t) \tag{109}$$

The approximation is equivalent to ignoring commutators[16,202] $[V(t), V(t')]$, and although it does not, outside the range of validity of first-order perturbation theory, reliably predict relative cross sections among Zeeman sublevels, it often does yield useful estimates of Zeeman-level averaged cross sections even when the interaction is strong.

4.1.7. Frequency Spectrum of Collision; Massey Parameter

According to Eq. (109) the phase shift η_{nm} is the Fourier transform or "frequency spectrum" of the interaction $V_{nm}(t)$ at the frequency ω_{nm} corresponding to the separation of the atomic levels n and m: The probability for a direct $n \leftrightarrow m$ transition is sizable only if the frequency spectrum of $V_{nm}(t)$ has a significant component at ω_{nm}, i.e., if the shortest time Δt over which $V_{nm}(t)$ changes significantly is small compared to ω_{nm}^{-1}. Putting $\Delta t = a/v$, where a is thus the smallest characteristic distance for changes in V_{nm}, we define a "Massey parameter"

$$\lambda \equiv \Delta t\, \omega_{nm} = \frac{aE_{nm}}{\hbar v} \tag{110}$$

Only when $|\lambda| \lesssim 1$ do we expect direct transitions $n \leftrightarrow m$ to be likely. For a typical thermal atomic collision with $a \simeq 1$ Å and $v \simeq 1$ km/s, we have $(\hbar v/a)(1/hc) \simeq 50$ cm^{-1}, so we expect a rather sharp falloff in thermal cross section as energy separations E_{nm} increase through 50 cm^{-1}. Indeed this is often found, as for example in alkali fine-structure transitions induced by collisions with noble-gas atoms[90] (see also Chapter 29 of this work).

The above discussion is rather simplistic in that the transition probability $|S_{nm}|^2$ depends generally on all the matrix elements of η. Indeed, $\mathbf{S}$ as given in Eq. (108) is usually found by determining the unitary transformation $\mathbf{A}$ that diagonalizes η and then using

$$\mathbf{S} = \mathbf{A}^{-1} \cdot \exp(2i\mathbf{A} \cdot \eta \cdot \mathbf{A}^{-1}) \cdot \mathbf{A} \tag{111}$$

In the two-state case, the result is analytic and one finds

$$S = e^{i(\eta_{11}-\eta_{22})} \begin{pmatrix} \cos \beta - i(\eta_{11}-\eta_{22})(\sin \beta/\beta) & 2i\eta_{12}(\sin \beta/\beta) \\ 2i\eta_{12}^{*}(\sin \beta/\beta) & \cos \beta + i(\eta_{11}-\eta_{22})(\sin \beta/\beta) \end{pmatrix} \tag{112}$$

where

$$\beta = [(\eta_{11}-\eta_{22})^2 + 4|\eta_{12}|^2]^{1/2} \tag{113}$$

In particular, the transition probability is

$$|S_{12}|^2 = 4|\eta_{12}|^2 (\sin^2 \beta)/\beta^2 \tag{114}$$

which clearly shows the importance of η_{11} and η_{22} as well as η_{12}.

4.1.8. Sudden Approximation

If $V(t)$ acts over times short compared to ω_{nm}^{-1}, η_{nm} can be approximated by setting $\omega_{nm} = 0$ in Eq. (109). This gives the "sudden" approximation.[205,206] If $V(t)$ is constant except for isolated times, the total S matrix can be built up of a product of S-matrix factors [Eqs. (108) and (109)].

4.1.9. Diabatic Basis

If instead of splitting the Hamiltonian H into $H_0 + V(t)$ we write $\mathbf{H} = \mathbf{H}_1(t) + \Delta\mathbf{V}(t)$, where $\mathbf{H}_1(t) = \mathbf{H}_0 +$ diagonal part of $\mathbf{V}(t)$ and $\Delta\mathbf{V}(t) =$ off-diagonal part of $V(t)$, then in place of Eq. (89) we find the diabatic evolution

$$i\hbar\dot{\mathbf{\Psi}} = \left\{ \exp\left[i \int_0^t dt' \mathbf{H}_1(t')/\hbar \right] \Delta\mathbf{V}(t) \exp\left[-i \int_0^t dt' \mathbf{H}_1(t')/\hbar \right] \right\} \cdot \mathbf{\Psi}(t) \tag{115}$$

The S matrix and its approximations may be found as before except in place of matrix elements $[\mathbf{V}_I(t)]_{nm} = e^{i\omega_{nm}t} V_{nm}(t)$ we have $\exp[i \int_0^t dt' \, v_{nm}(t')] \Delta V_{nm}(t)$ where $v_{nm}(t) = \omega_{nm} + V_{nn}(t) - V_{mm}(t)$. The first-order perturbation theory on Eq. (115) corresponds to the distorted-wave Born approximation. In the two-state Magnus approximation, $|S_{12}|^2$ [Eq. (114)] becomes simply

$$|S_{12}|^2 = \sin^2 \beta, \qquad \hbar\beta = -\int_{-\infty}^{\infty} dt \, e^{-i\int_0^t dt' \, v_{12}(t')} V_{12}(t) \tag{116}$$

The probability for transition between two states can be seen to be small unless the *diabatic* energy separation $\hbar v_{nm}$ becomes small compared to $\hbar \Delta t^{-1}$, where Δt is the time scale for changes in V_{nm}.

4.1.10. Landau–Zener Formula

If the diabatic levels cross $(\nu_{12} = 0)$ at some $t = \pm t_c$, the phase $i \int_0^t dt' \, \nu_{12}(t)$ is stationary there and the corresponding contribution to $|S_{12}|^2$ is large. Another approximation suggests itself in this case: $V_{12}(t)$ is replaced by $V_{12}(t_c)$ and $\nu_{12}(t)$ by $\dot{\nu}_{12}(t_c)(t - t_c)$. If $|\dot{\nu}_{12}(t_c)|t_c^2 \gg 1$, contributions to β [Eq. (116)] are well isolated and can simply be added. The Magnus approximation then yields

$$|S_{12}|^2 = \sin^2 \beta, \qquad \hbar\beta = -2 V_{12}(t_c) |\dot{\nu}_{12}(t_c)/\pi|^{-1/2} \tag{117}$$

Landau[207] and Zener[208] have shown that if V_{12} and $\dot{\nu}_{12}$ are constant, the transition probability between diabatic curves can be found exactly.[209] Assuming again that the crossings at $\pm t_c$ are well isolated, one obtains

$$|S_{12}|^2 = 2 \, e^{-\beta^2/2}(1 - e^{-\beta^2/2}) \tag{118}$$

where β is as in Eq. (117). The Magnus approximation is valid in this case only if $\beta^2 \ll 1$. If $\beta^2 \gg 1$ the Landau–Zener result [Eq. (117)], in contrast to the Magnus approximation, gives a vanishingly small $|S_{12}|^2$. The conditions that only two states be involved, that V_{12} and $\dot{\nu}_{12}$ be constant over the crossing region, and that $|\dot{\nu}_{12}(t_c)|t_c^2 \gg 1$ severely limit the number of cases for whch the Landau–Zener formula can be expected to give quantitative results. Nevertheless, it has found wide application, and numerous extensions have been discussed.[15,16]

4.1.11. Adiabatic Basis

The relation of scattering to adiabatic potentials is more direct in an adiabatic basis where the Hamiltonian $\mathbf{H}_0 + \mathbf{V}(t)$ is diagonalized by a time-dependent unitary matrix $\mathbf{A}(t)$ whose columns contain the coefficients of the adiabatic states:

$$\mathbf{A}^{-1} \cdot [\mathbf{H}_0 + \mathbf{V}(t)] \cdot \mathbf{A} = \boldsymbol{\varepsilon}(t) \tag{119}$$

where the elements of the diagonal matrix $\boldsymbol{\varepsilon}$ are the adiabatic potentials [see Eq. (32)]. The time dependence of states $\boldsymbol{\Psi}$ in an *adiabatic* basis follows directly from the Schrödinger equation. Using again the interaction picture we find

$$i\hbar\dot{\boldsymbol{\Psi}} = \left\{ \exp\left[i \int_0^t dt' \, \boldsymbol{\varepsilon}(t')/\hbar \right] \mathbf{V}_A(t) \exp\left[-i \int_0^t dt' \, \boldsymbol{\varepsilon}(t')/\hbar \right] \right\} \cdot \boldsymbol{\Psi} \tag{120}$$

where the adiabatic perturbation is the matrix representaion of $-i\hbar \, \partial/\partial t$:

$$\mathbf{V}_A(t) = -i\hbar \mathbf{A}^{\dagger}(t) \cdot \dot{\mathbf{A}}(t) \tag{121}$$

The matrix $\mathbf{A}$ can be written as a product of a time-dependent rotation by an angle θ in the collision plane and a transformation depending only on internuclear separation R. The perturbation can thus be conveniently expanded

$$-i\hbar \frac{\partial}{\partial t} = -i\hbar\dot{\theta}\frac{\partial}{\partial \theta} - i\hbar\dot{R}\frac{\partial}{\partial R} \qquad (122)$$

into rotational and radial coupling terms. The latter only couples states of the same Ω value, whereas the former can couple states differing in Ω values by ± 1. (Note from the rotational part of $\mathbf{A}$ that the matrix representation of $-i\hbar\,\partial/\partial\theta$ is simply the projection $\mathbf{j}\cdot\hat{\mathbf{n}}$ of the total internal angular momentum normal to the collision plane.)

In calculating an S matrix from the adiabatic basis, one must note that the final adiabatic basis is rotated an angle $\hat{\mathbf{n}}\int_{-\infty}^{\infty} dt\,\theta(t)$ with respect to the initial one. The inverse rotation needs to be applied before S is computed. The adiabatic basis is especially useful when nonadiabatic couplings are small and hence accurately reproduced by perturbation theory or the Magnus approximation. Applications have been discussed by Nikitin.[15]

4.1.12. Validity of Classical-Path Treatments

Even without approximations in the derivation of S from the Schrödinger equation [e.g., Eqs. (89) or (92)], the classical-path treatment is liable to be in error whenever trajectories calculated in the different adiabatic potentials of allowed channels differ significantly from each other or when quantum mechanical tunneling is important. The classical-path treatment is usually justified as an approximation to semiclassical theory valid when initial- and final-state trajectories are similar (see Section 4.2). Paths dependent on internal state amplitudes have been proposed as a means of generalizing the classical-path approach.[210,211] When orbiting collisions or quasibound resonances[72,73] substantially influence the scattering, however, the classical-path approach must be abandoned in favor of a semiclassical or preferably a full quantum treatment for any but qualitative results.

4.2. Semiclassical and JWKB Approximations

One-dimensional JWKB solutions to the elastic-scattering problem were considered in Section 2.1. There a single adiabatic potential $\varepsilon_0(R)$ sufficed to describe the motion and the internal (electronic) coordinates could be ignored. Here the extension of semiclassical methods to inelastic scattering is sketched.

The internal motion of the system may be either quantized or treated semiclassically. In the former case only the internuclear motion is subject to the JWKB approximation, whereas in the latter, the entire system— electronic (also vibrational and rotational for molecules) coordinates as well as internuclear ones—is handled semiclassically. Quantal treatments of internal motion, which of course are better justified and should lead to more accurate results, give scattering matrices $S_{nm}^{(l)}$ depending on sets $\mathbf{n}$ and $\mathbf{m}$ of discrete internal quantum numbers. The fully semiclassical theory yields S matrices as a function of continuous action variables $\mathbf{N}$ and $\mathbf{M}$ corresponding to $\mathbf{n}$ and $\mathbf{m}$.

When the internal degrees of freedom are quantized, the multichannel equations take the form

$$[\nabla_{\mathbf{m}}^2 + k_m^2 - U_{mm}(\mathbf{R})]\chi_m(\mathbf{R}) = -\sum_{\substack{n \\ n \neq m}} U_{mn}(\mathbf{R})\chi_n(\mathbf{R}) \qquad (123)$$

where $\hbar k_m = (2\mu E_m)^{1/2}$ is the wave vector, $U_{mm}(\mathbf{R})$ and $U_{mn}(\mathbf{R})$ are $(2\mu/\hbar^2)$ times the diagonal potential and the corresponding coupling elements in the chosen basis (e.g., diabatic or adiabatic), and $\chi_m(R)$ is the wave function for internuclear motion in the mth channel [see for example Eq. (35) in Section 3.1]. The wave function depends implicitly on the angular-momentum quantum numbers (see Section 3.2) J^2, M_J, and L^2 as well as on the electronic state. By writing $\chi_m(\mathbf{R})$ as a product of radial and angular parts

$$\chi_m(\mathbf{R}) = (1/R)\psi(R)Y_{L_m M_m}(\hat{\mathbf{R}}) \qquad (124)$$

and evaluating the angular averages

$$\bar{U}_{mn} \equiv \int_{4\pi} d\hat{\mathbf{R}} \, Y_{L_m M_m}^*(\hat{\mathbf{R}})U_{mn}(\mathbf{R})Y_{L_n M_n}(\hat{\mathbf{R}}) \qquad (125)$$

one arrives at the coupled radial equations

$$\left[\frac{d^2}{dR^2} + k_m(R)\right]\psi_m(R) = -\sum_m \bar{U}_{mn}(R)\psi_n(R) \qquad (126)$$

where

$$k_m(R) = [k_m^2 - L_m(L_m + 1)R^{-2} + \bar{U}_{mm}(R)]^{1/2}$$

If solutions that are regular at the origin are found in the form‡

$$\psi_m(R) = A_{mn}(R)k_m^{-1/2}(R)\exp\left[i\int^R dr\, k_m(r)\right]$$
$$+ \delta_{nm}k_n^{-1/2}\exp\left[i\int^R dr\, k_n(r)\right] \qquad (127)$$

‡ Airy functions Ai (x) and Bi (x) may easily be used in place of the normal exponential forms of the JWKB approximation.[229] They have the advantage that they are easily continued into the classically forbidden region.

i.e., as a linear combination of JWKB solutions, the coefficients $A_{mn}(\infty)$ are, within a phase factor, just the S-matrix elements S_{mn}.

In applications by Stueckelberg[212] to the two-state problem, the two coupled, second-order equations [Eq. (126)] are combined into a single fourth-order equation whose solution, written in the form $\exp\{\hbar^{-1}[S_0 + \hbar S_1 + \hbar^2 S_2 + \cdots]\}$, is found to order $\hbar$. To connect solutions at small and large R, the potentials involved are analytically continued into the complex R plane and phase integrals are calculated on complex trajectories. (Techniques similar to this "Zwaan-Stueckelberg method"[213] have also been used in the time-dependent classical-path formulation[15] and in a related momentum-space formulation.[16] Applications to line-broadening theory have been made as well.[80])

In the case of a Landau–Zener-type crossing of diabatic levels (see Section 4.1), the Zwaan–Stueckelberg method yields

$$|S_{12}|^2 = 4 \, e^{-\beta^2/2}(1 - e^{-\beta^2/2})\sin^2 \tau \tag{128}$$

which is simply $2\sin^2 \tau$ times the Landau–Zener result [Eq. (118)], where

$$\tau = \int^R dr[k_+(r) - k(r)] \tag{129}$$

and

$$2k_\pm^2(r) = k_1^2(r) + k_2^2(r) \pm \{[k_1(r) - k_2(r)]^2 + 4\bar{U}_{12}^2\}^{1/2} \tag{130}$$

In a random-phase approximation for τ, $\sin^2 \tau \approx 1/2$ and the "Stueckelberg–Landau–Zener" (SLZ) result reduces to that of Landau and Zener. The phase τ can be seen to arise from competing trajectories on adiabatic potentials. In comparisons of SLZ and more complete quantum calculations, the importance has been emphasized of having a well-defined and isolated crossing region.[214,215] Stueckelberg has also considered noncrossing potentials and pairs of potentials with more than one complex crossing point. Rederivations, corrections, and extensions of Stueckelberg's work have been widely published.[199,209,216–221] Unfortunately errors in calculations of the phase differences seem rather easy to make. Some of the pitfalls have been discussed by Crothers.[213,217]

Now we turn to the fully semiclassical theory, in which both relative nuclear motion and internal motion is treated semiclassically. One first seeks action variables **N** corresponding to internal quantum numbers n as well as the conjugate "angle" variables φ. Then, in one derivation, a JWKB-like function is substituted into the Schrödinger equation and its amplitude and phase are found by expanding in powers of $\hbar$.[222–229] In another derivation Feynman's path-integral formulation of quantum mechanics[230,231] is used to construct a semiclassical propagator[200,232–237] that is directly related to the S matrix. In both derivations trajectories are calculated classically, but phases accumulated along the trajectories can give rise to quantum interference phenomena.

The S matrix found must be averaged over initial values of the internal variables φ. Changes in these variables change the trajectory and the phase. For given initial and final "quantum numbers" (more precisely: values of the action variables $\mathbf{N}$) the phase will be stationary at some values of φ, and a stationary-phase or higher-order[222,223,225] ("uniform") approximation can be made in the integration over φ. If the transition investigated is classically forbidden, the value φ_c where the phase is stationary is complex and the transition probabilities decrease roughly expoenentially as $|\text{Im}\, \varphi_c|$ increases.

Most applications reported[16,200,232] of the fully semiclassical theory have treated only vibrational and rotational transitions. However, Miller[200] has discussed an extension to electronic transitions.[238] Using the path-integral approach, he has shown how one can consistently handle electronic motion quantally and other degrees of freedom semiclassically. One formally obtains a path integral over the product of the impact-parameter S matrix (Section 4.1) and the nuclear-motion phase factor both calculated for a given path. With a first-order perturbation approximation of electronic-transition phase contributions and a steepest-descent evaluation of the resultant path integral, Miller regains essentially the Stueckelberg results for the case of two electronic states. Miller's treatment[200] is attractive not only because it unifies various approaches, namely, time-dependent impact parameter, JWKB, and fully semiclassical approximations, but also because with it improvements and extensions of these approaches can be formulated.

4.3. Full Quantum Treatment

If the number of internal states to be considered is small, say equal to $N \lesssim 20$, it may be practicable to integrate the coupled radial equations [Eq. (126)] directly using perhaps a Numerov‡[239–242] or de Vogelaere[243,244]

---

‡ A Numerov-like algorithm for coupled second-order differential equations of the form $\mathbf{y}'' + \mathbf{G} \cdot \mathbf{y} = 0$ can be performed with no matrix inversions and with only one matrix multiplication per step, a considerable saving over the usual algorithms.[239–242] Let the indices $-$, 0, and $+$ indicate values at three successive positions, $x - h$, x, and $x + h$. A first approximation to $\mathbf{y}_+$, correct to order h^4, is $\mathbf{y}_+^{(1)} = 2\mathbf{y}_0 - \mathbf{y}_- - h^2 \mathbf{G}_0 \cdot \mathbf{y}_0$ which can be improved by a fourth-order correction $\mathbf{y}_+ = \mathbf{y}_+^{(1)} + (h^2/12)(2\mathbf{G}_0 \cdot \mathbf{y}_0 - \mathbf{G}_- \cdot \mathbf{y} - \mathbf{G}_+ \cdot \mathbf{y}_+^{(1)})$. The truncation error, although of order h^6, is 20 times larger than for standard Numerov. However, an additional iteration can reduce by a factor of 30 if desired. The fourth-order correction can be used to test the accuracy as, say, a criterion for adjusting the step size.

procedure, the "Gordon program"[245,246] or the "log-derivative" method.[247] (Also possible are related but less direct procedures,[248–250] which we do not consider here.) The set of equations (126) must be solved N times to obtain N linearly independent solutions. Equivalently, the matrix equation

$$\boldsymbol{\psi}'' + (\mathbf{k}^2 + \bar{\mathbf{U}}) \cdot \boldsymbol{\psi} = 0 \tag{131}$$

must be solved once, where $\boldsymbol{\psi}$ is an $N \times N$ matrix, each column of which is to contain a linearly independent wave function, $\bar{\mathbf{U}}$ is the matrix with elements $\bar{U}_{mn}$ [see Eq. (125)] and $\mathbf{k}^2$ is the diagonal matrix with elements $k_m^2(R)$.

The numerical integration is started near $R = 0$ with the condition $\boldsymbol{\psi}(R) \to 0$ as $R \to 0$. If only open channels are considered for the moment, the integration can proceed outward until $\bar{\mathbf{U}} \approx 0$ and $\mathbf{k}^2 = \text{const.}$ $\boldsymbol{\psi}$ can then be expressed in its asymptotic form

$$\boldsymbol{\psi}(R) \overset{R \to \infty}{\sim} \mathbf{j}(R)\,\mathbf{A} - \mathbf{n}(R) \cdot \mathbf{B} \tag{132}$$

where $\mathbf{j}(R)$ and $\mathbf{n}(R)$ are diagonal matrices with elements $k_m^{1/2} R j_{L_m}(k_m R)$ and $k_m^{1/2} R n_{L_m}(k_m R)$, respectively [compare Eq. (6) in Section 2.1]. The scattering matrix is then ‡

$$\mathbf{S} = (\mathbf{A} + i\mathbf{B}) \cdot (\mathbf{A} - i\mathbf{B})^{-1} \tag{133}$$

or equivalently in terms of the real symmetric "reactance" matrix $\mathbf{R} = \mathbf{B} \cdot \mathbf{A}^{-1}$

$$\mathbf{S} = (\mathbf{1} + i\mathbf{R}) \cdot (\mathbf{1} - i\mathbf{R})^{-1} \tag{134}$$

The solutions (i.e., columns of $\boldsymbol{\psi}$) must be linearly independent, otherwise $(\mathbf{A} - i\mathbf{B})^{-1}$ will be singular and $\mathbf{S}$ indeterminate. To ensure linear independence and a sufficiently small starting R value when off-diagonal elements of $(\mathbf{k}^2 + \bar{\mathbf{U}})$ are large, it may be necessary to diagonalize $(\mathbf{k}^2 + \bar{\mathbf{U}})$ near the starting point. If $\mathbf{C}^{-1} \cdot (\mathbf{k}^2 + \bar{\mathbf{U}}) \cdot \mathbf{C}$ is diagonal at R_s the columns of $\mathbf{C}$ are orthogonal adiabatic states of the effective potential (including any rotational coupling) at R_s. Of course the diagonal elements of $\mathbf{C}^{-1} \cdot (\mathbf{k}^2 +$

---

‡ The formulas here differ by a transposition from those given in Chapter 6 of Child's useful book[16] because, in spite of his statement at the bottom of p. 93 that *columns* of his matrix **M** are independent solutions to the coupled equations, his equations are consistently written for a matrix **M** whose *rows* (M_{ij}, $j = 1, 2, \ldots, N$) contain the components ψ_j of independent solutions. Compare, for example, Lester's paper.[244]

$\bar{U}) \cdot C$ must be sufficiently negative to ensure that the solutions found are regular at $R = 0$.

Numerical instabilities in the integration can arise for heavy-particle collisions at high energies because of the short de Broglie wavelength for internuclear motion. The approximate method of constructing piecewise analytic solutions[245] (in particular Gordon's program[246] using linear segments to approximate the effective potential and Airy-function solutions for the wave functions) are unhindered by small wavelength. The step size h is limited simply by the truncation error in the solutions (h^3 in the Gordon program for coupled equations compared to h^5 for uncoupled ones or to h^6 for coupled equations solved by a Numerov procedure) rather than by the wavelength. If classical turning points are widely separated, exponential growth of one or more components from error levels may cause problems that require the integration of the logarithmic derivative $\psi' \cdot \psi^{-1}$ rather than of ψ itself[247] or, alternatively, occasional orthogonalization of the columns of ψ.

The evaluation of the differential cross section $d\sigma_{nm}/d\Omega$ from S is straightforward for a spherically symmetric interaction potential, because the angular momenta L and j are individually conserved: Only channels with the same L, j, and m values are coupled, and one finds[16]

$$\frac{d\sigma_{nm}}{d\Omega} - k_m^{-2} \left| \sum_{L=0}^{\infty} (L + \tfrac{1}{2}) T_{nm}^{(L)} P_L (\cos \theta) \right|^2$$

where $i T^{(L)} = S^{(L)} - 1$. When the potentials are nonisotropic, only J and not L or j is conserved. The expressions are then much more complicated and of course depend on the formulation and the coupling involved. Both fixed-axis[251] and rotating-axis[252-255] formulations have been used,[16,256] and approximations to neglect part of the angular-momentum coupling have been proposed in order to reduce the number of coupled channels.[256-261] If closed channels are to be considered, asymptotic forms $\sim e^{-|k_m|r}$, where k_m is imaginary, must be used. The S or T matrix can be partitioned into closed–closed, closed–open, open–closed, and open–open parts, of which only the last affects cross sections.[249] Inward integration and matching of solutions is generally necessary.[262] Several original papers that have detailed their numerical full quantum calculations of inelastic atom–atom scattering can be recommended for reference.[263-268] They have provided standards against which semiclassical and impact-parameter calculations have been tested.

The approximations discussed in Section 4.1 used in time-dependent treatments generally have analogies in "full quantal" time-independent treatments. In particular Born and distorted-wave[199] approximations as well as their unitary exponential forms[269] have been popular. From the standpoint of accuracy—and for knowing, for example, whether a disagreement (or an agreement!) in calculated and measured cross sections gives valid data about the interatomic potentials assumed—the full numerical treatment, though tedious, time consuming, and less"intuitive" than semiclassical or classical path treatments, is obviously preferable. Of course it also invokes an approximation, the close-coupling approximation, since only a small number of states are included. The severity of the closed-coupling approximation is less with an adiabatic basis ("perturbed stationary states"[199] then with an atomic one. Adiabatic bases are rarely used, however, evidently because of difficulties in the evaluation of the nonadiabatic coupling elements[270] or in applying the asymptotic boundary conditions. However, as proposed by Levine and Johnson[271] and essentially as used earlier by Gordon,[246] a series of diabatic bases can be used, each of which is identical to the adiabatic basis at the centre of the small R interval over which it is applied. Then no awkward nonadiabatic matrices need be calculated and the coupling between states appears largely through connection matrices[246,271] joining solutions in neighboring intervals. Such a technique, perhaps using adiabatic solutions found numerically or a finite number of points, can combine "the best" of the adiabatic and diabatic approaches.

References

1. A. Dalgarno, *Rev. Mod. Phys.* **39**, 850, 858 (1967).
2. D. Mihalas, *Stellar Atmospheres*, W. H. Freeman, San Francisco (1970).
3. F. H. Mies, *Mol. Phys.* **26**, 1233 (1973).
4. C. W. Werner, E. V. George, P. W. Hoff, and C. K. Rhodes, *Appl. Phys. Lett.* **25**, 235 (1974).
5. A. Tam, G. Moe, and W. Happer, *Phys. Rev. Lett.* **35**, 1630 (1975).
6. U. Brinkmann, W. Hartig, H. Telle, H. Walther, *App. Phys.* **5**, 109 (1974).
7. G. Herzberg, *Spectra of Diatomic Molecules*, Van Nostrand, Princeton, New Jersey (1950).
8. J. A. Barker, in *Rare Gas Solids*, Eds M. L. Klein and J. A. Venables, Vol. I, Chap. 4, Academic Press, New York (1976).
9. A. L. Burgmans, J. M. Farrar, and Y. T. Lee, *J. Chem. Phys.* **64**, 1345 (1976).
10. P. M. Morse, *Phys. Rev.* **34**, 57 (1929).
11. I. M. Torrens, *Interatomic Potentials*, Academic Press, New York (1972).
12. J. Goodisman, *Diatomic Interaction Potential Theory*, Vols. I and II, Academic Press, New York (1973).
13. J. N. Bardsley, *Case Stud. At. Phys.* **4**, 299 (1974).
14. J. O. Hirschfelder, Ed., *Adv. Chem. Phys.* **12** (1967).
15. E. E. Nikitin, *Adv. Chem. Phys.* **28**, 317 (1975).
16. M. S. Child, *Molecular Collision Theory*, Academic Press, New York (1974).
17. U. Buck, *Adv. Chem. Phys.* **30**, 313 (1975).
18. U. Buck, *Rev. Mod. Phys.* **46**, 369 (1974).
19. H. Pauly, *Physical Chemistry, an Advanced Treatise*, Vol. 6B, p. 553, Academic Press, New York (1974).

20. J. P. Toennies, *Physical Chemistry, an Advanced Treatise*, Vol. 6A, p 227, Academic Press, New York (1974).
21. M. A. D. Fluendy and K. P. Lawley, *Chemical Applications of Molecular Beam Scattering*, Chapman and Hall Ltd., London (1973).
22. Ch. Schlier, *Ann. Rev. Phys. Chem.* **20**, 191 (1969).
23. E. A. Mason and L. Monchick, *Adv. Chem. Phys.* **12**, 329 (1967).
24. R. B. Bernstein and J. T. Muckerman, *Adv. Chem. Phys.* **12**, 389 (1967).
25. H. Goldstein, *Classical Mechanics*, Chap. 3, Addison-Wesley, Reading, Massachusetts (1959).
26. A. van Wijngaarden, E. J. Brimner, and W. E. Baylis, *Can. J. Phys.* **48**, 1835 (1970).
27. E. Everhart, G. Stone, and R. J. Carbone, *Phys. Rev.* **99**, 1287 (1955).
28. A. van Wijngaarden, B. Miremadi, and W. E. Baylis, *Can. J. Phys.* **50**, 1938 (1972).
29. F. T. Smith, in *Lectures in Theoretical Physics: Atomic Collision Processes*, Ed. S. Geltman, p. 95, Gordon and Breach, New York (1969).
30. M. Abramowitz and I. A. Stegun, Eds., *Handbook of Mathematical Functions*, National Bureau of Standards Applied Mathematics Series, No. 55, U.S. Government Printing Office, Washington, D.C. (1964).
31. R. G. Newton, *Scattering of Waves and Particles*, Chap. 20, McGraw Hill, New York (1966).
32. R. G. Newton, in *Mathematics of Profile Inversion*, Ed. L. Collin, NASA (USA) technical memorandum No. X-62 (1972).
33. P. C. Sabatier, in *Mathematics of Profile Inversion*, Ed. L. Collin, NASA (USA) technical memorandum No. X-62 (1972).
34. R. Klingbeil, *J. Chem. Phys.* **56**, 132 (1972).
35. W. G. Rich, S. M. Bobbio, R. L. Champion, and L. D. Doverspike, *Phys. Rev. A* **4**, 2253 (1971).
36. U. Buck, *J. Chem. Phys.* **54**, 1923 (1971).
37. W. H. Miller, *J. Chem. Phys.* **51**, 3631 (1969).
38. G. Vollmer, *Z. Phys.* **226**, 423 (1969).
39. G. Vollmer and H. Krüger, *Phys. Lett.* **28A**, 165 (1968).
40. N. Fröman and P. O. Fröman, *JWKB Approximation*, North-Holland, Amsterdam (1965).
41. E. A. Power, *Adv. Chem. Phys.* **12**, 167 (1967).
42. C. Schlier and R. Düren, *Disc. Faraday Soc.* **40**, 56 (1965).
43. U. Buck, H. O. Hoppe, F. Huisken, and H. Pauly, *J. Chem. Phys.* **60**, 4925 (1974).
44. R. B. Bernstein and T. J. P. O'Brien, *Disc. Faraday Soc.* **40**, 35 (1965).
45. R. B. Bernstein and T. J. P. O'Brien, *J. Chem. Phys.* **46**, 1208 (1967).
46. A. B. Brutschy, H. Haberland, H. Morgner, and K. Schmidt, *Phys. Rev. Lett.* **36**, 1299 (1976).
47. J. Bentley, J. L. Fraites, and D. H. Winicur, *J. Chem. Phys.* **65**, 653 (1976).
48. D. H. Winicur, J. L. Fraites, and J. Bentley, *J. Chem. Phys.* **64**, 1757 (1976).
49. D. H. Winicur and J. L. Fraites, *J. Chem. Phys.* **62**, 63 (1975).
50. C. H. Chen, H. Haberland, and Y. T. Lee, *J. Chem. Phys.* **61**, 3095 (1974).
51. H. Haberland, C. H. Chen, and Y. T. Lee, in *Atomic Physics 3*, Eds. S. J. Smith and K. Walters, p. 339, Plenum Press, New York (1973).
52. T. A. Davidson, M. A. D. Fluendy, and K. P. Lawley, *Disc. Faraday Soc.* **55**, 158 (1973).
53. R. Morgenstern, D. C. Lorents, J. R. Peterson, and R. E. Olson, *Phys. Rev. A* **8**, 2372 (1973).
54. E. W. Rothe, R. H. Neynaber, and S. M. Trujillo, *J. Chem. Phys.* **42**, 3310 (1965).
55. G. M. Carter, D. E. Pritchard, M. Kaplan, and T. W. Ducas, *Phys. Rev. Lett.* **35**, 1144 (1975).

56. R. Düren, H. O. Hoppe, and H. Pauly, *Phys. Rev. Lett.* **37**, 743 (1976).
57. R. Schieder, H. Walther, and L. Woste, *Opt. Commun.* **5**, 337 (1972).
58. G. M. Carter, D. E. Pritchard, and T. W. Ducas, *Appl. Phys. Lett.* **27**, 498 (1975).
59. C. Bottcher, *J. Phys. B* **9**, 3099 (1976).
60. C. Bottcher, T. C. Cravens, and A. Dalgarno, *Proc. R. Soc. London A* **346**, 157 (1975).
61. R. H. G. Reid, *J. Phys. B* **8**, L493 (1975).
62. S. Wofsy, R. H. G. Reid, and A. Dalgarno, *Astrophys. J.* **168**, 161 (1971).
63. R. W. Anderson, T. P. Goddard, C. Perravano, and J. Warner, *J. Chem. Phys.* **64**, 4037 (1976).
64. F. M. Kelly and M. S. Mathur, *Can. J. Phys.* **55**, 83 (1977).
65. F. M. Kelly, T. K. Koh, and M. S. Mathur, *Can. J. Phys.* **52**, 795, 1438 (1974).
66. W. E. Baylis, *Can. J. Phys.* **55**, 1924 (1977).
67. R. LeRoy and J. Van Kranendonk, *J. Chem. Phys.* **61**, 4750 (1974).
68. J. T. Vanderslice, E. A. Mason, W. G. Maisch, and E. R. Lippincott, *J. Mol. Spectry.* **3**, 17 (1959); **5**, 83 (1960).
69. R. H. Davies and J. T. Vanderslice, *J. Chem. Phys.* **45**, 95 (1966).
70. R. H. Davies and J. T. Vanderslice, *Can. J. Phys.* **44**, 219 (1966).
71. J. T. Vanderslice, R. Davies, and S. Weissman, *J. Chem. Phys.* **43**, 1075 (1965).
72. M. S. Child, *Mol. Spectrosc.* **2**, 446 (1974).
73. W. E. Baylis, *Phys. Rev. A* **1**, 990 (1970).
74. J. P. Toennies, W. Welz, and G. Wolf, *J. Chem. Phys.* **64**, 5305 (1976).
75. J. P. Toennies, W. Welz, and G. Wolf, *J. Chem. Phys.* **61**, 2461 (1974).
76. T. G. Waech and R. B. Bernstein, *J. Chem. Phys.* **46**, 4905 (1967).
77. R. J. LeRoy, *Mol. Spectrosc.* **1**, 113 (1973).
78. K. R. Way and W. C. Stwalley, *J. Chem. Phys.* **59**, 5298 (1973).
79. R. E. Smalley, D. A. Auerbach, P. S. H. Fitch, D. H. Levy, and L. Wharton, *J. Chem. Phys.* **66**, 3778 (1977).
80. J. Szudy and W. E. Baylis, *J. Quant. Spectrosc. Radiat. Transfer*, **15**, 641 (1975).
81. W. R. Hindmarsh and J. M. Farr, *Prog. Quantum Electron.* **2**, 141 (1972).
82. W. Behmenburg, *J. Quant. Spectrosc. Radiat. Transfer* **4**, 177 (1964).
83. G. York, R. Scheps, and A. Gallagher, *J. Chem. Phys.* **63**, 1052 (1975).
84. C. G. Carrington and A. Gallagher, *J. Chem. Phys.* **60**, 3436 (1974).
85. D. L. Drummond and A. Gallagher, *J. Chem. Phys.* **60**, 3426 (1974).
86. C. G. Carrington, D. L. Drummond A. Gallagher, and A. V. Phelps, *Chem. Phys. Lett.* **22**, 511 (1973).
87. R. E. M. Hedges, D. L. Drummond, and A. Gallagher, *Phys. Rev. A* **6**, 1519 (1972).
88. J. O. Hirschfelder, C. F. Curtis, and R. B. Bird, *Molecular Theory of Gases and Liquids*, Wiley, New York (1964).
89. J. Apt and D. E. Pritchard, *Phys. Rev. Lett.* **37**, 91 (1976).
90. L. Krause, *Adv. Chem. Phys.* **28**, 267 (1975).
91. L. Krause, in *Physics of Electronic and Atomic Collisions*, Eds. T. R. Grovers and F. J. De Heer, p. 65, North-Holland, Amsterdam (1972).
92. G. Nienhuis, *J. Phys. B* **9**, 167 (1976).
93. J. C. Slater, *Quantum Theory of Molecules and Solids*, Vol. I, *Electronic Structure of Molecules*, McGraw-Hill, New York (1963).
94. W. Kołos and L. Wolniewicz, *Rev. Mod. Phys.* **35**, 473 (1963).
95. P. Gombás and T. Szondy, *"Solutions of the Simplified Self-Consistent Field for All Atoms of the Periodic System from Z = 2 to Z = 92*, Adam Hilger, London (1970).
96. P. Gombás, *Pseudopotentiale*, Springer-Verlag, Berlin (1967).
97. P. Gombás and D. Kisdi, *Theor. Chim. Acta* **5**, 127 (1966).
98. P. Gombás, in *Handbuch der Physik*, Ed. S. Flugge, Vol. XXXVI, Springer-Verlag, Berlin (1956).

99. P. Gombás, *Die statistische Theorie des Atoms und ihre Anwendungen*, Springer-Verlag, Berlin (1949).
100. R. G. Gordon and Y. S. Kim, *J. Chem. Phys.* **56**, 3122 (1974).
101. V. K. Nikulin, *Zh. Tekh. Fiz.* **41** (1971) [*Sov. Phys. Tech. Phys.* **16**, 28 (1971)].
102. V. I. Gaydaenko and V. K. Nikulin, *Chem. Phys. Lett.* **7**, 360 (1970).
103. W. E. Baylis, *J. Chem. Phys.* **51**, 2665 (1969).
104. E. Fermi, *Z. Phys.* **48**, 73 (1928).
105. L. H. Thomas, *Proc. Cambridge Phil. Soc.* **23**, 542 (1926).
106. N. H. March, *Adv. Phys.* **6**, 1 (1957).
107. F. Seitz, *The Modern Theory of Solids*, p. 342, McGraw-Hill, New York (1940).
108. A. I. M. Rae, *Chem. Phys. Lett.* **18**, 574 (1973).
109. W. J. Carr, Jr., R. A. Coldwell-Horsfall, and A. E. Fein, *Phys. Rev.* **124**, 747 (1961).
110. W. J. Carr, Jr. and A. A. Maradudin, *Phys. Rev.* **133**, A371 (1964).
111. Y. S. Kim and R. G. Gordon, *J. Chem. Phys.* **60**, 4323 (1974).
112. J. P. Desclaux and Y.-K. Kim, *J. Phys. B* **8**, 1177 (1975).
113. J. P. Desclaux, *Comp. Phys. Comm.* **9**, 31 (1975).
114. J. Eichler and U. Wille, *Phys. Rev. A* **11**, 1973 (1975).
115. R. Latter, *Phys. Rev.* **99**, 510 (1955).
116. O. B. Firsov, *Zh. Eksperim. i Theor. Fiz.* **33**, 696 (1957) [*Sov. Phys. JETP* **6**, 534 (1958)].
117. J. Lindhard, V. Nielsen, and M. Scharff, *K. Dan. Vidensk. Selsk. Mat. Fys. Medd.* **36**,(10), 1 (1968).
118. A. D. Buckingham, *Adv. Chem. Phys.* **12**, 107 (1967).
119. B. W. Ninham and J. Mahanty, *Dispersion Forces*, Academic Press, New York (1975).
120. E. Merzbacher, *Quantum Mechanics*, 2nd ed, John Wiley, New York (1970).
121. A. Dalgarno, G. W. F. Drake, and G. A. Victor, *Phys. Rev.* **176**, 194 (1968).
122. J. Callaway, R. W. LaBahn, R. T. Pu, and W. M. Duxier, *Phys. Rev.* **168**, 12 (1968).
123. C. J. Kleinman, Y. Hahn, and L. Spruch, *Phys. Rev.* **165**, 53 (1968).
124. C. Bottcher, *J. Phys. B* **4**, 1140 (1970).
125. P. K. L. Drude, *The Theory of Optics*, Longmans Green, London (1933).
126. F. London, *Trans. Faraday Soc.* **33**, 8 (1937).
127. A. Dalgarno, *Adv. Chem. Phys.* **12**, 143 (1967).
128. K. T. Tang, J. M. Norbeck, and P. R. Certain, *J. Chem. Phys.* **64**, 3063 (1976).
129. M. B. Doran, *J. Phys. B* **7**, 558 (1974).
130. G. Starkschall and R. G. Gordon, *J. Chem. Phys.* **56**, 2801 (1972).
131. J. S. Cohen and R. T. Pack, *J. Chem. Phys.* **61**, 2372 (1974).
132. Y. S. Kim and R. G. Gordon, *J. Chem. Phys.* **61**, 1 (1974).
133. K. T. Tang and J. P. Toennies, *J. Chem. Phys.* **66**, 1496 (1977).
134. R. Ahlrichs, R. Penco, and G. Scoles, *Chem. Phys.* **19**, 119 (1977).
135. H. F. Schaefer, *The Electronic Structure of Atoms and Molecules*, Addison-Wesley, Reading, Massachusetts (1972).
136. W. G. Richards, T. E. H. Walker, and R. K. Hinkley, *Bibliography of ab initio Calculations*, Oxford University Press, London (1971).
137. J. C. Slater, *Quantum Theory of Atomic Structure*, Vols. I and II, McGraw-Hill, New York (1960).
138. A. Messiah, *Quantum Mechanics*, Vols. I and II, North-Holland, Amsterdam (1961).
139. D. D. Ebbing, *J. Chem. Phys.* **36**, 1361 (1962).
140. J. C. Browne and F. A. Matsen, *Phys. Rev. A* **135**, 1227 (1964).
141. R. McWeeny and B. T. Sutcliffe, *Methods of Molecular Quantum Mechanics*, Academic Press, New York (1972).
142. K. Ruedenberg and R. D. Poshusta, *Adv. Quantum Chem.* **7**, 1 (1972).

143. G. Berthier, in *Molecular Orbitals in Chemistry, Physics, and Biology*, Eds. P. O. Löwdin and B. Pullman, Academic Press, New York (1964).
144. C. C. J. Roothaan, *Rev. Mod. Phys.* **32**, 179 (1960).
145. B. Roos, *Chem. Phys. Lett.* **15**, 153 (1972).
146. A. F. Wagner, G. Das, and A. C. Wahl, *J. Chem. Phys.* **60**, 1885 (1974).
147. G. Das and A. C. Wahl, *Phys. Rev. A* **4**, 825 (1971).
148. P. Bertoncini and A. C. Wahl, *Phys. Rev. Lett.* **25**, 991 (1970).
149. E. S. Sachs, J. Hinze, and N. H. Sabelli, *J. Chem. Phys.* **62**, 3393 (1975).
150. D. McWilliams and S. Huzinaga, *J. Chem. Phys.* **63**, 4678 (1975).
151. M. J. Seaton and P. M. H. Wilson, *J. Phys. B* **5**, L1 (1972).
152. H. E. Saraph and M. J. Seaton, *Phil. Trans. R. Soc. London A* **271**, 1 (1971).
153. M. Cohen and R. P. McEachran, *Proc. Phys. Soc. (London)* **92**, 539 (1967).
154. M. J. Seaton and P. M. H. Wilson, *J. Phys. B* **5**, L175 (1972).
155. R. P. McEachran, C. E. Tully, and M. Cohen, *Can. J. Phys.* **46**, 2675 (1968); **47**, 875 (1969).
156. E. A. Hylleraas, *Z. Phys.* **54**, 347 (1929); **65**, 209 (1930).
157. E. A. Hylleraas and J. Midtdal, *Phys. Rev.* **109**, 1013 (1958).
158. C. S. Schwartz, *Phys. Rev.* **126**, 1015 (1962).
159. F. W. Byron, Jr. and C. J. Joachain, *Phys. Rev.* **157**, 1 (1967).
160. C. L. Pekeris, *Phys. Rev.* **112**, 1649 (1958).
161. H. M. James and A. S. Coolidge, *J. Chem. Phys.* **1**, 825 (1933).
162. W. Kołos and L. Wolniewicz, *J. Mol. Spectrosc.* **54**, 303 (1975).
163. W. Kołos and L. Wolniewicz, *J. Chem. Phys.* **41**, 3663 (1964).
164. W. Kołos and L. Wolniewicz, *J. Chem. Phys.* **41**, 3674 (1964).
165. J. Lennard-Jones and J. A. Pople, *Proc. R. Soc. London A* **210**, 190 (1951).
166. C. F. Bender, E. R. Davidson, and F. D. Peat, *Phys. Rev.* **174**, 75 (1968).
167. J. P. Desclaux, *At. Data Nucl. Data Tables* **12**, 311 (1973).
168. G. Das and A. C. Wahl, *J. Chem. Phys.* **64**, 4672 (1976).
169. J. C. Phillips and L. Kleinman, *Phys. Rev.* **116**, 287 (1959).
170. J. D. Weeks, A. Hazi, and S. A. Rice, *Adv. Chem. Phys.* **16**, 283 (1969).
171. B. J. Austin, V. Heine, and L. J. Sham, *Phys. Rev.* **127**, 276 (1962).
172. M. H. Cohen and V. Heine, *Phys. Rev.* **122**, 1821 (1961).
173. L. Szász and L. Brown, *J. Chem. Phys.* **65**, 1393 (1976).
174. C. S. Ewig and J. R. Van Wazer, *J. Chem. Phys.* **63**, 4035 (1975).
175. V. Bonifacic and S. Huzinga, *J. Chem. Phys.* **62**, 1507; 1509 (1975).
176. V. Bonifacic and S. Huzinga, *J. Chem. Phys.* **60**, 2779 (1974).
177. C. F. Melius, W. A. Goddard III, and L. R. Kahn, *J. Chem. Phys.* **56**, 3392 (1972).
178. L. R. Kahn and W. A. Goddard III, *Chem. Phys. Lett.* **2**, 667 (1968).
179. W. A. Goddard III, *Phys. Rev.* **174**, 659 (1968).
180. W. A. Goddard III, *Phys. Rev.* **169**, 120 (1968).
181. W. A. Goddard III, *Phys. Rev.* **157**, 73, 81 (1967).
182. G. Simons and A. Mazzotti, *J. Chem. Phys.* **52**, 2449 (1970).
183. L. Szász and G. McGinn, *J. Chem. Phys.* **48**, 2997 (1968).
184. L. Szász and G. McGinn, *J. Chem. Phys.* **47**, 3495 (1967).
185. L. Szász and G. McGinn, *J. Chem. Phys.* **45**, 2898 (1966).
186. H. Preuss, *Z. Naturforsch.* **10A**, 365 (1955).
187. M. Klapisch, C.R.Acad.Sc. (Paris), **265**, 914 (1967).
188. H. Hellmann, *J. Chem. Phys.* **3**, 61 (1935).
189. D. W. Norcross, *Phys. Rev. A* **7**, 606 (1973).
190. G. A. Victor and C. Laughlin, *Chem. Phys. Lett.* **14**, 74 (1972).
191. J. C. Weisheit, *Phys. Rev. A* **5**, 1621 (1972).

192. C. Bottcher, A. C. Allison, and A. Dalgarno, *Chem. Phys. Lett.* **11**, 307 (1971).
193. A. Dalgarno, C. Bottcher, and G. A. Victor, *Chem. Phys. Lett.* **7**, 265 (1970).
194. J. Pascale and J. Vandeplanque, *J. Chem. Phys.* **60**, 2278 (1974).
195. J. N. Bardsley, B. R. Junker, and D. W. Norcross, *Chem. Phys. Lett.* **37**, 502 (1976).
196. C. Bottcher and A. Dalgarno, *Proc. R. Soc. London A* **340**, 187 (1974).
197. J. N. Bardsley, *Chem. Phys. Lett.* **7**, 517 (1970).
198. R. D. Levine and R. B. Bernstein, *Molecular Reaction Dynamics*, Oxford University Press, New York (1974).
199. N. F. Mott and H. S. W. Massey, *The Theory of Atomic Collisions*, 3rd ed., Oxford University Press, London (1965).
200. W. H. Miller, *Adv. Chem. Phys.* **25**, 69 (1974).
201. F. Calogero, *Variable Phase Approach to Potential Scattering*, Academic Press, New York (1967).
202. P. Pechukas and J. C. Light, *J. Chem. Phys.* **44**, 3897 (1966).
203. J. Callaway and E. Bauer, *Phys. Rev. A* **140**, 1072 (1965).
204. W. Magnus, *Commun. Pure Appl. Math.* **7**, 649 (1954).
205. R. B. Bernstein and K. H. Kramer, *J. Chem. Phys.* **44**, 4473 (1966).
206. K. H. Kramer and R. B. Bernstein, *J. Chem. Phys.* **40**, 200 (1964).
207. L. D. Landau *Phys. Z. Sowjetunion* **2**, 46 (1932).
208. C. Zener, *Proc. R. Soc. London A* **137**, 696 (1932).
209. J. B. Delos and W. R. Thorson, *Phys. Rev. A* **6**, 728 (1972); **9**, 1026 (1974).
210. J. T. Muckerman, I. Rusinek, R. E. Roberts, and M. Alexander, *J. Chem. Phys.* **65**, 2416 (1976).
211. K. J. McCann and M. R. Flannery, *Chem. Phys. Lett.* **35**, 124 (1975).
212. E. C. G. Stuekelberg, *Helv. Phys. Acta* **5**, 369 (1932).
213. D. S. F. Crothers, *Adv. Phys.* **20**, 405 (1971).
214. S. A. Evans, J. S. Cohen, and N. F. Lane, *Phys. Rev. A* **4**, 2235 (1971).
215. J. S. Cohen, S. A. Evans, and N. F. Lane, *Phys. Rev. A* **4**, 2248 (1971).
216. D. F. Crothers, *J. Phys. B* **9**, 635 (1976).
217. D. S. F. Crothers, *J. Phys. B* **8**, L442 (1975).
218. J. Delos, *Phys. Rev. A* **9**, 1626 (1974).
219. G. V. Dubrovski and I. Fischer-Hjalmars, *J. Phys. B* **7**, 892 (1974).
220. W. R. Thorson, J. B. Delos, and S. A. Boorstein, *Phys. Rev. A* **4**, 1052 (1971).
221. M. S. Child, *Molec. Phys.* **20**, 171 (1971).
222. J. N. L. Connor, *Disc. Faraday Soc.* **55**, 51 (1973).
223. J. N. L. Connor, *Mol. Phys.* **25**, 181 (1973).
224. R. A. Marcus, *J. Chem. Phys.* **56**, 311, 3548 (1972).
225. J. N. L. Connor and R. A. Marcus, *J. Chem. Phys.* **55**, 5636 (1971).
226. R. A. Marcus, *J. Chem. Phys.* **54**, 3965 (1971).
227. R. A. Marcus, Chem. Phys. Lett. **7**, 525 (1970).
228. U.-I. Cho and B. C. Eu, *Mol. Phys.* **32**, 1 (1976).
229. B. C. Eu, *J. Chem. Phys.* **56**, 2507 (1972).
230. R. P. Feynman and A. R. Hibbs, *Quantum Mechanics and Path Integrals*, McGraw-Hill, New York (1965).
231. R. P. Feynman, *Rev. Mod. Phys.* **20**, 367 (1948).
232. W. H. Miller, *Adv. Chem. Phys.* **30**, 77 (1975).
233. W. H. Miller and T. F. George, *J. Chem. Phys.* **56**, 5637; **56**, 5668; **57**, 2458 (1972).
234. W. H. Miller, *J. Chem. Phys.* **53**, 1949; 3578 (1970).
235. P. Pechukas, *Phys. Rev.* **181**, 166; 174 (1969).
236. J. -T. Hwang and P. Pechukas, *J. Chem. Phys.* **65**, 1224 (1976).
237. F. J. McLafferty and T. F. George, *J. Chem. Phys.* **63**, 2609 (1975).

238. U.-I. Cho and B. C. Eu, *Molec. Phys.* **32**, 19 (1976).
239. N. F. Lane and S. Geltman, *Phys. Rev.* **160**, 53 (1967).
240. J. M. Blatt, *J. Comput. Phys.* **1**, 382 (1967).
241. I. H. Sloan, *J. Comput. Phys.* **2**, 414 (1968).
242. A. C. Allison, *J. Comput. Phys.* **6**, 378 (1970).
243. W. A. Lester, *J. Comput. Phys.* **3**, 322 (1968).
244. W. A. Lester, *Meth. Comput. Phys.* **10**, 211 (1971).
245. R. G. Gordon, *Meth. Comput. Phys.* **10**, 81 (1971).
246. R. G. Gordon, *J. Chem. Phys.* **51**, 14 (1969).
247. B. R. Johnson, *J. Comput. Phys.* **13**, 445 (1973).
248. D. G. Truhlar, J. Abdallah, Jr., and R. L. Smith, *Adv. Chem. Phys.* **25**, 211 (1974).
249. D. Secrest, *Meth. Comput. Phys.* **10**, 243 (1971).
250. W. N. Sams and D. J. Kouri, *J. Chem. Phys.* **51**, 4809 (1969).
251. A. M. Arthurs and A. Dalgarno, *Proc. R. Soc. London A* **256**, 540 (1960).
252. M. Jacob and G. C. Wick, *Ann. Phys. N.Y.* **7**, 404 (1959).
253. H. Klar, *J. Phys. B* **6**, 2139 (1973).
254. H. Klar, *Nuovo Cimento* **4A**, 529 (1971).
255. H. Klar, *Z. Phys.* **228**, 59 (1969).
256. D. Secrest, *J. Chem. Phys.* **62**, 710 (1975).
257. M. Tamir and M. Shapiro, *Chem. Phys. Lett.* **31**, 166 (1975).
258. D. Kouri, *Chem. Phys. Lett.* **31**, 599 (1975).
259. P. McGuire and D. J. Kouri, *J. Chem. Phys.* **60**, 2488 (1974).
260. D. J. Kouri and P. McGuire, *Chem. Phys. Lett.* **29**, 414 (1974).
261. R. T. Pack, *J. Chem. Phys.* **60**, 633 (1974).
262. K. Smith, *The Calculation of Atomic Scattering Processes*, Wiley-Interscience, New York (1971).
263. R. H. G. Reid, *J. Phys. B* **8**, 2255 (1975).
264. A. D. Wilson and Y. Shimoni, *J. Phys. B* **8**, 2393; 2415 (1975).
265. A. D. Wilson and Y. Shimoni, *J. Phys. B* **7**, 1543 (1974); **8**, 1392 (erratum).
266. R. H. G. Reid, *J. Phys. B* **6**, 2018 (1973).
267. F. H. Mies, *Phys. Rev. A* **7**, 942 (1973); **7**, 957 (1973).
268. J. C. Weisheit and N. F. Lane, *Phys. Rev. A* **4**, 171 (1971).
269. R. D. Levine, *Mol. Phys.* **22**, 497 (1971).
270. F. T. Smith, *Phys. Rev.* **179**, 111 (1969).
271. R. D. Levine and B. R. Johnson, *Chem. Phys. Lett.* **13**, 168 (1972).

II
Methods and Applications of Atomic Spectroscopy

7

New Developments of Classical Optical Spectroscopy

KLAUS HEILIG AND ANDREAS STEUDEL

1. Introduction

This article will be concerned with the effects of fine structure, hyperfine structure, and isotope shift in atomic spectra as studied by classical optical spectroscopy, and with reporting the progress that has been achieved mainly during the last decade.

Although spectral analysis is as old as atomic physics itself, there still remains a lot of work to be done, especially on complex spectra and the spectra of very highly ionized atoms. New spectrometers for the infrared region have become available and made possible considerable progress in the classification of complex spectra. The development of computers and data-handling systems has also contributed to the recent progress in fine-structure classification. Highly ionized spectra, which are now of great interest, can be studied by means of beam-foil spectroscopy and laser-produced plasmas, as well as by conventional light sources. Analysis of spectra has also been greatly promoted by theory; on the other hand, there are several recent examples where the development of the theory of atomic spectra was stimulated by new experimental results.

Conventional optical spectroscopy has found many applications in neighboring fields. The great importance of spectroscopy to chemistry is

KLAUS HEILIG AND ANDREAS STEUDEL • Institut A für Experimentalphysik, Technische Universität Hannover, Appelstrasse 2, D-3000 Hannover, Federal Republic of Germany.

well known. The sensitivity of spectroscopic methods and their nondestructive character are the essential reasons for their application in biology, medicine, criminology, geology, and archaeology. Important parts of our astronomical and astrophysical knowledge are based on optical spectroscopy. Recently, the elimination of the atmospheric obstruction by using space vehicles has opened up the wavelength region below 300 nm to astronomy and, therefore, new studies of spectra from highly ionized atoms have been carried out in laboratories in order to supply the basic knowledge needed to interpret the astronomical spectra. Furthermore, optical spectroscopy is widely used as a diagnostic tool in physics and technology, since it does not perturb conditions as other methods do. Many applications of this kind are found, for example, in plasma physics.

In high-resolution optical spectroscopy hyperfine structures and isotope shifts have for a long time been studied successfully by the Fabry–Perot technique. In recent years Fourier spectroscopy instruments have been designed especially for high-resolution studies in the near-infrared and visible regions, and excellent results have been reported. Many experiments have been considerably facilitated by the use of samples of enriched isotopes that have become available for any element. Thus complex blends of components belonging to different isotopes can be avoided.

Laser techniques have also been used in several different ways for reducing the experimental uncertainty and thus obtaining results of unrivaled accuracy (see Chapter 15).

Nuclear spins and nuclear moments have been evaluated from conventional optical hyperfine-structure investigations for more than five decades. Nowadays the determination of nuclear spins and nuclear magnetic dipole moments by conventional optical spectroscopy is of interest only in a few particular cases. One of the most important problems in the hyperfine-structure analysis, however, is still the evaluation of reliable values of nuclear electric quadrupole moments. The reliability of the spectroscopic quadrupole moments may be checked by making measurements in several configurations, a procedure for which optical spectroscopy is particularly suited.

In some cases there are now sufficient experimental data available for the interpretation of observed hyperfine structures by means of the relativistic effective operator formalism; it is then of interest to compare the experimental parameters obtained in this way with the corresponding ones calculated by theoretical methods, thereby allowing a check to be made on the validity of the single-electron approach and the significance of higher-order contributions. In particular, the comparison of experimental and calculated hyperfine parameters may be a suitable touchstone of many-body theories. Thus hyperfine-structure measurements are tools for the study of nuclear as well as electronic structure.

Like hyperfine-structure investigations, optical isotope shift studies have a nuclear and an electronic aspect. They have proved to be an important source of information on the change of the mean square nuclear charge radii between the isotopes of an element, and, more recently, they have often been used to study details of the electronic structure. For stable and long-lived isotopes nearly all the isotope shift data have been obtained by classical optical spectroscopy.

The study of hyperfine structure and isotope shift of a long sequence of isotopes of an element, which gives information on how nuclear parameters depend upon the attractive forces between the nucleons in the nucleus, is very valuable. Such a study usually requires making measurements on unstable nuclei. Conventional investigations on isotopes with lifetimes as short as 33 minutes have been reported.[1] The minimum time needed to fabricate a light source is 30 minutes. Much shorter-lived isotopes with half-lives in the range of minutes or seconds may be investigated when on-line techniques are used and when, for example, the fluorescence light after laser excitation is observed. Such a technique also allows the amount of substance required for an investigation to be reduced to less than 10^9 atoms in the resonance cell whereas, for conventional spectroscopy, at least 10^{12} atoms in the light source (electrodeless discharge lamp) are usually needed.

The quotations given in this chapter refer in many cases to typical examples and are in general not exhaustive. The chapter was completed in January 1977.

2. Experimental Techniques

2.1. Light Sources

This section is concerned with devices producing or absorbing discrete spectral lines particularly with reference to atomic spectroscopy. Many of the well-known light sources have been further improved in detail, but no spectacular development has been reported in the recent past.

Only light sources that have been used very frequently during the past five years are referred to here. The heat-pipe oven—an interesting device in many fields of physics—is also described. Beam-foil light sources are discussed elsewhere (see Chapter 20).

2.1.1. Sliding Spark

Roughly half of the fine-structure (fs) analyses of atoms ionized from about two to eight times were done by using sliding sparks as light sources. Still higher degrees of ionization can be produced and have been used in a few cases. Sliding sparks have been found to be highly effective in sorting

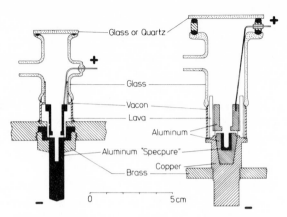

Figure 1. Examples of liquid-nitrogen-cooled hollow cathodes. Left: from Ref. 8. Right: from Ref. 9.

out spectra of different stages of ionization by variation of the peak spark current. Descriptions and further references to these long-known lamps are given in many publications on spectra in the vacuum ultraviolet (see, e.g., Refs. 2–7).

2.1.2. Hollow Cathode

The very reliable hollow cathode (HC)—which has been unchanged in principle for more than 40 years—is widely used as a light source in hyperfine-structure (hfs) investigations. Many types of HC have been constructed for discharge currents ranging from parts of a milliampere up to some 10 A. Low-current HC with liquid-nitrogen cooling are widely used in hfs investigations because of their small Doppler linewidth, their constant light intensity, and their simple construction. Figure 1 gives examples of recent types of liquid-nitrogen-cooled HC. The currents in the common commercial HC for fs investigations, for example the Fe/Ne–HC secondary wavelength standard,[10] lie between about 0.1 and 0.5 A. Either a medium- or high-current HC (up to 40 A[11]) can be used as a broad-line light source in absorption experiments or sometimes as a continuum source in the vacuum ultraviolet.[12]

2.1.3. Electrodeless Discharge Lamps

The main interest in electrodeless discharge lamps (EDL) comes from atomic fluorescence spectrometry, from investigations of emission lines under the influence of strong magnetic fields, and from their use as light sources for exciting highly radioactive samples and small amounts of substance in fs and hfs investigations.

The requirements for EDLs are in general high intensity and long life and—for nonphotographic work—great stability (no noise and no drift). In spite of numerous investigations these requirements cannot be met simultaneously. The main problems still lie in the difficulty of reproducing the preparation of the lamp, and in the fact that there are many mutually dependent parameters, whereas in most cases few of them have been carefully investigated for a small number of elements.[13,14]

Make and operation: The most common dimensions are 5–10 mm internal diameter and 2–5 cm length. The substance (very often chlorides and iodides) must have sufficient volatility and should not react with the quartz (or glass) walls. Argon seems to be the most suitable gas for filling the lamp. The optimum pressure has to be found by trial and error. Microwave energy (2450 MHz; 80–200 W, in few cases up to 1 kW) is used for excitation, radio frequency seldom being applied nowadays.[15] Modulation of the microwave power seems to be advisable in order to increase the lifetime and stability of the lamp. As in other light sources, the addition of another chemical element (like Na) may enhance the line intensity of the elements under investigation. Arc and spark spectra are predominant in EDL.

In order to obtain good stability the vaporization and excitation processes are sometimes separated: a conventional HC produces free atoms in a vapor that is excited by a high-frequency discharge.[16]

2.1.4. Heat-Pipe Oven

The heat-pipe oven is very useful for generating well-defined homogeneous metal vapors, for which the temperature and density are known. The essential feature of the oven is a heat-conductive element, which

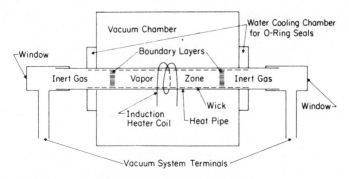

Figure 2. Schematic arrangement of the heat-pipe oven; from Ref. 17.

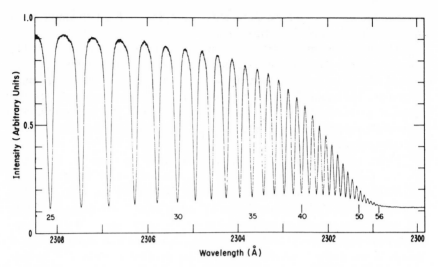

Figure 3. Lithium absorption spectrum of the principal series $2\,^2S-n\,^2P$ measured at a vapor pressure of 8.2 torr; from Ref. 17.

works on the principle that large amounts of heat can be transferred by evaporating a liquid, transporting the vapor, and condensing it again. For spectroscopic measurements both ends of the heat-pipe tube are replaced by windows, the tube is filled with an inert gas of about 1/100–10 torr pressure, and is heated in the middle (see Figure 2). The pumping action of the flowing vapor causes the inert gas to be completely separated from the vapor except for a short transition region. The pressure in the vapor then equals that of the inert gas.

This oven may be considered as a particular version of the well-known King furnace. Spectroscopic applications reported in the recent past are, for example: absorption measurements in atomic vapors (see Figure 3), the production of homogeneous mixtures of saturated and unsaturated vapors,[18] and the observation of resonance fluorescence spectra of molecules.[19] Emission spectra may be observed by inserting electrodes at both ends of the heat-pipe in order to maintain a discharge in the substance under investigation.[20]

2.1.5. Atomic Beam Oven

The reasons for using atomic beams are to reduce the Doppler width, to reduce the atomic collision rate, and, sometimes, to overcome technological problems (e.g., difficulties in handling the vapor in a cell). For details of the design of atomic beam ovens see Refs. 21–23 and Chapter 4. Constructions range from those which vaporize very refractory metals by direct electron bombardment[24] to crucibles which are heated directly, by electron bombardment, or by induction. Ovens exist that give atomic beams at very low rates of material consumption (a Ca atomic beam was constructed for isotope shift measurements which gave a rate of evaporation of only 0.5 mg per hour,[25] thus enabling the use of enriched isotopes in the atomic beam). Typical evaporation rates for spectroscopic applications are about 0.1–1 g/h. For the use of atomic beams together with laser scanning see Ref. 26.

2.1.6. Other Light Sources

There are some other spectroscopic light sources which are used mainly in fs investigations: *flame* (for spectrochemical analysis, mostly atomic absorption[27]); *King furnace* (for investigating spectra in different stages of excitation)[28,29]; *Geissler tube* (for investigation of permanent gases or volatile material, e.g., krypton-86 standard lamp); *dc arc* (for spectrochemical purposes, especially for detection of trace elements); *spark* (to excite second—"spark"—and higher spectra); *hot spark* (to produce highly ionized spectra in low-pressure gases); *laser-induced plasma* (for producing highly ionized atoms from the surface of a solid).[30]

2.2. Spectrometers

2.2.1. Principles

(*a*) *Fine Structure*. Most of the radiation from free neutral or singly ionized atoms is in the near infrared, visible, and ultraviolet region. Experimental methods of fs investigations in these regions widely use photographic recording, but recently Fourier spectrometers (see Section 2.2.3) have come into use, mainly in the near infrared. Photographic techniques allow the application of light sources of very different behavior, including pulsed or otherwise extremely unstable ones, whilst Fourier spectrometers require light sources of stable emission.

For photographic recording plane gratings are used in either the Ebert mounting[31] or the Czerny–Turner mounting,[32] or concave gratings in a Rowland circle in the Paschen–Runge mounting[33] (a famous spectrometer of this type is at the Argonne National Laboratory; its radius is 9.15 m[34]) or in the Eagle mounting.[35] A review on spectrometers in optical spectroscopy is given in Ref. 37.

A variety of photographic plates for spectroscopic purposes is currently manufactured by Kodak only.[36]

(*b*) *Hyperfine Structure and Isotope Shift*. The bulk of optical hfs and isotope shift investigations is done by means of the photoelectric recording

Fabry–Perot spectrometer. In its normal version it consists of a light source, monochromator, scanned Fabry–Perot interferometer, photomultiplier, amplifier, and a digital and analog data output (see, e.g., Refs. 38 and 39).

The *light source* has to emit narrow lines (Doppler broadening is the main reason for the observed linewidth) of constant intensity. Liquid-nitrogen-cooled hollow cathodes meet these requirements and are widely used (recording of reference light at the entrance of the interferometer is otherwise expedient). Neon and argon are the favorite carrier gases, but other noble gases are also used. Discharge currents lie between about 1 and 50 mA. The temperature in the discharge is about 50–150° C higher than the temperature of the cooled cathode (except for extremely low currents), so that cooling by liquid He or H_2 achieves very little further reduction of the Doppler width. In favorable cases, a fraction of a microgram is enough to give sufficient intensity for an investigation of some hours duration,[38] but usually 0.1–1 mg of substance is needed.

The *monochromator* will not be discussed in detail. Its task is to select the fs line under investigation. Depending on the number of lines per wavelength interval this can be done with interference filters, prism monochromators (with constant deviation), or grating monochromators.

The high resolution required is provided by a Fabry–Perot *interferometer*. Scanning is done by linearly changing the pressure of the gas (purified N_2 or other gases) between the mirrors (dielectric multilayer coatings) or piezoelectrically [see Section 2.2.5(a)]. Part of the central interference ring is seen through a diaphragm by a *photomultiplier* (cooled by liquid nitrogen or by Peltier elements to decrease the dark current). At low light intensities photon-counting techniques may be used,[40–45] but often the multiplier current is amplified and then split into two channels: an analogous recording for display and a digital data output (magnetic tape or punched paper tape). Either slow recording, with time constants of some seconds, or signal-averaging methods can be used when only low photon rates are available.

In isotope shift investigations the shift may be much smaller than the width of the line under investigation. Highly enriched isotope samples may then be used in different light sources—one isotope per light source—which are brought into the right position during the interferometer scan.[38] As the center of a symmetrical line can be determined with an accuracy of about 1/100 of the full width at half-maximum (FWHM), experimental uncertainties of measured isotope shifts down to about $0.1 \times 10^{-3} \text{cm}^{-1}$ ($\hat{=} 3 \text{ MHz}$) can be obtained. Unresolved hfs leads to somewhat larger uncertainties of isotope shift results. The methods used to evaluate isotope shifts or hfs from unresolved structures are described in Section 2.3.2.

Other light sources, like electrodeless discharge lamps, have been used for the investigation of either small amounts of radioactive isotopes or dangerous materials.

In some cases photography together with grating spectrometers of high dispersion has been used.[46] The main drawback of the photographic technique lies in the nonlinearity of the intensity scale, which prohibits the use of sophisticated data-handling methods. Hyperfine-structure investigations by Fourier transform spectroscopy were recently performed in the near infrared and in the visible region [see Section 2.2.3(c)].

2.2.2. Gratings

By 1950 the prism was being supplanted by the diffraction grating as a dispersing device, the mechanical exactness of grating production being combined with interferometric and electronic servo control. By 1967 the first holographically recorded gratings were made. Recent reviews can be found in Refs. 47–54.

Today problems in producing *ruled gratings* arise from failures in blank coatings, and from difficulties with diamonds.[47] Scattered light and other consequences of random spacing or periodic errors (ghosts) can be kept below 10^{-4}–10^{-5} times the intensity of the stronger spectral lines. Blaze angles can be obtained up to 85° with an accuracy to within a few minutes of arc.

Holographic gratings (= interference gratings) are produced by causing two halves of a coherent monochromatic light beam to interfere and to give rise to straight-line fringes in a grainless film, which is etched to yield grooves.[49,51–53,55] With holographic gratings, ghosts are avoided because of the absence of periodic errors, and stray light levels are about 10–20 times lower than for ruled gratings.

The holographic method seems to be best suited to the production of gratings with more than 1200 grooves/mm. Concave holographic gratings can be made easily and can be adjusted to be stigmatic for three wavelengths.[54] There are unrivaled possibilities for selecting the radius of curvature, and apertures up to $f/1$ are possible. Large concave holographic gratings have been produced yielding high throughput and high resolution (above 1,500,000).[52]

Some of the properties of ruled and of holographic gratings are compared in Table 1. Little or no difference exists with regard to aberrations, coatings, resistance to thermal shocks, humidity, vibrations, laser beams (for standard gratings), necessary care when handling (first surface mirror), and theoretical resolution. To sum up, it is not possible to draw general conclusions about which type of grating is better. In the infrared region and for echelles the ruled gratings seem to be more useful than

Table 1. Comparison between Ruled and Holographic Gratings

Property	Ruled plane grating	Holographic grating
Spectral region	uv, visible, ir	uv, visible, ir
Size (width)	up to 250 mm	up to 420 mm
Grooves/mm	20–3600	600–6000
Blaze	can be blazed at almost any angle	can be produced with very different groove forms (sinusoidal, rectangular, saw-tooth shaped)
Efficiency	60%–99% at the blaze wavelength λ_B; about 20% at 0.6 and at $2\lambda_B$	up to 70% if $0.8 < \lambda/g < 1.7$ (g grating constant); lower than that of ruled grating elsewhere
Ghost intensity	10^{-6}–10^{-4} of parent line intensity	no ghosts
Stray light	depends on kind of ruling	below that of best ruled gratings

holographic ones. In the visible and the near ultraviolet regions ruled gratings offer advantages in low dispersion apparatus, whereas holographic gratings seem to be better suited for high dispersion in addition to offering better spectral purity. In the vacuum ultraviolet and in the x-ray regions it appears that similar results can be obtained with both types of gratings.

2.2.3. Fourier-Transform Spectrometer

(a) *Principle.* Fourier-transform spectrometers (FTS) have been in use for about 25 years and have been described in detail in other review articles (see, e.g., Refs. 56–65). The light under investigation is split into two beams, the path difference between them is usually increased at a constant rate and after interfering of the beams the intensity is recorded (the *interferogram*). The interferometer is therefore used essentially as a mechanical heterodyning system which transforms the optical frequency range into an audio frequency range. To convert the interferogram into a spectrum its *Fourier transform* has to be computed.

In a detector-noise limited spectrometer, a gain of $n^{1/2}$ in the signal-to-noise ratio may be achieved when a spectrum consisting of n resolution elements is investigated with a FT spectrometer, rather than with a monochromator, provided both instruments have the same power throughout, imploy the same detector, and operate for the same length of time.* This, the "Fellgett advantage" or "*multiplex advantage*," is one major reason for using FT spectroscopy.

---

* For the particular case of analytic optical spectroscopy, signal-to-noise ratios for single-channel methods (sequential and multiplex) are compared with multichannel methods in Ref. 66.

The interferogram has two limitations which determine the quality of the spectrum: There is an upper limit, X, of the path difference, x, and the detector output is sampled at constant path intervals, Δx. The limited length of the interferogram results in a limited resolution of the spectrum (resolution $= \nu \cdot X$, where ν is the wave number) and the discrete spacing of sampling causes aliasing in the spectrum (free spectral range $= 1/\Delta x$).

Considerable *computer time* is needed for a direct calculation of the FT, since all points in the interferogram contribute to each point in the final spectrum and the number n of resolution elements may be of the order of 50,000 or more. Fortunately, however, the n^2 operations for an n-point spectrum can be reduced drastically by working on the individual "bits" of the numbers (fast Fourier transform, FTT).[67,68] The FTT algorithm gives a time reduction factor of 51 for $n = 2^{10}$ and 1089 for $n = 2^{15}$.

The power in the interferogram is often concentrated in the central part. This calls for a large dynamic range, which creates a problem if high resolution is wanted since, in this case, the upper limit in signal-to-noise ratio obtainable experimentally is given at present by ADCs with only about 15 bits and minicomputers with 12 to 16 bits word length. High-speed scanning with signal averaging in the computer can be used to overcome ADC problems.

(*b*) *Experimental Arrangements in General.* At the highest wave number of the spectral range under investigation, it is necessary to sample at mirror positions known to an accuracy of $1/10$ of a fringe. This accuracy may be achieved by using a laser (the red line of a He–Ne laser gives fringes with $\Delta x = 0.31 \ \mu m$) together with a beam of white light (which gives a sharp peak at $x = 0$).[69,70] This arrangement has the additional advantage (*Connes advantage*) that the wavelengths in the spectrum are inherently measured *ab initio* to an accuracy determined by the reference laser light.

The sensitivity of a FT spectrometer is enhanced by the *Jacquinot advantage*: a larger throughput can be obtained than in conventional spectrometers, since slits are not needed. The gain in intensity may amount to a factor of 50 and, for example, at low resolutions ($8 \ cm^{-1}$) the entire spectrum can be recorded in less than 1 sec.

In the infrared, emission spectra can be recorded from samples with temperatures only $20° \ C$ higher than that of the detector.

In *slow-scanning interferometers* the velocity of the mirror is sufficiently slow to ensure that all frequencies in the interferogram are lower than 1 Hz. To decrease the pickup of noise the light beam is modulated, either by means of a mechanical chopper (amplitude modulation) or by small periodic variations of path difference (phase modulation). In several interferometers the path difference is increased stepwise: in far infrared interferometers this is achieved by a stepper-motor, and in high-resolution interferometers by adding to the constant mirror velocity a saw-tooth movement of opposite sign and quick retracing.[69,70]

For slow-scanning interferometers, detectors with a long response time can be used, which may be important with infrared detectors; interferograms are usually symmetric, and real-time computations can be performed. If line-rich light sources are used, the dynamic range may be too high to be digitalized accurately, as was explained above.

In *rapid-scanning interferometers* the mirror velocity is sufficiently high to give detector frequencies in the audio range that can be easily amplified without any further modulation. By using a bandpass filter, noise of higher frequencies than the modulation frequency of the shortest wavelength in the spectrum is eliminated. The signal-to-noise ratio per scan is low, but it can easily be increased by use of signal-averaging techniques.

With rapid-scanning interferometers the use of rapid-response detectors is necessary, interferograms are usually asymmetric (so phase correction must be performed), and real-time computations are very difficult to carry out. Rapid-scanning interferometers are not suitable for high-resolution measurements.

(c) *High-Resolution Instruments.* Many high-resolution Fourier spectrometers have been constructed,[69-76] starting in 1966[69] with 0.1 cm^{-1} resolution, 5 samples per sec operation, and a maximum number of samples of about 50,000. Only the most recent developments will be described here, namely, the so-called interferometer IV[75] for near-infrared investigations and an interferometer for the visible and near-ultraviolet range.[76] Interferometer IV has been built for astronomical observations at a Coudé focus. Its specifications are as follows: spectral range 0.8–3.5 μm; maximum path difference 1 m; speed of operation normally 10^6 samples in the interferogram, recorded on magnetic tape; transform time 3 min; real-time special-purpose computer, which displays a 4096-sample slice of the spectrum to check resolution and signal-to-noise ratio; weight 200 kg.

An essential part of this interferometer is the carriage of the mirrors: Whereas in previous high-resolution systems oil bearings or air bearings were used, a new device, the so-called "slave carriage" has been constructed, which gives very accurate mirror movement. The retroreflector—a cat's eye—is supported by the carriage through an astatic parallelogram

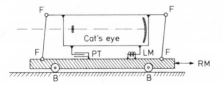

Figure 4. Slave carriage of the "interferometer IV,"[75] simplified drawing. F—flexure hinges; PT—position transducer; LM—linear motor; B—ball bearing; RM—rotary motor.

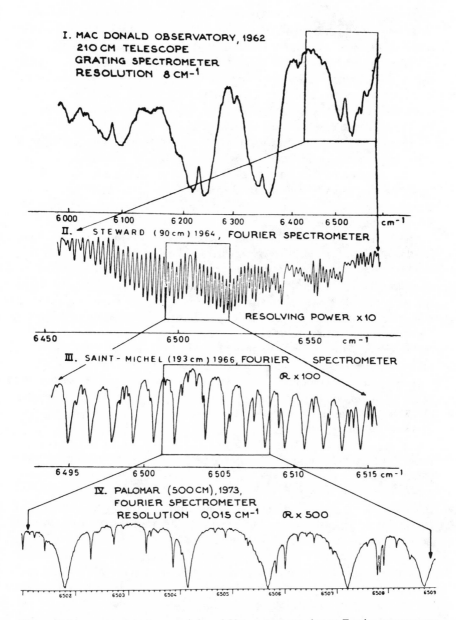

Figure 5. Improvement in the near-infrared Venus spectrum due to Fourier spectroscopy (same type detectors); from Ref. 75.

suspension (see Figure 4). Relative motion is measured with about 1 μm accuracy by a position transducer, the error signal of which is fed to the rotary motor through a suitable servo-loop. The carriage is thus compelled to follow the movement of the cat's eye and a frictionless support of the cat's eye is simulated. The first movement of the cat's eye and correction are both made by a short linear motor (a loudspeaker-type device). Different recording modes are possible, which are suitable for different light sources and conditions of operation.[75]

Figure 5 shows the improvements in resolution in near-infrared work obtained in the last ten years.

An instrument was constructed recently on the same principle as "interferometer IV" but adapted specially to the visible and near-ultraviolet region,[76] where the noise is mainly due to the signal itself. Specifications are as follows: maximum path difference of 32 cm; step $\Delta x = 1/(2 \, \nu_{laser})$, $\nu_{laser} = 15{,}798 \, cm^{-1}$; 10^6 steps in the interferogram, recorded in approximately 5 hr; resolving limit $\delta\nu = 16 \times 10^{-3} \, cm^{-1}$; free spectral range 13,000 cm^{-1}. Results are illustrated by Figure 6. As can be seen, the average noise far from intense lines is of the order of 5×10^{-4} of the most intense line, A_0, of the spectrum. Instability of the light source gives rise to parasitic components in the immediate vicinity of the intense lines, increasing noise by a factor of 10–100. Weak lines may easily be distinguished from noise by comparing spectra from two different interferograms.

First measurements on about 3000 lines were performed in praseodymium and in uranium (fs and hfs). When the signal-to-noise ratio of a line is greater than 200, the reproducibility of the difference of wave numbers calculated from the same interferogram is about $0.1 \times 10^{-3} \, cm^{-1}$, which corresponds to 1/300 of the half-width of a line. A signal-to-noise ratio of 10 gives a reproducibility of about $2 \times 10^{-3} \, cm^{-1}$. The difference in the absolute values of the wavenumbers calculated from two successive interferograms can amount to $2 \times 10^{-3} cm^{-1}$, i.e., to one part in 10^7; this is due to external effects (shifts in the light source due to pressure change, Stark effect, or self-absorption), whereas the instrument itself has an accuracy of the order of one part in 10^8.

The gain in the signal-to-noise ratio, compared to results achieved with a conventional scanning spectrometer (area of beam splitter and area of grating taken to be equal), was found in the particular case of uranium to be 14 for the throughput gain (Jacquinot advantage) and 5 for the multiplex gain (Fellgett advantage).[76] Multiplex gain vanishes for weak lines.

Two examples of hyperfine structures recorded by Fourier spectroscopy are given in Figure 7, illustrating the good performance obtained both in resolution and in line intensity measurements.

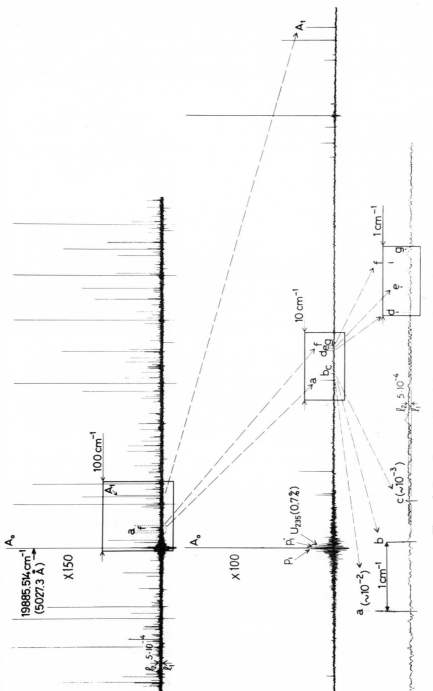

Figure 6. Part of a recording of the spectrum of natural uranium; from Ref. 76.

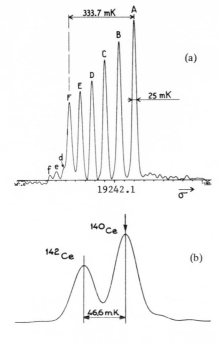

Figure 7. Examples of spectral lines highly resolved by Fourier transform spectroscopy; (a) hfs of a Pr I line ($n = 10^6$),[76] (b) shift between two Ce isotopes.[77]

The usefulness of the extension of Fourier transform spectroscopy in the visible range is no longer in doubt. On the other hand, high-resolution studies of unstable sources (including sources having short lifetimes) by means of Fourier transform spectroscopy are not yet possible. Also, when investigating only a few intense or medium lines, the conventional spectrometer remains the best instrument.

2.2.4. Other Spectrometers

(a) SISAM (spectromètre interférentiel à sélection par l'amplitude de modulation). This is a Michelson interferometer with the two mirrors replaced by identical blazed gratings at fixed distances from the beam splitter.[57,78,79,80] The two gratings have equal angles of incidence, and the optical path difference is equal to 0, $2\lambda_0$, $4\lambda_0$, . . . , so only light of the wavelength λ_0 can reach the detector. As both gratings are turned through the same angles, different wavelengths are transmitted. The intensity on the detector at the chosen wavelength is completely modulated by a periodic variation in the path difference (for example achieved by rotating the compensation plate in the Michelson interferometer).

After an instrument has been set up, its resolution has a fixed value equal to the theoretical resolution of one grating. This value can be

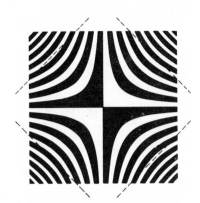

Figure 8. Principle of a grid in the grid spectrometer (actual size 3×3 cm^2, 0.05 mm smallest width of hyperbolas); the dashed lines give an example of how to cut off the edges in order to achieve apodization of the instrumental line function.

changed only by using another pair of gratings, thus hindering a quick change from high to low resolution.

The SISAM yields a radiant flux much higher than a conventional spectrometer (Jacquinot advantage). It has been used for investigation of near-infrared spectra of iodine and of some rare earths.[81–85]

(*b*) *Grid Spectrometer.* While Fourier transform spectrometers encode each resolution element of the spectrum into a sinusoidal structure (see Section 2.2.3), Hadamard transform spectroscopy[57,63,86,87] in its spectrometric imager version[88] uses moving coding masks on both the entrance and exit planes. One special case of this version—slightly modified—is the grid spectrometer.

In its general construction a grid spectrometer may be described as a conventional grating spectrometer where the entrance and exit slits have been replaced by a grid, a two-dimensional set of alternately transparent and nontransparent zones, the limit curves of which are regularly assembled equilateral hyperbolas (see Figure 8); one of their asymptotes is parallel to the slits of the conventional spectrometer.

The spectra are scanned by slowly turning the grating. For a fixed angle, α, of the grating only one wavelength, λ_0, exists that gives an exact superimposition of the entrance grid on the exit grid. Light of other wavelengths is half transmitted and half reflected by the exit grid. Both halves are measured. The intensity ratio of the halves jumps to a large value when, for the grating angle, α, light of wavelength λ_0 enters the spectrometer. This spectrometer was developed in 1960[89] and is used to investigate the infrared spectra of molecules and atoms (rare earths for the most part).[89,90]

Compared to conventional slit spectrometers, a gain in luminosity of about 150 can be achieved.[89] The resolution is nearly the same as can be obtained with a spectrometer equipped with a slit as small as the smallest distance on the grid.

2.2.5. Fabry–Perot Interferometer

The photoelectric, scanning Fabry–Perot interferometer (FPI) with plane plates and digital data output is still the device used most widely to obtain high resolution in optical spectroscopy. Both new experimental arrangements and improvements in the details of its construction are reported regularly. The following compilation is only a selection of the work done since about 1969.

(a) *Scanning FPI.* Pressure scanning[91–96] (i.e., changing the index of refraction of the gas between the interferometer plates) was the first method of scanning and is still the most popular. It is relatively simple and gives the highest finesse (up to 100). Linearity in wave number scale of about 0.2% is often achieved.[96] Digital scanning has even been used by changing the pressure in small discrete steps.[97,98] Pressure scanning is, however, a very slow method.

Other methods of scanning have been described, making use of the elastic deformation of springs,[99] thermal expansion,[100] magnetostrictive effect,[101] electrostrictive (piezoelectric) effect,[102] and of frictionless motion on air bearings over large distances.[103] Versatile FPI were recently constructed by combining coarse mechanical shifting from zero to some 10 cm distance with subsequent servo-controlled piezoelectric adjustment and sweep.[104,105] Besides the pressure-scanned FPI only the piezoelectric FPI is widely used, since it allows rapid scanning[106–113] (sweep times from some seconds down to 1 μs, the short sweep times requiring very intense light sources), stepwise scanning,[110,112] and easy adjustment and control.[104,114–118]

An alternative to these methods which is seldom used is to keep the etalon spacing fixed and change the inclination of rays with respect to the optical axes; this can be done by using an axicon (a prism of revolution).[119,120]

(b) *Multiple FPI.* If two or more FPI are used in a chain, the resolution of the system is of the same value or larger than that of the widest spaced etalon giving multiplication of the instrumental line shape functions. The free spectral range of a double FPI with spacers a and b, a/b rational and $a \geq b$, is given by $1/2b$, if the instrumental widths are smaller than the difference of the corresponding free spectral ranges, $1/2a$ and $1/2b$. When the widths increase, "ghosts" appear, as discussed in detail in Refs. 121 and 122. Similar results are valid for multiple FPI with spacers of rational ratio.[118] If this condition is not met, the free spectral range can be largely extended so that the usual predispersion by a prism or grating monochromator may be omitted[123] (PEPSIOS[95,80]).

In order to pressure scan a multiple FPI, the etalons may be placed in the same air-tight box, provided that the ratio $a:b:c \cdots$ has been fixed correctly (e.g., by means of piezoelectric spacers); if not, a constant pressure difference has to be maintained between the etalons.[108]

If the spacers of a multiple FPI are all exactly the same length, the free spectral range remains unchanged but the resolution and contrast are increased. In practice light passes two or three times through the same etalon by means of triple-prism mirrors and adequate optical decoupling (as is necessary in any multiple FPI). For this purpose the returning light passes through difference areas of the plates[124-126] or the light under investigation is linearly polarized and the plane of polarization is turned through 90° before repassing the FPI.[125,127-129]

(c) *Spherical FPI.* The confocal spherical FPI[130] was designed as an instrument capable of realizing higher resolving power, higher light flux, and lower requirements with regard to accuracy of adjustment than the plane FPI. Although the theory, design, and use of this instrument in optical spectroscopy has been described in recent texts,[21,131-133] experimental verifications are rare,[134,135] with the large-scale exception of the many uses with lasers and laser scanning (see Chapter 15).

Scanning is carried out in the same way as described in Section 2.2.5(a). With piezoelectric scanning an atomic beam passing through the interior of the interferometer can be used.[25,134,136]

The use of solid block FPI of optical glass shaped as a quasiconfocal resonator has also been described.[133]

(d) *Theory, Instrumental Function.* An analytic description of a FP spectrometer is given in Ref. 137. Resonances and fields in the resonator are dealt with in Ref. 138. Diffraction caused by finite apertures in long interferometers leads to a significant loss of signal, decreased fringe visibility, and a phase shift in the interference pattern.[139] Many texts have been published from the viewpoint of laser resonators (e.g., Ref. 140). They are not quoted here.

The instrumental line shape function of a FPI can be determined with the help of a light source that emits very narrow lines (mercury 198 lamp or thorium hollow cathode lamp[141]). Another method is to use a source emitting broad spectral lines but to illuminate the etalon under test with essentially monochromatic light using a second etalon with an instrumental profile much narrower than that of the first etalon.[142]

Studies have been made of the modification of transmission characteristics of a FPI due to surface defects on the plates.[143-145] Methods that flatten the surface by adding a deposit of thin films of appropriate size and thickness are described in Refs. 146-148.

Deconvolution techniques are presented in Refs. 149-155.

(e) *Multiplex FPI Systems.* Attempts have been made to use multiplex methods with FPI in order to overcome the overall loss of light, since the conventional FPI records only a small part of the spectrum at a time [see Section 2.2.3(a)].

One solution is the simultaneous FP,[119,156-158] in which up to 16 channels have been used, each with its own photomultiplier; this device has a very high time resolution. Another system using many channels but with a lower time resolution has been constructed,[159] which uses a rotating multizone disk (50 zones) placed in the focal plane of the interference fringes. Each of the 50 channels of the spectrum is then chopped at a different frequency and a Fourier transform recovers the spectrum.

Regularly spaced spectral lines can be simultaneously transmitted by a FPI when the free spectral range is equal to the frequency difference between adjacent spectral lines. The rotational spectra of some molecules have been investigated this way.[160,161]

2.3. Data Handling

2.3.1. Evaluation of Photoplates

The investigation of the fs in optical spectra usually starts with the recording of a spectrogram on a photographic plate (the simplest type of a signal-averaging system). In recent years, when line-rich spectra (like the lanthanide or actinide spectra) were examined, it became desirable to obtain high-precision wavelength data at the highest speed at which spectrograms could be measured and reduced. This demand promoted the development and improvement of photoelectric setting devices, automatic detections, and measuring of spectral lines on photographic plates using modern digital computers.

Automatic and semiautomatic comparators have been described in a number of papers (see for example references 1-7 in Ref. 162, 1-7 in Ref. 163, and 9-14 in Ref. 164). The instruments at the Argonne National Laboratory,[165] at the Los Alamos Scientific Laboratory,[164] and at the Zeeman Laboratorium Amsterdam[163,166,167] will be described here as examples.

The Argonne comparator uses a commercial type of instrument, supplemented by an oscilloscope which shows a stationary pattern corresponding to the transmission contour of the line under investigation.[165] To do this, the spectrum is projected onto a slit in front of a photocell, after being modulated laterally by means of a rotating octagonal prism. The prism provides an alternating direction of sweep by having alternated faces blackened. The synchronized oscilloscope sweep thus gives a density (intensity) versus distance trace of a small area of the photographic plate, successive traces being mirror images. These traces can be made to coincide by advancing the plate stage (asymmetrical lines can be set to cross each other at half-maximum intensity). The comparator drum is read at the moment at which the traces coincide. The settings can be reproduced to

0.5 μm. The comparator can be operated with a punched card output, or readout and storage is done directly by a computer. Wavelength comparison measurements (using thorium standards) can be made without shifting the plate. The dispersion curve is calculated for each plate separately.

The design of the Los Alamos automatic comparator began in 1960. While, in basic principle, similar to the preceding instrument, it is described[164] as being "not matched by other automatic comparators for position and intensity precision, or in ability to find and measure all the lines in a given spectrum." Absorption and emission spectra either together or separately may be processed as well as any set of photographic data with peaks or depressions.

Two automatic systems in use in Amsterdam have been described in the literature. The older one,[166] called COSPINSCA, uses an optical system similar to those described before, but the displacement of the plate by a screw has been modified into an automatic interference-fringe counting system. The interferometer is essentially a Michelson with reliable counting up to a limit of 16 cm shifting of the plate stage.

The other automatic Amsterdam comparator,[163,167] called ZELACOM, uses a simple microphotometer as the optical part, and a displacement measuring system consisting of a pair of transmission gratings, a lamp, and a photocell. The gratings are 25 cm and 1 cm long and are ruled with a grating constant of 10 μm. When one grating is fixed to the moving plate stage and the other one is stationary, the photocell receives an almost sinusoidal light signal.

In all the comparators described here the settings can be reproduced to 0.25–0.5 μm. The complete job of an operator can be done by the automatic comparators: output of position, wavenumber, or wavelength (including compensation of dispersion curve failures caused by unavoidable maladjustments of photoplates in the spectrograph), and recording the intensity. Even with slow scanning instruments (ZELACOM requires 2 h for a 25-cm length of plate) a large amount of data can be gained within a few days.

Although automatic comparators are well established in fs analysis of atomic spectra, the use of this technique for emission-spectrographic analysis in chemistry seems to be still in its infancy.[162]

2.3.2. Handling of Experimental Data from Hfs and Isotope Shift Measurements

Great efforts have been made to analyze recorded line structures by using computers, thus getting more information than would be possible by conventional methods of evaluation. Problems arise from the measured curves when (i) the hfs components are partly or completely unresolved, (ii) either components of unwanted isotopes (due to insufficiently enriched samples) or lines of the carrier gas in the light source are intruding, or (iii) small isotope shifts need to be investigated by mixing different isotopes in one light source rather than using the method of alternating light sources [see Section 2.2.1(b)].

Computerized analysis of the recorded patterns has been performed in many laboratories, but only very little detailed information can be found in the literature.[168-172]

The first step in the analysis usually involves finding the centers of intensity of separated components. In many cases the intensity ratios of the components are known from the abundance of the isotopes and from theoretical values of odd isotopes. By using estimated values for the splitting constants and the line profile a theoretical structure is simulated. The theoretical structure is compared with the experimental curve and an attempt is made to get good agreement between both by variation of suitable chosen parameters. Parameter variation is performed by different mathematical methods. Figure 9 gives an illustrative example of a computer analysis.

If the starting values of all free parameters (positions, intensity, underground, profile parameters) that are available are fairly good, then very good results are obtained. The calculations are affected by noise, but computational superimposition of several interference orders to enhance the signal-to-noise ratio of the experimental data is possible.

3. Fine Structure

3.1. Analysis of Spectra

There are many experimental reasons for the revival of interest during the past decade in the analyses of the spectra of all elements: projects of high-temperature plasma research and the opening up of the wavelength range from the near ultraviolet down to the x-ray region for astronomical observations. A large amount of new results has been published, relating to many stages of atomic ionization. The emphasis in this article will lie on neutral or lowly ionized atoms.

The analysis of atomic spectra implies finding the energy levels, the terms, electron configurations, and multiplets. As is well known, this can be achieved for many elements by exploring their spectra in a wide wavelength region with sufficient resolution. This method of classification is standard for light and medium heavy elements, but many spectra, mainly in the heavy elements, are so complex that they require the extension of

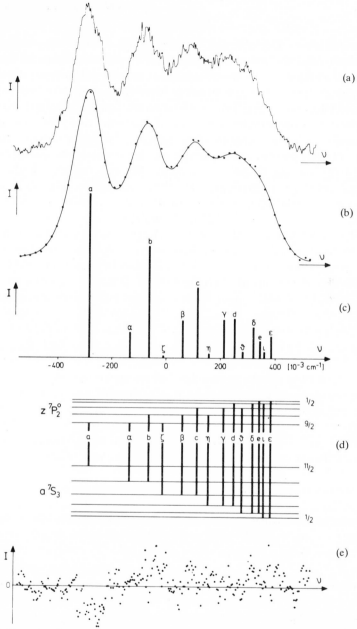

Figure 9. Example of a computer analysis of an unresolved hfs recording; Mn II, transition $3d^5 4s\, a\; {}^7S_3 - 3d^5 4p\, z\; {}^7P_2^0$, $\lambda = 2605$ Å.[125,173] (a) One interference order of the recorded hfs; (b) experimental data averaged over 25 interference orders and least-squares fit (solid line) – for sake of clarity only 1/5 of the experimental points have been printed; (c) calculated positions of the hfs components (intensity is taken from theory); (d) level scheme, evaluated from (c); (e) intensity differences of diagram (b), intensity scale 8 times increased (here all the measured values are used).

the investigation deep into the infrared region (beyond the photographic boundary), and additional information for classification must be found by means of Zeeman, hyperfine-structure, or isotope shift investigation.

The Zeeman effect is a means of identifying multiplets, but usually it gives no information about the configurations. Large grating spectrometers equipped with photoplates are required (currently many Zeeman spectra are furnished by the Argonne spectrometer[34]). The Zeeman effect has been so widely employed for a long time that it will not be discussed here.

In the spectra of odd Z elements hfs may heavily interfere with Zeeman splittings. In these cases hf splitting can help to identify levels and configurations (the nuclear spin I is nearly always known), as, for example, in Pu,[174,175] Am,[176] Tb,[177] Tm,[178] and Ho.[179]

The field shift part of an isotope shift (mostly outweighing the mass shift in the spectra of the rare earth or heavier elements; see Section 5) can be approximately predicted for different types of transitions and can thus lead to the identification of configurations. Recently experiments of this kind have been performed, sometimes combined with Zeeman effect studies, in Nd,[180] Sm,[181] Ce,[77] U,[182] and Cm.[183]

During the last years great progress has been made in the understanding of the spectra of the lanthanides and actinides, which had remained—at least as far as classification was concerned—nearly unexplored; for example, only in 1970 was the ground level of terbium established correctly.[184] The improvement in the classification of spectra was made possible by great experimental progress in the infrared region making use of Fourier spectrometers, SISAM, and grid spectrometers (see Sections 2.1.3 and 2.1.4), as well as by large efforts in theoretical and computational work. Infrared measurements led to the discovery of the connection between separated term systems, for example in Tb.[184]

An understanding of the level structure of the spectra of neutral and low ionized atoms also comes from the use of different types of light sources which strongly differ in intensity for different types of transitions.[185] (For example, it is well known from spectrochemical analysis that flame spectra and certain plasma spectra of rare earths are relatively simple.[186])

The most adequate light source for the study of rare earth and actinide spectra is the electrodeless discharge lamp.[179] Interesting problems still exist in the third and fourth spectra, the main features of which are described in Refs. 187 and 188. The systematics of the relative energies of some of the low-lying configurations in the lanthanide and actinide series are given in Ref. 189. The progress in the analyses of these spectra has allowed a large increase in the number of spectroscopically determined ionization potentials. Some of the values were derived directly from spectroscopic series, but most were obtained by more indirect methods relying on predictions based on the behavior of energy levels in neighboring elements.[190,191]

For a listing of compilations of fine-structure data, see the Appendix.

3.2. Forbidden Lines

The occurrence of forbidden transitions has been known for a long time.[192] The probability of a forbidden transition is about 10^5 times less than the probability for an electric dipole transition. Forbidden transitions may be of mixed type, that is, they are permitted for both magnetic dipole and electric quadrupole radiation. In this case the presence of interference effects has been predicted and observed. Isotope shift, hfs, and Zeeman splitting of mixed transitions were also investigated, and made possible the evaluation of the mixing coefficients.

Recent research has included, for example, transitions between levels within the ground configuration $3s^2\,3p^4$ of S I[193] (observed in the positive column of a gas discharge), transitions between the $3p$ ground state of Al I and the excited configurations $4p$–$8p$ and $4f$–$7f$[194] (using a laser produced plasma as light source), and hfs of the $^2P_{1/2}$–$^2P_{3/2}$ transition in the np^5 configuration of Br and I[195] (using an electrodeless discharge lamp).

4. Hyperfine Structure

4.1. Hyperfine-Structure Interactions

Armstrong[196] and Lindgren and Rosén[197] have recently given extensive reviews of the theory of hyperfine (hf) interaction in free atoms. Lindgren and Rosén have also discussed the principles involved in analyzing experimental hf data. Their comparisons between experimental and theoretical hf parameters are mainly concerned with atomic ground configurations which have been studied by atomic-beam magnetic resonance and are, therefore, not discussed here any further.

The magnetic dipole and electric quadrupole interaction can be represented by the operators (see, e.g., Refs. 196 and 197)

$$\mathbf{H}_\mu \propto \sum_{\substack{\text{all} \\ \text{electrons}}} [\mathbf{l}r^{-3} - (10)^{1/2}(\mathbf{s}\mathbf{C}^{(2)})^{(1)}r^{-3} + (8\pi/3)\mathbf{s}\,\delta(\mathbf{r})] \cdot \mathbf{I} \tag{1}$$

$$\mathbf{H}_Q \propto \sum_{\substack{\text{all} \\ \text{electrons}}} \mathbf{C}^{(2)}r^{-3} \cdot \mathbf{Q}^{(2)} \tag{2}$$

The last term in the dipole operator, the contact term, is caused by s

electrons only, while the remaining orbital and spin–dipole terms, as well as the quadrupole term, are due to non-s electrons. In the central-field model no contribution to the sum is obtained from the closed shells, since, in this model, closed shells are exactly spherical and therefore exhibit only the monopole Coulomb interaction with the nucleus. The simple central-field model, however, often predicts hf interactions that are in poor agreement with experiments. The reason for this is that polarization, correlation, and relativistic effects are not taken into account.

Polarization effects can be considered as distortions of the electron orbitals caused by perturbations from the nonspherical open shells so that the "closed shells" are no longer spherically symmetric. In the language of perturbation theory the polarization effects can be described by means of single-electron excitations (crossed-second-order effects[198,199] of the hf operator and the interelectronic electrostatic operator, sometimes referred to as far-configuration-mixing). Distortions of closed s shells are then interpreted as single excitations of s–s type, an effect which is usually called core polarization (see Section 4.3). Other types of polarization effects are represented by single excitations of the types p–p, d–d, s–d, p–f, ... Such excitations, for example, enter the quadrupole interaction and give rise to the Sternheimer correction (see, e.g., Refs. 200 and 201).

In contrast to the polarization effects, correlation effects require multiple excitation. They arise because the electrons do not move independently of each other.

It has been shown by Judd[198,199] and others[202,203] that all the types of configuration-interaction effects (polarization and correlation) that can be described by a one-body effective operator operating by definition only within the open shells are included in an effective Hamiltonian of the form

$$\mathbf{H}_{\mu}^{\text{eff}} \propto \sum_{\substack{\text{open} \\ \text{shells}}} [l\langle r^{-3}\rangle_l - (10)^{1/2}(\mathbf{s}\mathbf{C}^{(2)})^{(1)}\langle r^{-3}\rangle_{sd} + \mathbf{s}\langle r^{-3}\rangle_s] \cdot \mathbf{I} \tag{3}$$

$$\mathbf{H}_{Q}^{\text{eff}} \propto \sum_{\substack{\text{open} \\ \text{shells}}} \mathbf{C}^{(2)}\langle r^{-3}\rangle_q \cdot \mathbf{Q}^{(2)} \tag{4}$$

In contrast to Eqs. (1) and (2) the summation is now limited to open shells, and the radial functions are replaced by radial parameters, the values of which depend on the perturbations. The core polarization is represented in Eq. (3) by a "contact" term, even when there are no s electrons in the open shells. (In the case of an s electron, $\langle r^{-3}\rangle s$ is expressed by the density of the s electron at the nucleus, $|\psi(0)|_s^2$.) Other polarization and most of the correlation effects are included by allowing the radial factors of the orbital, spin-dipole and quadrupole terms to be different. Effective operators of the many-body type are not included in Eqs. (3) and (4); however, they are expected to be less important so that Eqs. (3) and (4) may be quite adequate in most cases.

Sandars and Beck[204] have shown that relativistic effects can be included in the dipole operator Eq. (3) without further modification, while in the quadrupole case two additional terms are required which have no nonrelativistic counterparts. In the dipole case it is therefore not possible to separate the relativistic and the configuration-interaction effects in the experimental data. When core polarization is present the hf anomaly may provide a tool for the discrimination of both effects (see Section 4.4).

The relativistic effective-operator formalism is frequently used in analyzing experimental hf data. In order to perform such an analysis, data have to be available for a sufficient number of levels of the same configuration. The first case in which optical hf measurements have been analyzed by this method seems to be the hyperfine structure (hfs) of levels of the configuration $4f^7(^8S)6s6p$ of Eu I.[205,206]

The original aim of hf measurements was the determination of nuclear spins and nuclear moments. Fuller and Cohen[207] have given in their compilation of nuclear spins and moments a table of values obtained by optical spectroscopy. (For updated values of the nuclear moments, see Ref. 208.) Although the spins of all stable isotopes now are known and magnetic moments can be measured directly with high accuracy, optical hf investigations are still useful for the determination of nuclear electric quadrupole moments, Q, which is one of the most important problems of the hfs analysis. The only remaining possibilities for the determination of nuclear spins and magnetic dipole moments by optical spectroscopy are in unstable isotopes. Some recent examples are given in Section 4.5.

In contrast to radio-frequency methods and level crossing, optical spectroscopy has the advantage that the hfs of levels belonging to different configurations in the atom as well as in the singly or even doubly ionized spectrum can easily be studied provided the splitting is large enough for sufficiently accurate measurements to be made. Such investigations are of particular interest for the determination of quadrupole moments, since they give good information about different Sternheimer corrections and allow the reliability of the evaluated quadrupole moments to be tested.

As a consequence of the theoretical development mentioned above, the aim of hf investigations now has somewhat shifted to the study of atomic structure, since the interaction between the electrons and the nucleus is very sensitive to the various perturbations, and valuable information on the atomic structure can therefore be obtained from a careful analysis of the hfs. If, for example, the hfs of a sufficient number of levels belonging to different configurations has been studied, a set of effective parameters can be obtained for each configuration. Comparison of these parameters with theoretical values can give information about the different types of configuration interactions present in the various configurations.

In many cases the magnetic hf splitting factors, A, have been used to check on calculated wave functions, which in turn are required to evaluate the quadrupole moment from the measured quadrupole coupling constants, B. The different procedures for obtaining $\langle r^{-3} \rangle_q$, which is needed for the evaluation of Q [See Eq. (4)], have been discussed critically in Refs. 209, 197, and 210. It should also be noted that the relativistic correction factors given by Kopfermann[211] ("Casimir factors") that are frequently used in such calculations are rather uncertain. New relativistic correction factors have therefore been calculated using relativistic self-consistent-field wave functions.[197] It turns out that the Casimir factors are particularly poor for d and f electrons (see also Refs. 212 and 213).

4.2. Parametric Treatment of the Hyperfine Structure

The first step in a modern parametric analysis of hf data is a theoretical parametric study of the fine-structure level energies in order to determine the wave functions in intermediate coupling while taking into account close-configuration-mixing. In this procedure the angular wave function for a particular state is written as a linear combination of basic states. In such an expansion LS states are normally used since most of the tensor-operator formalism is developed for these states. The coefficients in the expansion can be obtained from a least-squares fit of the theoretical level energies to the observed ones, treating the electrostatic integrals and spin-orbit and other constants as adjustable parameters. Usually g_J values are also evaluated in order to give a further check on the wave functions.

The angular part of the hfs Hamiltonian is then evaluated, and expressions for the hf constants in terms of certain radial parameters are obtained. Finally, the experimental values of these radial parameters are determined in a least-squares fit of the theoretical hf splitting constants to the measured ones.

As an example we shall consider the magnetic hf splitting in the configuration $4f^7(^8S)6s6p$ of Eu I.[214] In this case the effective tensor operator, $\mathbf{O}$, in the most general relativistic expression for the magnetic hf interaction $\mathbf{H}_\mu^{\text{eff}} = \mathbf{O} \cdot \mathbf{I}$ can be taken to be

$$\mathbf{O} = a_f \cdot \mathbf{S}_f + a_s \cdot \mathbf{s} + a_p \cdot \mathbf{s} + b_p \cdot \mathbf{l} - (10)^{1/2} c_p \cdot (\mathbf{s}\mathbf{C}^{(2)})^{(1)} \tag{5}$$

where $a_f \cdot \mathbf{S}_f$ is the contact term due to the half-filled $4f$ shell, a_s is the a value of the $6s$ electron, and the radial parameters a_p, b_p, c_p represent contact, orbital, and spin–dipole terms of the $6p$ electron, respectively.

Twelve energy levels belong to the configuration $4f^7(^8S)6s6p$; the hf splitting has been measured in 11 of them. Nine levels were studied by optical spectroscopy and two others ($z\,^6P_{7/2}$ and $y\,^8P_{7/2}$) by level crossing (see Table 2). Wave functions taking into account configuration mixing

Table 2. Theoretical and Experimental A Values and Best Values of the Radial Parameters in the Configuration $4f^7(^8S)6s6p$ of Eu I According to Lange $^{(214)a}$

Level		Theory	Expt.	Diff.	Ref.
z^6P	3/2	-76.04	—		
	5/2	-19.79	$-19.72(5)$	-0.07	205
	7/2	0.19	$-0.217(2)$	0.407	214
z^8P	5/2	-20.29	$-20.337(60)$	0.047	215
	7/2	-8.05	$-7.943(26)$	-0.107	215
	9/2	22.89	22.18(11)	0.71	205
$z^{10}P$	7/2	32.06	32.32(2)	-0.26	205
	9/2	34.13	34.138(8)	-0.008	205
	11/2	30.69	31.13(10)	-0.44	205
y^8P	5/2	-4.93	$-5.25(5)$	0.32	216
	7/2	-7.51	$-7.30(1)$	-0.21	217
	9/2	-8.24	$-7.68(11)$	-0.56	205

$$a_f = -2.4(4) \qquad b_p = 16.0(2.3)$$
$$a_s = 336(6) \qquad c_p = 27.5(5.0)$$

^{a} All values are for the isotope ^{151}Eu and are given in mK (10^{-3} cm^{-1}). References refer to experimental values.

with $4f^7(^8S)5d6p$ and $4f^6(^7F)5d6s^2$ were available. Since the hfs in $4f^7(^8S)6s6p$ is governed by the interaction of the $6s$ electron with the nucleus, the immediate contributions of the perturbing configurations to the hf splitting were neglected and only the influence of the perturbation on the coupling coefficients within the configuration $4f^7(^8S)6s6p$ was considered. Using these wave functions theoretical expressions for the A values were derived, and a least-squares fit to the experimental A values was carried out. Four of the five radial parameters in Eq. (5) were left free, while the ratio a_p/b_p was calculated by means of Casimir's relativistic correction factors. The results are given in Table 2. The agreement between theoretical and experimental A values is excellent. The r.m.s. error is 0.36 mK only, and the largest deviation amounts to 0.7 mK (1 mK = 10^{-3} cm^{-1}). The value found for a_f can reasonably be explained by core polarization (see Section 4.3), the result for b_p is in good agreement with the prediction $b_p = 17.4$ mK obtained from the spin–orbit coupling constant, and the ratio c_p/b_p agrees within the limits of error with the relativistic prediction $c_p = 1.34b_p$.

The wave functions applied for the interpretation of the A values were also used for the discussion of the observed B values.$^{(217)}$

The parametric method can usually be applied if a sufficiently large number of A or B values has been measured. A recent example is the study of the hfs in Tc I by optical spectroscopy,[218] where the hf splitting has been measured in 10 even levels belonging to $(4d + 5s)^7$ and in 21 odd levels belonging to $4d^5 5s 5p$ and $4d^6 5p$. Strong configuration mixing had to be taken into account and a new and more accurate value for the nuclear electric quadrupole moment of ^{99}Tc could be derived.

4.3. Core Polarization

Core polarization in the hfs of alkalilike atoms has been reviewed by Fischer.[219] The influence of core polarization on the hfs observed in $5d$-shell atoms has recently been discussed by Müller and Winkler.[220]

As an example, we shall briefly discuss the core polarization caused by the spins of the unpaired $4f^7(^8S)$ electrons in Eu. In this case optical hf measurements in $4f^7(^8S)6s6p$ of Eu I and $4f^7(^8S)6s$ and $4f^7(^8S)6p$ of Eu II are combined with results from radio-frequency spectroscopy on the hfs in $4f^7(^8S)6s^2$ of Eu I and $4f^7\ {}^8S$ of Eu III. If there is no core polarization the following five magnetic hf splitting factors would only be caused by relativistic effects of the $4f$ electrons and would then be approximately equal, since the wave functions of the $4f$ electrons are about the same in the various electronic structures: $A(4f^7(^8S)6s^2\ {}^8S_{7/2})$, $A(4f^7\ {}^8S_{7/2})$, and the a_f factors (contact terms) of the $4f^7(^8S)$ electrons in $4f^7(^8S)6s6p$, $4f^7(^8S)6s$, and $4f^7(^8S)6p$. The experimental results, however, are (for ^{151}Eu)

$$^{151}A(4f^7(^8S)6s^2) = -0.668\ \mathrm{mK}^{(221)}$$

$$^{151}a_f(4f^7(^8S)6s6p) = -2.4(4)\ \mathrm{mK}^{(214)}$$

$$^{151}a_f(4f^7(^8S)6s) = -2.32(8)\ \mathrm{mK}^{(206,222)}$$

$$^{151}a_f(4f^7(^8S)6p) = -3.20(20)\ \mathrm{mK}^{(222)}$$

$$^{151}A(4f^7\ {}^8S) = -3.430\ \mathrm{mK}^{(223)}$$

The A values of $4f^7(^8S)6s^2$ and $4f^7\ {}^8S$ will be discussed first. The difference between these A values indicates that different contributions of core polarization are present. The reason for this is that in Eu III inner s electrons can be (virtually) excited into the vacant $6s$ orbitals, i.e., $ns^2 \rightarrow ns6s$, with $n < 6$. In Eu I the $6s$ orbitals are occupied but the $6s$ electrons may be excited, i.e., $ns^2 6s^2 \rightarrow ns^2 6sn's$, with $n' > 6$. (In both cases, of course, excitation like $ns^2 \rightarrow nsn's$ are possible.)

Detailed calculations of core polarization show that the relation

$$a_f(4f^7(^8S)6s6p) = \tfrac{1}{2}[A(4f^7(^8S)6s^2) + A(4f^7\ {}^8S)]$$

should be valid under the simplifying assumption that all s electron wave functions, all $4f$ electron wave functions, and all excitation energies are the same for the three electronic structures.[206] The experimental values

$$-2.4 \text{ mK} \approx \tfrac{1}{2}(-0.668 - 3.430) \text{ mK} = -2.05 \text{ mK}$$

fulfill the theoretical prediction very well. Under the same assumptions

$$a_f(4f^7(^8S)6s6p) = a_f(4f^7(^8S)6s)$$

also should hold, because the two configurations have the same possibilities for the excitation of inner s electrons to vacant s orbitals. The experimental values agree excellently. For the same reasons also

$$a_f(4f^7(^8S)6p) = A(4f^7 {}^8S)$$

which again is confirmed by the experimental values.

The discussion of the A values of $4f^7(^8S)\ 6s^2$ and $4f^7\ ^8S$ reveals a different contribution of core polarization in both configurations but the size of these contributions cannot be decided. The consideration of hf anomaly (see Section 4.4) enables this to be done.

Recently the hfs in Tc I has been studied by optical spectroscopy,[218] and it has been shown that the observed A value of the ground state $4d^5 5s^2\ ^6S_{5/2}$ can be explained by core polarization. Because of the high configuration and Russell–Saunders purity of this level, as revealed by a parametric study of the wave functions, only the contact term of the $4d$ electrons, a_{4d}, has an appreciable importance in the interpretation, so that $A\ (4d^5 5s^2\ ^6S_{5/2}) = a_{4d}(4d^5 5s^2)$. A spin-polarized Hartree–Fock (HF) calculation has been performed, and, using the known value of the magnetic dipole moment of ^{99}Tc, the *ab initio* value $a_{4d}(4d^5 5s^2) = -3.43$ mK has been obtained and is in good agreement with the experimental value $A(4d^5 5s^2\ ^6S_{5/2}) = -3.64(3)$ mK.

In excited levels of Tc I much larger experimental a_{4d} values are found, for example $a_{4d}(4d^6 5s) = -10.5(5)$ mK and $a_{4d}(4d^5 5s 5p) = -12.33(26)$ mK. This can be explained by the spin-polarized HF calculation which shows that in the configuration $4d^5 5s^2$ the large polarization effect of the $5s^2$ subshell amost cancels the sum of the contributions of the other subshells, whereas this cancellation disappears in the excited configurations. This behavior of the core polarization is similar to that observed in Eu which has been discussed above.

Luc[224] has studied core polarization effects in the configuration $3d^5 4s 4p$ in Mn I and also compared experimental results with spin-polarized HF calculations.

4.4. Hyperfine Anomaly

An evaluation of the hf interaction must take into consideration that an s or $p_{1/2}$ electron can penetrate the nucleus and within the nuclear volume the distribution of magnetization is not uniform. A measurable quantity is the hf anomaly, defined for isotopes 1 and 2 of the same element as

$$_1\Delta_2 = \frac{A_1}{A_2}\frac{g_2}{g_1} - 1$$

where A_1, A_2 are the A values of a particular energy level and g_1, g_2 the nuclear g factors. A nonvanishing value of $_1\Delta_2$ is in fact expected if it is realized that the direct nuclear moment measurement in an external, homogeneous, magnetic field does not measure the same interaction as the hfs in which the nuclear magnetization interacts with the slightly inhomogeneous magnetic field produced by the electrons at the nucleus. The hf anomaly will be nonzero only if (i) the distribution of nuclear magnetism over the nuclear volume is different for the two isotopes and (ii) the A value contains a contribution from an s or a $p_{1/2}$ electron.

Hf anomalies have been tabulated by Fuller and Cohen.[225] Most hf anomaly values are based on atomic-beam magnetic resonance measurements. Optical spectroscopy may give an important contribution to the study of hf anomalies if excited levels of an atom or levels of ions are investigated provided the hf splitting is large enough so that sufficiently accurate measurements can be made. In the following we give a typical application of hf anomaly to the study of atomic structure.

The hfs of the two stable Eu isotopes ^{151}Eu and ^{153}Eu has been investigated in several resonance lines of the Eu II spectrum. From these measurements is derived the hf anomaly in the two levels of $4f^7(^8S)\,6s$, which is mainly due to the $6s$ electron (other influences may be neglected). The value obtained is $_{151}\Delta(s)_{153} = -0.61\,(15)\%$[222] where $\Delta(s)$ stands for a hf anomaly solely caused by s electrons. In three levels of the configuration $4f^7(^8S)6s6p$ in Eu I the hf anomaly also has been measured.[205] The $_{151}\Delta_{153}$ values are different in these levels because the contribution of the $6s$ electron to the A values is different. Since these contributions are known from theory, the hf anomaly $_{151}\Delta(s)_{153}$ due to the $6s$ electron can easily be obtained from $_{151}\Delta_{153}$ by multiplying by an appropriate factor.[206] The mean value, $_{151}\Delta(s)_{153} = -0.60\,(15)\%$, is in excellent agreement with the result obtained in Eu II.

Besides the optical investigations just mentioned, the ground states of Eu I $4f^7(^8S)6s^2\,{}^8S_{7/2}$[226] and Eu III $4f^7\,{}^8S_{7/2}$[223] have been studied by radio-frequency spectroscopy. The hf anomaly in $4f^7(^8S)6s^2$ turns out to be zero, thus indicating that the small A values observed in the ground state of Eu I are entirely caused by relativistic effects of the $4f$ electrons. On the other hand, in the ground state of Eu III a hf anomaly $_{151}\Delta_{153} = -0.53\%$ is found, which means that there is a contribution of s electrons to the A values of $4f^7\,{}^8S_{7/2}$ (core polarization, see Section 4.3), besides the contribution from the relativistic effects of the $4f$ electrons. Moreover, assuming that the relativistic contribution to the A value of $4f^7\,{}^8S_{7/2}$ in Eu III is approximately equal to $A(4f^7({}^8S)6s^2\,{}^8S_{7/2})$, it follows that the hf anomaly due to s electrons in $4f^7\,{}^8S_{7/2}$ is ${}_{151}\Delta(s)_{153} = -0.65\ (10)\%$,[206] which is in good agreement with the values found in Eu I and Eu II (the hf anomaly is independent of the principal quantum number of the s electrons). It may be concluded, therefore, that the hf anomalies observed in Eu fall into a consistent pattern.

4.5. Optical Hyperfine Investigations of Radioactive Isotopes

In the following we give four typical examples for the study of the hfs of unstable isotopes and isomers.

The knowledge about the nuclear properties of the odd–odd silver nuclides was recently completed by a study of the hfs of ^{108m}Ag and ^{110m}Ag ($T_{1/2} = 127y$ and $256d$, respectively) in the resonance lines of Ag I.[227] The isomers can be easily produced by neutron irradiation of the stable isotopes. Because of the large splitting of the isomers compared to the stable isotopes, the hf pattern could be well resolved, and no isotope separation was needed. The hollow cathode light sources were therefore prepared with the appropriate stable enriched isotope (^{107}Ag, ^{109}Ag, respectively) and afterwards irradiated in the reactor. In this way the nuclear magnetic dipole moment of ^{108m}Ag and the electric quadrupole moments of both isomers have been determined for the first time.

Champeau et al.[228] investigated the hfs of ^{169}Yb ($T_{1/2} = 32d$) in the intercombination line of Yb I. The isotope was produced by a (n, γ) reaction from a pure sample of ^{168}Yb. About 10 μg of ^{169}Yb were used in the hollow cathode for the recording of the hfs. The value of the spin $I({}^{169}$Yb$) = 7/2$, already expected from γ-ray spectroscopy results, was confirmed and the nuclear moments were determined. Since the Yb nuclei are strongly deformed, the intrinsic quadrupole moment Q_0 can be calculated from the spectroscopic one [Eq. (23)]. The value of Q_0 thus obtained is close to those of the neighboring isotopes of Yb.

Magnetic moments of nuclei in the vicinity of doubly magic numbers, and in particular those of bismuth, play an important part in nuclear-structure theory and in the observation of the effects of meson exchange on the nucleon g values. Stroke and his collaborators therefore studied the hfs of 30-year ^{207}Bi[229] and 15-day ^{205}Bi[230] in the 3067-Å resonance line $(6p^3\,{}^4S_{3/2}$–$6p^27s\,{}^4P_{1/2})$. The radioisotopes were produced in a cyclotron bombardment of natural lead, ^{207}Pb (p, n) ^{207}Bi at a proton-beam energy of 22 MeV and ^{207}Pb $(p, 3n)$ ^{205}Bi at 30 MeV, followed by mass separation. A procedure was developed for making low-impurity electrodeless spectral lamps using only nanogram quantities of a radioactive Bi isotope. The nuclear magnetic dipole moments of ^{205}Bi and ^{207}Bi were derived from the hf splitting. The nuclear spin of ^{207}Bi was determined by using a computer analysis of the hfs intensity profile of the ^{207}Bi line.[231] Calculations of magnetic moments of odd bismuth isotopes, including configuration mixing and pion-exchange contributions, are in agreement with the experimental results.[230]

The hfs of the einsteinium isotope ^{253}Es ($T_{1/2} = 20d$) has been studied by Worden et al.[232] together with an extensive investigation of the fine structure. Electrodeless discharge lamps filled with some μg of ^{253}Es were used as light sources. The nuclear spin was determined,[233] and the magnetic dipole and electric quadrupole moment were extracted from the hfs of the two lowest levels $5f^{11}7s\,^5I_8$ and $^3I_7$ of Es II. The results are in agreement with those obtained by atomic-beam magnetic resonance measurements on the ground state $4f^{11}7s^2\,^4I_{15/2}$ of Es I.

5. Isotope Shift

Optical isotope shift (IS) measurements in the heavier elements ($Z \gtrsim 38$) can provide accurate data concerning the variation in the nuclear charge distribution as a function of neutron number. The extraction of the quantity $\delta\langle r^2 \rangle$, the change in mean-square nuclear charge radius, is possible to a moderate precision (typically 10%–15%), while the relative values of $\delta\langle r^2 \rangle$ over a sequence of isotopes can be found to much greater accuracy.

The relation between nuclear structure and IS has been the essential reason for the interest in the IS of optical transitions for many years. This aspect is reflected in review papers by Brix and Kopfermann,[234] Stacey,[235] Golovin and Striganov,[236] and Heilig and Steudel.[237] During the past ten years, however, the interest has somewhat shifted towards the electronic problems of IS. This point of view is stressed in the recent review paper by Bauche and Champeau[238] on the theory of atomic IS.

5.1. General Features of Optical Isotope Shifts

5.1.1. Isotope Shift Formula

The isotope shift $\delta\nu_i^{AA'}$ between two isotopes with mass numbers A and A' observed in the wave number ν of an atomic spectral line i is the sum of a mass shift (MS) and a field shift (FS), also referred to as "volume shift":

$$\delta\nu_i^{AA'} = M_i(A' - A)/AA' + F_i\lambda^{AA'} \tag{6}$$

with $A' > A$. There are small corrections to Eq. (6) (see, e.g., Refs. 235

and 239), which can be neglected within the accuracy of the measurements.

The MS is caused by the change of the nuclear mass and results from the additional kinetic energy due to the motion of the nucleus. The factor M_i depends only on the wave functions of the electronic states of the transition; it is customarily divided into a normal mass shift (NMS) and a specific mass shift (SMS),

$$M_i = M_{i \text{ normal}} + M_{i \text{ specific}} \tag{7}$$

The NMS is due to the reduced mass correction, which can be calculated easily and is always positive ($M_{i \text{ normal}} = \nu_i / 1836.1$, where ν_i is the wave number in cm^{-1}; the IS in a spectral line is positive when the heavier isotope is shifted towards larger wave numbers).

The SMS originates from the influence of correlations in the motion of the electrons on the recoil energy of the nucleus as given by $(1/M) \sum_{k>j=1}^{N} \mathbf{p}_j \mathbf{p}_k$, where $\mathbf{p}_j$ are the electron momenta, M is the nuclear mass, and N is the number of electrons in the atom. The factor $M_{i \text{ specific}}$ can be positive or negative depending on the correlations between all the electrons in the atom. Although there is a factor $1/AA'$ as in the NMS [see Eqs. (6) and (7)], the SMS may drop off more slowly because of the increase in the possible number of electrons, N, contributing to the above sum. In particular transitions (d or f electron jumps, see Section 5.3) $|M_{i \text{ specific}}|$ may be up to 30 times larger than $M_{i \text{ normal}}$ (see, e.g., Ref. 240).

The FS originates from the change of the finite size and angular shape of the nuclear charge distribution when neutrons are added to the nucleus; consequently the binding energy of electrons that penetrate the nucleus differs for different isotopes.

It is customary to factorize F_i into an electronic factor E_i depending on the change of the total nonrelativistic electron-charge density at the (point) nucleus, $\Delta |\psi(0)|_i^2$, in the transition i:

$$E_i = \pi a_0^3 \Delta |\psi(0)|_i^2 / Z \tag{8}$$

and a known function $f(Z)$, which increases with Z and which includes the relativistic correction to E_i and also takes into account the finite nuclear charge distribution:

$$F_i = E_i f(Z) \tag{9}$$

For the determination of E_i the most reliable approach at present is to make use of empirical data as much as possible (see Section 5.4.3). The values of $f(Z)$ are often taken from the calculation of Babushkin.[241]

The FS depends precisely on a power series in the change of the radial nuclear charge moments

$$\lambda^{AA'} = \delta \langle r^2 \rangle^{AA'} + \frac{C_2}{C_1} \delta \langle r^4 \rangle^{AA'} + \frac{C_3}{C_1} \delta \langle r^6 \rangle^{AA'} + \cdots \tag{10}$$

Because the s-wave function at the nucleus is largely independent of the principal quantum number, the relative contributions of different moments to the FS are nearly equal for K x-ray and optical transitions; therefore the ratios C_n/C_1 can be obtained from the C_n values given for K x-ray shifts by Seltzer.[239] Since, to a good approximation, the electron wave function can be considered constant over the nuclear volume, the contributions of higher charge moments are small,[242,243] usually smaller than the errors in the evaluation of λ from optical measurements; therefore in the following

$$\lambda^{AA'} \approx \delta\langle r^2\rangle^{AA'} \qquad (11)$$

is used.

The dependence of FS on the moments of the nuclear charge distribution can alternatively be expressed by the change of a single moment $\delta\langle r^k\rangle$, where k is not an integer. According to Wu and Wilets,[244] optical electrons measure $\delta\langle r^{2\sigma'}\rangle$ with $\sigma' = 1 - 0.177\alpha^2 Z^2$. This result was confirmed by Boehm (see Ref. 245 and Figure 5 in Ref. 242), who calculated k by numerical integration using Seltzer's C values[239] (see also Ref. 246).

The product $f(Z)\delta\langle r^2\rangle = C$ is often referred to as the IS constant, first introduced by Brix and Kopfermann.[247]

FS, of course, can be observed only in transitions with $\Delta|\psi(0)|_i^2 \neq 0$. These comprise not only lines with an s-electron jump (or to a much smaller extent with a $p_{1/2}$-electron jump) but also transitions in which the screening of the inner closed s-electron shells is changed. For example, transitions of the type $nf^m - nf^{m-1}(n+1)d$ with $n = 4$ (rare earths) and $n = 5$ (actinides) show a large FS comparable to that of an $(n+2)s$-electron jump (see, e.g., Ref. 248).

5.1.2. Relative Isotope Shifts

If both MS and FS contribute to the observed IS in the spectral lines of an element, the relative contributions of MS and FS are usually different in various transitions [that is, $F_i/M_i \neq F_{i'}/M_{i'}$ in Eq. (6)] and therefore the relative isotope shift (RIS) often differs in different lines provided $\delta\langle r^2\rangle^{AA'}$ is not coincidentally proportional to $(A'-A)/AA'$.

In very light elements where the FS is negligibly small within the experimental errors the RIS is proportional to $(A'-A)/AA'$. Any deviation of the RIS from this proportionality must be interpreted as an influence of FS. For elements with $Z \lesssim 37$ where $|MS| \gg |FS|$, this is so far the only way of recognizing FS. At the present moment calcium is the lightest element in which FS were clearly observed in this manner.[25,249-251]

In medium-heavy elements ($Z = 38, \ldots, 57$) the contributions of MS and FS to the observed shifts are roughly of the same order; the FS are larger than the MS in favorably chosen lines, that is, lines involving the transition of electrons to or from low s states provided the $\delta\langle r^2 \rangle$ values are not extremely small, as is the case just below the closed neutron shells $N = 50$ and $N = 82$ (see Figure 12).

In heavier elements ($Z \gtrsim 58$) the observed shifts are mainly determined by FS, though the contributions of MS cannot be neglected, as is revealed by the fact that in many elements the RIS of some lines are quite different than for other lines. This was first clearly demonstrated by a careful investigation of the shift of the even isotopes of natural samarium by Striganow *et al.*[252] King[253] pointed out that the different RIS in Sm can be explained if SMS of an order of magnitude larger than the NMS are taken into account. In neodymium SMS were estimated to be up to 30 times larger than the NMS.[240] Even in plutonium the RIS does not seem to be constant.[254]

5.1.3. King Plots

The consistency of optical IS measurements with Eq. (6) can be checked easily when IS measurements have been made on at least four isotopes and the RIS is not proportional to $(A' - A)/AA'$. For two lines i and i' it follows from Eqs. (6) and (9) that

$$\delta\nu_i^{AA'} = \delta\nu_{i'}^{AA'} E_i/E_{i'} + [(A' - A)/AA'] \times (M_i - M_{i'} E_i/E_{i'}) \qquad (12)$$

Thus, if the "modified IS"

$$\zeta = \delta\nu_i^{AA'} \times AA'/(A' - A) \qquad (13)$$

is plotted for all possible pairs of values AA' against the same expression ξ for another line i', the points should lie on a straight line of slope $E_i/E_{i'}$ and intersect $M_i - M_{i'} E_i/E_{i'}$. This method is due to King[253]; it was slightly modified[255,256] in order to include the $1/AA'$ dependence of the MS, which is necessary when very accurate measurements are available. Figure 10 shows an example of a King plot. In all cases studied so far, no significant deviations of the experimental points from a straight line have been found for even isotopes. This is not the case for the odd isotopes which may deviate from the King line due to second-order hf perturbations (mixing of hf states having the same angular momentum F and the same parity and being sufficiently close together). Such a perturbation shifts the center of gravity of the hf splitting.

A striking example is the hf splitting of the $6s6d\ ^3D_1$ and $^1D_2$ levels in Hg I. The second-order hf interaction between these levels has been

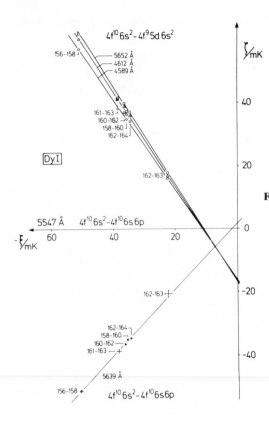

Figure 10. King graph for some Dy lines.[237] For the plot ζ and ξ [see Eq. (13)] were multiplied by $2/(162 + 164)$. At the points the mass numbers of the isotope pairs are given. For simplicity error bars are only shown on the lower line. They are nearly the same for the corresponding isotope pairs in the three other lines. IS data are from Ref. 257, Classifications of transitions from Ref. 258.

recently studied by using precision measurements of odd–even RIS (Gerstenkorn and Vergès[259]).

5.1.4. Separation of Mass and Field Shift

The unknown quantities F_i, $\lambda^{AA'} \approx \delta \langle r^2 \rangle^{AA'}$, and M_i cannot be determined separately from Eq. (6) even if measurements have been made for many lines and isotopes so that the number of measurements is larger than the number of unknowns. Since the equations are not independent, only ratios or other combinations of the unknowns can be derived. Thus additional information is needed in order to separate MS and FS.

The various possible methods for the separation of MS and FS have been reviewed extensively by Heilig and Steudel[237] and by Bauche and Champeau.[238] A critical comparison of some of these methods has been given recently for the case of neodymium.[240]

At present, the most reliable way to separate MS and FS seems to be to start with pure ns–np or ns^2–$nsnp$ transitions in which the SMS is small and can be estimated.[237] In all other transitions of any spectrum of the same element SMS and FS can then be derived by means of a King plot. The King plot for Dy I in Figure 10, for example, shows that the two $4f^{10}6s^2 - 4f^{10}6s6p$ transitions have approximately the same SMS. If the SMS in these transitions is taken to be zero, a large SMS, of about -16 mK for $\delta A = 2$, results for the three $4f^{10}6s^2 - 4f^9 5d6s^2$ transitions (the HF calculation[260] yields -30 mK for these transitions; the NMS is about 0.8 mK).

In principle, the comparison of optical IS with electronic x-ray shifts in a King plot would be an excellent method of separating MS and FS, since both optical IS and electronic x-ray shifts measure the same nuclear parameter, and, in the electronic x-ray shifts, the MS can be allowed for exactly. Unfortunately, this procedure still suffers from the insufficient accuracy of the x-ray data.

Whichever method is used, the separation of MS and FS usually causes a loss of precision. The parametric description of IS (see Section 5.2) is therefore of particular interest, since it does not require a separation of MS and FS.

5.2. Parametric Method

Considerable progress has been achieved in the phenomenological description of IS, also referred to as the "empirical" or "parametric" method, which is based on the central-field model.[238] This method is quite similar to the well-known parametric study of level energies, in which the radial integrals are taken as adjustable parameters and the angular wave functions of the levels are determined in intermediate coupling while allowing for possible configuration mixing. The IS of each level may correspondingly be written as a numerical expansion in terms of IS formal parameters. These parameters are determined by a least-squares fitting procedure, so that the experimental IS are reproduced in the most accurate way. An advantage of this method is that FS and MS need not necessarily be separated. The full accuracy of the measurements is therefore available for the least-squares fit.

The first parametric description of IS was published in 1959 by Stone,[261] who studied the low $np^5 (n+1)s$ configuration in Ne and Kr. The method was then fully developed by Bauche.[262]

5.2.1. Application of the Parametric Method to Xe

A remarkable example of a parametric study of IS is the recent work of Jackson et al.[263] on five even Xe isotopes. With Xe MS and FS are small and of the same order of magnitude. The discussion will be confined to only the odd levels that have been investigated: 11 levels of the configuration $5p^5 5d$ and 4 levels of $5p^5 6s$. Because of configuration mixing both configurations are treated together. The IS in these 15 levels has been described by six parameters. One parameter, a, with coefficient unity for levels of both configurations, and another parameter, d, whose coefficient for a given level is equal to the fraction of unity for which this level belongs to the second configuration, thus representing the well-known sharing rule (see below). The parameters a and d account at the same time for FS and for all the MS contributions which are equal for all levels of a configuration. For the description of other contributions the parameters $g^1(5p, 6s)$ and $g^1(5p, 5d)$ are required. They have exactly the same angular coefficients as the Slater integrals of rank 1, $G^1(5p, 6s)$ and $G^1(5p, 5d)$, which are necessary for expanding the total atomic energy of the relevant levels.[261,262] $g^1(5p, 5d)$ is a pure MS parameter, whereas $g^1(5p, 6s)$ also absorbs second-order FS contributions. Two further parameters, $z_{5p}(5p^5 5d)$ and $z_{5p}(5p^5 6s)$, having the same angular coefficients as the spin-orbit constants $\zeta_{5p}(5p^5 5d)$ and $\zeta_{5p}(5p^5 6s)$, are required for describing the observed IS. The z_{5p} parameters account essentially for all the FS contributions in the core, for example, those due to the $5p_{1/2}$ electrons; through the relativistic corrections they also contribute to the MS in the first order of perturbation.[264,265] The influence of $z_{5p}(5p^5 5d)$ is already clearly visible on the experimental IS, which in the configuration $5p^5 5d$ groups approximately into two classes: one for the levels built on $5p^5 \, J = 1/2$ and one for $5p^5 \, J = 3/2$.

The calculated IS agree satisfactorily with the observed values. In the case of Xe the limiting factor in the accuracy of the parametric results seems to be the quality of the wave functions.[263]

In very complex spectra the wave functions are often not accurate enough to enable a complete parametric analysis of the IS measurements to be made. In such cases only one main parameter per configuration can be determined and the shifts are then said to obey the sharing rule; that is, the IS δT of a state, whose wave function ψ results from the mixing of N configurations, can be written as

$$\delta T = \sum_{i=1}^{N} c_i^2 \delta T_i \qquad (14)$$

where δT_i is the IS of the pure configuration i and c_i^2 is the weight of this configuration in ψ ($\sum_{i=1}^{N} c_i^2 = 1$).

As an example, recent IS measurements in Dy II will be considered.[266] In this investigation the IS of ^{162}Dy–^{164}Dy has been studied in transitions from levels of the low-lying odd configurations $4f^9 5d6s$, $4f^9 5d^2$, and $4f^{10} 6p$ (which are strongly mixed), to the four lowest levels of the ground-state configuration, $4f^{10} 6s$. The mixture of the odd levels is known from a parametric study of the fine structure by Wyart.[267] Using the three IS between pure configurations $\delta\nu(f^{10}s - f^9 ds)$, $\delta\nu(f^{10}s - f^9 d^2)$, and $\delta\nu(f^{10}s -$

Table 3. Observed and Calculated IS in Some Dy II lines for ^{162}Dy–^{164}Dy and Best Values of IS between Pure Configurations[266]a

Line (Å)	Composition of upper level (%)[267]			$\delta\nu^{162,164}$ (mK)		
	f^9ds	f^9d^2	$f^{10}p$	Observed	Calculated	Difference
3385	25	13	62	−23.5	−22.1	−1.4
3393	13	40	47	−22.0	−28.0	6.0
3445	0	94	6	−29.7	−31.6	1.9
3523	2	24	74	−43.4	−37.9	−5.5
3531	0	24	76	−44.7	−39.4	−5.3
3836	2	79	19	−29.4	−31.8	2.4
3872	1	34	65	−32.6	−37.5	4.9
3873	8	90	2	−32.8	−26.1	−6.5
3898	2	79	19	−32.3	−31.8	−0.5
3968	41	29	30	−14.8	− 8.5	−6.3
4000	1	34	65	−31.5	−37.5	6.0
4078	2	69	29	−31.1	−32.9	1.8
4111	97	2	1	+32.6	+29.9	2.7

$$\delta\nu(4f^{10}6s - 4f^95d6s) = +32(4)\ \text{mK}$$
$$\delta\nu(4f^{10}6s - 4f^95d^2) = -31(3)\ \text{mK}$$
$$\delta\nu(4f^{10}6s - 4f^{10}6p) = -42(3)\ \text{mK}$$

^{a} For further explanation see text.

$f^{10}p$) as free parameters, calculated IS were fitted to the 13 observed ones. As can be seen from Table 3, satisfactory agreement is obtained which indicates that the composition of the odd levels given by Wyart is reasonable. Furthermore, the values of the IS between pure configurations obtained by the least-squares fit can be explained by comparing them with results in similar transitions and interpreting them in terms of MS and FS.

5.2.2. Crossed-Second-Order Effects

As in the theoretical interpretation of hfs, crossed-second-order (CSO) effects and relativistic effects have, in the last few years, proved to be fruitful in the interpretation of observed IS.

CSO effects of the electrostatic operator **G** and of any IS operator **O** may be written as

$$\sum_X \frac{(\psi_0|\mathbf{G}|X)(X|\mathbf{O}|\psi_0) + (\psi_0|\mathbf{O}|X)(X|\mathbf{G}|\psi_0)}{E_0 - E_X}$$

where E_0 is the zeroth-order energy of a level with wave function ψ_0 belonging to the configuration C_0 and the E_X are the zeroth-order energies of all states X of all other configurations C_X. Only the configurations C_X with sufficiently large energy distances E_0-E_X should be included in the sum over X (far-configuration-mixing effects), to enable rapid convergence of the perturbation expansion. Close-configuration-mixing effects should be treated in connection with the intermediate-coupling formalism.

CSO effects of the electrostatic operator, G, and the FS operator, F, can play an important role in the observed IS. The main contributions to this CSO effect come from excited configurations C_X in which an s electron is promoted to any other s subshell (singly occupied or empty, in the discrete spectrum or in the continuum). It can be shown that the FS remains constant in configurations containing no unpaired s electron (and no open p-subshell); in configurations of the type $nl^m n's$, however, the CSO effect of G and F has to be described by a parameter, $g^l(nl, n's)$, which has the same angular coefficient as the Slater integral $G^l(nl, n's)$.[238] The FS may thus be different for the various terms of the configuration $nl^m n's$.

The effect can, for example, be seen most clearly in the IS of the terms of the configuration $4d^5 5s$, ^7S and ^5S, in Mo I.[268] For the isotope pair ^{92}Mo–^{94}Mo the FS derived from the measured IS is

$$\delta T(^7S) = 22.1 \text{ mK} \quad \text{and} \quad \delta T(^5S) = 12.7 \text{ mK}$$

referred to $4d^5 5p$ ^7P. Since close-configuration-mixing effects are negligible, the difference in the FS of the two levels has to be attributed to CSO effects with the parametric description

$$\delta T(^7S) = d - 1/2 g^2(4d, 5s)$$
$$\delta T(^5S) = d + 7/10 g^2(4d, 5s)$$

where d represents the first order FS and g^2 the CSO effect. The resulting numerical values

$$d = 18.2 \text{ mK} \quad \text{and} \quad g^2 = -7.8 \text{ mK}$$

show that the CSO effect is of the same order of magnitude as the first-order FS. The values can reasonably be explained by HF calculations, which can be shown to account exactly for the CSO effect (see also Section 5.4.2).

The crossed perturbation of the SMS operator with the electrostatic operator can be accounted for by introducing one IS parameter per Russell–Saunders term, since the SMS operator is diagonal with respect to L and S. This procedure was first used in the interpretation of the IS in the configuration $2p^5 3p$ of Ne I.[269,270]

5.2.3. Relativistic Effects

Relativistic effects may produce a J dependence of the IS in the levels of a pure Russell–Saunders term. Since such effects are small, little experimental data are yet available on them.

J-dependent contributions to the non-Bohr MS were first shown to exist in Ne I by Bauche and Keller[269] and Keller.[270]

The IS in the ground multiplet $4f^6 6s^2 \, {}^7F$ of Sm I has been studied carefully by Bauche et al.[271] Small differences of the order of 1 mK for ${}^{144}Sm–{}^{152}Sm$ have been found between the IS of the 7 levels of ${}^7F$. These differences can be approximately described by the introduction of one parameter z_{4f}[238] (see Section 5.2.1.) and can be interpreted in two equivalent ways: (i) in the central-field scheme, as resulting from the crossed-second-order effect of the magnetic interactions and the field-shift operator, (ii) in the relativistic self-consistent-field scheme as resulting from the different screening effects of the $4f_{5/2}$ and $4f_{7/2}$ electrons.

5.3. Specific Mass Shifts

Considerable progress has been made by Bauche[260,272,273] in understanding and evaluating the SMS by HF techniques. This progress has been made possible by the development, during the sixties, of large computers and powerful computer codes for the calculation of atomic wave functions.

On the basis of Bauche's calculations two general conclusions can be drawn: (i) alkalilike $ns–np$ transitions between the lowest ns and np states in any optical spectrum display calculated SMS whose absolute values are rarely larger than the NMS values. The same is true for alkaline-earth-like $ns^2–nsnp$ transitions. (ii) The jump of an nd electron (in an nd series with $n = 3, 4,$ or 5) or of an nf electron (in an nf series with $n = 4$ or 5) results in absolute SMS values about ten times or more larger than the corresponding NMS values. The SMS is negative if the level with the larger number of nd (or nf) electrons is the lower one, and vice versa.

These findings are strongly supported by experimental evidence. Quantitative comparisons between experiment and theory, however, are often difficult because the configuration-mixing situation is unknown in many spectra or the SMS cannot be extracted from the measurements with sufficient accuracy.

More (details on SMS are given in Ref. 237 and particularly in Ref. 238.

Table 4 shows a comparison between calculated and experimental SMS values in Mo I and II. The experimental values have been extracted from the observed IS by two independent methods: (i) in the $5s \, {}^7S–5p \, {}^7P$ transition the SMS was assumed to be 0.3 ± 0.9 of the NMS[237]; (ii) the

Table 4. Experimental and Theoretical SMS and FS Values in Some
Lines of Mo I and Mo II[268] [a]

Transition	SMS		FS	
	Expt.	Theory	Expt.	Theory
$4d^55s\,^7S-4d^55p\,^7P$	1.4	2.0	-22.1	-22.4
$4d^55s\,^5S-4d^55p\,^5P$	1.2	-0.6	-11.3	-9.6
$4d^45s^2\,^5D-4d^55p\,^5P$	21.2	23.1	-50.4	-49.9
$4d^45s\,^6D-4d^45p\,^6F$	1.4	1.4	-30.9	-31.6

[a] Values in mK. For further explanation see text.

optical IS was compared in a King plot with the muonic x-ray shift which
was available for four isotope pairs. Both methods yield the same results
for the SMS. Corrections for close-configuration-mixing have been made
and the agreement between experimental and calculated SMS is very
satisfactory.

5.4. Field Shifts and $|\psi(0)|^2$

5.4.1. Screening Ratios

The FS in a spectral line is proportional to the change in the total
electron-charge density at the nucleus, $|\psi(0)|^2$, between the appropriate
initial and final states. The ratio of FS in different transitions of an element
is usually described by screening ratios introduced by Brix and Kopfermann.[274] These authors also showed that the screening ratios are
approximately the same for corresponding configurations in Gd and Hg,
i.e., in the $4f$ and $5d$ series.[234] Blaise and Steudel[248] compiled more
recent experimental data and extended the analogy to the $5f$ series. They
also pointed out that, because of the screening of the closed s electron
shells by an f electron, the FS in a transition $nf^{m-1}(n+1)d-nf^m$ is of the
same order of magnitude as the FS of an alkalilike s electron in $nf(n+2)s$
$(n=4,5)$.

The experimental values of screening ratios have an uncertainty of
about 10% and seem to be independent of Z within these limits. This
remarkable fact is of great use in the interpretation of observed FS and the
evaluation of $\delta\langle r^2\rangle$ values and may also be a considerable help in the
fine-structure classification of complex spectra.

Screening ratios, being purely electronic quantities, can also be used as
a check for *ab initio* calculations of $|\psi(0)|^2$. The first person to study
screening effects on $|\psi(0)|^2$ was Wilson,[275] who used nonrelativistic HF

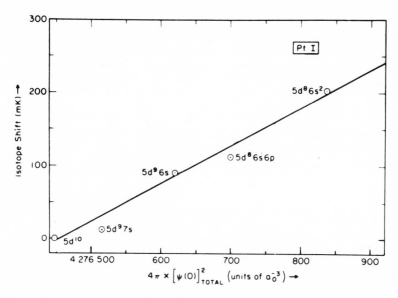

Figure 11. Measured IS versus $|\psi(0)|^2$ values obtained by HF calculations for Pt according to Wilson.[275]

calculations. Because of the simplifying approximations of the model, correct absolute values of $|\psi(0)|^2$ are not obtained; however, screening ratios can often be evaluated with reasonable certainty, as Wilson was able to show in the cases of Hg, Tl, and Pt. As an example, Figure 11 shows the IS in five configurations in Pt I plotted against the calculated $|\psi(0)|^2$ values. The proportionality between observed FS and calculated values of $|\psi(0)|^2$ is remarkably good, thus demonstrating the usefulness of such calculations. Wilson[275] was also able to explain the strong change in the screening of the closed shells when an f electron jump occurs. His HF calculations in Pu show that the charge cloud of the $5f$ orbital expands with increasing number of f electrons. This expansion causes the $5s$, $5p$, $5d$, $6s$, and $6p$ closed shell orbitals to contract since the nuclear shielding due to the $5f$ electrons decreases as they expand. The contraction of these closed-shell orbitals produces a corresponding increase in the values of $|\psi(0)|^2$ for the $5s$ and $6s$ orbitals. Other well-known screening effects could also be explained by Wilson's calculations, and he later extended them to Sm and Eu.[276] He stated that, as a general rule, "large changes in core s electron densities at the nucleus occur down to the n value of the core s electron which corresponds to the n value of the valence electron which is excited or ionized."

Relativistic HF calculations of the electron density at the nucleus were carried out by Coulthard[277] in Eu. In general, improved agreement with empirical screening ratios is obtained. This is due to the fact that the relativistic enhancement of the change in electron-charge density strongly depends upon whether s or p electrons or d or f electrons are excited.

5.4.2. Calculation of Crossed-Second-Order Effects

CSO field-shift effects have been described in Section 5.2.2. These far-configuration-mixing effects are exactly accounted for by the HF method, as was shown by Bauche and Klapisch[278] and Labarthe.[279]

Because of this property, HF calculations have also been used to explain CSO FS effects. As an example, experimental and theoretical results on the FS in some Mo I and II lines[268] are shown in Table 4. (The experimental values were corrected for close-configuration-mixing.) Since by the HF calculations correct absolute values of $|\psi(0)|^2$ are not obtained, the calculated values were multiplied by a normalization factor that was chosen so as to give the best overall agreement with the experimental data. The large difference between the FS of $4d^5 5s\ ^7S$ and ^5S due to CSO contributions was already mentioned in Section 5.2.2 (the difference in the FS between $4d^5 5p\ ^7P$ and ^5P is, on the contrary, very small). As can be seen from the table, the normalized HF values agree excellently with the experimental values, thereby demonstrating that HF calculations are suited to the explanation of observed CSO effects as well as screening ratios. Relativistic HF calculations carried out by Bauche[280] do not substantially change the nonrelativistic results in Table 4.

5.4.3. Evaluation of $\delta\langle r^2 \rangle$

For the extraction of $\delta\langle r^2 \rangle$ from the observed IS, spectral lines that are known to be pure ns–np transitions are preferred. This is done for two reasons: (i) In such transitions the SMS is small and can be estimated (see Section 5.3), and (ii) for an alkalilike ns electron the electronic charge density at the nucleus, $|\psi(0)|^2_{ns}$, may be determined from empirical data, either from the magnetic hf splitting due to the ns electron or from the fine-structure level scheme of the respective atom or ion by means of the Goudsmit–Fermi–Segrè formula (see, e.g., Refs. 211 and 281). Both procedures are believed to have an uncertainty of a few percent.

The total change of the electronic charge density at the nucleus, $\Delta|\psi(0)|^2_{ns-np}$, in the transition $(\text{core} + ns) - (\text{core}' + np)$, which is needed for the evaluation of $\delta\langle r^2 \rangle$, can then be written as

$$\Delta|\psi(0)|^2_{ns-np} = \beta|\psi(0)|^2_{ns} \tag{15}$$

with

$$\beta = \frac{|\psi(0)|^2_{core} + |\psi(0)|^2_{ns} - |\psi(0)|^2_{core'}}{|\psi(0)|^2_{ns}} \tag{16}$$

where the screening factor, β, takes into account the change in the screening of inner closed-shell electrons from the nuclear charge by the valence electron as it changes from ns to np. β is derived by HF calculations and has an error that is supposed to be a few percent, whereas HF calculations of absolute values of electron densities are believed to underestimate $\Delta|\psi(0)|^2_{ns-np}$ by 30%–40%.[282] β values for various spectra of several elements are compiled in Table A of Ref. 237. The extraction of $\delta\langle r^2\rangle$ from the IS in other than ns–np transitions is also discussed in Ref. 237.

Whenever possible $\delta\langle r^2\rangle$ should be derived from IS measured in various electronic transitions as well as in different spectra of an element. As an example, Table 5 shows the evaluation of $\delta\langle r^2\rangle$ from different transitions in Mo I and Mo II.[268] The three $\delta\langle r^2\rangle$ values obtained agree very well.

5.4.4. $\delta\langle r^2\rangle$ Values from Optical Isotope and Isomer Shifts

Extensive tables of relative and absolute $\delta\langle r^2\rangle$ values as extracted from optical isotope and isomer shifts are given in Ref. 237.

It is customary, and convenient, to consider the ratio of the experimental $\delta\langle r^2\rangle$ values to the values $\delta\langle r^2\rangle_{unif}$ which would be obtained if the nuclei were homogeneously charged spheres with radii $R = r_0 A^{1/3}$. In Figure 12, taken from Ref. 237, the ratios $\delta\langle r^2\rangle/\delta\langle r^2\rangle_{unif}$ are shown as a function of neutron number (Brix–Kopfermann diagram[274]). This simple and instructive way of presenting data is, of course, not a comparison with any theoretical expectation.

Three well-known features can be seen in Figure 12: (i) on the average, the experimental values are smaller than those predicted by the

Table 5. Evaluation of $\delta\langle r^2\rangle^{92,94}$ from the FS in Some Transitions of Mo I and II[268]a

Transition	FS_{5s-5p} (mK)	E_{5s}	β	$\delta\langle r^2\rangle^{92,94}$ (fm²)
$4d^5 5s\ ^7S - 4d^5 5p\ ^7P$	22.1	0.49	1.26	0.213
$4d^5 5s\ ^5S - 4d^5 5p\ ^5P$	11.3	0.30	1.01	0.222
$4d^4 5s\ ^6D - 4d^4 5p\ ^6F$	30.9	0.63	1.18	0.247

^{a} E_s was calculated by means of the Goudsmit–Fermi–Segrè formula. For further explanation see text.

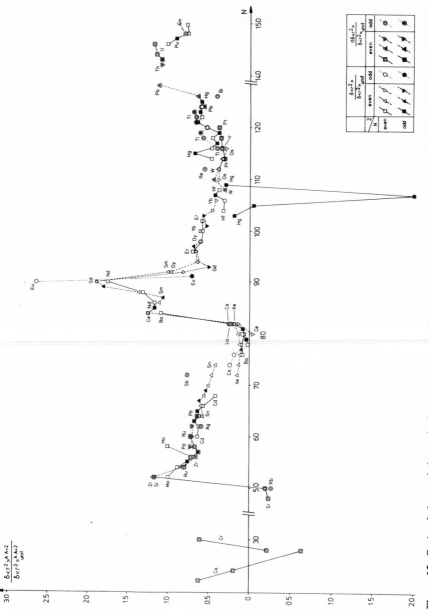

Figure 12. Ratio of observed change in the mean-square nuclear charge radius to the corresponding change for a uniform-density nucleus ($R = r_0 A^{1/3}$), $\delta\langle r^2\rangle^{A,A+2}/\delta\langle r^2\rangle^{A,A+2}_{unif}$, with $\delta\langle r^2\rangle^{A,A'}_{unif} = (2/5)r_0^2 A^{-1/3}\,\delta A$ and $r_0 = 1.20$ fm.[237] Points are plotted versus neutron number and are placed at the higher N value for each pair of isotopes.

$A^{1/3}$ law. This fact, usually referred to as IS discrepancy, is further discussed in Section 5.6; (ii) there are considerable deviations from the average value, which are mainly attributed to changes in nuclear deformation (see Section 5.5) and partly to shell effects (e.g., jumps in $\delta\langle r^2 \rangle$ at the magic neutron numbers $N = 28, 50, 82, 126$, when two neutrons are added to the closed shell); (iii) relative (and absolute) $\delta\langle r^2 \rangle$ values in neighboring elements are often approximately the same when isotonic nuclides are compared.

Kuhn *et al.*[283] have pointed out that the variations of the IS of the even–even isotopes in the medium-heavy elements are associated with particular values of $N - Z$ and concluded that there is evidence for some degree of α-particle structure in these nuclei.

A fourth feature cannot be seen in Figure 12. Odd-N nuclei often appear to be smaller than the mean size of the neighboring even-N nuclei (odd–even staggering). This is a kind of pairing effect which implies that an odd neutron does not participate as effectively in polarizing and expanding the (proton) core as does a pair of neutrons.[244] Blaise[284] suggested that the effect might be connected with the fact that the last neutron in the nucleus $N + 1$ (N even) generally has low angular momentum, compared with the last two neutrons in the nucleus $N + 2$, since pairing energy favors states of high angular momentum. The work of Stroke and his collaborators[285–287] supports these views. It was observed that near the closed neutron shell, $N = 126$, the odd–even staggering parameter $\gamma = 2\,\delta\nu_{FS}^{N,N+1}/\delta\nu_{FS}^{N,N+2}$ (N even) tends to increase with the odd-nucleon angular momentum for a limited mass number range, which means that the radial expansion of the odd nucleus increases. (See also Section 5.6.)

$\delta\langle r^2 \rangle$ values derived from various experimental sources (elastic electron scattering, muonic isotope and isomer shifts, electronic x-ray IS, optical IS, and Mössbauer isomer shifts) have been recently compiled in *Atomic Data and Nuclear Data Tables*, Vol. 14, No. 5–6 (1974). $\delta\langle r^2 \rangle$ values obtained from electronic x-ray IS are particularly suited to comparison with the $\delta\langle r^2 \rangle$ values extracted from optical IS, since both methods depend on the same nuclear parameter. Such a comparison was recently made by Heilig and Steudel[237]; fair agreement was found in most cases. Some $\delta\langle r^2 \rangle$ values obtained by different experimental methods are compared in Table 6. Results from muonic spectra and electron scattering are also included. In recent years ways of extracting model-independent $\delta\langle r^2 \rangle$ values from electron scattering measurements have been found[288] (Zr and Mo values are given in Table 6). The change of a model-independent equivalent radius can also be obtained from muonic IS measurements[289]; the conversion of this quantity into $\delta\langle r^2 \rangle$, however, is again model dependent.

Table 6. Comparison of $\delta\langle r^2\rangle$ Values for Some Selected Isotope Pairs

		$\delta\langle r^2\rangle\,(10^{-3}\,fm^2)$		
Elem. $A_1 - A_2$	Optical[237]	Electronic x-ray[245]	Muonic x-ray[289]a	Electron scattering[288]
Zr 90–92	302(60)[b]		306	302(50)[c]
92–94	209(56)[b]			183(50)[c]
Mo 92–94	226(19)[d]			346(41)[e]
92–96	419(18)[d]		588	590(57)[e]
92–100	796(20)[d]	1030(150)		1185(80)[e]
96–98	150(12)[d]		209	249(41)[e]
Ce 140–142	276(17)	274(10)	280	
Nd 142–144	277(12)		276	
142–146	534(13)	522(33)	540	526(36)
146–148	286(9)	275(34)	295	
146–150	667(20)	678(34)	686	670(40)
142–143	130(7)	116(21)	102	
144–145	111(6)	126(18)	98	
Yb 170–172	116(16)	163(19)	146	
172–174	92(15)	141(13)	120	
174–176	87(13)	103(12)	108	
170–171	41(10)	77(32)	67	
172–173	41(10)	50(27)	51	

[a] $\delta\langle r^2\rangle$ calculated for a Fermi distribution with $\delta t = 0$.   [d] From Ref. 268.
[b] $\beta\delta\langle r^2\rangle$ is quoted.   [e] From Ref. 291.
[c] From Ref. 290.

5.5. Field Shifts and Nuclear Deformation Effects

Two effects are generally assumed to give rise to changes in $\delta\langle r^2\rangle$ as neutrons are added: (i) The volume of the charge distribution increases, and (ii) the shape of the nucleus changes at constant volume and density and this will also alter $\delta\langle r^2\rangle$. This deformation effect was first suggested by Brix and Kopfermann[292,247] to account for the anomalously large IS observed between rare earth isotopes with neutron numbers $N = 88$ and $N = 90$ and was further developed by Wilets et al.[293] The reason for the effect is that a nonspherical shape appears to be more extended radially than spherical nuclei of the same volume.

The simple incompressible fluid model of the nucleus[294] is generally used to correlate $\delta\langle r^2\rangle$ with these two effects (see, e.g., Ref. 235). The surface of the uniformly charged nucleus may then be described by

$$R(\theta, \phi) = R_0\left[1 + \kappa + \sum_m \alpha_{2m} Y_{2m}(\theta, \phi)\right] \qquad (17)$$

where R_0 is the equilibrium charge radius of the drop and the quantities α_{2m} describe the shape of the drop, which may be time-dependent owing to vibrational motion. The requirement that $R(\theta, \phi)$ be real leads to the requirement $\alpha_{2m}^* = (-1)^m \alpha_{2-m}$, while the condition that the nuclear volume remains constant under deformation yields $\kappa = -\langle \beta^2 \rangle / 4\pi$, where

$$\langle \beta^2 \rangle = \sum_m \langle |\alpha_{2m}|^2 \rangle \qquad (18)$$

is the mean square deformation in the ground state. (A comparison of different deformation parameters used in the literature may be found in both Refs. 235 and 295, the latter being more complete.)

When nuclear vibration occurs the α_{2m} are the destruction and creation operators for a quadrupole vibration. For a statically deformed axially symmetric nucleus only $\alpha_{20} = \beta \neq 0$. Asymmetrically deformed nuclei are conveniently described in terms of the parameters β' and γ, where $\alpha_{20} = \beta' \sin \gamma$ and $\alpha_{22} = (1/2)^{1/2} \beta' \sin \gamma$. The asymmetry parameter, γ, gives the deviation from an axially symmetric shape with $\gamma = 0$ corresponding to a pure prolatelike deformed nucleus.

The mean square nuclear charge radius is

$$\langle r^2 \rangle = \frac{3}{5} R_0^2 \left(1 + \frac{5}{4\pi} \langle \beta^2 \rangle\right) \qquad (19)$$

up to terms of order α^2 and, since $\langle \beta^2 \rangle$ is usually small compared with

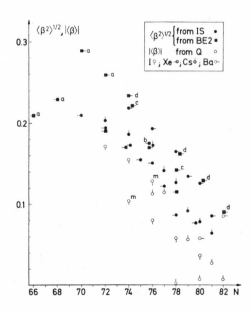

Figure 13. Deformation parameters for I, Xe, Cs, and Ba as a function of neutron number according to Ullrich and Otten.[296] (a) Ref. 297, (b) Ref. 298, (c) Ref. 299, (d) Ref. 300. Other $B(E2)$ values from Ref. 301. Q values from Ref. 302.

unity, the approximate result for $\delta\langle r^2\rangle$

$$\delta\langle r^2\rangle = \delta\langle r^2\rangle_{\text{sph}} + \frac{5}{4\pi}\langle r^2\rangle_{\text{sph}}\delta\langle\beta^2\rangle \tag{20}$$

is obtained where $\langle r^2\rangle_{\text{sph}} = 3R_0/5$; $\delta\langle r^2\rangle$ can thus be expressed ⸳ a sum of two terms, one depending on changes in nuclear volume and the other depending on changes in nuclear shape. If the approximation that $\langle\beta^2\rangle$ is small compared with unity is not made, a more cumbersome result is obtained, details of which can be found in Ref. 235. In this case the asymmetry parameter γ would enter Eqs. (19) and (20) via a term containing a factor $\langle\beta^3\rangle$ of lowest order. The use of a more realistic charge distribution, e.g., a deformed Fermi distribution, does not modify Eq. (20) significantly. As can be seen from Eq. (20), IS measurements cannot distinguish whether $\langle\beta^2\rangle$ stems from a static deformation or from vibration or from a mixture of both.

If in Eq. (20) $\delta\langle r^2\rangle_{\text{sph}}$ is equated to the change of the mean square charge radius of a homogeneously charged nuclear sphere, $\delta\langle r^2\rangle_{\text{unif}}$, the experimental $\delta\langle r^2\rangle$ values cannot be described properly since, on average, the charge radius does not increase as rapidly as $R = 1.2\,A^{1/3}$ fm but at a rate which is considerably lower; this can be seen in Figure 12. This result indicates that addition of neutrons expands the radius more slowly than addition of a mixture of neutrons and protons together. In order to account for this phenomenon an empirical factor

$$\eta = \delta\langle r^2\rangle_{\text{sph}}/\delta\langle r^2\rangle_{\text{unif}} \tag{21}$$

usually referred to as IS discrepancy, is introduced, and Eq. (20) is then written in the form

$$\delta\langle r^2\rangle = \eta\delta\langle r^2\rangle_{\text{unif}} + \frac{5}{4\pi}\langle r^2\rangle_{\text{unif}}\delta\langle\beta^2\rangle \tag{22}$$

Quantitative information on η may be obtained by fitting Eq. (22) to experimental $\delta\langle r^2\rangle$ values and $\langle\beta^2\rangle$ values extracted from reduced nuclear transition probabilities $B(E2, 0^+ \to 2^+)$ via the relation (valid in the limit of small deformations)

$$\langle\beta^2\rangle = B(E2, 0^+ \to 2^+)(3ZR^2/4\pi)^{-2} \tag{23}$$

with $R = 1.2\,A^{1/3}$ fm. [For $B(E2)$ values, see, e.g., the tables in Refs. 295, 300, and 301). It turns out that usually $\eta = 0.65 \pm 0.15$; however, there may be exceptions, as can be seen from the recent work by Fischer et al.[303] These authors have shown that, on the addition of two neutrons to the closed neutron shell $N = 82$, the Ce as well as the Nd nuclei undergo an unusually large increase in size resulting in $\eta = 1.03$ and $\eta = 1.09$, respectively, whereas between $N = 84$ and $N = 86$ the increase has the usual value.

Another empirical correction factor to Eq. (22) called "compressibility under deformation" was suggested by Fradkin.[304] This effect and other improvements to Eq. (22) have been reviewed in Ref. 235. A neutron skin of the nucleus has also been considered to explain the IS discrepancy.[305]

In Eq. (22) all the irregularities in the observed FS and therefore in $\delta\langle r^2 \rangle$ are assumed to be due to changes in nuclear deformation. This assumed behavior of nuclear matter in bulk has proved successful at points of sharp discontinuities of nuclear shape such as in the rare earth region[247] or in very light Hg isotopes[306] (see Section 5.5.2). This approach is, however, inadequate for investigating detailed variations of $\langle r^2 \rangle$ between neighboring isotopes, since finer effects of shell structure usually compete with the deformation effect. On the other hand, such local fluctuations may cancel out when the IS is taken over a long chain of isotopes, as in the following example on Xe, Cs, and Ba. This demonstrates that, by fitting Eq. (22) to the experimental data, $\delta\langle \beta^2 \rangle$ values and, provided $\langle \beta^2 \rangle$ for at least one isotope is known, $\langle \beta^2 \rangle$ values can be obtained that are in good agreement with those extracted from $B(E2)$ transition probabilities.

5.5.1. Nuclear Deformation in Xe, Cs, and Ba

As an example of the application of Eq. (22) recent IS studies made in the elements Xe, Cs, and Ba are considered (see Ref. 237 and the work of Ullrich and Otten[296]). As can be seen from Figure 12, these elements show FS a factor of 5–10 smaller than expected from the model of uniform charge distribution. This is a characteristic of the FS of elements in the region $Z > 52$, $N < 82$. The FS of such elements is of particular interest since the knowledge of changes in the nuclear charge distribution in nuclei near closed proton or neutron shells is helpful to the understanding of nuclear structure. Moreover, the large number of measured isotopes in Xe, Cs, and Ba offers an opportunity for the study of changes in the nuclear charge distribution over a large range of neutron numbers.

The $\delta\langle r^2 \rangle$ values have been derived from the measured IS and $\delta\langle \beta^2 \rangle$ values have been extracted using $\eta = 0.5$.[296] Nuclei having a closed neutron shell $N = 82$ are expected to have very small deformation parameters. ^{137}Cs and ^{136}Xe were therefore taken as reference isotopes and $\langle \beta^2 \rangle$ was assumed to be zero, whereas in the case of ^{138}Ba a reference value was available from $B(E2)$ measurements. The deformation parameters derived in this way from IS measurements are plotted in Figure 13 as a function of neutron number together with $\langle \beta^2 \rangle^{1/2}$ values derived from $B(E2)$ values. Within the limits of error (not shown in the figure) values from IS and

$B(E2)$ agree. At fixed neutron number the deformation is always found to grow with increasing Z. For comparison, β values calculated from spectroscopic quadrupole moments Q are also shown in Figure 13. These values were derived using the projection formula (Q_0 being the intrinsic quadrupole moment)

$$Q_0 = (I+1)(2I+3)/I(2I-1) \times Q \qquad (24)$$

and

$$\langle \beta \rangle = (5\pi)^{1/2} Q_0 / 3ZR^2 \qquad (25)$$

The projection formula, however, holds good only for a rigid axial symmetric rotor in strong coupling, conditions that are more or less violated in the region $Z > 52$, $N < 82$. Consequently the $|\langle \beta \rangle|$ values derived from Q are always smaller than the $\langle \beta^2 \rangle^{1/2}$ values obtained from IS and $B(E2)$. As the neutron number falls the $|\langle \beta \rangle|$ values clearly approach the $\langle \beta^2 \rangle^{1/2}$ field, thus confirming that the deformation will turn from a dynamic to a more static character with increasing distance from the closed neutron shell $N = 82$, as is expected.

5.5.2. Charge Radii in the Isotopic Series ^{181}Hg–^{205}Hg

The most extensively studied element is mercury and IS measurements have been made over a range of 20 isotopes and 4 nuclear isomers, from ^{181}Hg to ^{205}Hg, extending from nuclei far from the valley of nuclear stability to almost closed-shell nuclei. Only seven nuclides are stable, the other 17 having half-lives between $3.6\,s$ and $\geqslant 1.5\,y$. The IS measurements made in the intercombination line λ 2537 Å of Hg I have been converted to $\delta \langle r^2 \rangle$ and the changes of charge radii relative to ^{204}Hg, $\delta \langle r^2 \rangle^{A,204}$, are plotted in Figure 14 as a function of mass number.[307] These data were obtained by means of high resolution optical spectroscopy, level crossing, optical pumping, and $\beta(\gamma)$ radiation detected optical pumping.

The shift of the isotopes from ^{205}Hg to ^{187}Hg follows the general trend in this mass region: a fairly uniform decrease of $\langle r^2 \rangle$ with an absolute rate of about 0.5 of the $A^{1/3}$ value (dot-dashed line in Figure 14). From the measured $\delta \langle r^2 \rangle$ values $\delta \langle r^2 \rangle_{sph}$ can be derived by calculating $\delta \langle \beta^2 \rangle$ using deformation parameters from $B(E2)$ values and spectroscopic quadrupole moments Q (for odd-A isotopes). The dashed lines in Figure 14 give the upper and lower limits for $\delta \langle r^2 \rangle_{sph}$ obtained in this way. The IS discrepancy then turns out to be $\eta = 0.78(4)$.

As Figure 14 shows, a dramatic change in $\delta \langle r^2 \rangle$ occurs between ^{187}Hg and ^{185}Hg. The charge radius of ^{185}Hg jumps suddenly to that of ^{196}Hg.

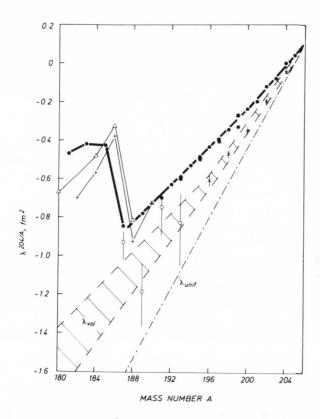

Figure 14. $\lambda \approx \delta\langle r^2\rangle$ of Hg isotopes (full dots) and isomers (stars) relative to ^{204}Hg according to Bonn et al.[307] Dashed–dotted line: $A^{1/3}$ dependence. The dashed lines give upper and lower limits for $\lambda_{vol} \approx \delta\langle r^2\rangle_{sph}$ [open circles from Q, crosses from $B(E2)$ values]. $\triangle$[308] and $+$[309]: Theoretical $\delta\langle r^2\rangle$ values, adjusted to experimental line at $A = 190$.

This was discovered in 1972 by Bonn et al.[306] using the technique of optical pumping and detection of the nuclear polarization via the angular asymmetry of the β decay. They interpreted this jump in $\delta\langle r^2\rangle$ as a sudden onset of nuclear deformation by analogy to a similar observation in the IS of rare earths between $N = 88$ and $N = 90$. The work of Bonn et al.[306] has stimulated a lot of theoretical and experimental publications (for references see 307), which are supporting and completing the original interpretation in the sense that the effect is most probably caused by a transition from a small oblate (^{187}Hg) to a strong prolate (^{185}Hg) deformation. Very recent results on the neutron-deficient Hg isotopes will be discussed in Chapter 17.

5.6. Calculations of $\delta\langle r^2 \rangle$

The phenomenon of IS discrepancy has been the subject of a number of theoretical investigations. It has been interpreted as due to finite nuclear compressibility by Wilets et al.[293]; this work has been extended to the nuclear liquid drop model by Bodmer.[310] Perey and Schiffer[311] related the IS discrepancy to the known increase of proton binding energy with increasing N along an isotopic series. These authors have taken into account the isospin-dependent term in the nuclear optical potential and have been able to explain the observed anomalous isotope shift in Ca by properly adjusting this term (see also Refs 312, 313, and 246). Brown[314] pointed out the importance of the density dependence of the effective interaction in explaining the IS discrepancy. Beiner and Lombard[315] have evaluated $\delta\langle r^2 \rangle$ for the case of the Hg isotopes with mass numbers 187–204 by a nuclear HF calculation using density-dependent forces and thereby introducing nuclear compressibility and obtained excellent agreement with the experimental average. This work seems to be the first quantitative success in the solution of the old problem of IS discrepancy. Calculations of $\delta\langle r^2 \rangle$ have been carried out based on Migdal's theory of finite Fermi systems.[316] Bunatyan and Mikulinskii[317] were able to give a qualitative explanation of the jumps in FS at the closed neutron shells with 82 and 126 neutrons; the FS of near-spherical nuclei has also been explained on this basis. This work has been extended by Bunatyan[318] and by Krainov and Mikulinskii,[319] who succeeded in lowering the theoretical $\delta\langle r^2 \rangle$ values below the liquid drop limit, but the course of the experimental data is not reproduced quantitatively, nor its differential changes. A somewhat better approach for even–even isotopes was obtained by Uher and Sorensen[320] using the pairing-plus-quadrupole model. Reehal and Sorensen[321] were able to reproduce qualitatively observed odd-even staggerings on the basis that $\langle \beta^2 \rangle$ associated with the zero-point motion of quadrupole vibration is greater for even than for odd nuclei.

Table 7. $\delta\langle r^2 \rangle$ Values from Optical IS[237] and Theory[322]

Isotope pair	$\delta\langle r^2 \rangle \,(10^{-3}\ \mathrm{fm}^2)$	
	Expt.	Theory
^{157}Gd–^{156}Gd	21 ± 13	29.5
^{158}Gd–^{156}Gd	135 ± 20	153.0
^{171}Yb–^{170}Yb	41 ± 10	64.9
^{172}Yb–^{170}Yb	116 ± 16	99.7
^{184}W–^{182}W	92 ± 16^a	68.7

^{a} $\beta\delta\langle r^2 \rangle$ is quoted.

The jumps in the IS between ^{187}Hg and ^{185}Hg (see Section 5.5.2) are fully reproduced (see Figure 14) by nuclear HF (Cailliau et al.[308]) and Strutinski-type calculations (Nilsson et al.[309]).

Extensive microscopic calculations in the region of the deformed rare earths nuclei have recently been performed by Zawischa and Speth[322] and for the first time satisfactory agreement between calculated and experimental $\delta\langle r^2 \rangle$ values has been obtained (see Table 7) with even the strong odd–even staggering in Gd being explained quantitatively.

Appendix: Compilations of Data

G. H. Harrison, *Wavelength Tables*, John Wiley, New York (1956). These tables contain wavelengths and intensities of about 110,000 lines between 2000 and 10,000 Å.

A. N. Zaidel, V. K. Prokofev, S. M. Raiskii, V. A. Slavnyi, and E. Ya. Shreider, *Tables of Spectral Lines*, Plenum Press, New York (1970).

C. E. Moore, *Atomic Energy Levels*, Vol. I ^{1}H–^{23}V (1949); Vol. II ^{24}Cr–^{41}Nb (1952); Vol. III ^{42}Mo–^{57}La, ^{72}Hf–^{89}Ac (1959); National Bureau of Standards Circular No. 467; reprint: National Standard Reference Data services, National Bureau of Standards 35 (1971); Vol. IV on rare earths in preparation.

C. E. Moore, *Bibliography on the Analysis of Optical Spectra*, Sec. 1 H–V, Sec. 2 Cr–Nb, Sec. 3 Mo–La, Hf–Ac, Sec. 4 La–Lu, Ac–Es; National Bureau of Standards, special publication No. 306 (1968/69).

W. F. Meggers, C. H. Corliss, and B. F. Scribner, *Tables of Spectral-Line Intensities*, Part I—Arranged by Elements; Part II—Arranged by Wavelengths, 2nd ed., National Bureau of Standards, monograph No. 145 (1975), revision of monograph No. 32, Parts I and II (1961). These tables contain wavelengths and relative intensities for nearly 39,000 lines (33,500 of them are classified) between 1936 and 9005 Å. These tables are also available as magnetic tape (NBS magnetic tape No. 10, September 1975, National Technical Information Service, Springfield, Virginia 22161).

L. Hagan and W. C. Martin, *Bibliography on Atomic Energy Levels and Spectra*, National Bureau of Standards, special publication No. 363 (1972).

C. E. Moore, *A Multiplet Table of Astrophysical Interest*, revised edition, Part I—Table of Multiplets, Part II—Finding List of all Lines in the Table of Multiplets, National Bureau of Standards, Technical note No. 36 (1959); Reprint: National Bureau of Standards Reference Data Series, NBS, Vol. 40 (rev.).

C. E. Moore, *An Ultraviolet Multiplet Table*, Sec. 1/2 spectra of H–V, Cr–Nb (1950/52), Sec. 3 spectra of Mo–La, Hf–Ra (1962), Sec. 4 finding list H–Nb (1962), Sec. 5 finding list Mo–La, Hf–Ra (1962); Circular National Bureau of Standards 488.

C. E. Moore, *Selected Tables of Atomic Spectra—Atomic Energy Levels and Multiplet Tables*, Sec. 1/2 Si I–IV (1970), Sec. 3 C I–IV (1970), Sec. 4 N IV–VII (1971), Sec. 5 N I–III (1975), Sec. 6 H I, D I (1972), Sec. 7 O I (1976); National Standard Reference Data Series, National Bureau of Standards, Vol. 3.

V. Kaufman, *Reference Wavelengths from Atomic Spectra in the Range from 15 Å to 25,000 Å*[323]; this compilation contains 5432 lines with wavelength uncertainties between 0.0001 and 0.002 Å.

R. Zalubas, "Energy Levels, Classified Lines, and Zeeman Effect of Neutral Thorium[324]";
this list contains about 9500 classified lines. A list of 35,000 Th I and Th II lines is to be
published.

C. E. Moore, *Ionization Potentials and Ionization Limits Derived from the Analyses of Optical
Spectra*, National Standard Reference Data Series, National Bureau of Standards 34
(1970).

J. Sugar and S. Reader[190,191] give a compilation on spectroscopically determined ionization
potentials in rare earths.

B. Eldén[325,326] gives a compilation of the present state of analysis of elements from H to Ni
together with an exhaustive list of recent references.

Very useful information about work to be published, in progress, or planned for the future is
given from time to time, together with a list of recent abstracts, by EGAS (European
Group for Atomic Spectroscopy, Section of the Atomic Division of European Physical
Society; present secretary; A. Dönszelmann, Zeeman Laboratory, Amsterdam).

M. W. Smith and W. L. Wiese[327] give a critical data compilation of forbidden lines for the iron
group elements with lines selected according to their astrophysical importance.

For further compilations see "Data Compilations," in *Physik Daten, Physics Data*, eds. H.
Behrens and G. Ebel, Zentralstelle für Atomkernenergie-Dokumentation, Eggenstein,
Germany, 3-1 (1976), 3-2 (1977).

References

1. D. Goorvitch, S. P. Davis, and H. Kleiman, *Phys. Rev.* **188**, 1897 (1969).
2. J. Sugar, *J. Opt. Soc. Am.* **53**, 831 (1963).
3. J. Reader, G. L. Epstein, and J. O. Ekberg, *J. Opt. Soc. Am.* **62**, 273 (1972).
4. W. Perrson and S. Valind, *Phys. Scr.* **5**, 187 (1972).
5. A. Borgström, *Ark. Fys.* **38**, 243 (1968).
6. B. Petersson, *Ark. Fys.* **38**, 157 (1968).
7. R. Poppe, *Physica* **40**, 17 (1968).
8. J. E. Hansen, A. Steudel, and H. Walther, *Z. Phys.* **203**, 296 (1967).
9. K. Heilig, unpublished.
10. H. M. Crosswhite, *J. Res. Natl. Bur. Stand., Sect. A* **79**, 17 (1975).
11. E. W. Weber, *Z. Phys.* **256**, 1 (1972).
12. N. Elander and H. Neuhaus, *Phys. Ser.* **10**, 130 (1974).
13. J. P. S. Haarsma, G. J. deJong, and J. Agterdenbos, *Spectrochim. Acta* **29B**, 1 (1974).
14. E. F. Worden, R. G. Gutmacher, and J. G. Conway, *Appl. Opt.* **2**, 707 (1963).
15. J. Reader, *J. Opt. Soc. Am.* **65**, 286 and 988 (1975).
16. H. G. C. Human, *Spectrochim. Acta* **27B**, 301 (1972).
17. C. R. Vidal and J. Cooper, *J. Appl. Phys.* **40**, 3370 (1969).
18. C. R. Vidal and M. M. Hessel, *J. Appl. Phys.* **43**, 2776 (1972).
19. C. R. Vidal, *J. Appl. Phys.* **44**, 2225 (1972).
20. P. Camus, *J. Phys. B* **7**, 1154 (1974).
21. D. A. Jackson, *Proc. R. Soc. London* **263A**, 289 (1961).
22. H. Lew, *Methods of Experimental Physics*, Ed. L. Marton, Vol. 4, Part A, pp. 155–198,
    Academic Press, New York (1976).
23. P. Kusch, in *Handbuch der Physik*, Ed. S. Flügge, Vol. 37/1, pp. 3–25, Springer, Berlin
    (1959).
24. S. Büttgenbach, G. Meisel, S. Penselin, and K. H. Schneider, *Z. Phys.* **230**, 329 (1970).
25. H.-W. Brandt, K. Heilig, H. Knöckel, and A. Steudel, *Phys. Lett.* **64A**, 29 (1977).

26. P. Jacquinot, in *Topics in Applied Physics*, Ed. K. Shimoda, Vol. 13, pp. 51–93, Springer, Berlin (1976).
27. J. D. Winefordner, J. J. Fitzgerald, and N. Omenetto, *Appl. Spectrosc.* **29**, 369 (1975).
28. A. S. King, *Astrophys. J.* **28**, 300 (1908).
29. F. S. Tomkins and B. Ercoli, *Appl. Opt.* **6**, 1299 (1967).
30. R. E. Honig and J. R. Woolston, *Appl. Phys. Lett.* **2**, 138 (1963)
31. H. Ebert, *Ann. Phys. (Leipzig) (Wiedemann's Ann.)* **38**, 489 (1889).
32. M. Czerny and A. F. Turner, *Z. Phys.* **61**, 792 (1930).
33. K. Runge and F. Paschen, *Abh. Kgl. Preuss. Akad. Wiss., Phys. Abh.* **1**, (1902).
34. F. S. Tomkins and M. Fred, *Spectrochim. Acta* **6**, 139 (1954); and *Appl. Opt.* **2**, 715 (1963).
35. A. Eagle, *Astrophys. J.* **31**, 120 (1910).
36. Eastman Kodak Company, *Plates and Films for Scientific Photography*, P-315, Rochester, New York (1973).
37. P. F. A. Klinkenberg, in *Methods of Experimental Physics*, Ed. L. Marton, Vol. 13, Part A, pp. 253–346, Academic Press, New York (1976).
38. K. Heilig, *Z. Phys.* **161**, 252 (1961).
39. J. Kuhl, A. Steudel, and H. Walther, *Z. Phys.* **196**, 365 (1966).
40. G. A. Morton, *Appl. Opt.* **7**, 1 (1968).
41. J. Rolfe and S. E. Moore, *Appl. Opt.* **9**, 63 (1970).
42. R. Jones, C. J. Oliver, and E. R. Pike, *Appl. Opt.* **10**, 1673 (1971).
43. F. Robben, *Appl. Opt.* **10**, 776 (1971).
44. J. D. Ingle, Jr., and S. R. Crouch, *Anal. Chem.* **44**, 785 (1972).
45. H. V. Malmstadt, M. L. Franklin, and G. Horlick, *Anal. Chem.* **44**, 63A (1972).
46. R. J. Hull, and H. H. Stroke, *J. Opt. Soc. Am.* **51**, 1203 (1961).
47. G. R. Harrison, *Appl. Opt.* **12**, 2039 (1973).
48. E. W. Palmer, M. C. Hutley, A. Franks, J. F. Verrill, and B. Gale, *Rep. Progr. Phys.* **38**, 975 (1975).
49. D. J. Schroeder, in *Methods of Experimental Physics*, Ed. L. Marton, Vol. 12, pp. 463–489, Academic Press, New York (1974).
50. W. R. Hunter, Rep. on Seminar, Washington 1975, *Appl. Opt.* **15**, 1365 (1976).
51. J. M. Moran and I. P. Kaminow, *Appl. Opt.* **12**, 1964 (1973).
52. A. Labeyrie and J. Flamand, *Opt. Commun.* **1**, 5 (1969).
53. M. C. Hutley, *J. Phys. E.* **9**, 513 (1976).
54. Jobin Yvon, *Diffraction Gratings, Ruled and Holographic*, Lonjumeau, France (1973).
55. D. Rudolph and G. Schmahl, *Optik* **30**, 475 (1970).
56. J. Connes, *Rev. Opt. (Paris)* **40**, 45, 116, 171, 231 (1961).
57. P. Jacquinot and B. Roizen-Dossier, in *Progress in Optics*, Ed. E. Wolf, Vol. III, pp. 31–188, North-Holland, Amsterdam (1964).
58. L. Mertz, *Transformations in Optics*, Wiley, New York (1965).
59. E. V. Lowenstein, *Appl. Opt.* **5**, 845 (1966).
60. H. A. Gebbie and R. Q. Twiss, *Rep. Progr. Phys.* **29**, 729 (1966).
61. G. A. Vanasse and H. Sakai, in *Progress in Optics*, Ed. E. Wolf, Vol. VI, pp. 261–332, North-Holland, Amsterdam (1967).
62. R. J. Bell, *Introductory Fourier Transform Spectroscopy*, Academic Press, New York (1972).
63. J. Strong, in *Essays in Physics*, Eds. G. K. T. Conn and G. N. Fowler, Vol. 5, pp. 197–224, Academic Press, New York (1973).
64. J. Cuthbert, *J. Phys. E* **7**, 328 (1974).
65. P. R. Griffith, "Chemical Infrared Fourier Transform Spectroscopy," in *Chemical Analysis*, Eds. P. J. Elving and J. D. Winefordner, Vol. 43, Wiley, New York (1975).

66. J. D. Winefordner, R. Avni, T. L. Chester, J. J. Fitzgerald, L. P. Hart, D. J. Johnson, and F. W. Plankey, *Spectrochim. Acta* **31B**, 1 (1976).
67. F. Abramovici, *J. Comput. Phys.* **11**, 28 (1973).
68. J. Connes, in Aspen International Conference on Fourier Spectroscopy, AFCRL special report No. 114, 83–115 (1971).
69. J. and P. Connes, *J. Opt. Soc. Am.* **56**, 896 (1966).
70. J. Pinard, *J. Phys. (Paris)* **28**, Suppl. C2 136 (1967).
71. J. Connes, G. Delouis, P. Connes, G. Guelachvili, J. P. Maillard, and G. Michel, *Nouv. Rev. Opt. App.* **1**, 3 (1970).
72. E. Luc-Koenig, C. Morillon, and J. Vergès, *Physica (Utrecht)* **70**, 175 (1973).
73. R. Beer, R. H. Norton, and C. H. Seaman, *Rev. Sci. Instr.* **42**, 1393 (1971).
74. R. A. Schindler, *Appl. Opt.* **9**, 301 (1970).
75. P. Connes and G. Michel, *Appl. Opt.* **14**, 2067 (1975).
76. S. Gerstenkorn and P. Luc, *Nouv. Rev. Opt. (Paris)* **7**, 149 (1976).
77. R.-J.-Champeau and J. Vergès, *Physica C (Utrecht)* **83**, 373 (1976).
78. P. Connes, Thesis Paris (1957), *Rev. Opt. (Paris)* **38**, 157, 416 (1959).
79. P. Jacquinot, *J. Phys. (Paris)* **28**, Suppl. C2 183 (1966).
80. J. Meaburn, *Appl. Opt.* **12**, 279 (1973).
81. J. Vergès, *Spectrochim. Acta* **24B**, 177 (1969).
82. P. Camus, G. Guelachvili, and J. Vergès, *Spectrochim. Acta* **24B**, 373 (1969).
83. J. Blaise, C. Morillon, M.-G. Schweighofer, and J. Vergès, *Spectrochim. Acta* **24B**, 405 (1969).
84. J. Blaise, J. Chevillard, J. Vergès, and Y. F. Wyart, *Spectrochim. Acta* **25B**, 333 (1970).
85. J. Blaise, J. Chevillard, J. Vergès, J. F. Wyart, and Th. A. M. van Kleef, *Spectrochim. Acta* **26B**, 1 (1971).
86. J. A. Decker, Jr., *Appl. Opt.* **10**, 510 (1971).
87. M. Harwit, P. G. Phillips, T. Fine, and N. J. A. Sloane, *Appl. Opt.* **9**, 1149 (1970).
88. M. Harwit, *Appl. Opt.* **10**, 1415 (1971); **12** 285 (1972).
89. A. Girard, *Opt. Acta* **7**, 81 (1960); *Appl. Opt.* **2**, 79 (1963).
90. C. Morillon, *Spectrochim. Acta* **25B**, 513 (1970); **27B**, 527 (1972).
91. P. Jacquinot and Ch. Dufour, *J. Rech. C.N.R.S.* **6**, 91 (1948).
92. H. Chantrel, *J. Rech. C.N.R.S.* **46**, 17 (1959).
93. J. E. Mack. D. P. McNutt, F. L. Roesler, and R. Chabbal, *Appl. Opt.* **9**, 873 (1963).
94. A. Steudel and H. Walther, *J. Phys. (Paris)*, Suppl. **28**, C2 255 (1967).
95. C. Vaucamps, J. P. Chabrat, P. Lozano, J. Rouch, and B. Serene, *J. Phys. E.* **6**, 970 (1973).
96. K. Heilig, Thesis, Heidelberg (1960).
97. V. G. Cooper, B. K. Gupta, and A. D. May, *Appl. Opt.* **11**, 2265 (1972).
98. V. Hansen and J. Thorsrud, *Appl. Opt.* **15**, 2418 (1976).
99. D. J. Bradley, *J. Sci. Instr.* **39**, 41 (1962).
100. J. Roig, *J. Phys. Radium* **19**, 284 (1958).
101. P. N. Slater, H. T. Betz, and G. Henderson, *Japan. J. Appl. Phys.* **4**, Suppl. 1, 440 (1965).
102. G. Hesse, *Feingerätetechnik* **14**, 535 (1965).
103. J. J. Barrett and G. N. Steinberg, *Appl. Opt.* **11**, 2100 (1972).
104. H.-W. Brandt, Private communication, Hannover (1976).
105. C. F. Bruce and R. M. Duffy, *Rev. Sci. Instr.* **46**, 379 (1975).
106. J. R. Greig and J. Cooper, *Appl. Opt.* **7**, 2166 (1968).
107. M. Herscher, *Appl. Opt.* **7**, 951 (1968).
108. J. Kuhl, A. Steudel, and H. Walther, *J. Phys. (Paris)*, **28**, Suppl. C2, 308 (1967).
109. P. Y. Brannon and F. M. Bacon, *Appl. Opt.* **12**, 142 (1973).
110. J. H. R. Clarke, M. A. Norman, and F. L. Borsay, *J. Phys. E* **8**, 144 (1975).

111. J.-M. Gagné, M. Giroux, and J.-P. Saint-Dizier, *Appl. Opt.* **12**, 522 (1973).
112. R. A. McLaren and G. I. A. Stegeman, *Appl. Opt.* **12**, 1396 (1973).
113. G. F. Kirkbright and M. Sargent, *Spectrochim., Acta* **25B**, 577 (1970).
114. J. V. Ramsay, *Appl. Opt.* **1**, 411 (1962); **5**, 1297 (1966).
115. G. Hernandez and O. A. Mills, *Appl. Opt.* **12**, 126 (1973).
116. T. R. Hicks, N. K. Reay, and R. J. Scaddan, *J. Phys. E* **7**, 27 (1974).
117. N. K. Reay, J. Ring, and R. J. Scaddan, *J. Phys. E* **7**, 673 (1974).
118. D. Drummond and A. Gallagher, *Rev. Sci. Instr.* **44**, 396 (1973).
119. J. Katzenstein, *Appl. Opt.* **4**, 263 (1965).
120. M. Platisa, *Fizika (Yugoslavia)* **5**, 83 (1973).
121. R. Chabbal, *Rev. Opt. (Paris)* **37**, 49, 336, 501, 608 (1958).
122. B. Moghrabi and F. Gaume, *Nouv. Rev. Opt.* **5**, 231 (1974).
123. F. L. Roesler and J. E. Mack, *J. Phys. (Paris)* **28**, Suppl. C2, 313 (1967).
124. C. Roychoudhuri and M. Hercher, *J. Opt. Soc. Am.* **62**, 1400 (1972).
125. M. Siese, private communication, Hannover 1976.
126. Commercial instrument: Burleigh multipass Fabry–Perot, Burleigh Instruments, East Rochester, New York 14445 (1974).
127. D. S. Cannell, J. H. Lunacek, and S. B. Dubin, *Rev. Sci. Instr.* **44**, 1651–1653 (1973).
128. D. Clarke, *Opt. Acta* **20**, 527 (1973).
129. G. Müller and R. Winkler, *Optik* **28**, 143 (1968/69).
130. P. Connes, *Rev. Opt.* **35**, 37 (1956); *J. Phys. Radium* **19**, 262 (1958).
131. M. Hercher, *Appl. Opt.* **7**, 951 (1968); **8**, 709 (1969).
132. J. R. Johnson, *Appl. Opt.* **7**, 1061 (1968).
133. A. Persin and D. Vukicevic, *Appl. Opt.* **12**, 275 (1973).
134. J. Kuhl, *Z. Phys.* **242**, 66 (1971).
135. D. Beysens, *Rev. Phys. Appl. (Paris)* **8**, 175 (1973).
136. H.-W. Brandt, Diplomarbeit, Inst. A. Exp. Physik, TU Hannover (1973).
137. G. Hernandez, *Appl. Opt.* **5**, 1745 (1966); **9**, 1591 (1970); **13**, 2654 (1974).
138. F. Pasqualetti and L. Ronchi, *Appl. Opt.* **10**, 2488 (1971); **11**, 1133 (1972).
139. W. Y. Tango and R. G. Twiss, *Appl. Opt.* **13**, 1814 (1974).
140. H. K. V. Lotsch, *Optik* **28**, 65, 328, and 555 (1968/69); **29**, 130, 622 (1969).
141. R. Bacis, *Appl. Opt.* **10**, 535 (1971).
142. D. N. Stacey, V. Stacey, and A. R. Malvern, *J. Phys. E* **7**, 405 (1974).
143. I. J. Hodgkinson, *Appl. Opt.* **8**, 1373 (1969).
144. G. Koppelmann, *Optik* **41**, 385 (1974).
145. G. S. Bhatnagar, K. Singh, and B. N. Gupta, *Nouv. Rev. Opt. (Paris)* **5**, 237 (1974).
146. R. P. Netterfield and J. V. Ramsay, *Appl. Opt.* **13**, 2685 (1974).
147. I. J. Hodgkinson, *Appl. Opt.* **10**, 396 (1971); **11**, 1970 (1972).
148. G. Koppelmann and K. Schreck, *Optik* **29**, 549 (1969).
149. A. F. Jones and D. L. Misell, *J. Phys. A* **3**, 462 (1970).
150. R. A. Day, *Appl. Opt.* **9**, 1213 (1970).
151. J. S. Hildum. *Appl. Opt.* **10**, 2567 (1971).
152. V. G. Cooper, *Appl. Opt.* **10**, 525 (1971).
153. V. N. Ivanov and I. S. Fishman, *Opt. Spectrosc.* **35**, 679 (1973); *Opt. Spektrosk.* **35**, 1175 (1973).
154. Y. F. Kielkopf, *J. Opt. Soc. Am.* **63**, 987 (1973).
155. K. Kaminishi and S. Nawata, *Japan, J. Appl. Phys.* **13**, 1640 (1974).
156. J. G. Hirschberg and P. Platz, *Appl. Opt.* **4**, 1375 (1965).
157. M. Daehler, *Appl. Opt.* **9**, 2529 (1970).
158. R. Behn and H. F. Döbele, *Appl. Opt.* **15**, 1850 (1976).
159. J. G. Hirschberg, W. I. Fried, L. Hazelton, Jr., and A. Wouters, *Appl. Opt.* **10**, 1979 (1971).

160. J. J. Barrett and S. A. Myers, *J. Opt. Soc. Am.* **61**, 1246 (1971).
161. P. J. Hargis, Jr., and R. A. Hill, *J. Opt. Soc. Am.* **65**, 219 (1975).
162. B. L. Taylor and F. T. Birks, *Analyst (London)* **97**, 681 (1972).
163. R. Hoekstra and R. Slooten, *Spectrochim. Acta* **26B**, 341 (1971).
164. D. W. Steinhaus, K. J. Fisher, and R. Engleman, Jr., *Chem. Instr.* **3**, 141 (1971).
165. F. S. Tomkins and M. Fred, *J. Opt. Soc. Am.* **41**, 641 (1951); *Appl. Opt.* **2**, 715 (1963).
166. R. Poppe, R. Hoekstra, and P. F. A. Klinkenberg, *Appl. Sci. Res.* B **11**, 293 (1964).
167. R. Hoekstra, *Appl. Opt.* **6**, 807 (1967).
168. H. Hühnermann, *J. Phys. (Paris)* **28**, Suppl. C2, 260 (1967).
169. G. Guthöhrlein, *Z. Phys.* **214**, 332 (1968).
170. H. Krause, Thesis, Berlin (1970).
171. G. Müller, Thesis, Berlin (1974).
172. H.-W. Brandt, Thesis, Hannover (1976).
173. H.-P. Clieves, private communication, Hannover (1977).
174. S. Gerstenkorn, *C. R. Acad. Sci.* **250**, 825 (1960).
175. J. Blaise, M. Fred, S. Gerstenkorn, and B. Judd, *C.R. Acad. Sci.* **255**, 2403 (1962).
176. M. Fred and F. S. Tomkins, *J. Opt. Soc. Am.* **47**, 1076 (1957).
177. E. Meinders, *Physica* **42**, 427 (1969).
178. J. Blaise and R. Vetter, *C.R. Acad Sci.* **256**, 630 (1963).
179. J. Blaise, P. Camus, G. Guelachvili, J. Vergès, and J.-F. Wyart, *C.R. Acad. Sci.* B **274**, 1302 (1972); **275**, 81 (1972).
180. J. Blaise, J. F. Wyart, R. Hoekstra, and P. J. G. Kruiver, *J. Opt. Soc. Am.* **61**, 1335 (1971).
181. A. Carlier, J. Blaise, and M. G. Schweighofer, *J. Phys. (Paris)* **29**, 729 (1968).
182. J. Blaise and L. J. Radziemski, Jr., *J. Opt. Soc. Am.* **66**, 644 (1976).
183. E. F. Worden and J. G. Conway, *J. Opt. Soc. Am.* **66**, 109 (1976).
184. P. F. A. Klinkenberg and Th. A. M. van Kleef, *Physica* **50**, 625 (1970).
185. N. Spector, *Phys. Scr.* **9**, 313 (1974).
186. V. A. Fassel, R. N. Knisely, and C. C. Butler, in *Analysis and Application of Rare Earth Materials*, NATO Advanced Study Institute, Ed. O. B. Michelson, Kjeller, Norway (1972); Univ.-forlaget, Oslo, pp. 71–86 (1973).
187. J. Blaise, and J. F. Wyart, *Rev. Chim. Min. (Paris)* **10**, 199 (1973).
188. J. F. Wyart, J. Blaise, and P. Camus, *Phys. Scr.* **9**, 322, 325 (1974).
189. K. L. Van der Sluis and L. J. Nugent, *J. Opt. Soc. Am.* **64**, 687 (1974).
190. W. C. Martin, L. Hagan, J. Reader, and J. Sugar, *J. Phys. Chem. Ref. Data* **3**, 771 (1974).
191. J. Sugar, *J. Opt. Soc. Am.* **65**, 1366 (1975).
192. S. Mrozowski, *Rev. Mod. Phys.* **16**, 153 (1944).
193. K. B. S. Eriksson, *J. Opt. Soc. Am.* **63**, 196 (1973).
194. P. S. P. Wei, *J. Chem. Phys.* **64**, 1531 (1976).
195. E. Luc-Koenig, C. Morillon, and J. Vergès, *Physica* **70**, 175 (1973).
196. L. Armstrong, Jr., *Theory of the Hyperfine Structure of Free Atoms*, Wiley-Interscience, New York (1971).
197. I. Lindgren and A. Rosén, *Case Stud. At. Phys.* **4**, 93 (1974).
198. B. R. Judd, *Proc. Phys. Soc. (London)* **82**, 874 (1963).
199. B. R. Judd, in *La Structure Hyperfine Magnétique des Atomes et des Molécules*, Eds. R. Lefebvre and C. Moser, Colloques Internationaux du C.N.R.S., No. 164, pp. 311–320, Paris (1967).
200. R. M. Sternheimer and R. F. Peierls, *Phys. Rev. A* **3**, 837 (1971).
201. R. M. Sternheimer, *Phys. Rev. A* **9**, 1783 (1974).
202. J. Bauche and B. R. Judd, *Proc. Phys. Soc. (London)* **83**, 145 (1964).

203. P. G. H. Sandars, *Adv. Chem. Phys.* **14**, 365 (1969).
204. P. G. H. Sandars and J. Beck, *Proc. R. Soc. London* **A 289**, 97 (1965).
205. W. Müller, A. Steudel, and H. Walther, *Z. Phys.* **183**, 303 (1965).
206. Y. Bordarier, B. R. Judd, and M. Klapisch, *Proc. R. Soc. London* **A 289**, 81 (1965).
207. G. H. Fuller and V. W. Cohen, *Nucl. Data Tables A* **5**, 433 (1969).
208. G. H. Fuller, *J. Phys. Chem. Ref. Data* **5**, 835 (1976).
209. A. Rosén, *Phys. Scr.* **8**, 159 (1973).
210. I. Lindgren, in *Atomic Physics* 4, Eds. G. zu Putlitz, E. W. Weber, and A. Winnacker, pp. 747–772, Plenum Press, New York (1975).
211. H. Kopfermann, *Nuclear Moments*, Academic Press, New York (1958).
212. A. Rosén and J. Lindgren, *Phys. Scr.* **8**, 45 (1973).
213. J. Lindgren and A. Rosén, *Phys. Scr.* **8**, 119 (1973).
214. W. Lange, *Z. Phys. A* **272**, 223 (1975).
215. H. D. Krüger and W. Lange, *Phys. Lett.* **42A**, 293 (1972).
216. J. Kuhl, *Z. Phys.* **242**, 66 (1971).
217. R. J. Champeau, E. Handrich, and H. Walther, *Z. Phys.* **260**, 361 (1973).
218. D. Wendlandt, J. Bauche, and P. Luc, *J. Phys.* **B10**, 1989 (1971).
219. W. Fischer, *Fortschr. Phys.* **18**, 89 (1970).
220. G. Müller and R. Winkler, *Z. Phys. A* **273**, 313 (1975).
221. P. G. H. Sandars and G. K. Woodgate, *Proc. R. Soc. London A* **257**, 269 (1960).
222. G. Guthöhrlein, *Z. Phys.* **214**, 332 (1968).
223. J. M. Baker and F. I. B. Williams, *Proc. R. Soc. London A* **267**, 283 (1962).
224. P. Luc, *Physica 62*, 239 (1972).
225. G. H. Fuller and V. W. Cohen, Oak Ridge National Laboratory report No. ORNL-4591 (1970).
226. L. Evans, P. G. H. Sandars, and G. K. Woodgate, *Proc. R. Soc. London A* **289**, 114 (1965).
227. W. Fischer, H. Hühnermann, and Th. Meier, *Z. Phys. A* **274**, 79 (1975).
228. R.-J. Champeau, J.-J. Michel, and H. Walther, *J. Phys. B: Atom. Molec. Phys.* **7**, L262 (1974).
229. R. Chuckrow and H. H. Stroke, *J. Opt. Soc. Am.* **61**, 218 (1971).
230. C. A. Mariño, G. F. Fülöp, W. Groner, P. A. Moskowitz, O. Redi, and H. H. Stroke, *Phys. Rev. Lett.* **34**, 625 (1975).
231. B. Buchholz, D. Kronfeldt, G. Müller, and R. Winkler, *Physica* **83C**, 247 (1976).
232. E. F. Worden, R. W. Lougheed, R. G. Gutmacher, and J. G. Conway, *J. Opt. Soc. Am.* **64**, 77 (1974).
233. E. F. Worden, R. G. Gutmacher, R. W. Lougheed, J. G. Conway, and R. J. Mehlhorn, *J. Opt. Soc. Am.* **60**, 1297 (1970).
234. P. Brix and H. Kopfermann, *Rev. Mod. Phys.* **30**, 517 (1958).
235. D. N. Stacey, *Rep. Progr. Phys.* **29**, 171 (1966).
236. A. F. Golovin and A. R. Striganov, *Sov. Phys. Usp.* **10**, 658 (1968); *Usp. Fiz. Nauk* **93**, 111 (1967).
237. K. Heilig and A. Steudel, *At. Data Nucl. Data Tables* **14**, 613 (1974).
238. J. Bauche and R.-J. Champeau, *Adv. At. Molec. Phys.* **12**, 39 (1976).
239. E. C. Seltzer, *Phys. Rev.* **188**, 1916 (1969).
240. W. H. King, A. Steudel, and M. Wilson, *Z. Phys.* **265**, 207 (1973).
241. F. A. Babushkin, *Sov. Phys. JETP* **17**, 1118 (1963); *J. Exp. Theor. Phys. USSR* **44**, 1661 (1963).
242. F. Boehm, Proceedings of the International Conference on Inner Shell Ionization Phenomena, Atlanta, 370–395, USEAC, Conf.-720404 (1972).
243. P. L. Lee and F. Boehm, *Phys. Rev. C* **8**, 819 (1973).

244. C. S. Wu and L. Wilets, *Ann. Rev. Nucl. Sci.* **19**, 527 (1969).
245. F. Boehm and P. L. Lee, *At. Data Nucl. Data Tables* **14**, 605 (1974).
246. R. C. Barrett, *Rep. Prog. Phys.* **37**, 1 (1974).
247. P. Brix and H. Kopfermann, *Z. Phys.* **126**, 344 (1949).
248. J. Blaise and A. Steudel, *Z. Phys.* **209**, 311 (1968).
249. R. Neumann, F. Träger, J. Kowalski, and G. zu Putlitz, *Z. Phys. A* **279**, 249 (1976).
250. R. Bruch, K. Heilig, D. Kaletta, A. Steudel, and D. Wendlandt, *J. Phys (Paris)* **30**, Suppl. C1-51 (1969).
251. G. L. Epstein and S. P. Davis, *Phys. Rev. A* **4**, 464 (1971).
252. A. R. Striganov, V. A. Katulin, and V. V. Eliseev, *Opt. Spectrosc.* **12**, 91 (1962);*Opt Spektrosk.* **12**, 171 (1962).
253. W. H. King. *J. Opt. Soc. Am.* **53**, 638 (1963).
254. F. S. Tomkins and S. Gerstenkorn, *C.R. Acad. Sci. (Paris) B* **265**, 1311 (1967).
255. J. E. Hansen, A. Steudel, and H. Walther, *Phys. Lett.* **19**, 565 (1965).
256. J. E. Hansen, A. Steudel, and H. Walther, *Z. Phys.* **203**, 296 (1967).
257. J. W. M. Dekker, P. F. A. Klinkenberg, and J. F. Langkemper, *Physica* **39**, 393 (1968).
258. J.-F. Wyart, Thèse, Orsay, France (1973).
259. S. Gerstenkorn and J. Vergès, *J. Phys. (Paris)* **36**, 481 (1975).
260. J. Bauche, *J. Phys. (Paris)* **35**, 19 (1974).
261. A. P. Stone, *Proc. Phys. Soc. (London)* **74**, 424 (1959).
262. J. Bauche, *Physica* **44**, 291 (1969).
263. D. A. Jackson, M.-C. Coulombe, and J. Bauche, *Proc. R. Soc. London A* **343**, 443 (1975).
264. A. P. Stone, *Proc. Phys. Soc. (London)* **77**, 786 (1961).
265. A. P. Stone, *Proc. Phys. Soc. (London)* **81**, 868 (1963).
266. P. Aufmuth, K. Heilig, and A. Steudel, Summaries of the Eighth EGAS Conference, No. 15, Oxford (1976).
267. J.-F. Wyart, *Physica* **61** 182 (1972).
268. P. Aufmuth, H. P. Clieves, K. Heilig, A. Steudel, D. Wendlandt, and J. Bauche, Summaries of the Eighth EGAS Conference, No. 14, Oxford (1976); *Z. Phys. A*, in press (1978).
269. J. Bauche and J.-C. Keller, *Phys. Lett.* **36A**, 211 (1971).
270. J.-C. Keller, *J. Phys. B: Atom. Molec. Phys.* **6**, 1771 (1973).
271. J. Bauche, R.-J. Champeau, and C. Sallot, *J. Phys. B: Atom. Molec. Phys.* **10**, 2049 (1977).
272. J. Bauche, Thèse, Université de Paris, France (1969).
273. J. Bauche and A. Crubellier, *J. Phys. (Paris)* **31**, 429 (1970).
274. P. Brix and H. Kopfermann, *Festschrift Akad. Wiss. Göttingen, Math.-Phys. Kl.*, 17 (1951).
275. M. Wilson, *Phys. Rev.* **176**, 58 (1968).
276. M. Wilson, *J. Phys. B: Atom. Molec. Phys.* **5**, 218 (1972).
277. M. A. Coulthard, *J. Phys. B: Atom. Molec. Phys.* **6**, 23 (1973).
278. J. Bauche and M. Klapisch, *J. Phys. B: Atom. Molec. Phys.* **5**, 29 (1972).
279. J.-J. Labarthe, *J. Phys. B: Atom. Molec. Phys.* **5**, L181 (1972).
280. J. Bauche, private communication.
281. H. G. Kuhn, *Atomic Spectra*, 2nd Ed., Longmans, London (1969).
282. W. Fischer, H. Hühnermann, G. Krömer, and H. J. Schäfer, *Z. Phys.* **270**, 113 (1974).
283. H. G. Kuhn, P. E. G. Baird, M. W. S. M. Brimicombe, D. N. Stacey, and V. Stacey, *Proc. R. Soc. London A* **342**, 51 (1975).
284. J. Blaise, *Ann. Phys. (Paris)* **3**, 1019 (1958).

285. W. J. Tomlinson III and H. H. Stroke, *Nucl. Phys.* **60**, 614 (1964).
286. G. F. Fulop, C. H. Liu, P. A. Moskowitz, O. Redi, and H. H. Stroke, *Phys. Rev. A* **9**, 593 (1974).
287. O. Redi and H. H. Stroke, *J. Opt. Soc. Am.* **65**, 1 (1975).
288. C. W. de Jager, H. de Vries, and C. de Vries, *At. Data Nucl. Data Tables* **14**, 479 (1974).
289. R. Engfer, H. Schneuwly, J. L. Vuilleumier, H. K. Walter, and A. Zehnder, *At. Data Nucl. Data Tables* **14**, 509 (1974).
290. H. Rothhaas, Contr. EPS International Conference on the Radial Shape of Nuclei, Cracow, June 22–25, pp. 3–4 (1976).
291. B. Dreher, *Phys. Rev. Lett.* **35**, 716 (1975).
292. P. Brix and H. Kopfermann, *Nachr. Akad. Wiss. Göttingen, Math.-Phys. Kl.*, 31–32 (1947).
293. L. Wilets, D. L. Hill, and K. W. Ford, *Phys. Rev.* **91**, 1488 (1953).
294. A. Bohr, *Mat. Fys. Medd. Dan. Vid. Selsk.* **26**, No. 14 (1952).
295. K. E. G. Löbner, M. Vetter, and V. Hönig, *Nucl. Data Tables* **A7**, 495 (1970).
296. S. Ullrich and E. W. Otten, *Nucl. Phys. A* **248**, 173 (1975).
297. W. Kutschera, W. Dehnhardt, O. C. Kistner, P. Kump, B. Povh, and H. J. Sann, *Phys. Rev. C* **5**, 1658 (1972).
298. P. F. Kenealy, G. B. Beard, and K. Parsons, *Phys. Rev. C* **2**, 2009 (1970).
299. C. W. Towsley, R. Cook, D. Cline, and R. N. Horoshko, *J. Phys. Soc. Jap. Suppl.* **34**, 442 (1973).
300. A. Christy and O. Häusser, *Nucl. Data Tables* **A11**, 281 (1973).
301. P. H. Stelson and L. Grodzins, *Nucl. Data A* **1**, 21 (1965).
302. F. Ackermann, E. W. Otten, G. zu Putlitz, A. Schenck, and S. Ullrich, *Nucl. Phys. A* **248**, 157 (1975).
303. W. Fischer, H. Hühnermann, K. Mandrek, Th. Meier, and D. C. Aumann, *Physica* **79C**, 105 (1975).
304. E. E. Fradkin, *Sov. Phys.—JETP* **15**, 550 (1962); *Zh. Eksp. Teor. Fiz.* **42**, 787 (1962).
305. W. D. Myers, *Phys. Lett.* **30B**, 451 (1969).
306. J. Bonn, G. Huber, H.-J. Kluge, L. Kugler, and E. W. Otten, *Phys. Lett.* **38B**, 308 (1972).
307. J. Bonn, G. Huber, H.-J. Kluge, and E. W. Otten, *Z. Phys.* **A276**, 203 (1976).
308. M. Cailliau, J. Letessier, H. Flocard, and P. Quentin, *Phys. Lett.* **46B**, 11 (1973).
309. S. G. Nilsson, J. R. Nix, P. Möller, and I. Ragnarsson, *Nucl. Phys. A* **222**, 221 (1974).
310. A. R. Bodmer, *Nucl. Phys.* **9**, 371 (1959).
311. F. G. Perey and J. P. Schiffer, *Phys. Rev. Lett.* **17**, 324 (1966).
312. L. R. B. Elton and A. Swift, in Proceedings of the Williamsburg Conference on Intermediate Energy Physics, pp. 731–741 (1966).
313. A. Swift and L. R. B. Elton, *Phys. Rev. Lett.* **17**, 484 (1966).
314. G. E. Brown, in *Facets of Physics*, pp. 141–150, New York, Academic Press (1970).
315. M. Beiner and R. J. Lombard, *Phys. Lett.* **47B**, 399 (1973).
316. A. B. Migdal, *Theory of finite Fermi systems and properties of atomic nuclei*, Nauka (1965).
317. G. G. Bunatyan and M. A. Mikulinskii, *Sov. J. Nucl. Phys.* **1**, 26 (1965); *J. Nucl. Phys. (USSR)* **1**, 38 (1965).
318. G. G. Bunatyan, *Sov. J. Nucl. Phys.* **4**, 502 (1967); *J. Nucl. Phys. (USSR)* **4**, 707 (1966).
319. V. P. Krainov and M. A. Mikulinskii, *Sov. J. Nucl. Phys.* **4**, 665 (1967); *J. Nucl. Phys. (USSR)* **4**, 928 (1966).
320. R. A. Uher and R. A. Sorensen, *Nucl. Phys.* **86**, 1 (1966).
321. B. S. Reehal and R. A. Sorensen, *Nucl. Phys. A* **161**, 385 (1971).

322. D. Zawischa and J. Speth, *Phys. Lett.* **56B**, 225 (1975).
323. V. Kaufman, *J. Phys. Chem. Ref. Data* **3**, 825 (1974).
324. R. Zalubas, J. Res. Natl. Bur. Stand. **80A**, 221 (1976).
325. B. Edlèn, *Phys. Scr.* **7**, 93 (1973).
326. B. Edlèn, in *Beam-Foil Spectroscopy*, Eds. I. A. Sellin and D. J. Pegg, Vol. 1, pp. 1–9, Plenum Press, New York (1975).
327. M. W. Smith and W. L. Wiese, *J. Phys. Chem. Ref. Data* **2**, 85 (1973).

8

Excitation of Atoms by Impact Processes

H. Kleinpoppen and A. Scharmann

1. Introduction

The wide variety of atomic excitation processes spans the simplest form of fire and the most sophisticated forms of highly intense laser radiation, synchrotron radiation, and beam-foil excitation, etc. The wide range of possible spectroscopic investigations requires the experimenter to choose the most suitable type of excitation process. Of course, simple types of commercially available discharge tubes or the classical light source for optical hyperfine structure studies, the Schüler hollow cathode, may still serve their purposes; alternatively it has become common that intense discharge tubes may be combined with lasers for multistep or multiphoton transitions in atoms, for example.

The excitation of atoms by impact with particles of uniform direction has been applied in such spectroscopic studies where other alternatives were neither possible nor suitable. Particularly successful were electron impact processes for production of metastable atoms and excited ions. Ion bombardment has proved to be extremely successful in connection with x-ray excitations (Chapter 30 of this work). The physical processes connected with particle impact excitation of atoms are normally of secondary importance in spectroscopic studies. The spectroscopist's main interest in impact light sources is related to problems of intensity, polarization (which indicates an imbalance in magnetic sublevel population), and also

H. Kleinpoppen • Institute of Atomic Physics, University of Stirling, Stirling, FK9 4LA, Scotland. A. Scharmann • I. Physikalisches Institut der Universität, Heinrich-Buff-Ring 16, 63 Giessen, West Germany.

the coherency of line radiation (which enables, for example, phenomena such as quantum beats of the emitted radiation to be studied). In connection with the spectroscopy of atomic compound states, collision methods have predominantly and successfully been applied for the configuration analysis or the "spectroscopy" of atomic compound states (see, e.g., Chapter 23 of this work).

In this article we restrict our review to impact excitation in electron and heavy-particle impact collisions. Neither part is comprehensive; rather, each part is selective with regard both to spectroscopic applications and to spectroscopic observables.

2. Electron Impact Excitation

2.1. Cross Sections and Excitation Functions

The method of producing spectral lines or metastable atoms by electron impact can be divided into two parts: (a) impact processes with intersecting beams of electrons and atoms (crossed-beam techniques) and (b) impact excitations by electron beams in a volume uniformly filled with the sample atoms.

The crossed-beam technique requires beam collimation both for electrons and atoms. An ideal intersection consists of identical rectangular cross sections for both the electron and the atomic beam.[1] If the atomic target density is n, the electron beam current density j, the total cross section Q_j for exciting a given state j, and if the branching ratio is f_{jk} for a line transition from the given state j to state k, the total number of photons emitted over 4π is

$$I = njQ_jf_{jk} \tag{1}$$

While in the early experiments of electron impact excitations relative "optical excitation functions" were mostly measured, in more recent experiments reliable absolute calibrations of excitation functions have been carried out. In the traditional experiments for measuring optical excitation functions, cylindrical symmetry is valid with reference to the direction of the electron beam. Assuming that electric dipole emission is the dominant radiative transition process, the radiation observed in any direction can be regarded as due to three independent dipole transitions with one of the axes of reference parallel to the electron beam and two axes perpendicular to the electron beam. The total excitation cross section Q_T is then given by

$$Q_T = Q_\parallel + 2Q_\perp \propto I_\parallel + 2I_\perp \tag{2}$$

with the intensity components $I_\parallel$ and $I_\perp$ polarized parallel or perpendicular to the electron beam, respectively. Before discussing examples of optical excitation cross sections in detail, we summarize the characteristics of excitation functions. The typical gross structure of an excitation function is characterized by the onset at threshold energy, a maximum (or several maxima) characteristic of the transition and the atom, and also by the monotonic fall of the excitation function at large electron energies. This gross structure of excitation functions has not changed much since the first quantitative measurements of optical excitation functions by Hanle,[2] Hanle and Schaffernicht,[3] and Bricout.[4] More recent measurements,[5] however have shown that the existence of detailed structure in excitation functions is very common close to threshold (Figures 1–4). Such structure can display several maxima or minima, an important feature for low-energy electron–atom excitations. The reason for most of the structure has been attributed to resonance in the inelastic electron scattering process (see Williams, Chapter 23).

Electron–atom cross sections are normally given in units of $\pi a_0^2 \simeq 0.7 \times 10^{-16}$ cm^2. Cross sections can vary over a wide range from approximately 10^{-12} cm^2 (e.g., excitation cross section for $2S \to 2P$ transition in atomic hydrogen) to cross sections $<10^{-20}$ cm^2 for lines of very highly excited states of atoms or ions (e.g., ion line excitations of He$^+$ ions from the neutral atom) at intermediate electron energies. The following figures illustrate examples of cross sections for various important line excitations. The hydrogen excitation cross sections show structure very close to the threshold energy and below the next inelastic channel. Such structure is related to compound states (Feshbach or type-I resonances) or shape resonances (type-II resonances) characterized by a given electron configuration. An analysis of the spectroscopic configuration of the resonances is indicated in Figure 1 (a more detailed discussion on the analysis of resonance structure can be found in Chapter 23). Resonance structure is a typical characteristic of excitation cross sections close to the excitation threshold and only a few striking examples of resonances are illustrated here. The study of the electron configuration of resonances leads to "level schemes" of atomic compound states which are similar to "normal" atomic level schemes. In Figure 2 we show some of the latest experimental data on the excitation function of mercury and we refer to a proposal by Heddle[10] for the level scheme of the resonance spectrum associated with the negative compound states of Hg$^-$ (Figure 17 of Chapter 23). As pointed out by Heddle,[10] the energy sequence of the Hg$^-$ ion is related to the $6s^2 6p$ and $6s6p^2$ configurations, of the thallium I isoelectronic sequence (Tl 1, Pb II, Bi III, ...) which supports the identification of the resonance structure in electron–mercury scattering. We note that according to the above interpretation, the two strong peaks in

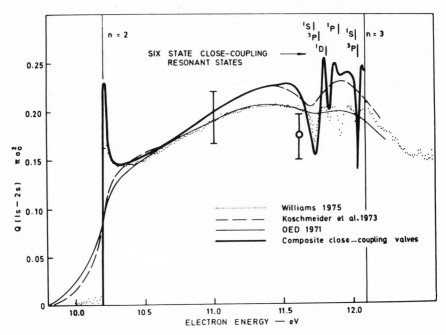

Figure 1. Total $1S \rightarrow 2S$ excitation cross section of atomic hydrogen by electron impact. Data point $\tilde{Q}$ from Kauppila *et al.*,[8] Koschmieder *et al.*,[6] Oed.[7] ^1S, ^3P, . . . mark the energy of the H′ compound states as predicted by theory[55] (three-state plus 20 correlation terms[55a] from 10.2 to 10.7 eV and six-state close-coupling values[55b] from 10.7 to 12.1 eV).

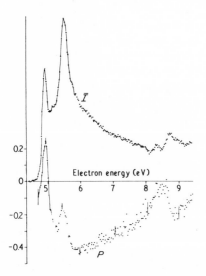

Figure 2. Optical excitation (top) and polarization P function (bottom) of the 6^3P_1–6^1S_0 2537 Å line of mercury (Ottley and Kleinpoppen[9]). E is the electron energy.

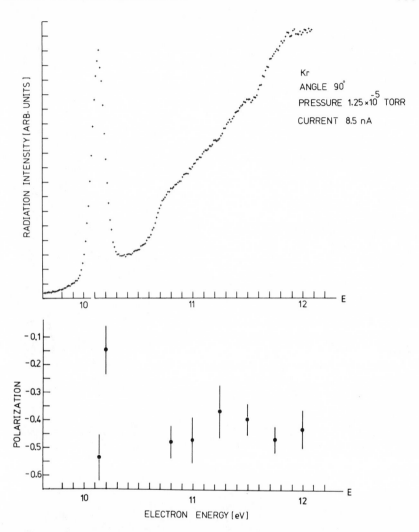

Figure 3. Excitation and polarization function of the two lowest excited states of krypton with the configurations $(4p^5 5s)^3 p$ and $(4p^5 5s)^1 p$ (Al-Shamma and Kleinpoppen[12]).

the excitation function of the 2537-Å mercury line are associated with $^2D_{3/2,5/2}$ negative ion compound states. Detailed studies of further mercury excitation functions (λ 3341 Å, 6^3P_2–8^3S_1; λ 3022/23/26 Å, 6^3P_2–7^3D_j; metastable states 6^3P_0 and 6^3P_2) by Shpenik et al.[47] also revealed interesting resonance structure.

Heddle and co-workers[11] have studied the excitation of helium in detail. By plotting $(-R/T)^{1/2}$, where T is the binding energy of a

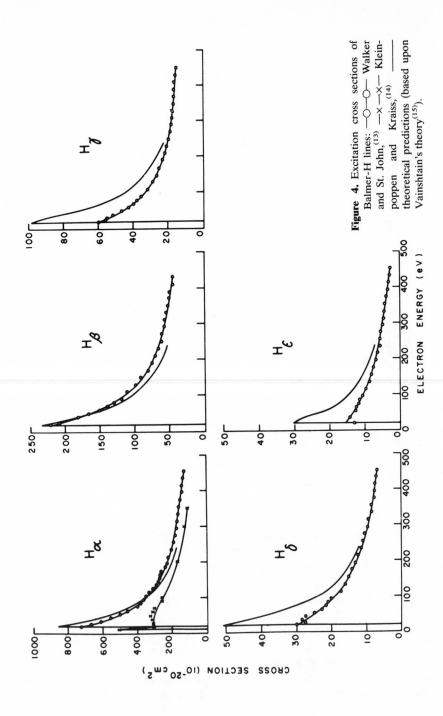

Figure 4. Excitation cross sections of Balmer-H lines: —○—○— Walker and St. John,[13] —×—×— Kleinpoppen and Kraiss,[14] ——— theoretical predictions (based upon Vainshtain's theory[15]).

resonance state, as a function of n as quantum number with integral values, Heddle[11] could observe the resonance series following a linear relationship between $(R/T)^{1/2}$ and n (see Chapter 23).

Another striking example of resonances in optical excitation functions is seen in the excitation of the two lowest excited states of krypton with the configurations $(4p^5 5s)^3 P$ and $(4p^5 5s)^1 P$. Figure 3 shows the combined excitation function of the 1235-Å and 1164-Å lines corresponding to transitions from these levels to the ground state of krypton. As pointed out by Swanson et al.[11a] the analysis of the krypton resonance structure is difficult because of the fact that intermediate coupling must be used to characterize the krypton states. In spite of such difficulties, Swanson et al.[11a] suggested the configuration $4P^5_{1/2} 5s^2 P_{1/2}$ for the strong krypton resonance just above threshold. Interesting resonance structure was also seen in the first resonance lines of xenon.[12]

Some of the more recent electron–atom excitation experiments were solely concerned with the gross behavior of cross sections over a wide energy range. In such experiments multistage electron guns were used instead of electron monochromators. Particularly interesting results were obtained for the Balmer lines and the first resonance lines of alkali and alkaline earth spectra. Figure 4 presents the results for the Balmer cross sections as obtained by Walker and St. John,[13] who used a highly intense beam of atomic hydrogen from a Wood's discharge tube. The relative contributions of the $3S$, $3P$, and $3D$ states to the Balmer-α line has been measured by modulating the electron beam and discriminating between the

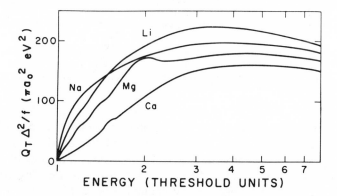

Figure 5. Comparison of cross sections for the excitation of the first resonance lines (total doublet intensity) of Li, Na, Mg, and Ca. Q_T is the absolute excitation cross section, Δ threshold energy, and f is the optical oscillator strength used for the Born normalization of the excitation function (from Leep and Gallagher[17]).

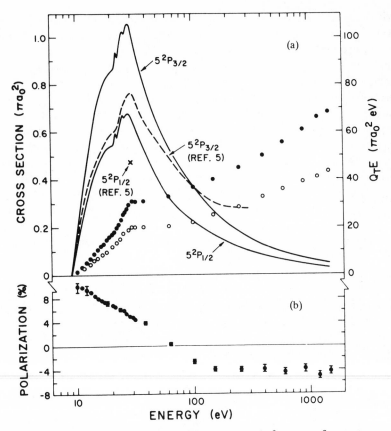

Figure 6. (a) Total excitation cross sections for the Sr^+ $5^2P_{3/2}$ and $5^2P_{1/2}$ levels;
(b) polarization for the 4078 Å line ($5^2P_{3/2} \rightarrow 5^2S_{1/2}$) excited from the
ground state of Sr. Solid curve, Chen *et al.*[18]; cross and dotted line,
Starodub *et al.*[19] The cross sections of Chen *et al.*[18] are also plotted on
the $Q_T E$ scale (cross section times electron energy, dots for $5^2P_{3/2}$, circles
for $5^2P_{1/2}$).

different contributions from the above states on the basis of their different
lifetimes for radiative decay after excitation (Mahan *et al.*[16]). Gallagher
and co-workers[17] recently measured the excitation functions of the first
resonance lines of Na, Ca, Li, Mg, and Sr and also of the ionizing excitation
of the Mg^+ $3^2P \rightarrow 3^2S$ resonance doublet and of Sr^+ $5^2P_{3/2,1/2} \rightarrow 5S_{1/2}$ transitions. The excitation cross sections of the resonance lines of Li, Na, Mg,
Ca are compared in Figure 5 with each other by plotting $Q_T \Delta^2/f$ versus
the energy (in threshold units), where Q_T is the absolute cross section
obtained by normalizing the excitation function to the Born approximation, Δ is the excitation threshold, and f is the optical oscillator strength

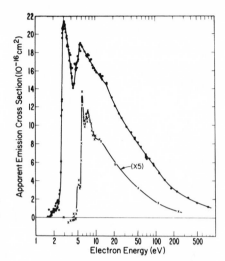

Figure 7. Excitation cross sections for 4554 Å line ($-\cdot-\cdot-$, $6^2P_{3/2} \to 6^2S_{1/2}$) and five times the cross-section for 4113+4166 Å lines ($-\times-\times-$, $6D_{5/2,3/2} \to 6P_{3/2}$) of Ba$^+$ excited by electron impact from the ground state of the Ba$^+$ ion (Crandall *et al.*[20]).

used for the Born normalization of Q_T. As an example for ionizing excitation we refer to Figure 6 for the Sr$^+$ $5^2P \to {}^2S$ transitions excited from the ground state of Sr.[18]

Recently, Trajmar and Williams[48] reviewed their measurements on electron–metal-atom cross sections (Li, K, Mg, Ba, Mn, Cn, Zn, Pb, and Bi).

Crossed-charge beam techniques for electrons and ions have recently been applied to the study of the excitation of Na$^+$, Ba$^+$, Ca$^+$, Hg$^+$, Ar$^+$, and He$^+$ (see references quoted in Grandell *et al.*[20]). Particular attention will only be given here to an example (Figure 7) of the careful absolute radiometry calibration measurements of excitation cross sections of Ca$^+$, Be$^+$, N$_2^+$, and Hg$^+$ which claim total uncertainties for the absolute cross sections ranging from just below 10% to about 40%.

2.2. Linear Polarization of Impact Radiation

Atomic line radiation excited by a collimated beam of electrons (or any collimated beam of exciting particles) will, in general, be polarized and the emitted photons will have an anisotropic angular distribution. This anisotropy appears to be a consequence of the geometrical anisotropy imposed on the excitation process by the incoming particle beam (cylindrical symmetry rather than spherical symmetry). With the incoming beam in the z direction the radiation can be considered as resulting from an incoherent superposition of three electric dipoles oscillating in the z direction and in the two coordinate directions perpendicular to the z direction. The total intensity of the radiation perpendicular to the z direction is given by $I(90°) = I_\parallel + I_\perp$ with $I_\parallel$, $I_\perp$ as polarized intensities parallel and perpendicular to z. The intensity contribution from a given dipole is proportional to $\sin^2 \theta$, with θ being the angle between the dipole axis and the direction of observation. The combined intensity at any direction $I(\theta)$ of the three dipoles is given by

$$I(\theta) = I_\parallel \sin^2 \theta + I_\perp (1 + \cos^2 \theta) \tag{3}$$

or with the "degree of polarization" $P = (I_\parallel - I_\perp)/(I_\parallel + I_\perp)$ we obtain

$$I(\theta) = I(90°)(1 - P \cos^2 \theta) \tag{4}$$

With the total intensity $I = 4\pi \bar{I}$ ($\bar{I}$ = mean intensity) integrated over all angles one obtains

$$I(\theta) = \bar{I} \frac{1 - P \cos^2 \theta}{1 - P/3} \tag{5}$$

The total excitation cross section follows directly from the total intensity or from a combined measurement of the degree of polarization and the intensity observed for any direction. The theoretical basis for impact line polarization lies with quantum mechanics. The theory is well formulated for cases when spin–orbit interactions between the projectile particle and the atom can be neglected.

The revival of interest in polarization studies of impact radiation since about the beginning of the sixties originates from the following facts. As illustrated in Figure 8, the threshold polarization of electron impact line radiation is governed by the selection rule $\Delta m_L = 0$. Accordingly, threshold polarization can easily be calculated for many simple cases, e.g., the threshold polarization of the excitation–deexcitation process $^1S \to {}^1P \to {}^1S$ should be 100%, because only the $m_L = 0$ sublevel of the P state is excited,

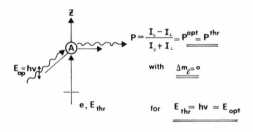

Figure 8. Equivalence of threshold impact excitation by unidirectional electrons and resonance excitation by linearly polarized light (z axis parallel to electron beam direction and the light vector) with respect to the selection rule $\Delta m_L = 0$. As a consequence, the threshold polarization P^{thr} of the electron impact excitation is equal to the polarization P^{opt} of the resonance fluorescence radiation. Note Table 1, which provides experimental evidence for $P^{thr} = P^{opt}$ for the first resonance lines of Li^6, Li^7, and Na^{23}.

and subsequently only the π transition to the S state occurs. There were several experiments in the twenties, particularly that of Skinner and Appleyard,[21] which were in complete disagreement with the threshold polarization required by angular momentum conservation. This startling discrepancy remained unexplained for several decades. The other important factor that stimulated renewed interest in polarization studies was the new theoretical approach to the problem mainly due to the papers by Percival and Seaton[22] and by Baranger and Gerjuoy[23] in 1958. The theory developed by Percival and Seaton overcame the deficiencies of the previous Oppenheimer[24] and Penney[25] theory by allowing for radiation damping or what is equivalent to taking the finite level width of the excited fine- and hyperfine-structure states into account. The particular importance of the paper by Baranger and Gerjuoy[23] is often referred to as the introduction of the atomic compound model for atomic collision processes. However, the most important applications and consequences of this model were related to the angular distribution and polarization of light emitted following excitation through atomic compound states (see Chapter 23).

Although it is expected that the existing theory on polarization of impact radiation correctly describes the experimental data in its range of application, there are still problems left that might call either for extensions or modifications of certain aspects of the theory, or alternatively, for an improvement in the experimental conditions. It therefore seems reasonable to review selected examples that illustrate cases of agreement and disagreement between theory and experiment.

The polarization of the line radiation depends on the unequal population of the magnetic sublevels excited during the collision and also on the relative transition probabilities for π and σ transitions of these substates into the magnetic sublevels of the state into which the atom decays. In order to discuss recent results on line polarization, we concentrate on examining applications for atoms in which the ground state has no orbital momentum. The intensity components polarized parallel or perpendicular to the direction of the incoming beam (z direction, see Figure 8) are then given by

$$I_{\parallel} \propto \sum_m \frac{A_m^{\pi}}{A} Q_m, \qquad I_{\perp} \propto \sum_m \frac{1}{2}\frac{A_m^{\sigma}}{A} Q_m \qquad (6)$$

(A is the total transition probability, A_m^{π} and A_m^{σ} refer to magnetic substates for π and σ transitions, Q_m is the excitation cross section for a magnetic substate with magnetic quantum m).

It then follows for the polarization of the line radiation (with $A = A_m^{\pi} + A_m^{\sigma}$; note that the total transition probability does not depend on m)

that

$$P = \frac{I_{\parallel} + I_{\perp}}{I_{\parallel} + I_{\perp}} = \frac{3K^{\pi} - K}{K^{\pi} + K} \tag{7}$$

where

$$K^{\pi} = \frac{1}{A} \sum_m A_m^{\pi} Q_m \quad \text{and} \quad K = \sum_m Q_m = Q \tag{8}$$

Note that in the above equations we do not specify the magnetic quantum number which could be taken for m_L, m_j, or m_F. The quantities K^{π} and K are proportional to the production rates of the total linear polarized and total unpolarized line intensity of the emission, respectively.

The recent emphasis on line polarization measurements according to the scheme in Figure 8 is mostly connected with some simple examples and arguments, which can best be illustrated as follows: Let us consider the excitation–deexcitation process ${}^1S \rightarrow {}^1P \rightarrow {}^1S$ with the total excitation cross sections, $Q(S \rightarrow P) = Q_{m_L=0} + Q_{m_L=+1} + Q_{m_L+-1} = Q_0 + 2Q_1$.* Only the Q_0 component results in the excitation of π light, therefore, with $K^{\pi} \propto Q_0$ and $K \propto Q$, the polarization of the above excitation process is given by

$$P = \frac{Q_0 - Q_1}{Q_0 + Q_1} = \frac{Q_0/Q_1 - 1}{Q_0/Q_1 + 1} \tag{9}$$

At threshold the excitation is governed by the selection rule $\Delta m_L = 0$ or $Q_1 = 0$ and $Q = Q_0$ and $P = 100\%$. A beautiful example, which proves that threshold polarization of the above process should be 100%, is revealed in the study of the polarization of the first resonance line of calcium investigated as a function of the electron energy by Ehlers and Gallagher[26] (Figure 9). The high-energy limit (E_{∞}) of the polarization should approach $P_{\infty} = -100\%$ because $Q_0(E_{\infty})/Q_1(E_1) \rightarrow 0$. King et al.[52] have made a study on the threshold selection rule $\Delta m_L = 0$ by means of an electron-photon coincident experiment (see Section 2.3). Electrons scattered in the forward direction do not transfer orbital angular momentum in that direction. Accordingly, the polarization of a line excited by forward-scattered electrons should be governed by a selection rule $\Delta m_L = 0$. This could be confirmed by the above authors for the He $3^1P \rightarrow 2^1P$, $\lambda = 5016$ Å transition.

Including fine and hyperfine structure interactions in the $S \rightarrow P \rightarrow S$ transitions, the theory has to be modified by taking into account the finite

---

* Parity invariance of the excitation process requires that the $m_L = \pm 1$ substates are excited with equal probability.

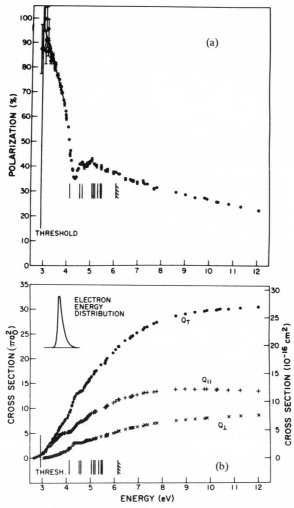

Figure 9. (a) Polarization of the 4^1P-4^1S, $\lambda = 4227$-Å radiations of calcium for low electron-bombardment energies (one-σ statistical error bars arising from counting statistics, Ehlers and Gallagher[26]). (b) Total cross section and the polarized components of the same line.

level width and also the jI recoupling of the excited states after the excitation. For the alkali resonance lines the polarization is given by the formula (Flower and Seaton[27])

$$P = \frac{3(9\alpha - 2)(Q_0 - Q_1)}{12Q_0 + 24Q_1 + (9\alpha - 2)(Q_0 - Q_1)} \qquad (10)$$

Table 1. Hyperfine Structure Coefficients α for Line Polarization

$2^2 P_{1/2,3/2}$ of atomic hydrogen	with hfs[a]	$\alpha = 0.441$
	without hfs	$\alpha = \frac{4}{9} = 0.445$
$2^2 P_{1/2,3/2}$ of Li^6	with hfs	$\alpha = 0.413$
$2^2 P_{1/2,3/2}$ of Li^7	with hfs	$\alpha = 0.326$
$3^2 P_{1/2,3/2}$ of Na^{23}	with hfs	$\alpha = 0.288$

[a] Hyperfine structure.

with

$$\alpha = \sum_{F,F'} \frac{\zeta(I,F,F')}{1+\varepsilon_{F,F'}^2}, \qquad \varepsilon_{F,F'} = \frac{2\pi \Delta\nu_{F,F'}}{A} \qquad (11)$$

and $h\Delta\nu_{F,F'}$ the energy separation of the hyperfine structure states and $\zeta(I,F,F')$ expressed in terms of Racah and vector coupling coefficients. The value of α depends largely on the ratios of the hyperfine-structure separations to the natural line width ratios, which can be calculated from the theory of Flower and Seaton[27] for cases of interest. These ratios are summarized in Table 1.

Table 2 gives results for theoretical and experimental threshold values and optical line radiation of the first resonance line of Li^6, Li^7, and Na^{23}, and also the Lyman-α impact polarization of atomic hydrogen. Notice the excellent agreement between theory and experiment of the alkali resonance lines. This proves that the relation $P_{opt} = P_{thr}$ is valid for the alkali resonance lines, at least for light alkalis. Figures 10 and 11 compare experimental and theoretical calculations of linear polarizations of the first resonance lines of Li^6, Li^7, and Na^{23} in an energy range from threshold to 300 eV.

Contrary to the above examples, there are quite a number of atomic lines that appear not to satisfy the expected threshold polarization required by the $\Delta m_L = 0$ selection rule. Important examples are certain helium lines,

Table 2. Threshold (P^{thr}) and Resonance Fluorescence Polarization (P^{opt}) of the First Resonance Line of Li^6, Li^7, Na^{23}, and H^a

Polarization	Na^{23}	Li^6	Li^7	H
P_{cal}^{thr}	14.1%	37.5%	21.6%	42.1%
P_{exp}^{thr}	$(14.8 \pm 1.8)\%$	$(39.7 \pm 3.8)\%$	$(20.6 \pm 3.0)\%$	13%(?)[34]
P_{cal}^{opt}	$(14.2 \pm 0.2)\%$	$(37.7 \pm 1.0)\%$	$(21.7 \pm 1.5)\%$	42.1%
P_{exp}^{opt}	$(14.0 \pm 0.8)\%$	$(37.5 \pm 1.7)\%$	$(23.1 \pm 1.3)\%$	—

[a] P_{cal}, P_{exp}: theoretical[27,28] and experimental[28] data.

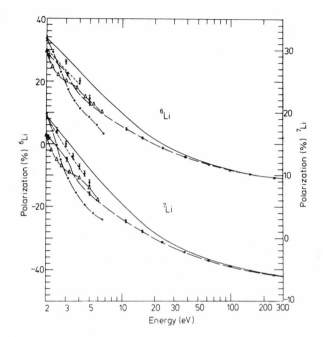

Figure 10. Percentage polarizations of the resonance line $2p \rightarrow 2s$ emitted from ^{6}Li and ^{7}Li for electron energies up to 250 eV. —— Kumar and Shrivastava,[29] —$\ominus$— experimental results of Leep and Gallagher,[30] —$\triangle$— close-coupling calculations of Burke and Taylor,[31] $---$ modified close-coupling calculations of Feautrier,[32] —$\bigcirc$— variational calculations of McCavert and Rudge,[33] $\ominus$ experimental data of Hafner and Kleinpoppen.[11]

the Lyman-α radiation[34] (Table 2), the $3^2D \rightarrow 2^2P$ transition in lithium,[28] and also certain mercury lines. Most of these lines show a considerable structure in the polarization close to threshold. Obviously, it has been suspected that structure in the polarization near threshold is related to the existence of resonances in the inelastic electron atom cross section near threshold. Fano and Macek[39] suggest that at a resonance the colliding electron and the one being excited in the atom remain strongly coupled for a time interval sufficient to allow extensive exchange of angular momentum between them with a decrease in alignment. The following examples to be discussed give clear evidence of the influence of resonances on the polarization of line radiation. We first discuss the mercury excitation–deexcitation processes $6^1S_0 \rightarrow 6^3P \rightarrow 6^1S_0$ ($\lambda = 2537$ Å) and $6^1S_0 \rightarrow 6^1P_1 \rightarrow 6^1S_0$ ($\lambda = 1849$ Å). For threshold excitation of the 6^1P state, only the $M_j = 0$ is excited because of the selection rule $\Delta m_L = 0$, for the excitation of

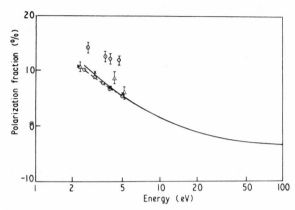

Figure 11. Polarization of resonance line of Na line. ——
Tripath *et al.*[35]; calculations of Moores and Norcross[36]
using a four-state exchange approximation. Experimental
data: $\dot{\bigcirc}$: data of Hafner and Kleinpoppen[11];
$\odot$ - - - $\odot$ data of Enemark and Gallagher[37]; $\triangle$: data of
Gould.[38]

the 6^3P_1 state, only $M_j = \pm 1$ states can be excited at threshold because of
the fact that the relevant Clebsch–Gordan coefficient for the $M_j = 0$ state
with $M_L = 0$ is zero: $C^{S=1,L=1,J=1}_{M_S=0,M_L=0,M_J=0} = 0$, while all the other angular
momentum coupling coefficients are nonzero. This has the consequence
that the threshold polarization of the above mercury lines should provide
the largest possible difference ($\pm 100\%$) for the line polarization, $P_{thr}(\lambda =
2537\text{ Å}) = -100\%$ and $P_{thr}(\lambda = 1849\text{ Å}) = +100\%$. Allowance for depar-
tures from LS coupling was made for the threshold polarization of the
2537-Å line by Penney.[25] He obtains $P_{thr} = -92\%$ for zero nuclear spin,
-53% for nuclear spin $I = \frac{1}{2}$, and -48% for $I = \frac{3}{2}$; with the normal isotope
mixture Penney's calculation for threshold polarization of the 2537-Å line
gives $P_{thr} = -80\%$.

    Precision studies of resonances in the excitation of mercury atoms
have been reported by Shpenik *et al.*[147] and Ottley and Kleinpoppen.[9]
The lower parts of Figure 2 and also Figure 12 show the latest results of the
experimental studies, with high resolution, of the mercury 2537-Å line
(Ottley and Kleinpoppen[9]). The resonance structure of the polarization
curve above threshold coincides with various other types of electron–atom
collision experiments (e.g., the transmission experiment carried out by
Burrow and Michejda,[40] the scattering and spin polarization experiment
of Düweke *et al.*[41]). Baranger and Gerjuoy[23] first developed a theory in
which they associated the failure of experimental data to approach the
required threshold polarization with the formation of atomic compound

states. They assumed that two negative-ion compound states are formed just above the excitation threshold of the 2537-Å line and made predictions for the polarization of the 2537-Å line in the resonance states of 60% and 0% in the center of resonances with total angular momenta $j = \frac{3}{2}$ and $j = \frac{1}{2}$, respectively. The calculations of Baranger and Gerjuoy[23] assumed zero orbital angular momentum of the outgoing electrons. The experimental data of the polarization curve do not confirm the theoretical predictions in the two resonance peaks at the moment. One must, however, bear in mind that the natural isotope composition and the remaining finite energy resolution of the electron beam still affect the polarization. It can be seen by comparing Figure 2 and Figure 12, that the polarization peak at the first resonance, close to the threshold, rises from about 25%, where the energy spread of the electron beam is about 140 meV, to almost 40%, where the energy spread of the electron beam is about 100 meV. A rough estimate shows that the predicted polarization of 60% in the $j = \frac{3}{2}$ resonance should be reduced to approximately 45%, taking into account the hyperfine-structure splitting for the normal isotope mixture of mercury. Fano and Cooper[42] have suggested the identification $^4P_{1/2,3/2,5/2}$ for the three lowest resonances associated with the compound state configuration $6S\,6P^2$. As discussed in Chapter 23, Heddle[10] offered a new interpretation for the classification of electron–mercury resonances. According to his suggestions the peaks or features in the polarization and excitation function of the $\lambda 2437$-Å line are due to Hg compound states with the configurations $^2D_{3/2}$ (at 4.92 eV), $^4P_{5/2}$ (at 5.23 eV, a weakly pronounced feature in the intensity components and the polarization), and $^2D_{5/2}$ (at 5.50 eV).

High-resolution studies of polarization curves near threshold for line transitions from the $4D$ states of helium were reported by Heddle et al.[43] The polarization of the line transition from the D states of helium can be

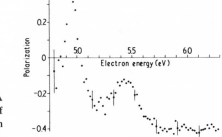

Figure 12. Polarization of the Hg 2537-Å intercombination line as a function of electron energy (energy resolution 100 meV, Ottley and Kleinpoppen[9]).

expressed as follows:

$$P(4^1D \to 2^1P) = \frac{3Q_0 + 3Q_1 - 6Q_2}{5Q_0 + 9Q_1 + 6Q_2}$$

$$P_{\text{thr}}(4^1D \to 2^1P) = \frac{3}{5} = 60\%$$

$$P(4^3D \to 2^3P) = \frac{213Q_0 + 213Q_1 - 426Q_2}{671Q_0 + 1271Q_1 + 1058Q_2}$$ (12)

$$P_{\text{thr}}(4^3D \to 2^3P) = \frac{213}{671} = 32\%$$

Figures 13 and 14 show excitation and polarization function lines associated with the above transition. Although the total error in the data in these figures is approximately 30%, close to threshold, the measurements agree well with the threshold predictions. The above authors suggest that the sharpness of the fall in the polarization immediately above threshold is associated with a $1s4d^2\,^2S$ resonance, which strongly decays into $m_L = |2|$ substates of the D state. Previous studies reveal structures in the excitation functions of the above line transitions that coincide with the energy posi-

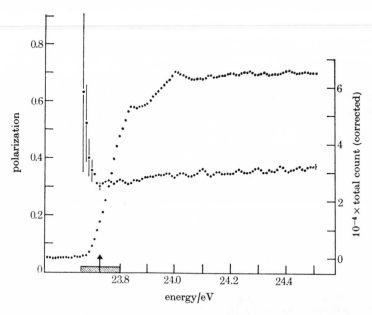

Figure 13. The excitation function and polarization curve of the $\lambda = 4922$-Å, $4^1D \to 2^1P$ line of helium as measured by Heddle et al.[43] after 120 h.

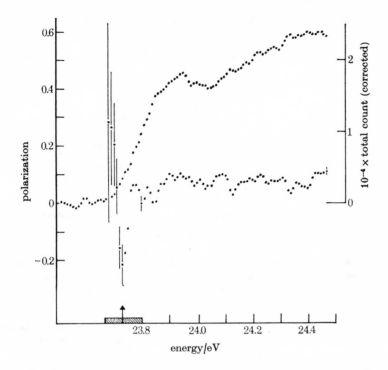

Figure 14. The excitation function and polarization curve of the 4472-Å, $4^3D \to 2^3P$ line of helium as measured by Heddle *et al.*[43] after 170 h.

tion of the polarization curves. The lowest possible polarization values occur if Q_2 is large compared with Q_0 and Q_1, which are -100% and -40% for the singlet and triplet transitions, respectively. The initial fall to -22% in the triplet transitions is followed by the return to a positive value. The true minimum polarization might be approached only with even higher-energy resolution, as demonstrated in the case of the mercury excitation described above. The double feature around 23.8 eV in the polarization of the helium triplet transitions can also be associated with structure in the excitation function.

Recently, Mumma *et al.*[44] determined the polarization of $n^1P \to 1^1S$ transitions of helium by measuring the angular intensity distribution of the photons. It was, however, not possible to separate the different lines of the $nP \to 1S$ transitions from each other in the vacuum ultraviolet spectral region. Figure 15 shows the results of the polarization data from such angular distribution measurements. Obviously, the lack of separation of the vacuum ultraviolet lines in the experiment of Mumma *et al.*[44] makes a comparison with theoretical predictions rather difficult; however, the

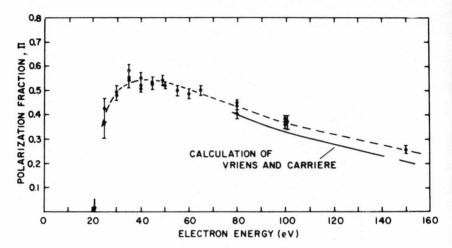

Figure 15. Polarization curve of the collective $\sum_n n^1P \to 1^1S$ transition of helium as investigated by Mumma *et al.*[44] and compared with the calculation of Vriens and Carriere.[45]

polarization of the collective $\sum_n n^1P \to 1^1S$ transitions might become independent of n at hither energies as indicated in comparing the experimental data with the theoretical predictions of Vriens and Carriere.[45]

Oscillatory structure, as seen in the polarization curves of helium and mercury, has also been detected in an electron ion excitation process. Figure 16 shows a striking feature in the polarization of the 4554-Å Ba^+ $2^2P_{3/2} \to 2^2S_{1/2}$ resonance line (Crandall *et al.*[28]).

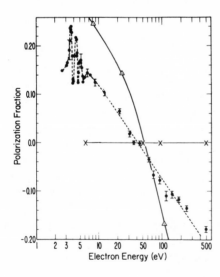

Figure 16. Polarization of Be^+ ion lines as observed by Crandall *et al.*[46] $\circleddash$: polarization of 4554-Å radiation $(6^2P_{3/2} \to 6^2S_{1/2})$; ×: polarization of 4934-Å radiation $(6^2P_{1/2} \to 6^2S_{1/2})$; $\triangle$: polarization of the mixture of 4131-Å radiation $(6^2D_{5/2} \to 6^2P_{3/2})$ with 4166-Å radiation $(6^2D_{3/2} \to 6^2P_{3/2})$. Bars shown are 1 rms error added to a small systematic error.

2.3. Electron–Photon Angular Correlations

2.3.1. Coherence Effects and Vector Polarization in Impact Excitation

Recently, a development in the physics of collisional excitation of atoms has emerged that can be characterized by the topics in the title of this chapter. It has been demonstrated in an experiment by Standage and Kleinpoppen[49] that quantities related to the above topics can lead to a complete analysis of electron–photon angular correlations from electron impact excitation of atoms. Part of the analysis of such experiments follows from the theoretical formulation of Fano and Macek[39] with regard to extracting alignment and orientation data from radiating atoms; the analysis of coherence parameters in conjunction with photon vector polarization is based upon the work of Wolf[50] and Blum and Kleinpoppen.[51]

Obviously, problems of coherence, orientation, and alignment go along with important applications in various subfields of spectroscopy. There are, however, significant differences between photon and electron impact excitation processes, which simply result from the difference in the nature of the two particles. Coherence effects in photon excitation of atoms occur in connection with a variety of processes such as level crossing (including zero-field level crossings, i.e., Hanle effect), pulsed and modulated photon excitation, quantum beat processes, etc. Such processes are well summarized (e.g., see Chapters 9 and 14). Photons with nonzero helicity can directly transfer orbital angular momentum to the excited atom parallel or antiparallel to the direction of the photon propagation. The excited atoms will, therefore, be orientated. In electron impact experiments with cylindrical symmetry around the electron beam axis, the atomic ensemble cannot be orientated in any direction because, on the one hand, the electron has no orbital angular momentum in its direction of travel (zero-helicity particle), and on the other hand, no orbital angular momentum transfer or orientation of the excited atom parallel to the electron beam direction is possible. Furthermore, it is expected that electron impact excitation with cylindrical symmetry is an incoherent process since the electrons are inelastically scattered all over 4π with excitation amplitudes depending on the scattering angle. However, coherence of the excitation process is conceivable with pulsed electron impact excitation and anisotropic scattering and phase distribution of the scattering amplitudes. The essence of this new type of electron impact excitation experiments based on electron–photon coincidences is the detection of completely coherent excitation, the direct measurement of quantum mechanical excitation amplitudes, the observation of circular and vector polarization, and also of the orientation of atoms (in the traditional type of impact excitation experiments, only alignment but no orientation can be produced).

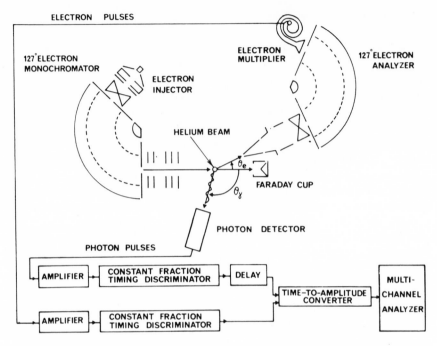

Figure 17. Schematic scheme of an apparatus for electron–photon coincidence measurements with the atomic beam (helium atoms) emerging perpendicular to the scattering plane.

In the following description of electron–photon coincidence experiments, we briefly refer to the analysis of scattering amplitudes, coherence correlation, and the connection between the anisotropy of the photon radiation and the atomic target parameters (orientation and alignment).

The experimental scheme for electron–photon measurements (see Figure 17) has been described in detail elsewhere (King *et al*,[52] Eminyan *et al*,[53] Standage and Kleinpoppen[49]). The method consists of crossing an atomic beam with an energy-selected electron beam and observing delayed electron–photon coincidences between electrons scattered inelastically at a particular angle and the photons resulting from the decay of the excited state. Those electron–photon pairs originating from a single excitation event are correlated in time, whereas uncorrelated events form a uniform background of random electron–photon coincidences.

The assumptions used in the analysis of electron–photon angular correlations from electron impact excitations of atoms are closely related to problems of coherency and the quantum mechanical theory of the measurement of interacting electrons and atoms.[73] Relevant basic concepts concerning these aspects may be found in Fano's paper[65] on density matrix formalism. (See also Chapter 2.) We assume that the electrons and the atoms are in "pure states" before the excitation. The state of the electron is then represented by the state vector $|P_0 m_0\rangle$, i.e., electrons with linear momentum P_0 and spin component m_0 (completely polarized electrons). The atoms are assumed to be in their ground state with orbital angular momentum $L = 0$, spin S_0, and atomic spin component M_0. The initial state is then characterized by $T_0 = S_0 M_0 P_0 m_0$; after the impact excitation the final state with the quantum numbers $Y_1 = L M_L S_1 M_1 P_1 m_1$ is represented by

$$|\psi(m_0 M_0)\rangle = \sum_{M_L M_1 m_1} a(T_0, T_1)|L M_L S_1 M_1\rangle|P_1 m_1\rangle \qquad (13)$$

The coefficients $a(T_0, T_1)$ depend on the variables of both the electron and the atom. We assume definite values for $\mathbf{P}_1$, L, and S_1. $|\psi(m_0 M_0)\rangle$ is the state vector of the combined system (inelastically scattered electron and excited atom) with definite initial values of m_0 and M_0. It is not possible to assign a single-state vector to either of the two ensembles having interacted with each other since, in general, only the combined system is in a pure state and can be characterized by a single-state vector.

In the electron–photon coincidence measurements carried out so far, unpolarized electrons and atoms were used. Accordingly, the initial state is not a pure quantum state and necessarily the final state is a mixed state as well, represented by a density matrix ρ_1 rather than by a linear superposition of states. Its elements are given by

$$\langle L M_L', S_1 M_1' ; m_1'|\rho_1|L M_L, S_1 M_1; m_1\rangle = \frac{1}{2(2S_0 + 1)} \sum_{m_0 M_0} a(\Gamma_0, \Gamma_1') a(\Gamma_0, \Gamma_1)^* \qquad (14)$$

The elements of the density matrix ρ_A describing only the atomic subensemble of combined system (with S_1) can be derived as follows:

$$\langle L M_L', S_1 M_1'|\rho_A|L M_L, S_1 M_1\rangle = \frac{1}{2(2S_0 + 1)(2S_1 + 1)} \sum_S (2S + 1) f_{M_L'}^{(S)} f_{M_L}^{(S)*} \delta_{M_1 M_1'} \qquad (15)$$

with electron amplitudes $f_{m_L}^{(S)}$ for the magnetic sublevels M_L in the different spin channels (S total spin). Following Eq. (15) ρ_A is diagonal in M_1, which means that the atomic ensemble is in an incoherent superposition of spin states. With the standard decomposition of amplitudes (see, e.g., Ref. 74) the state vector can be written as

$$|\psi(m_0, M_0)\rangle = \sum_{M_L S M_S} f_{M_L}^{(S)}|L M_L\rangle|P_1\rangle|S M_S\rangle \qquad (16)$$

where the atomic and spin functions are coupled to the total spin function $|SM_S\rangle$.

If the initial atom has spin zero ($S_0 = 0$), e.g., singlet or triplet excitation from the ground state, we have only one spin channel with $S = 1/2$. In this case the spin part of the state vector can be separated from the rest of the state vector

$$|\psi(m_0)\rangle = \left\{ \sum_{M_L} f^{(S)}_{M_L} |LM_L\rangle |P_1\rangle \right\} |S = \tfrac{1}{2}, M_0 = M_S\rangle \qquad (17)$$

Accordingly the orbital part of the wave function of the excited atoms can be written as a linear superposition of the states $|LM_L\rangle$ or, in other words, the atomic ensemble is "coherently excited" in M_L. In general, however, there are two different spin channels, $S = S_0 \pm \tfrac{1}{2}$, and because of the dependence of the scattering amplitudes on S there is only partial coherence in M_L. For this general case it clearly is not possible to write the atomic state as a coherent superposition of states $|LM_L S_1 M_1\rangle$.

We now describe the first experimental results which give evidence for completely coherent excitation of the ^1P state of helium. According to Eq. (17) the atomic ensemble in the excited ^1P state can be represented as coherent superposition of their magnetic sublevels under the same conditions for the excitation process, i.e., P_0 and P_1 have to be fixed. Practically, this means fixed electron energy and electron scattering angle: $|\psi(^1P)\rangle = \sum_{M_L} a_{M_L} \psi_{M_L}$. By normalizing $\langle \psi(^1P)|\psi(^1P)\rangle = \sigma(E_0, \theta)$ to the differential inelastic excitation cross section for exciting a ^1P state we associate the partial excitation amplitudes a_{M_L} with the partial cross sections $\sigma_0 = |a_0|^2$ and $\sigma_1 = |a_1|^2$, and with $a_1 = -a_{-1}$ (parity invariance), we have $\sigma = \sigma_0 + 2\sigma_1$. Coherence in the ^1P state is equivalent to the linear superposition of the magnetic substates in the above Ansatz, which requires a fixed phase relation between the two amplitudes a_0 and a_1.

The proof of the validity of the above theory of coherent substate excitation is based upon the following arguments.[54] If completely coherent excitation of the helium ^1P substates occurs, no phase randomness should exist between the excitation amplitudes a_0 and a_1. The fixed phase relation of the excitation amplitudes has the consequence that the photon radiation from the deexcitation process $^1P \to {}^1S$ should be governed by nonrandomness of the phase between the excitation amplitudes. The state or degree of coherency of the excitation process must then be related to the characteristics of the coherency of the photon radiation. In order to disentangle problems related to the coherence in electron impact, Standage and Kleinpoppen[49] carried out the following experiments. The characteristics of the photon radiation can be completely represented by the "coherence matrix" or, equivalently, by the density matrix of the photon.

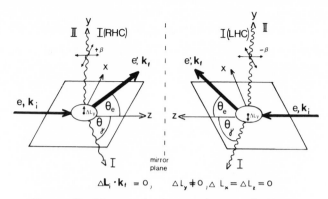

$$\Delta \mathbf{L}_i \cdot \mathbf{k}_f = 0, \qquad \Delta L_y \neq 0, \Delta L_x = \Delta L_z = 0$$

Figure 18. Scheme of electron–photon coincidence experiments including mirror symmetry operations. Positions I and II indicate, respectively, photon detection in or perpendicular to the scattering plane as defined by the wave vectors $\mathbf{k}_i$ and $\mathbf{k}_f$. Right-handed circularly polarized light I(RHC) changes to left-handed circularly polarized light I(LHC) in the mirror system. Parity invariance requires pseudoscalar $\Delta \mathbf{L}_i \cdot \mathbf{k}_f$ to vanish ($\Delta \mathbf{L}_i$ is the orbital angular momentum component of atom in excited state).

The state of coherence of the coincident photon radiation was investigated perpendicular to the scattering plane (Figure 18). The coherence matrix is then determined by the linearly polarized light vectors with components in the z and x directions ($E_z = E_{z0}\, e^{-i(\omega\tau - \phi_1)}$, $E_x = E_{z0}\, e^{-i(\omega\tau - \phi_2)}$):

$$\mathbf{J} = \begin{pmatrix} E_z E_z^* & E_z E_x^* \\ E_x E_z^* & E_x E_x \end{pmatrix} = \begin{pmatrix} J_{zz} & J_{zx} \\ J_{xz} & J_{xx} \end{pmatrix} \tag{18}$$

By normalizing the trace of the matrix to the total intensity of the light to unity, a "coherence correlation factor" can be defined as follows:

$$\mu_{zx} = |\mu_{zx}|\, e^{i\beta_{zx}} = \frac{J_{zx}}{(J_{zz}J_{xx})^{1/2}} \tag{19}$$

$|\mu_{zx}|$ is the "degree of coherence" and β_{zx} the "effective phase difference" of the photon radiation. For fixed phase relation between the two orthogonal light vectors $|\mu_{zx}|$ and β_{zx} become $|\mu_{zx}| = 1$ and $\beta_{zx} = \phi_1 - \phi_2$. Randomness in $(\phi_1 - \phi_2)$ will result in $|\mu_{zx}| = 0$ (completely incoherent radiation). As first shown and reported in detail by Standage and Klein-poppen,[49] the coherence correlation factor can be determined from the measurement of the Stokes parameter P_1, P_2, and P_3 (normalizing the total photon intensity to unity) as follows: $\mu = (P_2 - iP_3)/(1 - P_1^2)^{1/2}$.

The Stokes parameters are defined in connection with the photon radiation observed perpendicular to the scattering plane:

$$P_1 = I(0°) - I(90°) = J_{zz} - J_{xx}, \qquad P_2 = I(45°) - I(135°) = J_{zx} + J_{zx}$$

$$P_3 = I(\text{RHC}) - I(\text{LHC}) = i(J_{xz} - J_{zx})$$

where $I(\alpha = 0°, 45°, 90°, \text{or } 135°)$ are the linearly polarized light intensities at the angle α with reference to the z direction and where $I(\text{RHC, LHC})$ are the right- or left-handed circular polarized components of the photon radiation. Using the Stokes parameters a further quantity that characterizes the coherency of a photon beam is the "vector polarization" $\mathbf{P}$ with the three

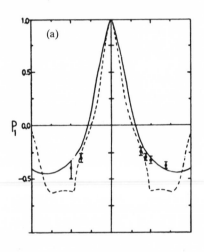

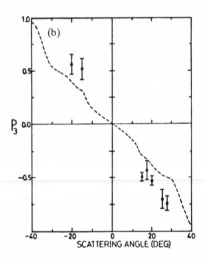

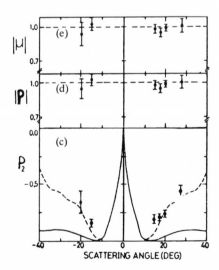

Figure 19. (a)–(c) Vector polarization components P_1, P_3, and P_2, respectively, of the He 3^1P-2^1S (5016-Å) coincident photons observed perpendicular to scattering plane (position II of Figure 1); experimental data with error bars (Standage and Kleinpoppen[49]) at 80 eV incident electron energy versus electron scattering angle. Solid line: first Born approximation (note that Born approximation predicts zero phase for χ and hence no orientation and circular polarization). Dashed line: multichannel eikonal $E4$ approximation (private communication from Professor McCann to Professors McDowell, Flannery, and McCann[59]). (d) Magnitude of vector polarization. (e) Degree of coherence.

Table 3. Various Data^a for the He $3\,^1P$ Excitations Electron Scattering Angles θ_e

θ_e (deg)	λ	$\lvert\chi\rvert$ (rad)	β (rad)	$\lvert O_{1-}^{col}\rvert$	$P_3/2\,(=O_{1-}^{col})$
−20	0.30±0.05	0.76±0.14	0.69±0.13	0.31±0.05	0.28±0.05
−15	0.35±0.02	0.49±0.07	0.55±0.09	0.23±0.03	0.26±0.05
15	0.38±0.02	0.58±0.07	−0.55±0.04	0.27±0.03	−0.25±0.04
	(0.38±0.01)	(0.55±0.02)		(0.25±0.01)	
17.5	0.36±0.02	0.62±0.06	−0.51±0.09	0.28±0.02	−0.22±0.05
20	0.34±0.02	0.63±0.05	−0.62±0.04	0.28±0.02	−0.27±0.02
	(0.36±0.02)	(0.64±0.05)		(0.29±0.02)	
27.5	0.36±0.02	0.98±0.07	−0.93±0.06	0.04±0.03	−0.38±0.04

^{a} Explained in text.
^{b} Incident electron energy 80 eV. Data in parentheses are from $3\,^1P-1\,^1S$ (537 Å) angular correlation measurements.[49] Note that $\lvert O_{1-}^{col}\rvert$ is calculable from λ and $\lvert\chi\rvert$ data (see, e.g., Ref. 49).

Stokes parameters as components of a three-dimensional vector $\mathbf{P} = (P_1, P_2, P_3)$. The magnitude of $\mathbf{P}$ is the "degree of polarization" of a photon beam: $\lvert\mathbf{P}\rvert = (\lvert P_1\rvert^2 + \lvert P_2\rvert^2 + \lvert P_3\rvert^2).^{1/2}$ A photon beam is completely coherent if and only if the degree of coherence and the degree of polarization are equal to unity.

A definite advantage in using the coherence correlation factor compared to the polarization vector results from the information obtained about the effective phase difference between two orthogonal light vectors. Standage and Kleinpoppen[49] determined μ and $\mathbf{P}$ for the helium $3\,^1P \rightarrow 2\,^1S$ (5016-Å) transition. Figure 19 presents data for P_1, P_2, P_3, $\mathbf{P}$, and μ. The fact that $\lvert\mathbf{P}\rvert = \lvert\mu\rvert = 1$ is within the error bars confirms that with the special conditions of the excitation under consideration, the photon radiation is completely coherent (100% degree of coherence!). Note the interesting fact in Figure 19 that the circular polarization changes sign as the electron scattering angle is changed from positive to negative. This is expected in accordance with parity invariance of the excitation process.

Table 3 displays the dependence of the effective phase β as a function of scattering angle for an incident electron energy of 80 eV. The fact is that $\lvert\mu\rvert = 1$ with a fixed phase between the two light vectors has to be interpreted as resulting from the excitation process of a single helium atom which appears as a source for emission of completely coherent line radiation ($\lambda = 5016$ Å). Taking up the model of coherent excitation of the magnetic sublevels by representing the excited ^1P state by a linear superposition of magnetic substates [according to Eq. (17)] the following conclusions can be drawn leading to further experimental tests:

By assigning a phase difference χ between the two excitation amplitudes a_0 and a_1 the probability of observing a photon in the scattering plane (position I in Figure 18) in the angular interval $\theta_\gamma\,d\Omega_\gamma$ for fixed

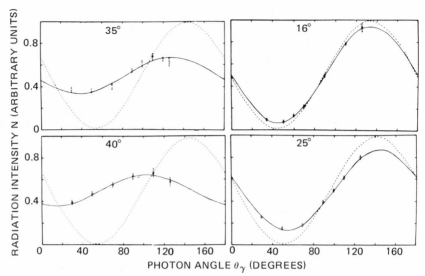

Figure 20. Electron–photon angular correlations from the helium excitation–deexcitation process $1^1S \to 2^1P \to 1^1S$ for the four different angles at 60 eV. Experimental data ($\dot{Q}$) with 1-rms error bars; full curve, least-squares fit of angular correlation function N [Eq. (20)] to experimental data; dashed curve, first Born approximation.

electron scattering angle is given by $dW = (3/8\pi)N \, d\Omega_\gamma$ with

$$N = \lambda \sin^2 \theta_\gamma + (1 - \lambda) \cos^2 \theta_\gamma - 2[\lambda(1 - \lambda)]^{1/2} \cos \chi \sin \theta_\gamma \cos \theta_\gamma \quad (20)$$

where N is the angular correlation function and $\lambda = \sigma_0/\sigma = |a_0|^2/(|a_0|^2 + 2|a_1|^2)$. Examples of typical measurements of electron–photon angular correlations are given in Figure 20. A least-squares fit of the angular correlation function to the data is also included in the figure, allowing the extraction of values for λ and χ. As can be seen from Table 3 the values for $|\chi|$ and β are identical within the error bars. It should, however, be reemphasized that the extraction of $|\chi|$ parameters from the angular correlation data in the scattering plane is based upon the assumption of coherent excitation, whereas data for β are obtained from a full set of measurements of Stokes parameters independent of any model of the atomic excitation process. Obviously the identity of $|\beta| = |\chi|$ for the excitation studied implies that the *phase χ between two quantum mechanical excitation amplitudes appears as a macroscopic measurable phase difference between two orthogonal light vectors.* Vector polarization and coherence parameters of the coincident radiation from the above excitation process provide direct and unambiguous tests for the coherent nature of the excitation of the helium atom at the given excitation energy. Very interesting coherence effects in the electron impact excitation of different l states in the $n = 3$ shell of atomic hydrogen have also been demonstrated by Mahan, Krotkov, Gallagher, and Smith.[68-71] By applying an axial electric field along the electron beam, the authors found that the Balmer-α line intensity, emitted perpendicular to the axis, has an asymmetric dependence with regard to the direction of the applied field. This electric field asymmetry effect on the Balmer-α line is interpreted in terms of coherent excitation of the different orbital angular momentum states (S, P, and D states). The observed asymmetry is the result of coherent excitation of states of coherent excitation of opposite parity, predominantly the pairs of states $P_{3/2,3/2}-D_{3/2,3/2}$, $P_{3/2,1/2}-D_{3/2,1/2}$, and $S_{1/2,1/2}-P_{1/2,1/2}$.

Morgan and McDowell[72] have dealt with the theory of coherent excitation of nP states of atomic hydrogen by which photons from deexcitations $nP \rightarrow nS$ are detected in coincidence with the inelastically scattered electrons. The authors showed that electron–photon coincidence measurements (e.g., Lyman-α photons in coincidence with the inelastically scattered electrons) of such processes do not allow a measurement of the relative phase of the m_L components of the nP state unless the singlet and triplet scattering can be separated from each other by using spin-polarized electrons and atoms.

2.3.2. Substate—Cross Sections, Phases of Excitation Amplitudes, and Target Parameters (Alignment and Orientation) in Impact Excitation

Having established, by the measurement of the coherence parameters, the electron–photon coincidence experiment as a feasible method, its observables and derived quantities represent a potentially more thorough test of electron–atom collision theory than traditional tests using total or differential cross sections only. It is possible to extract $\lambda = \sigma_0/\sigma$ and χ data for different electron energies and scattering angles from angular correlation measurements as displayed in Figure 20. Examples for a variety of λ and χ data are given in Figure 21.

By using the equations $\lambda = \sigma_0/\sigma$ and $\sigma = \sigma_0 + 2\sigma_1$ and by applying absolute cross-section measurements for the differential cross sections of the 2^1P and 3^1P excitation of helium, Chutjian[62] translated the λ data into absolute partial cross sections σ_0 and σ_1. For comparison we present σ, σ_0, $\sigma_1' = 2\sigma$, and $|\chi|$ for the $1^1S \rightarrow 3^1P \rightarrow 2^1S$ excitation/de-excitation process at 80 eV (Figures 22–24). We do not intend to discuss or evaluate the quality of the different theoretical approximations (discussions on such topics may be found in Refs. 62 and 58, for example) but we note by comparing Figure 22 with Figures 23–24 that partial differential substate cross sections and phases between the amplitudes a_0 and a_1 appear to be

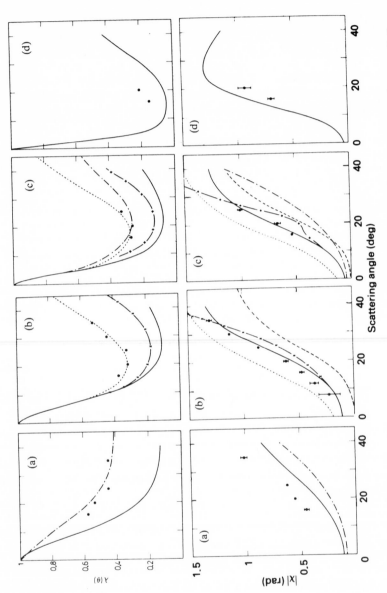

Figure 21. Angular correlation parameter $\lambda = \sigma_0/\sigma$ (top) and phase magnitude $|\chi|'$ (bottom) between substate excitation amplitudes a_0 and a_1 for the 2^1P excitation of helium by (a) 50-eV, (b) 80-eV, (c) 100-eV, and (d) 200-eV electron energy. Full curves: based upon impact parameter approximation of Berrington $et\,al.$;[57] chain curves: distorted-wave approximation of Bransden and Winter;[58] full curves with crosses: eikonal approximation of Flannery and McCann;[59] broken curves: eikonal distorted-wave approximation of Joachain and Vanderpoorten;[60] dotted curves: McCann;[59] broken curves: eikonal distorted-wave approximation of Joachain and Vanderpoorten;[60] dotted curves: [61] Experimental data points.[53]

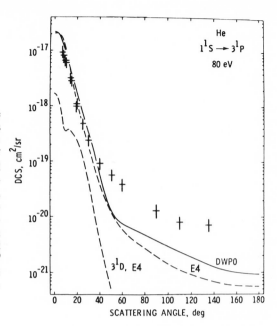

Figure 22. Experimental absolute differential cross sections (crosses) for the helium $1^1S \rightarrow 3^1P$ transition at 80 eV (Chutjian[62]). Theoretical multichannel data from Flannery and McCann[59] (E4, multichannel eikonal approximation—private communication from Professor McCann to Professor McDowell) and from Scot and McDowell[63] (DWPO, distorted-wave polarized orbital approximation).

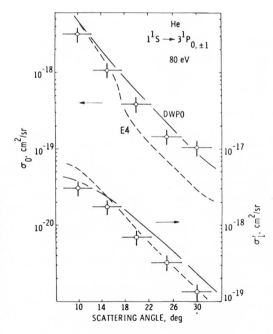

Figure 23. Experimental magnetic substate cross sections (crosses) for the helium $1^1S \rightarrow 3^1P$ transition at 80 eV (13,23), σ_0 is for $m_L = 0$ excitation, σ_1' is twice the cross section for $m_L = |1|$ excitation. Theoretical approximations as in Figure 22.

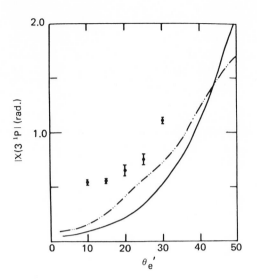

Figure 24. Magnitude of phase shift χ between excitation amplitudes a_0 and a_1 for helium $3\,^1P$ excitation at 80 eV as function of scattering angle. Experimental data with error bars,[56] $- \cdot\cdot - \cdot\cdot -$ E4 eikonal approximation[59] (private communication from Professor McCann to Professor McDowell), —— distorted-wave polarized orbital approximation,[63] (DWPO, I).

more sensitive test quantities for theoretical approximations than total differential or integral cross sections.

We would like to illustrate how the electron–photon coincidence method is capable of providing more detailed information not only on scattering parameters but also on target parameters. By "target parameters" we mean the state in which an atom is, when excited by electron impact under the condition that all atoms have experienced the same excitation and scattering processes (see the above discussion). Of course, as proved above, the state of the excited atom is a "pure" state with a_0 and a_1 as coefficients of the coherent superposition of the magnetic substates. In addition, however, to the description of the target atom by the characteristics of a pure state, further useful target parameters are the Fano–Macek[39] orientation and alignment parameters or the multipole moments of the excited state.[51,64]

We restrict our discussion to applications of orientation (O_{1-}^{col}) and alignment parameters (A_{q+}^{col}) which are defined and calculable from λ and χ or from the Stokes parameters (P_1, P_2, P_3) as follows:

$$A_0^{col} = \tfrac{1}{2} < 3L_z^2 - L^2 \rangle = (1 - 3\lambda)/2 - (1 + 3P_1)/4$$

$$A_{1+}^{col} = \tfrac{1}{2}\langle 3L_xL_z + L_zL_x \rangle = \{\lambda(1 - \lambda)\}^{1/2} \cos \chi = -P_2/2$$

$$A_{2+}^{col} = \tfrac{1}{2}\langle L_x^2 - L_y^2 \rangle = \tfrac{1}{2}(\lambda - 1) = (P_1 - 1)/4 \qquad (21)$$

$$O_{1+}^{col} = \tfrac{1}{2}\langle L_y \rangle = -\{\lambda(1 - \lambda)\}^{1/2} \sin \chi = -P_3/2$$

A_{q+}^{col} are the components of an alignment tensor of rank 2, O_{1-}^{col} is an orientation vector which is directly related to the orbital angular momentum component transferred to the atom by the colliding electron. The relation between the orientation and alignment parameters on the one hand and the quantum mechanical expectation values for orbital angular momentum quantities ($\langle L_y \rangle, \cdots$) is based upon definition [see Eqs. (21)]. In Figures 25 and 26 we present examples of the alignment and orientation parameters. Note that orientation of magnitude 0.5 means that the atom is completely orientated, which is identical with 100% circular polarization of photons emitted parallel to the scattering normal.

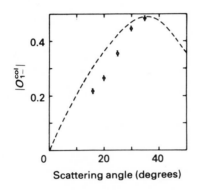

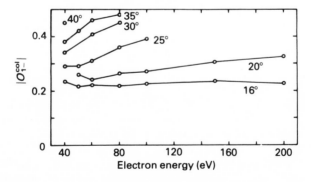

Figure 25. Variation of the Fano–Macek orientation parameter (in units of $\hbar/2$) with electron energy and different scattering angles (bottom), and with the scattering angle and fixed energy (80 eV) for the helium $1^1S \to 2^2P$ excitation process. The bottom diagram only includes the experimental data,[53] whereas the top diagram also displays theoretical predictions (dashed curve) for 78 eV by Madison and Shelton.[61]

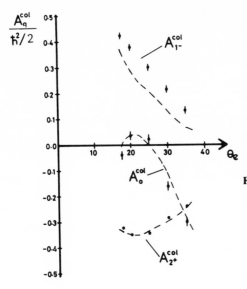

Figure 26. Alignment parameters A_0^{col}, A_1^{col}, and A_2^{col} (in units of $\hbar^2/2$) calculated from experimental λ and χ data[53] and from the distorted-wave approximation[61] (dashed curve) for the 2^1P excitation of helium at 80-eV excitation energy.

Finally, we note from the complete analysis of the polarization of the photons that a polarization vector of magnitude unity ($|P| = 1$) implies that "maximum knowledge" can be obtained for the state vectors of the photons and the excited atoms (see, e.g., Fano,[65] Macek,[66] or Blum[67]). Thus the electron–photon coincidence experiment for the helium ^1P excitation represents a strikingly simple example of the quantum mechanical linear superposition principle of pure states.

3. Heavy-Particle Impact Excitation

By heavy-particle excitations of atoms we will mean (a) direct excitation of atoms by impact of ions or atoms, (b) excitation of projectile atoms or ions in collisions with heavy particles (atoms, ions), (c) charge exchange processes of ions with neutral atoms or molecules.

The excitation processes listed above have successfully been applied in spectroscopy. For example, in the experiments on electric-field-induced quantum beats[73] of metastable atomic hydrogen the metastable atoms were produced by charge exchange of protons fired through a gas cell containing molecular hydrogen. In the first experiment[74] for detecting in coincidence the two-photon decay of hydrogen atoms in the $2S$ state, the metastables were produced by a resonant charge exchange process of protons passing through a cesium cloud. In connection with the excitation processes listed above, phenomena like polarization of line radiation or even coincident simultaneous excitations of colliding atoms have been observed recently. Another interesting phenomenon is the theoretical prediction that spin-polarized hydrogen atoms may be produced by charge exchange of protons with xenon atoms.[75] It will be by no means surprising if future investigations reveal further interesting phenomena connected with atomic excitation by heavy-particle collisions.

Although the goal of a unified theoretical understanding of heavy particle–atom collisions lies in the area of atomic collision processes, the development of "collisional spectroscopy" in atom–atom collisions shows characteristics of spectroscopy, e.g., structure in line intensities, line polarization, population ratio, curve crossings of atomic states, etc.

In the following sections we will only select examples displaying some typical characteristics of excitation processes by heavy-particle impact. We also will report the powerful two-step excitation process in which heavy-particle excitation is combined with laser excitation.

3.1. Excitation of Atomic Hydrogen and Helium Ions

Excitation of atomic hydrogen by heavy-particle impact has been successfully used in atomic spectroscopy for investigations in which either the total density of the population of excited states or where fast time resolution of the photon emission from the populated states was significant. In this section we first report on the near-resonant charge-exchange process between protons and cesium atoms resulting in the excitation of the metastable state of atomic hydrogen. This process is based upon the charge-exchange reaction $p + Cs \rightarrow H(2S) + Cs - 0.49$ eV. Following the historically first-discussed charge transfer theory[76] maximum cross section is expected when the transfer energy between the initial and final state is zero; this is nearly fulfilled in the above reaction for producing metastable hydrogen atoms (-0.49 eV is the difference between the excitation energy of the hydrogen $2S$ state and the ionization energy of cesium). Figure 27 shows the cross section of the above reaction in comparison with the total capture cross section for any state of the hydrogen atom. The fraction for the production of the metastables was determined by quenching the metastable $2S$ atoms in a static electric field and by measuring the Lyman-α quench radiation. It is worth noting that the above charge-exchange cross section of protons with cesium is substantially larger than those with atomic or molecular hydrogen or with the rare-gas atoms.

Polarization effects of the Lyman-α radiation in collisions of H^+ or D^+

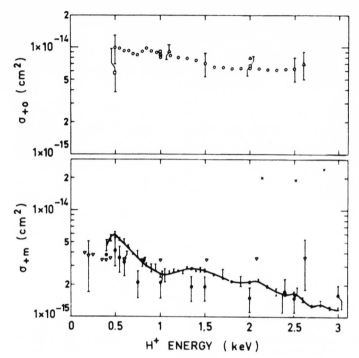

Figure 27. Cross sections (σ_{+m}) for H($2S$) production in p + Cs collisions with total electron capture cross section (σ_{+0}) for protons. ●: Tuan et al;[77] ○: Spiers[78]; □: Schlachter[79]; ×: Sellin and Granoff[80]; ▽: Donnally[81]; ─●─: Roussel.[82]

ions with rare-gas atoms have been reported by Gailey et al.[83] In their experiment the degree of polarization P was obtained from angular distribution measurements of the L_α photons. According to the connection of the Lyman-α polarization with the sublevel excitation cross sections Q_0 (partial cross section for magnetic sublevel excitation with $m_L = 0$) and Q_1 (partial cross section for magnetic sublevel $m_L = 1$) $P = \{Q_0 - Q_1\}/\{2.3575Q_0 + 3.749Q_1\}$ with the total capture cross section $Q = Q_0 + 2Q$. In Figure 28, data on polarization, sublevel, and total excitation cross sections for the P + Ne → H($2P$) + Ne$^+$ charge exchange excitation process are presented. A substantial difference in the population of the two types of magnetic sublevel occurs around 5 keV.

Remarkable progress has recently been made in exciting hydrogen atoms into states with high values of the quantum number n. Of particular interest is the experiment by Koch et al.[84] by which firstly a fast H($2S$)

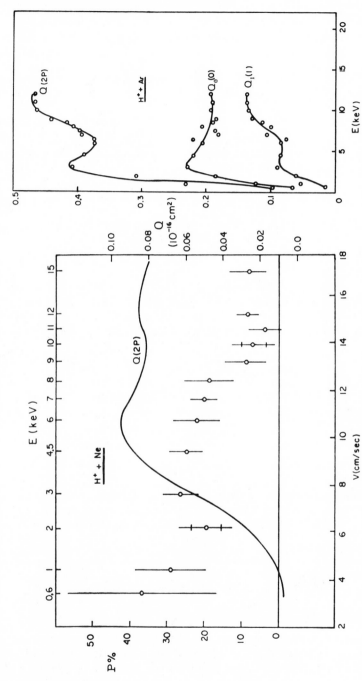

Figure 28. Polarization of Lyman-α radiation (Gailey et al.[83]) produced in proton–neon collisions (upper part). $Q(^2P)$ is the total cross section for the L_α production in the ^2P capture process; Q_0 and Q_1 are magnetic substate populations of $Q(^2P)$.

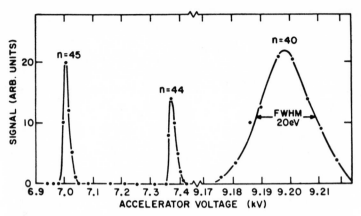

Figure 29. Ionization signal of the Doppler tuned laser excitation spectrum to highly excited states of atomic hydrogen. Charge $[H(2S) + h\nu_D \to H(n)$, Koch et al.[84]].

beam was produced by charge exchange collisions of a proton beam passing through an H_2 gas target and secondly the uv ArIII laser line at 3637.89 Å induced $H(2S) \to H$ (high n) transitions. With the laser beam propagating parallel to the keV-$H(2S)$ atomic beam, the wavelength of the laser line can be "tuned" over a wide range of exciting high n states simply by using the Doppler effect of the fast atomic $H(2S)$ beam. A "Doppler scan" of exciting states with $n = 40$, 44, and 45 (obtained by varying the proton energy before the charge exchange process takes place) is illustrated in Figure 29. The highly excited states were detected by tunnel-effect ionization of an electric field in which ionization of the highly excited state is achieved; in other words the signal in Figure 29 is proportional to the number of protons produced. Koch and Bayfield[85] performed experiments with hydrogen atoms of quantum numbers $4 < n < 80$ based upon the above production method of combining charge exchange processes with laser excitation. At $n = 65$ the electric field tunnel method of detecting highly excited states was absolute in the quantum number n with an uncertainty of, at most, $\Delta n = \pm 5$ (see also Chapter 16).

An interesting scaling law for the excitation cross section of the nS states in atomic hydrogen has been found which appears to be valid even as low as $n = 1$. Figure 30 gives results for the product $n^3\sigma(nS)$ for the excitation process $p + He \to H(nS) + He^+$. Note the independence of that product on n at large proton energies. Similar n-dependence studies for the direct excitation processes $p + H \to H^* + p$ at larger proton energies have been reported by Park et al.,[91] of which Figure 31 gives an example. They separated the excitation into the different n states by analyzing the energy loss of the protons with a $127°$ monochromator for protons. As seen from

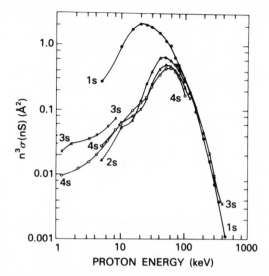

Figure 30. Charge transfer excitation cross section for $p +$ He $\rightarrow$ H$(nS)+$He$^+$ weighted by the factor n^3. The figure is taken from Bayfield.[86] Low-energy data from Dawson and Lloyd,[87] intermediate-energy data from Dawson and Lloyd,[87] high energy data from Ford and Thomas.[90]

Figure 31, the data for exciting atomic hydrogen to $n = 2$, 3, and 4 are very similar in shape with a peak at about 60 eV. By normalizing to the Born approximation, these data have been converted to absolute cross sections. At 200-keV proton energies such absolute cross sections were reported as $\sigma(n = 2) = (6.637 \pm 0.35)\, 10^{-17}$ cm^2, $\sigma(n = 3) = (1.41 \pm 0.2)\, 10^{-17}$ cm^2, and $\sigma\,(n = 4) = (0.77 \pm 0.3)\, 10^{-17}$ cm^2.

An interesting case of charge exchange excitation is the reaction

$$\text{He}^{++} + \text{H}(1S) \rightarrow \text{He}^+(2S) + \text{H}^+$$

for which the energy defect is zero; therefore, this reaction would appear to be a simple example of resonant charge transfer under exact conditions. However, as pointed out by Gilbody,[94] the Coulomb interaction between

Figure 31. Relative cross sections σ_2, σ_3, and σ_3 for excitation of atomic hydrogen to the $n = 2$, 3, and 4 states (Park *et al.*[91]). All experimental and theoretical data have been set equal to 1 at 200 keV. σ_2 and σ_3 relative Glauber approximation (Franco and Thomas[92]), $C7_2$ relative seven-state close-coupling calculation.[93] See text on absolute cross sections at 200 keV.

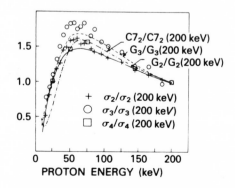

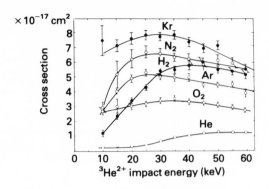

Figure 32. Cross sections for one-electron capture into the He^+ $(2^2S_{1/2})$ state by $^3He^{2+}$ ions passing through various gases (Shah and Gilbody[97]).

the charged collision products might be responsible for a "nonresonant" behavior of this reaction for which rather low cross sections have been measured not exceeding 10^{-16} cm^2 in the energy range from about 10 to 100 keV (Shah and Gilbody[95]).

Cross sections σ_{21}^* for $He^+(2S)$ formation by the process $He^{2+} + X \rightarrow He^+(2S) + X^+$ were determined together with total cross sections σ_{21} for one-electron capture $He^{2+} + X \rightarrow He^+ + X^+$ and σ_{20} for two-electron capture $He^2 + X \rightarrow He + X^{2+}$ (Shah and Gilbody[96,97]). Figures 32 and 33 present results for a large variety of atoms and molecules. As pointed out by the authors, one of the most striking features of the results for σ_{21}^* and σ_{21} is that the energy dependence of the cross section in various gases is not greatly different in spite of the large difference in the energy defects of the above processes. It is suspected that the markedly different energy depen-

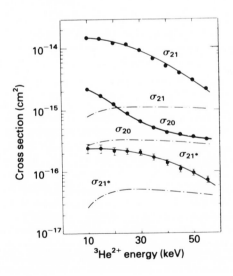

Figure 33. Cross sections σ_{21}^*, σ_{21}, and σ_{20} (see text) for formation of $H^+(2S)$, He^+, and He in collisions of $^3He^{2+}$ with potassium ($-\bullet-$)[96] and argon ($-\cdot-$).[97]

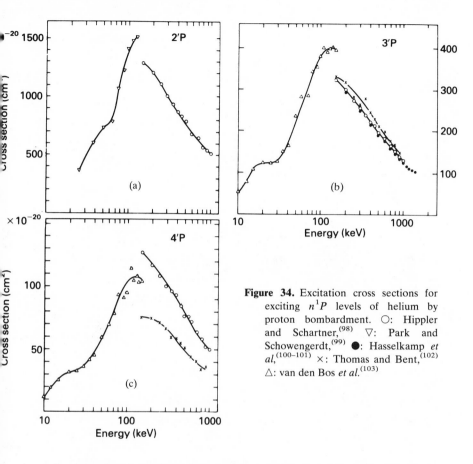

Figure 34. Excitation cross sections for exciting n^1P levels of helium by proton bombardment. ○: Hippler and Schartner,[98] ▽: Park and Schowengerdt,[99] ●: Hasselkamp *et al*,[100–101] ×: Thomas and Bent,[102] △: van den Bos *et al*.[103]

dence in potassium is attributable, in part, to the influence of pseudocrossings of the potential curves involved in the process.

As pointed out by Shah and Gilbody,[97] formation and destruction processes of fast $He^+(2S)$ ions may well play an important role in producing a nuclear polarized $^3He^+$ ion source (or also an electron-spin polarized $^4He^+$ source). It has been found, by these authors, that the cross section for conversion of $^3He^+(2S)$ ions into $^3He^{2+}$ (which may provide polarized nuclei through level-crossing effects according to the Lamb-shift method[11]) is considerably larger than that for $He^+(1S)$ in gases like H_2, He, Ar, N_2, and O_2.

3.2. Ion Impact Excitation of Other Atoms

While heavy-particle impact excitation processes have been important in particular in connection with spectroscopic applications of hydrogen

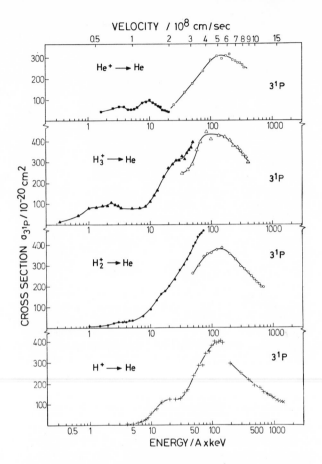

Figure 35. Excitation cross section for the 3^1P state of helium
excited by various ions. The abscissa is plotted in units of
$V = 4.377 \ 10^7 \ (E/A)^{1/2}$ with E in keV for projectile
energy and A in atomic mass units. $\times \bigcirc \ \triangle$: Hasselkamp;[104] +: van den Bos *et al.*[131] (up to 150 keV),
Hasselkamp *et al.*[100] (200 keV–1 MeV); $\square$: Hasselkamp
et al.[132]; ● ▲: van den Bos *et al.*[133]; ■: de Heer and
van den Bos.[134]

atoms, a large amount of experimental data is now available for heavy-
particle impact excitation of other than hydrogen atoms.* In this section we
select only examples that are significant in connection with spectroscopic
problems (line intensity and line polarization, for example). In a series of

---

* A review up to 1971 was given by E. W. Thomas (*Excitation in Heavy Particle Collisions*,
New York, Wiley-Interscience, 1972).

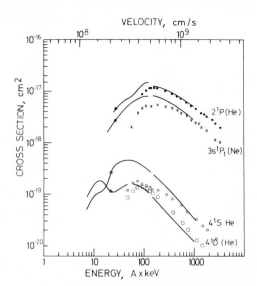

Figure 36. Comparison of proton and electron impact excitation of helium and neon; — · —, —O— proton impact data, other data from electron impact (figure courtesy of K.-H. Schartner[105]).

figures we concentrate first on the excitation of the n^1P states of helium atoms. Figure 34 presents excitation cross sections for the states 2^1P, 3^1P, and 4^1P by proton impact. Note the fact that maxima of the cross sections occur at rather high energies (>100 keV). Hippler and Schartner[98] also applied the Bethe–Fano plot for these transitions (i.e., $QE \propto f_n \ln \alpha E$, where Q is the total cross section, f_n is the optical oscillator strength, E is the energy of incoming protons, and α is a constant). Although the linear relationship between QE and $\ln \alpha E$ was quite well approached by their data, the f_n values extracted from their data appeared to be somewhat too high, and for $n = 3$ they were in disagreement with the theoretical values of oscillator strength.

An interesting aspect of collisional excitation with heavy particles is the fact that the velocity rather than the energy is the decisive quantity for the dependence of the cross section at large velocities. The velocity dependence of cross sections displays similar features for different ions (see Figure 35) or even for ions like protons in comparison to electron impact excitation (Figure 36). With decreasing velocities, charge exchange processes become increasingly important (Figure 37). At low velocities the excitation functions often show more or less regular oscillations, which can in simple cases be explained by the formation of transient molecules.

An interesting fact should be known to the spectroscopist concerning the excitation of atoms with no spin–orbit interaction by bare nuclei. In these cases it follows from the so-called "Wigner spin rule" that the multiciplicity of the atomic state before the collision must be maintained during the excitation process.[138] This has the important consequence that

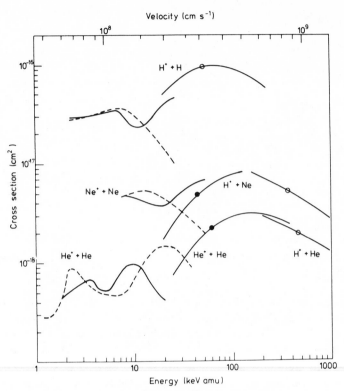

Figure 37. $Ne^+ + Ne$: Emission cross section for the target excitation
(——) and the projectile charge exchange excitation (– – –) of the
$Ne(3s\ ^1P_1)$ level as a function of the impact velocity. $H^+ + Ne$:
Excitation cross $(3s\ ^1P_1)$ for the target excitation of the $(3s\ ^1P_1)$
level of Ne excited by H^+ impact. —●—: York *et al.*[142] —○—:
Hippler and Schartner.[143] $H^+ + H$: Excitation cross section for
target excitation (——) and projectile charge-exchange excitation
(– – –) of the $H(2p)$ level excited in $H^+ + H$ collisions (Morgan *et
al.*[144]). —○—: Park *et al.*[145] $He^+ + He$: Excitation cross section for
the target excitation (——) and the projectile charge exchange excitation (– – –) of the $He(3^1P)$ level excited in $He^+ + He$ collisions
(Wolterbeek-Muller and de Heer,[116] de Heer and van den
Bos;[146] —●—: Hasselkamp *et al.*[147] $H^+ + He$: Excitation cross
section for the target excitation of the $He(3^1P)$ level of H^+ impact:
—○—: Hasselkamp *et al.*[147] (The above figure is taken from Schartner *et al.*[148])

the excitation from the ground state to the triplet system of helium by H^+
or He^{2+} ions is formally forbidden (Gray *et al.*[139]).

Extensive studies on the polarization of heavy-particle impact processes have been reported. The formal description of such polarization

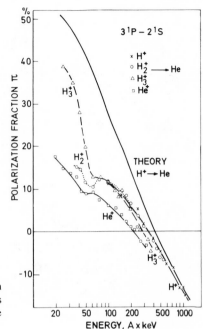

Figure 38. Polarization of 3^1P–2^1S transition in helium for different projectile particles (Hasselkamp,[104] theory[105–107]); A is the atomic mass number.

phenomena can be based on the theoretical approach by Percival and Seaton,[22] taking into account incoherent excitation of the magnetic sublevels, which, in cases of unequal population of the sublevels, can lead to polarization of lines emitted from states with $l \neq 0$. Figure 38 gives an example for the polarization of the helium $3^1P - 2^1S$ transition excited by various ions. Note that the highest possible degree of polarization would be 100% when only the sublevel $m_L = 0$ is excited.

Amongst the many further examples and features of heavy-particle atom collisions, the oscillatory structure of excitation cross sections at low energy particularly attracts attention. A typical example of this feature for the excitation of helium by helium ions at low energy is presented in Figure 39. Such oscillations in the cross sections can at least qualitatively be understood at present. The phenomena are related to crossings and pseudocrossings between the molecular potentials of the collision partners (Figures 39 and 40). Transitions from the ground to the excited state take place near such crossings, which provide the primary mechanism of excitation of these collisions. Such nonadiabatic effects were first studied by Landau,[109] Zener,[110] and Stuekelberg[111] in connection with two potential curves crossing each other (Figure 40a). In the low-energy collisional processes of $He^+ + He$ the primary mechanism for populating excited states

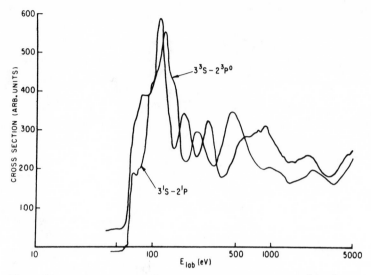

Figure 39. Relative cross section for the excitation of the He $3^3S_1-2^3P$ and $3^1S_0-2^1P$ transition by He$^+$ impact (Dworetsky *et al.*[(108)]).

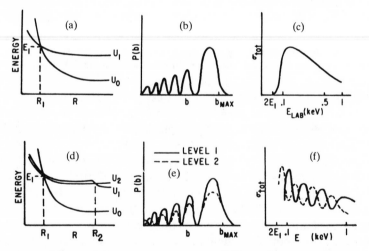

Figure 40. Diagrammatic representation of single-crossing (a) and two-crossing models (d) for ion–atom excitation (Rosenthal–Foley model). (b) Excitation probability at fixed energy and based upon single-crossing model as a function of impact parameter. (c) Total excitation probability as a function of incident energy for single-crossing model. (e) Excitation probabilities for two-crossing model at fixed energy as a function of impact parameter. (f) Total excitation probabilities for two-crossing model as a function of incident energy.

is provided through a pseudocrossing of the excited-state $^2\Sigma_g$ potentials of He_2^+ with the ground-state $^2\Sigma_g$ potential. As pointed out by Lichten,[112] the $^2\Sigma_g$ ground-state potential diabatically crosses all inelastic potentials. Rosenthal and Foley[113] explained the oscillatory behavior of the $He^+ +$ He excitations by a phase interference between the inelastic and the elastic potential at the crossing point of the two potentials of the $He + He^+$ system. Such processes can be treated with the Landau–Zener–Stueckelberg theory, which, however, provided evidence for oscillatory structure only in the differential cross section; in other words, oscillatory structure does not show up in the total cross section (Figure 40). Rosenthal and Foley,[113] however, developed a model that reproduced the oscillatory structure in the total cross section. Their basic assumption was that two inelastic channels are involved in a pseudocrossing with the elastic channel of the potential curves (Figure 40a). In that case the two amplitudes of the inelastic channels are coherently mixed, and this mixing manifests itself as an interference effect in the final populations of the excited states. The two inelastic levels are first populated by the crossings with the elastic potential near $R \approx R_1$ (Figure 40d). Hence, as the collision partners separate from each other, they pseudocross at R_2 (leading to Landau–Zener–type nonadiabatic transitions) and the amplitudes of the two inelastic channels are coherently mixed. Accordingly, the critical phase of this interference effect is the phase difference between the two inelastic amplitudes at the distance R_2 between the two collision partners. Based upon a model calculation in the WKB approximation by Rosenthal[114] the physical reason that the interference mechanism leads to an oscillatory total cross section is the fact that the phase difference of the two inelastic amplitudes is almost independent of the impact parameter. This phase is a function of the kinetic energy of the internuclear motion. While the internuclear separation R_1 is crossed twice, it is only as the collision partners separate that the interference takes place, since the inelastic amplitudes are zero on the way in. The Rosenthal–Foley model not only predicts structure in the cross section but also reproduces the anticoincidence for maxima and minima of the 3^3S and 3^1S cross section of helium (Figure 39). It is also notable that the spacing of the oscillations is explained by the model: Experimentally the spacing of the successive minima and maxima of the excitation cross sections is linear when plotted against v^{-1}, the reciprocal of the velocity in the outgoing channel. As far as nonoscillatory cross sections were observed (e.g., for the 2^1P and 2^3P excitation[108] of He by He^+ impact) the outer-crossing model correctly predicts such a feature inasmuch as the excitation process operates through channels in which no efficient mixture of inelastic amplitudes occurs beyond the inner crossing region (i.e., either where there is no outer pseudocrossing or where such crossing is diabatic).

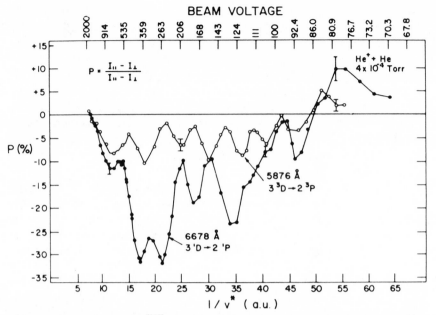

Figure 41. Linear polarization[118] of light emitted perpendicular to helium ion beam exciting helium atoms as a function of $1/v^*$ $[v = 2^{1/2}(E_{on} - Q)/\mu$ in atomic units]. $Q = 23.07$ eV, average excitation energy of $3^{3,1}D$ states, μ reduced mass of He–He$^+$.

Strong interference effects in the line polarization of the emitted light from He$^+$ excitations of the He $3^{1,3}D$ states has been observed[118] (Figure 41) for which the Rosenthal–Foley analysis appears to be applicable in order to explain the observed oscillatory structure of the line polarization (pseudocrossing of the $3^{1,3}D$ states at large internuclear distance). Oscil-

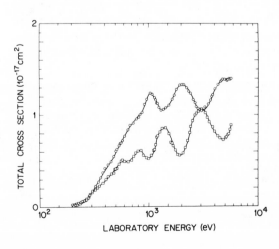

Figure 42. Absolute population cross section for Ne* (3P) levels (—□—□—) and Na* (3P) levels (—○—○—) (Tolk et al.[119]) in collisions between Na$^+$ ions and neon atoms.

latory behavior in the emission cross section for the 4 ^1S-state of He, excited in collisions with Ne$^+$, He$^+$, and H$^+$, has been reported by various authors and is summarized and compared by Hasselkamp *et al.*[117]

Further systematic studies of interference effects in optical excitation cross sections and in line polarization from collisions between various atoms and ions have been carried out by Tolk *et al.*[122] (Na$^+$ on Ne), by Anderson *et al.*[120] (all ions ranging from Li$^+$ to Al$^+$ on Ne), by Dworetsky *et al.*[108] (K$^+$ on Ar), by Zavilopulo *et al.*[141] (Mg$^+$ on Rb) and by Bobashev and Kritskii[140] and Tolk *et al.*[119] (Na$^+$ on Ne). Highly regular oscillatory structure found in the energy dependence of the emission cross sections for ten NeI ($3P \rightarrow 3S$) optical transitions is observed to be in antiphase with similar structure in the NaI ($3P \rightarrow 3S$) optical emission cross sections (see Figure 42). The radiation observed in this Na$^+$–Ne experiment arises from the excitation of both Ne and Na in the $3P$ electronic states as a result of direct and charge-exchange collision processes:

$$Na^+ + Ne \begin{array}{l} \nearrow Na^+ + Ne^*(3p) \\ \searrow Na^*(3p) + Ne^+ \end{array}$$

The analysis of the data by Tolk *et al.*[119] supports a model in which the oscillatory behavior of the excitation cross sections arises predominantly from the coherently excited molecular states $\pi(\Omega = \pm 1)$ and $^3\pi(\Omega = \pm 2)$ of

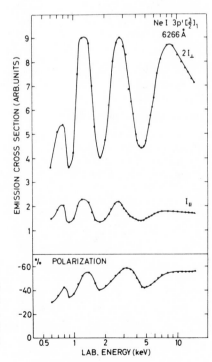

Figure 43. The two polarization components ($I_\perp$ and $I_\parallel$) of the total emission cross section and the polarization of the 6226-Å NeI radiation from the $3p'[3/2]_1$ level of neon in collisional Na$^+$–Ne excitation (Anderson *et al.*[120]).

the Na$^+$–Ne system, which are populated at small internuclear separations and which evolve independently as the collision partners recede and finally interact via a charge-transfer mechanism at large internuclear separations. According to this interpretation the oscillations in the excitation cross section are due to quantum mechanical interference between direct excitation of neon and charge exchange collisional excitation of sodium. Anderson *et al.* showed that the strong oscillating structure in the emission cross section of the Ne $3p^1$ (3/2) level is due almost entirely to the polarization component perpendicular to the ion beam direction (Figure 43). In the above study of Anderson *et al.* for collisional excitation of neon by ions from Li$^+$ to Al$^+$ strong oscillatory systems are N$^+$–Ne, O$^+$–Ne, Ne$^+$–Ne, and Mg$^+$–Ne; these systems show interference between direct excitation and charge-exchange collision excitation in the target (neon) and the projectile (the ions). The Li$^+$–Ne, Be$^+$–Ne, and Al$^+$–Ne systems all show a tendency towards weak oscillations in the emission cross sections for the $3P$ states of neon, but no regular oscillations occur. For the B$^+$–Ne, C$^+$–Ne, F$^+$–Ne or Ne$^+$–Ne systems, the emission cross sections in neon were similar and no tendency towards an oscillatory structure in the emission cross section for neon was observed. A discussion of different theoretical models for the interpretation of the oscillatory structure in the total cross sections has been given by Anderson *et al.*[120] Ankundinov *et al.*[121] attempted a quantitative theory based upon the Rosenthal and Foley[113] original model. On the basis of this model, Tolk *et al.*,[122] for instance, interpreted the oscillations in the excitation curve of the Na$^+$–Ne system by applying a

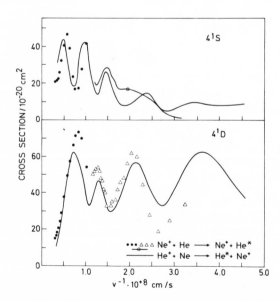

Figure 44. Excitation cross section for direct and charge transfer processes of the helium 4^1S and 4^1D state as a function of $1/v$ (v is the velocity Muller of the projectile particle). He$^+$ + Ne: — — Wolterbeek Muller and de Heer,[116] Ne + He: $\cdots$ Hasselkamp *et al.*,[108] $-\bigcirc-$, $\triangle\triangle\triangle$.[106]

molecular orbital–energy diagram, based upon the promotion model of Barat and Lichten.[123]

The presence of nonadiabatic outer crossings in excited states of atom–ion systems has led to a variety of studies of potentials between atomic collision partners. The whole field of this research is, however, very much in flux. As has been pointed out by Schartner,[105] the excitation of atoms by ion bombardment in the medium-energy range may need applications of molecular many-channel models combined with two-state atomic models to explain the v^{-1} dependence of the cross sections for exciting the 4^1S and 4^1D state in helium (Figure 44). For example, Larsen[115] used a model in which he applied a two-state approximation with a constant nonzero interaction during an interaction period and with no interaction outside this period.

3.3. Excitation from Atom–Atom Collisions

Considerable progress has recently been made in the study of impact excitation of colliding neutral atoms. Such measurements led to observations of excitation cross sections, line polarization, and even coherence effects in photon–photon coincidences from the emission of both the projectile and the target atom.

3.3.1. Linear Polarization of Line Radiation

In experiments on impact excitation of colliding atoms, monoenergetic projectile atoms are normally obtained by charge exchange of monoenergetic ions with atoms of a gaseous target. The excitation and polarization function or cross sections of atoms by impact with atoms can be obtained by a method similar to that used in the common experiments on impact excitation: The intensity of a given emission line is monitored as a function of the energy of the projectile atom. From many possible examples we select some of the results of Kempter and co-workers. These authors, particularly, investigated excitation cross sections and polarization fractions of various spectral lines in collisions between a target of rare-gas atoms and monoenergetic beams of potassium atoms. Figure 45 presents an example of the polarization of the $4^2P_{3/2} \rightarrow 4^2S_{1/2}$ transition of potassium atoms. Contrary to the majority of examples for electron impact excitation of atoms, the polarization in Figure 45 starts from very low values. For the colliding system K + He the polarization rises from about zero at 200 eV to its maximum within approximately 2 eV; at low energy the line polarization varies rapidly with the impact energy while it decreases monotonically at higher energies. It can be shown[124] that the

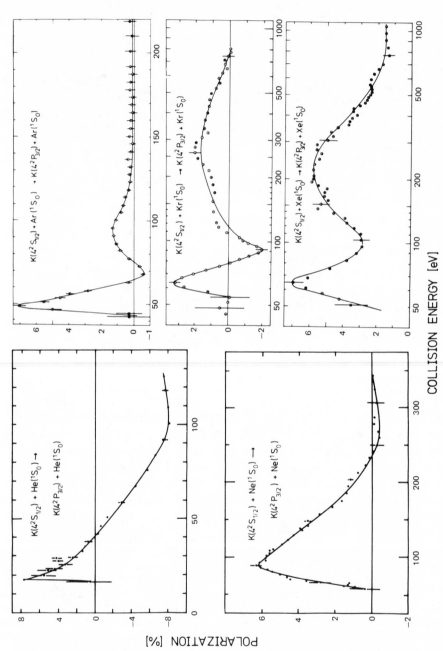

Figure 45. Linear polarization of the line transition $(4\,^2P_{3/2} \rightarrow 4\,^2S_{1/2})$ in collisional excitation of potassium with various rare-gas atoms (Alber *et al.*[124]).

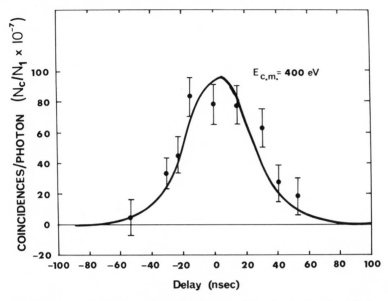

Figure 46. Coincidence peak from the photons of the simultaneously excited
states of neon[129] transition $3p'[1/2]_0 \to 3s'[1/2]_1$ at 5852 Å and
$3p'[3/2]_1 \to 3s'[1/2]_0$ at 6266 Å.

shape of the polarization function depends mainly on the relative population of the molecular states $A^2\pi$ and $B^2\Sigma$ which are correlating with the
atomic states $K(4^2P_{3/2}) + $ rare gas atoms $(^1S_0)$.

3.3.2. Coincidence Measurements

Studies of simultaneous excitations of both the colliding atoms have
recently been successful using coincidence methods. Differential cross-
section measurements have already shown[125] that, in collisions between
neon atoms, simultaneous excitation of the target and projectile atoms to
$Ne(2P^53P)$ is the dominant process for certain collision energies and
scattering angles. Martin *et al.*[126] observed coincident photons propagating in the opposite direction from the collision center as a result of the
deexcitation of the two excited neon states $Ne(3p'[3/2]_1)$ and $Ne(3p'[1/2]_0)$.
The simultaneous emission of the photons was revealed as a time correlation
effect in the delayed coincidence spectrum. Figure 46 shows an example of
such a time correlation from the simultaneous excitation of the above neon
states.

Figure 47 demonstrates a combination of the measurement of total
cross section $Q(p'[1/2]_0)$ of neon with that for the measurements of the cross

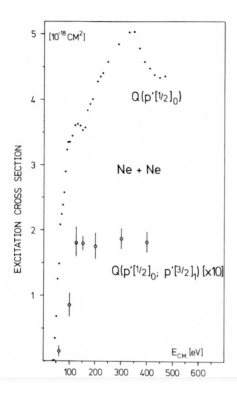

Figure 47. Cross sections for simultaneous excitation of projectile and target neon atoms to the states $p'[1/2]_0$ and $p'[3/2]_1$, $Q(p'[1/2]_0$, $p'[3/2]_1$ and the $p'[1/2]_0$, $Q(p'[1/2]_0$ for Ne-Ne collisions versus collision energy (Martin et al.[126]).

section for simultaneous excitation of the Ne states $p'[1/2]_0$ (or $2P_1$ state in Paschen notation) and $p'[3/2]_1$ (or $2P_5$ state in Paschen notation) for Ne–Ne collision versus center-of-mass energy. By selecting appropriate combinations of interference filters Kempter and his co-workers[127] were able to measure excitation cross sections for simultaneous excitation of (a) one $2p_n$ ($n = 1 - 10$ in Paschen's notation for all Ne $(2p^5 3p)$ states) in neon and all other Ne $(2p^5 3p)$ states (denoted by Σ), $Q(2p_n/\Sigma)$, and (b) of two individual Ne $(2p^5 3p)$ states $Q(2p_n/2p_n)$. From these measurements of simultaneous excitation $Q(2p_n/\Sigma)$ for the states $n = 1, 5, 6$, and 10, it could be estimated that approximately 25% of the total excitation to Ne $(2p^5 3p)$ occurs through simultaneous excitation of both the target and projectile atom at a collision energy of 300 eV.

Finally, we would like to mention the new type of measurements in which coherence effects can be detected in atom–atom excitation processes. Martin et al.[128] have studied the excitation processes $K(4^2S) + Y \rightarrow K(2^2P) + Y$ for $Y = Ar$, Xe in such a way that the photon emitted from the decay of $K(4^2P)$ and the scattered potassium atoms in the ground state were measured by both a delayed coincidence system and a TAC–MCA system (time-to-amplitude conversion plus multichannel analyzer; see Section 2.3). Interesting results from such measurements are summarized in Figure 48, and Figure 48a gives the excitation probability $P(b)$ for the above excitation process with argon as target as a function of the impact

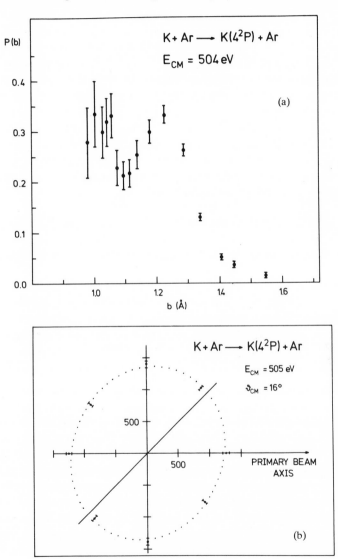

Figure 48. (a) Probability $P(b)$ for excitation of $K(4^2P)$ in collisions K–Ar as a function of the impact parameter b, and (b) number of photon-atom [K(4S)] coincidences versus polarizer orientation for K–Ar collisions[128]).

parameter b. The quantity measured in the coincidence experiment is

$$\frac{N_c(\theta)}{N_T} = \frac{\text{number of coincidences}}{\text{total number of scattered atoms}}$$

which is proportional to $\delta_{\text{in}}(\theta)/\delta_{\text{tot}}(\theta)$, the excitation probability $P(\nu)$ for the above process [e.g., $K(4^2S) + Ar \rightarrow K(4^2P) + Ar$] in comparison to all possible elastic and inelastic processes.[129] In the relevant range of internuclear distances, the functional connection between the scattering angle θ and the impact parameter b is known, and accordingly $N_c(\theta)/N_T \propto P(b)$.

On the assumption that the magnetic sublevels $m_l = 0, \pm 1$ of the $K(^2P_{3/2,1/2})$ states are coherently excited in the above collisional excitation process, the angular correlation parameters $\lambda = \delta_0/\delta$ and the phase χ between the excitation amplitudes a_0 and a_1 of the magnetic sublevels for notations λ and χ (see also Section 2.3) can be determined based upon the following equation for the coincidence rate:

$$I(\beta) = \tfrac{1}{3}CS\{1 + \tfrac{1}{4}h^{(2)}[3(1 - \lambda)\cos 2\beta + 3(\lambda(1 - \lambda))^{1/2}\cos \chi \sin 2\beta - 1]\}$$

β is the polarizer angle (angle between polarization filter and incoming potassium beam) of the polarization filter (linear polarizer) in front of the photon detector observing photons perpendicular to the scattering plane. $h^{(2)} = -0.297$ is a factor that depends on the fine-structure splitting of the $K(^2P)$ state and C,S are constants given by Eriksen *et al.*[130] Figure 48b presents a measurement of $I(\beta)$ for the above collisional excitation process with Ar as target. The dotted curve in Figure 48b is a least-squares fit of $I(\beta)$ to the data. From that follows $\delta_0/\delta_1 = 1.20 \pm 0.1$ and $\chi = (122 \pm 16)°$.

4. Conclusions

Excitations of atoms by collisional impact processes are of interest both in spectroscopy and in collision physics. Phenomena such as linear polarization and vector polarization of line radiation, coherence effects, orientation and alignment, and interatomic potentials between excited atoms can only be studied and fully understood by taking into consideration knowledge of both the field of atomic spectroscopy and collision physics. Impact excitation by electrons has been most comprehensively investigated with regard to coherence effects, alignment, and orientation. It may be enlightening to draw attention to the difference between observing coherence effects in excitations by photons and by particle impact. In resonance absorption of photons with given polarization, all atoms of the ensemble can be excited in the same substate or in different substates with equal distribution and fixed phase relations for all excitation amplitudes. In electron impact excitation by monoenergetic electrons, the substate excitation depends not only on the energy of the electrons but also on the collision dynamics. This has the consequence that atoms can only be coherently excited in a complete way, by electron impact, if the observation of the photons is restricted to those excitation processes from electrons scattered inelastically in one given direction. Any electron (or any other colliding particle) having excited a discrete energy state and being scattered in a given direction is characterized by the amount of energy it has lost and also by the scattering angle. Accordingly, all those electrons scattered in a given direction and detected in coincidence with photons emitted from excited atoms in a given state have experienced the same scattering process, and alternatively all the photons observed in coincidence result from the emission of atoms in the excited states of equal conditions (same substate population with fixed phase relations for the excitation amplitude due to the fact that the conditions of the inelastic scattering process of the impact particles are the same). In this way, electron impact excitation studied in electron–photon coincidences matches excitation by photons. The excitation of atoms by resonance absorption of photons is simplified by the fact that the incoming photon is completely absorbed, whereas in particle impact excitation the particle can be scattered in all possible directions leaving the excited state of the atoms in an infinite variety of conditions of the excited state unless the excitation process is restricted by selecting a given scattering angle. However, all criteria for the observation of coherence phenomena in excitation processes with photons (e.g., Hanle effect,[135] level crossings,[136] modulation effects[137]) are expected to occur in impact excitation if observation occurs under conditions of particle–photon coincidences. Of course, these possibilities have not yet been fully exploited, but they are obvious extensions of this advanced technology of coincidence measurements in impact excitations of atoms.

Excitation by heavy particles adds a wealth of new information relevant to the spectroscopist. While at large impact velocities the excitation functions as well as the polarization of radiation show similar features for different ions (and electrons), effects unknown in electron impact experiments are observed at low and intermediate velocities. These include the formation of fast excited atoms through charge exchange and the occurrence of molecular effects which often give rise to velocity-dependent oscillations in the excitation functions and the polarization of radiation. Only recently, coherence effects as well as simultaneous excitation of projectile and target atoms have been observed, a field that promises to give new and interesting results in the future.

Acknowledgments

We gratefully acknowledge the helpful discussions with Dr. D. Hasselkamp and Professor K.-H. Schartner. We also appreciate very much receiving information from Dr. V. Kempter prior to publication.

References

1. B. Bederson and L. J. Kieffer, *Rev. Mod. Phys.* **43**, 601 (1971).
2. W. Hanle, *Z. Phys.*, **54**, 848 (1929).
3. W. Hanle and W. Schaffernicht, *Ann. Phys. Leipzig* **6**, 905 (1930).
4. P. Bricout, *Compt. Rend.* **185**, 1029 (1927).
5. J. Williams, in *The Physics of Electronic and Atomic Collisions*, Eds. J. S. Risley and R. Geballe, IXth I.C.P.E.A.C. Invited Papers, p. 139 (1975).
6. H. Koschmieder, V. Raible, and H. Kleinpoppen, *Phys. Rev. A* **8**, 1365 (1973).
7. A. Oed, *Phys. Letters*, **34A**, 435 (1971).
8. W. E. Kauppila, W. R. Ott, and W. L. Fite, *Phys. Rev. A* **1**, 1099 (1970).
9. T. W. Ottley and H. Kleinpoppen, *J. Phys. B* **8**, 621 (1975).
10. D. W. Heddle, *J. Phys. B* **8**, L33 (1975).
11. D. W. O. Heddle, *Contemp. Phys.* **17**, 443 (1976); also in *Electron and Photon Interactions with Atoms*, Eds. H. Kleinpoppen and M. R. C. McDowell, p. 671, Plenum Press, New York (1976).
11a. N. Swanson, G. W. Cooper, and C. E. Kuyatt, *Phys. Rev. A* **8**, 1825 (1973).
12. S. H. Al-Shamma and H. Kleinpoppen, Xth ICPEAC, Paris, 1977, Abstracts, p. 518.
13. J. D. Walker, Jr., and R. M. St. John, *J. Chem. Phys.* **61**, 2394 (1974).
14. H. Kleinpoppen and E. Kraiss, *Phys. Rev. Lett.* **20**, 361 (1967).
15. L. A. Vainshtein, *Opt. Spectrosc.* **18**, 538 (1965).
16. A. H. Mahan, A. Gallagher, and S. J. Smith, *Phys. Rev. A* **13**, 156 (1976).
17. D. Leep and A. Gallagher, *Phys. Rev. A* **13**, 148 (1976).
18. S. T. Chen, D. Leep, and A. Gallagher, *Phys. Rev. A* **13**, 947 (1976).
19. V. P. Starodub, I. S. Aleksakhin, I. I. Garga, and I. P. Zapesochnyi *Opt. Spectrosc.* **35**, 1037 (1973).
20. D. H. Crandall, P. O. Taylor, and G. H. Dunn, *Phys. Rev.* **10**, 141 (1974).
21. H. W. B. Skinner and E. T. Appleyard, *Proc. R. Soc. London A* **117**, 224 (1927).
22. I. C. Percival and M. J. Seaton, *Phil. Trans. R. Soc., London, Ser. A* **251**, 113 (1958).
23. E. Baranger and E. Gerjouy, *Proc. Phys. Soc. London* **73**, 326 (1958).
24. J. R. Oppenheimer, *Z. Phys.* **43**, 27 (1927); *Proc. Natl. Acad. Sci.* **13**, 800 (1927); *Phys. Rev.* **32**, 361 (1928).
25. W. G. Penney, *Proc. Natl. Acad. Sci.* **18**, 231 (1932).
26. V. J. Ehlers and A. C. Gallagher, *Phys. Rev. A* **7**, 1573 (1973).
27. D. R. Flower and M. J. Seaton, *Proc. Phys. Soc. London* **91**, 59 (1967).
28. H. Hafner, H. Kleinpoppen, and H. Krüger, *Phys. Lett.* **18**, 270 (1965); H. Hafner and H. Kleinpoppen, *Z. Phys.* **198**, 315 (1967).
29. S. Kumar and M. K. Shrivastava, *J. Phys. B* **9**, 1911 (1976).
30. D. Leep and A. Gallagher, *Phys. Rev. A* **10**, 1082 (1974).
31. P. G. Burke and A. J. Taylor, *J. Phys. B* **2**, 869 (1969).
32. N. Feautrier, *J. Phys. B* **3**, L152 (1970).
33. P. McCavert and M. R. H. Rudge, *J. Phys. B* **3**, 1286 (1970).
34. W. R. Ott, W. E. Kauppila, and W. L. Fite, *Phys. Rev. Lett.* **19**, 1361 (1963).

35. A. N. Tripathi, K. C. Mathur, and S. K. Joshi, *J. Phys. B* **6**, 1431 (1973).
36. D. L. Moores and D. W. Norcross, *J. Phys. B* **5**, 1482 (1972).
37. E. A. Enemark and A. Gallagher, *Phys. Rev. A* **6**, 192 (1972).
38. G. Gould, Ph.D. Thesis, University of South Wales (1970).
39. U. Fano and J. H. Macek, *Rev. Mod. Phys.* **45**, 553 (1973).
40. P. D. Burrow and J. A. Michejda, in Book of Abstracts of Stirling Symposium on Electron and Photon Interactions with Atoms (1974).
41. M. Düweke, N. Kirchner, E. Reichert, and S. Schön, *J. Phys. B* **9**, 1915 (1976).
42. U. Fano and J. W. Cooper, *Phys. Rev. A* **138**, 400 (1965).
43. D. W. O. Heddle, R. G. W. Keesing, and R. D. Watkins, *Proc. R. Soc. London A* **337**, 443 (1974).
44. M. J. Mumma, M. Misakian, W. M. Jackson, and J. L. Faris, *Phys. Rev. A* **9**, 203 (1974).
45. L. Vriens and J. D. Carriere, *Physica* **49**, 517 (1970).
46. D. H. Crandall, P. O. Taylor, and G. H. Dunn, *Phys. Rev. A* **10**, 141 (1974).
47. O. B. Shpenik, V. V. Souter, A. N. Zavilopulo, I. P. Zapesochny, and E. E. Kontosh, *Sov. Phys. JETP* **42**, 23 (1976).
48. S. Trajmar and W. Williams, Invited Lecture SPIG 76, VIIth International Summer School and Symposium on the Physics of Ionized Gases, Dubrovnik, Yugoslavia (1976).
49. M. Standage and H. Kleinpoppen, *Phys. Rev. Lett.* **36**, 577 (1976).
50. E. Wolf, *Nuovo Cimento* **13**, 1165 (1959); see also M. Born and E. Wold, in *Principles of Optics*, 5th ed., Pergamon Press, New York (1975).
51. K. Blum and H. Kleinpoppen, *J. Phys. B* **8**, 922 (1975); and *Int. J. Quant. Chem. Symp.* **10**, 231 (1976).
52. G. C. M. King, A. Adams, and F. H. Read, *J. Phys. B* **5**, 254 (1972).
53. M. Eminyan, K. B. MacAdam, J. Slevin, and H. Kleinpoppen, *Phys. Rev. Lett.* **31**, 576 (1972); also *J. Phys. B* **7**, 1519 (1974).
54. H. Kleinpoppen, K. Blum, and M. Standage, in *The Physics of Electronic and Atomic Collisions*, IXth ICPEAC Invited Lectures, Review Papers and Progress Reports, Eds. J. S. Risley and R. Geballe, pp. 641–659, University of Washington Press (1976).
55. (a) A. J. Taylor and P. G. Burke, *Proc. Phys. Soc. London* **92**, 336 (1967); (b) P. G. Burke, S. Ormonde and H. Whitaker, *Proc. Phys. Soc. London* **92**, 319 (1967).
56. M. Eminyan, K. B. MacAdam, J. Slevin, M. C. Standage, and H. Kleinpoppen, *J. Phys. B* **8**, 2058 (1975).
57. K. A. Berrington, B. H. Bransden, and J. P. Coleman, *J. Phys. B* **6**, 436 (1973); P. S. Nichols and K. H. Winter, *J. Phys. B* **6**, L250 (1973).
58. B. H. Bransden and K. H. Winter, *J. Phys. B* **9**, 1115 (1976).
59. M. R. Flannery and K. J. McCann, *J. Phys. B* **8**, 1716 (1975).
60. C. J. Joachain and R. Vanderpoorten, *J. Phys. B* **7**, L528 (1974).
61. D. H. Madison and W. N. Shelton, *Phys. Rev. A* **7**, 449 (1973).
62. A. Chutjian, *J. Phys. B* **9**, 1749 (1976).
63. T. Scott and M. R. C. McDowell, *J. Phys. B* **8**, 1851 (1975); *J. Phys. B* **9**, 2235 (1976).
64. J. Macek and I. V. Hertel, *J. Phys. B* **7**, 2173 (1974).
65. U. Fano, *Rev. Mod. Phys.* **29**, 74 (1957).
66. J. Macek, in *The Physics of Electronic and Atomic Collisions*, Invited Lectures of IXth ICPEAC, Eds. J. S. Risley and R. Geballe, p. 627, University of Washington Press (1975).
67. K. Blum, Chapter 2 in Part I of this book.
68. A. H. Mahan, Ph.D. Thesis, University of Colorado (1974).
69. S. J. Smith, in *Electron and Photon Interactions with Atoms*, Eds., H. Kleinpoppen and M. R. C. McDowell, p. 365, Plenum Press, New York (1976).

70. A. H. Mahan, R. V. Krotkov, A. C. Gallagher, and S. J. Smith, *Bull. Am. Phys. Soc.* **18**, 1506 (1973).
71. R. V. Krotkov, *Phys. Rev. A* **12**, 1793 (1975).
72. L. A. Morgan and M. R. C. McDowell, *J. Phys. B* **8**, 1073 (1975).
73. A. van Wijngaarden, E. Goh, G. W. F. Drake, and P. S. Farago, *J. Phys. B* **9**, 2017 (1976).
74. D. O'Connell, K-J. Kollath, A. J. Duncan, and H. Kleinpoppen, *J. Phys. B* **8**, L214 (1975).
75. J. Macek and R. Shakeshaft, *Phys. Rev. Lett.* **27**, 1487 (1971).
76. H. S. W. Massey and R. A. Smith, *Proc. R. Soc. London A* **142**, 142 (1933).
77. V. N. Tuan, G. Gautherin, and A. S. Schlachter, *Phys. A* **9**, 1242 (1974).
78. G. Spiers, A. Valance, and P. Pradel, *Phys. Rev. A* **6**, 746 (1972).
79. A. S. Schlachter, P. J. Bjorkholm, D. H. Loyd, L. W. Anderson, and W. Haeberli, *Phys. Rev.* **177**, 184 (1969).
80. I. A. Sellin and L. Granoff, *Phys. Lett.* **25A**, 484 (1967).
81. B. L. Donnally, T. Clapp, W. Sawyer, and M. Schultz, *Phys. Rev. Lett.* **12**, 502 (1964).
82. F. Rousel, Thesis, University of Paris (1973).
83. T. D. Gaily, D. H. Jaecks and R. Geballe, *Phys. Rev.* **167**, 81 (1968).
84. P. M. Koch, L. D. Gardner, and J. E. Bayfield, IVth International Conference on Beam-Foil Spectroscopy, Gatlinburg, Tennessee, 1975.
85. P. M. Koch and J. E. Bayfield, Abstract IVth ICAP, Heidelberg (1974).
86. J. E. Bayfield, *Atomic Phys* 4, Eds. G. zu Putlitz, E. W. Weber, and A. Winnacker, p. 397, Plenum Press, New York (1974).
87. H. R. Dawson and D. H. Lloyd, *Phys. Rev. A* **9**, 166 (1974).
88. R. H. Hughes, C. A. Stigers, B. M. Doughty, and E. D. Stokes, *Phys. Rev. A* **1**, 1424 (1970); R. H. Hughes, H. R. Dawson, B. M. Doughty, D. B. Kay, and C. A. Stigers, *Phys. Rev.* **146**, 53 (1966); R. H. Hughes, H. R. Dawson, and B. M. Doughty, *Phys. Rev.* **164**, 166 (1967).
89. J. F. Williams and D. N. F. Dunbar, *Phys. Rev.* **149**, 62 (1966); F. J. De Heer, J. Schutten, and H. Moustafa, *Physica* **32**, 1766 (1966).
90. J. C. Ford and E. W. Thomas, *Phys. Rev. A* **5**, 1694 (1972).
91. J. T. Park, J. E. Aldag, J. M. George, and J. L. Peacher, *Phys. Rev. A* **14**, 608 (1976).
92. V. Franco and B. K. Thomas, *Phys. Rev. A* **4**, 945 (1971).
93. D. Rapp and D. Dinwiddie, *J. Chem. Phys.* **57**, 4919 (1972).
94. H. B. Gilbody, Proceedings of Invited Lectures, 7th Yugoslav Symposium and Summer School on the Physics of Ionized Gases, Ed. V. Vujnovic, p. 167 (1974).
95. M. B. Shah and H. B. Gilbody, *J. Phys. B* **7**, 630 (1974).
96. M. B. Shah and H. B. Gilbody, *J. Phys. B* **7**, 637 (1974).
97. M. B. Shah and H. B. Gilbody, *J. Phys. B* **7**, 256 (1974).
98. R. Hippler and K-H. Schartner, *J. Phys. B* **7**, 618 (1974).
99. J. T. Park and F. D. Schowengerdt, *Phys. Rev.* **185**, 152 (1969).
100. D. Hasselkamp, R. Hippler, A. Scharmann, and K.-H. Schartner, *Z. Phys.* **248**, 254 (1971).
101. D. Hasselkamp, A. Scharmann, and K.-H. Schartner, VIIIth ICPEAC, Institute of Physics, Beograd, p. 651 (1973).
102. E. W. Thomas and G. D. Bent, *Phys. Rev.* **164**, 143 (1967).
103. J. van den Bos, G. J. Winter, and F. J. de Heer, *Physica* **40**, 357 (1968).
104. D. Hasselkamp, Ph.D. thesis, Giessen (1976).
105. K.-H. Schartner, Habilitationsschrift, Giessen (1976).
106. J. van den Bos, *Physica* **42**, 245 (1969).

107. L. Vriens and J. D. Carriere, *Physica* **49**, 517 (1970).
108. S. Dworetsky, R. Novick, W. Smith, and N. Tolk, *Phys. Rev. Lett.* **18**, 939 (1967).
109. L. D. Landau, *Phys. Z. Sowjetunion* **2**, 46 (1932).
110. C. Zener, *Proc. R. Soc. London A* **137**, 696 (1932).
111. E. C. G. Stueckelberg, *Helv. Phys. Acta* **5**, 370 (1932).
112. W. Lichten, *Phys. Rev.* **131**, 229 (1963); **164**, 131 (1967).
113. H. Rosenthal and H. M. Foley, *Phys. Rev. Lett.* **25**, 1480 (1969).
114. H. Rosenthal, *Phys. Rev. A* **4**, 1030 (1971).
115. H. B. Larsen, Diplomarbeit, Kopenhagen (1976).
116. L. Wolterbeek-Müller and F. J. de Heer, *Physica* **48**, 345 (1970).
117. D. Hasselkamp, A. Scharmann and K-H. Schartner, *IXth ICPEAC, Abstracts of Papers*, p. 733, University of Washington Press, Seattle (1974).
118. D. A. Clark, J. Macek, and W. W. Smith, *IXth ICPEAC, Abstracts of Papers*, p. 731, University of Washington Press, Seattle (1974).
119. N. H. Tolk, J. C. Tulley, C. W. White, J. Krauss, A. A. Monge, D. L. Simms, and M. F. Robbins, *Phys. Rev. A* **13**, 969 (1976).
120. T. Andersen, A. Krikegaard Nielsen, and K. J. Olsen, *Phys. Rev. A* **10**, 2174 (1974).
121. V. A. Ankundinov, S. V. Bobashev, and V. I. Perel, *Zh. Eksp. Teor. Fiz* **60**, 906 (1971) [English transl. *Sov. Phys.—JETP* **33**, 490 (1971)].
122. N. H. Tolk, C. W. White, S. H. Neff, and W. Lichten, *Phys. Rev. Lett.* **31**, 671 (1973).
123. M. Barat and W. Lichten, *Phys. Rev. A* **6**, 211 (1972).
124. H. Alber, V. Kempter, W. Mecktenbrauch, *J. Phys. B* **8**, 913 (1975).
125. J. C. Brenot, D. Dhuicq, J. P. Gauyacq, J. Pommier, V. Sidis, M. Barat, and E. Pollack, *Phys. Rev. A* **11**, 1245 (1975).
126. P. J. Martin, G. Riecke, L. Zehnle, and V. Kempter, Book of Abstracts, Xth ICPEAC, Paris (1977).
127. V. Kempter, G. Riecke, F. Veith, and L. Zehnle, *J. Phys. B* **9**, 3081 (1976).
128. J. P. Martin, E. Clemens, V. Kempter, and L. Zehnle, Contributed Paper to IXth ICPEAC, Paris (1977).
129. V. Kempter, private communication.
130. F. J. Eriksen, D. H. Jaecks, W. de Rijk, and J. Macek, *Phys. Rev. A* **14**, 119 (1976).
131. J. van den Bos, G. J. Winter, and F. J. de Heer, *Physica* **40**, 357 (1968).
132. D. Hasselkamp, R. Hippler, A. Scharmann, and K.-H. Schartner, *Z. Phys.* **257**, 43 (1972).
133. J. van den Bos, G. J. Winter, and F. J. de Heer, *Physica* **44**, 143 (1969).
134. F. J. de Heer and J. van den Bos, *Physica* **31**, 365 (1965).
135. W. Hanle, *Ergeb. Exakten Naturwiss.* **4**, 214 (1925).
136. W. Happer and R. Gupta, Chapter 9 of this book.
137. G. W. Series, in *Physics of One- and Two-Electron Atoms*, p. 268, North-Holland, Amsterdam (1968).
138. E. Wigner, *Nachr. Akad. Wiss. Göttingen, Math. Phys.*, 375 (1927).
139. R. C. Gray, H. H. Haselton, D. Krause Jr., and E. A. Soltysik, *VIIth ICPEAC, Abstracts of Papers*, p. 831 North-Holland, Amsterdam (1971).
140. S. V. Bobashev and V. A. Kritskii, *Zh. Eksp. Teor. Fiz. Pis'ma Red.* **12**, 280 (1970) [English transl. *JETP Lett.* **12**, 189 (1970)].
141. A. N. Zavilopulo, I. P. Zapesochnyi, G. S. Panev, O. A. Skalko, and O. B. Shpenik, *Zh. Eksp. Teor. Fiz. Pis'ma Red.* **18**, 417 (1973) [English transl. *JETP Lett.* **18**, 245 (1973)].
142. G. W. York, Jr., J. T. Park, V. Pol, and D. H. Crandall, *Phys. Rev. A* **6**, 1497 (1972).
143. R. Hippler and K.-H. Schartner, *J. Phys. B* **7**, 618 (1974).
144. T. J. Morgan, J. Geddes, and H. B. Gilbody, *J. Phys. B* **6**, 2118 (1973).
145. J. F. Park, J. E. Aldag, and J. M. George, *Phys. Rev. Lett.* **34**, 1253 (1975).

146. F. J. de Heer and J. van den Bos, *Physica* **31**, 365 (1965).
147. D. Hasselkamp, R. Hippler, A. Scharmann, and K.-H. Schartner, *Z. Phys.* **248**, 254 (1971).
148. K.-H. Schartner, R. Hippler, and H. F. Beyer, *J. Phys. B* **10**, 93 (1977).

9

Perturbed Fluorescence Spectroscopy

W. HAPPER AND R. GUPTA

1. Introduction

One might imagine that a free atom is a simple and well-understood system, since it is held together by electromagnetic forces, which are the most completely understood forces of nature. In one sense this is true, and as yet there is no evidence that forces other than those of electromagnetism play any role in atomic physics. However, except for hydrogen, and to a lesser extent helium, the many-body effects of the different electrons of an atom on each other are so complex that *a priori* calculations cannot even predict the correct sign of some important atomic properties. This is particularly true of small effects like atomic fine structure and hyperfine structure. Many a newly educated undergraduate physicist has found to his or her dismay that a thorough knowledge of Maxwell's equations is not sufficient for repairing a television receiver. The situation is much the same with atoms, and if one really wants to know precise information about multielectron atoms it is necessary to resort to experiment.

The most precise information about atoms is obtained from spectroscopic measurements, from experiments that are designed to measure the intervals between different energy levels of an atom. The intervals between different electronic states of an atom can be measured by the methods of optical spectroscopy, and extensive optical spectroscopic data have been accumulated for many years. Optical spectroscopic methods are also precise enough to give useful data about the fine-structure and

W. HAPPER AND R. GUPTA • Columbia Radiation Laboratory, Department of Physics, Columbia University, New York, New York 10027.

hyperfine-structure splittings of various electronic states. However, particularly in the case of hyperfine structure, the resolution of optical spectroscopy is often too poor to allow one to make good measurements of very-small-energy splittings.

We may compare an atom to an unknown country and the electronic energy levels of the atom to the cities of that country. The distances between the cities of this country (i.e., the optical energy intervals of the atom) have already been mapped out, but the maps of the cities themselves (i.e., the fine and hyperfine structure of the atoms) are still being surveyed. These surveys are, however, proceeding at a very rapid rate now, as is illustrated in Figure 1.

This rapid progress has been made possible by the application of lasers and other new excitation techniques like beam-foil excitation to a very old type of experiment, which, for want of a better name, we shall call perturbed fluorescence experiments. There are three steps to such an experiment, as we have indicated in Figure 2. The atoms are excited by

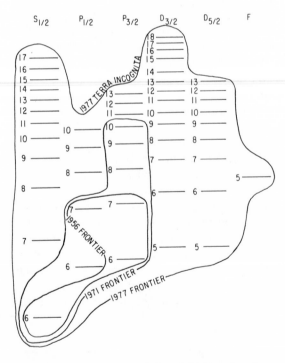

Figure 1. Expansion in knowledge of the hyperfine structures of cesium in the last two decades due to perturbed atomic fluorescence spectroscopy.

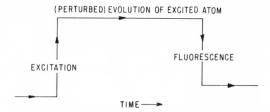

Figure 2. Schematic diagram of a perturbed fluorescence experiment.

some mechanism, for example by resonant light or by an electron beam, the excited atoms evolve for a mean time τ in the excited state, and they finally fluoresce. Various internal and external perturbations may influence the evolution of the excited atoms, and by observing how these perturbations affect the fluorescence, the experimentalist is able to make very precise measurements of various properties of the excited state.

Perturbed fluorescence experiments are very similar to perturbed angular correlation experiments[1] in nuclear physics. However, since optical photons can be more easily analyzed for polarization and frequency than gamma rays, one often observes the polarization or some property other than the angular distribution in "angular correlation" experiments of atomic and molecular physics. We have therefore chosen the name "perturbed fluorescence" to designate these experiments.

Perturbed fluorescence experiments are not affected by the Doppler broadening of the fluorescence, and the resolution of the experiment is usually limited by the natural radiative width of the excited atoms. Thus, perturbed fluorescence experiments have always been "Doppler free," and this explains why recently developed, purely optical techniques like saturated absorption spectroscopy and two-photon spectroscopy have found only limited application in atomic physics.

The first step toward the development of these techniques was made in 1923 when Wood and Ellet[2] showed that the 2537-Å resonance fluorescence of mercury vapor was almost completely linearly polarized when the vapor was excited by linearly polarized light. They also noticed that very weak magnetic fields (of the order of a few gauss) could strongly modify the polarization of the fluorescent light.

Wood and his contemporaries were struck by the fact that the polarization of the fluorescent light could be destroyed by magnetic fields that were about a thousand times weaker than those required to produce a noticeable Zeeman splitting of the optical emission lines of the atom.

Upon hearing of the experiment of Wood and Ellet, W. Hanle[3] carried out a series of independent experiments with resonance radiation from mercury vapor, and he was soon able to construct a model for the magnetic depolarization of resonance radiation that was in complete accord with all of the experimental facts. Although this explanation was based on the naive classical picture of an elastically bound electron within the atom, the formulas Hanle derived were essentially correct. Hanle showed that the ease with which an atom is depolarized by a magnetic field is proportional to the lifetime of the excited atom, and he was able to infer a value for the lifetime of the lowest excited state of mercury that was not greatly different from the presently accepted value. To this day magnetic depolarization of resonance radiation or the "Hanle effect" has remained one of the simplest and most precise ways to measure the lifetimes of excited atoms.

Although Hanle's classical model was essentially correct, the early experiments on the magnetic depolarization of resonance radiation soon revealed a number of puzzling discrepancies between measurements and the classical theory. For instance, careful experimental measurements showed that the actual zero-field polarization of the 2536 Å resonance radiation of mercury was only about 80% instead of the 100% predicted by the classical theory. Eventually the zero-field depolarization was found to occur only in the odd isotopes of mercury Hg^{199} and Hg^{201}. The zero-field polarization of the even isotopes was 100%, as expected from simple theory. In the odd isotopes, the nonzero magnetic moments of the nucleus produced internal magnetic and electric fields that partially depolarized the atomic electrons. An excellent review of the gradual unraveling of the factors that influence the polarization of resonance radiation can be found in the classic book by Mitchell and Zemansky, *Resonance Radiation and Excited Atoms.*[4]

In 1933 Breit[5] published a systematic quantum mechanical theory of atomic fluorescence. Much of the earlier theoretical work was found to be rigorously correct within the framework of a quantized radiation field. Even Hanle's formulas, which were based on the "unrealistic" model of an elastically bound electron, turned out to be correct in many, though not all, instances.

During the 1930s experimentalists began to use other aspects of atomic fluorescence besides magnetic depolarization to measure properties of excited atoms. Particularly noteworthy was the work of Heydenburg and Ellet[6] on the hyperfine structure of excited atoms. They recognized that the partial loss of polarization of atomic fluorescence from an atom with hyperfine structure could be eliminated by decoupling the nucleus from the electrons with strong external magnetic fields. The magnitude of the field required to effect this decoupling was a measure of the hyperfine structure, and Heydenburg and his co-workers were able to use these methods to measure a number of small excited-state hyperfine structures with accuracies of 20% or better.

Work on resonance radiation and atomic fluorescence seemed to be drawing to a close in the 1940s with the upsurge of interest in nuclear physics, solid state physics, and other newly fashionable fields. This lull in activity was only temporary, however, for in 1952 Brossel and Bitter[7] published the details of their newly discovered "optical double resonance" technique. The name is due to the fact that in these experiments an atom is subject to two resonant electromagnetic fields—resonance light, which produces excited atoms, and a radio-frequency field, which causes transitions between the sublevels of the excited atoms. The radio-frequency transitions can be detected because they cause changes in the polarization or angular distribution of the fluorescent light.

The Brossel–Bitter experiment was the first perturbed atomic fluorescence experiment of the modern era, and it incorporated a number of important advances over the earlier work. In almost all of the previous experiments, observations of the fluorescence were made visually or with photographic plates. The impressive achievements of Wood, Hanle, and other early workers called for excellent eyesight as well as uncommon ingenuity. The use of photomultiplier tubes by Brossel and Bitter and later workers brought the sensitivity of the experiments close to the theoretical, photon-counting limit. Also, the magnetic resonance method introduced by Brossel and Bitter has fewer sources of systematic error than most of the earlier methods. For instance, the spectral profile of the lamp and isotopic composition of the sample need not be known with high precision in magnetic resonance work, but they are of utmost importance in Hanle-effect experiments and in Heydenburg's decoupling experiments. Other important technical improvements included better optical filters (interference filters, Schott colored glass filters, etc.) and the use of phase-sensitive detection (lock-in amplifiers) for signal averaging.

One very important stimulus to the further development of perturbed fluorescence techniques was the work on various nuclear models during the 1950s. One of the key quantities that could serve to distinguish between the merits of the nuclear shell model and collective models was the electric quadrupole moment of the nucleus. No experiments have yet been devised to measure a nuclear quadrupole moment directly, because one must somehow subject the nucleus to a well-defined electric field gradient, an as yet unsolved experimental problem. What is done in practice is to measure the interaction energy of the nucleus with the natural electric field gradients that exist within atoms, molecules, or solids. Since these gradients must be *calculated*, it seemed reasonable that calculations would be most reliable for a free atom. Even here very subtle many-body shielding and antishielding effects must be taken into account, as was first done by Sternheimer.[8] Although the beautiful atomic beam magnetic resonance techniques of Rabi[9] and his co-workers had been used to obtain quadrupole interactions of many ground-state atoms, these methods could not be applied to many whole columns of the periodic table, for example alkali atoms, alkaline earth atoms, and noble-gas atoms, where the electron charge distribution in the ground state was spherically symmetric. To obtain quadrupole interaction energies for such atoms it was necessary to measure hyperfine interaction energies in nonspherically symmetric excited states.

In 1955 Perl et al.[10] succeeded in measuring the hyperfine structures of the lowest $^2P_{3/2}$ states of a number of alkali atoms by an atomic beam magnetic resonance method with optically excited atoms. At about the same time Ritter and Series[11] and Sagalyn[12] showed that the same sort of information could be obtained much more conveniently by the optical double-resonance method of Brossel and Bitter. A very productive experimental period ensued and much of that work has been reviewed by zu Putlitz.[13]

In 1959 Colegrove et al.[14] discovered the striking influence of high-field level crossings on atomic fluorescence. While investigating the 1.08 μ fluorescence from the 2^3P state of helium, they observed a sharp resonance in the intensity of the polarized fluorescent light at a field of 300 g. They pointed out that on simple theoretical grounds, similar resonances were to be expected whenever two energy sublevels of an excited atom crossed. By measuring the external field required to make two levels cross (as determined by the resonant change in the fluorescent light), one could deduce various properties (e.g., fine structure, hyperfine structure) of the excited atom.

Level-crossing spectroscopy turned out to have much in common with radio-frequency spectroscopy (e.g., optical double resonance) in that one could obtain sharp resonances whose positions were relatively insensitive to the spectral profile of the exciting light and whose shape was much simpler than the shape of Heydenburg's decoupling curves. It was a particularly attractive technique since it worked well for very-short-lived excited states where very high rf powers would have been required for radio-frequency spectroscopy. It was soon realized that a Hanle-effect experiment could be thought of as a level-crossing experiment at zero field, and in retrospect it is puzzling that the discovery of high-field level crossing came over 30 years after the discovery of the Hanle effect.

Soon after the discovery of high-field level-crossing resonances, Eck and Wieder[15] discovered the phenomenon of high-field anti-level-crossing resonances. These resonances, which occur when a pair of energy levels go through an avoided crossing or anticrossing, can be thought of as high-field versions of Heydenburg's decoupling experiments.

During the 1960s many new applications of perturbed fluorescence techniques were made. Novick and his co-workers[16] were able to use optical double-resonance and level-crossing techniques to measure quadrupole interactions in many excited atomic states of radioactive isotopes; also the use of new excitation mechanisms, electron excitation, beam-foil excitation, heavy-particle beam excitation, etc. was investigated.

It is probably safe to predict that the decade of the 1970s will be regarded as the decade of the tunable laser for perturbed fluorescence experiments. The intensity and tuning range of these devices has opened up many new possibilities for spectroscopic work on hitherto intractable but very interesting problems.

It is impossible in an article of this nature to give an exhaustive review of all aspects of perturbed fluorescence experiments. We have therefore tried to survey the main features of the field in order to give the reader some flavor of the problems and successes of the field. We have adopted a rather utilitarian view, and we have discussed mainly those techniques that can be used to measure new properties of excited atoms, properties that cannot be calculated with any confidence *a priori*. We shall also ignore the many important studies of collisional effects that have been made with perturbed fluorescence techniques. These topics are treated elsewhere in this volume in articles by Elbel and by Baylis. Our article is naturally biased by our own prior experience, and we hope we have not done too much injustice to topics with which we have less familiarity than some of our colleagues.

2. Theory

2.1. The Production of Fluorescent Light by Polarized Excited Atoms

In all of the experiments discussed in this paper the basic piece of observational data is the measured intensity of fluorescent light. The fluorescent intensity is usually measured as a function of some external electric or magnetic field that "perturbs" the fluorescent intensity. In some cases, we may regard the internal fields of the nuclear moments as the perturbing influence, but, of course, we do not have much control over such fields. In order to extract the maximum amount of information from perturbed atomic fluorescence experiments, we must have a precise theory that relates the measured fluorescent intensity to the properties of the excited atom.

Suppose an excited atom described by the wave function $|\psi\rangle$ decays to a ground-state sublevel $|\mu\rangle$. According to Fermi's "golden rule," the intensity of light emitted into the solid angle $\Delta\Omega$ is

$$\langle \mathcal{L} \rangle \Delta\Omega = (2\pi/\hbar)|\langle\psi|V|\mu\rangle|^2 \Gamma(E)\, \Delta\Omega\, \hbar\omega \tag{1}$$

where $\Gamma(E)$ is the density of final photon states at the resonant frequency

ω, and the interaction energy for electric dipole radiation is

$$V = C\hat{\mathbf{u}} \cdot \mathbf{p} \tag{2}$$

Here $\hat{\mathbf{u}}$ is the polarization of the detected fluorescent light, $\mathbf{p}$ is the momentum operator of the atomic electrons, and C is a constant, which we need not discuss in more detail here.

We must modify (1) somewhat to account for the conditions that prevail in most real experiments. First, in most experiments the observed fluorescence corresponds to the total intensity from many excited atoms, each of which may have a different wave function. Let us denote the wave function of the ith excited atom by $|\psi_i\rangle$. Then we should average (1) over all N excited atoms of the vapor. Secondly, the atom can often decay to many different ground-state sublevels $|\mu\rangle$. It is usually impossible to tell experimentally in which of several ground-state sublevels the atom was left after it emitted the fluorescent photon. We must therefore sum (1) over a complete set of final states $|\mu\rangle$ for which detectable photons are produced. With these considerations in mind, the mean fluorescent intensity is

$$\langle \mathscr{L} \rangle \propto \frac{1}{N} \sum_{i=1}^{N} \sum_{\mu} \langle \mu|\hat{\mathbf{u}}^* \cdot \mathbf{p}|\psi_i\rangle\langle\psi_i|\hat{\mathbf{u}} \cdot \mathbf{p}|\mu\rangle \tag{3}$$

One can easily verify that Eq. (3) can be written as the trace of the product of two excited state operators, which we denote by $\mathscr{L}$ and ρ:

$$\langle \mathscr{L} \rangle = \mathrm{Tr}\,[\mathscr{L}\rho] = \sum_{mm'} \mathscr{L}_{mm'}\rho_{m'm} \tag{4}$$

Here the summation extends over any complete set $|m\rangle$ of excited-state sublevels. The operator ρ is defined by its matrix elements

$$\rho_{mm'} = \frac{1}{N} \sum_{i=1}^{N} \langle m|\psi_i\rangle\langle\psi_i|m'\rangle \tag{5}$$

and is called the *excited-state density matrix*.

The operator $\mathscr{L}$ is defined by the matrix elements

$$\mathscr{L}_{m'm} = \sum_{\mu} \langle m'|\hat{\mathbf{u}} \cdot \mathbf{p}|\mu\rangle\langle\mu|\hat{\mathbf{u}}^* \cdot \mathbf{p}|m\rangle \tag{6}$$

and is called the *light-intensity matrix*. The density matrix describes the degree of polarization of the excited state. The light-intensity matrix describes the way in which the excited-state polarization affects the polarization and angular distribution of the fluorescent light.

If we denote the number of different excited-state sublevels by G, then we shall call

$$\rho_{mm} - 1/G \tag{7}$$

the *population imbalance* of atoms in the sublevel G. Furthermore we call the off-diagonal matrix element of ρ between different excited state sublevels m and n

$$\rho_{mn} \tag{8}$$

the *coherence* between the sublevels m and n. We will always assume that ρ is normalized such that

$$\text{Tr}\,[\rho] = \sum_m \rho_{mm} = 1 \tag{9}$$

Perturbed fluorescence experiments are possible only if at least some of the population imbalances (7) or coherences (8) are nonzero. We shall see that it is usually possible to classify perturbed fluorescence experiments as those that depend on population imbalances (e.g., rf spectroscopy, decoupling spectroscopy, anti-level-crossing spectroscopy) or those that depend on coherence (e.g., level-crossing spectroscopy, light-modulation spectroscopy).

We may discuss the light-intensity matrix (6) in much the same way. If $\mathscr{L}_{mn}$ is nonzero for $m \neq n$ we say that we have a *coherent detection system*, that is, a system capable of giving a signal proportional to the coherence ρ_{mn} of the density matrix. If $\mathscr{L}_{mm} \neq \mathscr{L}_{nn}$ the detection system will be sensitive to the population difference $\rho_{mm} - \rho_{nn}$ between the excited-state sublevels m and n.

The light-intensity matrix depends chiefly on the polarization of the detected light, and one can see from the form of (6) that if the quantum numbers m and m' are azimuthal quantum numbers then the matrix element $\mathscr{L}_{mm'}$ will be zero unless

$$\Delta m = |m - m'| \leqslant 2 \tag{10}$$

2.2. Evolution of the Excited Atoms

It is customary to describe the small splittings between the sublevels of an atomic state by as few parameters as possible, and these parameters are normally chosen as coefficients of an *effective Hamiltonian operator* of the atom. The effective Hamiltonian is a truncated version of the true (very intractable) Hamiltonian of the atom. For example the effective Hamiltonian of an atom in an electronic state of total angular momentum J and with a nuclear spin of I is

$$\mathscr{H} = hA\mathbf{I} \cdot \mathbf{J} + hB \left\{ \frac{3(\mathbf{I} \cdot \mathbf{J})^2 + \frac{3}{2}(\mathbf{I} \cdot \mathbf{J}) - I(I+1)J(J+1)}{2I(2I-1)J(2J-1)} \right\}$$

$$+ g_J\mu_B \mathbf{H} \cdot \mathbf{J} + g_I\mu_B \mathbf{H} \cdot \mathbf{I} - \frac{1}{2}a_0 E^2$$

$$- \frac{1}{2}\alpha_2 \left\{ \frac{3(\mathbf{E} \cdot \mathbf{J})^2 - J(J+1)E^2}{J(2J-1)} \right\} \tag{11}$$

Table 1

Quantity	Name	Physical meaning	Typical units
A	Magnetic dipole hfs constant	Magnetic field at nucleus $\times$ nuclear magnetic dipole moment	MHz
B	Electric quadrupole hfs constant	Electric field gradient at nucleus $\times$ nuclear electric quadrupole moment	MHz
g_J	g value for J	Ratio of electronic magnetic moment to angular moment	$-\mu_B/\hbar$
g_I	g value for I	Ratio of nuclear magnetic moment to angular momentum	$-\mu_B/\hbar$
α_0	Scalar polarizability	Ratio of the mean induced dipole moment to the electric field	cm^3
α_2	Tensor polarizability	Ratio of the m_J-dependent induced dipole moment to the electric field	cm^3
τ	Lifetime	Mean lifetime of the excited atom	nsec

Here the magnitudes of external electric and magnetic fields are denoted by E and H, respectively, and additional terms that describe normally small effects (for example, the diamagnetic susceptibility) have been omitted. Except for the ground state of an atom, all atomic states also have a lifetime τ, which is usually due to radiative decay, but which may also include effects due to autoionization, field ionization, collisions, or other mechanisms. The parameters that characterize the effective Hamiltonian (11) and their physical meanings are listed in Table 1. The goal of a perturbed atomic fluorescence experiment is to measure one or more of these parameters.

Different effective Hamiltonians must be defined for the different excited states of an atom, since the various physical properties represented by the parameters of Table 1 are different for different atomic states. For example, highly excited electrons, since they are far away from the nucleus for a large fraction of time, produce relatively small electric and magnetic fields at the site of the nucleus, and therefore they have only small contributions to A and B. However, the highly excited electrons are relatively easily displaced from the nucleus by external electric fields and the polarizabilities α_0 and α_2 will be large.

The matrix elements of the effective Hamiltonian (11) between different sublevels of an atomic state are very nearly the same as the matrix elements of the true Hamiltonian. However, the effective Hamiltonian will not be an adequate description of an atom when the external fields E or H

are so large that substantial amounts of other atomic states are mixed into the state of interest. For example, it is not unusual to do experiments in magnetic fields that are so large that the total electronic angular momentum J is no longer a good quantum number. More elaborate effective Hamiltonians can be constructed to account for such state mixing.

The effective Hamiltonian plays a key role in perturbed fluorescence experiments, since it determines the evolution of the excited-state density matrix that was defined in (5). The Liouville equation of evolution (the analog of the Schrödinger equation) for the excited-state density matrix is

$$i\hbar\frac{d}{dt}\rho = [\mathcal{H}, \rho] - i\hbar\gamma\rho \qquad (12)$$

where $\gamma = \tau^{-1}$ is the natural decay rate of the excited atoms.

2.3. Excitation of Atoms

We now turn to the initial step of a perturbed fluorescence experiment, the process of excitation. Since there are many ways to produce excited atoms for a perturbed fluorescence experiment, we shall discuss a few of the more important techniques in more detail in Section 3. In nearly all of these excitation mechanisms we may assume that polarized excited atoms are suddenly produced, and that they then evolve freely under the influence of the effective Hamiltonian of the excited state. We shall therefore represent the excitation by a source matrix S in the Liouville equation, which now becomes

$$\frac{d}{dt}\rho = \frac{1}{i\hbar}[\mathcal{H}, \rho] - \frac{1}{\tau}\rho + S \qquad (13)$$

An excitation mechanism is useful for perturbed fluorescence experiments only if some polarization is generated in the excited state. That is, we shall only be interested in excitation processes for which S is not a multiple of the unit matrix, or, to state the requirement more physically, we shall be interested in excitation mechanisms that populate the different sublevels of an excited atom with different efficiencies, or excitation mechanisms that produce nonzero coherence, as defined by (8).

3. Excitation Mechanisms

3.1. Direct Optical Excitation

The oldest and simplest excitation method for perturbed fluorescence experiments is direct optical excitation. The basic components of such a

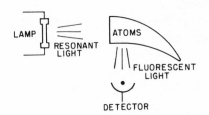

Figure 3. Schematic illustration of direct optical excitation.

system are indicated schematically in Figure 3. Only one light source is necessary for direct optical excitation, and in favorable cases the fluorescence can be seen by the naked eye. The apparatus can be extremely simple, especially when the atoms can be contained in a transparent, sealed-off cell.

Direct optical excitation also has a number of limitations, which have encouraged experimentalists to look for alternate methods of excitation. Perhaps the most serious drawback is that only a limited number of states can be populated by direct optical excitation of atoms in their ground states or low-lying thermally populated metastable states. For instance, in the case of alkali atoms, the $^2S_{1/2}$ ground state can be efficiently excited only into $^2P_{1/2}$ or $^2P_{3/2}$ excited states by direct optical excitation. Although forbidden single quantum transitions to the $^2D_{3/2}$ and $^2D_{5/2}$ state can be excited, these transitions are so weak that they have never been used for precision spectroscopic work. Also the oscillator strengths for excitation of the very highly excited P states of an alkali atom are so small that direct optical excitation without a laser is not of much practical use. Similar difficulties are encountered for atoms from all groups of the periodic table.

Another serious disadvantage of direct optical excitation is that the excitation wavelengths may be so far in the ultraviolet that appropriate light sources or optical materials are not available.

A major advantage of optical excitation is that the source matrix of the excited state is very simply related to the polarization $\hat{e}$ of the light and to the ground state density matrix ρ_g. Barrat and Cohen-Tannoudji[17] have shown that

$$S = (2\pi/\hbar^2)u\hat{e} \cdot \mathbf{D}\rho_g\hat{e}^* \cdot \mathbf{D} \tag{14}$$

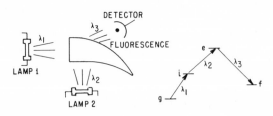

Figure 4. Typical stepwise excitation experiment. The ground-state (g) atoms are excited to the desired excited state e by two successive excitations via an intermediate state i. Fluorescent decay of the excited state e to some final state f is observed.

where u is the energy density of the light, and $\mathbf{D}$ is the electric dipole moment of the atom. For an unpolarized ground state (14) implies that one can only generate coherence with $\Delta m \leqslant 2$. The form of (14) is completely analogous to the form of the fluorescent light operator (6), and both can be conveniently discussed in terms of irreducible tensors.[18]

3.2. Stepwise Optical Excitation

Stepwise optical excitation is one way to populate states that cannot be produced by direct optical excitation. A very schematic diagram of such an experiment is shown in Figure 4. One of the earliest investigations of the use of stepwise excitation for perturbed fluorescence experiments was that of Kibble and Pancharatnam.[19] A serious difficulty with stepwise excitation is that, for conventional light sources, only a small fraction of the ground-state atoms are transferred to the intermediate state i by the light from the first lamp. Therefore, even fewer atoms are transferred to the excited state of interest e by light from the second lamp, and the fluorescent light is exceedingly weak. Only a few serious measurements have been done with conventional lamps, and the ingenious measurements of Smith and Eck[20] on the D states of lithium and Tam[21] in helium are some of the most impressive examples.

A completely different situation obtains when one of the lamps is replaced by an intense laser. Dye lasers are so intense that it is not difficult to saturate an optical transition. Thus, very efficient stepwise excitation can be obtained with a dye laser and a conventional lamp, as was first demonstrated by Svanberg et al.[22]

Another advantage of stepwise excitation is the fact that one can often make use of "cross fluorescence," i.e., fluorescence that results when the excited atom decays to a final state f that is different from the intermediate state i. In this case the wavelength λ_3 of the fluorescent light may be sufficiently different from the exciting wave length λ_1 and λ_2 that optical filters can be used to eliminate instrumentally scattered light of wavelengths λ_1 and λ_2 from the detector. This is especially important when a laser supplies one of the exciting light beams, because even a small amount of scattered laser light can overwhelm the relatively weak fluorescent signal.

3.3. Cascade Optical Excitation

Cascading is another way to populate states that cannot be produced by direct optical excitation. For example, the branching ratios for some of the lower states of the rubidium atom are shown in Figure 5. The P states can be populated by direct optical excitation, and they in turn populate lower-lying S and D states by spontaneous decay. Fairly large amounts of atomic polarization can also be transferred by cascading, so perturbed fluorescence experiments are possible.[23]

It is not possible to observe strong high-field level-crossing signals in most cascade experiments. As Tai *et al.*[24] have shown, strong high-field level-crossing signals can only be observed when level crossings occur at almost the same value of the magnetic (or electric) field in both the excited state and the branch state *b*. This is quite an unlikely occurrence. However, zero-field level-crossings always occur in all atomic states with non zero angular momenta, so that Hanle-effect signals can be observed in cascade. Cascade Hanle-effect signals can be used (e.g., Ducloy *et al.*[25] and Bhaskar and Lurio[25])to measure the lifetimes of the branch states in a cascade experiment.

Cascade experiments are particularly well adapted to rf spectroscopy, since population imbalances can be transferred with fairly good efficiency from state to state by cascading.[26] In practice only a few stages of cascading can be used to reach a state because branching ratios are not too large and some polarization is lost in each stage of cascading. Lam *et al.*[27] have made systematic use of triple cascades to measure the hyperfine structures of the lowest S and D states of alkali atoms, and a discussion of some of the difficulties involved in such experiments can be found in their paper.

Some of the major advantages of a cascade experiment are that only one light source is needed, and the fluorescence is usually at a sufficiently different wavelength from the exciting light that optical filters can be used

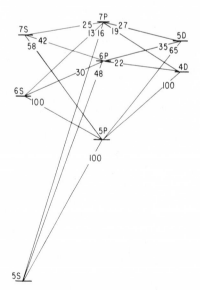

Figure 5. Energy levels involved in a typical cascade excitation of the low-lying S and D states of Rb. Atoms are excited to a relatively highly excited P state by resonance light. The P-state atoms subsequently decay to all lower-lying S and D states. The numbers in this diagram indicate the branching ratios in percent.

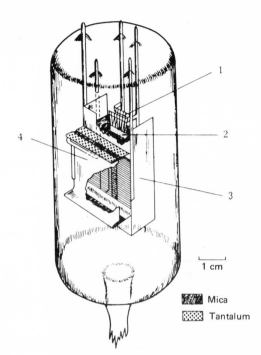

Figure 6. For electron excitation a cell of the type shown here is used by Pebay-Peyroula and co-workers.[32] Parts of the diagram are as follows: (1) heaters, (2) cathode, (3) grid, and (4) plate.

1 cm

▨ Mica
▨ Tantalum

to eliminate instrumental scattering. A serious disadvantage is that the fluorescent wavelengths tend to lie so far toward the infrared end of the spectrum that quantum efficiencies of photomultipliers may be very small. Also, in magnetic resonance experiments the spectrum of the feeding states appears superimposed on the spectrum of the state of interest. While this may be an inconvenience if the spectra of several different states overlap, it also has some advantages for calibration of the apparatus and encouragement of the experimentalist.

3.4. Particle Beam Excitation*

Atoms may be excited by collisions with charged or neutral particles. A typical experimental arrangement for electron excitation is shown in Figure 6. The selection rules for excitation by particles are much less severe than those for optical excitation. Consequently, with particle beams one can directly excite many states that could not be produced by direct optical excitation. For example, some of the cross sections for electron excitation of alkali atoms[28] are shown in Figure 7.

A particle beam has a preferred direction, and the excited atoms are therefore often found to be aligned parallel to the beam direction. The

* For more details, see Chapter 8 (Kleinpoppen and Scharmann) of this work.

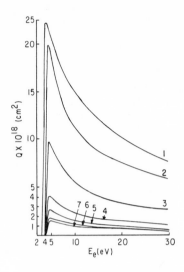

Figure 7. Cross sections for excitation of Rb atoms by slow electrons, plotted as a function of the electron energy.[28] The cross sections shown in this diagram are for the diffuse series $5^2P_{3/2} \to n^2D_{5/2}$, and the numbers 1–7 correspond to $n = 5$–11, respectively.

excited atoms may then emit polarized light,[29] and this makes it possible to perform perturbed atomic fluorescence experiments with particle beam excitation. The fluorescent polarization is usually a maximum for energies just above the excitation threshold, as is illustrated in Figure 8.

The polarizations that can be obtained by particle beam excitation are usually smaller than those that can be obtained by direct optical excitation. Another problem is that one has very low selectivity in particle beam excitation, and the beam excites many different states of the atom. The state of interest is therefore partially populated by cascades from higher-lying states, a fact that can seriously complicate the analysis of experimental data. For example, one may have to make extensive corrections to account for cascade effects on lifetime measurements.

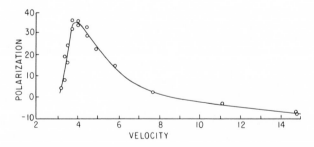

Figure 8. Polarization (%) of the 4347 Å (2^1P_1–4^1D_2) light from Hg atoms as a function of the electron velocity in units of $eV^{1/2}$.[29]

To produce maximum transverse coherence for a level-crossing experiment it is necessary to direct the particle beam at right angles to the external magnetic (or electric) field. Thus, it is often practically impossible to use charged particles to excite atoms for a high-field level-crossing experiment since the particles are bent too strongly by the field. High-energy beams of H_2^+ ions have been used by Kaul[30] to excite atoms for high-field level-crossing work. The heavy ions have much larger radii of curvature than low-energy electrons, but they produce very small atomic polarization. Neutral beam excitation has been utilized by Lhuillier *et al.*[31] to perform high-field level-crossing experiments. Unfortunately the signals generated by a neutral beam are very small.

Electron beam excitation is well adapted to rf spectroscopy, and Pebay-Peyroula[32] has used this method to measure the properties of many highly excited states of mercury. Because of the free electrons and ions in these experiments, resonant effects associated with electron cyclotron resonance are frequently observed and they interfere with the desired signals from the excited atoms. Another difficulty is that the relative cross sections for the population of the different atomic sublevels are seldom known, and this makes it very difficult to analyze the lineshapes and intensities with any precision.

Lombardi[33] and his co-workers have shown that one can use an rf discharge at sufficiently low pressures to produce polarized excited atoms. If the electron collision rate is less than the rf frequency, the electrons will tend to oscillate along the direction of the rf electric field and they will therefore tend to align the excited atoms with the direction of the rf electric

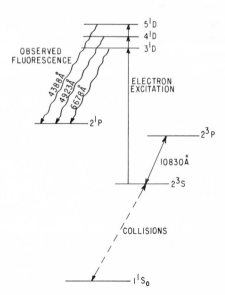

Figure 9. Scheme used by Pavlović and Laloë[34] to produce polarized excited ^1D states of He by electron excitation of the optically pumped 2^3S states.

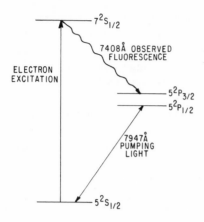

Figure 10. Polarized $7^2S_{1/2}$ state of Rb were produced by the electron excitation of the optically pumped ground state by Chang and Happer.[35]

field. Such experiments have the advantage that they do not require cells with internal electrodes and a simple sealed-off glass cell will suffice.

It is also possible to use electron excitation with randomly directed electrons if the target atoms are polarized. The most impressive experiments of this type have been carried out by Laloë and his collaborators[34] with He^3. The He^3 nuclei are polarized by optical pumping of the 2^3S metastable state, and the electron excited states of the singlet and triplet series are all found to be partially polarized since the nuclear spin of the $^1S_0$ ground state is polarized. A sketch of the states involved in this work is

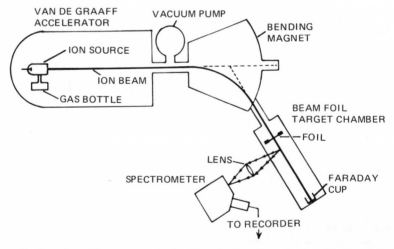

Figure 11. Experimental arrangement of a Van de Graaff accelerator, bending magnet, beam-foil target, and spectrometer for beam-foil experiments (From Ref. 94).

shown in Figure 9. Chang *et al.*[35] have shown that excited-state polarization of alkali atoms can be produced by isotropic electron excitation of spin-polarized optically pumped alkali atoms, as shown in Figure 10.

3.5. Beam-Foil Excitation

One of the most promising new excitation mechanisms, beam-foil excitation, was recently discovered by Bashkin and his collaborators.[36] In this type of experiment, which is illustrated in Figure 11, a beam of ions from a Van de Graaff accelerator passes through a thin foil of carbon or some other refractory material. The particles emerging from the other side are found to be in various states of ionization and excitation. Furthermore, since the excitation mechanism is anisotropic the excited atomic or ionic states are found to be polarized by an amount that depends in a complicated and still not totally understood way on the angle of incidence of the beam on the foil and on many other experimental parameters. Fortunately, the exact mechanism and degree of polarization is not too important for applications of beam-foil techniques in perturbed fluorescence spectroscopy. Beam-foil excitation experiments are discussed in more detail by Andrä in Chapter 20 of this work.

3.6. Collisions with Excited Polarized Atoms*

Another very interesting and potentially useful excitation mechanism is excitation transfer, where the atoms of interest are excited by collisions with another excited polarized atom or ion. For example, Schearer *et al.*[37] have used Penning ionization of zinc atoms by metastable helium atoms to produce polarized excited states of singly ionized zinc, as shown in Figure 12. Excitation transfer experiments are discussed in more detail in the

* For more details, see Chapter 18 (Schearer and Parks) of this work.

Figure 12. Schematic diagram of the experimental arrangement used by Schearer and co-workers[37] for producing polarized excited states of Zn^+ by Penning ionization. Zn is evaporated by the oven as shown. The metastable He atoms are produced by microwaves and polarized by optical pumping.

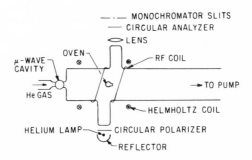

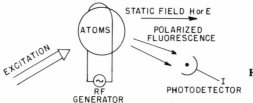

STATIC FIELD H or E

ATOMS

POLARIZED
FLUORESCENCE

EXCITATION

RF
GENERATOR

PHOTODETECTOR

I

Figure 13. Sketch of the experimental setup for a rf spectroscopy experiment.

article by Elbel, Chapter 29 of this work. So far this method has not been systematically used in perturbed fluorescence work.

4. Basic Spectroscopic Methods

In this section we shall discuss the experimental methods of perturbed fluorescence that have proven most successful in practice or that appear to have the best prospects for further development. Most of these methods can be used with all of the excitation mechanisms discussed in the last section, although certain combinations are more felicitous than others.

4.1. Radio-Frequency Spectroscopy (Optical Double Resonance)

The most versatile and reliable way to measure small energy splittings in an excited atom is to use a radio-frequency (rf) field to cause transitions between the sublevels. The energy splitting ΔE is then related to the resonant rf frequency ω by the Bohr frequency condition

$$\Delta E = \hbar\omega \tag{15}$$

The key idea for the development of practical methods of rf spectroscopy for excited atoms is due to Kastler and Brossel.[38] They suggested that the

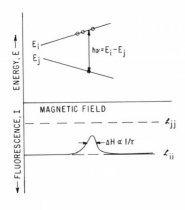

ENERGY, E →

E_i
E_j

$h\nu = E_i - E_j$

MAGNETIC FIELD

FLUORESCENCE, I →

$\mathscr{L}_{jj}$

$\Delta H \propto 1/\tau$

$\mathscr{L}_{ii}$

Figure 14. For rf spectroscopy experiments, unequal population distribution of the atoms among different sublevels i and j of the excited state is produced by some means. Transitions induced between these two sublevels by the rf field of frequency $\nu = (E_i - E_j)/h$ are detected as a change in the intensity of the fluorescence.

rf transitions could be detected by observing the changes produced in the angular distribution or polarization of the fluorescent light from the excited atoms. The first experimental demonstration of this new method was made by Brossel and Bitter[7] in 1952. The basic components of their apparatus, and of many analogous experiments that have been done since then, are shown in Figure 13. The atoms are selectively excited by some means (for example, by resonant light or by an electron beam) and the atomic fluorescence, of an appropriate polarization and wavelength, is detected, preferably with a photomultiplier tube. The fluorescence may exhibit resonant changes in intensity as the external field or the rf frequency is varied.

The basic physics of rf spectroscopy is illustrated in Figure 14. Suppose that atoms are produced in the excited-state sublevel i, for example by direct optical excitation, and the sublevel j is not populated. From the discussion of Section 2.1 we know that the fluorescence intensity of an atom in state i is $\mathscr{L}_{ii}$ and the fluorescent intensity of an atom in state j is $\mathscr{L}_{jj}$. The quantities $\mathscr{L}_{ii}$ and $\mathscr{L}_{jj}$ depend on the polarization of the fluorescent light, and by an appropriate choice of polarization we can usually ensure that $\mathscr{L}_{ii} \neq \mathscr{L}_{jj}$. Now suppose the atoms are subject to a magnetic field that oscillates at a radio frequency ω. The evolution equation (13) would then be modified by adding to the effective Hamiltonian $\mathscr{H}$ a term

$$V = g_J \mu_B \mathbf{H}_1 \cdot \mathbf{J}\, e^{-i\omega t} + \text{c.c.} \tag{16}$$

Here $\mathbf{H}_1$ is the amplitude of the oscillating field. When the frequency ω is not close to the resonance frequency $(E_i - E_j)/\hbar$ the oscillating field will have little effect on the evolution of the excited atom and the fluorescent intensity will be proportional to $\mathscr{L}_{ii}$, the intensity characteristic of atoms in the state i. However, if the frequency ω is close to the resonance frequency $(E_i - E_j)/\hbar$, transitions will be induced between the states i and j, and for sufficiently intense rf fields, equal numbers of atoms will be in states i and j. The fluorescent intensity will then be proportional to $\frac{1}{2}(\mathscr{L}_{ii} + \mathscr{L}_{jj})$, the mean fluorescent intensity from the states i and j. The actual line shape of the rf resonance can be shown[39] to be

$$\Delta I \propto \frac{\omega_1^2}{\omega_1^2 + \gamma^2 + (\omega - \omega_{ij})^2} \frac{\mathscr{L}_{ii} - \mathscr{L}_{jj}}{2} \tag{17}$$

where ω_1 is proportional to the matrix element of the rf interaction between the states i and j, and γ is the natural radiative decay rate of the states i and j (for simplicity we assume that both states decay at the same rate). The frequency of the rf field is ω and the resonance frequency is $\omega_{ij} = (E_i - E_j)\hbar$. We note that the limiting linewidth for vanishing rf power is just γ, the natural width of the excited state.

One is sometimes concerned with atoms that have many closely spaced energy levels. In that case many different rf transitions may overlap and the composite line shape may be too complicated to analyze. Fortunately, exact line-shape functions can be calculated for the important case of a Zeeman multiplet such as the one shown in Figure 15. The line shape corresponding to (17) is

$$\Delta I = C_1 y_1 + C_2 y_2 \tag{18}$$

Here C_1 and C_2 are appropriate linear combinations of the fluorescent detection efficiencies and of the excitation rates, while the line-shape factors are

$$y_1 = \frac{\omega_1^2}{\omega_1^2 + \gamma^2 + (\omega - \omega_0)^2} \tag{19}$$

and

$$y_2 = \frac{3\omega_1^2[\gamma^2 + \omega_1^2 + 4(\omega - \omega_0)^2]}{[\gamma^2 + 4\omega_1^2 + 4(\omega - \omega_0)^2][\gamma^2 + \omega_1^2 + (\omega - \omega_0)^2]} \tag{20}$$

The line shapes y_1 and y_2 are associated with the orientation and alignment of the excited state,[41] and their functional form is the same for all spins. However, there is no alignment resonance for a multiplet of angular momentum $1/2$. Since the orientation is associated with circular polarization the simple line shape y_1 can be observed only when the detected light has some degree of circular polarization and if the mean angular momentum of the excited atoms is not zero. It is interesting to note that the line shape y_2 develops a double peak at sufficiently high rf powers, as shown in Figure 16. For large rf powers the resonance lines can be seriously distorted and shifted, and some care is needed to avoid erroneous

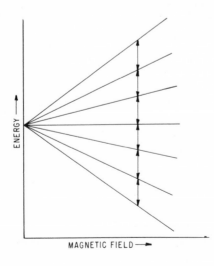

Figure 15. Linear Zeeman splittings of an atomic state in an external magnetic field.

ENERGY →

MAGNETIC FIELD →

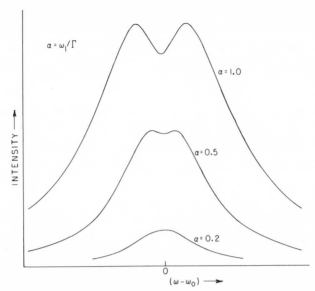

Figure 16. Resonance line-shape function y_2 plotted against the frequency ω of the rf. ω_1 is the amplitude of the rf field and γ is the decay rate of the atoms.

interpretation of the experimental data. A good discussion of some of the pitfalls of rf power shifts can be found in a recent article by Pegg.[42]

Multiple quantum transitions and a large number of other resonant effects in high rf fields have been studied by many investigators, and some of these phenomena are discussed by Stenholm in Chapter 3 of this book. Some of these effects can be used to make precision measurements of rf field amplitudes. In practical rf spectroscopic measurements of previously unknown atomic parameters, simple single quantum rf transitions (or in the case of Zeeman multiplets, orientation or alignment resonances) are nearly always used.

A major advantage of rf spectroscopy is the fact that the resonance frequency (i.e., the center of the "peak" on the recording apparatus) is quite insensitive to the details of the excitation mechanism. These details (e.g., the spectral profile of the exciting lamp) are usually not too well known, and they can be a major source of uncertainty in decoupling spectroscopy. One can also use rf spectroscopy on states that do not have high-field level crossings. Coherent excitation is not necessary for rf spectroscopy, and this is a decisive advantage in cases where cascade excitation or particle beam excitation at high magnetic fields is used. Neither excitation mechanism is well suited to producing transverse excited-state

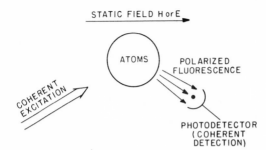

STATIC FIELD H or E

ATOMS

POLARIZED
FLUORESCENCE

COHERENT
EXCITATION

PHOTODETECTOR
(COHERENT
DETECTION)

Figure 17. Sketch of an experimental setup for a level-crossing experiment.

coherence, but both can produce sizeable population differences for rf spectroscopy.

A serious disadvantage of rf spectroscopy is the fact that for atoms with very short lifetimes or for systems with very small transition moments (e.g., molecules in high rotational states) it is difficult or impossible to obtain sufficient rf power to saturate the transitions. For such cases some other technique of perturbed fluorescence spectroscopy like high-field level-crossing or modulated excitation can often be used to good advantage.

4.2. Level-Crossing Spectroscopy

Level-crossing experiments are the oldest technique of perturbed fluorescence spectroscopy, since the Hanle effect, which we discussed in the Introduction, is due to the crossing of Zeeman sublevels at zero field. It was not until 1958 that Colegrove et al.[14] discovered that similar level-crossing effects could also be observed at high magnetic fields. The basic experimental arrangement for a level crossing experiment is sketched in Figure 17.

A simple example of an atomic state with level crossing is shown in Figure 18. Here we have sketched the hyperfine-structure sublevels of a $^2P_{3/2}$ state as a function of the magnetic field H. There are several zero-field level crossings which cause a large Hanle effect, and there are also two high-field level crossings. The qualitative behavior of the fluorescence is sketched in the lower part of Figure 18.

The theory of level-crossing experiments follows directly from the basic equation of evolution (13). If we assume that the source matrix is independent of time then the steady-state solution of (13) is

$$\rho_{ij} = \frac{S_{ij}}{\gamma + i\omega_{ij}} \tag{21}$$

The intensity of fluorescent light associated with the coherence ρ_{ij} is

$$I_{ij} = \rho_{ij}\mathscr{L}_{ji} = \frac{S_{ij}\mathscr{L}_{ji}}{\gamma + i\omega_{ij}} \tag{22}$$

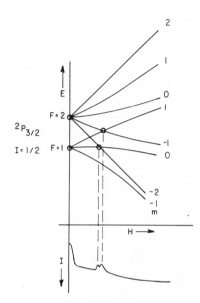

Figure 18. Hyperfine structure sublevels of an atom (with nuclear spin $I = 1/2$) in a $P_{3/2}$ state placed in an external magnetic field. With a suitable choice of excitation and detection scheme, level crossings (shown with circles) may be detected as a change in the intensity of the fluorescence.

Since both ρ and $\mathscr{L}$ are Hermitian, only the real part of (22) contributes to the total fluorescent intensity.

We see that there will be a resonant change in I_{ij} whenever ω_{ij} passes through zero, that is, whenever the levels i and j cross. In contrast to the resonances for rf spectroscopy, which are usually symmetrical about the center of the line, the level-crossing resonances may be either symmetric (absorptive) or antisymmetric (dispersive) about the line center. The line shape is determined by the phase of the complex number $S_{ij}\mathscr{L}_{ji}$ in the numerator of (22). The phase of $\mathscr{L}_{ji}$ depends on the polarization of the detected light, and it can often be adjusted to yield the desired line shape by placing an appropriate polarizer between the atoms and the photodetector.

From (22) we see that level-crossing signals can be observed only if both S_{ij} and $\mathscr{L}_{ji}$ are nonzero. From the definition of $\mathscr{L}$ in Section 2.1 we know that $\mathscr{L}_{ij}$ will be zero unless the difference Δm in the azimuthal quantum numbers of the states i and j is less than 2, i.e., we must have $\Delta m \leqslant 2$ for an observable level crossing. Therefore no resonance associated with the $\Delta m = 3$ level crossing of Figure 18 has been indicated.

The chief advantage of level-crossing spectroscopy is the simplicity of the experimental apparatus. No radio-frequency generators are required, nor are there rf coils to get in the way of the exciting or detected light. Also, fast photodetectors are not required since one observes the time-averaged value of the fluorescent intensity. For well-resolved high-field level crossings the resonance line shape is not very sensitive to the details of the excitation mechanism (e.g., the spectral profile of the lamp is not as important as it is for decoupling experiments). In contrast to rf spectroscopy, it is no more difficult to measure level-crossing resonances in very short-lived excited states than in longer-lived states. There are also no rf power shifts to worry about.

There are some serious disadvantages to level crossing experiments too. The resonance magnetic fields typically give information about the ratio of hyperfine- or fine-structure coupling constants to certain g values (e.g., A/g_J), and Hanle-effect linewidths give products of g values with excited-state lifetimes (e.g., $g_J\tau$). Thus it is necessary to find some other independent technique to measure the g values (or the A values or τ values, etc.). Some excited states have no high-field level crossings. Finally, the necessity of coherent excitation is often an insuperable obstacle for the application of the level-crossing technique. In particle beam excitation of optically inaccessible excited states it is difficult to keep the particle trajectories transverse to high magnetic fields (transverse components of the particle velocity are needed for coherent excitation). In cascade excitation much of the high-field coherence generated in the primary step of excitation is lost before the state of interest is generated. In spite of these drawbacks level-crossing spectroscopy has proved to be only slightly less fruitful than rf spectroscopy in the production of new results.

4.3. Decoupling and Anti-Level-Crossing Spectroscopy

Anti-level-crossing spectroscopy and decoupling spectroscopy are related to each other in much the same way as high-field level-crossing spectroscopy and the Hanle effect. Like the Hanle effect, decoupling spectroscopy was developed very early by Heydenburg and Ellet in 1933.

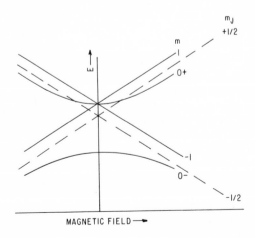

MAGNETIC FIELD ⟶

Figure 19. Energy levels of an atom (without hyperfine structure) in a $J = 1/2$ state placed in an external magnetic field (dotted lines). In the presence of a hyperfine structure ($I = 1/2$), energy levels as shown by the solid lines result. 0^+ and 0^- levels anticross due to the hyperfine interaction.

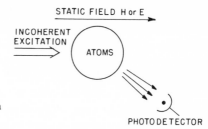

STATIC FIELD H or E

INCOHERENT
EXCITATION

ATOMS

Figure 20. Sketch of an experimental setup for a
decoupling experiment.

PHOTODETECTOR

We can think of decoupling and anti-level-crossing spectroscopy as a
peculiar kind of zero-frequency rf spectroscopy induced by the internal
fields of the atom.

We shall begin with a discussion of decoupling spectroscopy. Some
atomic systems have no high-field level crossings and only zero-field level
crossings exist. A typical example is a $^2P_{1/2}$ state with a nuclear spin of
$I = 1/2$. The energy-level diagram for such a state is shown in Figure 19.
The dotted lines represent the energies $E = g_J\mu_B H m_J$ that the atom would
have in the absence of any hyperfine interaction, and the solid lines
represent the energies of the atom when a magnetic dipole hyperfine
interaction $A\mathbf{I} \cdot \mathbf{J}$ is included in the Hamiltonian.

Since the atom does have zero-field level crossings, the Hanle effect
(with circularly polarized excitation and detection) can be used to measure
the excited-state lifetime (more precisely, $g\tau$). However, the Hanle effect is

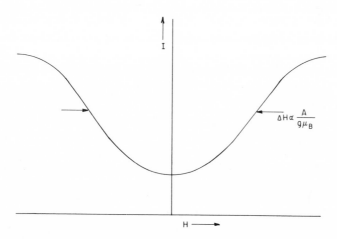

I

$\Delta H \propto \dfrac{A}{g\mu_B}$

H $\longrightarrow$

Figure 21. Typical shape of a decoupling signal. Width of the
decoupling curve is proportional to the hyperfine structure.

very insensitive to the hyperfine structure of the excited state as long as $A\tau \gg 1$. Nevertheless, as Heydenburg and Ellet showed in 1933, it is possible to obtain information about the hyperfine structure by a decoupling experiment, the basic components of which are sketched in Figure 20.

In the case of a $P_{1/2}$ excited atom the excitation might be circularly polarized light propagating along the magnetic field and the circularly polarized fluorescence is detected in a direction as nearly parallel or antiparallel to H as possible. The intensity of the fluorescence is then found to vary with the magnetic field as sketched in Figure 21. A curve somewhat reminiscent of the Hanle effect is obtained, but the width of the curve is found to be proportional to the hyperfine coupling constant A rather than to the natural width γ of the excited state. The constant of proportionality can be calculated from theory and the experimental width can therefore be used to determine A (or A/g).

The theory of anticrossing or decoupling spectroscopy is similar to that of rf spectroscopy. If the atom of Figure 19 had no hyperfine structure we could cause an rf transition between the unperturbed energy levels with an rf field, as we discussed in Section 4.1.

Now consider the hyperfine interaction energy

$$V = A\mathbf{I} \cdot \mathbf{J} = \tfrac{1}{2}A\mathbf{I} \cdot \mathbf{J}\, e^{-i0t} + \text{c.c.} \tag{23}$$

If we compare (16) and (23) we see that the hfs interaction is formally equivalent to an rf field of zero frequency and amplitude $\mathbf{H}_1 = A\mathbf{I}/2g_J\mu_B$. Consequently, the hfs interaction can cause transitions between the electronic states $i(m_J = 1/2)$ and $j(m_J = -1/2)$ whenever the energy difference between these states is equal to the "rf frequency," i.e., whenever the states i and j cross. In analogy to the case of rf spectroscopy we expect a fluorescent signal of the form (17)

$$\Delta I \propto \frac{\omega_1^2}{\omega_1^2 + \gamma^2 + \omega_{ij}^2}(\mathscr{L}_{ii} - \mathscr{L}_{jj}) \tag{24}$$

where the "Rabi frequency" is

$$\omega_1 \propto (1/\hbar)\langle i|A\mathbf{I} \cdot \mathbf{J}|j\rangle \tag{25}$$

The magnitude of the hyperfine interaction constant A must be comparable to or greater than the natural decay rate γ of the excited state to saturate the decoupling signal.

Since the hyperfine interaction conserves the total angular momentum $\mathbf{F} = \mathbf{I} + \mathbf{J}$ of the atom, the hyperfine interaction causes transitions between sublevels with the same total azimuthal quantum numbers, $m = m_I + m_J$ (the levels labeled 0+ and 0− in Figure 19). If we include the hyperfine interaction $A\mathbf{I} \cdot \mathbf{J}$ in the unperturbed Hamiltonian we obtain the energies

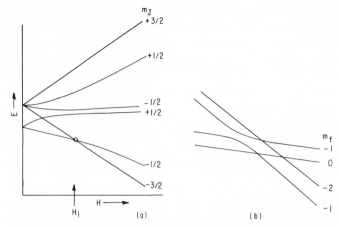

Figure 22. (a) In the absence of hyperfine structure, two Zeeman sublevels of an atom in the ^2P state with $\Delta m = 1$ cross at a field H_1. (b) In the presence of a hyperfine structure ($I = 1/2$), each of the sublevels involved in the "crossing" splits into two sublevels, as shown, and the sublevels with the same total magnetic quantum number m_F anticross.

indicated by the solid lines on Figure 19. We see that the sublevels between which transitions are induced, actually anticross at zero frequency, while the two sublevels with the quantum numbers $m = \pm 1$ only suffer a uniform upward shift of $A/4$. The anticrossing of a pair of sublevels is therefore equivalent to a zero-frequency rf transition between the sublevels.

The same anticrossing phenomena can occur at high magnetic fields, as Eck[15] first demonstrated. For example, the energy levels of a ^2P state of an excited atom are sketched in Figure 22. In Figure 22a the hyperfine structure has been ignored in order to show that the fine-structure levels have both a $\Delta m = 1$ and a $\Delta m = 2$ high-field level crossing. In Figure 22b the behavior of the fine-structure levels is sketched in the neighborhood of the $\Delta m = 1$ high-field level crossing and the hyperfine sublevels for a nuclear spin of $I = 1/2$ have been included. The two sublevels with azimuthal quantum numbers m and $-m$ anticross near the field H_1; that is, in the neighborhood of the field H_1 the hyperfine interaction causes a transition between the electronic states of azimuthal quantum numbers $m_J = -1/2$ and $m_J = -3/2$. Of course the sublevels $m_J = -1/2$ and $m_J = -3/2$ must be populated at different rates by the excitation mechanism, and the fluorescent detection efficiency for atoms in the states $m_J = -1/2$ and $m_J = -3/2$ must be different for an observable signal to result.

Anticrossings may also be induced by the electric quadrupole interaction for $\Delta m_j \leqslant 2$ provided that the quadrupole interaction is nonzero.[43]

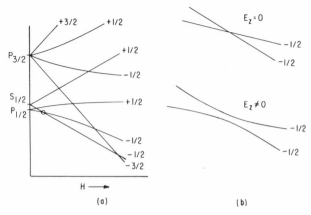

Figure 23. (a) Zeeman sublevels of a hydrogenic atom in an external magnetic field. As shown in (b), two sublevels with $m_J = -1/2$ cross at a magnetic field H_0. However, an external electric field can make these sublevels anticross.

A kind of artificial anticrossing spectroscopy can be induced with external electric fields.[44] For example, the magnetic field dependence of the ($m = 1$) electronic sublevels of a hydrogenic atom are shown in Figure 23a. In Figure 23b, the crossing sublevels in the neighborhood of the magnetic field H_0 are shown. A very small longitudinal electric field E_z is sufficient to make the levels anticross (i.e., the electric field causes zero-frequency rf transitions). Electric-field-induced anticrossings have been used by Kleinpoppen and his collaborators[45] to measure Lamb shifts and fine-structure intervals in hydrogenlike atoms.

It is interesting to note that at a high-field level crossing in a magnetic field, the matrix element of the transverse magnetic moment operator is zero, as was pointed out by Kocher.[46] This prevents one from using magnetic resonance to cause transitions between sublevels in the neighborhood of a high-field level crossing.

Decoupling experiments or anti-level-crossing experiments are very simple since they require no rf equipment, and the signals can be quite large for decoupling experiments. However, the theoretical analysis of decoupling curves can be quite involved, and for atoms with complicated hyperfine structure, elaborate computer programs may be necessary to extract the desired information from the data. The line shapes may also be extremely sensitive to the spectral profile of the exciting light and to other details of the excitation mechanism that may not be under good experimental control. For this reason decoupling experiments are most useful for providing rough preliminary values of previously unknown atomic parameters. These approximate values can greatly simplify subsequent, more precise measurements by rf or level-crossing spectroscopy. It is also important to note that decoupling experiments can provide information about the signs of excited-state parameters.[47] Radio-frequency spectroscopy often determines only the magnitude of the coupling constants.

One would seldom be tempted to use high-field anticrossing resonances in cases where high-field level-crossing experiments are feasible. However, in some special cases, for example, cascade excitation or particle excitation at very high magnetic fields, it may be impossible to provide the necessary coherent excitation for a level-crossing experiment, and an anticrossing experiment may be the only recourse. High-field anticrossing resonances have widths that are proportional to the strengths of the various interactions (e.g., hyperfine coupling constants) which cause the anticrossing. If these interactions are too large the signals may be too wide to be easily detectable, while if the interactions are smaller than the natural widths of the levels, the anticrossing signal amplitude may be too small to be detected. Because of these limitations anticrossing and decoupling spectroscopy are not as widely used as rf or level-crossing spectroscopy.

4.4. Light-Modulation Spectroscopy

By light-modulation spectroscopy we mean experiments in which one observes the index of modulation and phase of intensity-modulated fluorescent light. The modulation is produced by some kind of driving frequency, for example an rf field or an intensity-modulated exciting lamp, and pronounced changes in the phase and amplitude of the modulated components of the fluorescent light are observed when the modulation frequency is nearly equal to the characteristic energy splittings (times h^{-1}) of the excited state.

Light modulation experiments were pioneered by Series and his collaborators.[48] Series pointed out that when one induces an rf transition between sublevels i and j of an excited atom in an experiment like that sketched in Figures 13 and 14, one also induces a coherence

$$\rho_{ij} = \frac{S_{ii} - S_{jj}}{\gamma} \frac{\gamma - i(\omega_{ij} - \omega)}{\gamma^2 + \omega_1^2 + (\omega_{ij} - \omega)^2} \cdot \frac{i\omega_1}{2} e^{-i\omega t} \qquad (26)$$

between the sublevels i and j. This is associated with a modulated component of the detected light

$$\Delta I = \rho_{ij} \mathscr{L}_{ji} \qquad (27)$$

It is worth noting that the magnitude of the coherence reaches a maximum when $\omega_1^2 \simeq \gamma^2 + (\omega - \omega_{ij})^2$ and the coherence tends to zero for higher rf powers. This means that there is an optimum rf power for rf-induced light modulation experiments. The case where many sublevels of a Zeeman multiplet are coupled by an rf field has been discussed by Series[49] and others. It is possible to identify rf-induced light-modulation signals that are characteristic of various components of orientation or alignment. Light modulation induced by rf fields has been of considerable academic interest but has not proved to be a popular experimental tool.

A closely related type of light-modulation experiment does show promise of becoming an important tool in spectroscopy. In these experiments, which were pioneered by Aleksandrov,[50] atoms are excited by modulated light, and the modulated component of the fluorescent light is observed as indicated in Figure 24. The index of modulation of the fluorescent light is found to undergo resonant changes whenever the rf frequency ω is equal to certain coherence frequencies of the excited atom. The basic theory of this phenomenon can be understood if we represent the modulated excitation by a modulated term $S_{ij}(\omega)\,e^{-i\omega t}$ in the source matrix for the excited-state density matrix. Then the density matrix evolves as

$$\frac{d}{dt}\rho_{ij} = -i\omega_{ij}\rho_{ij} - \gamma\rho_{ij} + S_{ij}(\omega)\,e^{-i\omega t} \tag{28}$$

If we assume a steady-state solution to (28) of the form

$$\rho_{ij} = \rho_{ij}(\omega)\,e^{-i\omega t} \tag{29}$$

we find that the amplitude $\rho_{ij}(\omega)$ is

$$\rho_{ij}(\omega) = \frac{S_{ij}(\omega)}{\gamma + i(\omega_{ij} - \omega)} \tag{30}$$

That is, the oscillating excitation produces an oscillating coherence which has a resonant peak at the frequency $\omega = \omega_{ij}$.

The most convenient way to produce modulated excitation is by

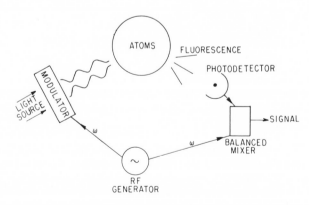

Figure 24. Sketch of an experimental setup for a light-modulation spectroscopy experiment.

intensity-modulating a beam of resonant light, although in principle other modulated excitation schemes, for example, modulated electron beams, could be used. Modulated excitation spectroscopy has many of the advantages of radio-frequency spectroscopy, since it can be used for atoms without level crossings, and by varying the modulating frequency one can "track" the resonances from a simple limiting low-field or high-field region to more informative but much more complicated intermediate-field regions. In contrast to rf spectroscopy, there is no rf power broadening of the resonance. Furthermore, there is no danger of rf-induced breakdown in the sample cells, so modulated excitation spectroscopy might prove to be particularly interesting for plasmas. A further advantage is that the states i and j of (30) may be coupled very weakly by rf, for example in a molecule, but they may nevertheless have strong modulated excitation resonances. The basic disadvantage of modulated excitation spectroscopy is the necessity to produce modulated light and detect modulated fluorescence. Good light modulators are particularly difficult to find, and fast photomultiplier tubes add additional problems, as will be discussed in Section 5. Nevertheless we expect that as these technical difficulties are overcome, modulated excitation spectroscopy will be much more widely used. Some of the nicest recent work with this technique has been done by Lehmann and his collaborators.[51]

It is worth noting that the same kind of arguments that lead to (30) imply that if the state i is excited at a rate

$$S_{ii} = s(1 + m \cos \omega t) \tag{31}$$

then the population of the state i will be

$$\rho_{ii} = \frac{s}{\gamma} \left[1 + \frac{\gamma m}{(\gamma^2 + \omega^2)^{1/2}} \cos (\omega t - \varphi) \right] \tag{32}$$

That is, the modulated component of the population (and therefore the modulated component of the fluorescence) lags the excitation by a phase φ which is

$$\tan \varphi = \omega / \gamma \tag{33}$$

This is the basis of the well-known phase-shift method[52] of measuring atomic lifetimes, since if both ω and φ can be determined experimentally γ follows from (33).

4.5. Transient Light-Modulation Spectroscopy*

As in many other branches of physics it is possible to study excited atoms either with sinusoidal excitation of the type discussed in Section 4.4

---

* See also Chapter 14 (Dodd and Series) of this work.

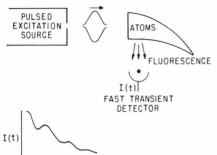

Figure 25. Sketch of an experimental setup for a transient light-modulation spectroscopy experiment.

or by pulsed excitation. A typical experiment with pulsed excitation is sketched in Figure 25. Under the proper experimental conditions the transient exponentially decaying fluorescence will be modulated at frequencies corresponding to the energy splittings of the excited states. One of the first experiments of this type was performed by Dodd.[53]

The theory of transient light-beat experiments is straightforward. Just after the atoms have been excited at time $t = 0$, the density matrix evolves according to Eq. (12). The solution to (12) is

$$\rho_{ij} = \rho_{ij}(0)\, e^{-(\gamma + i\omega_{ij})t} \tag{34}$$

The corresponding fluorescent intensity $\mathscr{L}_{ji}\rho_{ij}$ will therefore "ring" at the difference frequency between the sublevels i and j, and it will also damp at the natural decay rate γ of the excited atom. One can therefore infer the energy splittings of the excited state by determining the ringing frequencies of the transient.

It is instructive to note that one can think of the signal in a level-crossing experiment as the time average of the signal in a transient light-beat experiment. The level-crossing resonances are simply zero-frequency light beats that do not average out with time. A major advantage of transient light-beat experiments over level-crossing experiments is that one can carry out a transient light-beat experiment at zero field, where constructive interference of many coherences of the same frequency takes

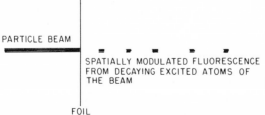

Figure 26. Spatially modulated fluorescence is observed in beam-foil excited atoms.

place. In principle, one could gain the same advantage in rf spectroscopy by sweeping the frequency of the rf source while maintaining the atomic sample at zero magnetic field. However, at least in the case of magnetic dipole rf transitions where high rf fields are required to saturate the resonances, experimental difficulties usually discourage one from sweeping the rf frequency and one must resort to a fixed rf frequency and a swept, nonzero magnetic field.

Another advantage of transient light-beat experiments is the fact that the atoms can decay freely without any external perturbations like rf fields. However, in the case of pulsed laser excitation it is necessary to keep the density of excited atoms small enough to avoid superradiant coupling of the atoms with each other, as was shown by Haroche.[54]

Transient light-beat experiments are particularly appropriate for beam-foil excitation, since the temporal light beats of the atoms are transferred into spatial modulation of the light as the atoms move away from the foil. A typical beam-foil light-beat experiment is shown in Figure 26.

5. Experimental Considerations

In this section we discuss some of the characteristic practical problems that are encountered in perturbed fluorescence experiments. The discussion is necessarily brief and intended to give only a general idea of the experimental considerations that make the difference between a successful experiment and a frustrating waste of time.

5.1. Light Sources

The success of an experiment with optical excitation depends to a large measure on the efficiency of the available light sources. Since atomic or molecular absorption lines are quite narrow it is advantageous to use a light source with an intense, narrow spectral output. This will maximize the atomic scattering while minimizing instrumental scattering.

The most straightforward way to obtain monochromatic light is to construct a lamp in which the light is produced by an excited vapor of the same kind of atoms as those that are to be investigated. For example, a simple and effective mercury resonance lamp can be made from a small sealed-off quartz cell containing a droplet of mercury and a small amount of noble gas. The lamp can be excited by a high-radio-frequency electromagnetic field or by microwaves, as is illustrated in Figure 27. Similar lamps can be constructed for the heavier alkali metals and certain other elements. Two such lamps that have been used successfully for heavier alkali metals are shown in Figure 27.

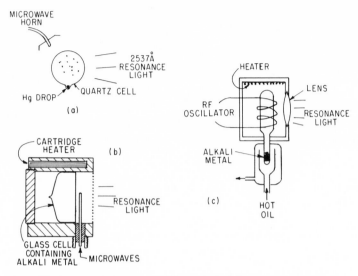

Figure 27. Some typical lamps used in perturbed fluoresence spectroscopy
experiments. (a) Hg and (b, c) heavy alkali metals are contained in a
quartz or a glass cell. Discharge is produced by microwaves or rf.

Some elements (lithium for instance) are so corrosive that it is almost
impossible to find a transparent material that is not destroyed by the
atomic vapor. Also it may be impossible to obtain a sufficient vapor
pressure of the more refractory elements in a simple lamp like the ones in
Figure 27. Such materials may be handled in a "flow lamp," which operates
in the manner sketched in Figure 28. With proper operation, lamps like
those in Figures 27 and 28 are stable enough that they contribute no noise
to an experiment except for the shot noise of the detected photons. A

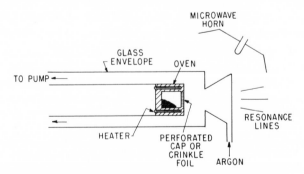

Figure 28. Flow lamp for refractory or corrosive elements.
Argon flow keeps the element from depositing on the
window.

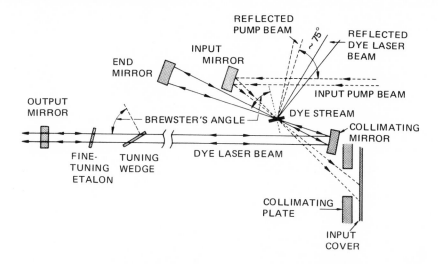

Figure 29. Schematic diagram of a jet stream cw dye laser (courtesy of Spectra-Physics). Dye lasers of this kind are usually pumped by an Ar^+ laser.

review of various lamps suitable for perturbed atomic fluorescence experiments is given by Budick *et al.*[55]

The most promising recent development for optical excitation has been the tunable laser. For example the basic components of a cw tunable dye laser are sketched in Figure 29. The approximate tuning range of a cw dye laser with various dyes is shown in Figure 30. These lasers can furnish hundreds of milliwatts of quasimonochromatic light, while the useful output of a good conventional lamp seldom exceeds a few milliwatts. The frequency stability of the laser is not too important as long as the frequency jitter is less than the width of the absorption line.

One of the most important properties of a lamp is its spectral profile. Some typical spectral profiles are shown in Figure 31. The broad line profile whose spectral intensity does not vary in the neighborhood of the absorption line (Figure 31a) is particularly useful since the details of the spectral profile are not very important for an analysis of the experimental data. Approximate broad line profiles can be obtained with lamps like those of Figures 27 and 28 if the vapor density in the lamp is not too high and if the absorption of cooler layers of vapor near the edges of the plasma does not cause self-reversal of the line. Self-reversed lines, such as those in Figure 31b are usually to be avoided since they have few photons of the correct frequency to cause optical excitation of the atoms but they have many photons in the wings of the line which can contribute to instrumental scattering.

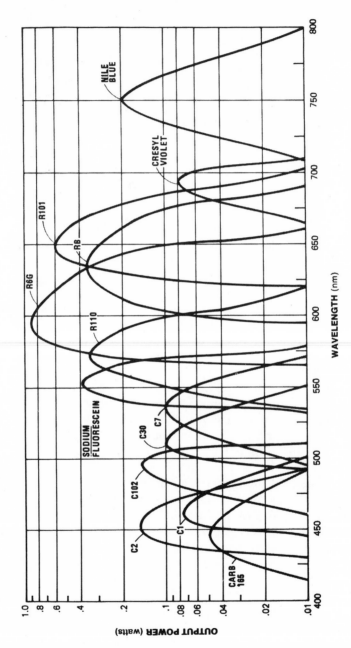

Figure 30. Typical tuning range of a cw dye laser (courtesy of Spectra-Physics).

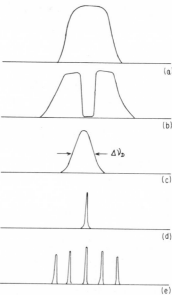

Figure 31. Typical spectral profiles of various light sources. (a) Broad-line profile obtained from the lamps of the type shown in Figures 27 and 28. (b) If these lamps are run with a very high vapor density, self-reversal of the lines can result. (c) Narrow line profile that is obtained by certain types of lamps, e.g., hollow cathode lamps. The spectral width of a line from such a lamp is comparable to the Doppler width of the absorption line $\Delta\nu_D$. (d) Spectral profile of a single mode dye laser, and (e) that of a multimode dye laser.

Perhaps the most common line shape is the narrow-line profile of Figure 31c, which may be obtained, for example, from a hollow cathode lamp. This profile has a width that is comparable to that of the absorption line of the atoms of interest, and it may be offset in frequency because of isotope shifts or various complicated broadening mechanisms in the lamp. Lamps with narrow-line spectral profiles often are very efficient in exciting the atoms of interest and they are quite satisfactory for radio-frequency spectroscopy and high-field level-crossing spectroscopy. However, the analysis of Hanle-effect and decoupling experiments with narrow-line excitation is exceedingly complicated, and it is seldom that the spectral profile of the lamp is known with sufficient accuracy to make precision measurements with these techniques.

An extreme case of a narrow-line profile is the quasimonochromatic profile, Figure 31d, which can be obtained from a single-mode dye laser. The spectral width of the laser profile is determined by the jitter of the mode. In addition to the problems associated with narrow-line excitation from a conventional lamp, such lines may be so intense that effects of nonlinear excitation, like light shifts and broadening of the atomic levels, can be important. In more common use are the multimode dye lasers which have spectral profiles of the type shown in Figure 31e. The width of each mode is again determined by the jitter of the mode and there are a number of modes separated by the free spectral range of the laser cavity (typically a few hundred MHz). Usually only a few laser modes overlap the absorption line.

5.2. Detectors

In a few favorable cases it is possible to observe the fluorescence of excited atoms by eye. The yellow resonance line of sodium atoms and the many fluorescent lines of iodine molecules are good examples that are suitable for demonstration. However, visual observations are not suitable for quantitative work, and various photoelectric devices are the most widely used detectors today.

Simple photodetection without electron multiplier chains can be used in experiments with very high light levels. Photocells are not very sensitive, and except for very high light levels the noise is determined by the temperature of the measuring apparatus, i.e., the noise power in the bandwidth Δf is $4kT\,\Delta f$.

The best general-purpose detector is a photomultiplier tube. Detailed descriptions of the characteristics and operating procedures for photomultiplier tubes are usually furnished by the manufacturer. A photomultiplier tube can be operated either as an analog or digital device, that is, one can either count pulses (photons) or measure the current through the anode. Pulse counting has the advantage of great stability and freedom from drift, but it requires more equipment than an analog arrangement. Furthermore, the light levels are often so high that photon counting is not feasible.

With a good photomultiplier tube it is frequently possible to obtain signal-to-noise ratios that are limited only by the shot noise of the detected photon. That is, the "noise" is equal to the square root of the number of detected photons.

For very low light levels where dark current is a problem it is possible to significantly improve the signal-to-noise ratios of photomultiplier tubes by cooling them with dry ice or liquid nitrogen. Except for $S1$ photosurfaces (Ag CsO) dry ice (solid CO_2) temperatures are adequate and no

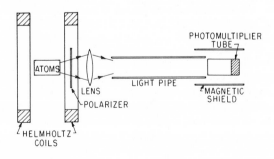

Figure 32. It is desirable to keep the photomultiplier tube outside the region of large magnetic field. Large solid angle of detection may still be obtained by use of a light pipe.

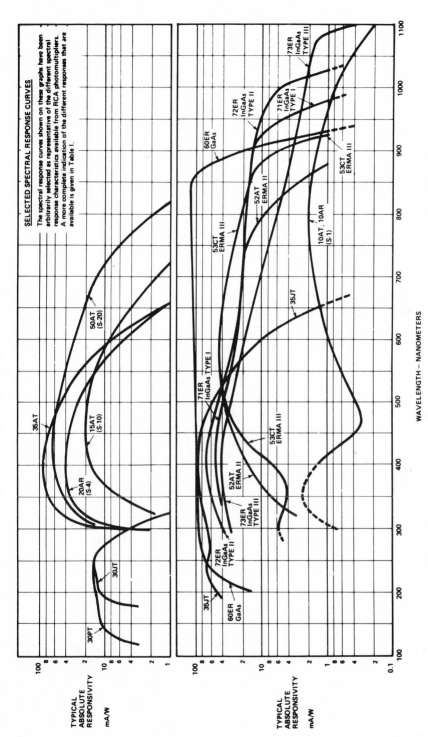

Figure 33. Spectral response characteristics of various photosurfaces (courtesy of RCA Corporation).

significant gain in signal-to-noise ratio is achieved by cooling to liquid-nitrogen temperatures. Some caution is required to prevent condensation of atmospheric moisture on the window of the tube.

Photomultiplier tubes are very sensitive to magnetic fields and it is necessary to move them outside of the region of the magnetic field and to provide magnetic shielding. A typical arrangement is shown in Fig. 32. The light pipe consists of a cylindrical glass tube whose diameter is about the same as that of the window of the photomultiplier tube. The interior of the pipe is coated with highly reflecting aluminum. Since the multiple reflections in the light pipe scramble the polarization of the light, it is necessary to analyze the polarization of the fluorescence before the light enters the light pipe. Additional magnetic shielding can be provided by one or more cylindrical, high-permeability ferromagnetic shields which are wrapped around the P.M. tube.

The basic limit on the time resolution of a photomultiplier tube is determined by the spread in the transit times of the electrons through the dynode chain. The fastest commercially available tubes can be used to detect steady-state modulated light at frequencies up to about 100 MHz, and a few experimental tubes are available that operate at frequencies as high as 10,000 MHz. For good high-frequency response the effective RC time constant of the anode load must be kept as small as possible, and if long coaxial cables are needed it may be necessary to build impedance-matching preamplifiers into the base of the photomultiplier tube.

The practical wavelength range of photomultiplier tubes extends from about 10,000 to about 2000 Å. The short-wavelength limit is determined by the transparency of the windows of the tube, and windowless photomultiplier tubes are available for use at wavelengths much shorter than 2000 Å. The long-wavelength limit of photomultiplier tubes is set by the quantum efficiency of the photo surface. Experimental tubes with gallium arsenide and indium antimonide are available with useful quantum efficiencies at 11,000 Å. The quantum efficiencies of several representative photo surfaces are shown in Figure 33.

Other detectors have been used for special situations. For instance, Bhaskar and Lurio have used "channeltron" detectors, which can be

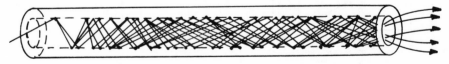

Figure 34. Schematic illustration of a channeltron electron multiplier. Multiplication of photoelectrons is achieved by applying a potential difference across the two ends of a ceramic tube.

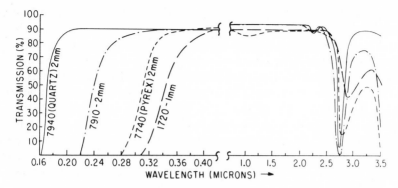

Figure 35. Optical transmission characteristics of various glasses and quartz. These curves are essentially flat in the 0.4–1.0 μ range. The numbers (e.g., 7740) specify the code used by Corning Glass Works.

operated in a vacuum and which have extremely low background counting rates. A schematic diagram of a channeltron is shown in Figure 34.

In the near-infrared region of the spectrum beyond about 10,000 Å it is possible to use solid-state detectors. Silicon photovoltaic devices and lead sulfide photoconductive devices have been most widely used. Since these devices have no electron multiplication they are much less sensitive than photomultiplier tubes and they can be used only with the strongest sources of fluorescence.

5.3. Sample Containers

For many elements the atomic vapor is contained in a glass or a quartz cell. The element is generally distilled under high vacuum into the cells.

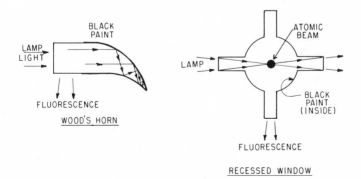

Figure 36. Instrumentally scattered light can, in general, be minimized by an arrangement of the type shown here.

Typical transmission characteristics of various common glass and quartz are shown in Figure 35.

The vapor cells have the advantage of great convenience, and of economy for expensive separated isotopes. However, some elements are either too refractory or are too reactive at elevated temperatures to be contained in glass or quartz cells. In such cases atomic beams must be used.

If fluorescent light of the same wavelength as the exciting light is detected, great pains must be taken to eliminate instrumental scattering of the exciting light. In many early experiments a Wood's horn was used to suppress instrumental scattering. A few of the experimental arrangements used to suppress instrumental scattering of the excited light are shown in Fig. 36.

5.4. Optical Filters, Polarizers, Retardation Plates

The success of an experiment often depends critically on the choice of appropriate optical filters. Two types of filters are in general use, colored glass filters and interference filters.

Colored glass filters owe their properties to special heat treatment which results in the precipitation of fine particles in the glass or to the presence of ions of transition or rare-earth metals within the glass matrix. These filters are inexpensive and are usually available in any desired size. They are quite insensitive to the solid angle of the incident light and they are capable of giving tremendous discrimination between widely separated wavelengths. Extensive tabulations of the properties of colored glass filters

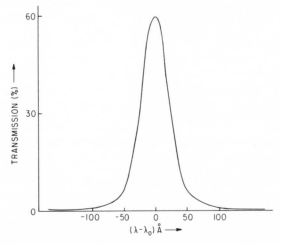

Figure 37. Transmission characteristics of a typical interference filter.

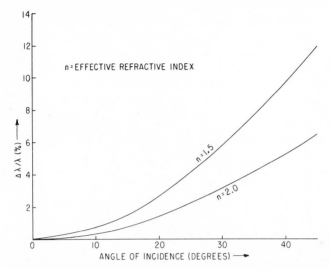

Figure 38. The transmission peak of an interference filter shifts toward short wavelength side for light that is not incident normally. This figure shows the fractional shift in the peak transmission as a function of the angle of incidence for two different (and typical) *effective* refractive indices of the filter.

have been published by the Schott and Corning glass companies. Some care is necessary when using the colored glass filters since some of them fluoresce at longer wavelengths.

The most serious disadvantage of colored glass filters is their poor resolution for closely spaced wavelengths. Very high resolution can be

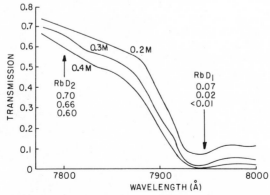

Figure 39. Transmission characteristics of a 1-cm thickness of neodymium chloride solution of 0.2M, 0.3M, and 0.4M concentration (from Ref. 56).

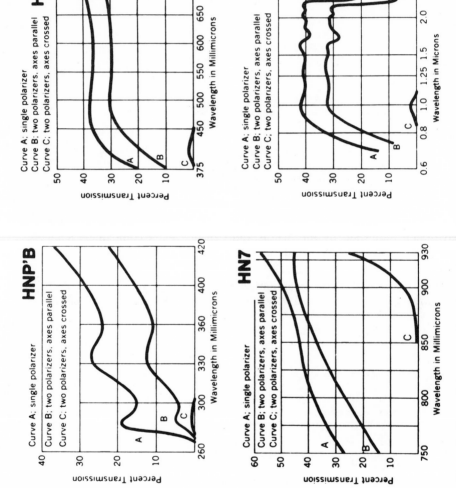

Figure 40. Transmission curves for a few of the many types of polarizer sheets that are commercially available (courtesy of Polaroid Corporation).

obtained with interference filters, which operate as a closely spaced, fixed frequency Fabry–Perot interferometer. The transmission curve of a typical interference filter is shown in Figure 37.

A serious problem with interference filters is their sensitivity to the angle of incidence of the light. The shift of the peak transmittance wavelength of an interference filter as a function of the angle of incidence is shown in Figure 38. This shift is always toward shorter wavelength. For proper operation of an interference filter it is necessary to collimate the light so that angle of incidence does not exceed the limits set by the bandpass of the filter.

Various liquid solutions generally of transition metals or rare earth and certain vapors (for example I_2 and Cl_2) have been used as optical filters. These are inexpensive and they can be prepared in any size or thickness. For example, the transmission curve of NdCl solution[56] is shown in Figure 39. This solution can be used, for example, to filter out the 7947-Å line of the Rb resonance doublet.

The most convenient polarizers for the visible, near infrared, and near ultraviolet region of the spectrum are the plastic sheets developed by the Polaroid Corporation. Representative extinction data are shown in Figure 40. Some care must be taken to select the appropriate plastic polarizer for the spectral region of interest.

Polarizers for the ultraviolet region of the spectrum can be prepared by depositing certain dyes on rubbed quartz substrates, as described by McDermott and Novick.[57] "Polacoat" polarizers of this type are available commercially.

Because of their great expense and their limited acceptance area and solid angle Nicol polarizers and similar quartz crystal polarizers have not been widely used in fluorescence experiments.

Plastic phase-retardation plates for the visible and near infrared region of the spectrum are available from the Polaroid Corporation. These

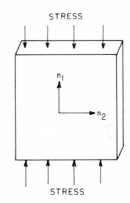

Figure 41. A uniformly stressed fused quartz plate acts as a tunable quarterwave plate.[95] These plates are very sensitive to thermal drifts.

are inexpensive and can be obtained with very large areas. For the ultraviolet region of the spectrum it is still necessary to use mica phase-retardation plates or crystalline quartz plates.

Adjustable phase retardation plates can be made from plates of fused quartz which are uniformly stressed to produce the desired degree of birefringence. The basic principle is illustrated in Figure 41. The indices of refraction are different for light that is polarized parallel or perpendicular to the stress axis. Such plates will operate at wavelengths down to about 2000 Å.

5.5. Magnetic Fields

The magnetic fields required for many fluorescence experiments are in the zero to few hundred gauss range and are most conveniently produced by a pair of Helmholtz coils. It is usually necessary to water-cool or oil-cool the coils if fields of a few hundred gauss are required. Helmholtz coils give sufficiently uniform fields and have the advantage that the atomic vapor is accessible from all sides for excitation and fluorescence detection, and large solid angles can be used. Moreover, the relationship between the field and the current through the coils is linear and it is necessary to calibrate the coils only once. A very precise and convenient way to calibrate the coils is by the optical pumping of the ground state.

For fields in excess of a kilogauss iron core magnets are usually required. Such magnets make it much more difficult to get light into or out of the sample container, and the space between the pole faces is usually quite limited. Another point to keep in mind is that fields of the order of a kilogauss are usually sufficient to scan the absorption line of an atom by several Doppler widths. Narrow-line lamps may fail to excite atoms in such high fields. Because of hysteresis effects, such fields must be continuously monitored by some kind of probe, and nuclear magnetic resonance probes are most widely used. Good current regulation is essential for both Helmholtz coils and iron core magnets.

5.6. Electric Fields

Electric fields are much more difficult to work with than magnetic fields. It is not wise to apply external electric fields to glass cells, since the cells walls can easily accumulate stray electrostatic charges and distort the field inside the cell. Internal electrodes are necessary, and in practice it is difficult to obtain fields much in excess of 60,000 V/cm, although Marrus[58] has shown that larger fields can be obtained with heated glass electrodes. Sharp edges must be avoided on leads and electrodes. Alkali metals may form a conducting film within a glass cell and completely prevent the use of internal electrodes. In such cases an atomic beam can be used. The useful volume and homogeneity of electric fields are usually much worse than those of magnetic fields.

5.7. Radio-Frequency Sources

Fairly large radio-frequency fields are required to produce magnetic dipole transitions between the different Zeeman sublevels of an excited atomic state, and the radio-frequency equipment in such an experiment is a major part of the experimental apparatus. Hundreds of watts of rf power may be required to cause transitions in a short-lived excited state. Experiments in which electric dipole transitions are used require very small amounts of power, and it is often possible to obtain sufficient microwave power with crystal mixers.

The basic components of the rf system are illustrated in Figure 42. Fairly inexpensive rf generators and amplifiers can often be obtained from firms that cater to the amateur radio market. The source should be capable of generating some tens and preferably hundreds of watts of rf power. The power should be available at a load impedance that matches that of the transmission line and a power meter should be included in the transmission line to monitor the transmitted and reflected power. To avoid reflected rf power a matching device should be attached to the rf loop. Both the primary and secondary circuits are tuned to resonate at the rf frequency and the mutual coupling is adjusted by varying the distance between the primary and secondary turns of the transformer until no power is reflected. For frequencies above a few hundred MHz cavity resonators are more convenient. A typical cavity for operation at 150 MHz is shown in Figure 43. At high frequencies and high rf powers it is important to avoid discharges in the cells. Since the discharge is initiated by the rf electric field, one way to avoid discharges is to place the sample in a region of maximum magnetic field and minimum electric field as indicated in Figure 43. It is

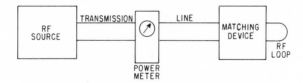

Figure 42. Typical setup for feeding rf into a loop. In order
to avoid standing waves in the transmission line, it is
necessary that the effective impedance of the loop
(plus the matching network) be equal to the characteristic impedance of the transmission line and the
impedance of the rf generator.

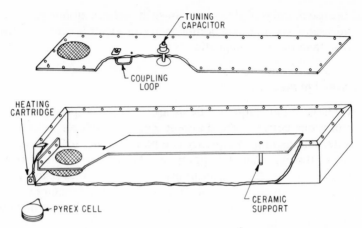

Figure 43. For frequencies on the order of 10^8 Hz and above, lumped circuits are not particularly convenient. However, a $\lambda/4$ rf cavity may be conveniently used. The central conductor (a flat strip) is approximately $\lambda/4$ long and the sample cell is placed near the shorted end of the cavity where the rf magnetic field is maximum but the rf electric field is minimum.

also helpful to keep the cell thickness as thin as possible to avoid multipacting breakdown.[26]

Completely enclosed cavity resonators like those of Figure 43 are also helpful in preventing rf pickup in the detection electronics. Pickup is particularly troublesome in light-modulation experiments where modulated components of the light intensity must be detected at the frequency of the rf field. Good rf shielding, the avoidance of large standing-wave ratios in the coaxial cables, and modulation of the magnetic field can be used to eliminate most of the serious pickup problems.

5.8. Signal-Processing Equipment

To obtain the optimum signal-to-noise ratio in a perturbed fluorescence experiment it is necessary to use proper signal-averaging equipment.

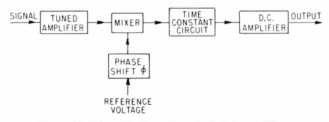

Figure 44. Schematic illustration of a lock-in amplifier.

Certainly the most versatile device is the lock-in amplifier or phase-sensitive detector which is illustrated in Figure 44. In order to use phase-sensitive detection, the signal should be modulated at some reference frequency. The amplifier shifts the phase of a reference voltage by ϕ and mixes it with the signal. The mixed signal (a dc level with noise on it) is filtered by an RC circuit of time constant τ and the output is used to drive a chart recorder or other recording device. When used with a photomultiplier tube the voltage signal-to-noise ratio of the lock-in amplifier is essentially the same as the statistical signal-to-noise ratio one would obtain by counting pulses from the phototube for a time τ.

The signal from the phototube can be modulated in several ways, three of them being most commonly used. In rf spectroscopy experiments it is generally best to modulate the rf (though not always the most convenient). In certain situations, polarization modulation is most effective, and this can be achieved as shown in Figure 45. Finally, magnetic field modulation has been most commonly used, particularly for level-crossing signals. With magnetic field modulation the signal shape is drastically changed. For example, a signal that is Lorentzian in dc detection is observed as a dispersion-shaped signal with magnetic field modulation. Care must be used in analyzing signals obtained with magnetic field modulation. The width of the observed signals depends on the amplitude of modulation, and any asymmetries or nonlinearities in the modulation can cause a shift of the center. Wahlquist[59] and Isler[60] have discussed the effects of modulation in detail.

In many cases the signals are so small that long integration times are necessary in order to obtain good signal-to-noise ratio. It is not very practical to use time constants longer than 30 sec on the lock-in amplifiers owing to the drifts in the light sources or other instruments. In these cases minicomputers or multichannel analyzers have been used to do signal averaging.

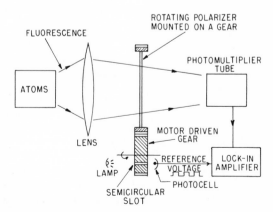

Figure 45. A typical arrangement for polarization modulation. A linear polarizer (or a $\lambda/4$ plate for circular polarization) is mounted on a rotating gear. The reference frequency is generated by a incandescent lamp, a semicircular slot in the gear, and a photocell as shown in the diagram.

6. A Case Study: Rubidium

A large number of results have been obtained by perturbed fluorescence techniques in a wide variety of atoms. The atomic parameters that have generally been measured are listed in Table 1. Not all of the techniques that we have discussed in the previous sections have been used on any one kind of atom. However, the scope of this article does not permit us to discuss results for all atomic species. We shall therefore confine ourselves to experiments on rubidium atoms only, as an example. We have chosen Rb as an example because Rb atoms (and all alkali metal atoms) have a single valence electron which determines the gross spectroscopic properties as shown in the Grotrian diagram of Figure 46. However, as we shall discuss later, the core electrons do have a profound influence on the fine and hyperfine structures of these atoms, and perturbed fluorescence techniques have proved to be most fruitful in the investigation of these structures.

6.1. The P States

We will discuss the results for the P states of Rb atoms first because these states were the first ones to be investigated by perturbed

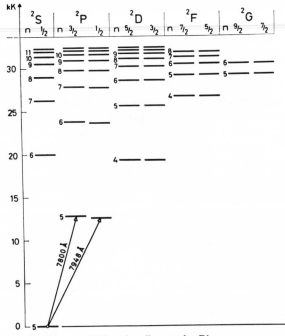

Figure 46. Grotrian diagram for Rb.

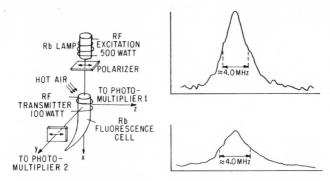

Figure 47. Experimental setup and the results of the first perturbed fluorescence experiments on Rb by Meyer-Berkhout.[61] The curves are the optical double resonance curves for the $6^2P_{3/2}$ states of Rb^{85} and Rb^{87}.

fluorescence techniques, since P states can be optically excited directly from the ground state. The first measurement of the hfs in a P state of Rb was done by Meyer-Berkhout[61] in 1955 on the $6^2P_{3/2}$ state by optical double resonance, and the values of the nuclear quadrupole moments of Rb^{85} and Rb^{87} were deduced* (Rb^{85} and Rb^{87} are the two stable isotopes of Rb). Figure 47 shows Meyer-Berkhout's experimental arrangement and typical results. At that time magnetic dipole moments of the Rb^{85} and Rb^{87} nuclei were already known from atomic-beam magnetic-resonance experiments in the ground $5^2S_{1/2}$ state. However, the nuclear electric quadrupole moments could not be derived from the hfs measurements in the ground state because the quadrupole interaction constant B is zero for states where $J \leq 1/2$ owing to the spherical symmetry of the charge distribution. The determination of the nuclear electric quadrupole moments (Q) was a big triumph for the perturbed fluorescence techniques.

Subsequent to Meyer-Berkhout's measurement of the hfs in the $6^2P_{3/2}$ state, several other measurements have been performed, and Table 2 shows the final values of A and B that have been obtained. In this table and in the subsequent tables in this section, we have usually listed values that we consider representative of the most accurate existing measurements and we have also referenced the first measurement of the quantity. A critical review of all experimental measurements can be found in the recent paper by Arimondo *et al.*[63]

Although measurements of the hfs are invaluable for the determination of nuclear moments, they are also very sensitive tests of the

---

* Almost simultaneously with Meyer-Berkhout, Senitzky and Rabi also measured the electric quadrupole moments of Rb^{85} and Rb^{87} by atomic-beam magnetic-resonance techniques. Their measurement, however, was on the $5^2P_{3/2}$ states (Ref. 62).

Table 2. Results of Measurements by the Perturbed Fluorescence Spectroscopy in the P States of Rb

State	A (MHz)		B (MHz)		τ (nsec)	g_J	α_2 [MHz/(KV/cm)^2]
	Rb^{85}	Rb^{87}	Rb^{85}	Rb^{87}			
$5\,^2P_{3/2}$	25.029(16)[a]	84.853(30)[b]	26.032(70)[a]	12.611(70)[b]	25.5(5)[c]	1.3362(6)[d]	
$6\,^2P_{3/2}$	8.16(6)[e]	27.63(10)[e]	8.199(40)[f]	4.000(39)[f]	111(3)[d]	1.3337(10)[g]	−0.521(21)[h]
$7\,^2P_{3/2}$	3.71(1)[i]	12.57(1)[i]	3.68(8)[i]	1.768(8)[d]	233(10)[d]	1.3359[d]	−3.2(2)[j]
$8\,^2P_{3/2}$	1.99(2)[k]	6.747(14)[d]	1.98(12)[k]	0.933(20)[d]	400(80)[k]		
$9\,^2P_{3/2}$		4.05(3)[l]		0.55(3)[l]		1.3335(15)[l]	
$10\,^2P_{3/2}$		2.60(8)[l]				1.3332(20)[l]	
$5\,^2P_{1/2}$					30(3)[m]		
$6\,^2P_{1/2}$	39.11(3)[n]	132.56(3)[n]			114(3)[n]	0.6659(3)[n]	
$7\,^2P_{1/2}$	17.68(8)[n]	59.92(9)[n]				0.6655(5)[n]	
$8\,^2P_{1/2}$		32.12(11)[o]					

[a] Reference 82, 96. [b] Reference 83. [c] Reference 84. [d] Reference 85. [e] Reference 86, 88. [f] Reference 87. [g] Reference 87. [h] Reference 89. [i] Reference 90. [j] Reference 91. [k] Reference 92. [l] Reference 93. [m] Reference 77. [n] Reference 65. [o] Reference 67.

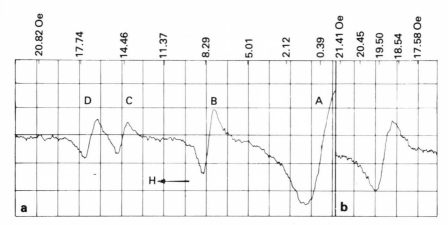

Figure 48. (a) Level-crossing signals in the $6^2P_{3/2}$ state of Rb^{85} observed by Bucka *et al.*[64] A is the zero-field level crossing (Hanle effect) while B, C, and D are the high-field level crossings. Dispersion shape of the signal results from the magnetic field modulation. (b) High-field level-crossing signal in the $6^2P_{3/2}$ state of Rb^{87}.

accuracy of computational methods in atomic physics. For this reason the hfs in many excited states has been measured as shown in Table 2. Typical zero-field (Hanle-effect) and high-field level-crossing signals in the $6^2P_{3/2}$ state of Rb, obtained by Bucka *et al.*,[64] are shown in Figure 48. Feiertag and zu Putlitz[65] have measured hfs in several $P_{1/2}$ states in order to study the core-polarization effects in Rb, and some of their results are shown in Figure 49. Although P states are accessible by direct optical excitation of the ground-state atoms by resonance radiation, the oscillator strengths for higher-resonance lines are so low that good signal-to-noise ratios cannot be obtained by this method for highly excited P states. Therefore, the highly excited states have recently been measured by rf spectroscopy in

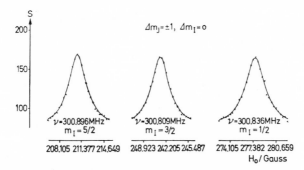

Figure 49. Optical double-resonance signals for the $6^2P_{1/2}$ state of Rb^{85} observed by Feiertag and zu Putlitz.[65]

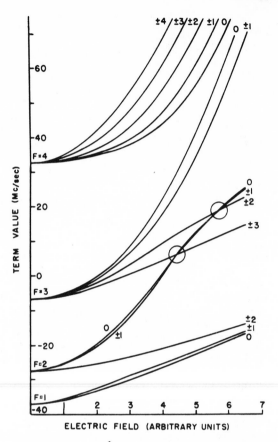

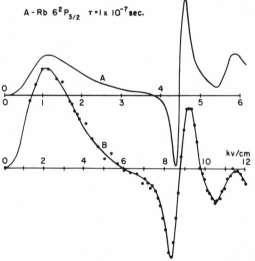

ELECTRIC FIELD (ARBITRARY UNITS)

A - Rb $6^2P_{3/2}$ $\tau = 1 \times 10^{-7}$ sec.

Figure 50. Top: Dependence of the Rb[85] hyperfine-structure sublevels on the applied electric field E. A common downward shift of all levels proportional to E^2 has been suppressed. Two level crossings are shown by circles. Bottom: Level-crossing signals. Curve A is the theoretical derivative curve for the natural isotope mixture of 72.15% Rb[85] and 27.85% Rb[87]. Curve B is the phase-sensitive amplified output signal as a function of the applied field observed by Khadjavi *et al.*[68]

conjunction with a multiple-excitation scheme involving a cw dye laser and both stepwise and cascade excitation.[66,67]

In 1966 Kadjavi et al.[68] showed that it is possible to observe level-crossing signals when sublevels of an atom cross in a *pure electric* field as shown in Figure 50. Using the method of pure electric field level crossing, they were able to measure the tensor Stark polarizability α_2 for the $6^2P_{3/2}$ state of Rb.[69] The tensor polarizability α_2 is a sensitive test of the radial part of the electronic wave functions.

Many radiative lifetimes and g_J values have also been measured by the fluorescence techniques as shown in Table 2.

Recently, zu Putlitz and collaborators[70] used the fluorescence techniques for the determination of nuclear moments in the short-lived radioactive isotopes of Rb.

6.2. The S States

The hyperfine structure of the ground state $(5^2S_{1/2})$ of Rb is much larger than the Doppler width of the spectral lines and it was measured quite early by conventional optical spectroscopy. Later, precision measurements were performed by several investigators by optical pumping and atomic-beam magnetic-resonance techniques. However, accurate measurements of the hfs in the excited S states were not performed till 1971. The difficulty, of course, had been that those states could not be populated by direct optical excitation of the ground states. The first perturbed fluorescence measurements in the excited S states were performed by Chang et al.,[71] who used decoupling spectroscopy with cascade excitation to investigate the hfs. The cascade decoupling curves for the $7^2S_{1/2}$ state of Rb^{85} and Rb^{87} are shown in Figure 51. Later, more accurate measurements in the low-lying excited S states were performed by cascade radio-frequency spectroscopy[47] and typical data for the $7^2S_{1/2}$ state of Rb^{87} are shown in Figure 52. Cascade excitation is very efficient in transferring excitation and polarization from state to state. For example, it has been possible to measure the hfs of the $6^2S_{1/2}$ state of Rb by a multiple cascade scheme[26] as shown in Figure 53. Although cascade excitation is an efficient process it is not terribly useful for the investigation of highly excited S states because the oscillator strengths for excitation of high P states are very low and the wavelengths are in the ultraviolet, beyond the cutoff of most glasses. Tsekeris and Gupta[72] and Farley et al.[73] have used stepwise excitation, with a cw dye laser used in the second step of the excitation, as shown in Figure 54, to reach the highly excited S states. Radio-frequency spectroscopy has been used to measure the hyperfine structures. We have compiled the results for the excited S states of Rb in Table 3.

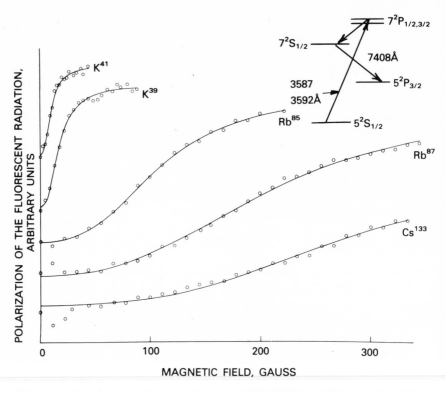

Figure 51. Decoupling signals for the $7^2S_{1/2}$ states of Rb^{85} and Rb^{87} (along with those for the second excited states of K^{39}, K^{41}, and Cs^{133}) observed by Gupta *et al.*[23] The insert shows the excitation and detection scheme. The width of the decoupling curve increases systematically from K^{41} to Cs^{133}, as the hyperfine structure increases.

Coherent excitation is transferred efficiently at zero magnetic field in the cascade process as pointed out in Section 3.3. Bulos *et al.*[74] have used cascade Hanle effect to measure the lifetime of the $7^2S_{1/2}$ state of Rb, and their results are shown in Figure 55. Although the cascade Hanle effect is a very useful technique for measuring the lifetimes of the optically inaccessible states, its fundamental limitation is that lifetimes (and g_J values) of all the intermediate states involved in the cascade process must be known reliably. Measurement of the excited S-state lifetimes by the Hanle effect in the laser stepwise excited atoms has, to our knowledge, not been attempted owing to the uncertainty associated with the spectral profile of the exciting laser light. Either the spectral profile of the exciting light should be essentially flat over the absorption width of the atoms, or it should be known with sufficient reliability that a detailed theoretical

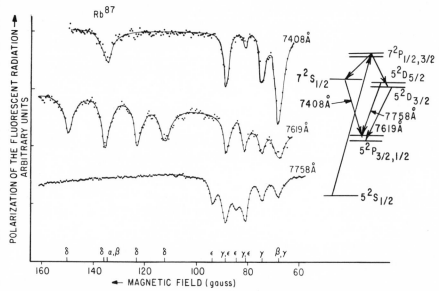

Figure 52. Cascade rf spectroscopy signals in the $7^2S_{1/2}$ (α), $5^2D_{3/2}$ (δ), and $5^2D_{5/2}$ (ε) states of Rb^{87} at $\nu = 147.5$ MHz as observed by Gupta et al.[47] Rf resonances in the $7^2P_{3/2}$ state (γ) are also observed, while those in the $7^2P_{1/2}$ state (β) are too small to be seen here. Excitation and detection scheme is shown on the right.

analysis of the Hanle curve is possible. None of the above two conditions is easily attained in a multimode laser.

6.3. The D States

The first hyperfine structure measurements in the D states of Rb were performed by cascade radio-frequency spectroscopy in 1972.[47] The rf resonances in the $5^2D_{3/2}$ and the $5^2D_{5/2}$ states of Rb^{87} are shown in Figure 52. These resonances have been used to determine accurate magnitudes of

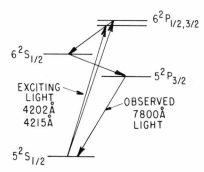

Figure 53. $6^2S_{1/2}$ state of Rb has been investigated in a multiple cascade scheme by Gupta et al.[26] $6^2S_{1/2}$ state fluoresces in the infrared, where photomultiplier tubes cannot be used. Therefore, rf resonances in the $6^2S_{1/2}$ state as observed in the 7800-Å fluorescent light.

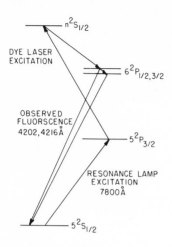

Figure 54. Stepwise excitation scheme for the highly excited S states used by Tsekeris and Gupta.[72] In order to get away from the instrumentally scattered laser light, direct fluorescence of the $n^2S_{1/2}$ atoms is not observed, rather, 4202- and 4216-Å fluorescent light is observed.

the hfs in the 5^2D states, as shown in Table 4. Rf spectroscopy, however, does not conveniently yield the sign of the hyperfine interaction constants. Decoupling spectroscopy, on the other hand, is quite sensitive to the sign of the hfs. Figure 56 shows the decoupling signal for $5^2D_{3/2}$ and the $5^2D_{5/2}$ states of Rb^{87} and the theoretical fit to these data. In generating the theoretical curves the magnitude of the magnetic dipole coupling constant A has been taken to be that given by the radio-frequency spectroscopy. The results of these fits are extremely surprising. For the $5^2D_{3/2}$ state the positive value of A fits the data very well but the $5^2D_{5/2}$-state data are consistent with a negative value of A only. Inverted hfs in the $D_{5/2}$ state is unexpected, especially since the hfs is normal for the P states, which have a larger overlap with the core electrons than the D states. It is worth pointing out here that the fine-structure intervals in the D states of Rb are also anomalous. The fine-structure intervals in the low-lying D states are much

Table 3. Results of the Measurements by Perturbed Fluorescence Spectroscopy in the S States of Rb

	A (MHz)		
State	Rb^{85}	Rb^{87}	τ (nsec)
$6^2S_{1/2}$	$239.3(12)^a$	$809.1(50)^a$	
$7^2S_{1/2}$	$94.00(64)^a$	$318.1(32)^a$	$91(11)^d$
$8^2S_{1/2}$	$45.5(20)^a$	$159.2(15)^b$	
$9^2S_{1/2}$		$90.9(8)^b$	
$10^2S_{1/2}$		$56.3(2)^c$	
$11^2S_{1/2}$		$37.4(3)^c$	

^{a} Reference 26. ^{b} Reference 72. ^{c} Reference 73. ^{d} Reference 74.

Table 4. Results of the Measurement by Perturbed Fluorescence Spectroscopy in the D States of Rb

State	A (MHz)		B (MHz)		τ (nsec)	g_J	α_2 [MHz/(KV/cm)^2]
	Rb^{85}	Rb^{87}	Rb^{85}	Rb^{87}			
$4^2D_{3/2}$	7.3(5)[a]	25.1(9)[a]					
$5^2D_{3/2}$	4.18(20)[b]	14.43(23)[b]	<5[b]	<3.5[b]	205(40)[b]		
$6^2D_{3/2}$	2.28(6)[e]	7.84(5)[c]		0.53(6)[c]			-0.105(7)[e]
$7^2D_{3/2}$	1.34(1)[e]	4.53(3)[c]		0.26(4)[c]			4.95(25)[e]
$8^2D_{3/2}$	0.84(1)[e]	2.840(15)[d]		0.17(2)[d]			27.01(1.4)[e]
$9^2D_{3/2}$		1.900(10)[d]		0.11(3)[d]			
$4^2D_{5/2}$	-5.2(3)[a]	-16.9(6)[a]					
$5^2D_{5/2}$	-2.12(20)[b]	7.44(10)[b]	<5[b]				
$6^2D_{5/2}$	-0.95(20)[e]	-3.4(5)[e]					0.94(5)[e]
$7^2D_{5/2}$	-0.55(10)[e]	-2.0(3)[e]					11.90(60)[e]
$8^2D_{5/2}$	-0.35(7)[e]	-1.20(15)[e]				1.1998(15)[d]	56.9(30)[e]
$9^2D_{5/2}$		(-)0.80(15)[d]				1.1995(15)[d]	180.3(90)[d]

[a] Reference 27. [b] Reference 24. [c] Reference 76. [d] Reference 77. [e] Reference 75.

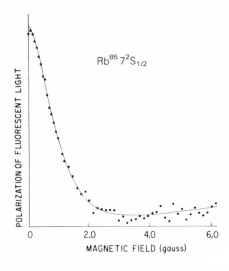

Figure 55. Cascade Hanle-effect data for the $7^2S_{1/2}$ state of Rb^{85} observed by Bulos et al.[74]

smaller than those predicted by the Landé formula, and the fine structure is actually inverted for the first excited D state (4^2D).

An obvious question one might ask is, what happens in the highly excited D states where the valence electron is very far away from the core? Does the hyperfine structure continue to be inverted in the $D_{5/2}$ states or does it become normal? Svanberg and collaborators[75–77] have carried out experiments to answer this question. They have used stepwise excitation with a resonance lamp in the first excitation step and a cw dye laser in the second step to reach the highly excited D states, as shown in Figures 57 and 58. They have performed level-crossing spectroscopy on the $D_{3/2}$ states and radio-frequency spectroscopy on the $D_{5/2}$ states, and some of their results for Rb^{87} are shown in Figures 57 and 58. They have also been able to deduce some electric quadrupole coupling constants B from the level-crossing signals in the $D_{3/2}$ states. Decoupling spectroscopy is not particularly suitable for determining the sign of the hfs coupling constants when dye laser excitation is used because of the uncertainty in the knowledge of the spectral profile of the dye laser. Hogervorst and Svanberg[75] have therefore used Stark-effect measurements to determine the signs of A. They have performed level-crossing experiments with parallel electric and magnetic fields, and from the shift in the position of the level-crossing signal in the presence of the electric field they determined the tensor polarizability α_2 for several $D_{3/2}$ and $D_{5/2}$ states as shown in Table 4. Some of their data are shown in Figure 59. The measured values of α_2 show that A is positive for all $D_{3/2}$ states that they have measured and it is negative for all the $D_{5/2}$ states that they have measured, as shown in Table 4.

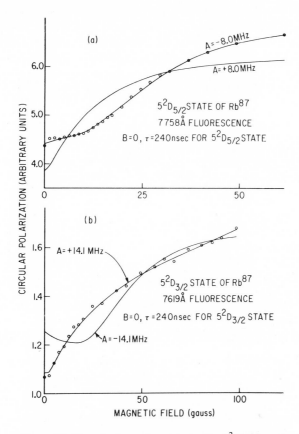

Figure 56. Cascade decoupling data for the 5^2D states of Rb[87] taken by Gupta *et al.*[(47)] Circles are the data points while the lines are the theoretical fit to these data with the indicated values of A (magnitude of A obtained by rf spectroscopy). These data show that the hfs in the $5^2D_{5/2}$ state is inverted.

   In order to gain a better understanding of the anomalous hfs in the D states, Liao *et al.*[(43)] have performed an anticrossing experiment on the 4^2D state of Rb at a sufficiently large magnetic field ($\sim$4 kg) such that J is no longer a good quantum number. Experiments at such high magnetic fields yield information about the hfs that cannot be obtained at low fields (where J is a good quantum number), as described below.

   The Hamiltonian that describes the fine and hyperfine structure of an atom placed in an external magnetic field H is

$$\mathcal{H} = \mathcal{H}_{\text{hfs}} + \zeta \mathbf{L} \cdot \mathbf{S} + \mu_B (g_l \mathbf{L} + g_s \mathbf{S} - g_I \mathbf{I}) \cdot \mathbf{H} \qquad (35)$$

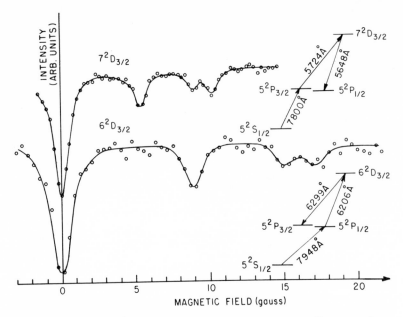

Figure 57. Zero-field and high-field level-crossing signals in the $6^2D_{3/2}$ and $7^2D_{3/2}$ states of Rb^{87} observed by Svanberg *et al.*[22] Their stepwise excitation scheme is shown in the inserts. First step of excitation is produced by a resonance lamp while the second one is produced by a cw dye laser.

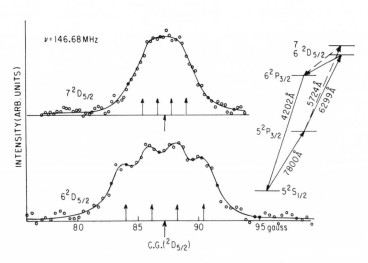

Figure 58. Rf spectroscopy signals in the $6^2D_{5/2}$ and the $7^2D_{5/2}$ states of Rb^{87} obtained by stepwise excitation by Svanberg and Tsekeris.[76] 4202-Å fluorescence was observed.

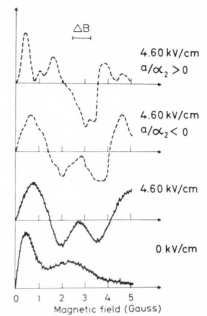

Figure 59. Level-crossing signals in the $6^2D_{5/2}$ state of Rb^{85} at zero electric field and at $E = 4.60$ kV/cm observed by Hogervorst and Svanberg.[75] Theoretical curves for two possible signs of A/α_2 are also shown. It can be concluded that $A/\alpha_2 < 0$. α_2 is calculated to be positive, indicating that the hfs in the $6^2D_{5/2}$ state is negative.

Here $\zeta = \delta w/(L + \frac{1}{2})$, δw being the zero-field fine-structure interval, and $\mathcal{H}_{\text{hfs}}$ is the hyperfine-structure Hamiltonian. The magnetic dipole and electric quadrupole contribution to the hyperfine-structure Hamiltonian of a hydrogenic atom can be written as

$$\mathcal{H}_{\text{hfs}} = \mathcal{H}_l + \mathcal{H}_d + \mathcal{H}_c + \mathcal{H}_q \tag{36}$$

The electric quadrupole contribution is

$$\mathcal{H}_q = \frac{2LQe^2}{2L+3}\langle r^{-3}\rangle_q \frac{3(\mathbf{L} \cdot \mathbf{I})^2 + (3/2)(\mathbf{L} \cdot \mathbf{I}) - L(L+1)I(I+1)}{2L(2L-1)I(2I-1)} \tag{37}$$

where Q is the nuclear quadrupole moment. The orbital contribution is

$$\mathcal{H}_l = \frac{2\mu_B\mu_I}{I}\langle r^{-3}\rangle_l \mathbf{L} \cdot \mathbf{I} \tag{38}$$

and the dipole contribution is

$$\mathcal{H}_d = \frac{4\mu_B\mu_I}{I}\langle r^{-3}\rangle_d \frac{L(L+1)}{(2L-1)(2L+3)}\left[\mathbf{S} \cdot \mathbf{I} - 3\left\{\frac{\mathbf{S} \cdot \mathbf{LL} \cdot \mathbf{I} + \mathbf{I} \cdot \mathbf{LL} \cdot \mathbf{S}}{2L(L+1)}\right\}\right] \tag{39}$$

The contact interaction is

$$\mathcal{H}_c = \frac{\mu_B\mu_I}{I}g_s\frac{8\pi}{3}|\Psi(0)|^2\mathbf{S} \cdot \mathbf{I} \equiv \frac{2\mu_B\mu_I}{I}\langle r^{-3}\rangle_c \mathbf{S} \cdot \mathbf{I} \tag{40}$$

For a nonrelativistic hydrogenic atom $\langle r^{-3}\rangle_q$, $\langle r^{-3}\rangle_l$, and $\langle r^{-3}\rangle_d$ are identical and are the inverse cube of the distance of the electron from the nucleus. The parameter $\langle r^{-3}\rangle_c$, which is defined by (40), is introduced for convenience of notation and is a measure of $|\Psi(0)|^2$ rather than the inverse cube of the distance of the electron from the nucleus.

We may also describe the hyperfine structure of the 4^2D state of rubidium with the Hamiltonian (36). However, since configuration interactions modify the hyperfine interactions, we must regard the quantities $\langle r^{-3}\rangle_i (i = q, l, d, \text{ and } c)$ as convenient independent parameters that characterize the strengths of the interactions (37)–(40). The parameters $\langle r^{-3}\rangle_i$ $(i = q, l, d, \text{ and } c)$ do not necessarily represent, even approximately, the inverse cube of the distance of the valence electron from the nucleus.

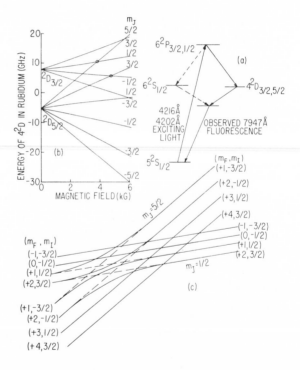

Figure 60. (a) Excitation and detection scheme in the anticrossing experiment of Liao et al.[43] (b) Zeeman splittings of the fine-structure levels in the 4^2D state of Rb in an external magnetic field. (c) The hyperfine levels involved at the 3800-G crossing in the 4^2D state of Rb[87]. Circles indicate the positions of the anticrossings.

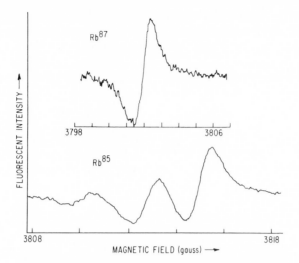

Figure 61. Typical anticrossing signals with σ exciting light observed by Liao et al.[43]

At low magnetic fields where $J = |\mathbf{L} + \mathbf{S}|$ is a good quantum number, the hfs Hamiltonian (36) reduces to

$$\mathscr{H}_{\text{hfs}} = hA_J \mathbf{J} \cdot \mathbf{I} + \mathscr{H}_q \tag{41}$$

where, for $L = 2$ and $S = \frac{1}{2}$,

$$A_{3/2} = \frac{2\mu_B\mu_I}{hI}\left(\frac{6}{5}\langle r^{-3}\rangle_l + \frac{2}{5}\langle r^{-3}\rangle_d - \frac{1}{5}\langle r^{-3}\rangle_c\right) \tag{42}$$

$$A_{5/2} = \frac{2\mu_B\mu_I}{hI}\left(\frac{4}{5}\langle r^{-3}\rangle_l - \frac{4}{35}\langle r^{-3}\rangle_d + \frac{1}{5}\langle r^{-3}\rangle_c\right) \tag{43}$$

$A_{3/2}$ and $A_{5/2}$ are the magnetic dipole coupling constants in the $D_{3/2}$ and the $D_{5/2}$ states. Low-field measurement of the magnetic hfs yields only two quantities, $A_{3/2}$ and $A_{5/2}$, while there are three independent parameters, $\langle r^{-3}\rangle_l$, $\langle r^{-3}\rangle_d$, and $\langle r^{-3}\rangle_c$, to be determined.

The experiment of Liao et al.[43] is shown schematically in Figure 60. The 4^2D state is populated by cascade excitation and 7947-Å (D_1 line) fluorescence is observed. The figure also shows the fine-structure magnetic sublevels in an external field, and there is a level crossing at about 3800 G. If a nuclear spin is present each "crossing" consists of a large number of crossings and anticrossings, as shown in Figure 60c for the case of Rb^{87}. The anticrossings are due to the hyperfine coupling of levels with the same value of the total magnetic quantum number $m_F = m_I + m_J$, and they are observed as a change in the intensity of the 7947-Å fluorescent light at the

Table 5. Results of the Fine- and Hyperfine-Structure Measurements in the 4^2D State of Rb by Liao et al.[43]

Parameter	Results
ζ	$-5344.12(28)\,\text{MHz}$
$\langle r^{-3}\rangle_c$	$-4.22(10)\times 10^{24}\,\text{cm}^{-3}$
$\langle r^{-3}\rangle_l$	$+0.195(19)\times 10^{24}\,\text{cm}^{-3}$
$\langle r^{-3}\rangle_d$	$-0.318(78)\times 10^{24}\,\text{cm}^{-3}$
$\langle r^{-3}\rangle_q$	$+0.82(57)\times 10^{24}\,\text{cm}^{-3}$

anticrossing points. Liao et al. have observed several anticrossings in Rb^{85} and Rb^{87} and some of their data are shown in Figure 61. Their anticrossing results along with the low-field results of $A_{3/2}$ and $A_{5/2}$ for Rb^{85} and Rb^{87} overdetermine the four unknown parameters $\langle r^{-3}\rangle_i$ ($i = l$, c, d, and q). Results of their least-squares fit are shown in Table 5. These results are extremely surprising. The effective value of $\langle r^{-3}\rangle$ for the dipole contribution is *negative* and the core polarization term is very large.

A determination of the total hfs Hamiltonian in this fashion gives a great deal of insight into the mechanisms responsible for anomalous hfs, and the results of Liao et al. have stimulated several theoretical attempts to explain these anomalies.[78-80] Unfortunately, total hfs Hamiltonians have not been determined for the other D states, primarily because of the problems associated with the very large magnetic fields that are needed to decouple L and S in these states. Calculations by Lindgren et al.[80] indicate that $\langle r^{-3}\rangle_l$ should indeed be negative.

6.4. The F States

The hyperfine structure in the F states is smaller than the natural width of the levels and therefore it cannot be measured easily. The fine structures in F states of Rb, however, are reminiscent of the $D_{5/2}$-state hyperfine structures. It has been known for some time from the conventional spectroscopic measurements that the fine-structure intervals in the F states are inverted; however, their magnitudes were very poorly known. Because of the difficulty in populating the F states, the perturbed fluorescence techniques were not applied for an accurate measurement of the fine structures in the F states till very recently. Farley and Gupta[81] have used radio-frequency spectroscopy in conjunction with a multistep excitation scheme involving a cw dye laser to measure the fine structures of the 6^2F and the 7^2F states of Rb. Their excitation scheme and some of the data are shown in Fig. 62 and the results are listed in Table 6.

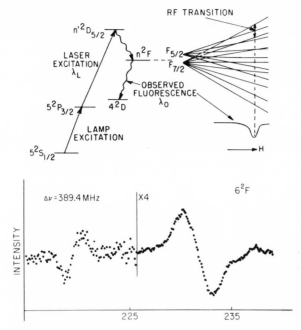

Figure 62. n^2F states are populated by the spontaneous decay of $n'^2D_{5/2}$ states. The $D_{5/2}$ states are excited by stepwise excitation and rf transitions are observed in the $n^2F \to 4^2D$ fluorescence. Typical data for the 6^2F state of Rb obtained by Farley and Gupta[81] is shown. The large resonance on the right is due to the $8^2D_{5/2}$ state, while the resonance on the left is due to the 6^2F state.

6.5. The G and Higher Angular Momentum States

No measurements have yet been carried out in the G and higher angular momentum states owing to the difficulty in populating these states by optical methods. The hyperfine structures are expected to be unresolved, but fine-structure measurements would be very interesting. It would be especially interesting to find out if the fine structures in these states are normal or inverted. Judging from the rapid progress that has

Table 6. Results of the Fine-Structure Measurements in the F States of Rb[81]

State	6^2F	7^2F
Fine-structure interval (MHz)	−486(4)	−347.6(10)

been made in the last few years, we can look forward to these measurements in the very near future.

Acknowledgments

Supported in part by the Joint Services Electronics Program (U.S. Army, U.S. Navy, and U.S. Air Force) under contract No. DAAB07-74-C-0341, and also by a grant from the Alexander von Humboldt Foundation.

References and Notes

1. E. Karlson, E. Matthias, and K. Siegbahn, *Perturbed Angular Correlations*, North-Holland Publishing Co., Amsterdam (1964).
2. R. W. Wood and A. Ellet, *Phys. Rev.* **24**, 243 (1924).
3. W. Hanle, *Z. Phys.* **30**, 93 (1924).
4. A. Mitchell and M. Zemansky, *Resonance Radiation and Excited Atoms*, Cambridge University Press, Cambridge (1961).
5. G. Breit, *Rev. Mod. Phys.* **5**, 91 (1933).
6. A. Ellet and N. P. Heydenburg, *Phys. Rev.* **46**, 583 (1934); N. P. Heydenburg, *ibid* **46**, 802 (1934).
7. J. Brossel and F. Bitter, *Phys. Rev.* **86**, 308 (1952).
8. R. M. Sternheimer, *Phys. Rev.* **80**, 102 (1950); **84**, 244 (1951); **86**, 316 (1952).
9. H. Kopfermann, *Nuclear Moments*, Academic Press, New York (1958).
10. M. L. Perl, I. I. Rabi, and B. Senitzky, *Phys. Rev.* **98**, 611 (1955).
11. G. J. Ritter and G. W. Series, *Proc. Phys. Soc. London A* **68**, 450 (1955); *Proc. R. Soc. London A* **238**, 473 (1956).
12. P. L. Sagalyn, *Phys. Rev.* **94**, 885 (1954).
13. G. zu Putlitz, *Ergeb. Exakt. Naturw.* **37**, 105 (1965).
14. F. D. Colegrove, P. A. Franken, R. R. Lewis, and R. H. Sands, *Phys. Rev. Lett.* **3**, 420 (1959); P. A. Franken, *Phys. Rev.* **121**, 508 (1961).
15. T. G. Eck, L. Foldy, and H. Wieder, *Phys. Rev. Lett.* **10**, 239 (1963); H. Wieder and T. G. Eck, *Phys. Rev.* **153**, 103 (1967).
16. M. N. McDermott and R. Novick, *Phys. Rev.* **131**, 707 (1963).
17. J. P. Barrat and C. Cohen Tannoudji, *J. Phys. Radium* **22**, 329 (1961); **24**, 443 (1961).
18. W. Happer, *Rev. Mod. Phys.* **44**, 169 (1972).
19. B. P. Kibble and S. Pancharatnam, *Proc. Phys. Soc. London* **86**, 1351 (1965).
20. R. L. Smith and T. G. Eck, *Phys. Rev. A* **2**, 2179 (1970).
21. A. Tam, *Phys. Rev.* **12**, 539 (1976).
22. S. Svanberg, P. Tsekeris, and W. Happer, *Phys. Rev. Lett.* **30**, 817 (1973).
23. R. Gupta, S. Chang, and W. Happer, *Phys. Rev. A* **6**, 529 (1972).
24. C. Tai, W. Happer, and R. Gupta, *Phys. Rev. A* **12**, 736 (1975).
25. M. Ducloy, *J. Phys. (Paris)* **31**, 533 (1970); N. Bhaskar and A. Lurio, *Phys. Rev. A* **13**, 1484 (1976).
26. R. Gupta, W. Happer, L. K. Lam, and S. Svanberg, *Phys. Rev. A* **8**, 2792 (1973).
27. L. K. Lam, R. Gupta, and W. Happer, *Phys. Rev. A* (to be published).

28. I. P. Zapesochnyi and L. L. Shimon, *Opt. Spectrosc.* **20**, 525 (1966).
29. H. W. B. Skinner and E. T. S. Appleyard, *Proc. R. Soc. London A* **117**, 224 (1928).
30. R. D. Kaul, *J. Opt. Soc. Am.* **58**, 429 (1968).
31. C. Lhuillier, P. Riviére, and J. P. Faroux, *C.R. Acad. Sci.* **276B**, 607 (1973).
32. J. C. Pebay-Peyroula, *J. Phys. Radium* **20**, 669 (1959); **20**, 721 (1959); in *Physics of the One- and Two-Electron Atoms*, Eds. F. Bopp and H. Kleinpoppen, North-Holland (1969).
33. M. Lombardi, *J. Phys. (Paris)* **30**, 631 (1969).
34. M. Pavlović and F. Laloë, *J. Phys. (Paris)* **31**, 173 (1970).
35. S. Chang, Ph.D. Dissertation, Columbia University (1972); R. Gupta, S. P. Chang, and W. Happer, *Bull. Am. Phys. Soc.* **15**, 1508 (1970).
36. S. Bashkin, *Nucl. Inst. Methods* **28**, 88 (1964).
37. L. D. Schearer and W. C. Holton, *Phys. Rev. Lett.* **24**, 1214 (1970); L. D. Schearer, *Phys. Rev. Lett.* **22**, 629 (1969).
38. A. Kastler and J. Brossel, *Compt Rend.* **229**, 1213 (1949).
39. G. W. Series, *Rep. Prog. Phys.* **22**, 280 (1959).
40. J. N. Dodd, G. W. Series, and M. J. Taylor, *Proc. R. Soc. London A* **273**, 41 (1963).
41. W. Happer and S. Svanberg, *Phys. Rev. A* **9**, 508 (1974).
42. D. T. Pegg, *J. Phys. B. Atom. Molec. Phys.* **2**, 1104 (1969).
43. K. H. Liao, L. K. Lam, R. Gupta, and W. Happer, *Phys. Rev. Lett.* **32**, 1340 (1974).
44. T. G. Eck and R. J. Huff, *Phys. Rev. Lett.* **22**, 319 (1969).
45. H.-J. Beyer and H. Kleinpoppen, *J. Phys. B* **4**, L129 (1971).
46. C. A. Kocher, *Phys. Rev. A*, **6**, 35 (1972).
47. R. Gupta, S. Chang, C. Tai, and W. Happer, *Phys. Rev. Lett.* **29**, 695 (1972).
48. G. W. Series, in *Physics of the One- and Two-Electron Atoms*, Eds. F. Bopp and H. Kleinpoppen, North-Holland, Amsterdam (1969).
49. J. N. Dodd, G. W. Series, and M. J. Taylor, *Proc. R. Soc. London* **273**, 41 (1963).
50. E. B. Aleksandrov, *Opt. Spektrosk.* **14**, 436 (1963) [English transl. *Opt. Spectrosc.* **14**, 232 (1963)].
51. M. Broyer, J.–C. Lehmann, and T. Vigue, *J. Phys. (Paris)* **36**, 235 (1975).
52. O. Osberghaus and K. Ziock, *Z. Naturforsch.* **11a**, 762 (1956).
53. J. N. Dodd, W. J. Sandle, and D. Zissermann, *Proc. Phys. Soc. London* **92**, 497 (1967).
54. M. Gross, C. Fabre, P. Pillet, and S. Haroche, *Phys. Rev. Lett.* **36**, 1035 (1976).
55. B. Budick, R. Novick, and A. Lurio, *Appl. Opt.* **4**, 229 (1965).
56. A. H. Firester, *Am. J. Phys.* **36**, 366 (1968).
57. M. N. McDermott and R. Novick, *J. Opt. Soc. Am.* **51**, 1008 (1961).
58. R. Marrus, E. Wang, and J. Yellin, *Phys. Rev. Lett.* **19**, 1 (1967).
59. H. Wahlquist, *J. Chem. Phys.* **35**, 1708 (1961).
60. R. C. Isler, *J. Opt. Soc. Am.* **59**, 727 (1969).
61. U. Meyer-Berkhout, *Z. Phys.* **141**, 185 (1955).
62. B. Senitzky and I. I. Rabi, *Phys. Rev.* **103**, 315 (1956).
63. E. Arimondo, M. Inguscio and P. Violino, *Rev. Modd. Phys.* **49**, 31 (1977).
64. H. Bucka, B. Grosswendt, and H. A. Schüssler, *Z. Phys.* **194**, 193 (1966).
65. D. Feiertag and G. zu Putlitz, *Z. Phys.* **261**, 1 (1973).
66. G. Belin and S. Svanberg, *Phys. Lett.* **47A**, 5 (1974).
67. P. Tsekeris, J. Farley, and R. Gupta, *Phys. Rev. A* **11**, 2202 (1975).
68. A. Khadjavi, W. Happer, and A. Lurio, *Phys. Rev. Lett.* **17**, 463 (1966).
69. A. Khadjavi, A. Lurio, and W. Happer, *Phys. Rev.* **167**, 128 (1968).
70. F. Ackermann, I. Platz, and G. zu Putlitz, *Z. Phys.* **260**, 87 (1973).
71. S. Chang, R. Gupta, and W. Happer, *Phys. Rev. Lett.* **27**, 1036 (1971).
72. P. Tsekeris and R. Gupta, *Phys. Rev. A* **11**, 455 (1975).

73. J. Farley, P. Tsekeris, and R. Gupta, *Phys. Rev. A* **15**, 1530 (1977), (to be published, April, 1977).
74. B. R. Bulos, R. Gupta, and W. Happer, *J. Opt. Soc. Am.* **66**, 4261 (1976).
75. W. Hogervorst and S. Svanberg, *Phys. Scr.* **12**, 67 (1975).
76. S. Svanberg, P. Tsekeris, and W. Happer, *Phys. Rev. Lett.* **30**, 817 (1973); S. Svanberg and P. Tsekeris, *Phys. Rev. A* **11**, 1125 (1975).
77. G. Belin, L. Holmgren, and S. Svanberg, *Phys. Scr.* **13**, 351 (1976).
78. H. M. Foley and R. M. Sternheimer, *Phys. Lett.* **55A**, 276 (1976).
79. T. Lee, J. E. Rodgers, T. P. Das, and R. M. Sternheimer, *Phys. Rev. A* **14**, 51 (1976).
80. I. Lindgren, J. Lindgren, and A.-M. Martensson, *Z. Phys.* **A279**, 113 (1976); *Phys. Rev. A* (to be published).
81. J. Farley and R. Gupta, *Phys. Rev. A* **15**, 1952 (1977).
82. H. A. Schüssler, *Z. Phys.* **182**, 289 (1965).
83. H. Bucka, H. Kopfermann, M. Rasiwala, and H. Schüssler, *Z. Phys.* **176**, 45 (1963); other measurements are from Refs. 82 and 85.
84. R. W. Schmieder, A. Lurio, W. Happer, and A. Khadjavi, *Phys. Rev.* **2**, 1216 (1970); other measurements are from Refs. 85 and 93.
85. G. Belin and S. Svanberg, *Phys. Scr.* **4**, 269 (1971); references to others may be found in this paper.
86. Reference 61, and other measurements from Refs. 64, 82, and 87.
87. H. Bucka, H. Kopfermann, and A. Minor, *Z. Phys.* **161**, 123 (1961); other measurements are from Refs. 82, 61, 64, 85, and 88.
88. G. zu Putlitz and A. Schenck, *Z. Phys.* **183**, 428 (1965).
89. Reference 69, and also measured by L. A. Volikova, V. N. Grigoriova, G. I. Khvostenko, and M. P. Chaika, *Opt. Spectrosc.* **30**, 88 (1971); **34**, 712 (1973).
90. H. Bucka, G. zu Putlitz, and R. Rabold, *Z. Phys.* **213**, 101 (1968); other measurements from Ref. 85.
91. S. Svanberg, *Phys. Scr.* **5**, 132 (1972).
92. G. zu Putlitz and K. V. Venkataramu, *Z. Phys.* **209**, 470 (1968); other measurements from Ref. 85.
93. J. D. Feichtner, J. H. Gallagher, and M. Mizushima, *Phys. Rev.* **164**, 44 (1967).
94. W. S. Bickel, *Surf. Sci.* **37**, 971 (1973).
95. W. Happer and E. B. Saloman, *Phys. Rev.* **160**, 23 (1967).
96. Reference 82; other measurements by E. Arimondo and M. Krainska-Miszczak, *J. Phys. B* **8**, 1613 (1975).

10
Recent Developments and Results of the Atomic-Beam Magnetic-Resonance Method

SIEGFRIED PENSELIN

1. Introduction

The atomic-beam magnetic-resonance (ABMR) method, devised by Rabi and co-workers[1] in 1938 and described in detail in many review articles,[2-7] has profited during recent years from a variety of new technical developments, which have opened up new fields of application for this method and provided a broad spectrum of new information concerning the electronic structure of free atoms and the properties of nuclei. These recent developments and their results will be reviewed in this article.

Most of the experiments performed with the ABMR method, in all its forms, represent a measurement of the hyperfine structure (hfs) of free atoms. They are, therefore, experiments in the field of atomic spectroscopy. On the other hand, however, each hyperfine interaction depends on properties both of the electrons and of the nuclei of the atoms under study. By measuring hyperfine interactions, therefore, information is obtained about the electrons and the nuclei of the atoms. Experiments emphasizing either the atomic or nuclear side of hyperfine interactions must, therefore, be considered in this review.

Until recently, there was a whole class of elements that, with very few exceptions, could not be studied by the ABMR method; namely, the highly

SIEGFRIED PENSELIN • Institut für Angewandte Physik der Universität Bonn, Bonn, Germany.

refractory elements. After the development of a universal method of producing intense atomic beams of these elements,[8,9] hfs measurements were performed on most of them. The bulk of the refractory elements are transition elements with an unfilled $4d$ or $5d$ electron shell, and because of this feature of their electronic structure they tend to have many low-lying metastable electronic states, which are occupied, in the atomic beam, by a usable number of atoms. In many cases, therefore, the hfs can be studied in a number of different states of the same atom, providing a severe test of fine details of the theory of hfs. On the other hand, these experiments provide much new information on the properties of the nuclei of these elements. This is especially interesting in the case of the $5d$-shell elements, since their nuclei tend to be highly deformed. These experiments will be reviewed in Section 2.

The populations of the metastable atomic states decrease rapidly with increasing energy of the state, and consequently classical ABMR studies become extremely difficult for metastable states more than $10{,}000 \text{ cm}^{-1}$ above the ground state. A new method, described in Section 3, for studying the hfs of such states was developed recently.[10] It is very similar to the ABMR method, replacing the two inhomogeneous deflection magnets of an ABMR apparatus by two laser beams. The laser beams serve only as a means of detecting transitions between adjacent hfs states by optical pumping and resonance fluorescence, and so the line widths of the resonances are the same as those obtained with a conventional ABMR apparatus and are not influenced by the spectral width of the laser beams. This allows the same high accuracy to be obtained for the hfs measurements as for the usual ABMR measurements.

The availability of commercial rf equipment of a very high accuracy (frequency decades, oscillators, frequency stabilizers, quartz frequency standards) has made it possible to measure the frequency of rf transitions using the ABMR method, in special cases, with a relative accuracy of better than one part in 10^9. This means that in favorable cases, even the small interaction between the nuclear magnetic dipole moment and an external magnetic field of several thousand oersted can be determined to better than one part in 10^6.[11] This allows, for example, the determination of nuclear magnetic dipole moments to the same accuracy, provided that special care has been taken to avoid all possible systematic errors.[12] As a result of this improvement in precision, a topic which will be described in Section 4, the search for such small effects as a hexadecapole interaction as part of the hfs interaction has become feasible.

For more than twenty years the ABMR method has also been applied to the study of the hfs of radioactive isotopes. Most of the spins and moments of the nuclear ground states of radioactive nuclei with lifetimes above 20 min have been measured with the ABMR method. Considerable progress has been made in recent years in the measurement of these parameters for nuclei with shorter lifetimes. By putting an ABMR apparatus on-line with the on-line isotope separation facility ISOLDE, at the 600-MeV proton synchro-cyclotron at CERN, it was possible to measure more than 30 isotope spins with lifetimes down to about 20 sec.[13] Many of these neutron-deficient nuclei lie very far from the region of stable isotopes and are therefore of special interest with respect to their nuclear structure. This and other aspects of progress in the study of radioactive isotopes will be reported in Section 5.

Finally, it should be mentioned that the ABMR method has also made substantial contributions to severe tests of quantum electrodynamics and to the search for violations of P or T invariance in atomic systems. These achievements have been reviewed recently by Hughes[14] and Sandars.[15]

In view of the remarkable progress of the ABMR method in recent years, it can be said that it has retained a very impressive liveliness, despite the fact that it was invented almost 40 years ago, and it can be expected that many interesting results will be obtained from its use in the future.

2. Hyperfine-Structure Measurements of Highly Refractory Elements

2.1. Beam Production

Atomic beams suitable for study with the ABMR method are normally produced by evaporating the element to be studied from a crucible through a narrow rectangular slit (for apparatus with two-pole deflecting fields) or through a small circular hole (for apparatus with six-pole deflecting fields).[16] Normally, resistance heating of the crucible is employed, but, for higher evaporation temperatures, electron impact may be used. A fairly common difficulty in the production of suitable beams of certain elements is that of finding a crucible material that does not react with the element to be evaporated. In most cases such reactions destroy the crucible and prevent the production of an atomic beam. These difficulties become almost insurmountable for highly refractory elements with evaporation temperatures above 2.800 K. There is only one known example of the production of an atomic beam with this method above this temperature, viz., the evaporation of carbon from a crucible of tantalum carbide.[17]

Consequently, several other methods for the production of atomic beams of some highly refractory elements have been developed. Doyle and Marrus[18] produced atomic beams of Ta, Wo, Re, and Ir by sublimating atoms from thin wires heated to just below melting point. The resultant beam intensity was sufficient for the determination of the spins of seven radioactive isotopes of these elements with the ABMR method, it being the case that the detection efficiency for radioactive isotopes is much better than for stable isotopes. In the case of Re, Schlecht et al.[19] were also able to measure the hfs separations of of the two isotopes ^{186}Re and ^{188}Re in the electronic ground state. Armstrong and Marrus[20] even managed to apply, in this case, the triple-resonance method[21] in order to detect transitions with $\Delta m_J = 0$, $\Delta m_I = \pm 1$ between Zeeman levels of the ground state in magnetic fields of up to 500 Oe. In this way, they were able to measure the interaction between the nuclear magnetic moment and the external field, thus obtaining a direct measurement of the nuclear dipole moments of the two isotopes with a relative accuracy of about 3×10^{-3}. Lindgren and co-workers at Göteborg modified slightly the method of beam production[22] and were able to determine the spins of 20 radioactive isotopes of the elements Nb, Mo, Tc, Rh, Ta, Re, Os, Ir and Pt[23] with half-lives longer than 2.3 h.

Another method for the production of atomic beams of stable highly refractory atoms has been developed by Pendlebury et al.[24,25] and was used successfully for Mo, Ta, and Wo. They fed a thin (0.36-mm-diameter) wire of the material to be evaporated down through a vertical guiding cylinder. At the end of the wire they produced a drop of molten material by heating this end above the melting point by electron impact. The drop is suspended at the end of the solid wire and atoms evaporate from it. The beam cross section is defined by a slit of 0.15 mm width positioned in front of the drop. The beam intensity is typically about 1/50 of that obtained by conventional evaporation from a crucible. This low intensity severely limits the application of this method.

For a long time, the ABMR method could not be employed, with the few exceptions mentioned, in the study of stable isotopes of the highly refractory elements owing to these very serious difficulties. The hfs of many of these elements is, however, of special interest, firstly, from the standpoint of nuclear physics, since many of these elements have highly deformed nuclei (mainly those beyond the rare-earth elements), their nuclear properties being not yet well known, and secondly from the standpoint of atomic physics since most of the refractory elements have unfilled $4d$ or $5d$ shells, and have several low-lying fine-structure multiplets. Some of these multiplets result from different electron configurations, which at the evaporation temperatures are well enough populated to permit ABMR studies. They provide, therefore, suitable sources of measurement of the hfs of the same isotope in several fine-structure states allowing the determination of fine details of the hfs such as configuration interactions and relativistic effects.

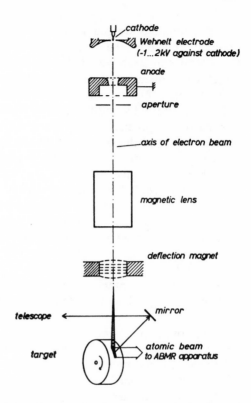

Figure 1. Arrangement for the production of atomic beams of refractory elements (schematic).

Finally, a universal method for the production of atomic beams of stable refractory elements was developed by the ABMR group at Bonn.[8] The principle of the method is shown in Figure 1. An area equivalent in its dimensions to the slit of a conventional atomic beam crucible is heated locally on the mantle surface of a cylindrical target consisting of a solid piece of the element to be studied. The local heating above the evaporation temperature is achieved by a well-collimated beam of electrons generated by the specially adapted electron gun of a commercial electron-welding apparatus. The accelerating voltage for the electrons of 100 kV produces at currents from 4 to 12 mA a power density of between about 300 and 1000 W/mm^2 on the target. The electron beam can be focused by a magnetic lens to a few tenths of a millimeter. In order to obtain an evaporation zone of rectangular shape with dimensions of about 0.3×6 mm, this being the required cross section of the atomic beam, the electron beam is deflected linearly back and forth in time in one direction by a deflection magnet with a frequency of 25 kHz. The geometrical shape of the atomic beam running through the apparatus is thus defined by the evaporation zone, the collimating slit in the middle of the apparatus, and the entrance slit of the detector.

The temperature required for the local evaporation of the beam is normally well above the melting point of the material. The surroundings of the evaporating area would therefore melt, causing deterioration of the target. Consequently the target is rotated at a slow angular speed of about 1 revolution in 2–5 sec, allowing the material to cool down below melting point soon after it has left the area hit by the electron beam. By very slowly moving the target back and forth in the direction of the axis the whole mantle surface of the target is gradually used for the evaporation process. As shown in Figure 2, the target is mounted, together with a suitable heat-shielding arrangement made of tantalum, onto a copper block which can be water-cooled from the inside by an axial water-cooling system, while it rotates together with the target and a suitable vacuum feedthrough. For shielding against the intense x-rays produced by the electron bombardment of the target, the electron gun assembly, the oven chamber, and the diffusion pumps are enclosed in a tight lead shielding of 5 mm thickness.

Atomic beam intensity fluctuations of up to 20% are typical in the case of an evaporation zone that is directly visible from the apparatus, a much larger fluctuation than in the case of a beam coming through a slit from the buffering vapor volume inside a crucible. The background of a conventional flop-in signal of an ABMR rf resonance fluctuates at the same relative rate as the atomic beam, and so an on-line computer PDP-8 is used to detect the rf resonances in the presence of this large background noise.[9] The rf power producing the rf transitions is turned on and off periodically by the computer with a period of, typically, 50 msec, the differences of the counting rates of atoms at the detector for rf "on" and "off," for preselected frequencies of the rf, being stored in different memory cells of the computer. The frequency of the rf signal is also varied, by the computer, back and forth through a series of preselected frequencies in a range where the resonance is expected. This frequency variation is achieved by steering, from the computer, a remotely controllable frequency decade, to which the rf oscillator is stabilized by a suitable regulating device. Preselected frequencies within a given range can thus be turned on and off within about 1 msec with a relative accuracy, in the frequency, of better than one part in

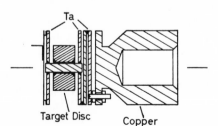

Ta

Target Disc   Copper

Figure 2. Cross section of the rotating target mounting.

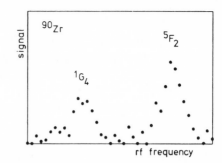

Figure 3. Zeeman transitions in the $4d^2 5s^2\ ^1G_4$ (8057 cm^{-1}) and $4d^3 5s\ ^5F_2$ (5023 cm^{-1}) levels of ^{90}Zr at 1700 Oe.

10^9. Figure 3 shows the recording of extremely weak rf transitions between adjacent Zeeman levels at 1700 Oe in the states $4d^2 5s^2\ ^1G_4$ and $4d^3 5s\ ^5F_2$ of ^{90}Zr.[26] The states are 8057 and 5023 cm^{-1}, respectively, above the ground level. To obtain the curve required about 10 min of data collection. The difficulties arising from the beam fluctuations can thus be overcome by a suitable data-handling system making use of an on-line computer.

The method of beam production described is completely universal. Targets composed of different refractory elements behave quite differently with regard to slow deformations during several hours of continuous use, but despite this, satisfactory beam production was possible with all of them.

2.2. Results of Measurements Using the Rotating-Target Method*

Using the rotating-target method, hfs and Zeeman-effect measurements have been performed on the following refractory elements: C,[27] Zr,[28] Nb,[29,30] Mo,[31] Ru,[32,33] Hf,[34–36] Ta,[37,38] W,[39] and Ir.[40] These experiments provided new information concerning the electronic structure of the atoms as well as the properties of their nuclei. There follows a brief report.

2.2.1. Electronic Structure of the Atoms

Except for C, all refractory elements studied using the rotating-target method belong to the 4d and 5d series of the periodic system. Because all these elements have fairly low-lying metastable electronic states, it was possible to perform, on nearly all these elements, measurements of the hfs in several fine-structure states, the only exception being Mo, on which, so far, only measurements in the ground state have been made. It was therefore possible to study the influences of relativistic and configuration interactions effects on the hfs of these elements. In order to take into account

* The author gratefully acknowledges the help of Dr. Büttgenbach, who prepared the material for this section, partly prior to publication of his results.

these effects, the method of effective Hamilton operators developed by Sandars and Beck[41] can be used. According to this method, the operator of the magnetic dipole interaction can be written for configurations of the type nl^N and $nl^N s$, to which the low-lying metastable states of the $4d$ and $5d$ elements belong, as follows:

$$\mathbf{H}(M1) = \sum_{i=1}^{N} [a_l^{01}\mathbf{l}_i - (10)^{1/2}a_l^{12}(\mathbf{s}\mathbf{C}^{(2)})_i^{(1)} + a_l^{10}\mathbf{s}_i]h\mathbf{I} + a_s^{10}\mathbf{s}_{N+1}h\mathbf{I}$$

$\mathbf{s}_i$ and $\mathbf{l}_i$ are the spin and orbital angular momentum operators of the electron j and $\mathbf{C}^{(2)}$ is a second-rank spherical tensor operator proportional to the spherical harmonic Y_{2q}. $\mathbf{I}$ is the nuclear spin operator. The a's can be regarded as independent parameters to be fitted to the experimental data in order to take into account relativistic and configuration-interaction effects. For nonrelativistic approximations, and without configuration interactions, the parameters $a_l^{k_s k_l}$ are related to the usual interaction constants a_{nl}:

$$a_l^{01} \rightarrow a_{nl} = \frac{2\mu_B\mu_N}{h}\frac{\mu_I}{I}\langle r^{-3}\rangle_{nl}$$

$$a_l^{12} \rightarrow a_{nl}, \qquad a_l^{10} \rightarrow 0$$

Usually the parameters a are expressed by effective expectation values of $\langle r^{-3}\rangle$ as follows:

$$a_l^{k_s k_l} = \frac{2\mu_B\mu_N}{h}\frac{\mu_I}{I}\langle r^{-3}\rangle_{k_s k_l}$$

whereas the parameter a_s is correlated to the density of the s electron at the nucleus by

$$a_s = \frac{16\pi}{3}\frac{\mu_B\mu_N}{h}\frac{\mu_I}{I}|\psi_s(0)|^2$$

If the eigenvector of a fine-structure state is adequately known for intermediate coupling and if, also, the magnetic moment of the nucleus is known with adequate precision, the expectation value of the magnetic dipole interaction operator and thus also the A factor of the state can be written as a linear combination of the four quantities $\langle r^{-3}\rangle_{01}$, $\langle r^{-3}\rangle_{12}$, $\langle r^{-3}\rangle_{10}$ and $|\psi_s(0)|^2$. If theoretical values for the three different expectation values of $\langle r^{-3}\rangle$ and for $|\psi_s(0)|^2$ can be calculated by relativistic self-consistent-field methods,[42] then the remaining differences between the experimental and theoretical values can be regarded as a measure of the influences of configuration interactions and of how good the intermediate coupling wave functions are.

With regard to this fairly sophisticated status of hfs theory, it is very interesting to perform hfs measurements in at least three or—for the case of configurations with an additional single s electron—four different fine-structure states, in order to determine the three or four different effective parameters $\langle r^{-3}\rangle_{k_s k_l}$ and $|\psi(0)|^2$. Measuring the hfs of still more fine-structure states of a given configuration will, of course provide a highly desirable experimental check on the internal consistency of the parametrization. Many such measurements have been performed in recent years with the ABMR method, mainly by Childs and Goodman at Argonne National Laboratory, on suitable nonrefractory elements, these being mainly elements with open d shells, and rare-earth elements. Very detailed and comprehensive reviews of the measurements and their comparison with theory have been published by Childs[43] and Lindgren and Rosén.[42]

For the case of the refractory elements, Table 1 shows the elements, configurations, and fine-structure multiplets on which successful experiments for the determination of effective radial parameters have been performed. In Figure 4 the experimental results for the three effective radial parameters are compared with theoretical predictions for the $4d^N 5s$ series. The solid lines indicate the results of relativistic Hartree–Fock calculations by Lindgren and Rosén.[42,46] The experimental results for Rh are taken from Chan et al.[47] The results for $\langle r^{-3}\rangle_{01}$ are in fairly good agreement with the theoretical values, whereas there is appreciable disagreement for $\langle r^{-3}\rangle_{12}$. This seems to indicate that $\langle r^{-3}\rangle_{01}$ is fairly insensitive to configuration interaction, but that $\langle r^{-3}\rangle_{12}$ is not. On the other hand the "experimental" values for $\langle r^{-3}\rangle_{12}$ are strongly dependent on the intermediate coupling of the wave functions, so it is fairly difficult to differen-

Table 1. Configurations and Multiplets of Refractory Elements, on which Experiments for the Determination of Effective Radial Parameters Have Been Performed

Isotope	Configuration	Multiplet	Ref.
^{91}Zr	$4d^2\,5s^2$	^3F	28, 44
	$4d^3\,4s$	^5F	
^{93}Nb	$4d^3\,5s^2$	^4F	29, 30
	$4d^4\,5s$	^6D	
^{99,101}Ru	$4d^7\,5s$	^5F	32, 33
^{177,179}Hf	$5d^2\,6s^2$	^3F	
^{181}Ta	$5d^3\,6s^2$	^4F	
^{183}W	$5d^4\,6s^2$	^5D	44, 45
^{191,193}Ir	$5d^7\,6s^2$	^4F	
	$5d^8\,6s$	^4F	

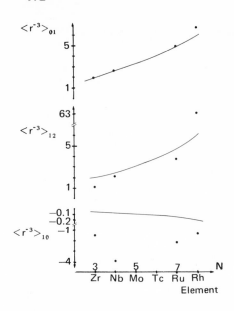

Figure 4. Comparison between experiment and theory for the radial parameters of the $4d^N 5s$ series ($\langle r^{-3} \rangle$ in units of a_0^{-3}).

tiate between these two influences on $\langle r^{-3} \rangle_{12}$. That the value of $\langle r^{-3} \rangle_{12}$ for Rh is too big by an order of magnitude is certainly caused by deficiencies of the eigenfunctions being used. The disagreement of an order of magnitude for the contact parameter $\langle r^{-3} \rangle_{10}$ shows that the strong contact part of the interaction cannot be explained by relativistic effects and must be caused

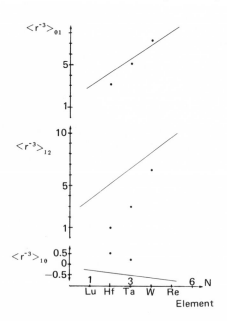

Figure 5. Comparison between experiment and theory for the radial parameters of the $5d^N 6s^2$ series ($\langle r^{-3} \rangle$ in units of a_0^{-3}).

by core polarization. The same situation has been established already for the $3d$-shell elements.[48]

The evaluation of the measurements performed so far on the $5d$-shell elements Hf, Ta, W, and Ir has not yet been finished; therefore Figure 5 shows only preliminary experimental results for Hf, Ta, and W. The theoretical values are available for Lu, Hf, Ta, and Re, again from Lindgren and Rosén.[42] The comparison between experiment and theory shows the same tendency as described for the $4d$-shell elements.

In order to resolve the discrepancies, it seems, now, very desirable to improve the eigenvectors of the single configurations under study and to perform explicit calculations of the influences of configuration interactions on the radial parameters using many-body perturbation theory. Enough experimental data have been obtained to justify renewed theoretical efforts.

2.2.2. Nuclear Properties

A complete review of the results relevant to the nuclear properties is beyond the limits of this report; therefore only a few typical examples will be described.

(a) *Nuclear Magnetic Dipole Moments.* In order to determine the nuclear magnetic dipole moments of the nuclei with the ABMR method independently of the interaction between the nucleus and the atomic electrons, one has to extract from adequate hfs measurements the direct interaction between the nuclear magnetic dipole moment and the external magnetic field applied in the C-field region of the ABMR apparatus, where the induced rf transitions take place. This was possible for the cases of Ru and Hf by applying the triple resonance method,[21] which allows the detection of rf transitions $\Delta m_J = 0$ by inducing two auxiliary transitions in separate homogeneous magnetic fields in front of and behind the C field. The results[33,36]

$$\mu_I(^{177}\text{Hf}) = 0.7836(6)\,\mu_N, \qquad \mu_I(^{99}\text{Ru}) = -0.6381(51)\,\mu_N$$

$$\mu_I(^{179}\text{Hf}) = -0.6329(13)\,\mu_N, \qquad \mu_I(^{101}\text{Ru}) = -0.7152(60)\,\mu_N$$

show that this method allows a very accurate determination of the nuclear moments. In the case of Ru it was possible to perform direct determinations of the nuclear moments of both isotopes in two different fine-structure states. The results agreed within the limits of error for both states. This represents a very severe test of the accuracy with which the hfs theory allows the extraction of the direct interaction between the nucleus and the external field from the measurements.[33]

(b) *Electric Quadrupole Moments.* For the electric quadrupole interaction an effective Hamilton operator, taking into account relativistic effects, can be derived that is similar to the one for the magnetic dipole interaction. In this case, three effective radial parameters, which are proportional to the electric quadrupole moment Q of the nucleus and to the expectation values of r^{-3}, are obtained:

$$b^{k_s k_l} = (e^2/h)Q\langle r^{-3}\rangle_{k_s k_l}, \qquad k_s k_l = 02, 13, 11$$

In most of the experiments using the rotating target method the electric quadrupole interaction could also be determined, and it was possible to obtain experimental values for b^{02}, b^{13}, and b^{11} in many cases, but a direct comparison between these results and theoretical calculations of $\langle r^{-3}\rangle_{02}$, $\langle r^{-3}\rangle_{13}$, and $\langle r^{-3}\rangle_{11}$ is difficult, because direct measurements of the electric quadrupole moments Q are not possible. It follows that only the ratios of the $\langle r^{-3}\rangle_{k_s k_l}$ can be determined. It seems that the purely relativistic contributions b^{13} and b^{11} to the quadrupole interaction can be explained only qualitatively by the theoretical calculations. In order, therefore, to obtain experimental values for the nuclear quadrupole moments of the nuclei, the unrelativistic approximation of the radial parameters is often used; that is, the quadrupole moments are determined from the formula

$$Q = \frac{2\mu_B \mu_N}{e^2} \frac{\mu_I}{I} \frac{b_{nl}}{a_{nl}}$$

Table 2 shows the results for some deformed nuclei of the $5d$ shell. The table also shows values, Q', for the quadrupole moments calculated from the internal quadrupole moments Q_0 as determined by methods of nuclear physics,[49] using the well-known projection factors between the internal and the spectroscopic quadrupole moments.[4] The ratio Q/Q' can be regarded as a measure of the influence of shielding effects and thus represents an "experimental" Sternheimer factor, which is, of course, fairly speculative because of the severe deficiencies of the theoretical assumptions being used.

Table 2. Comparison of Some Quadrupole Moments Determined by Spectroscopic and Nuclear Methods for Some $5d$ Elements

Isotope	Q (spectr.) (barns)	Q' (from Q_0) (barns)	Q/Q'
^{177}Hf	5.0 (5)	3.55 (12)	1.41 (14)
^{179}Hf	5.7 (6)	3.82 (13)	1.49 (15)
^{181}Ta	4.1 (4)	3.15 (6)	1.30 (13)
^{191}Ir	0.96 (27)	0.87 (5)	1.10 (30)
^{193}Ir	0.87 (24)	0.80 (3)	1.09 (26)

3. rf Transitions in the hfs of Metastable Atomic States Detected by Laser Fluorescence

There are several reasons for studying the hfs of excited metastable atomic states with the ABMR method: (1) There is a class of elements, typically the noble gas elements and the alkaline-earth elements, with diamagnetic ground states having total angular momentum $J = 0$. The atoms of these elements cannot be deflected by the inhomogeneous A and B magnets of the ABMR apparatus thus preventing ABMR studies in the ground states of these elements. Excitation of the atoms to higher metastable states would make possible hfs studies of these elements. (2) As explained in Section 2, experimental checks of the modern highly sophisticated theory of hfs require hfs measurements in as many states as possible of a given configuration.

For these reasons, considerable effort has been made, in the past, to produce beams of metastable atoms that are intense enough to allow ABMR experiments to be performed with them. There is one very elegant solution to this problem: by increasing the detection efficiency for the neutral atomic beam at the end of the ABMR apparatus, the thermal population of at least the low-lying metastable states, created by the thermodynamic equilibrium in the oven at the evaporation temperature, can be used. Very high detection efficiencies were achieved by Childs and Goodman at Argonne National Laboratory[43] and the ABMR group at Bonn[9,50] by using, for the universal electron bombardment detector, a mass spectrometer of very high luminosity and, even more important, an efficient data collection system with multichannel analyzer (Argonne) or an on-line computer (Bonn) allowing long integration times. These methods were highly successful and widely used with numerous experiments, but an upper limit seems to be an excitation energy for the metastable states of around $10,000\ \mathrm{cm}^{-1}$, since beyond that the population of the states becomes too low. The population is probably also increased considerably above the thermal equilibrium value by electron impact excitation from the electrons being used to heat the atomic beam ovens. The very dense beam from the electron gun used in the rotating target method is apparently most effective in this respect (described in Section 2).

In order to populate higher metastable states, several devices have been developed. Metastabilization by electron bombardment of the atomic beam in the region between the oven and the A magnet was successfully used in several experiments, especially by Faust and McDermott,[51] Lurio et al.,[52–54] and others.[55] The low efficiency of the electron bombardment (typically well below 1%) can be compensated for by a selective detector, which only registers the metastable atoms. This can be achieved by measuring the electron emission caused by the metastable atoms, when they hit a metal surface with very low work function (preferably lower than the excitation energy of the metastable state under study). For the noble gases especially the excitation can be produced by a gas discharge.[56,57] A considerable increase in the metastable population was achieved by Brinkmann et al.,[58] who ran an electric discharge in the metallic vapor inside a Ca oven, thus producing beams of metastable Ca atoms with population rates of 2%–12%, according to the particular metastable state. Because of the high population rates Aydin et al.[59] were able to adapt this source for use with an ABMR apparatus using a standard electron-bombardment detector without selective sensitivity for metastable atoms. They measured, in this way, the g_J factors of several metastable states of Ba and Sr. Similar population methods were used by Brink and Hull[60] and Schmelling.[61]

With the advent of tunable dye lasers it has become possible to achieve effective selective excitation of metastable states by optical excitation. Meisel and co-workers[62] were able to populate the metastable states $3d^2 4s\ {}^4F_{5/2,7/2}$ of ^{45}Sc, about $11,600\ \mathrm{cm}^{-1}$ above the ground state, by using a single-mode tunable dye laser to excite the atomic beam between the oven and the A field. They measured the hfs of these states using a conventional ABMR apparatus without selective detection of the metastable atoms.

The selective detection of metastable atoms is being improved at present at Göteborg[63] by exciting the metastable atoms at the end of the apparatus to suitable higher states using a tunable laser. The reemitted fluorescent light can then be observed and serves to indicate the presence of metastable atoms. Using this technique it was possible to measure the hfs of the $5d6s\ {}^3D$ states of Ba[64] and of the $6p^2\ {}^3P$ states of Pb using the ABMR method.

A completely new design for the detection of rf transitions in the hfs of ground and metastable atomic states has been developed at Bonn by Ertmer and Hofer[65] in 1976. It drastically changes the fundamental design of the ABMR apparatus by making full use of the advantages of laser beams. A similar method for use with molecular beams was developed at the same time independently.[66] The principle of the method is shown in Figure 6. The inhomogeneous magnetic field of the conventional ABMR method has been replaced by an optical pumping region, where atoms of the beam are excited, by the beam of a tunable single-mode cw dye laser, from one particular hfs level of the ground state (or of a metastable state) to a particular hfs level of a suitable excited state, which is not metastable. The optical pumping by the laser beam can be very efficient. So the lower level, from which the excitation takes place, can be almost completely depopulated in the beam. The remaining, very low, population of the level can then be probed by resonance fluorescence in a

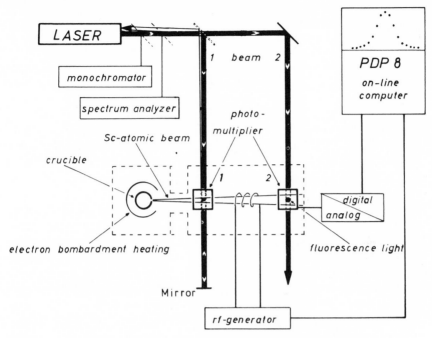

Figure 6. Schematic arrangement of the apparatus for atomic-beam magnetic-resonance detected by laser-induced resonance fluorescence (ABMR-LIRF).

second interaction region between the atomic beam and the same laser beam, this second interaction region being equivalent to the inhomogeneous magnet B of the ABMR method. Any change in the population of the depopulated hfs level, which takes place between the two interaction regions, e.g., by rf transitions from neighboring hfs states, can be detected by the change of the resonance fluorescence signal observed in the interaction region 2.

The first application of this new method (ABMR, detected by laser-induced resonance fluorescence: ABMR-LIRF) was the measurement of the hfs of the metastable fine-structure multiplet $3d^2 4s$ ^4F ($J = 3/2, 5/2, 7/2,$ and $9/2$) of ^{45}Sc with an excitation energy of about $11,600 \text{ cm}^{-1}$.[65] The relevant part of the Sc spectrum is shown in Figure 7.

In order to perform optical pumping and resonance fluorescence detection of the rf transitions, a rough knowledge is needed of the hfs intervals of both fine-structure states, which are connected by the spectral line used for pumping and detection. This was achieved by conventional high-resolution fluorescence spectroscopy performed at one of the two interaction regions. Because the laser beam crosses the well-collimated atomic beam at an angle of 90°, the Doppler width of the fluorescence

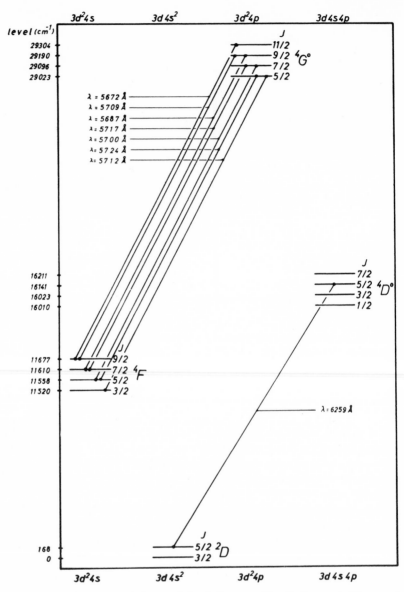

Figure 7. Relevant part of the atomic level diagram of Sc with the spectral lines used.

signal is negligible. By continuous tuning of the laser frequency the hfs of all seven lines connecting the multiplets ^4F and $^4G^0$ could be measured. Figure 8 shows the hfs of the line 5712 Å obtained this way. All components of the line are completely resolved, and the resonance

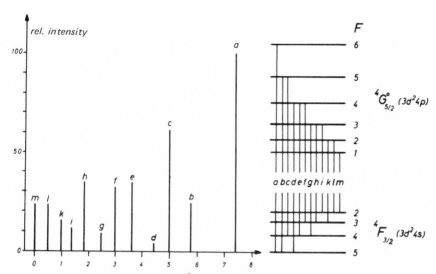

Figure 8. Hfs splitting of the line 5712 Å of Sc as measured by fluorescence spectroscopy, and the corresponding level diagram.

fluorescence of the single components could even be seen with the naked eye. These fluorescence experiments yielded values for the hfs intervals to an accuracy of about 3%. This was accurate enough to find the rf resonances easily. Figure 9 shows the recording of a typical rf resonance. The linewidth of the rf resonance of about 55 kHz is caused only by the finite interaction time between the rf field in the rf loop and the atoms of the beam, as is usual for ABMR experiments. This means that the linewidth of the rf resonances is independent of the spectral width of the laser light. The laser light is used only for the detection of the rf resonances, and so its spectral width has only to be narrow enough to resolve the single hfs components. This, nevertheless, makes single-mode operation of the laser necessary.

There are several special advantages of the ABMR-LIRF method compared with the conventional ABMR method. Because the distance between the oven and the second interaction region is relatively small (about 50 cm) and a collimation ratio for the atomic beam of about 1 : 100 is completely sufficient, the number of atoms available for the experiment is much higher than in conventional ABMR machines. Furthermore, the detection of the atoms by the laser light is very sensitive, since detection of about one photon per atom is easily achievable. Electron impact ionization detectors, however, which are normally used for ABMR experiments, detect generally only one in about 10^4 atoms. The rf transitions in Sc were performed at zero magnetic field, but of course it is possible to install a homogeneous magnetic field, in which the rf transitions between the two

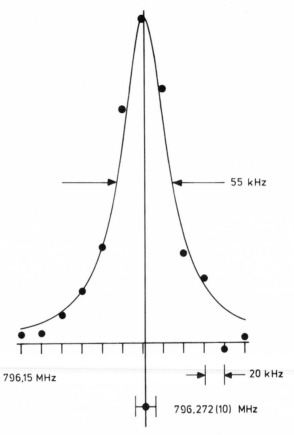

Figure 9. Line shape of the transition $3d^2 4s\ {}^4F_{3/2}(F = 4) \leftrightarrow (F = 5)$ of ^{45}Sc.

optical interaction regions can be induced. It will then be possible to measure $\Delta F = 1$ transitions between single Zeeman sublevels at higher magnetic fields and also $\Delta F = 0$ transitions, if the levels are separated by more than the width of the laser light, at the intersection with the laser beam. Then in principle all possible transitions $\Delta m = 0, \pm 1$ can be detected, and no change of sign of m_J is required as is the case with the normal ABMR method. This fact would increase the possibility of selection of transitions. Atomic ground states as well as metastable states can be examined by this method, but only few elements have spectral lines, involving the ground state, within the range of cw dye lasers, equipped with the dyes available today and without the use of frequency doubling. In the near future the main application of this method will be to elements having metastable states that can be excited to levels of opposite parity within the spectral range of about 4000–8000 Å. Most of these metastable states cannot be studied with the conventional ABMR method, because their thermal population is too low, whereas the very high sensitivity of the new method allows the detection of rf transitions in these states.

An ideal field of application of this new method is in the measurement of the hfs of metastable states of the configurations $3d^N 4s^2$ and $3d^N 4s$ of the 3d-shell elements. Many states of these configurations are metastable, since the lowest odd states lie fairly high in most of the spectra, and many states of the configurations $3d^N 4s$ are too high to be studied by the conventional ABMR method. On the other hand there are detailed theoretical predictions by Bauche-Arnoult[67,68] about the ratios α of the radial parameters $\langle r^{-3} \rangle_{01}$ and $\langle r^{-3} \rangle_{12}$ ($\alpha = \langle r^{-3} \rangle_{01} / \langle r^{-3} \rangle_{12}$) for configurations of neighboring elements in the 3d series, taking into account also the configuration interactions. In the meantime, therefore, a program has been started at Bonn to measure as many of these radial parameters as possible, in order to check the theoretical predictions. Results have already been obtained for the multiplets $3d^6 4s\ ^6D_{1/2,3/2,5/2,7/2,9/2}$ of ^{55}Mn (about 17,000 cm^{-1} above the ground state)[69] and $3d^7 4s\ ^5F_{2,3,4,5}$ of ^{57}Fe (about 7000–8000 cm^{-1} above the ground state).[70] Table 3 shows these results, whereas Table 4 gives all results for the ratios of the radial parameters known so far experimentally for the configurations $3d^N 4s^2$ and $3d^N 4s$.

Table 3. Results of the hfs Measurements with the AMBR–LIRF Method on ^{55}Mn and ^{57}Fe

^{55}Mn	$A(^6D_{1/2}) = 882.056\ (12)$ MHz	
$3d^6 4s$	$A(^6D_{3/2}) = 469.391\ (7)$ MHz	$B(^6D_{3/2}) = -65.091\ (50)$ MHz
Ref. (69)	$A(^6D_{5/2}) = 436.715\ (3)$ MHz	$B(^6D_{5/2}) = -46.769\ (30)$ MHz
	$A(^6D_{7/2}) = 458.930\ (3)$ MHz	$B(^6D_{7/2}) = 21.701\ (40)$ MHz
	$A(^6D_{9/2}) = 510.308\ (8)$ MHz	$B(^6D_{9/2}) = 132.200\ (117)$ MHz

$$\langle r^{-3} \rangle_{01}^d = 2.9143\ (40)\ a_0^{-3} \qquad \langle r^{-3} \rangle_{02}^d = 2.909\ a_0^{-3}$$
$$\langle r^{-3} \rangle_{12}^d = 2.855\ (32)\ a_0^{-3}$$
$$\alpha = \langle r^{-3} \rangle_{01}^d / \langle r^{-3} \rangle_{12}^d = 1.021\ (12)$$

^{57}Fe	$A(^5F_2) = 55.994\ (7)$ MHz
$3d^7 4s$	$A(^5F_3) = 69.632\ (5)$ MHz
(Ref. 70)	$A(^5F_4) = 78.435\ (4)$ MHz
	$A(^5F_5) = 87.246\ (3)$ MHz

$$\langle r^{-3} \rangle_{01}^d = 3.97 a_0^{-3} \qquad \langle r^{-3} \rangle_{10}^d = -1.42 a_0^{-3}$$
$$\langle r^{-3} \rangle_{12}^d = 3.89 a_0^{-3} \qquad |\psi_s(0)|^2 = 3.60 a_0^{-3}$$
$$\alpha = \langle r^{-3} \rangle_{01}^d / \langle r^{-3} \rangle_{12}^d = 1.02$$

Table 4. Experimental Results for the Ratios $\alpha = \langle r^{-3}\rangle_{01}/\langle r^{-3}\rangle_{12}$ of the Configurations $3d^N 4s^2$ and $3d^N 4s$ [a]

Element	Configuration	Multiplet	Average excited state energy (cm^{-1})	α	Ref.
Sc	$3d^1 4s^2$	$^2D_{3/2,5/2}$	ground state	1.123	73
Ti	$3d^2 4s^2$	$^3F_{2,3,4}$	ground state	1.032	74
V	$3d^3 4s^2$	$^4F_{3/2\cdots 9/2}$	ground state	1.10	75, 76
Cr	$3d^4 4s^2$	$^5D_{0\cdots 4}$	8,000	*	
Mn	$3d^5 4s^2$	$^6S_{5/2}$	ground state		
Fe	$3d^6 4s^2$	$^5D_{0\cdots 4}$	ground state	0.94	77
Co	$3d^7 4s^2$	$^4F_{3/2\cdots 9/2}$	ground state	0.924	78
Ni	$3d^8 4s^2$	$^3F_{2,3,4}$	1,100	1.026	79, 80
Cu	$3d^9 4s^2$	$^2D_{3/2,5/2}$	12,200		
Ca	$3d^1 4s$	$^3D_{1,2,3}$	20,300		
Sc	$3d^2 4s$	$^4F_{3/2\cdots 9/2}$	11,600	1.15	65, 81
Ti	$3d^3 4s$	$^5F_{1\cdots 5}$	6,700	*	
V	$3d^4 4s$	$^6D_{1/2\cdots 9/2}$	2,200	1.11	75, 76
Cr	$3d^5 4s$	$^5G_{2\cdots 6}$	20,500	*	
Mn	$3d^6 4s$	$^6D_{1/2\cdots 9/2}$	17,300	1.021	69
Fe	$3d^7 4s$	$^5F_{1\cdots 5}$	7,500	1.02	70
Co	$3d^8 4s$	$^4F_{3/2\cdots 9/2}$	4,000		
Ni	$3d^9 4s$	$^3D_{1,2,3}$	1,000		

[a] For the pairs of elements connected by a bracket $\langle$, the α's should be the same according to Bauche-Arnoult. For the α's marked by * experiments are in preparation at Bonn.

This table also includes the many results obtained by Childs and Goodman using the conventional ABMR method. As can be seen from the table, the predictions of Bauche-Arnoult are fairly well fulfilled for the pairs of elements Fe–Co and Mn–Fe, whereas there is a distinct discrepancy for the pair Sc–Ti. In a recent paper Bauche and Bauche-Arnoult[71] tried, therefore, to take into account the influence of far-configuration-mixing effects[72] on the off-diagonal hfs matrix elements between the configurations $d^N s^2$ and $d^{N+1} s$ of Sc and Ti. Their results for the theoretical α's of Sc and Ti are now $\alpha(Sc, 3d4s^2) = 1.13$ and $\alpha(Ti, 3d^2 4s^2) = 1.06$. This improves the agreement between theory and experiment considerably for Sc, but for Ti the discrepancy becomes slightly bigger. Altogether the results seem to indicate that the theory is making good progress towards explaining these very fine details of the hfs of excited states, details which have now become measurable with remarkable accuracy. As indicated in Table 3, there are more results to be expected in the near future, and it seems worthwhile to study the same effects for the $4d$ series as well. Several members of that series can be studied by the ABMR-LIRF method.

4. Precision Measurements of the hfs of Stable Isotopes

4.1. Elimination of Systematic Errors and Precision hfs Measurements of Alkali Atoms

The ABMR method has the advantage of dealing with free atoms; thus atomic properties may be studied in the absence of external perturbations. The precision attainable seems to be limited only by the width of the resonance lines and the number of detected atoms. Experiments that try to measure small effects in the hfs of free atoms, like the interaction between the nuclear magnetic moment and the external magnetic field or higher-order multipole interactions, very often require the determination of the resonance frequencies to within a very small fraction of the linewidth. Usually, however, there are shifts and distortions of the line shape causing the resonance frequencies to change by, typically, a few percent of the linewidth. Well-known examples of such shifts are the Millman effect,[82] caused by a nonconstant direction of the rf field within the transition region and—when Ramsey's technique of separated oscillatory fields[83] is used—the slight distortion of the Ramsey pattern caused by small phase differences in the rf fields of the two rf transition regions. From a very thorough analysis Böklen[12] derived a general relationship by which the appropriate experimental methods for determining, and thus correcting for, the shifts could be found without a detailed knowledge of the actual line shapes. Also included in the formalism are effects caused by small inhomogeneities of the C field and shifts due to the presence of neighboring states. Böklen and his co-workers have successfully applied his results[11] in precision experiments, in which frequency shifts of 10^{-2}–10^{-5} of the line width were eliminated. These experiments, performed on several alkali isotopes, combined Ramsey's method of separated oscillatory fields with the triple resonance method[21] in order to detect transitions with $\Delta m_J = 0$ and $\Delta m_I = \pm 1$ in external magnetic fields of about 3 kOe. These transitions have the special advantage that for them only the orientation of the nuclear magnetic moment is changed with respect to the

Table 5. Results of Precision Measurements of the hfs of Some Alkali Atoms

Isotope	$-g_I/g_J \times 10^3$	$-g_I \times 10^3$	$\Delta\nu$ (MHz)
^{6}Li	0.223 569 78 (10)	0.447 654 0 (3)	228.205 259 0 (30)
^{7}Li	0.590 427 19 (10)	1.182 213 0 (6)	803.504 086 6 (10)
^{23}Na	0.401 844 06 (40)	0.804 610 8 (8)	1771.626 128 8 (10)
^{39}K	0.070 886 13 (6)	0.141 934 89 (12)	461.719 720 2 (14)
^{41}K	0.038 908 37 (4)	0.077 906 00 (8)	254.013 872 0 (20)

external field. They are especially sensitive therefore to the interaction between the nuclear magnetic moment and the external magnetic field. By measuring an appropriate pair of such transitions in the same external magnetic field, the ratio g_I/g_J of the gyromagnetic ratios of the nuclear and electronic ground state can be determined. The results are shown in Table 5. Several appropriate pairs of frequencies were measured at various magnetic fields, and all experiments agreed within the limits of error. This represents a very severe and successful test of the method for eliminating all systematic errors with an accuracy of up to 10^{-5} of the linewidth. Because all the g_J's are known with at least the same precision, one can calculate, from the g_I/g_J ratios, values for the nuclear g_I factors. In several cases these have a relative accuracy of better than 10^{-6}. These values are also given in Table 5.

In order to calculate the experimental values for g_I/g_J from the measurements, the hfs separations $\Delta\nu$ of the $^2S_{1/2}$ ground states at zero external magnetic field must be known accurately. These were measured using the same technique, and the results can be seen in Table 5. The degree of experimental errors in these measurements lies between 1 and 3 Hz. This represents, in the case of ^{23}Na, a relative accuracy of about 6 parts in 10^{10}.

In order to achieve this kind of accuracy, several other conditions had to be fulfilled. The homogeneous C magnet had to be extremely well stabilized. The current regulation of the C-field supply was stable enough to allow a short-term stability of about 3×10^{-7}/min. In order to achieve an equivalent long-term stability, the C field was regulated by a strongly field-dependent resonance of Ag. For this purpose a second atomic beam of Ag atoms ran parallel with the alkali beam. Both beams were passing the apparatus simultaneously. A sketch of the apparatus is shown in Figure 10. In order to make full use of the excellent signal-to-noise ratio of the alkali resonances, an elaborate data-handling system using a PDP-8 on-line computer was used.

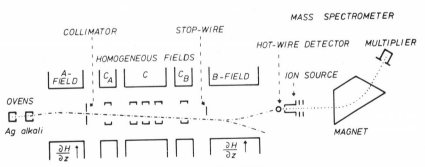

Figure 10. Sketch of the ABMR apparatus.

By comparing the g_I factors of Table 5, which were obtained from experiments with free atoms, with measurements of the same quantity performed with the NMR method in aqueous solutions, the chemical shifts of NMR frequencies caused by the hydrate surrounding of the alkali ions in the NMR probe can be determined.

To see how important the measurements contained in Table 5 are for the general system of natural constants of the alkali atoms, these constants being widely used for precision and calibration experiments in other fields, reference can be made to a comprehensive review article that has just been published.[84]

4.2. Hexadecapole Interaction in the hfs of Free Atoms

It can be seen from Table 5 that hfs separations at zero magnetic field can be measured with an absolute accuracy of a few hertz. This offers the possibility of looking for an electric hexadecapole interaction in the hfs of free atoms. In 1975 the hexadecapole moment of ^{165}Ho was determined from muonic x-rays. The result is $Q_{40} = 0.52(10)\,\text{eb}^2$[85] and is of the theoretically expected order but disagrees with a previous value obtained from an ABMR measurement.[86] A series of precision experiments was therefore started by Böklen and co-workers to try to detect hexadecapole moments larger than about 0.3 eb^2 in the hfs of rare-earth elements. The first experiment was performed on ^{141}Pr and yielded the result that the hexadecapole interaction constant A_4 is smaller than 0.34 Hz.[87] From this value it can be deduced that the hexadecapole moment Q_{40} must be smaller than 0.4 eb^2. Further experiments on other rare-earth elements are in progress at the time.[88]

4.3. g_J Factor of Li

Another challenge, at present, to precision ABMR measurements is presented by the theoretical predictions of Hegstrom[89] for the g_J factor of lithium in its ground state. They are based in the concept, which was developed previously[90] and tested in experiments with hydrogenic atoms[91] and helium[92] including correction terms of order α^3 and $\alpha^2 m/M$ (where α is the fine-structure constant and m/M is the electron nuclear mass ratio). Unlike He, there is a small uncertainty of about 10^{-7} in the theoretical calculation due to an uncertainty in the wave function for the Li ground state. If an experimental accuracy for g_J of about 10^{-8} could be achieved, this would test both the concept of atomic structure represented by the wave function and the bound state radiative correction of α^3 included in the Hamiltonian. In order to achieve this accuracy it is necessary to apply Ramsey's technique of separated oscillatory fields to strongly field-dependent lines at fields of several kOe. This has been done so far, only twice in experiments with He,[93,94] but not with the accuracy that is aimed at here. In order to achieve this accuracy Böklen[95] built a coil-shim device for the ABMR apparatus at Bonn, allowing him to homogenize an area of 104×5 mm in the C field of the apparatus to better than $3 \times 10^{-8}H$ for $H > 2$ kOe. In this field he was able to detect perfectly symmetric Ramsey patterns of strongly field-dependent Li resonances at a magnetic field of about 2 kOe. In a first experiment he measured the ratio of the g_J factors of ^{6}Li and ^{7}Li to be

$$g_J(^6\text{Li})/g_J(^7\text{Li}) = 1 + 3(70) \times 10^{-10}$$

A measurement of the ratio $g_J(^7\text{Li})/g_J(^{87}\text{Rb})$ is in progress at present. Because the ratio $g_J(^{87}\text{Rb})/g_J(H\ 1^2S_{1/2})$ is known with an accuracy of 6×10^{-10},[96] this measurement will yield a value for $g_J(^7\text{Li})$ in units of $g_J(H\ 1^2S_{1/2})$ and will thus test Hegstrom's predictions.

5. hfs Measurements of Radioactive Isotopes (Off-Line and On-Line)

Continual efforts have been made, during recent years, to measure spins and moments of radioactive nuclei using the ABMR technique, especially, at Göteborg, by Lindgren and co-workers, who have studied a large number of isotopes mainly in the rare-earth, the tin, and the refractory groups of elements.

These impressive results, together with all the references, are contained in the "Summary of Results from Measurements of Nuclear Spins and Moments" (updated in November 1976) by the Göteborg–Uppsala atomic-beam group. A detailed review of these measurements is beyond the scope of this report. Special reviews of the interpretation of the nuclear structure information obtained for the Lu isotopes[97] and for the elements between Hf and Ir[98] have been published by Ekström and co-workers. A new ABMR apparatus, for the study of radioactive isotopes, has been built recently at Amsterdam,[99] and first results have been obtained for the nuclear spins of some As isotopes.

During the last few years considerable efforts have been made to measure, using the ABMR method, spins and moments of nuclei that are far from stability. Because these isotopes are normally very short-lived, their study is only possible by adequate on-line techniques for the production of the isotopes, a method first developed by Ames and co-workers at Princeton.[100] Ekström, Lindgren, and co-workers installed an ABMR apparatus at the ISOLDE facility at CERN. The isotope separator ISO-

LDE is connected, on-line, with the CERN 600-MeV synchrocyclotron. The ion beam from the separator is focused directly into the oven of the ABMR apparatus, from which the radioactive isotopes are evaporated continuously as free atoms into the conventional ABMR apparatus. More than 30 spins of isotopes of the elements Rb, Cs, Au, and Fr, with lifetimes down to about 20 sec, have been determined so far. A review of the experimental technique and a systematic discussion of the nuclear structures in the regions around the elements Rb, Cs, and Au was given in a report by Ekström and co-workers[101] at the Third International Conference at Cargèse in 1976 on Nuclei Far from Stability. As more varieties of isotopes become available at ISOLDE, a wealth of new and interesting data will become available in the future concerning short-lived isotopes and especially concerning long chains of isotopes of the same element.

An interesting new atomic-beam optical-resonance method was developed recently at Orsay. The method is based on the principle[102] of detecting optical pumping, produced by a cw tunable dye laser in an atomic beam of Na, and by subsequent deflection of the atomic beam in a hexapole magnet. The laser beam crosses the atomic beam at an angle of 90° shortly after the oven. Its linewidth is narrow enough to resolve the hfs of the Na D lines. If the laser is tuned to a hfs component, which changes, by optical pumping, the sign of the quantum number m_J of some atoms, then these atoms will be deflected differently by the hexapole magnet, which is positioned behind the optical-pumping region. This change of deflection can be detected by a mass spectrometer, which registers all atoms focused onto its entrance hole on the axis of the atomic beam behind the hexapole magnet. This scheme allows the detection of the hfs of optical-resonance lines. The first experiments were carried out on $^{21-25}$Na produced from a molten aluminium target by spallation with 150-MeV protons from the Orsay synchrocyclotron.[103] In addition to a confirmation of the nuclear parameters of these nuclei, it was possible also to measure the isotope shift which turned out to be a pure mass shift. A review of the first experiments with this method can be found in Ref. 104. At present these investigations are being extended to the very neutron-rich isotopes,[104] which can be produced by the reaction of high-energy protons from the CERN Proton Synchrotron with heavy elements such as uranium.[105]

References

1. I. I. Rabi, J. R. Zacharias, S. Millman, and P. Kush, *Phys. Rev.* **53**, 318 (1938).
2. K. F. Smith, Molecular Beams, Methuen, London (1955).
3. N. F. Ramsey, *Molecular Beams*, Oxford University Press, London (1956).
4. H. Kopfermann, *Nuclear Moments*, trans. E. E. Schneider, Academic Press, New York (1958).

5. P. Kusch and V. W. Hughes, "Atomic and Molecular Beam Spectroscopy," in *Handbuch der Physik*, ed. S. Flügge, Vol. 37/1, pp. 1–172, Springer–Verlag, Berlin (1959).

6. W. A. Nierenberg, "The Measurement of the Nuclear Spins and Static Moments of Radioactive Isotopes," in *Annual Review of Nuclear Science*, Vol. 7, pp. 349–406, Annual Reviews Inc., Palo Alto, California (1957).

7. P. G. H. Sandars (Notes prepared by B. Dodsworth), "Hyperfine Structure in Atomic Beam Resonance," in *Hyperfine Interactions*, Eds. A. J. Freeman and R. B. Frankel, pp. 115–139, Academic Press, New York (1967).

8. S. Büttgenbach, G. Meisel, S. Penselin, and K. H. Schneider, *Z. Phys.* **230**, 329 (1970).

9. S. Büttgenbach and G. Meisel, *Z. Phys.* **244**, 149 (1971).

10. W. Ertmer and B. Hofer, *Z. Phys.* **A276**, 9 (1976).

11. A. Beckmann, K. D. Böklen, and D. Elke, *Z. Phys.* **270**, 173 (1974).

12. K. D. Böklen, *Z. Phys.* **270**, 187 (1974).

13. Annual Report 1976 for the Research Group of Atomic Physics, Department of Physics, Chalmers University of Technology, University of Gothenburg, Göteborg, Sweden.

14. V. W. Hughes, in *Atomic Physics 3*, Eds. S. J. Smith and G. K. Walters, pp. 1–32, Plenum Press, New York (1973).

15. P. G. H. Sandars, in *Atomic Physics 4*, Eds. G. zu Putlitz, E. W. Weber, and A. Winnacker, pp. 71–92, Plenum Press, New York 1975.

16. H. Lew and E. Lipworth, in *Methods of Experimental Physics*, Eds. V. W. Hughes and H. L. Schultz) Vol. 4, Part A, pp. 155–256, Academic Press, New York (1967).

17. G. Wolber, H. Figger, R. A. Haberstroh, and S. Penselin, *Phys. Lett.* **29A**, 461 (1969); *Z. Phys.* **236**, 337 (1970).

18. W. M. Doyle and R. Marrus, *Nucl. Phys.* **49**, 449 (1963).

19. R. G. Schlecht, M. B. White, and D. W. McColm, *Phys. Rev.* **138B**, 306 (1965).

20. L. Armstrong Jr. and R. Marrus, *Phys. Rev.* **138B**, 310 (1965).

21. W. A. Nierenberg and G. O. Brink, *J. Phys. Radium* **19**, 816 (1958); P. G. H. Sandars and G. K. Woodgate, *Nature* **181**, 1395 (1958).

22. H. Rubensztein, I. Lindgren, L. Lindström, H. Riedl, and A. Rosén, *Nucl. Instr. Methods* **119**, 269 (1974).

23. H. Rubensztein and M. Gustafsson, *Phys. Lett.* **58B**, 283 (1975).

24. J. M. Pendlebury, D. B. Ring, and K. F. Smith, in *Actes du Colloque International du CNRS sur "La Structure Hyperfine Magnétique des Atomes et des Molécules,"* pp. 71–72, Edition du CNRS, Paris (1967).

25. J. M. Pendlebury and D. B. Ring, *J. Phys.* **B5**, 386 (1972).

26. S. Büttgenbach, R. Dicke, and H. Gebauer, *Phys. Lett.* **58A**, 56 (1976).

27. A. Beckmann, K. D. Böklen, G. Bremer, and D. Elke, *Z. Phys.* **A272**, 143 (1975).

28. S. Büttgenbach, R. Dicke, and H. Gebauer, *Phys. Let.* **58A**, 56 (1976).

29. S. Büttgenbach, R. Dicke, H. Gebauer, M. Herschel, and G. Meisel, *Z. Phys.* **A275**, 193 (1975).

30. S. Büttgenbach and R. Dicke, *Z. Phys.* **A275**, 197 (1975).

31. S. Büttgenbach, M. Herschel, G. Meisel, E. Schrödl, W. Witte, and W. J. Childs, *Z. Phys.* **266**, 271 (1974).

32. S. Büttgenbach, M. Herschel, G. Meisel, E. Schrödl, W. Witte, and W. J. Childs, *Z. Phys.* **269**, 189 (1974).

33. S. Büttgenbach, R. Dicke, H. Gebauer, and M. Herschel, *Z. Phys.* **A280**, 217 (1977).

34. S. Büttgenbach and G. Meisel, *Z. Phys.* **250**, 57 (1972).

35. S. Büttgenbach, M. Herschel, G. Meisel, E. Schrödl, and W. Witte, *Phys. Lett.* **43B**, 479 (1973).

36. S. Büttgenbach, M. Herschel, G. Meisel, E. Schrödl, and W. Witte, *Z. Phys.* **260**, 157 (1973).
37. S. Büttgenbach, G. Meisel, S. Penselin, and K. H. Schneider, *Z. Phys.* **230**, 329 (1970).
38. S. Büttgenbach and G. Meisel, *Z. Phys.* **244**, 149 (1971).
39. S. Büttgenbach, R. Dicke, H. Gebauer, R. Kuhnen, and F. Träber, *Z. Phys.* **A283**, 303 (1977).
40. S. Büttgenbach, M. Herschel, G. Meisel, E. Schrödl, W. Witte, and W. J. Childs, *Z. Phys.* **263**, 341 (1973).
41. P. G. H. Sandars and J. Beck, *Proc. R. Soc. London A* **289**, 97 (1965).
42. I. Lindgren and A. Rosén, *Case Stud. At. Phys.* **4**, 93 (1974).
43. W. J. Childs, *Case Stud. At. Phys.* **3**, 215 (1973).
44. S. Büttgenbach, R. Dicke, H. Gebauer, R. Kuhnen, and F. Träber: *Z. Phys.* **A283**, 303 (1977).
45. S. Büttgenbach, R. Dicke, H. Gebauer, R. Kuhnen, and F. Träber: to be published.
46. A. Rosén, private communication.
47. Y. W. Chan, W. J. Childs, and L. S. Goodman, *Phys. Rev.* **173**, 107 (1968).
48. W. J. Childs, in *Atomic Physics 4*, Eds. G. zu Putlitz, E. W. Weber, and A. Winnacker, pp. 731–745, Plenum Press, New York (1975).
49. K. E. G. Löbner, M. Vetter, and V. Hönig, *Nucl. Data Tables* **A7**, 495 (1970).
50. A. Beckmann, K. D. Böklen, and D. Elke, *Z. Phys.* **270**, 173 (1974).
51. W. L. Faust and M. N. McDermott, *Phys. Rev.* **123**, 198 (1961).
52. A. Lurio, *Phys. Rev.* **126**, 1768 (1962).
53. A. G. Blachmann, D. A. Landmann, and A. Lurio, *Phys. Rev.* **150**, 59 (1966).
54. A. G. Blachmann, D. A. Landmann, and A. Lurio, *Phys. Rev.* **161**, 60 (1967).
55. T. Hadeishi, O. A. McHarris, and W. A. Nierenberg, *Phys. Rev.* **138A**, 983 (1965).
56. V. Hughes, G. Tucker, E. Rhoderick, and G. Weinreich, *Phys. Rev.* **91**, 828 (1953).
57. H. Kuiper, *Z. Phys.* **165**, 402 (1961).
58. U. Brinkmann, A. Steudel, and H. Walther, *Z.f. Angew. Phys.* **22**, 223 (1967).
59. R. Aydin, R. Aldenhoven, L. Degener, H. Gebauer, and G. Meisel, *Z. Phys.* **A273**, 233 (1975).
60. G. O. Brink and R. J. Hull, *Phys. Rev.* **179**, 43 (1969).
61. S. G. Schmelling, *Phys. Rev. A* **9**, 1097 (1974).
62. W. Zeiske, G. Meisel, H. Gebauer, B. Hofer, and W. Ertmer, *Phys. Let.* **55A**, 405 (1976).
63. Annual Report 1976 for the Research Group of Atomic Physics, Chalmers University of Technology and University of Gothenburg, Göteborg, Sweden.
64. M. Gustavsson, I. Lindgren, G. Olsson, A. Rosén, and S. Svanberg, *Phys. Lett.* **62A**, 250 (1977).
65. W. Ertmer and B. Hofer, *Z. Phys.* **A276**, 9 (1976).
66. S. D. Rosner, R. A. Holt, and T. D. Gaily, *Phys. Rev. Lett.* **35**, 785 (1975).
67. C. Bauche-Arnoult, *Proc. R. Soc. London A* **322**, 361 (1971).
68. C. Bauche-Arnoult, *J. Phys.* **34**, 301 (1973).
69. J. Dembczyński, W. Ertmer, B. Hofer, U. Johann, and P. Stinner, private communication, to be published.
70. J. Dembczyński, W. Ertmer, B. Hofer, U. Johann, and P. Stinner, private communication, to be published.
71. J. Bauche and C. Bauche-Arnoult, *J. Phys. B: Atom. Molec. Phys.* **10**, L125 (1977).
72. B. R. Judd, *Proc. Phys. Soc. London* **82**, 874 (1963).
73. H. Gebauer, R. Aldenhoven, and R. Aydin, *Phys. Lett.* **51A**, 417 (1975).
74. K. H. Channappa and J. M. Pendlebury, *Proc. Phys. Soc. London* **86**, 1145 (1965).
75. W. J. Childs and L. S. Goodman, *Phys. Rev.* **156**, 64 (1967).

76. W. J. Childs, *Phys. Rev.* **156**, 71 (1967).
77. W. J. Childs and L. S. Goodman, *Phys. Rev.* **148**, 74 (1966).
78. W. J. Childs and L. S. Goodman, *Phys. Rev.* **170**, 50 (1968).
79. W. J. Childs and L. S. Goodman, *Phys. Rev.* **170**, 136 (1968).
80. W. J. Childs and B. Greenebaum, *Phys. Rev. A* **6**, 105 (1972).
81. C. Bauche-Arnoult, private communication.
82. S. Millman, *Phys. Rev.* **55**, 628 (1939).
83. N. F. Ramsey, *Phys. Rev.* **76**, 996 (1949).
84. E. Arimondo, M. Inguscio, and P. Violino, *Rev. Mod. Phys.* **49**, 31 (1977).
85. R. J. Powers, F. Boehm, P. Vogel, A. Zehnder, T. King, A. R. Kunselman, P. Robertson, P. Martin, G. H. Miller, R. E. Welsh, and D. A. Jenkins, *Phys. Rev. Lett.* **34**, 492 (1975).
86. W. Dankwort, J. Ferch, and H. Gebauer, *Z. Phys.* **267**, 229 (1974).
87. K. D. Böklen, T. Bossert, W. Foerster, H. H. Fuchs, and G. Nachtsheim, *Z. Phys.* **A274**, 195 (1975).
88. K. D. Böklen, private communication.
89. R. A. Hegstrom, *Phys. Rev. A* **11**, 421 (1975).
90. R. A. Hegstrom, *Phys. Rev.* **184**, 17 (1969); R. Faustov, *Phys. Lett.* **33B**, 422 (1970); F. E. Close and H. Osborne, *ibid.* **34B**, 400 (1971); H. Grotch and R. A. Hegstrom, *Phys. Rev. A* **4**, 59 (1971); **8**, 1166 (1973).
91. J. S. Tiedeman and H. G. Robinson, in *Atomic Physics 3* , Eds. S. J. Smith and G. K. Walters, pp. 85–91, Plenum Press, New York (1973); F. G. Walter, W. D. Phillips, and D. Kleppner, *Phys. Rev. Lett.* **28**, 1159 (1972); D. J. Larson and N. F. Ramsey, *Phys. Rev. A* **9**, 1548 (1974).
92. C. W. Drake, V. W. Hughes, A. Lurio, and J. A. White, *Phys. Rev.* **112**, 1627 (1958); E. Aygün, B. D. Zak, and H. A. Shugart, *Phys. Rev. Lett.* **31**, 803 (1973); G. M. Keiser and H. G. Robinson, *Phys. Rev. Lett.* **35**, 1223 (1975); B. E. Zundell and V. W. Hughes, *Bull. Am. Phys. Soc.* **21**, 625 (1976).
93. C. W. Drake, V. W. Hughes, A. Lurio, and J. A. White, *Phys. Rev.* **112**, 1627–1637 (1958).
94. B. E. Zundell and V. W. Hughes, *Bull. Am. Phys. Soc.* **21**, 625 (1976).
95. K. D. Böklen, W. Foerster, H. H. Fuchs, G. Nachtsheim, and W. Nitschke, *Z. Phys.* **A282**, 249 (1977).
96. W. M. Hughes and H. G. Robinson, *Phys. Rev. Lett.* **23**, 1209 (1969).
97. C. Ekström, *Phys. Scr.* **13**, 217 (1976).
98. C. Ekström, H. Rubinsztein, and P. Möller, *Phys. Scr.* **14**, 199 (1976).
99. H. A. Helms, W. Hogervorst, G. J. Zaal, and J. Blok, *Phys. Scr.* **14**, 138 (1976).
100. O. Ames, E. A. Phillips, and S. S. Glickstein, *Phys. Rev.* **137 B**, 1157 (1965).
101. C. Ekström, S. Ingelman, G. Wamberg, and M. Skarestad, CERN report No. 76–13, pp. 193–205 (1976).
102. H. T. Duong and J.-L. Vialle, *Opt. Commun.* **12**, 71 (1974).
103. G. Huber, C. Thibault, R. Klapisch, H. T. Duong, J.-L. Vialle, J. Pinard, P. Juncar, and P. Jacquinot, *Phys. Rev. Lett.* **34**, 1209 (1975).
104. G. Huber, R. Klapisch, C. Thibaut, T. H. Duong, P. Juncar, S. Liberman, J. Pinard, J.-L. Vialle, and P. Jacquinot, CERN report No. 76–13, pp. 188–192 (1976).
105. R. Klapisch, C. Thibault, A. M. Poskanzer, R. Prieels, C. Rigaud, and E. Roeckl, *Phys. Rev. Lett.* **29**, 1254 (1972).

11

The Microwave–Optical Resonance Method

WILLIAM H. WING AND KEITH B. MACADAM

1. Introduction

1.1. Method

The spectroscopic method of *microwave–optical resonance* is one of a class of stimulated resonant transition techniques which, over the past 40 years, have exhibited both a remarkable propensity to differentiate and mutate into new forms as details of particular experimental situations have demanded, and a great fertility in producing precise experimental data on atomic and molecular energy levels. The class may be said to have originated with the *molecular-beam radio-frequency resonance* experiments of Rabi and co-workers at Columbia.[1] Its range of applicability was extended greatly by the development in 1949–50 of the *optical double resonance* method by Brossel and Kastler[2] and by Bitter[3] and the microwave–optical resonance method by Lamb and Skinner.[4,5] The various techniques have the following steps in common (Figure 1):

(1) The production of atoms or molecules in the initial states a and b with unequal populations n_a and n_b.

(2) The stimulation of transitions between a and b with electromagnetic radiation (rf) whose angular frequency ω lies near the resonant frequency $\omega_0 = |E_a - E_b|/\hbar$; a net transfer of population to the state with lower initial population results.

WILLIAM H. WING • Department of Physics and Optical Sciences Center, University of Arizona, Tucson, Arizona 85721. KEITH B. MACADAM • Department of Physics, University of Arizona, Tucson, Arizona 85721.

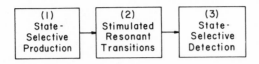

(1) State- Selective Production	(2) Stimulated Resonant Transitions	(3) State- Selective Detection

Figure 1. Basic principle of resonance spectroscopic methods.

(3)   The detection of the atoms or molecules in a manner that favors systems in one of the two states, so that the population transfer may be noted. The population transfer signal, plotted against ω, exhibits a maximum at a frequency ω_m, which, sometimes with the aid of systematic corrections, is taken equal to the true resonance frequency ω_0. Either the particles themselves or their absorption or emission of radiation may be detected.

In beam experiments these three steps occur in sequence along the beam axis (in regions designated successively A, C, and B in the molecular-beam literature). In microwave–optical resonance and optical double resonance, which have been termed "bottle" methods, the steps take place at the same point in space and either concurrently or in rapid time sequence, allowing ready application to the study of short-lived excited states. Both bottle methods are variants of the same basic technique (Figure 2), and differ principally in the means of excitation used in step (1). Optical double resonance employs a light source (usually a resonance lamp or laser) which is spectrally intense at the excitation frequency $\nu = E/h$ of at least one of the excited levels. In practice it is limited by the availability of sources to levels having excitation energies E less than (roughly) 10 eV, and also to those levels for which electric-dipole selection rules ($\Delta L = \pm 1$) allow population from the ground state by absorption of resonance radiation or by cascade from energetically higher levels. Thus it cannot easily be used to study the simplest atoms, hydrogen and helium, as well as most atomic ions, whose lowest excitation energies exceed 10 eV.

These restrictions are largely eliminated in the microwave–optical resonance method by the use of electron-impact excitation, since electron beams of sufficient kinetic energy to excite any atomic level of interest, and with sufficient current, can be produced readily, and since electric-dipole selection rules do not apply rigorously in electron–atom collisions. Experiments of this type have therefore produced a substantial body of data not obtainable by optical double resonance. The cost, however, is a

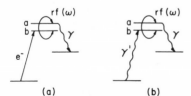

Figure 2. Comparison of "bottle" excited-state resonance spectroscopic methods. The fluorescent quanta γ are detected. (a) Microwave–optical resonance; (b) optical double resonance.

more complex environment. The excited atoms suffer perturbations from both macroscopic space charge and electrode fields and microscopic electric fields from ion and electron collisions. Electron-beam-excited experiments are usually also operated at higher gas pressures than optically excited experiments because of the relatively much lower electron-impact cross sections, resulting in increased rates of collision between excited atoms and ground-state neutrals.

The principal advantage that all types of microwave or radio-frequency resonance methods hold relative to classical spectroscopy appears in the measurement of small energy level differences, or fine structure. In classical spectroscopy, fine details are frequently obscured by the Doppler broadening of the spectral lines, which results from random thermal motion of the sample atoms. The Doppler shift in the measured frequency for an atom and a photodetector having relative velocity v along the line of sight is $(v/c)\nu$, which amounts to typically $10^{-5}\nu$ or $10^{-6}\nu$ for a sample of light atoms at ordinary temperatures. The Doppler breadths for microwave and for optical spectral measurements are in the ratio $\nu_{\text{microwave}}/\nu_{\text{optical}} \approx 10^{10}\,\text{Hz}/10^{15}\,\text{Hz} = 10^{-5}$. In most microwave experiments the Doppler broadening is small compared to the natural broadening due to the finite lifetimes of the excited states. In experiments at long microwave or radio wavelengths λ, the atoms may be confined to a volume of dimensions small compared to $\lambda/4$, in which case the Doppler broadening effectively disappears altogether, as shown by Dicke.[6]

The literature of radio-frequency and microwave resonance experiments in excited states is voluminous and has been reviewed previously by several authors.[7–9] Most reviews have stressed optically excited experiments employing magnetic-dipole rf transitions, and the atomic hyperfine structure, nuclear moments, fine structure, and g factors, and the molecular rotational and fine-structure spectra obtained thereby. In this chapter we shall treat principally electron-impact-excited experiments employing electric-dipole rf transitions. Further, we shall emphasize work on hydrogen, hydrogenlike singly ionized helium, and neutral helium in quasi-hydrogenlike Rydberg states, because of the evolution of our own work.

1.2. Classification of Fine Structure

Two general types of fine structure can be distinguished in atomic and molecular systems. The first is *relativistic fine structure*, in which we include spin–spin, spin–orbit, and quantum-electrodynamic contributions. The second is *electrostatic fine structure*. This term covers inner-electron screening, electron exchange, and core-polarization effects. Rotational structure in molecules also fits into this category. In making this distinction we have attempted to separate those fine-structure effects that properly originate in a treatment according to relativistic quantum mechanics (e.g., spin), or in a quantum-electrodynamic treatment, from those that are present when the system is treated as a collection of charged point masses interacting via the Coulomb field and obeying Schrödinger's equation. The hydrogen atom has no electrostatic fine structure because of the degeneracy of the Schrödinger solution with respect to orbital angular momentum. Other atoms do, however, because the electrostatic potential felt by an outer electron does not vary simply as $1/r$. Spectral features such as the sodium D lines and the infrared $2P$–$2S$ helium emission lines can be regarded as electrostatic fine structure. The doublet structure of the D lines, however, is relativistic in origin.

Microwave transitions in low atomic states usually involve relativistic fine–structure intervals, since electrostatic intervals are too large to be so spanned. In highly excited states, however, both types of structure are accessible via microwave experiments, which have provided much detailed information on these delicate levels.

1.3. Rate-Equation Treatment of Line Shape

1.3.1. Two Excited Levels

We follow here the treatment of Lamb and Sanders.[10] We assume that a and b are the only excited states, and that their lifetimes $\tau_a = 1/\gamma_a$ and $\tau_b = 1/\gamma_b$ are short enough that atoms in either state decay entirely within view of the photodetector. Atoms are excited to these states at rates r_a and r_b, and the microwave or radio-frequency field causes stimulated transitions between a and b at specific rates $W_{ab} = W_{ba} \equiv W$, as shown in Figure 2(a) or 2(b). No particular characteristics of the excitation process are assumed except that it does not induce ensemble-wide correlations between the excited states. Plausible differential equations for the populations n_a and n_b of excited atoms are then

$$\dot{n}_a = -\gamma_a n_a + W(n_b - n_a) + r_a$$
$$\dot{n}_b = -\gamma_b n_b + W(n_a - n_b) + r_b \tag{1}$$

The decay rates γ_a, γ_b are the total Einstein A coefficients for the states a, b, and the stimulated transition rate W is closely related to the Einstein B coefficient for the transitions $a \leftrightarrow b$. That rate equation ideas can be applied in the case of a monochromatic perturbation, having infinite spectral intensity at frequency ω, can be justified in the density matrix treatment.[10–12]

In the case of constant excitation rates r_a and r_b, setting $\dot{n}_a = \dot{n}_b = 0$ in Eqs. (1) yields the steady-state excited populations n_a and n_b. The excited populations decay by emitting quanta at rates $\gamma_a n_a$ and $\gamma_b n_b$, of which fractions f_a and f_b are detected. The rate of detection is

$$S(W) = f_a \gamma_a n_a + f_b \gamma_b n_b \tag{2}$$

The "signal" is the rf-power dependent part of the detected emission:

$$S = S(W) - S(0) = -(f_a - f_b)\left(\frac{r_a}{\gamma_a} - \frac{r_b}{\gamma_b}\right)\left[\frac{W}{1 + W(\gamma_a + \gamma_b)/(\gamma_a \gamma_b)}\right] \tag{3}$$

This equation results from carrying out the indicated operations on Eqs. (1) and (2). Equation (3) shows the two requirements for a definite effect: (i) the detection of emitted quanta of the two types with unequal probability (the first factor); and (ii) the production of excited atoms with unequal unperturbed steady-state populations (the second factor). In the usual situation (i) is made possible by selection rules that place the two types of emitted quanta in quite different parts of the optical spectrum or give them different polarizations; similarly (ii) results from dependences of optical or electron-impact excitation rates on polarization or on excited-state quantum numbers.

Another common case is that of repetitive nonconstant excitation, perhaps by a periodic short pulse. The signal expression for this "quasi-stationary" situation is very similar to the steady-state case if the observed signal is the *time-averaged* rf-induced emission rate. Equations (1) and (2), being linear, can be time-averaged; the left-hand sides of (1) again become zero. When the derivation is carried through, the signal expression is found to be identical to (3) except that r_a and r_b are replaced by time-averaged excitation rates

$$\langle r_a \rangle = F \int_0^{1/F} r_a(t)\, dt$$
$$\tag{4}$$
$$\langle r_b \rangle = F \int_0^{1/F} r_b(t)\, dt$$

where F is the repetition rate. The detection system is here assumed to be too slow to respond to transient interference phenomena.

If the excitation pulse is intense enough to deplete the ground-state population significantly, but short compared with τ_a and τ_b, Eq. (3) still holds if the factor $(r_a/\gamma_a - r_b/\gamma_b)$ is replaced by $F(n_{a0} - n_{b0})$, where n_{a0} and n_{b0} are the excited populations at the end of the pulse. The equation does not, however, predict coherent transients and related phenomena which are sometimes observed under such excitation conditions.

Both the density matrix treatment and quantum-mechanical time-dependent perturbation theory give for W, in the rotating-wave approximation:

$$W = \frac{\frac{1}{4}(\gamma_a + \gamma_b)|V|^2}{(\omega - \omega_0)^2 + \frac{1}{4}(\gamma_a + \gamma_b)^2} \tag{5}$$

where ω is the rf circular frequency, $\omega_0 = |E_a - E_b|/\hbar$ is the resonance circular frequency, and V is the matrix element of the perturbation amplitude operator. For electric-dipole transitions induced by the classical rf field $\mathbf{E} \cos \omega t$, we have $V = \langle a|e\mathbf{E} \cdot \mathbf{r}|b\rangle/\hbar$. When Eq. (5) is substituted into Eq. (3) the signal expression becomes

$$S = \frac{-\frac{1}{4}(f_a - f_b)(r_a/\gamma_a - r_b/\gamma_b)(\gamma_a + \gamma_b)|V|^2}{(\omega - \omega_0)^2 + \frac{1}{4}(\gamma_a + \gamma_b)^2[1 + |V|^2/(\gamma_a\gamma_b)]} \tag{6}$$

This expression represents a Lorentzian with full width at half-maximum (FWHM)

$$\Delta\omega_{1/2} = (\gamma_a + \gamma_b)(1 + |V|^2/\gamma_a\gamma_b)^{1/2} \tag{7}$$

and peak height

$$S_0 = -(f_a - f_b)\left(\frac{r_a}{\gamma_a} - \frac{r_b}{\gamma_b}\right)\left(\frac{\gamma_a\gamma_b}{\gamma_a + \gamma_b}\right)\left[\frac{|V|^2/(\gamma_a\gamma_b)}{1 + |V|^2/(\gamma_a\gamma_b)}\right] \tag{8}$$

At low rf power levels the peak height is linear in $|V|^2$, and hence in rf power. At high powers it saturates with a maximum intensity equal to $-(f_a - f_b)(r_a/\gamma_a - r_b/\gamma_b)[\gamma_a\gamma_b/(\gamma_a + \gamma_b)]$. The minimum linewidth, when $|V|^2 \ll \gamma_a\gamma_b$, is $\gamma_a + \gamma_b$. Lamb and Sanders have defined an "optimum" perturbation level $|V|^2 = \gamma_a\gamma_b$, at which the signal strength is 50% of the saturated value and the linewidth is $2^{1/2}$ times its minimum value. The linewidth can be seen to increase as the rf *power* at low rf intensity, but as the rf *field* (square root of power) at high rf intensity.

1.3.2. Many Excited Levels

When the rf perturbation or perturbations couple many excited levels the treatment is substantially more difficult. Two simplifying cases may be mentioned. First, Wilcox and Lamb[11] have shown that the rate equation approach can be used if it is possible to find a time-dependent diagonal unitary transformation matrix $U(t)$ such that the matrix

$$\Omega = \frac{1}{2}\Gamma + iU^{\dagger}HU - \dot{U}^{\dagger}U \tag{9}$$

is independent of time, where $H(t)$ is the atomic Hamiltonian $H_0 +$ time-dependent perturbation $V(t)$, in units of rad/sec. Γ is the diagonal matrix of decay rates λ_j, and $U^{\dagger}U = 1$. The populations n_j of the excited levels

$j = 1, 2, \ldots, N$ then obey the set of coupled differential equations

$$\dot{n}_j = -\gamma_j n_j + \sum_{k \neq j} W_{jk} n_k - n_j \sum_{k \neq j} W_{kj} + r_j \tag{10}$$

Here r_j is the excitation rate for state j and W_{jk} is an element of the matrix

$$W = \Gamma - K^{-1} \tag{11}$$

where

$$K_{jk} = \int_0^\infty |[\exp(-\Omega s)]_{jk}|^2 \, ds \tag{12}$$

One possibility leading to a time-independent Ω is that the oscillatory perturbation coupling any given pair of levels j, k has only one significant frequency component ω_α. Then in the rotating-wave approximation the off-diagonal matrix elements of $V(t)$ are $\frac{1}{2} V_{jk\alpha} e^{i\omega_\alpha t}$, where $V_{jk\alpha} = \langle j|e\mathbf{E}_\alpha \cdot \mathbf{r}|k\rangle/\hbar$ (diagonal elements being zero).

The diagonal unitary transformation required is

$$U(t) = [e^{-i\omega_1' t}, \ldots, e^{-i\omega_{N-1}' t}, 1] \tag{13}$$

where $\omega_j' - \omega_k' - \omega_\alpha = 0$. Since there are at most $N - 1$ frequencies ω_α, it is always possible to set at least one element of U equal to 1. In this case any number of levels may be close to resonance, but interference phenomena resulting from multiple transition paths between levels are excluded.

Secondly, the case of N levels, of which only a and b are in resonance, may be treated by time-dependent perturbation theory. Transitions between a and b may occur either directly or through off-resonant intermediate states n. The oscillatory components of the n-state amplitudes that are pertinent to the problem at hand are assumed to be small. Several off-diagonal perturbations may be present, with frequencies ω_α.

The time-dependent Schrödinger equation for the amplitude a_j of state j is

$$i\dot{a}_j = \sum_{k=1}^N \left(H_{jk} - \frac{i}{2} \gamma_j \delta_{jk} \right) a_k \tag{14}$$

where $H_{jk} = \omega_j \delta_{jk} + \Sigma_\alpha V_{jk\alpha} \cos \omega_\alpha t$. Written out with only the dominant terms retained in the equations for levels n, (14) becomes

$$i\dot{a} = (\omega_a - i\gamma_a/2)a + \tfrac{1}{2} \sum_\alpha \left[\left(V_{ab\alpha}b + \sum_n V_{an\alpha}n \right)\left(e^{i\omega_\alpha t} + e^{-i\omega_\alpha t} \right) \right] \tag{15a}$$

$$i\dot{b} = \text{same as above but with } a \text{ and } b \text{ interchanged} \tag{15b}$$

$$i\dot{n} = \omega_n n + \tfrac{1}{2} \sum_{k=a,b} \sum_\alpha V_{nk\alpha}(e^{i\omega_\alpha t} + e^{-i\omega_\alpha t})k \tag{15n}$$

With $V(t)$ a small perturbation, we expect the results $a \approx a_0\, e^{-i\omega_a t}$, $b \approx b_0\, e^{-i\omega_b t}$, where a_0 and b_0 vary slowly with time. Using the relation

$$i\dot{n} - \omega_n n = i\, e^{-i\omega_n t}\frac{d}{dt}(n\, e^{i\omega_n t})$$

we rearrange Eq. (15n), integrate, and obtain

$$n = \frac{1}{2}\sum_{k=a,b}\sum_{\alpha} V_{nk\alpha}\left(\frac{e^{i\omega_\alpha t}}{\omega_k - \omega_n - \omega_\alpha} + \frac{e^{-i\omega_\alpha t}}{\omega_k - \omega_n + \omega_\alpha}\right)k \tag{16}$$

which, when inserted into Eqs. (15a) and (15b), gives

$$i\dot{a} = (\omega_a - i\gamma_a/2)a + \left[\frac{1}{4}\sum_{n,\alpha,\beta} V_{an\alpha}V_{na\beta}\left(\frac{e^{i\omega_\beta t}}{\omega_a - \omega_n - \omega_\beta} + \frac{e^{-i\omega_\beta t}}{\omega_a - \omega_n + \omega_\beta}\right)\right.$$
$$\left. \times (e^{i\omega_\alpha t} + e^{-i\omega_\alpha t})\right]_a a$$
$$+ \left[\sum_\alpha\left\{\left[\frac{1}{2}V_{ab\alpha} + \frac{1}{4}\sum_{n,\beta}V_{an\alpha}V_{nb\beta}\left(\frac{e^{i\omega_\beta t}}{\omega_b - \omega_n - \omega_\beta} + \frac{e^{-i\omega_\beta t}}{\omega_b - \omega_n + \omega_\beta}\right)\right]\right.\right.$$
$$\left.\left. \times (e^{i\omega_\alpha t} + e^{-i\omega_\alpha t})\right\}\right]_b b \tag{17a}$$

$$i\dot{b} = \text{same with } a \text{ and } b \text{ interchanged} \tag{17b}$$

These lengthy expressions simplify in practical cases. Terms with frequencies far from $-\omega_a(-\omega_b)$ will have little effect on the population of level $a(b)$ on the average, and may be neglected. Accordingly we may retain only those terms in $[\ ]_a$ for which $\pm\omega_\beta \pm \omega_\alpha \approx 0$, and those in $[\ ]_b$ for which $\pm\omega_\alpha - \omega_b + \omega_a \approx 0$ (if they include $V_{ab\alpha}$)* or $\pm\omega_\beta \pm \omega_\alpha + \omega_a - \omega_b \approx 0$ (if they include $V_{an\alpha}V_{nb\beta}$). These three relations are the expression of the rotating-wave approximation for this problem. The last two give the resonance condition. The terms retained in $[\ ]_a$ constitute an rf Stark (Zeeman) shift in the effective energy of state a when $V(t)$ is an electric (magnetic) perturbation. The terms retained in $[\ ]_b$ constitute a modified effective perturbation matrix element which may be substituted for $\frac{1}{2}V$ in the two-level rate-equation treatment of the previous section.

In the simplest instance, direct transitions between a and b are forbidden and there is only a single perturbation frequency ω. Then we have for the rf power shift in $\omega_0 = \omega_a - \omega_b$

$$\delta\omega_0 = \frac{1}{4}\sum_n\left[|V_{an}|^2\left(\frac{1}{\omega_a - \omega_n - \omega} + \frac{1}{\omega_a - \omega_n + \omega}\right)\right.$$
$$\left. - |V_{bn}|^2\left(\frac{1}{\omega_b - \omega_n - \omega} + \frac{1}{\omega_b - \omega_n + \omega}\right)\right] \tag{18}$$

* If this is done carefully the Bloch–Siegert shift,[13] which does not depend on the off-resonant states n, will be obtained.

and for the effective matrix element (assuming $\omega_a - \omega_b > 0$)

$$\tfrac{1}{2} V_{\text{eff}} = \frac{1}{4} \sum_n V_{an} V_{nb} \left(\frac{1}{\omega_b - \omega_n + \omega} \right) \tag{19}$$

The resonance condition now becomes $\omega_a - \omega_b \approx 2\omega$, exhibiting a "two-quantum" resonance.

2. Experiments in Lowly Excited States

2.1. Fine Structure of Hydrogen and Hydrogenlike Atoms

The history and theory of hydrogen fine-structure measurements and their role in verifying the predictions of quantum electrodynamics have been reviewed in numerous places[14–18] and need not be described again here. Our restricted purpose will be to describe those measurements that have been done by the method that is the subject of this paper, in order to display the evolution of the technique.

2.1.1. n = 2 Singly Ionized Helium

The microwave–optical resonance method was first used by Lamb and Skinner[4,5] to measure the $2^2S_{1/2}$–$2^2P_{1/2}$ interval in hydrogenlike singly ionized helium (Figures 3 and 4). In 1950 they reported the value $14{,}020 \pm 100$ MHz for this interval, which was 1.4% higher than the theoretical value then available. The Lamb–Skinner apparatus is shown in Figure 5. Its similarity to more recent versions to be described will be readily apparent. A hot filament G and plate H formed the cathode and anode of an electron gun whose beam passed through the K band (18–26 GHz) waveguide section C, which could also be removed, allowing operation in the lower frequency range (8–12 GHz) of the X-band waveguide L. These components were mounted within and between the polepieces A and B of an electromagnet. Of the $n = 2$ He$^+$ ions formed by simultaneous electron-impact ionization and excitation, those in the 2^2P states decayed quickly with a natural lifetime $\tau = 1.0 \times 10^{-10}$ sec. The 304-Å decay photons from the transition $2^2P \to 1^2S$ produced a current in a photoelectric cell consisting of photocathode D, electron collector E, and aperture J. Ions in the $2^2S_{1/2}$ state, by contrast, are metastable, with a lifetime of 1.9×10^{-3} sec resulting from two-quantum electric dipole emission. In an electric field the $2S$-state lifetime can be substantially shortened by coupling to the $2P$ states. Nonetheless, with a mean thermal velocity around 10^5 cm/sec, most of the 2^2S ions reached an apparatus wall in a time much shorter than τ_{2S}, to deexcite there nonradiatively.

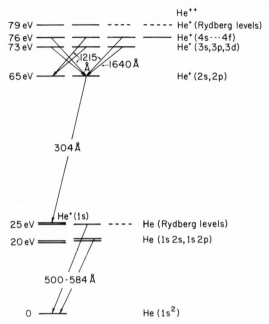

Figure 3. Some energy levels of the He, He$^+$ system.

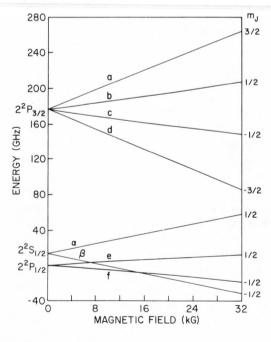

Figure 4. Zeeman energies of He$^+$, $n = 2$. To within the accuracy of the figure, the $n = 2$ H-atom energies can be obtained by multiplying both axis scales by the factor $1/16$, then increasing the $2^2S_{1/2}-2^2P_{1/2}$ displacement (Lamb shift) by the factor 1.21.

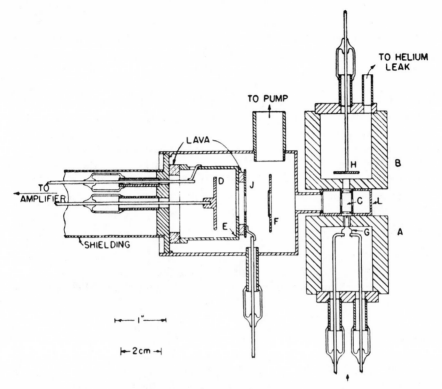

TO HELIUM
LEAK

TO PUMP

LAVA

TO
AMPLIFIER

SHIELDING

H   B

D   J

C   L

E

F

G   A

|← — 1″ — →|

|← 2 cm →|

Figure 5. The Lamb–Skinner apparatus.[5]

Because of the great lifetime disparity, the $n = 2$ state population was overwhelmingly $2^2S_{1/2}$. Thus when microwave radiation was introduced at the $2^2S_{1/2}$–$2^2P_{1/2}$ transition frequency, atoms were transferred into the $2^2P_{1/2}$ state, producing an increase in 304-Å intensity that was used to detect the resonance. A similar increase would have occurred at the $2^2S_{1/2}$–$2^2P_{3/2}$ transition frequency. This, however, is approximately 161.5 GHz, an experimentally difficult frequency even today; this experiment has not been done.

Initially Lamb and Skinner were plagued by a background signal about 70 times larger than their peak resonance signal, resulting principally from 20–25 eV photons emitted by neutral helium and from metastable 2^1S and 2^3S helium atoms; ions and electrons, to which the photocell might also have responded, were trapped by the magnetic field. These sources were greatly attenuated by the interposition of a collodion film filter F before the photodetector. At a thickness of 7 μg/cm^2 it both eliminated the metastable atoms and transmitted the 41-eV He$^+$ radiation preferentially to the lower-energy neutral helium emission.

The signal expression in this experiment differs from that given by the derivation of Section 1.3.1 because atoms in the $2^2S_{1/2}$ state pass from view of the detector before decaying. This inhomogeneous loss mechanism (i.e., depending on the past history of the excited atom) cannot be accounted for in the steady-state solution of Eqs. (1). Instead the following model is used. The population of $2^2P_{1/2}$ atoms is assumed to be small. The $2^2S_{1/2}$ atoms may be converted to $2^2P_{1/2}$ atoms by rf at a constant rate W, and by other processes (static electric-field perturbations, collisions, etc.) at a constant rate λ. Alternatively they may travel a distance L at a velocity v and be lost from view. If the $2^2S_{1/2}$ production rate is r, the rate for the second process is $re^{-(W+\lambda)L/v}$; that for the first process is $r(1-e^{-(W+\lambda)L/v})$. The detected light results from the first process. The rf transition rate W is given by Eq. (5). The signal expression is thus

$$S(W)-S(0)=fr\,e^{-\lambda L/v}(1-e^{-WL/v}) \qquad (20)$$

where f is the overall efficiency for detecting the emitted photons.

For low rf intensity (small W) the exponential may be expanded, yielding

$$S(W)-S(0)\approx fr(L/v)\,e^{-\lambda L/v}W \qquad (21)$$

Consequently the resonance signal is a Lorentzian with FWHM equal to γ_{2P}. The same result holds when both states a and b are short-lived, with FWHM $=\gamma_a+\gamma_b$. At higher intensities both linewidths increase. The shapes, however, differ considerably between the two cases.

An important practical difficulty in plotting out the resonance arises from the rapid decay of the $2^2P_{1/2}$ state, which gives the resonance a width of 1600 MHz. To cover this frequency interval Lamb and Skinner would have required several klystron oscillators. The authors therefore took the alternate course of fixing the microwave frequency and varying the atomic resonance frequency via the magnetic field Zeeman effect of the energy levels, as shown in Figure 4. Measurements were made on the αf transition near 2000 G. A correction for the Zeeman shift was then added to the observed resonance frequency to yield the true $2^2P_{1/2}-2^2S_{1/2}$ frequency. This tuning method has the significant advantage that the microwave electric field experienced by the atoms remains constant across the resonance. It has the disadvantages, however, of increasing the number of distinct resonances between sublevels and of introducing a motional electric field $\mathbf{E}=(\mathbf{v}/c)\times\mathbf{H}$, which can both contribute magnetic-field-dependent Stark shifts and couple $2S$ to $2P$, causing the nonresonant quenching rate λ to vary over the resonance. Lamb and Skinner's accuracy was not high enough for their results to be significantly affected by these phenomena; they have, however, troubled later workers.

Later experiments on the $2^2S_{1/2}-2^2P_{1/2}$ interval in $^4He^+$ were carried out by Novick *et al.*,[19] Lipworth and Novick,[20] and Narasimham and Strombotne.[21] All were similar in principle to the Lamb–Skinner experiment, but included technical improvements and refinements in data-collection technique. Novick *et al.* used pulsed excitation, which allowed rejection by electronic gating of most background radiation, which resulted from nonresonant short-lived states. At the same time, however, ion speeds and consequently motional Stark effects were increased, since at the end of the exciting pulse a cloud of positive ions was left, which then spread rapidly by "Coulomb explosion." The authors studied the αf resonance near 8000 G, where the β-state population was greatly reduced by motional-electric-field coupling at the βe crossing. To compensate for variations in production rate, detection probability, and S-state nonresonant quenching rate with magnetic field, the signal was taken as the ratio of a low-power and a high-power measurement; this procedure eliminates dependence on f, r, and λ for a line shape of the form of Eq. (20).

The last two experiments mentioned used steady-state excitation and lock-in detection, and studied the αe resonance at ≈ 29.3 GHz while operating near the βf crossing at ≈ 16 kG. The α and β sublevels are quenched unequally by motional electric fields; this choice of operating point minimized αe-resonance uncertainties from βf-resonance overlap, since the two are most distant at that magnetic field. Although the βe resonance is closer in frequency, it is excited by microwave electric fields polarized at right angles to the magnetic field (σ type, $\Delta m_J = \pm 1$), whereas αe and βf are excited by parallel electric fields (π type, $\Delta m_J = 0$). Thus by carefully designing the static magnetic and rf electric fields to be parallel, βe could be rendered unimportantly weak.

The experiment of Narasimham and Strombotne, the most recent and most accurate in this sequence, yielded the result $\nu(2^2S_{1/2}-2^2P_{1/2}, \, ^4He^+) = 14{,}046.2 \pm 2.0$ MHz, in agreement with theory as of its publication date. The accuracy quoted corresponds to $1/800$ of the $2^2P_{1/2}$-state natural breadth.

2.1.2. $n = 2$ Hydrogen

The majority of experiments on $n = 2$ hydrogen atoms have been carried out using other techniques.[14–16] One microwave–optical resonance experiment, however, was performed over several years by Kaufman *et al.*[22–24] The result was the value 9911.377 ± 0.026 MHz for the $2^2S_{1/2}-2^2P_{3/2}$ interval, with the level positions defined as the centers of gravity of their respective hyperfine sublevels. The experiment differed from its predecessors principally in that the excited atoms were produced directly by electron impact on H_2 molecules. The apparatus included a long, thin, hollow ellipsoidal light pipe to carry the 1216-Å vacuum-ultraviolet decay light to a NO-filled photoionization cell situated well outside the magnet. With this arrangement at most one surface reflection occurred, resulting in high light-collection efficiency. Because of the extremely high quoted accuracy (1/3800 of the 99.7-MHz $2^2P_{3/2}$ natural breadth) which the strong signals made statistically possible, much effort was expended on data collection and on analysis of possible line-center shifts such as space-charge- and magnetic-field-dependent Stark quenching, quenching caused by charged and neutral particle collisions, and overlapping resonances.

The velocity distribution of excited $2S$ atoms born from dissociating molecules, which determines the motional Stark shift, was measured in separate experiments.[25,26] The fact that atoms with a variety of flight times L/v participated was accounted for empirically by using as a line shape function the sum of two terms of the form of Eq. (20), whose relative contributions and individual L/v values could be determined from the data. The low-power/high-power ratio method of processing resonance data was used.

This measurement is still the most precise for the $2^2S_{1/2}-2^2P_{3/2}$ interval; however, it disagrees with others, for reasons that are not known.* Speculations by the authors mentioned the possible influence of microwave fields on the molecular dissociation process, microwave transitions in molecular hydrogen, residual electric fields, and the sufficiency of the theory of the Zeeman effect. We shall speculate on another possibility, that microwave transitions in long-lived hydrogen-atom Rydberg levels could have contributed to the 1216-Å fluorescence by cascade processes. Kaufman et al. assumed that the low rf power levels that were sufficient to induce $2S$–$2P$ transitions would not substantially affect populations of higher nonmetastable levels. Furthermore, they assumed that transitions in higher states would occur at lower frequencies, since the zero-magnetic-field fine structure scales approximately as $1/n^3$. However, Rydberg transitions can be induced at very low power levels because the electric dipole matrix elements increase rapidly with n. For the same reason, all fine-structure sublevels are coupled strongly by motional electric fields above a certain n. This leads to rf resonances of high frequency ($\omega/2\pi B \gg$ 1.4 MHz/G), which would have overlapped the resonances studied by Kaufman et al. Such "bumps" probably would have been observed in the raw data since the resonance height was measured at only 10 magnetic field

---

* It also disagrees with the current[17] theoretical value $\nu_{theor}(2S_{1/2}-2P_{3/2})=$ 9911.116±0.019 MHz by 0.261±0.032 MHz or 8 standard deviations. Here the uncertainty in ν_{theor} includes both that from theory and that from the fine-structure constant α.

points. An example of this phenomenon observed in helium by Mader and Wing is shown in Figure 12.

2.1.3. n = 2 Doubly Ionized Lithium

An experiment on hydrogenlike Li^{++} in the $n = 2$ state has been done by Leventhal and Havey.[27] This work was technically very difficult because of the short (135 Å) wavelength of the detected 2^2P-1^2S photoemission, the large magnetic field range required for tuning, and the problems of handling lithium vapor. The experimenters used a collimated lithium vapor stream from a heated molybdenum oven. A 2500-Å-thick Be filter kept both lithium and radiation outside its 100–300-Å bandpass from reaching the windowless photomultiplier detector. A small fraction of the ions formed by bombardment with a 6-μsec electron pulse was left in the doubly charged $n = 2$ state. The ions were trapped radially by the magnetic field and axially by two suitably biased reflector electrodes, and subjected to a 4-μsec rf pulse, during which photons were collected. The cycle was completed, in a time much shorter than the 167-μsec $2S$ lifetime, by removing the reflector bias to clear the trap. Alternate low-rf-power and high-rf-power cycles allowed ratio normalization. After several years of labor this experiment has yielded the result $\nu(2S_{1/2}-2P_{1/2}, {}^6Li^{++}) = 62,765 \pm 21$ MHz, in agreement with current theory. The quoted uncertainty corresponds to $1/400$ of the 8040-MHz $2P$ natural breadth.

2.1.4. n = 3 and 4 Hydrogenlike Levels

Several microwave-optical experiments on $n = 3$ and $n = 4$ H and He^+ fine structure have been performed. In such experiments both resonating states are short-lived enough to decay within view of the detector; therefore the theory of Section 1.3 applies. In principle it is still necessary to average over the distribution of rf intensity in the experimental module, weighted by the local density of excited atoms and the light collection efficiency; in practice the simple Lorentzian of Eq. (6) is sufficiently close to reality to produce accurate line-center values, which depend primarily on the symmetry of the resonance line shape and of its modeling function.

The first of these experiments was performed by Lamb and co-workers[10,11] on deuterium, whose relatively smaller nuclear magnetic moment makes accurate interpretation of hyperfine effects less critical than in hydrogen. The experiments were done in a baked and sealed-off glass tube, with an internal electron gun and external optical detection. $3S_{1/2}-3P_{1/2,3/2}$, $3P_{3/2}-3D_{3/2}$, and $4S_{1/2}-4P_{1/2}$ resonances were studied. The detection factor $(f_a - f_b)$ in Eq. (6) was rendered nonzero by the fact that the S or D states decayed principally to the $2P$ state with the emission of

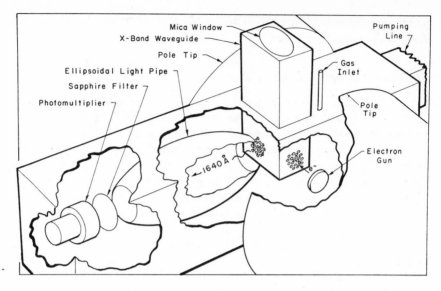

Figure 6. Modern microwave–optical resonance apparatus (Mader *et al.*[31]).

observed Balmer radiation (H_α or H_β), whereas the P states decayed principally to $1S$, accompanied by unobserved Lyman-series emission. Excepting unfortunate combinations of excitation cross sections and natural lifetimes, the population factor $(r_a/\gamma_b - r_b/\gamma_b)$ is quite generally nonzero, as it was here.

These authors included careful treatments of the resonance process from both the rate equation and density matrix viewpoints. They also treated[11] three-level resonances, which in their case resulted when rf transitions took place between two levels, one of which was also significantly coupled to a third level by static electric fields. The rate-equation treatment is successful in this case. The solution can be transformed into that of two perturbations of arbitrary frequency.

A succession of microwave–optical measurements has been made on $n = 3$ and $n = 4$ He$^+$ fine structure by Lamb and co-workers.[28–31] Similar experiments were done in $n = 4$ by Hatfield and Hughes[32] and by Beyer and Kleinpoppen,[33] and in $n = 3$ and 5 by Baumann and Eibofner[34,35] and by Eibofner.[36,37] The experiments of Jacobs *et al.*,[30] Mader *et al.*,[31] and Kaufman *et al.*[24] used a common apparatus, a later revision of which has been used by the present authors in their work on fine structure of Rydberg helium atoms, to be described in Section 3. The Jacobs and Kaufman experimental modules were substantially identical, since Jacobs also detected 1215-Å fluorescence, in his case the He$^+$ $n = 4$ to $n = 2$ emission line. In Mader's work, the 1640-Å $n = 3$ to $n = 2$ fluorescence was

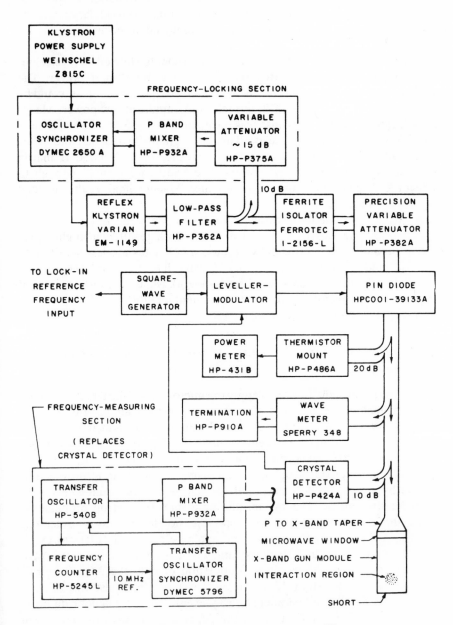

Figure 7. Typical microwave system (Jacobs *et al.*[30]).

detected by a sapphire filter and photomultiplier combination as shown in Figure 6. The microwave generation, stabilization, and modulation circuitry used by Jacobs, which is typical of that employed in recent experiments, is shown in Figure 7.

Both the Mader and the Jacobs experiments showed a mysterious dependence of computer-fitted resonance linecenters on rf power. The frequencies were linear in rf power at high power ($|V|^2 \gtrsim \gamma_a\gamma_b$), as would be expected from an rf Stark shift, but fell away rapidly at low power. Mader was able to show that the cause was resonances in $n = 5$ to 7 levels, seen in the $n = 3$ fluorescence by cascade. They were greatly broadened and hence essentially "flat" at high power levels because of their large rf matrix elements ($|V|^2 \propto n^4$). At low powers their widths were small enough to affect the shape of the main $n = 3$ resonance. Mader's final results included a correction for this effect.

These experiments have given values for both the $nS_{1/2}-nP_{1/2}$ and $nS_{1/2}-nP_{3/2}$ intervals ($n = 3, 4$). The $n = 3$ work resulted in higher precision than did $n = 4$, principally because $n = 3$ states are less sensitive to electric field perturbations. The $n = 3$ work had a quoted accuracy of $1/1000$ of the total natural breadth ($\gamma_{3S} + \gamma_{3P}$), while the most accurate of the $n = 4$ experiments, Jacobs' measurement of $\nu(4S_{1/2}-4P_{3/2})$, had an accuracy of $1/200$ of ($\gamma_{4S} + \gamma_{4P}$).

2.2. Relativistic Fine Structure of Helium, Other Atoms, and Molecules

Lamb proposed[38] a way to apply the microwave–optical method to the measurement of helium fine structure, and with Maiman and Wieder performed[39] measurements of 2^3P and 3^3P intervals by inducing magnetic-dipole transitions between Zeeman sublevels. The result was 300 times more accurate than the spectroscopic value for the interval.

Although the states belong to the same electrostatic term, electron impact by a collimated beam populates various sublevels unequally. The basis for this can be seen by considering a spinless electron propagating along the z axis, which excites a spinless, nondegenerate one-electron atom to a P state. At threshold the outgoing electron can carry away no angular momentum, and so the symmetry of the total initial-state wave function about the z axis allows population of only the $M_L = 0$ atomic sublevel. When the atom decays to the $1S$ state, which also has $M_L = 0$, only radiation polarized parallel to z can be emitted. If a magnetic dipole rf transition populates a P state having $M_L = \pm 1$, the ensuing decay will result in radiation polarized along x and y, with a different angular distribution. Either the polarization or the angular distribution change can be used to detect the resonance, just as in optical double resonance experiments. In a more realistic excitation model all sublevels will be populated, but unequally in general. Lamb's paper[38] treated the realistic helium case and also discussed the relativistic fine structure in detail.

Miller and Freund[40,41] have dubbed the Lamb–Maiman method MOMRIE (microwave–optical magnetic resonance induced by electrons). They have used it to measure the fine structure and Zeeman effect in the 4^3P and 5^3P states of ^{4}He. They have also used MOMRIE in a prolific study of fine structure in various excited states of H_2 and D_2.[42-52]

The rf– or microwave–optical resonance method with electron bombardment excitation has been used for accurate measurements on a number of other atoms and molecules in relatively low states of excitation. Pebay-Peyroula, Brossel, and their associates have studied magnetic dipole resonances in excited states of Hg, Na, Cs, Zn, Zn^+, Cd, Cd^+ (Refs. 53–63) as well as ^{4}He (Ref. 64) and ^{3}He (Refs. 65, 66). Lifetimes, g factors, and hyperfine structures have been reported. In general the oscillator frequency was held constant at less than a few hundred MHz, and the resonances were scanned by varying a magnetic field.

3. Experiments in Highly Excited States

3.1. Electrostatic and Relativistic Fine Structure of Helium Rydberg States

The nonrelativistic l-degeneracy of electronic energy levels in a pure Coulomb field, i.e., in H I, He II, Li III, etc., is lifted by relativistic and radiative level shifts. In helium and all other nonhydrogenlike atoms the electrostatic fine structure further separates the various terms and levels having a given principal quantum number n. Among complex atoms, the electrostatic fine structure is smallest in heliumlike atoms.

When a single atomic electron is excited to a state nl far above the states occupied by the core electrons, i.e., having high n, the atom is said to be in a Rydberg state because the Rydberg formula[67]

$$E_n = -hcR/(n - \delta_l)^2 \tag{22}$$

then provides a useful empirical rule for unifying the wealth of excited-state data. E_n is the binding energy of the excited electron, and the quantum defect δ_l is a quantity characteristic of the atom and term series that varies only slowly with n. The quantum defect expresses the non-Coulombic nature of the potential near the ionic core. Table 1 gives approximate scaling factors for many of the properties of highly excited

Table 1. Approximate Scaling Factors for Properties of Highly Excited
States

Property	Scaling factor
Binding energy	$1/n^2$
Mean atomic radii Electric dipole matrix elements ($\Delta n = 0$)	n^2
Geometric collision cross section	n^4
r.m.s. velocity of excited electron	$1/n$
Classical period of electron motion	n^3
Radiative decay rates Electron-impact excitation rates Level spacings	$1/n^3$
Lifetimes	n^3
Stark shift in electric field $\mathscr{E}$	$\mathscr{E}^2 n^7$
Quadratic Zeeman effect in magnetic field $\mathscr{H}$	$\mathscr{H}^2 n^4$
rf power required to produce "optimum" perturbation in $\Delta n = 0$ electric dipole transition	$1/n^{10}$

atoms in terms of n.* In excited atoms having complex cores, the regular progression of Rydberg series is often perturbed by series associated with different core states. The analysis of such series in terms of quantum defects can be complex.[70–72] However, the excited states and fine structure of ^{4}He are completely regular and provide an attractive example for the study of Rydberg atoms.

In high singly excited configurations $1snl$ of helium, the excited nl electron moves in the field of the He$^+$ core. Orbitals having high angular momentum l correspond to the nearly circular and nonpenetrating orbits of the Bohr–Sommerfeld model[68]; for these states the deviation from hydrogenic behavior is small. As l decreases, however, the states display ever greater core polarization and distortion, electron exchange and nonadiabatic effects until for S, P, and D states ($l = 0, 1, 2$) helium can be said to exist in two nearly independent species, orthohelium (total electron spin $S = 1$, "triplets") and parahelium ($S = 0$, "singlets"), having substantially different energy levels and properties. For the S, P, and D terms Russell–Saunders LS coupling is an excellent model for all n; the terms are described by "good" quantum numbers LSJ as $(1snl)^{2S+1}L_J$ with $l = L$ and $\mathbf{J} = \mathbf{S} + \mathbf{L}$. LS coupling is appropriate for $L \leq 2$ because the Coulomb-exchange interaction is far stronger than the relativistic spin–orbit interaction. Furthermore, the incomplete screening of the nuclear charge ($Z = 2$) by the $1s$ orbital and the strong polarization of the $1s$ core by the nl orbital depress the singlet–triplet barycenter of a $1snl$ configuration

* For general reviews of the properties of Rydberg atoms, see Refs. 68 and 69.

substantially below the hydrogenic Bohr level, $E_n = -hcR/n^2$. In states of $L \geq 3$ (F, G, H, etc.), screening and exchange become equal to or smaller than the relativistic fine structure, but core polarization still causes lower-l states to lie lower in energy. The terms converge rapidly to the Bohr level from below as $L \to L_{max} = n - 1$. Figure 8 shows the electrostatic fine structure of helium states $n = 10$ and demonstrates the effect of screening and polarization, and, for S, P, and D states, the effect of exchange. The displacement of levels from the Bohr level decreases by a factor of 0.25–0.5 for each increase of L by one unit for $L \geq 3$.

The calculation of electrostatic and relativistic fine structure in helium is complex if extremely high accuracy is required. Very detailed calculations have been carried out for a range of S and P states.[73] Chang and Poe[74,75] have used Brueckner–Goldstone perturbation theory to calculate electrostatic and relativistic fine structure of D and F states. Parish and Mires[76] have calculated fine structure using modifications of the method due to Araki.[77] For states of higher L electrostatic fine structure can be modeled with fair success by simpler polarization theories[78–82] and the relativistic fine structure by the Breit–Bethe theory.[14]

In a polarization theory, the non-Coulombic behavior of the ionic core is described by dipole and quadrupole polarizabilities; small corrections may be made for nonadiabatic effects. Some approaches are inherently unable to account for exchange because the electrons are supposed to remain distinguishable. For $L \leq 3$ this is a serious shortcoming, but for $L \geq 4$ the exchange effects make a negligible contribution to fine structure owing to the nonpenetration of the core by the outer electron.

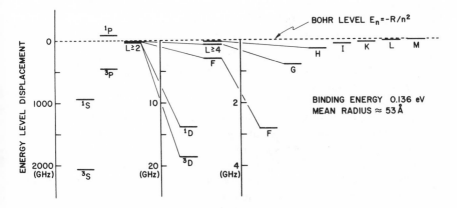

Figure 8. Electrostatic fine structure in helium $n = 10$, demonstrating the wide range of magnitudes for low and high values of orbital angular momentum (note the changes of vertical scale) and the rapid convergence to the Bohr level for high L.

Table 2. Nonzero Matrix Elements in the Breit Operators H_3^{SO}, H_3^{SOO}, and H_5 in the Breit–Bethe Approximation[a]

S	S'	J	J'	$(H_3^{SO})_{SS'JJ'}$	$(H_3^{SOO})_{SS'JJ'}$	$(H_5)_{SS'JJ'}$
0	0	L	L	0	0	0
1	1	$L+1$	$L+1$	$(Z-1)L$	$-2L$	$2L/(2L+3)$
1	1	L	L	$-(Z-1)$	$+2$	-2
1	1	$L-1$	$L-1$	$-(Z-1)(L+1)$	$2(L+1)$	$(2L+2)/(2L-1)$
1	0	L	L	$-(Z-1)[L(L+1)]^{1/2}$	$-2[L(L+1)]^{1/2}$	0

[a] These matrix elements when multiplied by $(2cR)(\alpha^2/4)\langle(a_0/r_2)^3\rangle$ will be given in Hz. The quantity $\langle(a_0/r_2)^3\rangle$ is obtained from hydrogenic wave functions and equals $2(Z-1)^3/[n^3(2L+1)(L+1)L]$. R is the Rydberg for helium (cm^{-1}), c the speed of light, α the fine-structure constant, and a_0 the Bohr radius.

The Breit–Bethe theory[14] of helium relativistic fine structure proceeds from the Pauli approximation of the relativistic Breit equation. Certain small terms that do not lift the degeneracy in an $\{SLJM\}$ manifold are dropped, and an approximation is made to the spin–orbit and spin–spin terms that remain. The assumptions are made that (a) $r_1 \ll r_2$, where r_i $(i = 1, 2)$ are the radial coordinates of the $1s$ and nl electrons; (b) consequently, $r_{12} = |\mathbf{r}_1 - \mathbf{r}_2|$ may be replaced by r_2; (c) the screening of the nucleus by the inner electron is complete, therefore the wave function of the outer electron is approximately hydrogenic with nuclear charge $Z_{eff} = Z - 1$; and (d) *unsymmetrized* product wave functions may be used. The assumptions become more reasonable as l and therefore n increase. The operators of the Breit Hamiltonian then take simple forms for the calculation of matrix elements diagonal in n and L, listed in Table 2. H_3^{SO} accounts for the spin–orbit interaction of the nl electron's spin magnetic moment with the magnetic field in its rest frame caused by orbital motion through the electric fields of the nucleus and the $1s$ electron; H_3^{SOO} is the spin–other-orbit interaction of the $1s$ electron's spin with the magnetic field generated by orbital motion of the nl electron. H_5 is the magnetic dipole spin–spin interaction of the two electrons.

In solving the secular determinant for the fine-structure splittings in a configuration $1snl$, the exchange integral must be included in the diagonal matrix elements in Table 2, $+K$ for $S = 0$ states, $-K$ for $S = 1$ states; K is positive. The off-diagonal matrix elements of the spin–orbit interactions mix the two $J = L$ basis states in a given configuration so that their pure singlet or triplet character is compromised. The $J \neq L$ states remain pure triplet. The degree of singlet–triplet mixing thus occurring depends on the magnitude of K in relation to the relativistic interactions and is described by a singlet–triplet mixing coefficient Ω,[76,83–85]

$$|^3L_{J=L}\rangle = (1 + \Omega^2)^{-1/2}[\Omega|S\rangle + |T\rangle] \tag{23}$$

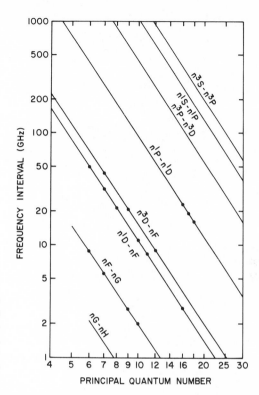

Figure 9. Electrostatic fine-structure intervals in helium $|\Delta L| = 1$. The deviation from straight lines having slope -3 is imperceptible on this scale. Dots indicate transitions measured by microwave-optical resonance.[87-93]

$$|^1 L_{J=L}\rangle = (1+\Omega^2)^{-1/2}[|S\rangle - \Omega|T\rangle] \qquad (24)$$

where here we use the notation $|^{1,3}L_{J=L}\rangle$ to refer to the energy eigenstates; $|S\rangle$ and $|T\rangle$ are the pure singlet and triplet $J = L$ basis states in Table 2. This somewhat confusing notation is consistent with past practice; hereafter it must be taken for granted that $J = L$ eigenstates given in LS-coupling notation are, to a greater or lesser extent, mixtures of singlet and triplet basis states. In a given configuration, triplets lie below singlets.[86] The $3P$ and $3D$ fine-structure levels form inverted triplets, i.e., the levels from lowest to highest energy are $J = L+1$, L, and $L-1$. The strong singlet-triplet repulsion in F, G, etc. states ($L \geqslant 3$) depresses the triplet $J = L$ level below the $J = L+1$ level to form irregular triplets. The levels from lowest to highest energy are then $J = L$, $L+1$, and $L-1$.

In Figure 9 the $\Delta L = 1$ electrostatic fine-structure intervals in ^{4}He are shown for a range of principal quantum numbers. The relativistic fine structure is too small to appear on the scales of Figures 8 and 9. Figure 10 shows the relativistic fine structure for $10F$, G, H, etc. states and lists the singlet-triplet mixing coefficients calculated in the Breit–Bethe theory.

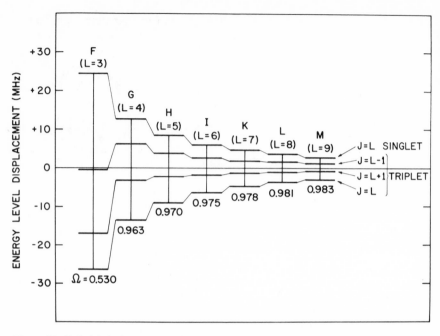

Figure 10. Relativistic fine structure in helium $n = 10$. The singlet–triplet mixing coefficients Ω, as calculated in the Breit–Bethe theory, are indicated.

3.2. Apparatus

A resonance module used for recent microwave–optical studies in helium[87] is shown in Figure 11. It consists of a simple electron gun, an electrostatically shielded bombardment and resonance region and microwave horns for irradiation of the excited atoms. The resonance region has a high geometrical transparency to aid in light collection. Periodically, in the course of an experiment, the wires that form the "cage" (Figure 11) can be heated to incandescence to remove insulating contaminant layers that might charge electrostatically under electron bombardment. The module is enclosed in a vacuum chamber continuously pumped to a base pressure of 2×10^{-7} Torr by a turbomolecular pump. Helium is leaked into the vacuum continuously to maintain a pressure in the range 10^{-5}–10^{-2} Torr. Fluorescence is collected by an ellipsoidal light pipe and focused at the entrance slit of a small monochromator. Microwave power of controlled frequency and amplitude irradiates the excited atoms and is chopped for purposes of synchronous detection of the fluorescence intensity.

The gas that fills the module is bombarded continuously by electron currents of 10–$500\ \mu A/cm^2$ at kinetic energies from the excitation

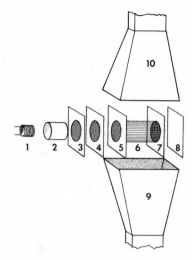

Figure 11. Exploded view of the resonance module used in Ref. 87: (1) heater; (2) cathode; (3) accelerating grid G1; (4) grid G2; (5) grid B1; (6) the "cage," consisting of wires joining the perimeters of grids B1 and B2; (7) grid B2; (8) collector C; (9) and (10) X-band waveguide horns. The grid supports are nonmagnetic stainless-steel plates, 2.54 cm square. The cage is 2.54 cm long × 1.6 cm diameter.

threshold to several hundred electron volts. The fluorescence count rates vary from 1 to 100 kHz depending on excitation conditions, the state being excited, the method of light detection, etc. The magnetic field in the resonance region is less than 100 mG.

3.3. Experimental Results in Helium

Preliminary results on fine-structure transitions in highly excited states of ^{4}He were obtained by Wing, Mader, and Lamb[88,89] and by Wing, Lea, and Lamb.[90] In some of this work a broadband vacuum-uv detector was used to monitor the fluorescence intensity. Magnetic field tuning with fixed microwave frequencies was used. A wealth of unidentified transitions was seen, some with very high Zeeman slopes indicating high angular momentum (Figure 12). All were affected by velocity-dependent Stark shifts owing to the passage of excited atoms through the magnetic field. In our own subsequent precision work[87,91–93] the magnetic field was carefully set to zero and the resonances scanned by varying the microwave frequency. The precision measurements have covered a wide range of ^{4}He transitions having $\Delta n = 0$ and $|\Delta L| = 1$ or 2.

The n's conveniently covered for particular L's by microwave sources in the range 5–100 GHz are listed in Table 3. The single-quantum electric dipole transitions satisfy the selection rules $\Delta L = \pm 1$ and $\Delta J = \pm 1, 0$. Because of the large degree of singlet–triplet mixing in F, G, etc. states ($L \geq 3$), transitions such as $n\,^1D_2$–$n\,^3F_3$ or $n\,^3D_3$–$n\,^1F_3$ are allowed, although superficially they would appear to violate the spin selection rule $\Delta S = 0$ of electric-dipole transitions. The reduced dipole matrix element connecting

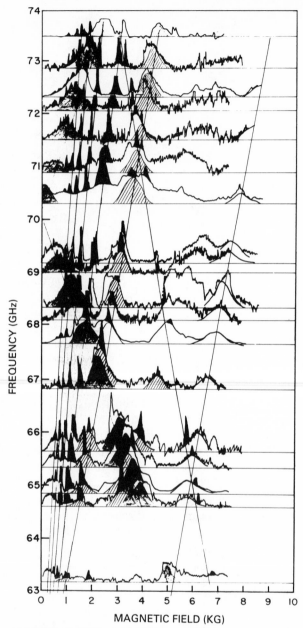

Figure 12. A "forest" of Rydberg state resonances seen in
500–584 Å helium decay light. The usual selection rule
$|\Delta L| = 1$ has been broken by the combination of strong rf
and motional $(\mathbf{v}/c \times \mathbf{H})$ electric fields. Zeeman slopes
$d\nu/dH$ as high as $8\mu_B/h$ appear.[90]

Table 3. Principal Quantum Numbers n Giving Rise to Transitions at Microwave Frequencies 5–100 GHz in ^{4}He

Transition series[a]	Range of n	Observed members[b]
$n^1S_0-n^1P_1$	22–58	—
$n^3S_1-n^3P_{J'}$	25–67	—
$n^1P_J-n^1D_2$	10–26	16–18
$n^3P_J-n^3D_{J'}$	17–44	—
$n^1D_2-n^{1,3}F_3$	5–13	6–8,10,11
$n^3D_J-n^{1,3}F_{J'}$	6–14	7,9,12
$n^{1,3}F_J-n^{1,3}G_{J'}$	5–7	6
$n^1D_2-n^{1,3}G_4$	5–11	7,9
$n^3D_J-n^{1,3}G_{J'}$	5–12	10
$n^1P_1-n^{1,3}F_3$	8–20	16

[a] The D–G and P–F series ($\Delta L = 2$) are observed by two-quantum transitions: The transition energy corresponds to twice the frequency of an oscillator in the 5–100 GHz range.
[b] References 87–93.

eigenstates of mixed spin multiplicity ($J = L$) can be written in terms of the reduced matrix elements for pure-spin basis states as

$$\langle^3L_{L'}'\|\mu\|^1L_L\rangle = [(1+\Omega'^2)(1+\Omega^2)]^{-1/2}(\Omega'\langle S'\|\mu\|S\rangle - \Omega\langle T'\|\mu\|T\rangle) \quad (25)$$

Here, Ω' and Ω are the singlet–triplet mixing coefficients for the primed and unprimed states, respectively; $|S\rangle$ and $|T\rangle$ (or $|S'\rangle$ and $|T'\rangle$) represent the pure singlet and triplet basis functions for the unprimed (or primed) eigenstates. Similarly the matrix elements for superficially spin-conserving transitions are modified by the singlet–triplet mixing,

$$\langle^1L_{L'}'\|\mu\|^1L_L\rangle = [(1+\Omega'^2)(1+\Omega^2)]^{-1/2}(\langle S'\|\mu\|S\rangle + \Omega'\Omega\langle T'\|\mu\|T\rangle) \quad (26)$$

$$\langle^3L_{L'}'\|\mu\|^3L_L\rangle = [(1+\Omega'^2)(1+\Omega^2)]^{-1/2}(\Omega'\Omega\langle S'\|\mu\|S\rangle + \langle T'\|\mu\|T\rangle) \quad (27)$$

In principle, it would be possible to deduce Ω and Ω' from the saturation characteristics of microwave resonances; in practice this has not yet been accomplished.

We have made accurate measurements of transition frequencies for the principal and angular momentum quantum numbers listed in Table 3. The measurements, comprising 46 different transitions (August 1976), are described in detail in the references. The observed resonances ranged in frequency from 8 to 49 GHz and had minimum widths of 1–4 MHz. The line centers could be determined by fitting the data with one or more Lorentzian functions using the method of least squares. Line-center uncertainties were 10–500 kHz. After consideration of systematic effects, and allowing for the statistical scatter among repeated line-center measurements, center uncertainties (one standard deviation, SD) could be

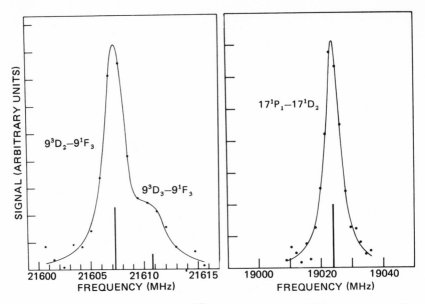

Figure 13. Resonances in ^{4}He n = 9, 17.[93] The points represent 100-sec integration of the lock-in amplifier output. The curves are computer least-squares fits.

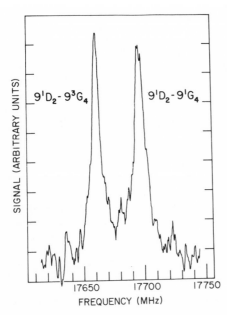

Figure 14. Chart recorder trace of the two-quantum resonances $9^1D_2–9^{1,3}G_4$ in ^{4}He.[87] Sweep rate 0.4 MHz/sec, lock-in time constant 2 sec.

reported from 0.03 to 0.70 MHz. Figures 13 and 14 show resonance line shapes for some of the better measurements.

We make here several observations of a general nature about the resonance measurements as a whole.

(a) *Search Problems.* When no member of a transition series had been observed previously, an estimate of resonance frequency based on other experimental[71,94] and theoretical work was required. Because the experimental linewidths are only a few MHz, whereas typical accuracy in optical spectroscopy is 0.01 cm^{-1} (300 MHz), the initial microwave–optical measurements of each series required scans over frequency ranges of hundreds of linewidths. However, simple accurate scaling laws could be used to predict subsequent members of a series after the first one or two had been located (see Section 3.4).

(b) *Signal Strengths.* As shown in Section 1, the signal strength depends on the population difference and the difference of fluorescence branching ratios for the states connected by the microwave transition. The observed n^1D-nF resonances had very good signal-to-noise (SN) ratios, on the order of 10–50 in 100 sec integration time. Signal strengths in the n^1P-n^1D resonances for $n = 16$–18 were a bit poorer than those in 11^1D-11F: An order-of-magnitude advantage gained in n^1P cross sections[95] relative to n^1D was offset by the higher n's required for work in a given frequency range and by greater sensitivity to electric fields [see (e) below]. SN was an order of magnitude poorer in n^3D resonances than n^1D. In this case, the reason was that cross sections are generally smaller for the triplet terms and that the term populations are diluted by relativistic fine-structure splittings. Furthermore, the F and G states receive population at the higher pressures by collisional excitation transfer (CET)[95–98] out of the readily excited singlet states, principally n^1P; n^3D states, being purely triplet to a good approximation, can receive such a contribution only from the less readily excited triplet states n^3S or n^3P. Therefore, at otherwise convenient operating pressures, resonances from n^3D states may be weak because of small population differences with nF or nG states. The SN was very poor in the $6F-6G$ transitions owing to small cross sections for electron bombardment, dilution by fine structure, canceling contributions from CET, and collisional angular momentum mixing[99] in the $L \geqslant 3$ states.

(c) *rf Power Levels.* The single-quantum transitions require only very small microwave power levels because of the large $\Delta n = 0$ matrix elements in Rydberg atoms (see Table 1). In our apparatus the single-quantum power levels were on the order of 1 μW/cm^2.[91,92] This level is so low that there is a danger of broadening a resonance beyond recognition by the inadvertent use of excessive power. The two-quantum resonances require much more power, depending on the energy mismatch of the nonresonant intermediate states (see Section 1). We used $1-100 \text{ mW/cm}^2$ for the $n = 7, 9$, and 10 $D-G$ two-quantum resonances.[87]

(d) *Oscillator Disposition.* In the case of two-quantum or multiple-quantum resonances, the experimenter has one or more degrees of freedom in selecting oscillator frequencies that sum to the overall transition energy. Provided that energy mismatches with intermediate states are not too unfavorable, a single oscillator at frequency ω can be used to produce n-quantum resonances having total transition energy $n\hbar\omega$. Because of the rapid convergence of nL levels in helium to the Bohr level at high L (Figure 8), such an arrangement may require rather high rf power levels. An alternative is to use two (or more) oscillators so that the frequencies may be better matched to the intermediate states,[100] thereby reducing the power requirements. Stepwise *resonant* transitions can in fact be produced by matching the oscillators exactly to the intermediate states; in this case, the analysis of line shape of Section 1.3.2 is required. Our two-quantum transitions have so far used a single X-band oscillator (8.0–12.4 GHz) except in the case $7^1D_2-7^{1,3}G_4$, where the X-band oscillator and a 26-GHz reflex klystron were used together.

(e) *Distortion by Stark Effect.* The highly excited atoms are very sensitive to electric fields. The quadratic Stark effect, in second-order perturbation theory, scales as n^7 for states of given LSJ. An ensemble of excited atoms in a region of small random electric fields will give rise to a resonance that is distorted and displaced according to the square of the electric fields. Fields on the order of a few volts per centimeter are readily produced by contact potentials, electronic or ionic space charge, charging of insulating surfaces, or penetration of fields into the interaction region from elsewhere. Such field magnitudes can cause distortions and shifts on the order of a natural linewidth. The worst effects of this type were encountered in our measurements of 10^3D-10G, 12^3D-12F, and $n = 16-18$ $^1P-^1D$ resonances.[87,93] In the latter case, in order to maximize signal strength by allowing all atoms of the excited ensemble to participate in the resonance, the resonances were power-broadened to a greater degree than the broadening by Stark-effect distortions. It was then necessary to correct the apparent line centers by 1–2 MHz, because the quadratic Stark shifts for individual atoms are unidirectional.

(f) *Collisional Excitation Transfer.* As mentioned above,[95–98] CET provides a mode of excitation of the high angular momentum states that competes favorably, at pressures of 10^{-3} Torr and above, with direct excitation by electron bombardment. The effects of CET are clearest in cases where only one of a pair of resonating states can readily receive such population. The resonance polarity, determined by the sign of the population difference, reverses for certain of these resonances as helium pressure is increased[87,92] owing to the competition between CET and direct electron-bombardment excitation. The dependence of resonance heights on pressure and bombarding voltage can in principle be exploited to study the rate$\mathcal{s}$ of CET processes.

(g) *Cascade Detection.* The microwave–optical resonances are detected by intensity changes in the fluorescence from the resonating states. Because the lower principal quantum numbers cannot support high orbital angular momentum, direct decays from high-L Rydberg states of neutral atoms have inconveniently long wavelengths. Fluorescence from F and G states in helium lies in the near and intermediate ir region of the spectrum where low-noise photon detectors are less readily available. For the 6F–6G resonances, therefore, we used a method of cascade detection.[87] The $6F$ state decays only to $5^{1,3}D$, $5G$, $4^{1,3}D$, and $3^{1,3}D$, whereas the $6G$ state decays to $5F$ and $4F$ and thence to $4^{1,3}D$ and $3^{1,3}D$. The D states fluoresce at convenient visible wavelengths. By observing emission from the 5^1D or 5^3D states, we could maintain the required difference of branching ratios, which would to some degree have been lost by observing $4^{1,3}D$ or $3^{1,3}D$ light; conventional photomultipliers could therefore be used. More generally, this approach provides a detection method for resonances in virtually any Rydberg state by observation of fluorescence from nearly any convenient lower state, because the branching ratios into the observed fluorescence channel from the two resonating states would be unlikely ever to cancel exactly.

3.4. Series Regularities

To a first approximation, the displacements of atomic levels from the Bohr levels due to internal interactions scale as $1/n^3$. Therefore the dominant term in a scaling law for fine-structure intervals is also $1/n^3$. Chang[101] has shown on theoretical grounds that in a certain approximation an appropriate scaling law for electrostatic or relativistic displacements and intervals ($L \geqslant 2$) is

$$\nu_n = A/n^3 + B/n^5 + C/n^7 \qquad (28)$$

This scaling behavior has been thoroughly corroborated in our experimental work.[87,90,92,93] With it the infinite task of measuring fine structure in helium becomes more manageable, because only a few members of each transition series need be measured. With the scaling formula an entire series can be predicted on the basis of a few measurements; in fact, because accurate experiments become more difficult as n increases, the accuracy of predictions based on low-n measurements can exceed that of values measured at higher n. The scaling law can also be used to reveal data that are suspect and to narrow the search for resonances not yet measured. We have analyzed our data for each tran-

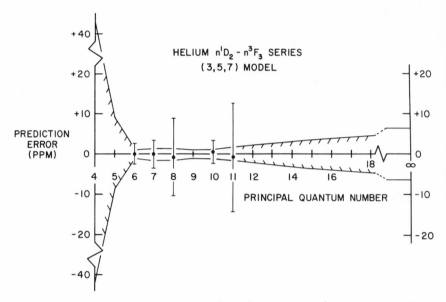

Figure 15. Error bands for predictions of $n^1D_2-n^3F_3$ intervals in ^{4}He. The hatched lines represent the one-standard-deviation uncertainties for predictions based on the fits of microwave–optical resonance data[87,91–93] to Eq. (28). The data points indicate the discrepancies between the measurements and predictions based upon them.

sition series by using the method of weighted least squares to deduce the coefficients A, B, and C.[87,93]

The terms in the scaling formula are highly nonorthogonal over the positive integers n. Hence the parameters A, B, and C determined by least squares are strongly correlated. This fact makes it pointless to give statistical uncertainties of the parameters without reporting the entire variance–covariance matrix. The greatest benefit may be derived from the uncertainties of experimental fine-structure intervals by using the variance–covariance matrix to propagate the parameter errors[102] back to predictions based on the scaling formula. The nonorthogonality of terms in the formula is then manifested as strong correlations between predictions of series members for nearby n. If M_A is the variance–covariance matrix of the parameters for a particular set of weighted data, then

$$M_\nu = TM_AT'$$ (29)

is the variance–covariance matrix of the predictions. T is a rectangular matrix whose rows are the derivatives of Eq. (28) with respect to each parameter in turn, evaluated at each n for which a prediction is required. Estimates of the standard deviations of the predictions are given by the square roots of the diagonal elements of M_ν. Figures 15 and 16 show

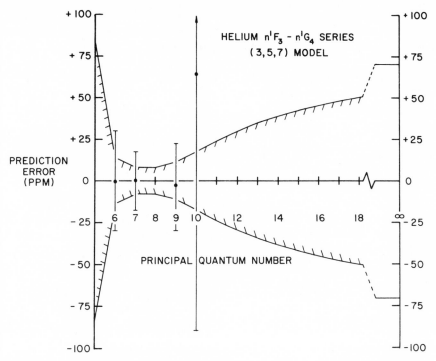

Figure 16. Error bands for predictions of $n^1F_3-n^1G_4$ intervals in ^{4}He.[87] See caption to Figure 15.

prediction error bands for two representative transition series calculated in this way.

In principle, basis functions for a scaling formula could be chosen that are orthogonal over a given set of integers n. In this case the required basis functions would be different for each set of integers n and for each choice of weights. Each of three orthogonal basis functions would require specification of three numerical coefficients in their definitions.

The usefulness of the scaling formula has been demonstrated many times in our work, in which resonances under study have been predicted within 1–2 MHz of their actual centers on the basis of previous results.

3.5. Measurements in Other Atoms

Wing *et al.*[90] measured an electric dipole transition in neon, $(^2P_{3/2})7d_2-(^2P_{3/2})7f_{1,2}$ (Paschen $7d''_1-7X$) at 38.44 ± 0.10 GHz. Fluorescence of the $7d''_1$ state at 4636 Å was used to detect the resonance. An electron beam current 0.3 mA at 75 V was used for excitation. There are many other intervals suitable for electric dipole transitions at microwave frequencies in the noble gases, although in general the energy-level schemes are much more complicated than in one- or two-electron atoms and the optical spectral line density is greater. Theoretical calculations of fine structure in the noble gases (except He) are more difficult, thereby rendering the search problems greater. There is no fundamental reason why extensive measurements as successful as those in ^{4}He cannot be performed.

Our own measurements on fine structure of Rydberg states in ^{4}He and Ne constitute the only rf spectroscopic studies of highly excited atoms using electron-bombardment excitation. However, in this connection certain other experimental programs deserve mention.

Happer and co-workers have studied hyperfine structure and lifetimes in many excited states of alkalis by rf cascade spectroscopy. The atoms are excited by stepwise excitation with a resonance lamp and a cw dye laser.[9] Svanberg et al.[103–105] have used optical double resonance (and level-crossing) techniques to measure fine and hyperfine structures and Stark effects in alkali states as high as $n = 18$. Gallagher et al.[100] have measured fine structure and d–f–g–h intervals in highly excited states of Na. In their work, one-, two-, and three-quantum rf electric dipole transitions were observed. The atoms, up to $n = 17$, were excited stepwise with two pulsed dye lasers pumped by a N_2 laser.

Brandenberger et al.[106] have measured several microwave electric dipole transitions in highly excited states of He$^+$ and He[107] by using a fast beam technique. Fast ions pass through a gas target and emerge after charge exchange in excited atomic or ionic states. They subsequently pass through a region of microwave fields. Their spontaneous decays are monitored optically farther downstream, and microwave resonances are detected by the change of intensity. These measurements exhibit significant transit-time broadening at present, although in principle this limitation can be overcome. The fact that measurements are carried out in a beam with spatially separated excitation, resonance, and detection regions does not alter the fact that apart from transit-time effects the method is that described in this chapter.

Acknowledgment

We are grateful to the U.S. National Science Foundation for partial support of this work.

References

1. I. Rabi, S. Millman, P. Kusch, and J. Zacharias, *Phys. Rev.* **55**, 526 (1939).
2. J. Brossel and A. Kastler, *C.R. Acad. Sci.* **229**, 1213 (1949).

3. F. Bitter, *Phys. Rev.* **76**, 833 (1949).
4. M. Skinner and W. E. Lamb, Jr., *Phys. Rev.* **75**, 1325A (1949).
5. W. E. Lamb, Jr. and M. Skinner, *Phys. Rev.* **78**, 539 (1950).
6. R. H. Dicke, *Phys. Rev.* **89**, 472 (1953).
7. G. W. Series, *Rep. Prog. Phys.* **22**, 280 (1959).
8. G. zu Putlitz, in *Ergebnisse der Exakten Naturwissenschaften*, Ed. E. A. Niekisch, Vol. 37, pp. 105–149, Springer-Verlag, Berlin (1965).
9. W. Happer and R. Gupta, Chapter 9 of this work.
10. W. E. Lamb, Jr., and T. M. Sanders, Jr., *Phys. Rev.* **119**, 1901 (1960).
11. L. R. Wilcox and W. E. Lamb, Jr., *Phys. Rev.* **119**, 1915 (1960).
12. U. Fano, *Rev. Mod. Phys.* **29**, 74 (1957).
13. F. Bloch and A. Siegert, *Phys. Rev.* **57**, 522 (1940).
14. H. A. Bethe and E. E. Salpeter, *Quantum Mechanics of One- and Two-Electron Atoms*, Springer-Verlag, Berlin (1957).
15. K. R. Lea, in *Atomic Masses and Fundamental Constants 4*, Eds. J. H. Sanders and A. H. Wapstra, pp. 355–372, Plenum Press, New York (1972).
16. B. N. Taylor, W. H. Parker, and D. N. Langenberg, *Rev. Mod. Phys.* **41**, 375 (1969).
17. E. R. Cohen and B. N. Taylor, *J. Phys. Chem. Ref. Data* **2**, 663 (1973).
18. H.-J. Beyer, Chapter 12 of this work.
19. R. Novick, E. Lipworth, and P. F. Yeargin, *Phys. Rev.* **100**, 1153 (1955).
20. E. Lipworth and R. Novick, *Phys. Rev.* **108**, 1434 (1957).
21. M. A. Narasimham and R. L. Strombotne, *Phys. Rev. A* **4**, 14 (1971).
22. S. L. Kaufman, W. E. Lamb, Jr., K. R. Lea, and M. Leventhal, *Phys. Rev. Lett.* **22**, 507 (1969).
23. S. L. Kaufman, W. E. Lamb, Jr., K. R. Lea, and M. Leventhal, in *Precision Measurement and Fundamental Constants*, Eds. D. N. Langenberg and B. N. Taylor, National Bureau Standards (U.S.), special publication No. 343 (August 1971).
24. S. L. Kaufman, W. E. Lamb, Jr., K. R. Lea, and M. Leventhal, *Phys. Rev. A* **4**, 2128 (1971).
25. M. Leventhal, R. T. Robsicoe, and K. R. Lea, *Phys. Rev.* **158**, 49 (1967).
26. S. J. Czuchlewski, Ph.D. Thesis, Yale University, 1973 (University Microfilms, Ann Arbor, Michigan).
27. M. Leventhal and P. E. Havey, *Phys. Rev. Lett.* **32**, 808 (1974).
28. M. Leventhal, K. R. Lea, and W. E. Lamb, Jr., *Phys. Rev. Lett.* **15**, 1013 (1965).
29. K. R. Lea, M. Leventhal, and W. E. Lamb, Jr., *Phys. Rev. Lett.* **16**, 163–165 (1966).
30. R. R. Jacobs, K. R. Lea, and W. E. Lamb, Jr., *Phys. Rev. A* **3**, 884 (1971).
31. D. L. Mader, M. Leventhal, and W. E. Lamb, Jr., *Phys. Rev. A* **3**, 1832 (1971).
32. L. L. Hatfield and R. H. Hughes, *Phys. Rev.* **156**, 102 (1967).
33. H.-J. Beyer and H. Kleinpoppen, *Z. Phys.* **206**, 177 (1967).
34. M. Baumann and A. Eibofner, *Phys. Lett.* **34A**, 421 (1971).
35. M. Baumann and A. Eibofner, *Phys. Lett.* **43A**, 105 (1973).
36. A. Eibofner, *Z. Phys.* **249**, 58 (1971).
37. A. Eibofner, *Z. Phys.* **249**, 73 (1971).
38. W. E. Lamb, Jr., *Phys. Rev.* **105**, 559 (1957).
39. W. E. Lamb, Jr. and T. H. Maiman, *Phys. Rev.* **105**, 573 (1957); I. Wieder and W. E. Lamb, Jr., *Phys. Rev.* **107**, 125 (1957).
40. T. A. Miller and R. S. Freund, *Phys. Rev. A* **4**, 81 (1971).
41. T. A. Miller and R. S. Freund, *Phys. Rev. A* **5**, 588 (1972).
42. R. S. Freund and T. A. Miller, *J. Chem. Phys.* **56**, 2211 (1972).
43. T. A. Miller and R. S. Freund, *J. Chem. Phys.* **56**, 3165 (1972).
44. T. A. Miller and R. S. Freund, *J. Chem. Phys.* **58**, 2345 (1973).
45. R. S. Freund and T. A. Miller, *J. Chem. Phys.* **58**, 3565 (1973).
46. R. S. Freund and T. A. Miller, *J. Chem. Phys.* **59**, 4073 (1973).

47. T. A. Miller and R. S. Freund, *J. Chem. Phys.* **59**, 4093 (1973).
48. R. S. Freund, T. A. Miller, and B. R. Zegarski, *Chem. Phys. Lett.* **23**, 120 (1973).
49. T. A. Miller, R. S. Freund, and B. R. Zegarski, *J. Chem. Phys.* **60**, 3195 (1974).
50. R. S. Freund and T. A. Miller, *J. Chem. Phys.* **60**, 4900 (1974).
51. T. A. Miller and R. S. Freund, *J. Chem. Phys.* **62**, 2240 (1975).
52. R. S. Freund, T. A. Miller, and B. R. Zegarski, *J. Chem. Phys.* **64**, 4069 (1976).
53. J. C. Pebay-Peyroula, *C.R. Acad. Sci.* **244**, 57 (1957).
54. J. C. Pebay-Peyroula, J. Brossel, and A. Kastler, *C.R. Acad. Sci.* **245**, 840 (1957).
55. J. P. Descoubes and J. C. Pebay-Peyroula, *C.R. Acad. Sci.* **247**, 2330 (1958).
56. J. C. Pebay-Peyroula, *J. Phys. Radium* **20**, 669 (1959).
57. J. C. Pebay-Peyroula, *J. Phys. Radium* **20**, 721 (1959).
58. A. D. May, *C.R. Acad. Sci.* **250**, 3616 (1960).
59. M. Barrat and J. C. Pebay-Peyroula, *C.R. Acad. Sci.* **251**, 56 (1960).
60. A. D. May, *C.R. Acad. Sci.* **251**, 1371 (1960).
61. Y. Archambault, J. P. Descoubes, M. Priou, A. Omont, and J. C. Pebay-Peyroula, *J. Phys. Radium* **21**, 677 (1960).
62. J. C. Pebay-Peyroula, in *Physical Sciences: Some Recent Advances in France and the U.S.*, Eds. H. P. Kallmann, S. A. Korff, and S. G. Roth pp. 93–102, New York University Press, New York (1962).
63. A. D. May, in *Topics in Radiofrequency Spectroscopy*, Ed. A. Gozzini, pp. 254–258, Academic Press, New York (1962).
64. B. Decomps, J. C. Pebay-Peyroula, and J. Brossel, *C.R. Acad. Sci.* **251**, 941 (1960).
65. B. Decomps, J. C. Pebay-Peyroula, and J. Brossel, *C.R. Acad. Sci.* **252**, 537 (1961).
66. J. P. Descoubes, B. Decomps, and J. Brossel, *C.R. Acad. Sci.* **258**, 4005 (1964).
67. J. R. Rydberg, *K. Sven. Vetenskapsakad. Handl.* **23**, No. 11 (1889); *Philos. Mag.* **29**, 331 (1890); *C.R. Acad. Sci.* **110**, 394 (1890); *Z. Phys. Chem. (Leipzig)* **5**, 227 (1890).
68. I. C. Percival, in *Atoms and Molecules in Astrophysics*, Eds. T. R. Carson and M. J. Roberts, pp. 65–83, Academic Press, New York (1972).
69. R. F. Stebbings, *Science* **193**, 537 (1976).
70. M. J. Seaton, *Proc. Phys. Soc. London* **88**, 801 (1966).
71. M. J. Seaton, *Proc. Phys. Soc. London* **88**, 815 (1966).
72. G. Herzberg, *Atomic Spectra and Atomic Structure*, Dover, New York (1944).
73. Y. Accad. C. L. Pekeris, and B. Schiff, *Phys. Rev. A* **4**, 516 (1971).
74. T. N. Chang and R. T. Poe, *Phys. Rev. A* **10**, 1981 (1974).
75. T. N. Chang and R. T. Poe, *Phys. Rev. A* **14**, 11 (1976).
76. R. M. Parish and R. W. Mires, *Phys. Rev. A* **4**, 2145 (1971).
77. G. Araki, *Proc. Phys. Math. Soc. (Japan)* **19**, 128 (1937).
78. B. Edlén, in *Encyclopedia of Physics*, Ed. S. Flügge, Vol. 27, Secs. 20 and 33, Springer, Berlin (1964).
79. C. Deutsch, *Phys. Rev. A* **2**, 43 (1970); **3**, 1516 (1971).
80. W. C. Martin, *J. Res. Natl. Bur. Stand. Sect. A* **74**, 699 (1970).
81. A. Temkin and A. Silver, *Phys. Rev. A* **10**, 1439 (1974).
82. C. Deutsch, *Phys. Rev. A* **13**, 2311 (1976).
83. G. Araki, M. Ohta, and K. Mano, *Phys. Rev.* **116**, 651 (1959).
84. G. W. F. Drake, *Phys. Rev.* **181**, 23 (1969).
85. R. K. van den Eynde, G. Wiebes, and T. Niemeyer, *Physica (Utrecht)* **59**, 401 (1972).
86. R. P. Messmer and F. W. Birss, *J. Phys. Chem.* **73**, 2085 (1969).
87. K. B. MacAdam and W. H. Wing, *Phys. Rev. A* **15**, 678 (1977).
88. W. H. Wing, D. L. Mader, and W. E. Lamb, Jr., *Bull. Am. Phys. Soc.* **16**, 531 (1971).
89. W. E. Lamb, Jr., D. L. Mader, and W. H. Wing, in *Fundamental and Applied Laser Physics: Proceedings of the Esfahan Symposium*, Eds. M. S. Feld, A. Javan, and N. Kurnit, pp. 523–548, Wiley, New York (1973).

90. W. H. Wing, K. R. Lea, and W. E. Lamb, Jr., in *Atomic Physics 3*, Eds. S. J. Smith and G. K. Walters, pp. 119–141, Plenum, New York (1973).
91. W. H. Wing and W. E. Lamb, Jr., *Phys. Rev. Lett.* **28**, 265 (1972).
92. K. B. MacAdam and W. H. Wing, *Phys. Rev. A* **12**, 1464 (1975).
93. K. B. MacAdam and W. H. Wing, *Phys. Rev. A* **13**, 2163 (1976).
94. W. C. Martin, *J. Phys. Chem. Ref. Data* **2**, 257 (1973).
95. B. L. Moiseiwitsch and S. J. Smith, *Rev. Mod. Phys.* **40**, 238 (1968).
96. H. S. W. Massey, *Electronic and Ionic Impact Phenomena. III. Slow Collisions of Heavy Particles*, 2nd. ed., pp. 1757–1767, Oxford, London (1971).
97. J. D. Jobe and R. M. St. John, *Phys. Rev. A* **5**, 295 (1972).
98. A. F. J. van Raan and J. van Eck, *J. Phys. B* **7**, 2003 (1974).
99. T. F. Gallagher, S. A. Edelstein, and R. M. Hill, *Phys. Rev. Lett.* **35**, 644 (1975).
100. T. F. Gallagher, R. M. Hill, and S. A. Edelstein, *Phys. Rev. A* **14**, 744 (1976).
101. T. N. Chang, *J. Phys. B* **7**, L108 (1973).
102. W. C. Hamilton, *Statistics in Physical Science*, Ronald, New York (1964).
103. W. Hogervorst and S. Svanberg, *Phys. Scr.* **12**, 67 (1975).
104. G. Belin, L. Holmgren, I. Lindgren, and S. Svanberg, *Phys. Scr.* **12**, 287 (1975).
105. G. Belin, L. Holmgren, and S. Svanberg, *Phys. Scr.* **13**, 351 (1976).
106. J. R. Brandenberger, S. R. Lundeen, and F. M. Pipkin, *Phys. Rev. A* **14**, 341 (1976).
107. S. R. Lundeen, private communication.

12

Lamb-Shift and Fine-Structure Measurements on One-Electron Systems

H.–J. BEYER

1. Introduction

1.1. Scope of the Article

It has been known for about a century that many atomic emission lines consist of several close-lying components, thus indicating that the transitions in fact occur between a number of more or less degenerate states. This fine structure has been the subject of numerous experimental and theoretical investigations that eventually led to the picture of atomic structure we have today. Naturally, interest focused in particular on the fine structure of the most fundamental hydrogenic systems, where calculations and (for some time now) measurements can be carried out with great accuracy. During the last 30 years, the study of the hydrogenic fine structure was dominated to a great extent by the continuing interest in quantum electrodynamical effects (Lamb shifts). Yet, at the same time, other fine-structure intervals were also measured to high precision and played an important role in the determination of fundamental constants like the fine-structure constant, α, or the Rydberg constant, R. It is the aim of this article to review the fine-structure measurements of hydrogenic systems which have been carried out with a wide variety of techniques. Muonium, positronium, and other exotic but essentially hydrogenic systems will be excluded. They are dealt with in Chapters 31 and 32 of this work. A review of the theory will also be found in Chapter 4 of this work.

H.-J. BEYER • Institute of Atomic Physics, University of Stirling, Stirling, Scotland.

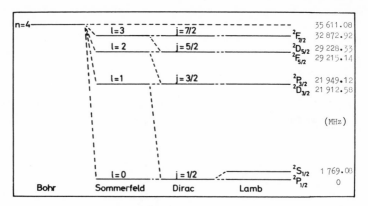

Figure 1. History of the fine structure of the $n = 4$ state of $^4\text{He}^+$. The energy values shown were calculated by Erickson.[198] Intervals with $\Delta l = 0$ and $\Delta j = 1$ (like $P_{1/2}$–$P_{3/2}$) depend little on QED effects and are commonly called fine-structure intervals, while the QED intervals with $\Delta l = 1$ and $\Delta j = 0$ (like $S_{1/2}$–$P_{1/2}$) are commonly referred to as Lamb-shift intervals.

1.2. The Fine Structure of Hydrogenic Systems

Figure 1 shows the historical development of our knowledge of the hydrogenic fine structure on the example of the $n = 4$ states of $^4\text{He}^+$ (which has no hyperfine structure). On the basis of the relativistic motion of the electron in the Coulomb field, Sommerfeld[1] calculated the fine-structure energies and obtained states (four for $n = 4$) labeled with $k = 1, \ldots, n$ ($l = 0, \ldots, n - 1$). Their energies agreed quite well with the experiments, but there appeared to be line components violating the $\Delta l = \pm 1$ selection rule. This problem was solved in the late twenties with the introduction of the electron spin, which added j as a further quantum number. The energies remained unchanged, but the number of fine-structure states increased to seven ($n = 4$) with some states degenerate. This is borne out by the fine-structure energies obtained on the basis of the Dirac equation for a state $|n, l, j\rangle$[2]:

$$E(n, l, j) = -\frac{RZ^2}{n^2}\left[1 + \frac{(\alpha Z)^2}{n}\left(\frac{1}{j + \frac{1}{2}} - \frac{3}{4n}\right) + \cdots\right] \tag{1}$$

l does not appear explicitly in Eq. (1), and states with the same j but different l are degenerate.

The agreement between Eq. (1) and the experiment was good, although measurements in the late thirties cast some doubt on the positions of the $^2S_{1/2}$ and $^2P_{1/2}$ states. The matter was decided in 1947 by the famous experiment of Lamb and Retherford,[3] who proved that $^2S_{1/2}$ and $^2P_{1/2}$ are in fact *not* degenerate. Encouraged by this experimental result, theory quickly managed to set aside the problems connected with the introduction of a quantized electromagnetic field and explained the "Lamb shift" as a quantum electrodynamical (QED) effect. Ever since, the Lamb shift has provided the foremost testing ground for the QED theory in atomic physics.

Including the QED shifts, all degeneracies are lifted, as can be seen in Figure 1, which also shows that the QED shift of $^2S_{1/2}$ is by far the largest. Although the Lamb shift is part of the fine structure, it is common to distinguish between Lamb-shift intervals ($\Delta l = 1$, $\Delta j = 0$) and fine-structure intervals ($\Delta l = 0$, $\Delta j = 1$). The latter are little affected by QED effects and are essentially described by Eq. (1). The most important Lamb-shift intervals $^2S_{1/2}-^2P_{1/2}$ are usually denoted by $\mathscr{S}$ and the most common fine-structure intervals $^2P_{1/2}-^2P_{3/2}$ by ΔE.

The theory of the Lamb shift is discussed in Chapter 4 of this work. It will, therefore, suffice to mention only a few general points. The Lamb shift is made up of several contributions which are the result of (i) the zero-point oscillations of the electromagnetic field (self-energy); (ii) the vacuum polarization; (iii) the anomalous magnetic moment of the electron (acting through the LS coupling); and, (iv) nuclear size and nuclear structure effects. Contributions based on (i) and (ii) require the electron wave function at the origin (nucleus) to be nonzero. To an appreciable extent this is the case only for S states, which thus show by far the largest shift (as shown in Figure 1) of which the dominant contribution comes from effect (i). The displacements of other than S states are much smaller and mainly caused by (iii) with some addition from (i) and (ii). Contributions from (iv) are always small though not negligible.

The calculation of the Lamb shift presents considerable problems and is achieved by evaluating the various contributions in orders of α, $(Z\alpha)$, and m/M. The calculated terms were discussed and tabulated by Erickson and Yennie[4] in 1965 and, including more recent calculations, by Appelquist and Brodsky[5] in 1970 and by Ermolaev in Chapter 4 of this book. A general survey of all fields of QED theory by Lautrup *et al.*[31] (1972) also contains an extended section on Lamb shift, and Erickson[8] (1977), in providing an up-to-date list of energy levels of one-electron systems, also discusses the present state of the Lamb-shift theory.

The largest contribution to the Lamb shift of S states comes from the one-photon self-energy term (of power α) which has been calculated explicitly in orders of $(Z\alpha)$ up to terms with $(Z\alpha)^6$. For large values of $Z(\geqslant 10)$ the $(Z\alpha)$ expansion does not converge very well, and recently Erickson[6] and Mohr[7] independently calculated the expansion for higher Z without truncation. In turn this allows the estimation of the uncalculated high-order terms for low Z, thus removing some of the uncertainty previously associated with these contributions. The results of the two cal-culations[6,7] differ slightly—not enough to allow the best present experi-mental results to decide between them, but by more than their estimated uncertainties (for hydrogen the situation is shown in Figure 29 in Section 4).

The Z dependence of the $S_{1/2}-P_{1/2}$ Lamb-shift intervals is somewhat slower than the Z^4 dependence of the Dirac fine structure of Eq. (1) since the leading term is proportional to $\alpha(Z\alpha)^4\{\ln[1/(Z\alpha)^2]-C\}$ with $C \simeq \frac{1}{4}$ of the ln term for $Z = 1$. This dependence is slightly modified by contributions of higher order in $(Z\alpha)$, which become rapidly more pro-minent with increasing Z.* To test the theory (including the nontruncated calculations[6,7]) it is, therefore, very important to extend the measure-ments to as high values of Z as possible. The n dependence follows very closely the $1/n^3$ dependence of the Dirac fine structure, and within the experimental accuracy this is quite well established by the measurements. The small Lamb shifts with $j > 1/2$ closely follow both the Z^4 and $1/n^3$ dependence of the Dirac fine structure, but there are only a few, not very accurate experimental results available (cf. Table 4 in Section 5).

The ultimate experimental accuracy attainable for Lamb-shift and fine-structure intervals hinges on the width of the observed signals, since even a well-known signal shape allows the signal position to be determined only to a certain fraction of the width. Thus, when striving for the highest possible precision, one should select Z and n values that provide the smallest signal width in comparison with the interval to be investigated. The signal width of modern experimental methods is determined by the sum of the natural widths of the states or at best by a (not too small) fraction of it. Unfortunately, the Z and n dependences of the natural width are almost exactly counterbalanced by the Z and n dependences of cor-responding Lamb-shift and fine-structure intervals, so that not much can be gained in this way. Since the signal-to-noise ratio is best for low Z and n and since low-n states are also less sensitive to external perturbations, the most accurate measurements were done on the $n = 2$ states of H, D, and He$^+$. A number of recent measurements showed that intervals like S–D, S–$F, \ldots$, avoiding the short-lived P states, can be derived from cor-respondingly narrow (higher-order) signals and stand a good chance of being determined to high precision even though the higher l values involved require these intervals to be studied at increased n values.

Numerous measurements of Lamb-shift and fine-structure intervals have been carried out over the last 30 years, encouraged, in the case of the

* For instance, the term with $\alpha(Z\alpha)^5$ contributes 0.7% of the total Lamb shift $\mathcal{S}$ in H ($Z = 1$) compared with 7.1% in C^{5+} ($Z = 6$) and the corresponding contributions from $\alpha(Z\alpha)^6$ are 0.04% for H and 1.0% for C^{5+}. Therefore, for C^{5+} measurements with an accuracy of approximately 1% are already very useful.

Lamb shift, by the somewhat unsatisfactory basis of the theory and by recurring discrepancies between theory and experiments and between various experiments. A very high level of accuracy has now been reached, and, with few exceptions, there is satisfactory agreement among the experiments and with theory.

1.3. Structure of the Article

There has been no comprehensive review of Lamb-shift and fine-structure measurements for some time, but early experiments and several aspects of more recent developments have been reviewed by Lamb,[9] Series,[10] Robiscoe,[11] Lea,[32] Taylor et al.,[12] and Cohen and Taylor.[13] The latter two articles[12,13] were concerned with a least-squares adjustment of the natural constants and provide a critical review of experimental values and their accuracy. The aim of the present article is mainly to discuss the various methods that have been employed to investigate Lamb-shift and fine-structure intervals. Thus, the division into sections is made in the first place according to the method used, and in the second place according to the intervals studied. Three basic methods will be distinguished and discussed in Sections 2–4.

Section 2 will deal with optical investigations. The method of resolving spectral lines and measuring the relative positions of their components was used extensively in the early days of fine-structure and Lamb-shift investigations, but it is limited by Doppler broadening, which has only recently been overcome in some cases by absorption spectroscopy with tunable lasers. Optical investigations cannot compete in accuracy with direct measurements and are now applied mainly to intervals that cannot be investigated otherwise, like the ground-state Lamb shift of hydrogenic systems or the Lamb shift of two-electron systems.

Section 3 will deal with slow-beam and bottle-type investigations. Most precision Lamb-shift and fine-structure investigations fall into this group of resonance experiments using radio-frequency, anticrossing, and level-crossing methods. The excited states are usually populated by electron impact, and a magnetic field is applied to tune the energies of substates through resonance. The signal width is not determined by the Doppler broadening but by the natural widths of the states.

The results rely on accurate knowledge of the Zeeman effect, which has been calculated with high precision by Brodsky and Parsons[14] for the $n = 2$ states of H and D. Nevertheless, the measured and implied Lamb-shift intervals of $n = 2$ of H appeared to depend on the magnetic field used for the measurements.[15] Even though this was probably an accidental dependence (the results for D did not show it), some doubt remains, not so much regarding the accuracy of the theory, but regarding the ultimate freedom from systematic effects when using a magnetic field. A prominent culprit is the motional electric field experienced by the atoms in a magnetic field. It can result in a velocity- and magnetic-field-dependent shift and distortion of the desired signal. It is, therefore important to obtain independent results with other methods, and such methods are discussed in Section 4.

Section 4 will deal with fast-beam investigations. The slow-beam and bottle-type experiments are limited to intervals with low Z, partly because the electron excitation process is not very efficient in producing hydrogen-like ions for $Z > 2$, partly because the intervals become too large to be tuned to resonance with common radio-frequency sources or electromagnets, and partly even because the signals become so broad (as a result of the rapidly decreasing lifetime of P states when Z is increased) that ordinary electromagnets cannot tune the full signal width. Such highly ionized states can be created with reasonable efficiency by passing a fast ion beam in the keV to MeV region through a thin foil or a gaseous target. At the same time, the fast-beam technique opens new ways of investigating fine-structure and Lamb-shift intervals.

The light emitted by the fast excited ions is observed further along the beam (even short-lived states travel some distance within their lifetime), and the separation between the position of excitation and the position of observation provides a useful time scale. Thus, time-dependent effects like the natural decay or quantum beats can readily be observed as spatial effects without pulsing techniques, provided the position of excitation and the beam velocity are well defined. Various aspects of the fast-beam technique have been explored in recent years in order to measure fine-structure and Lamb-shift intervals, usually without application of a magnetic field. With some notable exceptions, the fast-beam experiments have not yet reached the high relative accuracy of the slow-beam and bottle-type investigations, but they are providing independent results for lower Z and have allowed the extension of the range of measurable Lamb-shift and fine-structure intervals to higher Z values, where, as pointed out in Section 1.2, a lower level of accuracy is sufficient to provide a meaningful test of the calculated Lamb-shift values.

In Section 5, the numerical results will be compiled and compared with theory.

2. Optical Investigations

2.1. Interferometric Work

Before Lamb and Retherford[3] in 1947 made their famous radio-frequency experiment on the $n = 2$ states of H, the hydrogenic fine struc-

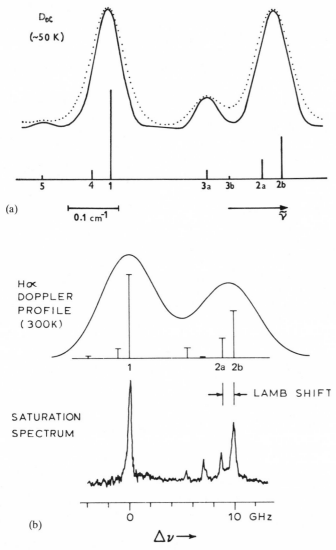

Figure 2. (a) Interferogram of the D_α line obtained by Kibble *et al.*[24] from an rf discharge cooled in liquid helium. The dotted curve represents the original measurement from which the solid curve is derived by deconvolution of the instrumental function. The theoretical fine-structure pattern is shown underneath the experimental curve. (b) (From Ref. 199) Upper part: Theoretical fine structure of the H_α line (as in Figure 2a) and the Doppler-broadened structure at 300 K reducing the fine structure to the much-investigated "doublet." Lower part: Saturated absorption spectrum of the H_α line obtained by Hänsch *et al.*[25] at ~300 K. As indicated, the Lamb shift of $n = 2$ is completely resolved.

ture could only be investigated by resolving the components of the spectral lines with high-resolution optical methods. Such measurements were carried out mainly on the lines of the Balmer series of H and D and on the blue Fowler-α line ($n = 4 \rightarrow n = 3$) of He$^+$. However, the fine-structure intervals are very small, and the Doppler width is large for these light elements (excited in a weak discharge) so that the width of the line components is of the order of or greater than the structure to be resolved (cf. Figure 2). As a further complication, each line complex contains the fine structure of the upper and of the lower state simultaneously.

Nevertheless, by about 1940, several (though not all) experiments had shown shifts of line components from the positions predicted by the Dirac theory, and Pasternack[33] concluded in 1938 that these shifts would be consistent with an upwards displacement of the $S_{1/2}$ states. Several proposals for modifications of the theory had also been made, but the resulting corrections were either too small or in the wrong direction or the calculations presented insurmountable problems. Thus, neither the experiment nor the theory managed to produce conclusive results. For a review of these attempts (and of more recent optical work), see Refs. 9 and 10.

After 1947, many spectroscopic investigations were again carried out, taking all possible steps towards reducing the Doppler width by carefully cooling the discharges or even using atomic beams[16] and by making full use of new developments in the interferometric technique. With these efforts it became possible to resolve more components of the lines, to support the findings of radio-frequency spectroscopy, and to extend the measurements to higher values of n, which were not immediately covered by radio-frequency investigations. Since the Doppler width is smaller for He$^+$ than for H and D lines, many of these studies were done on He$^+$, the most recent by Herzberg[17] on the line $n = 3 \rightarrow n = 2$, by Larson and Stanley[16] and by Berry and Roesler[18] on the line $n = 4 \rightarrow n = 3$, by Berry[19] on the line $n = 5 \rightarrow n = 3$, and by Kessler and Roesler[20] on the lines $n = 5 \rightarrow n = 4$ and $n = 6 \rightarrow n = 4$. Similar work on D was carried out by Kuhn and Series.[21,22] As an example, a recent interferometric recording of the D$_\alpha$ line is shown in Figure 2(a).

In spite of all the care taken in these experiments, fine-structure intervals could only be determined with an accuracy of 50–100 MHz and, within these limits, the results are now generally in agreement with the QED theory (cf. Table 3 in Section 5). In view of these limitations, recent optical studies put more emphasis on the investigation of line intensities and discharge characteristics or concentrated on the absolute measurement of single-line components to determine the Rydberg constant.[23,24]

2.2. Saturation Spectroscopy

The advent of narrow-band tunable dye lasers made it possible to overcome the problem of the Doppler width in optical absorption experiments. The method, described in more detail in Chapter 15 of this work, was developed for hydrogen by Hänsch et al.[25,26] and called saturation spectroscopy. The output of a tunable dye laser is divided into a strong and a weak beam traveling in opposite directions through the same region of a weak gas discharge used to populate the $n = 2$ states of hydrogen. The transmission of the weak beam is recorded when the laser is tuned through the Balmer-α line complex. Since the bandwidth of the laser is small compared with the Doppler width of the Balmer-α line components, each laser beam interacts only with a small proportion of the $n = 2$ atoms and independently scans the Doppler absorption profile (starting from different sides since the beams travel in opposite direction). Only at the center of the Doppler profile both beams interact with the same atoms, those with near zero velocity with respect to the beam directions. This affects the absorption recorded with the weak beam since the strong beam partly depletes the $2S$ states of the same set of atoms. As a result, an almost Doppler-free absorption spectrum can be obtained as shown for $H_\alpha^{(25)}$ in Figure 2(b). The two components marked 2a and 2b are essentially separated by the Lamb shift of the $n = 2$ state and were completely resolved here for the first time. In spite of the good resolution, the accuracy of Lamb-shift and fine-structure intervals derived with this method is unlikely to exceed that of the direct radio-frequency or anticrossing measurements. However, the method applied to the Balmer-α line of deuterium allowed the determination of the Rydberg constant with greatly improved accuracy.[26]

2.3. Ground-State Lamb Shift

Apart from hyperfine splitting, the $1S$ ground state of hydrogenic systems is a single state and its Lamb shift from the Dirac energy can thus only be derived from absolute wavelength measurements of spectral lines of the Lyman series or from an absolute measurement of the ionization potential. A general comparison of theoretical and experimental wavelength values of the Lyman series is provided by Boya.[27] Specific measured results were reported by Herzberg,[28] who made an absolute wavelength measurement of the Ly_α line of D and obtained 7.9 ± 1.1 GHz for the Lamb shift of $1S$, and by Hänsch et al.,[29] who induced Doppler-free two-photon transitions from $1S$ to $2S$ with a tunable dye laser and obtained $1S$ Lamb shifts of 8.3 ± 0.3 GHz for D and 8.6 ± 0.8 GHz for H in agreement with the theoretical values[8] of 8.172 GHz for D and 8.149 GHz for H. Since the publication of these values, the two-photon method has been improved further and new results are expected to be at least an order of magnitude more accurate.[30]

3. Slow-Beam and Bottle-Type Investigations

This section comprises fine-structure investigations based on electric-field-induced radio-frequency and anticrossing signals and on magnetic-field-induced radio-frequency signals and level-crossing signals. The first group of signals is appropriate for the investigation of Lamb-shift intervals since the electric field couples states with different l, while the second group should be suitable for the investigation of fine-structure intervals with $\Delta l = 0$. In practice, however, electric-field-induced signals are employed almost exclusively. These are created much more readily in hydrogenic systems, and it is common to derive intervals with $\Delta l = 0$ from two complementary measurements with $\Delta l = \pm 1$. In fact, there is only one fine-structure experiment of the second group, a level-crossing measurement of the $2^2P_{1/2}-2^2P_{3/2}$ (ΔE) interval of H. This will be described separately in Section 3.3.5, while the rest of this section deals with electric-field-induced signals.

Only the modern (post-1947) measurements will be described in this section, but it may be interesting to mention briefly some attempts made in the thirties to induce radio-frequency (rf) transitions between fine-structure states of hydrogen. Using the then only available very weak spark gap oscillators, Betz[34] recorded an increased absorption of the rf sent through a hydrogen gas discharge at the rf wavelengths of 3, 9, and 28 cm and no such effect on N_2 or O_2 discharges. This was tentatively ascribed to fine-structure transitions in $n = 2$ and 3. Even though there was another apparently positive experiment,[35] the results were not entirely convincing,[36] and Haase,[37] in a further measurement at wavelengths between 6 and 31 cm, failed to detect any absorption peaks and estimated that the effect would in fact be too small to be observable under the conditions used.[38] The more sensitive (modern) way of detecting rf transitions from intensity changes of spectral line components was also considered and apparently tried without success.[39]

3.1. Signals and Signal Shapes

The common principle of the electric-field-induced radio-frequency and anticrossing investigations is to subject atoms or ions, which are usually excited at a steady rate by electron impact, to a perturbation by an alternating or static electric field and to apply a magnetic field to tune the energies of appropriate Zeeman substates (differing in l) until their energy

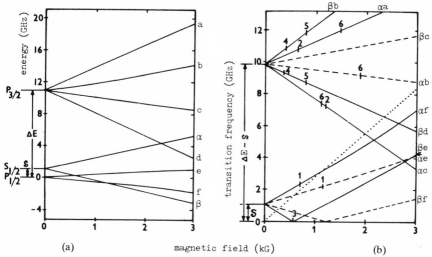

Figure 3. (a) Fine structure and Zeeman effect of the $n = 2$ states of H and D neglecting the
hyperfine structure. Lamb's notation of substates is used as explained in Table 1 (p. 546).
(b) Transition frequencies for electric dipole transitions between substates of S and P of
Figure 3a. σ transitions ($\Delta m = \pm 1$) are shown as solid lines. They require the perturbing
static or rf field to be perpendicular to the magnetic field while π transitions ($\Delta m = 0$),
shown as dashed lines, are induced by electric fields parallel to the magnetic field. The
dotted line represents the electron cyclotron frequency. The working points of all
precision experiments on $n = 2$ of H and D are indicated by vertical bars and numbers.
(1) Reference 51 (H, D), cf. Section 3.2.1; (2) Reference 52 (D), cf. Section 3.2.1; (3)
Reference 56 (H), Ref. 58, (D), cf. Section 3.2.2(a); (4) Reference 61 (H), cf. Section
3.2.2(b); (5) Reference 55 (H), cf. Section 3.2.2(b); (6) Reference 75 (H), cf. Section
3.3.2(b).

separation is in resonance with the frequency of the electric field (cf. Figure
3). At and near resonance, which is at the crossing of the substates if the
electric field is static, the substates are coupled by the perturbation, and
transitions are induced resulting in a change of the populations, if these are
different outside the resonance. The magnetic field position of the
resonance can be located by monitoring either the population change (for
instance in a beam experiment) or the change of the decay of the substates.
Since the total population of the two substates remains unchanged, the
detection process must discriminate at least in part between the substates.
From the magnetic field position of the resonance center and the known
frequency of the electric field, the zero-field fine-structure interval is
derived using the theoretical Zeeman splitting of the substates. The
Doppler broadening, being proportional to the working frequency, is
negligible for these signals since the rf is so much lower than the light
frequency of optical experiments. Thus, the resolution is now limited by the
natural width of the states.

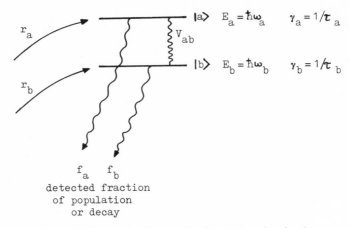

Figure 4. The two-level system for rf or anticrossing signals.

3.1.1. Signals between Short-Lived States

To find an expression for the radio-frequency and anticrossing signals, two states $|a\rangle$ and $|b\rangle$ are considered (as indicated in Figure 4) with energies E_a and E_b and decay rates γ_a and γ_b. The states may be coupled by the perturbation V_{ab}, which, in the case of hydrogenic fine-structure investigations, is provided by an alternating or static electric field. $|a\rangle$ and $|b\rangle$ are excited at the steady rates r_a and r_b, respectively. The mode of detection adopted may record the fraction f_a of excited states $|a\rangle$ ($1-f_a$ remains unrecorded) and the fraction f_b of excited states $|b\rangle$.

Using time-dependent perturbation theory, the steady-state resonance signal for such a two-level system was calculated by Lamb and Sanders[40] for rf perturbation and, in a very general way, by Wieder and Eck[41] for a static perturbation. Wieder and Eck also took into account coherence effects and so obtained a complex equation covering level crossing, anticrossing, and mixed signals. However, only the "noncoherent" part 5 of their Eq. (1) is normally encountered when investigating fine-structure intervals between states with $\Delta l \neq 0$, and this is identical with the resonance signal derived by Lamb and Sanders[40] if only the appropriate *rotating* part of the radio-frequency field is considered:

$$S = C - \frac{(f_a - f_b)(r_a/\gamma_a - r_b/\gamma_b)\frac{1}{4}(\gamma_a + \gamma_b)|2V_{ab}|^2}{(\Delta\omega - \omega)^2 + \frac{1}{4}(\gamma_a + \gamma_b)^2 + \frac{1}{4}[(\gamma_a + \gamma_b)^2/\gamma_a\gamma_b]|2V_{ab}|^2} \tag{2}$$

Here C represents the total intensity detected outside the resonance, originating from all substates including those not affected by the perturbation; the resonance curve itself is of Lorentzian shape centered at $\Delta\omega = \omega$, where ω is the circular frequency of the rf field and $\Delta\omega = |\omega_a - \omega_b|$; r_a/γ_a and r_b/γ_b

are the steady-state populations of the unperturbed states $|a\rangle$ and $|b\rangle$, and no resonance signal is observed if they are equal. No signal can be detected either (even though there might be a change of populations) if $f_a - f_b = 0$, i.e., if the detection efficiencies for $|a\rangle$ and $|b\rangle$ are identical.

The width and the amplitude of the signal depend on the strength of the perturbation, represented by V_{ab}. For a weak perturbation, $V_{ab}/(\gamma_a\gamma_b) \ll 1$, the signal width (FWHM) increases slowly from the sum of the natural width of $|a\rangle$ and $|b\rangle$, $(\gamma_a + \gamma_b)/(2\pi)$, and the signal amplitude rises from zero proportional to $|V_{ab}|^2$. For a strong perturbation, $V_{ab}/(\gamma_a\gamma_b) \gg 1$, the width increases further proportional to $|V_{ab}|$ while the amplitude reaches a saturation value of

$$A(V_{ab} \to \infty) = -(f_a - f_b)(r_a/\gamma_a - r_b/\gamma_b)\gamma_a\gamma_b/(\gamma_a + \gamma_b) \qquad (3)$$

The perturbation for the signals of interest here is provided by an external electric field and can thus be varied in strength so that width and amplitude of the signals can be controlled.* For signals between states with $\Delta l = \pm 1$, the perturbation element is (in circular frequency units)

$$V_{ab} = (1/\hbar)\langle a|e\mathbf{r}|b\rangle \cdot \mathbf{F} \qquad (4)$$

where $\langle a|e\mathbf{r}|b\rangle$ is the electric dipole matrix element between $|a\rangle$ and $|b\rangle$ and $\mathbf{F}$ the static or the rotating electric field strength. In higher order the electric field can also couple states with $\Delta l > 1$. This requires a multistep dipole coupling equivalent to multiphoton processes in the case of a radio-frequency field. For instance, the coupling element for $\Delta l = 2$ is

$$V_{ab} = (1/\hbar)\sum_c \langle a|e\mathbf{r} \cdot \mathbf{F}|c\rangle\langle c|e\mathbf{r} \cdot \mathbf{F}|b\rangle/(E_a - E_c) \propto F^2 \qquad (5)$$

where $|c\rangle$ are intermediate states coupling in first order with both $|a\rangle$ and $|b\rangle$.

No resonance signal can be observed following Eq. (2) if one of the states does not decay[42] because the slightest perturbation will then drive the signal into complete saturation. This is important since the most interesting state of hydrogenic systems, $2S$, is metastable ($\tau_{2S} \simeq \frac{1}{6}$ sec for H). However, the metastable atoms escape from the observation region long before they can decay naturally, and if this removal follows an exponential (decay) law with a mean lifetime τ_r, we can replace $\gamma_a \simeq 0$ in Eq. (2) by the effective decay rate $\gamma_r = 1/\tau_r$ so that a resonance or anticrossing signal is obtained as before, if only the removal process is fast enough.

According to Eq. (2), the signal amplitude is largest if only one of the substates is excited (r_a or r_b zero). However, if $|a\rangle$ and $|b\rangle$ have very

---

* As a compromise between large signal amplitude and small signal width, the perturbation is usually chosen to create an amplitude of around 50% of the saturation value when the signal width becomes (in frequency units) $2^{1/2}(\gamma_a + \gamma_b)/(2\pi)$.

different lifetimes (or effective lifetimes), it is important to choose the long-lived state for the exclusive excitation.* In hydrogenic systems this is the S state (compared with P), and investigations of intervals $S–P$ rely on an overpopulation of S. Because of the longer lifetime of S, a steady-state overpopulation of S is easily achieved even if other states are excited as well. In an atomic beam setup in particular, a beam containing only $2S$ and ground-state atoms is readily produced by letting the beam travel some (short) distance after excitation until all $2P$ states have decayed.

3.1.2. Signals Involving Long-Lived States

In measurements involving the metastable $2S$ states, the effective lifetime determined by the escape of excited atoms from the interaction region does not usually follow an exponential decay function as required for Eq. (2). In this case it is more appropriate to calculate first the change of the lifetime caused by the perturbation and from that the change of the decay over the period of the effective lifetime. Consider as before (Figure 4) two states $|a\rangle$ and $|b\rangle$ coupled by an electric field. The perturbed states are superpositions of $|a\rangle$ and $|b\rangle$ and if the natural decay rates are different, the coupling results in a change of the decay rates. In particular the decay constant of a metastable state ($|a\rangle$) changes drastically even if the interaction V_{ab} with the short-lived state $|b\rangle$ is small. The modification of the decay constant from $\gamma_a \simeq 0$ to γ_{St} as a function of the energy difference $\Delta\omega$ and the electric field frequency ω is given by the Bethe–Lamb theory[43,44,53]:

$$\gamma_a \to \gamma_{St} = \gamma_a + \gamma_b |V_{ab}|^2 / \{(\Delta\omega - \omega)^2 + \tfrac{1}{4}\gamma_b^2\} \qquad (6)$$

Equation (6) is not valid for very large values of the perturbation, and $\gamma_{St} \to \gamma_b/2$ for $V_{ab} \to \infty$:

If N metastable atoms in state $|a\rangle$ are subjected to the perturbation V_{ab} for the time τ, ΔN of them decay according to Eq. (6) through the coupled state $|b\rangle$ and the number of surviving metastables drops by the same amount†:

$$\Delta N = N(1 - e^{-\gamma_{St}\tau}) \qquad (7)$$

---

* If only the short-lived state $|b\rangle$ is excited ($r_a = 0$, $\gamma_a \ll \gamma_b$), the saturation amplitude of the signal [Eq. (3)] is $A = (f_a - f_b) r_b \gamma_a / \gamma_b$. Being governed by the ratio γ_a / γ_b, the amplitude of the resonance signal represents only a very small fraction of the originally excited $|b\rangle$ states, most of which continue to decay as $|b\rangle$ states even in resonance. For this reason exclusive (resonance light) excitation of the short-lived P states is not normally employed in measurements of hydrogenic $S–P$ intervals, and only one investigation on this basis appears to exist [Ref. 82; cf. footnote in Section 3.3.3(b) below].

† Any effect related to a sudden onset of the perturbation (unlikely in a slow-beam set up) is neglected, and it is further assumed that $\tau \gg \tau_b$, so that τ does not contribute noticeably to the signal width through the uncertainty relation.

Usually, not all atoms are subjected to the perturbation for the same time τ. This is taken into account by introducing the time-dependent fraction, $\alpha(t)$, of the atoms subjected to the perturbation at time $t = 0$ and still present at time t. Thus Eq. (7) becomes more generally

$$\Delta N = \gamma_{St} N \int_0^\infty \alpha(t) e^{-\gamma_{St} t} \, dt \tag{8}$$

In some experiments, the perturbation is switched on for a short time interval. This can be accounted for by adjusting the time interval of the integral. In a steady-state experiment with continuous excitation, perturbation, and observation, the resonance signal is [using the definitions of Figure 4 and Eq. (2)]

$$S' = C' - r_a(f_a - f_b)\frac{\Delta N}{N} = C' - r_a(f_a - f_b)\gamma_{St} \int_0^\infty \alpha(t) e^{-\gamma_{St} t} \, dt \tag{9}$$

Equation (9) is based on exclusive excitation of the long-lived state $|a\rangle$. Following the discussion in the first footnote on page 542, any contribution to the resonance signal from the short-lived state (if it is excited) can be neglected because of the lifetime disparity. Since γ_{St} of Eq. (6) shows a Lorentzian resonance dependence on the electric field frequency ω, the signals of Eqs. (7)–(9) are also nearly Lorentzian.

3.1.3. Comparison of the Signals of Sections 3.1.1 and 3.1.2

It is interesting to compare the signals of Eqs. (2) and (9) for appropriate identical conditions and $C = C'$, $r_b = 0$, $\tau = \tau_a \gg \tau_b$, where τ is the effective lifetime of $|a\rangle$ in Eq. (9) and taken to be the same for all excited atoms [i.e., $\alpha(t) = 1$ for $0 \leqslant t \leqslant \tau$, $\alpha(t) = 0$ for $t > \tau$]. In this case, both equations predict the same saturated signal amplitudes, and below saturation the signal shapes are similar, as is apparent from Figure 5. However, the signal based on Eq. (9) shows a higher degree of saturation for the same perturbation V_{ab} (e.g., 63% instead of 50% in Figure 5) and at the same time its width is still below that of the signal based on Eq. (2). This behavior is, of course, the result of the different distribution of the lifetimes τ and τ_a (although their mean values are identical).* This provides a simple example of how the line shape can be influenced by time biasing (see Chapter 14 of this work).

---

* Equation (9) does in fact transform into Eq. (2) if $\alpha(t)$ follows an exponential time dependence, $\alpha(t) = e^{-\gamma t}$. Taking $\gamma = \gamma_a$, Eq. (9) becomes

$$S' = C' - r_a(f_a - f_b)\frac{\gamma_{St}}{\gamma_a + \gamma_{St}} = C' - \frac{(f_a - f_b)(r_a/\gamma_a)(\gamma_b/4)|2V_{ab}|^2}{(\Delta\omega - \omega)^2 + \frac{1}{4}\gamma_b^2 + (\gamma_b/4\gamma_a)|2V_{ab}|^2}$$

This is identical with Eq. (2) when setting $r_b = 0$ and $\gamma_a \ll \gamma_b$.

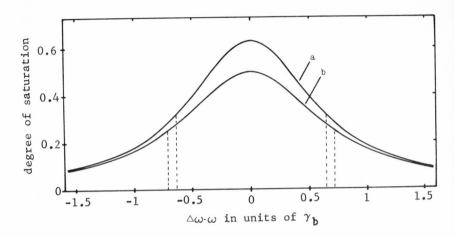

Figure 5. Theoretical resonance curves following Eq. (9) [curve (a)] and Eq. (2) [curve (b)]. Both curves are based on the exclusive excitation of the long-lived state, $|a\rangle(r_b = 0)$ and on the same *mean* lifetime τ_a, which, in the case of Eq. (9), is taken to be the same for *all* excited atoms [$\alpha(t)$ a step function] and, in the case of Eq. (2), is the mean value of an exponential distribution. Both curves are for the same perturbation $V = 0.5(\gamma_a \gamma_b)^{1/2}$. Note that because of the different lifetime distributions, curve (a) is narrower than curve (b) even though it shows a higher degree of saturation. The half-maximum positions are indicated.

3.1.4. Corrections to the Signal Shapes

In precision experiments a number of corrections to the above line shapes and to the measured resonance centers have to be considered, some of which are listed below. They may differ in importance in the various experiments.

1. Overlapping with other resonances, with cascading resonances and with "forbidden" resonances induced in second order by a combination of internal (e.g., hyperfine) and external (electric field) perturbations.

2. Radio-frequency power shift.[45]

3. Magnetic field inhomogeneity together with a finite size of the interaction region resulting in some transitions taking place at a magnetic field different from the "center" field.

4. Zeeman curvature of substates.

5. Variation with the magnetic field of the electric dipole matrix element between the substates due to $L–S$ decoupling.

6. Stark shift of the resonance center due to stray and motional electric fields.

7. The velocity distribution (in beam experiments) or the distribution of the time spent in the interaction region enters Eqs. (7)–(9).

8. The velocity distribution (of metastable atoms) is altered by motional electric fields since faster-moving atoms experience a larger motional field and are, therefore, quenched preferentiaily. The change of the velocity distribution affects the signal.[46]

3.2. Slow-Beam Experiments on H, D, n = 2

The breakthrough in the investigation of the hydrogenic fine structure came in 1947 when Lamb and Retherford[3] succeeded in inducing and detecting rf transitions between $2^2S_{1/2}$ and $2^2P_{1/2}$ on a thermal beam of H atoms and so established the existence of the Lamb shift beyond doubt. On this basis, a first series of precision measurements of $\mathscr{S}$ and $\Delta E - \mathscr{S}$ of H and D was carried out around 1950 by the same group. With some improvements of the technique, a second series of measurements was reported between 1965 and 1971.

3.2.1. Lamb's Measurements 1947–1953

The development of the experimental technique and of the data analysis from the first crude to the final precise measurements was laid down by Lamb and his co-workers, Retherford, Dayhoff, and Triebwasser in a series of papers commonly referred to as HI–HVI.[47]–[52] The basic experimental setup for these measurements is shown in Figure 6. A beam of H or D atoms is created by thermal dissociation of H_2 or D_2 in a

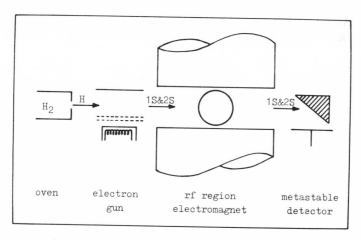

Figure 6. Schematic experimental arrangement for slow-beam investigations of the $n = 2$ states of H and D. The original setup of Lamb and Retherford[3,47] is shown. Later versions have all components inside the magnetic field.

high-temperature tungsten oven and crossed by an electron beam which excites some of the atoms, only slightly deflecting and diffusing the atomic beam through recoil effects. All but the metastable $2S$ states decay before the beam enters the rf field region inside the air gap of an electromagnet (magnetic field perpendicular to the beam). Here, transitions are induced by the rf field to the rapidly decaying $2^2P_{1/2}$ or $2^2P_{3/2}$ states when the energy separation between suitable Zeeman substates of S and P is tuned by the magnetic field to match the fixed frequency of the rf field. The surviving metastable $2S$ atoms of the beam are selectively detected by a metastable detector consisting of a tungsten surface which is indifferent to ground-state atoms but ejects electrons when hit by metastable atoms. The electrons are collected and the collector current is recorded as a function of

Table 1. Notation of Magnetic Substates of Hydrogenic Systems Following Lamb[a]

State	\multicolumn{10}{c}{m}									
	9/2	7/2	5/2	3/2	1/2	−1/2	−3/2	−5/2	−7/2	−9/2
$G_{9/2}$	A'' 4/+	B'' 3/+	C'' 2/+	D'' 1/+	E'' 0/+	F'' −1/+	G'' −2/+	H'' −3/+	I'' −4/+	J'' −4/−
$G_{7/2}$		K'' 4/−	L'' 3/−	M'' 2/−	N'' 1/−	O'' 0/−	P'' −1/−	Q'' −2/−	R'' −3/−	
$F_{7/2}$		A' 3/+	B' 2/+	C' 1/+	D' 0/+	E' −1/+	F' −2/+	G' −3/+	H' −3/−	
$F_{5/2}$			I' 3/−	J' 2/−	K' 1/−	L' 0/−	M' −1/−	N' −2/−		
$D_{5/2}$			A 2/+	B 1/+	C 0/+	D −1/+	E −2/+	F −2/−		
$D_{3/2}$				G 2/−	H 1/−	I 0/−	J −1/−			
$P_{3/2}$				a 1/+	b 0/+	c −1/+	d −1/−			
$P_{1/2}$					e 1/−	f 0/−				
$S_{1/2}$					α 0/+	β 0/−				

[a] At low magnetic field, m corresponds to m_j. At high magnetic field (Paschen–Back effect of the fine structure), m is made up by m_l and m_s as noted below each substate, + and − denoting $m_s = \frac{1}{2}$ and $m_s = -\frac{1}{2}$, respectively. In H with the nuclear spin $I = \frac{1}{2}$ a superscript + or − is used in the text to distinguish states with $m_I = +\frac{1}{2}$ and $m_I = -\frac{1}{2}$ (in the Paschen–Back effect of the hyperfine structure).

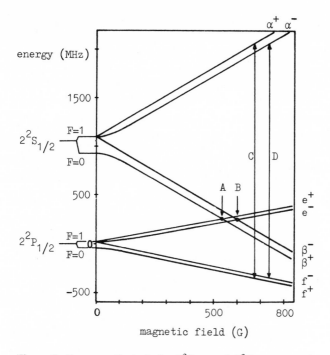

Figure 7. Zeeman effect of the $2^2S_{1/2}$ and $2^2P_{1/2}$ states of hydrogen including hyperfine structure ($A_{HFS} = 177.56$ MHz for $S_{1/2}$ and 55.19 MHz for $P_{1/2}$). The electric dipole selection rules for σ transitions are $\Delta l = \pm 1$, $\Delta j = 0, \pm 1$, $\Delta m_j = \pm 1$, $\Delta m_I = 0$ (assuming Paschen–Back effect for the hyperfine structure). The measuring positions are marked for the rf transitions $\alpha^+ f^+$ (C) and $\alpha^- f^-$ (D) employed by Triebwasser et al.,[51] cf. Section 3.2.1, and for the anticrossing signals $\beta^+ e^+$(A) and $\beta^- e^-$(B) investigated by Robiscoe and Cosens;[53,54] cf. Section 3.2.2(a). The hyperfine signal components are not resolved unless one component is suppressed.

the magnetic field: The current decreases near the resonance when, following Eqs. (7)–(9), the number of surviving metastables is reduced.

The four transitions αe, αf, αa, αc (the notation for the substates is explained in Table 1) were studied at the frequencies marked 1 and 2 in Figure 3b. The β substate which remains close to $P_{1/2}$ substates within the magnetic field range used, is almost totally quenched by the motional electric field seen by the atoms when traveling perpendicular to the magnetic field and no β signals were observed. To avoid corrections due to the β states, steps were taken in the final measurements to quench this component completely.

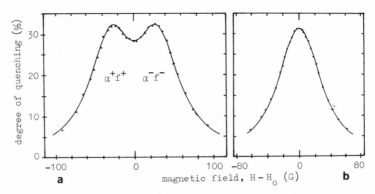

Figure 8. Resonance curves αf obtained by Triebwasser *et al.*[51] at a frequency of 2395 MHz. The dots are experimental points, the line is calculated. (a) Hydrogen, $H_0 = 703.77$ G. The hyperfine signal components overlap strongly. This curve may be compared with the hyperfine state selected signal in Figure 10. (b) Deuterium, $H_0 = 704.57$ G. The hyperfine structure of D is much smaller than for H so that the components overlap to a single only slightly broadened signal.

On the basis of the natural lifetime of the $2P$ states of 1.6×10^{-9} sec corresponding to a level width of 100 MHz, a width of the resonance curves of about 120 MHz would be expected according to Eq. (7) for 50% signal saturation. Unfortunately, in hydrogen the hyperfine structure of the $2S$ state is of the same order, so that the two components allowed for each transition as a result of the hyperfine structure (cf. Figure 7) cannot be fully resolved and form a double peaked signal. This is shown in Figure 8a. Figure 8b shows the corresponding signal in D, which has a much smaller hyperfine structure than H so that the composite signal comes much nearer to a single-component situation. D atoms also travel more slowly because of their larger mass so that velocity-dependent effects on the signal are reduced. Thus, the necessary corrections to the final results (cf. the list in Section 3.1.4) were considerably smaller for D than for H, and the precision measurement of the $2^2S_{1/2}$–$2^2P_{3/2}$ interval, for instance, was only carried out for D.

The accuracy of the final results, representing approximately three times the average deviation of the mean, was ± 0.1 MHz corresponding to 1/1000 of the natural linewidth of 100 MHz. The total corrections applied were between -0.3 MHz (D) and -3.2 MHz (H) in the case of the Lamb-shift intervals[51] and about 0.1 MHz in the $2^2S_{1/2}$–$2^2P_{3/2}$ interval.[52]

3.2.2. Measurements 1965–1971

While retaining the basic beam setup of Lamb and co-workers,[47–52] important new features were introduced in this second series of measure-

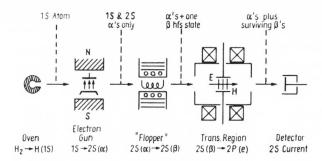

Figure 9. (From Ref. 11.) Schematic experimental arrangement for slow-beam anticrossing measurements of $n = 2$ states of H and D including hyperfine state selection.

ments, mainly to reduce the amount of correction required for the results. Firstly, the metastable beam was hyperfine-state-selected before entering the principal interaction region so that the signals were no longer superpositions of several hyperfine transition signals. Secondly, the magnetic field in the interaction region was applied parallel to the beam axis, thus reducing the motional electric field experienced by the atoms and its effect on the resonance signal (cf. the list of corrections in Section 3.1.4). In addition, the Lamb-shift experiments employed the anticrossing method replacing the rf field by a static electric field.

(a) *Anticrossing Measurements.* The method was introduced by Robiscoe[53] to remeasure the $n = 2$ Lamb shift of H, and Figure 9 shows his arrangement in schematic form. It differs from Lamb's setup in Figure 6 in the direction of the magnetic field, now parallel to the beam axis and created by Helmholtz coils, and in the hyperfine state selector. With the rf field replaced by a static electric field ($\perp H$) a resonance signal is obtained at ~575 G, where the substates β and e cross*(cf. Figure 3).

Including the hyperfine structure (cf. Figure 7) the anticrossing βe consists of the two allowed components $\beta^+ e^+(A)$ at 538 G and $\beta^- e^-(B)$ at 605 G in the same way as the rf signals did. One of these components can, however, be suppressed if a state-selected beam of β^+ or β^- is produced. This is achieved by first applying a magnetic field of 575 G (near the

---

* The static electric field mixes the states near degeneracy and causes the levels to repel each other so that the actual crossing is removed. The resulting signals are commonly known as anticrossings in contrast to the level-crossing signals, where the crossing persists but the substates become indistinguishable near the crossing with the result of possible interference effects in the excitation–decay process. If the natural lifetimes of the states differ, the levels do actually still cross as long as the electric field perturbation remains small ($V \lesssim |\gamma_a - \gamma_b|/4$).[42] This is usually the case in the measurements discussed here, but nevertheless the signals belong to the group of anticrossings.

crossing points A and B) perpendicular to the atomic beam axis in the excitation region so that all β states are quickly quenched by the motional electric field. The α branches remain unquenched and these $2S$ atoms pass on into the flopper where the magnetic field is brought to zero in such a way that a redistribution of the population of the states takes place (nonadiabatic Majorana transitions). Since the branches α^+, α^- and β^- degenerate in zero magnetic field to the $F = 1$ state, β^- is recreated in this way, while β^+, separated from the others by the hyperfine interval of ~ 177 MHz, is not. Hence, the $2S$ beam entering the second magnetic field and the interaction region consists only of α^+, α^-, and β^- so that the anticrossing component $B(\beta^- e^-)$ is observed without interference from A. It is also possible to produce a state-selected beam of $\beta^{+(54)}$ by applying a magnetic field of a few gauss (instead of zero) in the flopper and inducing rf transitions from α^- to $\beta^+(F = 1, m_F = 0 \to F = 0, m_F = 0)$. β^+ states thus created by adiabatic transitions were actually used in preference to β^- states whose creation is sensitive to the velocity distribution and external magnetic fields.[55] The basic signal shape is given by Eq. (8) and a

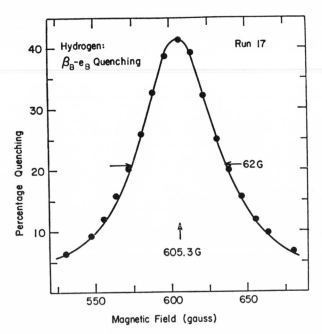

Figure 10. (From Ref. 53.) Hyperfine state selected anticrossing signal $\beta^- e^-$ (marked B in Figure 7) in hydrogen obtained by Robiscoe.[53] Without state selection the signal would resemble that of Figure 8a.

measured example is shown in Figure 10. Without state selection the signal shape would resemble that of Figure 8a.

Employing the state selection and avoiding motional fields considerably reduced the corrections to be applied to the results. However, the anticrossing signals are very sensitive even to small background static electric fields, such as motional fields caused by a misalignment of the main magnetic field with respect to the atomic beam axis. Even though this was estimated to be $< 1°$, the corresponding motional fields caused a slight asymmetry of the signals mainly through point (8) of the list of corrections in Section 3.1.4. This, and a measurement of the velocity distribution of the metastables[56,57] resulted in two revisions[46,56,57] of the original results[53,54] but, even including this, the corrections applied to the results were only of the order of -0.3 MHz, much less than in the rf measurements of Triebwasser et al.[51] Yet, in spite of this improvement, the accuracy of the final result (± 0.1 MHz) did not exceed that quoted by Triebwasser et al.[51] Similar measurements on deuterium* were carried out by Cosens[58,59] with comparable accuracy.

(b) *rf Measurements.* The interval $\Delta E\text{-}\mathscr{S}$ of H was remeasured by two groups[60,61,59;62,63,55] using the technique described in Section 3.2.2(a) and Figure 9 with the exception that an rf field was employed again since the suitable crossings ac and βd require too large magnetic fields to be achieved by coils. The transitions and frequencies used are marked 4 and 5 in Figure 3b. Cosens and Vorburger[59–61] introduced an additional refinement of the technique in order to distinguish the overlapping signal αa from the desired signal βd. This was done with a further quench region after the interaction region which (working at 575 G) essentially allowed the modulation of the β component while leaving the α component unaffected. The corrections and the final uncertainties were near 0.05 MHz in both sets.

3.3. Bottle-Type Experiments

3.3.1. Introduction

In bottle-type experiments, the spatial separation of the regions of excitation, resonance transition, and detection is removed as shown schematically in Figure 11. The electron beam, directed parallel to the magnetic field excites a thin atmosphere of gas (H_2, He, . . .) inside the rf or static electric field region and the light emitted by excited atoms or ions from that region is registered. It varies in intensity and polarization if

---

* Deuterium with nuclear spin 1 has more hyperfine components, but the same principle of state selection as for H can be applied.

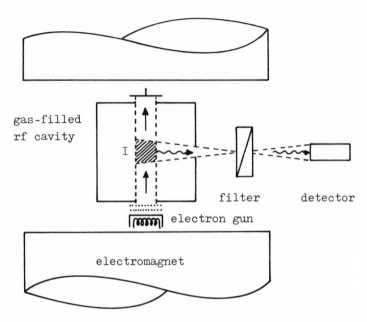

Figure 11. Schematic arrangement of bottle-type rf experiments (if
anticrossing signals are observed, the cavity is replaced by Stark
plates to apply a static electric field across the interaction region). As
interaction region we define the shaded region, I, which is shared by
all, the electron beam, the rf or static field, and the observation
cone. The rf field strength as well as the density of the excited atoms
and the detection sensitivity may vary over the excitation region,
and in precision experiments this has to be considered as a source of
line-shape distortion.

suitable fine-structure intervals are tuned through resonance with the rf
field by the magnetic field.

The method is suitable both for fine-structure intervals involving a
metastable state ($n = 2$), where it competes with the thermal beam method,
and for intervals involving two short-lived states ($n > 2$), where no thermal
beam can be formed. The advantage over the slow-beam method lies in the
higher density of excited states attainable in the interaction region which is
of particular importance if the excitation cross sections are small as for the
simultaneous ionization–excitation process of $He^{+(65,66)}$ and Li^{++} states.
The disadvantages of the new method are that discrimination is required
against the strong background light emitted by other excited states created
at the same time, and that the experiment has to be carried out inside the
excitation region, an environment usually disturbed by electric fields,
which are difficult to control and to estimate.

Bottle-type experiments, as far as rf transitions are employed, are also discussed as part of Chapter 11 of this book, and more detailed descriptions of some of the following experiments can be found there.

3.3.2. Experiments with Metastable States

(a) $He^+, n = 2(\mathcal{S})$. The first of these bottle-type experiments was carried out around 1949 by Lamb and Skinner[67,68] to measure the Lamb shift of the $n = 2$ state of He^+. It also represented the first extension of the direct Lamb-shift measurements beyond $n = 2$ of H and D. No attempt was made to form a beam of metastable He^+ ions, partly because an ion beam would be deflected by the magnetic field perpendicular to the beam* and partly because there were no suitable detectors to distinguish $2S$ ions from ground-state ions. Most of the $2S$ ions which, by virtue of their long lifetime, build up a much larger steady-state density than other excited states, escape from the interaction region before they decay naturally or as a result of (nonresonant) quenching or collisions. Thus, a resonant rf field induces transitions $2S \rightarrow 2P$ and so reduces the number of escaping $2S$ ions causing an increased emission of $2^2P \rightarrow 1^2S$ photons at 304 Å. Lamb and Skinner had considerable problems in distinguishing this increase from background radiation since, even with a thin-film filter in front of the detector, only partial isolation of the 304 Å line was achieved.

Further investigations of this Lamb-shift interval[69–72] were based on the same method with various refinements. A very effective way of reducing the background signal was introduced by Novick et al,[69] who employed pulsed excitation, followed by a delay period during which most of the short-lived states decayed. Only then were the rf field and the detection system switched on for a short period. However, the setup was complicated and much measuring time was lost in the low duty cycle. Also, there were problems connected with the fast-expanding cloud of positive ions left behind by the exciting electron pulse. The most recent experiments[70–72] therefore returned to continuous excitation and distinguished between the rf signal and the background by modulating the rf field and using the lock-in detection technique.

The signals are governed by Eq. (8) or (9) and consist of single components since the dominant He isotope has nuclear spin zero. The transition αe was investigated in the latest two experiments[70,72] and the rf frequency was chosen such that the signal coincided with the βf crossing position. This minimized asymmetries caused by the β branch which cannot be suppressed easily as in beam experiments. With such steps to

---

* The width of the He^+ signals is 16 times that of corresponding H signals and it would be difficult with coils to create a magnetic field parallel to the beam sufficient to sweep the full resonance curves.

avoid distortions and after consideration of a number of corrections (cf. the list of Section 3.1.4) the Lamb-shift values almost reached the relative accuracy of the H and D experiments. The latest measurement by Narasimham and Strombotne[71,72] carries an uncertainty of ± 2 MHz (1/800 of the signal width) with total corrections of only ± 1.2 MHz. The interval $2^2S_{1/2}-2^2P_{3/2}$ of He$^+$ at 161.5 GHz, which would determine an independent value of the fine-structure constant, has not yet been investigated.

(b) H, $n = 2(\Delta E - \mathscr{S})$. Complementing the slow-beam investigations discussed above (Section 3.2), Kaufman et al.[73-75] carried out a bottle-type measurement of the $2^2S_{1/2}-2^2P_{3/2}(\Delta E - \mathscr{S})$ interval of H using transitions αa, αb, (αc) as indicated in Figure 3b above. An excellent signal-to-noise ratio was obtained by using direct electron impact excitation of H_2 molecules and a very effective Ly$_\alpha$ light-collecting system. Even though no hyperfine state selection was possible, the strong signals allowed the resonance centers to be determined with unprecedented accuracy. The signals are based on Eq. (9), and the empirical velocity distribution of the $2S$ (Leventhal et al.[76]) was used. Several disturbing effects could be eliminated by adopting the difference between signals at high and low rf power as normalized signal, others were investigated experimentally, and few corrections entered the final value, which carries an uncertainty of ± 0.026 MHz corresponding to 1/3800 of the natural width. However, the result disagrees with the present theoretical value and also with the other experimental values, as shown in Figure 33 in Section 5.3. A number of suggestions have been put forward to explain the discrepancy (Refs. 15, 74, 75; and Chapter 11 of this work) but the precise reason is not known. Since bottle-type experiments work with a less "clean" environment than beam experiments, one would be inclined to search for a missed out environmental effect there, and indeed it will be seen that other bottle-type experiments also experienced difficulties of this sort.

(c) Li^{++}, $n = 2(\mathscr{S})$. The measurement of the Lamb-shift interval of the $n = 2$ state of Li^{++}, based on electron impact excitation and rf signals, is extremely difficult. The cross section for simultaneous double ionization and excitation is small, the Lamb-shift interval is large (~ 63 GHz), and the width of the resonance signals (~ 8 GHz) requires quite a powerful magnet for the tuning. Technically, lithium vapor is awkward to handle and the 135 Å Ly$_\alpha$ photons are not in a favorable spectral range. Nevertheless, one such experiment has been carried out by Leventhal and Havey.[77,78] They used a crude beam of Li vapor to reduce the contamination of the system and a Be filter in front of the open photomultiplier to improve the spectral selectivity. The pulsed excitation–detection method of Novick et al.[69] [cf. Section 3.3.2(a)] was employed with additional ion trapping to enhance the signal. βe transitions were induced at 35.8 GHz and readings were taken

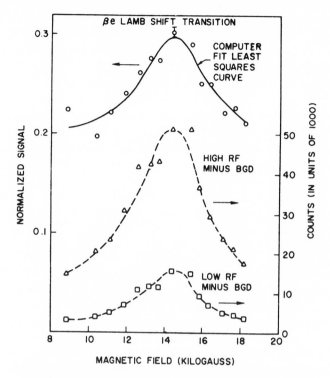

Figure 12. (From Ref. 78.) Raw resonance curves (bottom two curves) and normalized difference signal (top curve) for the βe transition of Li^{++}, $n = 2$ obtained by Leventhal and Havey.[77,78] Radio frequency 35.8 GHz, resonance center $\simeq 14.4$ kG, signal width (FWHM) $\simeq 4$ kG.

for high and low rf power so that a number of effects causing asymmetries could be eliminated as in earlier experiments[69,75] by forming the normalized signal

$$S_n = \frac{(\text{low-power counts}) - (\text{background counts})}{(\text{high-power counts}) - (\text{background counts})}$$

This is illustrated in Figure 12. The signal shape is based on Eq. (8) with appropriate time intervals. The hyperfine structure is not resolved. Considering the complexity of the experiment, the quoted accuracy (one standard deviation) is very remarkable with ± 21 MHz or 330 ppm, corresponding to 1/400 of the natural width. Slightly less accurate results for this interval have been obtained in fast-beam experiments [cf. Sections 4.1.2(a) and 4.5.2(c)].

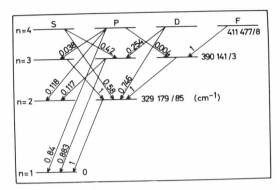

Figure 13. Branching ratios of one electron systems for electric dipole radiation averaged over all polarizations (Section 63 of Ref. 2). The energies are not to scale. Energy values shown are for ^{4}He$^+$.

3.3.3. First-Order Experiments with Short-Lived States ($n > 2$, $\Delta l = \pm 1$)

(a) *General Remarks.* Bottle-type experiments of the type shown in Figure 11 also allow the measurement of Lamb-shift and fine-structure intervals between short-lived states with $n > 2$. For all practical purposes the excited states can now be considered as decaying where they are created, and the signal is given by Eq. (2) in Section 3.1.1. The required unequal population of states (differing in l) is promoted by the disparity of the corresponding natural lifetimes. In particular the comparatively long-lived S states are much more populated than all other states of the same n. The discrimination between interacting states in the decay ($f_a \neq f_b$) is less obvious since both states contribute to the observed light, which is made up of all spectral line components of the transition $n \rightarrow n'(n > n')$. Yet, considering for instance the decay from $n = 3$, Figure 13 shows that differences in the branching ratios provide discrimination for any two states of $n = 3$ with $\Delta l = \pm 1$ irrespective of whether H$_\alpha$ or Ly$_\beta$ light is observed. Such differences in the branching ratios occur in practically all other state combinations and can further be influenced by the choice of the light polarization. In general they tend to decrease with increasing n at the same time as the absolute populations and the population differences decrease as well. Thus, the signal-to-noise ratio deteriorates towards higher n, and this is aggravated by the rapidly increasing sensitivity to electric fields ($\sim \propto n^7$) which are always present as space charge and motional fields and which shift and distort the resonance curves. Another problem is posed by the many possible signals occurring at n values higher than that under investigation. These are saturated at much lower rf or static electric field strengths and can affect the observed signals through cascades [cf. Section 3.3.3(d) and Figure 15 below]. Cascade signals have even been used to investigate the fine structure of the cascading state.[79,186] No hyperfine state selection has been attempted in any of the following experiments, so that the H and D signals are broadened superpositions of several components. In spite of such restrictions the measurements provide an indispensable test of the n dependence of the Lamb-shift and fine-structure intervals, and it will be seen that the best of them are in fact of similar accuracy as the $n = 2$ studies.

(b) *H, D, n = 3.* The first investigation of fine-structure intervals between short-lived states was carried out by Lamb and Sanders[80] on $n = 3$ of H, covering $\mathscr{S}$, $\Delta E - \mathscr{S}$, and some $P_{3/2}-D_{5/2}$ transitions. Later, the same group[40,81] gave a detailed account of the theoretical background of these experiments and obtained more accurate results of $\mathscr{S}$ and $\Delta E - \mathscr{S}$ on D ($\approx 1/40$ of the natural width) inducing rf transitions in a sealed-off glass system which provides a very stable gas pressure. On H, the Lamb shift was measured with improved accuracy by Kleinpoppen*[82] and again by Glass-Maujean and Descoubes,[83] who employed anticrossing signals induced by a static electric field and also observed a number of $P-D$ anticrossings. The observation of rf transitions $S_{1/2}-P_{1/2,3/2}$ and $P_{1/2,3/2}-D_{3/2,5/2}$ was reported by Richardson and Hughes.[84,85] The current most accurate values of the Lamb shift were obtained on a fast-beam setup [cf. Sections 4.5.2(a) and 4.6.2(a)].

(c) *H, D, n = 4.* The $n = 4$ states are extremely sensitive to electric fields, and only very weak rf signals had been detected in D by Wilcox and Lamb.[81]

Nevertheless, a very successful measurement of $\mathscr{S}$ and $\Delta E - \mathscr{S}$ of H was carried out by Brown and Pipkin,[86,87] who used pulsed electron excitation and observed the Balmer-β intensity (in a continuous rf field) during a time interval starting some time after the excitation. As a result of this time biasing towards the longer-lived $4S$ states the line shape is somewhat complicated and differs from Eq. (2) but the signal is enhanced with respect to the background and the disturbing electric fields are reduced by the absence of the electrons. Three transitions (αf, αa, αc) were studied each at the same two values of the magnetic field to ensure identical environmental conditions. The static electric field component

---

* This measurement combines characteristics of both slow-beam and bottle-type experiments in using Ly$_\beta$ resonance light to populate $3P$ states in a beam of H ground-state atoms. About 10% of the $3P$ states decay to $2S$ and so create metastable atoms in the beam which were detected in a metastable detector. The number of these metastables dropped when resonance transitions were induced in the excitation region from $3P$ to $3S$. $3S$ decays exclusively to the ground state through the short-lived $2P$ state.

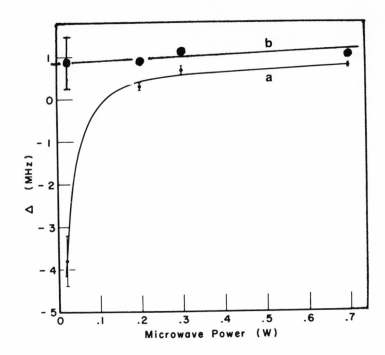

Figure 14. (From Ref. 89.). Curve (a): rf power shift of the He^+ $3^2S_{1/2}-$ $3^2P_{1/2}$ Lamb shift measured by Mader *et al.*[89] employing *ae* transitions at 10 GHz (9 point runs). If the contribution from the cascading signal $7\beta c$ (cf. Figure 15) is included in the data analysis, the corrected rf dependence of curve (b) is obtained. The energy scale is relative to the theoretical value of the Lamb shift [the current value (Table 3) would shift the zero point to +2.11 MHz on this scale].

perpendicular to the magnetic field is assumed to be the dominant disturbing factor and displaces the three transitions by different amounts so that the Lamb-shift and fine-structure results of the three transitions should be consistent only if a certain value of the electric field is assumed for correction. This was so determined for that particular experiment to $1.46 \pm^{0.19}_{0.26}$ V/cm. Unfortunately, the shift of the transitions by the electric field is such that a more accurate determination of the field strength was not possible, and this limited the final accuracy of the Lamb shift to about 0.5% ($\simeq 1/20$ of the natural width) and that of the fine-structure intervals to 230 ppm ($\simeq 1/50$ of the natural width).

    (*d*)   *He⁺*, $n = 3$. Radio-frequency measurements on this state were carried out by Leventhal *et al.*,[88] Mader *et al.*[89] ($\mathscr{S}, \Delta E - \mathscr{S}$), and by Baumann and Eibofner[90,91] ($P_{3/2}-D_{5/2}$), who also investigated anticrossing signals $S_{1/2}-P_{1/2}$ and $P_{3/2}-D_{5/2}$.[92,91] All these studies used the line at

1640 Å ($n = 3 \rightarrow n = 2$) for the detection. The most accurate results of this series were obtained by Mader et al.,[89] who, in view of the good signal-to-noise ratios, were able to make a direct investigation of disturbing effects like pressure, current, and rf power shifts. In the case of the αe transition, used to determine the Lamb-shift interval, a pronounced rf power shift was noticed, especially at low rf power as shown in Figure 14. A close inspection of the signal at low rf power revealed the structure shown in Figure 15 and made it possible to attribute the power shift to overlapping with nearby cascading resonance signals. Even though these signals are small, their saturation at extremely low rf power makes them comparable in size with the $n = 3$ signal in the power range where the latter just sets in. The effect was possibly amplified by the use of only 3–9 data points to fit the resonance. When, however, the $7\beta c$ signal was included in the data analysis, most of the

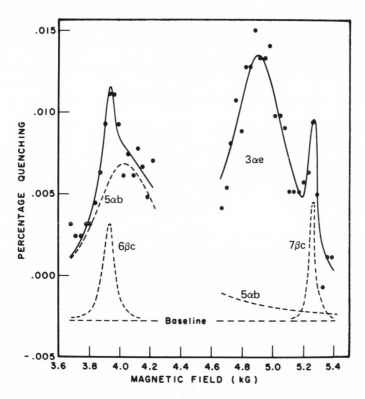

Figure 15. (From Ref. 89.) Recording of rf signals in the region of the $3\alpha e$ resonance taken at 9 GHz and very low rf power (0.5 mW) so that the $3\alpha e$ signal is only 0.1% saturated. The cascading $7\beta c$ signal is clearly shown as the main cause for the observed rf power shift (cf. Figure 14).

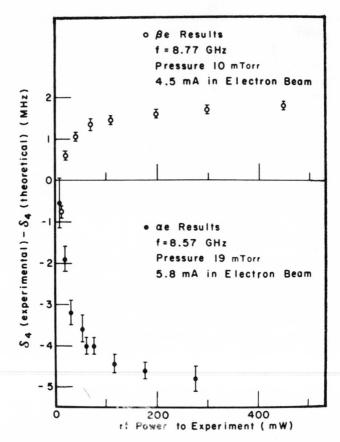

Figure 16. (From Ref. 97.) Radio-frequency power shift of the He$^+$ $4^2S_{1/2}$–$4^2P_{1/2}$ Lamb-shift results derived from βe (open circles) and αe transitions (full circles). The shifts are probably again caused by cascading signals, but these could not be identified. An extrapolation analogous to Figure 14 could not be applied since this would have led to contradicting results for the Lamb shift. The current theoretical Lamb-shift value (Table 3) would shift the zero point of the energy to +0.81 MHz in the given scale.

rf power shift disappeared, as indicated by the solid circles in Figure 14, and the intervals could be determined to an accuracy of ±0.5 MHz or 1/1000 of the combined natural level widths.

(e) *He$^+$, n = 4.* The fine structure of $n = 4$ of He$^+$ has been very popular among experimentalists mainly because of the convenient situation of the decay line $n = 4 \rightarrow n = 3$ at 4686 Å. The first attempt to detect rf signals was made by Series and Fox[93] in a glow discharge, but the experiment proved inconclusive. Later investigations used electron impact excitation in bottle-type experiments, and rf measurements of the Lamb shift were reported by Lea et al.,[94] Hatfield and Hughes,[95]* Beyer and Kleinpoppen,[96] Jacobs et al.,[97] and Eibofner.[98] Other intervals were measured by Jacobs et al.[97] $(\Delta E - \mathscr{S})$ and by Eibofner $(\Delta E - \mathscr{S}$,[99] $D_{3/2} - F_{5/2}$[100]).

The two Lamb-shift transitions αe and βe investigated by Jacobs et al.[97] were troubled by a rf power shift similar to that observed in $n = 3$.[89] This is shown in Figure 16. The most likely culprits are again cascade effects from high n signals, but it was not possible in this case to identify the cause positively. Since corrections in the manner indicated in Figure 14 for $n = 3$ would result in two distinct Lamb-shift values, the uncertainty was chosen to encompass the whole rf power shift. Eibofner[98] investigated the αe transition in another frequency range and experienced similar problems but nevertheless managed to improve the accuracy to ± 2 MHz or $1/100$ of the natural width. The $S_{1/2} - P_{3/2}$ transitions did not show these effects, and Eibofner[99] quotes an accuracy of ± 0.8 MHz ($1/270$ of the natural width), slightly exceeding the accuracy estimated by Jacobs et al.[97]

Another way to measure Lamb-shift intervals was tried out on $n = 4$ of He^+ by Hadeishi[101,102] following a proposal[42] and comments[103] by Series. It is basically an anticrossing method but involves modulated electron excitation and observation of the amplitude of the corresponding modulated decay light. For a first experiment[101] a relatively slow modulation of approximately 3 MHz was used and a reduction of the modulated light was observed at the βe and βf crossing positions (coupling was provided by static electric fields present in the interaction region). The signals were attributed to the electric-field-induced variation of the decay constants in the vicinity of the crossings when the signal amplitude should reflect the amount of state mixing which depends on the separation of the states and is not complete even at the crossing position if the electric field perturbation is small enough ($|V| < \frac{1}{4}|\gamma_s - \gamma_p|$). However, since the modulation period is large compared with the natural lifetime of both states, the signal could also be an electric-field-induced anticrossing signal as observed in dc conditions,[95,104] and it was estimated by Eck and Huff[104] that the dc contribution would be approximately 60% of the observed signal. In a second experiment[102] much higher modulation–detection frequencies of around 400 and 1700 MHz were used. In this case the modulation frequency is related to the energy separation between the substates, and the intensity change of the modulated light, observed when the two frequencies match on either side of the crossing position, cor-

---

* In this experiment pulsed electron impact excitation and delayed observation was used in a similar way as described in connection with the later experiment on H, $n = 4$ by Brown and Pipkin [Refs. 86, 87; cf. Section 3.3.3(c) above].

responds to the quantum beat phenomenon observed in fast-beam spectroscopy (cf. Section 4.3). No numerical results were derived.

(f) $He^+, n > 4$. Bottle-type experiments beyond $n = 4$ were carried out by Baumann and Eibofner, who measured the $S_{1/2}-P_{3/2}$ intervals for $n = 5^{(105,106,99)}$ and $n = 6.^{(106,99)}$ Radio-frequency transitions αc were investigated in the vicinity of the αc crossing, and systematic effects were reduced by taking signals at symmetric points on either side of the crossing.

3.3.4. Higher-Order Experiments with Short-Lived States ($n > 2$, $\Delta l > 1$)

(a) *Method.* It has been seen in the preceding section (Section 3.3.3) that there are few bottle-type measurements of $\Delta l = 1$ intervals other than $S-P$ such as $P-D, D-F, \ldots$. The reason can be found in the comparatively low accuracy attainable in such investigations, which is the result of several factors. Firstly, signals involving higher l values require correspondingly high n values where the (electron impact) excitation cross sections and the branching ratios are less favorable (resulting in weak signals) while the sensitivity to distortions increases rapidly. Moreover, the steady-state

Table 2. Characteristic Data for rf and Anticrossing Signals in $n = 4$ of He^+

Quantity	S	P	D	F
Lifetime[a] (nsec)	14.4	0.775	2.28	4.56
Natural width[b] (MHz)	11.1	205	69.8	34.9
Calculated electric field strength required for 50% signal saturation[c] (V/cm)	$S-P$ 4.9^d (βe) 6.9^d (αc)	$P-D$ 18.6^d (bD) 32.1^d (aC)	$D-F$ 5.5^d (JN') 17.3^d (GK')	
		$S-D$ (second order) 90^e (αJ) 41^e (αE)		
		$S-F$ (third order) 130^e ($\alpha G'$) 159^e ($\beta H'$)		

[a] Section 63 of Ref. 2.
[b] $1/(2\pi\tau)$, τ lifetime.
[c] The transitions shown were used in experimental investigations of this fine-structure system ($P-D$ transitions only for $n = 3$ of He^+). If an rf field is employed the field strength corresponds to the rotating field component.
[d] Calculated in (j, mj) representation (at zero magnetic field).
[e] Calculated at the magnetic field of the crossing position of the substates using matrix diagonalization in combined electric and magnetic fields. The calculated field values agree with the measured results (Ref. 107).

population differences between the states $P, D, F, \ldots$ are in general much smaller than between S and $P, D, F, \ldots$. Without considering the excitation cross sections in detail, this can be inferred from the relative lifetimes of the states like those of $n = 4$ of He^+ shown in Table 2. Finally the risk of signal distortions is increased by the large width of the signals (in particular P–D) and by the increased rf or static electric field perturbation required to induce them. The latter depends not only on the coupling element V but also on the lifetimes, and following Eq. (2) 50% signal saturation is reached for $|2V|^2 = \gamma_a \gamma_b$. Typical values of the field strength required for 50% saturation are given in Table 2 for a number of first-order transitions observed in bottle-type experiments.

It is obvious from these arguments and the data in Table 2 that strong *and* narrow signals would result from transitions S–D, S–$F, \ldots$. These require multistep dipole or higher multipole transitions. The multistep dipole coupling is dominant in hydrogenic systems, whose many nearly degenerate fine-structure states readily offer suitable intermediate states in "near resonance" conditions. The higher-order coupling elements V are thus of the form given by Eq. (5) in Section 3.1.1 and are not as small as might at first be expected. The comparatively long lifetimes of the states connected help further to keep the required electric field strength down. Typical values for 50% signal saturation are shown in Table 2.

The multistep or multiphoton character of the coupling process is not seen directly in anticrossing investigations, but it could be established there by comparing the theoretical and experimental electric field dependence of the signal amplitudes (Ref. 107; see also Chapter 13 of this work). In rf experiments the multiphoton character becomes immediately obvious since the frequency required for resonance is $(E_a - E_b)/(h\Delta l)$. The higher-order signals are well described by Eq. (2) with appropriate coupling elements V obtained independently. *Ab initio* calculations are very complex since at least $\Delta l + 1$ states have to be taken into account (usually more because of the summation over *all* intermediate states). Glass-Maujean and Descoubes[108,109] carried out such a calculation for second order S–D anticrossing signals in $H, n = 3$ using a three-level system, i.e., without summation over all intermediate states.

An important problem encountered when deriving fine-structure intervals from such measurements is the considerable Stark shift of the signals caused by the widespread coupling of fine-structure states in the comparatively strong static or rf field required to induce the signals. An obvious example is shown in Figure 17. In this respect, anticrossing studies have a slight advantage over the rf measurements since the dc Stark effect is better understood, so that the necessary extrapolation to zero electric field can easily be checked against the theoretical dependence. In spite of the Stark shift, the small signal widths and the large energy gaps spanned

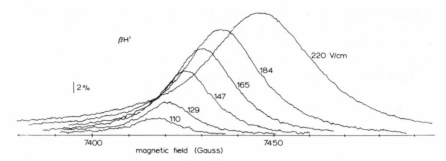

Figure 17. Recordings of the third-order anticrossing signal $4\beta H'$ $(4S\text{--}4F)$ of He^+ taken for several values of the static electric field F $(F \perp H)$ applied to provide the coupling. The steep rise of the signal amplitude ($\propto V^2 \propto F^6$ for a low degree of saturation) and the pronounced (quadratic) Stark shift of the signal position are typical for higher-order signals. The curves represent the change of the light intensity of the line 4686 Å ($n = 4 \to n = 3$) in σ polarization; the bar indicates a change of approximately 2% (Ref. 107).

suggest that signals of this type may provide very accurate fine-structure results in the future.

(*b*) *Measurements.* Static electric fields of the magnitude shown in Table 2 can easily be applied in bottle-type experiments provided they are directed perpendicular to the magnetic field, and anticrossing signals were indeed studied first. Radio-frequency fields of this magnitude tend to cause electrical breakthrough, but the field requirements can be reduced by going up in n or replacing He^+ by H, and such investigations have been carried out more recently both on fast beam [cf. Sections 4.5.3 and 4.6.2(c)] and on bottle-type arrangements. The anticrossing measurements are discussed in more detail in Chapter 13 of this book and will only be summarized.

Eck and Huff[110,104] first introduced higher-order signals with the observation of anticrossing signals $S\text{--}D$ and $S\text{--}F$ in $n = 4$ of He^+. A more detailed investigation of these intervals was carried out by Beyer and Kleinpoppen[112,113] and included the extrapolation of the signal positions to zero electric field, which at the same time allowed the investigation of the Stark effect.[111] A strong Stark shift of the signals is obvious from Figure 17, which shows the $S\text{--}F$ anticrossing signal $\beta H'$ for various values of the electric field strength. Similar measurements of the $S\text{--}D$, $S\text{--}F$,[114] and $S\text{--}G$[79] intervals in $n = 5$ of He^+ were carried out by the same group. All these experiments used sealed glass systems containing the electron gun in an atmosphere of about 10^{-2} Torr of He. The $n = 4$ signals, detected as a change of the intensity and polarization of the line 4686 Å ($n = 4 \to n = 3$), provide a very good signal-to-noise ratio, which will certainly allow a further improvement of the accuracy. At present the results of $n = 4$[112,113] are only quoted to the order of $1/10$ of the combined natural widths of the states, which, nevertheless, corresponds to less than 550 ppm of the respective intervals. Similar anticrossing measurements of $S-D$ intervals were carried out on H, $n = 3^{(115,109)}$ and $n = 4^{(116,109)}$ by Glass-Maujean. Here the motional fields of the excited atoms, produced by electron impact on H_2, were sufficient to induce the signals and no external electric field was applied.

The interval $6^2S_{1/2}-6^2D_{5/2}$ of He^+ was investigated by Eibofner[117] in the only bottle-type higher-order rf experiment so far. In line with the recent trend no magnetic field was applied and the radio frequency tuned instead. No finally corrected result is given, but the accuracy of the interval should be of the order of 200 ppm.

(c) *Leventhal's Anticrossing.* Second-order coupling (but of a different type) also formed the basis for the anticrossing signal αd observed by Leventhal[197] in H, $n = 2$. It is obvious that the states $\alpha(S_{1/2}, m_j = 1/2)$ and $d(P_{3/2}, m_j = -3/2)$, crossing at ~ 2360 G (cf. Figure 3a), are not connected in dipole interaction since they differ by $\Delta m_j = 2$. However, if the magnetic hyperfine coupling is included (the situation and the nomenclature are shown in Figure 18), there is a weak internal magnetic coupling between α^- and β^+ and between d^+ and c^-, f^- while α^+ and d^- remain undisturbed. As a result of this state mixing, electric dipole coupling becomes possible between $\{\alpha^-(\beta^+)\}$ and $\{d^+(c^-, f^-)\}$, and an electric field applied perpendicular to the magnetic field converts the $\alpha^- d^+$ crossing into an anticrossing. This causes a shortening of the lifetime of the α^- state which was observed by Leventhal via an increase of the Ly_α radiation in a bottle-type experiment. The interesting point about this experiment is that inherently it produced hyperfine state selection at a time (1966) when only the first steps had been made by Robiscoe[53] towards reducing the hydrogen signals to single components [cf. Section 3.2.2(a)]. In spite of this obvious advantage no precision measurement was carried out, probably because the internal coupling and therefore the signal were considered to be too weak (cf. Section 3.3.1 in Chapter 13 of this work).

Figure 18. Amplified section of Figure 3a showing the crossing αd of H, $n = 2$. The hyperfine structure is included (+ and − on the states denote $m_I = +\frac{1}{2}$ and $-\frac{1}{2}$ respectively), and hyperfine state mixing is indicated on the states. The only crossing that can be converted into an anticrossing by applying an electric field ($\perp H$) is marked.

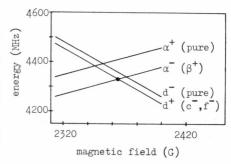

3.3.5. Level-Crossing Measurements of ΔE

Instead of combining measurements of $S_{1/2}-P_{1/2}$ and $S_{1/2}-P_{3/2}$ to determine $\Delta E(P_{1/2}-P_{3/2})$, this interval has also been investigated directly in level-crossing experiments. No external perturbation is required to induce a coupling at the crossing point of states with the same parity, but in order to be able to observe a level-crossing signal the two states must be excited coherently and the decay channel chosen for observation must be open to both states, e.g., decay to a common lower state. In this case interference effects result in a spatial redistribution of the radiation in the region of the crossing which may be detected as a change of the light intensity (Refs. 118, 119; see also Chapters 9 and 13 of this work). Coherent excitation of appropriate substates by electron impact is not possible if the electron beam is parallel to the magnetic field (which in the cases considered here is too strong to allow an electron beam to travel in any other direction). Light excitation is suitable, but all resonance lines of hydrogenic systems are in the vacuum ultraviolet region.

Theoretical aspects of $P_{1/2}-P_{3/2}$ level crossings were discussed by Rose and Carovillano[120] and by Himmell and Fontana,[121] and three

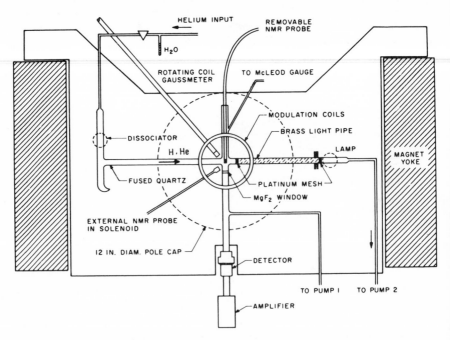

Figure 19. (From Ref. 124.) Diagrammatic view of the apparatus used to investigate the *ed* level-crossing signal in the Ly_α resonance fluorescence of H ($n = 2$). The magnetic field is at right angles to the paper plane.

measurements were carried out on $n = 2$ of H by Baird *et al.*,[122–124] Wing,[125] and by Kollath and Kleinpoppen.[126] The former two aimed at a precision measurement of ΔE and the most accurate result so far was obtained by Baird *et al.*[124] using the arrangement shown in Figure 19. Atomic hydrogen is produced in a high-frequency gas discharge dissociator and drifts in an He atmosphere through the interaction region in the center of the magnetic field where it is irradiated with Ly_α light. The resonance scattered Ly_α light is detected at right angles to both the exciting light and the magnetic field. Its intensity drops with a Lorentzian-type line shape when the magnetic field is varied through the crossing *ed* at approximately 3500 G. The signal width corresponds to twice the natural width of the P state. Yet in spite of this large width of ~200 MHz, the good signal-to-noise ratio and the direct investigations of many possible systematic effects led to a quoted uncertainty of ±0.1 MHz, which is mostly made up of the statistical uncertainty. This measurement represents a key contribution to the determination of the fine-structure constant α (cf. Section 5.3), and a very detailed discussion of the uncertainties is given by Taylor *et al.*,[12] who in particular point out the need to determine systematic effects to the accuracy of the main data if one wants to make sure that the systematic uncertainty is less than the statistical uncertainty.

4. Fast-Beam Investigations

4.1. Measurement of the Stark Quenching Lifetime of 2S

4.1.1. Method

The most important method of investigating Lamb-shift intervals $2^2S_{1/2}$–$2^2P_{1/2}$ beyond $Z = 2$ employs the measurement of the quenching lifetime of $2S$ hydrogenic ions in a (nonresonant) static electric field. Considering only two levels, $S_{1/2}$ and $P_{1/2}$, the $2S$ lifetime in a static electric field is given by Eq. (6) in Section 3.1.2 setting $\omega = 0$. It is a function of the electric field strength (contained in V_{ab}) and of the (Lamb-shift) interval $\Delta\omega$ if γ_b and the dipole matrix elements are taken to be known. The strong correlation between the Lamb shift and the lifetime is obvious from Figure 20 which shows the electric field dependence of the $2S$ lifetime in H as calculated by Lüders[127] for $\Delta\omega = 0$ and $\Delta\omega = \mathscr{S}$ and verified by Sellin[128] in a fast-beam experiment (for $\Delta\omega = \mathscr{S}$). In turn this correlation can be used to determine the Lamb-shift interval from the lifetime measured in a given electric field. Even though lifetimes cannot be measured with the same accuracy as the positions of resonance signals, the method has proved to be extremely useful outside the range available to resonance experiments.

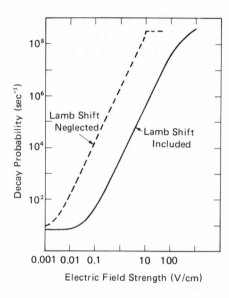

Figure 20. (From Ref. 129.) Decay probability of the $2^2S_{1/2}$ state of H in a static electric field (Stark quenching) as calculated by Lüders.[127]

The precise knowledge of the theoretical quenching lifetime, which was not too important for resonance experiments, is crucial now. As indicated in Section 3.1.2, Eq. (6) does not hold for large electric field perturbations, but it represents a good approximation for much of the field region shown in Figure 20. For accurate measurements higher-order terms as well as the effect of the $P_{3/2}$ state were calculated by Fan et al.[129] and in more general terms by Holt and Sellin.[130] The influence of nP states with $n > 2$ was considered by Drake and Grimley,[131] who also pointed out that

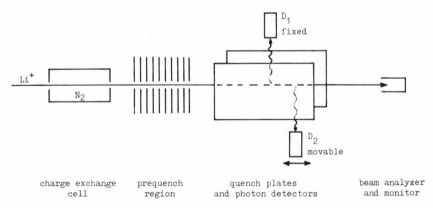

Figure 21. Schematic arrangement to measure the Stark quenching lifetime on a fast (~MeV) beam of 2S ions. The detector D_2 scans the decay curve when moved along the beam. Li^{++} [129] is taken as example.

relativistic effects would have to be calculated for high values of Z. Finally the influence of the hyperfine structure was also considered, e.g., for $Li^{(129)}$ (most other species studied like C, O, Ar have nuclear spin zero).

The experiments as developed by Fan et al.[132,129] make use of a fast (MeV) beam of $2S$ ions produced by beam-foil or charge exchange excitation. Assuming that the beam velocity is known and uniform, the time dependence of the quench radiation in a static electric field perpendicular to the beam is measured by recording the intensity of the quench radiation as a function of the detector position along the beam (cf. Figure 21). The strength of the quench field is usually chosen such that one or two decay lengths correspond to a detector motion of approximately 5 cm and the field has to be increased rapidly with Z. The Lamb shift is derived from the lifetime thus determined following Eq. (6) with appropriate corrections. A review of this type of measurement and of the problems encountered was given by Leventhal.[133]

4.1.2. Measurements

(a) Li^{++}. A schematic diagram of the experimental setup used by Fan et al.[129] to measure the Lamb shift of Li^{++} is shown in Figure 21. A 3-MeV beam of $^6Li^+$ is sent through a N_2-filled charge exchange cell which results in an approximate charge equilibrium of 70% Li^{3+}, 29% Li^{++} (some in the $2S$ state), and 1% Li^+. The long-lived $2S$ ions can be quenched in a prequench region of alternating 12 kV and earth potential thus allowing the $2S$ signal to be distinguished from the background. In the main quenching region a static electric field is applied perpendicular to the beam direction, and the photon intensity is observed by one fixed and one movable detector. Subsequently, the beam is analyzed and its intensity monitored. Measured exponential quench decay curves are shown in Figure 22 for two quench voltages. The analysis of the decay included higher-order terms, the $P_{3/2}$ contribution, and the relativistic time dilation. Hyperfine-structure effects were found to be negligible. It should be noted that the uncertainty of the beam velocity contributed only marginally (0.05%) to the overall uncertainty of ±0.5% which in the terms used in Section 3 corresponds to ~1/25 of the natural width of the $2P$ state.

(b) $^{12}C^{5+}$. The $^{12}C^{5+}$ Lamb shift has been studied by the group of Kugel, Leventhal, and Murnick.[134–136] The beam energy was increased to between 25 and 35 MeV, and instead of using the gas cell to strip and excite the ion beam, which would have resulted in a whole range of different ions, bare C^{6+} nuclei were created first and then allowed to pick up an electron in the gas cell. This has the advantage of producing predominantly C^{5+} ions (some in the $2S$ state) and few other metastable or long-lived ionic states, the decay of which would alter the apparent quench

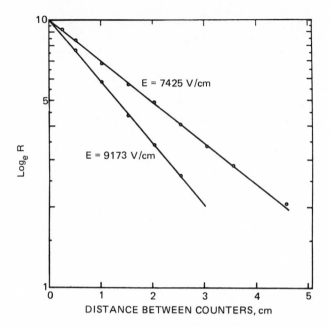

Figure 22. (From Ref. 129.) Semilog plot of the counting rates of the $2S$ Stark quenching radiation of Li^{++} (exponential decay curves) obtained by Fan *et al.*[129] for two values of the quench field. The normalized count rates of the movable counter are shown as a function of the separation from the fixed counter (cf. Figure 21).

lifetime. Another modification was to apply a magnetic field perpendicular to the beam direction and to use the motional electric field experienced by the fast ions to create the strong electric quench field required.

Additional corrections to those applied in the Li^{++} experiment had to be considered. The application of a magnetic field (of the order of 2500 G) results in some Zeeman splitting and thus in a ~2% difference of the lifetimes of the $2S$ substates α and β. Another effect of the strong magnetic field is the deflection of the beam away from or towards the detector. This results in a variation of the segment length of the beam seen by the detector when moved along, in a change of the solid angle of radiation reaching the detector (both effects are counteracting) and in a change of the path length compared with the undeflected beam. All these effects considered,[136] the overall correction was given as -0.29% with single components up to 1.2% and the result was quoted with an accuracy of $\pm 1.0\%$ corresponding to ~$1/16$ of the natural width of the $2P$ state of C^{5+}.

(c) O^{7+}. The Lamb shift of O^{7+} was measured by Leventhal et al.[137] and by Lawrence et al.[138] using an experimental technique very similar to that employed for C^{5+}. However, Lawrence et al. used a different detector setup and treated the background signal and the corrections in a different way. As the result of the increased quench field required (again created as motional field by a magnetic field), the beam deflection became more serious even though the beam energy was raised to around 40 MeV. It was also found necessary to correct for the natural two-photon decay rate of $2S$ which increases more rapidly with $Z(\propto Z^6)$ than the dipole decay rate of $2P(\propto Z^4)$. Both experimental results are in agreement and accurate to approximately 0.5% or $\sim 1/40$ of the natural width of $2P$.

(d) F^{8+}. The technique used for C^{5+} and O^{7+} was extended further to F^{8+} by Murnick et al.[139] but no numerical result was given for the Lamb shift. For an investigation of the interval $S_{1/2}-P_{3/2}$ using laser-induced transitions see Section 4.7.

(e) Ar^{17+}. Gould and Marrus[140] reported the investigation of Ar^{17+} using a 340-MeV argon beam and (motional) electric quench fields of up to 900 kV/cm. The authors claim to have overcome most of the systematic difficulties including beam bending and geometry effects. The experimental value, quoted by Erickson,[8] agrees with the theory to within the accuracy of $\sim 3\%$ corresponding to $\sim 1/10$ of the width of the $2P$ state.

4.2. Anisotropy of the 2S Quenching Radiation

The Ly_α quenching radiation emitted when hydrogenic $2S$ states are subjected to an electric field is polarized as was first measured by Fite et al.[141,142] This finding was surprising at first, since the $P_{1/2}$ state, which decays with unpolarized light, is 10 times nearer to the S state than $P_{3/2}$ and therefore dominates the state mixing, which is $\propto |\langle 2S|er|2P\rangle F|/|E_{2S}-E_{2P}|$ (F is the electric field strength). Yet the small $P_{3/2}$ mixing contribution has nevertheless a very noticeable effect on the decay as a result of interference terms. Through the energy dependence of the state mixing the polarization depends on the fine and hyperfine intervals of the system, and for hydrogen it is about 32%. In a detailed theoretical analysis of the quenching decay Drake and Grimley[131,143] noted that the polarization is equivalent to a spatial anisotropy of the quenching radiation (summed over all polarizations), which is much easier to measure and, being approximately proportional to the Lamb-shift interval, could be used to measure the Lamb shift.

Such measurements were carried out on H and D by the group of van Wijngaarden, Drake, and Farago,[144–146] who, in the latest version of the experiment, used an arrangement as shown in Figure 23. A $2S$ beam of about 10-keV energy is created from a proton beam by charge exchange in

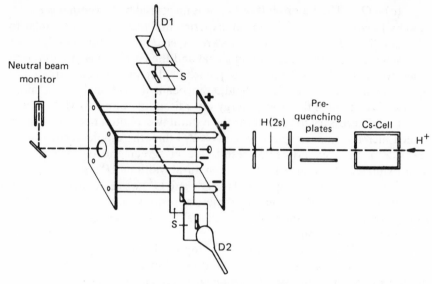

Figure 23. (From Ref. 145.) Apparatus used by Drake *et al.*[145] to measure the anisotropy of the 2S Stark quenching radiation of H and D. The quench field is applied through a quadrupole arrangement of rods, allowing the field direction to be changed without moving the detectors. A grounded conducting cylinder surrounding the quadrupole rods is not shown.

a Cs cell with subsequent removal of the remaining ions in a weak prequench field. The main electric quench field in the observation region is provided by four rods arranged symmetrically with respect to the beam axis and with pairs of adjacent rods kept at the same potential as indicated in Figure 23. The Ly_α light emitted in the directions parallel and perpendicular to the electric field is measured by two fixed slit–detector assemblies. The electric field direction at the beam can be changed electrically, and this eliminates, very effectively, instrumental asymmetries like differing acceptance angles and efficiencies of the detectors as well as effects from motional electric fields created by stray magnetic fields. No movement of a detector, which could introduce uncertainties, is required, an important advantage over the lifetime experiments described in Section 4.1.

Before the anisotropy is deduced from the measured count rates, the isotropic background radiation, which is measured by applying a strong electric field on the prequench plates to destroy all metastables, has to be deducted. Only minor corrections for the acceptance angles of the detectors are necessary since the radiation intensities change very slowly near the angles used. The variation of the anisotropy itself with the strength of the quench field is very slight since the anisotropy depends not so much on the actual mixing, which is proportional to the electric field, but on the ratio of $P_{3/2}$ to $P_{1/2}$ mixing, which does not alter much at reasonably low electric fields. Furthermore, as a result of small differences in the lifetimes of the various hyperfine components of $2S$ in an electric field, the anisotropy depends somewhat on the mean time the atoms spend in the quench field before radiating. Both effects can be eliminated by a simple extrapolation to zero electric field. Any time oscillations of the anisotropy as a result of a sudden onset of the quench field can be neglected in this setup.

The accuracy of single measured anisotropy points[145] was limited by the statistical uncertainty of (at best) ±0.15%. All corrections made, Lamb-shift values for H and D, $n = 2$ were obtained to about ±0.1%, approximately 10 times less accurate than the slow-beam results in Section 3. The same method was also used by Bentz et al.,[147] who reported a value accurate to ±0.04% for H. The results may not be able to compete with investigations based on resonance signals, but with the proof that anisotropy measurements can provide Lamb-shift intervals to better than 0.1%, there is every hope that this method can be successfully applied to higher-Z Lamb shifts (in particular $Z > 3$), even though there is a slow decrease of the anisotropy with increasing Z.[131,148] In this way many of the problems connected with the measurement of the quenching lifetime (discussed in Section 4.1) could be overcome.

4.3. Quantum Beat Experiments

Details of the quantum beat method will be discussed in Chapter 20 of this work and have been part of an earlier paper by Andrä.[149] It is therefore sufficient to give only a brief outline.

If two or more excited states with slightly different energies are prepared in such a way that they decay partly or wholly coherently to a common lower state, the intensity of the emitted radiation fluctuates at the difference frequency (superimposed on the natural decay). In order to be able to observe these beats the whole ensemble of atoms has to be excited simultaneously and the detection system must be able to resolve the beat frequencies involved. The beam-foil light source is particularly suited for such experiments since, even though it is a continuous light source, the excitation time is only spread over the short period the fast ions take to travel through the thin foil. On the detection side the time scale is provided by the distance from the foil and the beam velocity, and the time resolution is only restricted by the size of the detector moving along the beam.

In some cases the excitation process may be sufficient to provide the coherent superposition of states. In other cases the fine or hyperfine coupling subsequent to the sudden excitation forms the basis for the beats. Furthermore, external fields with a sudden (nonadiabatic) onset for the passing beam (or present already at the foil position) may induce beats between states that do not otherwise decay to a common lower state. In particular, quantum beats between states with different parity ($\Delta l = 1$) can only be observed if an external electric field is applied to mix these states.

Since the beat frequency is a direct measure of the energy separation of the interfering states, it can be employed to investigate fine structure and (applying an electric field to couple S and P) also Lamb-shift intervals. The frequency range that can be covered is limited on the high side by the spatial resolution of the detector, and on the low side by the given range of detector motion along the beam and the natural lifetimes of the states. Within limits, the beam velocity may be adjusted to increase the range. Frequencies between a few hundred and over 10,000 MHz have been resolved. Often there are more than two or even many fine-structure states involved, so that many beat frequencies may become superimposed. Considerable progress in disentangling such superpositions has been made in recent years by calculating Fourier transforms of the measured beat structure.

Quite a number of quantum beat experiments have been carried out on hydrogen and He^+, usually applying a static electric field to couple states with different parity. Yet most investigations were more concerned with the effect itself,[150–157,210] with the electric field dependence of the beat frequency which can be used to measure the low-field Stark effect[154,158–161,210] or with the excitation amplitudes of the various states.[157,158,162,163] The investigations of spectral lines from $n \geqslant 3 \to n \geqslant 2$ in H,[150,152,153,155] He^{+},[151,159,161] and Li^{2+} [159] are complicated by the multitude of beating fine and hyperfine states whose decay is observed simultaneously. However, the Ly_α[152,154,156–158,210,164,160] and Ly_β[162,163,210] lines show a less complicated frequency spectrum, and some of these measurements would allow the evaluation of fine-structure and Lamb-shift intervals with reasonable accuracy.

Andrä[158] observed and measured the beats between $2^2S_{1/2}$ and $2^2P_{1/2}$ of H in Ly_α radiation applying a static electric field of up to 400 V/cm parallel or perpendicular to the beam axis. The measured beats are shown in Figure 24. As expected, there are no beats for zero electric field (no S–P coupling). The mixing and thus the beat amplitude rise with increasing electric field and so does the beat frequency (as a result of the electric field, which causes the $S_{1/2}$ and $P_{1/2}$ levels to repel each other). The change of the beat frequency allows the nonlinear low-field Stark effect to be investigated and, extrapolated to zero electric field, the Lamb shift could have been measured to $\sim \pm 50$ MHz and probably better with some effort. Later, Andrä et al.[164] also resolved the 11-GHz $2^2P_{1/2}$–$2^2P_{3/2}$ quantum beats for which no electric field is required. Adding an electric field, the Fourier transform of the whole beat structure showed three

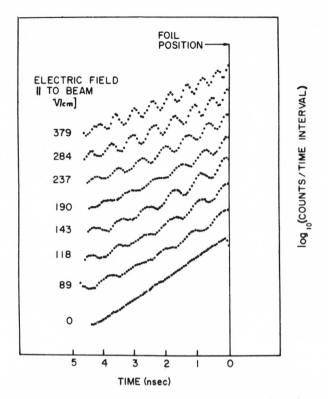

Figure 24. (From Ref. 158.) Decay curves of the Ly$_\alpha$ intensity (semilog plot) and superposed quantum beats between $2^2S_{1/2}$ and $2^2P_{1/2}$ of H observed by Andrä[158] using foil excitation of a fast (~2 MeV) beam. Without electric field only the $2P$ decay is observed. With an electric field applied from the foil position outward, coupling occurs between $S_{1/2}$ and $P_{1/2}$ resulting in beats which increase with the electric field in amplitude and frequency (Stark shift of the states).

frequencies related to the $S_{1/2}-P_{1/2}$, $S_{1/2}-P_{3/2}$, and $P_{1/2}-P_{3/2}$ intervals. This would, for the first time, allow the determination of the various intervals simultaneously. However, a precision measurement is made difficult by the presence of electric, and in this case, magnetic fields which split up the levels, and also by the hyperfine structure. Recently, van Wijngaarden et al.[160] carried out a refined study of Ly$_\alpha$ quantum beats observed after sudden application of a static electric field to a beam of $2S$ atoms (the influence of fringing fields in this experiment is discussed by Drake[196]). At an electric field of 170 V/cm they were able to resolve five components, three of which still consist of several transitions. No attempt was made to extrapolate to zero electric field, and the accuracy was approximately ± 30 MHz, mainly resulting from systematic effects. Again dealing mainly with the Stark effect, similar investigations of higher n states were reported by Pinnington et $al.$[159] on He^+ ($n = 5$–8) and Li^{++} ($n = 7$–10) and by Bourgey et $al.$,[161] who resolved a considerable number of beat frequencies in the Fourier transform of the beats of the $n = 4 \rightarrow n = 3$ line of He^+ in an electric field.

Sokolov[165,166] made the only attempt to carry out a precision measurement of the Lamb shift on this basis, but did not in fact observe quantum beats. A well-focused $2S$ beam is produced by charge exchange of 20-keV protons in H_2. The beam is allowed to drift over about 2 m while most excited states other than $2S$ decay, and remaining protons are removed by a small magnetic field perpendicular to the beam. The beam then passes through a constant static electric field either parallel or perpendicular to the beam designed such that the $2S$ atoms experience a nonadiabatic onset and offset of the field at the edges (at least with respect to the $2P_{1/2}$ state). Instead of observing the quantum beats in the electric field region the $2P$ and $2S$ contents of the beam are analyzed in the field-free region following the electric field condenser. This is done by measuring the Ly_α decay of $2P$ at a fixed point near the end of the electric field

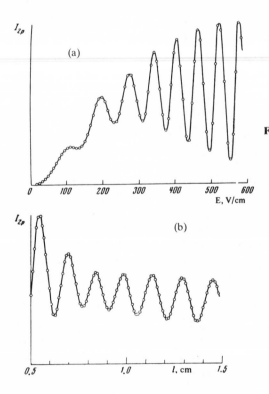

Figure 25. (From Ref. 166.) Analysis of the $2P$ population recorded by the Ly_α intensity, after the beam has left the static electric field region with nonadiabatic boundaries (Sokolov, Ref. 166). (a) $2P$ flux following a static electric field region of fixed length and variable field strength. The mixing between $2S$ and $2P$ in the field region and thus the amplitude of the intensity variation increases with the electric field. At the same time the period of the oscillations reduces as the $S_{1/2}$–$P_{1/2}$ separation increases by Stark shift. (b) $2P$ flux following a static electric field region of fixed field strength but variable length.

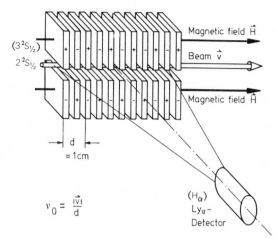

$$v_0 = \frac{|\vec{v}|}{d}$$

Figure 26. (From Ref. 149.) Experimental arrangement
for rf resonance Lamb-shift measurements using a
simulated rf field whose frequency is determined by
the beam velocity and the separation of the static
electric field plates. The resonance is scanned either
by an external magnetic field (shown here) or by varying the beam velocity in zero magnetic field.

and the $2S$ quench radiation further down the beam in another electric
field. The intensities of $2P$ and $2S$ depend on the superpositions of these
states at the point where the atoms leave the electric field, i.e., on the time
spent in the field, on the fine and hyperfine intervals involved, and on the
electric field strength which modifies these intervals by Stark effect. The
intensities of $2S$ and $2P$ fluctuate (180° out of phase) when either the time
spent in the field or the electric field strength is varied. This is shown in
Figure 25. The analysis included the influence of the hyperfine structure,
but not the effects of the $P_{3/2}$ state. On this basis, a Lamb-shift value for H
of 1058.3 MHz was obtained with a standard deviation of only 0.04 MHz.
However, this does not allow for systematic errors resulting from the
uncertainty of the beam velocity, which was measured to 10^{-4}, the fractional energy spread of the beam of 10^{-4}, the uncertainty of the electric
field strength both in the central region and near the edges, the effects of
the $P_{3/2}$ states, and remaining effects of the hyperfine structure. These
effects would probably increase the uncertainty by at least one order of
magnitude. Even then the accuracy remains remarkably high, and the
author claims that it could be improved further by choosing a field
configuration that allows an exact theoretical treatment of the signal
including the $P_{3/2}$ states and by using a hyperfine state selected $2S$
beam.

4.4. Quenching by Simulated rf Fields

Somewhere in between the static electric field quenching experiments and the rf experiments falls the group of Lamb-shift investigations based on sending a fast beam of excited atoms through a region of electrostatic fields which change polarity periodically in space along the beam. Owing to their motion the atoms experience an oscillating field with a frequency determined by their velocity and the spatial periodicity of the field. The basic experimental setup is shown in Figure 26.

The method was first applied to the $3^2S_{1/2}-3^2P_{1/2}$ interval in hydrogen by Hadeishi et al.[167,168] with the aim of testing its applicability to heavy hydrogenlike ions. Following the beam-foil excitation, the lifetime disparities quickly reduced the beam to S states (at a velocity of $\sim 5.10^8$ cm/sec the decay length of $3S$ atoms is ~ 80 cm compared with ~ 2.5 cm for $3P$ and ~ 8 cm for $3D$). The frequency of the oscillating field, i.e., the beam velocity, was held constant and a magnetic field was applied parallel to the beam to tune the substates. Resonance transitions $S \rightarrow P$ were found to cause a reduction in the intensity of the H_α radiation in agreement with the theory given[167,168] even though a simple estimate indicates that the signal should in fact be positive and very small.* On the other hand, the signal polarity changes if the decay length of S (outside resonance) is considerably shortened by Stark mixing with P, and sufficient $S-P$ mixing is indeed likely to occur since only 1° misalignment of the magnetic field axis with respect to the beam results in a motional field of 9 V/cm (which may also cause Stark shifts of the signal).

Corresponding experiments on the $2^2S_{1/2}-2^2P_{1/2}$ interval of hydrogen were carried out by Hadeishi,[168] by Andrä,[169] and by Newton et al.[170] Andrä used a pair of coaxial equally spaced helical springs at opposite potentials so that the atoms experienced a rotating electric field which, depending on the direction of the magnetic field, induced only $\Delta M = +1$ or $\Delta M = -1$ transitions. He also studied the signal for different detection time intervals. No numerical results were given, but the measurements appear to be accurate to about ±10 MHz.

In the recent experiment by Newton et al.[170] the magnetic field tuning was abandoned and the resonance scanned by varying the beam velocity between 10 and 30 keV. Hyperfine state selection of the $2S$ beam was incorporated by selectively quenching the $S_{1/2}$ ($F = 1$) component as

---

* N resonance transitions $3S \rightarrow 3P$ result in an approximate change of H_α radiation (observed over a distance l along the beam) of $\Delta N = -f_s N(1-e^{-l/d_s}) + f_p N(1-e^{-l/d_p})$ with branching ratios $f_s = 1$, $f_p = 0.117$. Using $l = 2.5$ cm and the decay lengths corresponding to the natural lifetimes, $d_p = 2.5$ cm, $d_s = 80$ cm, one obtains: $\Delta N = +0.043N$. If the decay length d_s is reduced (e.g., by Stark quenching) to 20 cm, ΔN changes to $\Delta N' = -0.044N$. For $l \gg d_s$ (full decay observed as in bottle-type experiments) ΔN becomes $\Delta N_{max} = (f_p - f_s)N = -0.887N$.

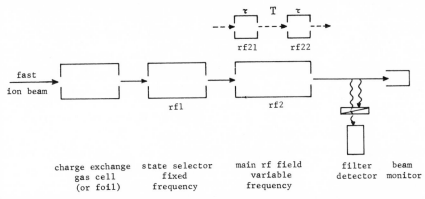

Figure 27. Typical experimental setup (schematic) for fast-beam Lamb-shift and fine-structure measurements using an rf field of variable frequency (rf2) to scan the resonance. No magnetic field is applied. A second rf field (rf1) is employed in most hydrogen experiments to provide hyperfine state selection as described in Figure 28. In separated rf field experiments (cf. Section 4.6), rf2 is replaced by two rf field sections, rf21 and rf22, where the atoms spend time τ, separated by a field-free region where the atoms spend time T. Under suitable conditions this allows the extraction of Ramsey pattern interference signals with a width smaller than that corresponding to the natural lifetimes of the states.

described in Figure 28 in Section 4.5 before the beam entered the simulated rf field region consisting of 32 equally spaced disks with a 0.5-cm hole in the center. The electric field distribution of this configuration was calculated and the corresponding signal shape was found to agree well with the measured variation in the number of surviving $2S$ atoms. The accuracy was limited by the uncertainty of the beam velocity (± 12 eV), which directly determines the simulated frequency, but this, it is claimed, could be bettered by actually measuring the velocity in a time-of-flight setup. Even so, the $2^2S_{1/2}-2^2P_{1/2}$ interval of H was determined with an uncertainty of ± 0.26 MHz, which is only three times less accurate than the slow-beam results (of Section 3) and ten times less than the best results [cf. Sections 4.5.2(d) and 4.6.2(b)].

4.5. rf Experiments

4.5.1. Method

The progress in microwave techniques since the time of the early rf measurements makes it possible now to maintain a constant power level in the interaction region even when sweeping the frequency over quite a wide range. This has led to a series of precision Lamb-shift and fine-structure investigations in which the microwave frequency was varied and no magnetic field applied. In principle, this method can be applied to slow-beam and bottle-type investigations, but in general fast-beam arrangements were used in such measurements. This has the additional advantage (at least over bottle-type experiments) of allowing the separation of the excitation and interaction regions even for short-lived states so that the interaction region can be kept free of disturbing electric fields. The typical experimental arrangement is shown in Figure 27. It is analogous to that of slow-beam rf experiments. Resonance transitions induced by the rf field of variable frequency (rf2) are detected further along the beam from the change of the number of excited states passing and radiating in front of the photomultiplier. In S–P experiments, these are mainly the longer-lived S states (any P states populated in the rf region having decayed long before), and in the case of $n = 2$ a quench field is applied to derive the number of surviving metastables from the quench Ly_α radiation. A second rf field (rf1) with fixed frequency, acting as a hyperfine state selector, is usually applied before or after rf2 if states with large hyperfine structure like S states of H are involved. This allows the selective quenching of the $S_{1/2}$ ($F = 1$) state as explained in Figure 28. The actual resonance signal is then reduced to a single component $S_{1/2}$ ($F = 0$)–$P_{1/2}$ or $P_{3/2}$ ($F = 1$). If a metastable $2S$ beam is investigated, a drift region is normally provided between the excitation and the first rf region to allow short-lived states to decay. Since the beam usually travels parallel to the direction of the rf field, the rf resonance signals are shifted by first-order Doppler effect, but this can easily be investigated and corrected by reversing the direction of the rf field.

The basic signal shape for these experiments has been calculated by

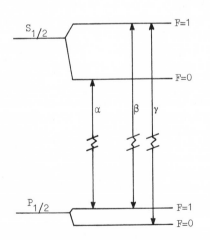

Figure 28. Hyperfine state selection of a beam of hydrogen nS states ($n = 2, 3, \ldots$) which, because of their long lifetimes, dominate the beam in comparison with the corresponding nP states. The $S_{1/2}$ ($F = 1$) state is quenched (in zero magnetic field) by resonant rf transitions β or γ to either of the short-lived P states. In fact both transitions can be driven by a single (fixed-frequency) rf field since the hyperfine separation of the P states is comparable to their natural width. The hyperfine splitting of the S state is larger, however, so that the resonance α is far enough separated from the resonances β and γ for most of the $S_{1/2}$ ($F = 0$) states to survive the state selecting field, while the $S_{1/2}$ ($F = 1$) states are almost completely depleted. As a result of this, only the single resonance α (there are no transitions with $\Delta F = 0$ for $F = 0$) is scanned by the main rf field with variable frequency and recorded.

Fabjan and Pipkin,[171,172] and various aspects have been treated in detail in a series of papers by Silverman and Pipkin.[173,174] Even though the process is very similar to the slow-beam rf experiments in creating a beam of excited S states (for S–P intervals) and recording the number of surviving S states, the equations for the resonance signal of the slow-beam experiments (cf. Section 3.1) are not readily applicable since the time τ spent by the excited atoms in the rf region is now comparable with or even shorter than the lifetime of both resonating states. Following uncertainty principle considerations, this results in an increase of the signal width beyond that determined by the natural level width and rf power broadening, thus impairing the resolution. Also, the natural decay of the S states (for $n > 2$) must now be taken into account as well as possible beat-type oscillations resulting from a sudden onset of the rf field at the boundaries.

4.5.2. Measurements (First-Order Transitions)

(a) H, $n = 3$–5. The fast-beam method with variable rf was developed by Fabjan and Pipkin[175] and first applied to the $3^2S_{1/2}$–$3^2P_{1/2}$ Lamb-shift interval of H[175,171,172] using a beam energy of typically 30 keV. State selection as described in Figure 28 was also investigated for the first time and proved very effective. A measured signal is shown in Figure 30a. In the final experiment[172] the time spent by the atoms in the rf region was 31 nsec, which, being much larger than the $3P$ lifetime of 5.4 nsec, caused little broadening of the resonance signal. With a standard error of about 200 ppm (1/500 of the combined natural widths) the single-field Lamb-shift value was considerably more accurate than previous measurements on $n = 3$ of H or D.[81–83]

Fabjan et al.[176] extended the investigations to the hydrogen Lamb-shift intervals of $n = 4$ and 5, which are very difficult to measure in bottle-type experiments [cf. Section 3.3.3(c)]. At the same time a large number of other $\Delta l = \pm 1$ S–P, P–D, and D–F fine-structure intervals of $n = 3, 4, 5$ were measured. In view of the high sensitivity of these states to electric field perturbations, which prevented most of them from being investigated before, these measurements represent quite an achievement, even though the accuracy, naturally, is not up to that of the $n = 3$ results.

(b) He$^+$, $n = 4$ to 7. A similar investigation in He$^+$ of the Lamb-shift intervals for $n = 4$ to $n = 7$ and of F–G and G–H fine-structure intervals for $n = 6$ was carried out by Churassy et al.[177] at somewhat higher beam energies of about 400 keV. No state selection is required since He has nuclear spin zero. The accuracy was of the order of $\pm 2\%$. Again the high n values reached prove the freedom from disturbing electric fields in this method.

(c) Li^{++}, $n = 2$. Dietrich *et al.*$^{(178,179)}$ investigated the $2^2S_{1/2}-2^2P_{1/2}$ Lamb shift of Li^{++} using a 4.5-MeV beam excited in a gas target. The rf region (58 GHz) is surrounded by an iron-clad solenoid producing a magnetic field parallel to the beam to tune the $S-P$ interval through resonance. The surviving metastables are recorded in a quench field further down the beam. Only preliminary results have been given so far.

(d) H, $n = 2$ (*Precision measurement.*) The most accurate measurement to date of the $2^2S_{1/2}-2^2P_{1/2}$ Lamb shift of H was carried out by Andrews and Newton,$^{(180)}$ equalled only by the measurement of Lundeen and Pipkin$^{(181)}$ to be discussed later [cf. Section 4.6.2(b)]. An H beam energy of 21 keV was used, slow enough for the metastables created by charge exchange in H_2 gas, to experience an adiabatic switch-on and switch-off of the rf field when passing through the microwave region. This was of slab line geometry, the field distribution of which was obtained analytically, thus allowing an accurate calculation of the resonance line shape. State selection as described in Figure 28 was employed and almost completely quenched the $F = 1$ state of $S_{1/2}$. The number of surviving

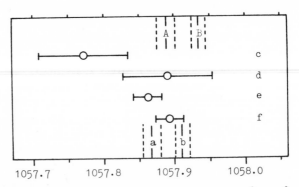

Figure 29. Comparison of experimental and theoretical values of the $2^2S_{1/2}-2^2P_{1/2}$ Lamb shift of hydrogen. Theory: (a) Mohr (Ref. 200); proton radius 0.80 ± 0.02 f; (b) Erickson (Ref. 8); proton radius 0.80 ± 0.02 f; (A) Mohr (Ref. 201), (a) but using the RMS proton radius 0.87 ± 0.02 f of Ref. 191; (B) Erickson, calculated from (b) for the RMS proton radius of Ref. 191 using Eq. (4.10b) of Ref. 4. Experiment: (c) Triebwasser *et al.* (Ref. 51) as quoted in Ref. 12. (d) Robiscoe and Shyn (Ref. 56), uncertainty as quoted in Ref. 12. (e) Andrews and Newton (Ref. 180). (f) Lundeen and Pipkin (Ref. 181). The proton radii used for the theoretical values a, b and A, B differ by approximately 10% and cause a variation of the theoretical Lamb shift $2^2S_{1/2}-2^2P_{1/2}$ of the order of the accuracy of the best current measured values. If the QED contributions are taken to be correct, precision Lamb-shift measurements could thus now be used to extract information on the nuclear structure. Andrews *et al.*$^{(213)}$ have in fact done this and derived an independent value of the proton radius of 0.845 ± 0.050 f.

metastables was recorded further down the beam by Stark quenching and Lyman-α detection. Seventeen points at different frequencies of the rf field were used to scan the resonance signal, and the position of the α transition (cf. Figure 28) was determined by a computer fit which included spurious contributions from the β and γ transitions. The measurements were taken for both directions of the rf field and individually corrected for Bloch–Siegert shift[182,183] by about 90 ppm (90 kHz). Further corrections amounted to only +27 ppm (29 kHz), the largest contribution being +19 ppm (20 kHz). The overall uncertainty (1 standard deviation) is made up to the larger part by systematic uncertainties and is given as ±19 ppm (20 kHz) corresponding to 1/5000 of the natural width of the P state. The comparison of precision Lamb-shift data for hydrogen, $n = 2$ in Figure 29 shows that the result agrees with the other measurements and the theory by Mohr[7] but differs somewhat from Erickson's[8] value.

4.5.3. Higher-Order (Multiphoton) Transitions

The principle of higher-order signals between hydrogenic states has been discussed in Section 3.3.4 in connection with bottle-type experiments. Most investigations of this type employed the anticrossing technique, but, more recently, a number of groups succeeded in inducing and detecting higher-order rf transitions which, naturally, show the multistep character of these signals more directly. A fast-beam arrangement is best suited for these investigations since it separates the interaction region from the "dirty" excitation region thus reducing the tendency to start an rf discharge at the elevated rf field strength necessary for the experiment. Also, the clean environment introduces few perturbations so that higher n values can be chosen for the measurement with a corresponding reduction in the required rf field strength.

The first double-photon rf signals were reported by Kramer et al.,[184] who observed $3^2S_{1/2}$–$3^2D_{5/2}$ transitions in $n = 3$ of H using two separated oscillatory fields [cf. Section 4.6.2(c)]. Multiphoton transitions between the $2^2S_{1/2}$ and $2^2P_{1/2}$ states of H were studied by Andrews and Newton[185] using an experimental arrangement similar to that in Figure 27. The coupling scheme in this case (for example in third order) is $2S \to 2P \to 2S \to 2P$ and it is obvious that only signals with uneven orders can be expected, which is confirmed by the observation. Since the Lamb-shift interval can be measured better in first order, the value of this study lies more in a test of the theoretical rf power shift, which forms a small, but important, correction in first-order experiments and can be studied here over a wider field range.

Intervals involving states up to $l = 5$ $(H_{11/2})$ in He^+, $n = 6, 7, 9$ were investigated by Brandenberger et al.[186] using a fast-beam setup of the type shown in Figure 27. All signals were detected as cascade effects in the He^+ line at 1640 Å $(n = 3 \to n = 2)$. In contrast to earlier higher-order anticrossing measurements [cf. Section 3.3.4(b)], no use was made of the long lifetime and the strong population of S states, and most signals were of the type $\{P_{3/2}–G_{9/2}, D_{3/2}–H_{9/2}\}$, which cannot be separated. In common with all higher-order signals, a considerable rf power shift was observed and the signals were extrapolated to zero rf power. The experimental results of the four intervals are accurate to between 0.3% and 0.6%.

4.6. Separated rf Field Experiments

4.6.1. Method

All resonance experiments described so far in Sections 3 and 4 were limited in their resolution by the natural widths of the states under investigation, and it has been seen that further improvements of the best available fine-structure measurements are extremely difficult since this requires the determination of the signal position to something like 1/1000 of the signal width, or even less in the case of the Lamb shift of $n = 2$ of H. In recent years, successful attempts have been made to shape resonance curves by appropriate selection of interaction and detection time intervals. A full discussion of the effects involved and of relevant experiments will be found in Chapter 14 of this book. Of particular interest are attempts to reduce the width of the resonance signal to below the combined natural widths of the states by biasing the observed sample in favor of states that have survived longer than the natural lifetime. Of course, this has to be paid for with a corresponding loss of intensity (i.e., signal-to-noise ratio), and thus biasing is usually applied only when other methods fail to resolve closely spaced signals. After the introduction of hyperfine state selection this is not the problem any more in hydrogenic fine-structure investigations. Yet, with the extremely good signal-to-noise ratio achieved in some of the recent experiments, a reduction of the signal width even at the cost of a considerable reduction of the signal amplitude is a challenging proposition, and such a technique was developed and successfully applied by Fabjan and Pipkin.[187]

To introduce time biasing, Ramsay's[188] well-known arrangement of two separated rf fields, which for some time has provided narrow magnetic resonance signals of stable states, was extended to short-lived states. A fast-beam arrangement was used as shown in Figure 27, now with the main rf field (rf2) split into two sections rf21 and rf22 oscillating coherently and separated by a variable gap. The states may spend the time τ in each section and the time T in the gap. The resulting resonance signal (e.g., S–P) is dominated by $S \to P$ transitions induced in the single rf field sections with a width determined (as discussed in Section 4.5.1) by the combined natural widths of S and P, the rf power broadening and the broadening from the short interaction time τ. Superimposed on this broad signal is an oscillatory pattern, the amplitude and width of which are determined by the time T spent between the rf field sections. This signal can be understood as an interference term in the intensity of S states detected behind the rf fields which is based on the two amplitudes of outgoing S states (1) having spent the time T between the rf field sections as S states and (2) having spent the time T as P states. Since the P states decay rapidly the interference term also decreases rapidly with increasing T. The signal has been calculated in general terms by Fabjan and Pipkin.[172] The form of the interference part is essentially given by[189,190]

$$S'_{S,\text{out}} = k\, e^{-\gamma_p T/2} \cos\left[(\omega_0 - \omega)T + (\delta_2 - \delta_1) + f_-(\omega_0 - \omega)\right] \qquad (10)$$

$\gamma_p (\gg \gamma_s)$ is the decay constant of the P state; ω_0 and ω are the angular frequencies of the interval and the applied rf field, respectively; δ_1 and δ_2 are the phases of the rf fields in the field sections 1 and 2 (rf21 and rf22), and f_- is a geometry-dependent correction function. The exponential factor shows that the amplitude of the interference signal goes down with increasing T (as fewer P states survive), while at the same time the number of oscillations increases and their width reduces (the signal corresponding to the single rf sections forms an envelope for the oscillations). The interference signal S' can be separated from the total signal S by recording the difference signal $S(0)$–$S(180)$ ($\delta_2 - \delta_1 = 0°$ and $180°$).

4.6.2. Measurements

(a) *Hydrogen*, $n = 3$. The first measurement with separated oscillatory fields was carried out by Fabjan and Pipkin[187,171,172,189] on the $3^2S_{1/2}$–$3^2P_{1/2}$ interval of hydrogen and compared with single rf field results [cf. Section 4.5.2(a)] obtained with the same apparatus. Hyperfine state selection as described in Figure 28 was used in both cases. Measured signals are compared in Figure 30 showing a considerable reduction in width for the separated field case. However, as expected on the basis of the reduced signal strength and the more complicated signal shape, the accuracy of the separated field results did not fully reflect the reduction in signal width. Still, with a standard deviation (before applying the final corrections) of ± 108 ppm, the Lamb-shift interval obtained with separated fields was more accurate than that for a single field (± 158 ppm). Combin-

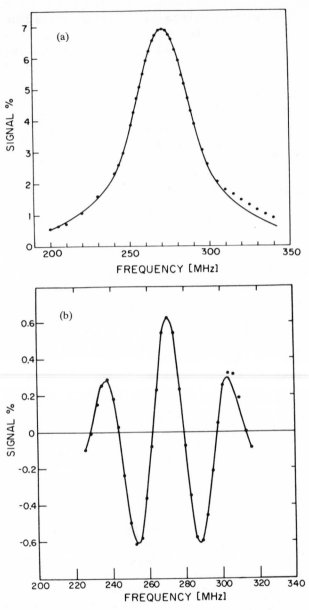

Figure 30. (From Ref. 172.) Theoretical (solid line) and measured (dots) resonance profile for rf transitions $3^2S_{1/2}$ $(F = 0) \to 3^2P_{1/2}$ $(F = 1)$ in H. Hyperfine state selection is applied (cf. Figure 28). (a) Single rf field configuration. Time spent in the rf field $2\tau = 2 \times 15.5$ nsec. A slight remaining trace of the β and γ resonances is visible in the wing. (b) Separated rf field configuration ($\tau_1 = 15.5$ nsec, $T = 18$ nsec, $\tau_2 = 15.5$ nsec), otherwise conditions as in (a). $S(0)$–$S(180)$ is recorded.

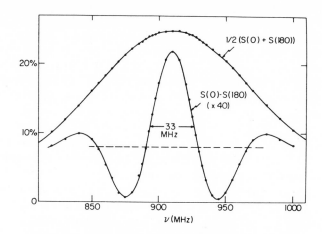

Figure 31. (From Ref. 181.) Theoretical and measured resonance curves of the $2^2S_{1/2}$ $(F=0) \rightarrow 2^2P_{1/2}(F=1)$ transition of H obtained by Lundeen and Pipkin.[181] The broad curve represents the single rf field resonance. Owing to the conditions chosen, it is about $\frac{1}{3}$ wider than corresponding signals used in other investigations (Ref. 180, Figure 8b, Figure 10). The narrow curve represents the interference signal for separated oscillatory fields showing a reduction of the resonance width to $\frac{1}{3}$ of the natural width of the P state.

ing both results, a final accuracy of ± 150 ppm was achieved corresponding to $1/630$ of the combined natural widths. This represents the highest accuracy obtained so far for this interval.

(b) *Hydrogen, $n = 2$ (Precision Measurement)*. Lundeen and Pipkin[181] applied a separated rf field method to the $2^2S_{1/2}-2^2P_{1/2}$ interval of hydrogen and in 1975 reported the most accurate result obtained so far for any Lamb-shift interval. {In the meantime Andrews and Newton[180] [cf. Section 4.5.2(d)] showed that the same accuracy can also be reached in a single rf field investigation.} $2S$ atoms of 50–100 keV energy were passed through two coherently oscillating rf field sections consisting of pairs of balanced rf plates designed to ensure uniformly polarized rf fields. The number of surviving $2S$ atoms was measured as a function of the frequency. Both the single rf field signal $\frac{1}{2}[S(0)+S(180)]$ and the reduced width interference signal $S(0)-S(180)$ are shown in Figure 31. Hyperfine state selection was employed as shown in Figure 28. Possible phase errors between the rf sections were eliminated by interchanging the plates. Doppler shifts were also eliminated. The signal centers were determined from symmetrical points with additional checks on the wider line profile. A number of further sources of corrections and uncertainties were carefully investigated.

Total corrections of the raw data range from $+64$ ppm (58 kHz) to $+149$ ppm (136 kHz) depending on the beam energy and the rf field spacing, with an estimated uncertainty of ±20 ppm (18 kHz). Including the statistical uncertainty of ±10 kHz the overall accuracy (1 standard deviation) with respect to the Lamb-shift interval is ±19 ppm (20 kHz) corresponding to 1/5000 of the natural width of the $2P$ state or about 1/1800 of the (reduced) signal width. The comparison with other Lamb-shift results in Figure 29 shows that the result is compatible with both theoretical values and with other recent measurements.

(c) *Higher-Order Transitions.* Higher-order signals were discussed in Section 3.3.4 and for fast-beam experiments with single rf fields in Section 4.5.3. Kramer et al.[184] used two separated oscillatory fields to investigate double quantum transitions $3^2S_{1/2}-3^2D_{5/2}$ in hydrogen [corresponding anticrossing signals were studied by Glass-Maujean,[115] (cf. Section 3.3.4)]. The experimental setup resembled that of the other separated rf field investigations and included hyperfine state selection so that mainly the transition $3^2S_{1/2}(F=0) \to 3^2D_{5/2}(F=2)$ was observed. The total signal was recorded showing the interference pattern superimposed on a broad resonance line (Figure 32). The interference pattern alone could not be extracted since in the case of double quantum transitions this would have required rf phase switching by 90°. In view of the relatively long lifetime of both the S and D states the time spent in and between the rf fields was so short that even the narrowed signals were still wider than corresponds to the combined natural width, but it is claimed that a reduction to 1/3 of the linewidth would be feasible. No numerical results were given at this stage. Further progress of this experiment was reported by Clark et al.[192]

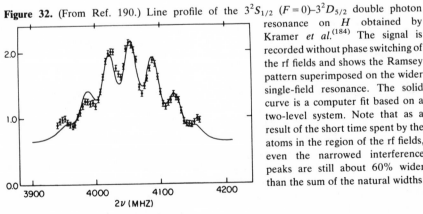

Figure 32. (From Ref. 190.) Line profile of the $3^2S_{1/2}$ $(F=0)$–$3^2D_{5/2}$ double photon resonance on H obtained by Kramer et al.[184] The signal is recorded without phase switching of the rf fields and shows the Ramsey pattern superimposed on the wider single-field resonance. The solid curve is a computer fit based on a two-level system. Note that as a result of the short time spent by the atoms in the region of the rf fields, even the narrowed interference peaks are still about 60% wider than the sum of the natural widths.

4.7. Laser Experiments

For high Z values, the fine-structure intervals of hydrogenic ions get into the region covered by lasers, which may thus be used to induce resonant transitions and to retain the high inherent accuracy of rf experiments. Tunable dye lasers are, of course, best suited to sweep the resonances, and on this basis a measurement of the $2^2S_{1/2}-2^2P_{3/2}$ interval of the $(\mu^- {}^4He)^+$ muonic ion was carried out by Bertin et al.[193] Yet, a fixed-frequency laser radiating near the fine-structure resonance frequency may be sufficient when used in conjunction with a fast beam of excited states. Tuning can then be accomplished by varying the angle between the laser and the ion beam (Doppler tuning). Such an experiment was reported by Kugel et al.[194,195] for the $2^2S_{1/2}-2^2P_{3/2}$ interval of hydrogenic ${}^{19}F^{8+}$. Metastable ions of 64 MeV energy were created by stripping and subsequent electron pickup [cf. Section 4.1.2(b)] and crossed with a pulsed HBr laser beam centered at 2382.52 cm^{-1}, near the $S_{1/2}-P_{3/2}$ fine-structure separation of $\sim$2296 cm^{-1}. The 826-eV Ly$_\alpha$ radiation emitted as a result of induced $S \to P$ transitions was detected with proportional counters. Their resolution allowed only partial discrimination against background radiation, but the background could be reduced by observing in coincidence with the laser pulses. The laser light consisted of several lines causing an asymmetric resonance line shape. This could be corrected in the data analysis since the positions and relative intensities of the lines were known. Effects of the hyperfine structure ($I = \frac{1}{2}$) could not be eliminated and a broadened signal of three unresolved components was observed, the center of which, however, should not be shifted in lowest order. The preliminary result given for ΔE-$\mathscr{S}^{(194)}$ carries a standard deviation of $\pm 0.05\%$. Using the theoretical value for ΔE, the Lamb shift is obtained to $\pm 1.0\%$, which is of the same order as results of the $2S$ quenching lifetime measurements for comparable values of Z (cf. Section 4.1). However, the authors claim that another order of magnitude may be gained over the present result.

5. Results

5.1. Lamb-Shift and Fine-Structure Measurements

Table 3 attempts to give a complete list of Lamb-shift and fine-structure measurements including a selection of optical results. It is (with some additions and alterations) based on material kindly provided by Professor Erickson,[211] who also calculated the theoretical values. The measured and the theoretical results are given with uncertainties estimated

Table 3. Lamb-Shift and Fine-Structure Measurements

Atom	Interval	Calculated values[a] ($\pm 1\sigma$)	Measured value ($\pm 1\sigma$)	Difference[b]	Method[c]	Text (section number)	Ref.,[a] Notes
H	$1S_{1/2}-1S_{1/2}$[p]	8149.44 ± 0.080 MHz	8600 ± 800 MHz	-0.6	two-photon absorption	2.3	29
	$2S_{1/2}-2P_{1/2}$	1057.911 ± 0.010 MHz	1057.862 ± 0.020 MHz	$+2.2$	fast beam, rf	4.5.2(d)	180
			1057.893 ± 0.020 MHz	$+0.8$	fast beam, separ. rf	4.6.2(b)	181
			1057.90 ± 0.10 MHz	$+0.1$	slow beam, ac	3.2.2(a)	56
			1057.772 ± 0.063 MHz	$+2.2$	slow beam, rf	3.2.1	51, corr. 12
			1058.05 ± 0.26 MHz	-0.5	fast beam, simul. rf	4.4	170
			1057.85 ± 0.44 MHz	$+0.1$	fast beam, anisotropy	4.2	147
			1057.3 ± 0.9 MHz	$+0.7$	fast beam, anisotropy	4.2	145
			1058.3 ± 0.5 MHz	-0.8	fast beam, modified quantum beats	4.3	166[j]
	$2P_{3/2}-2S_{1/2}$	9911.129 ± 0.011 MHz	9911.173 ± 0.042 MHz	-1.0	slow beam, rf	3.2.2(b)	61
			9911.250 ± 0.063 MHz	-1.9	slow beam, rf	3.2.2(b)	55
			9911.377 ± 0.026 MHz	-8.8	bottle, rf	3.3.2(b)	75
	$2P_{3/2}-2P_{1/2}$	10969.040 ± 0.005 MHz	10969.13 ± 0.10 MHz	-0.9	bottle, level crossing	3.3.5	124[d]
			10969.6 ± 0.7 MHz	-0.8	bottle, level crossing	3.3.5	125[d]
	$3S_{1/2}-3P_{1/2}$	314.898 ± 0.003 MHz	314.819 ± 0.048 MHz	$+1.6$	fast beam, separ. rf	4.6.2(a)	172
			315.11 ± 0.89 MHz	-0.2	fast beam, rf	4.5.2(a)	176
			314.9 ± 0.9 MHz	00	bottle, ac	3.3.3(b)	109, 83
			313.6 ± 2.9 MHz	$+0.4$	slow beam, Ly_β, rf	3.3.3(b)	82
	$3P_{3/2}-3S_{1/2}$	2935.191 ± 0.003 MHz	2933.5 ± 1.2 MHz	$+1.4$	fast beam, rf	4.5.2(a)	176
	$3D_{3/2}-3S_{1/2}$	2929.859 ± 0.003 MHz	2930.3 ± 0.8 MHz	-0.6	bottle, ac (2nd order)	3.3.4(b)	115
	$3D_{3/2}-3P_{1/2}$	3244.757 ± 0.002 MHz	3255.6 ± 8.4 MHz	-1.3	fast beam, rf	4.5.2(a)	176
	$3D_{5/2}-3S_{1/2}$	4013.197 ± 0.003 MHz	4013.8 ± 0.8 MHz	-0.8	bottle, ac (2nd order)	3.3.4(b)	115
	$3D_{5/2}-3P_{3/2}$	1078.0059 ± 0.0007 MHz	1080.3 ± 2.9 MHz	-0.8	fast beam, rf	4.5.2(a)	176
			1078.0 ± 1.1 MHz	$+0$	bottle, ac	3.3.4(b)	109

Transition	Theory	Experiment	Corr.	Method	Section	Ref.
$4S_{1/2}-4P_{1/2}$	133.0858±0.0013 MHz	132.53±0.58 MHz −0.78	+1.0	bottle, rf, pulsed	3.3.3(c)	87
		133.15±0.59 MHz	−0.2	fast beam, rf	4.5.2(a)	176
$4P_{3/2}-4S_{1/2}$	1238.044±0.001 MHz	1237.79+0.29 MHz −0.26	+0.9	bottle, rf, pulsed	3.3.3(c)	87
$4D_{3/2}-4S_{1/2}$	1235.756±0.001 MHz	1235.9±1.3 MHz	+1.6	fast beam, rf	4.5.2(a)	176
		1235.0±2.1 MHz	+0.4	bottle, ac (2nd order)	3.3.4(b)	109
$4D_{3/2}-4P_{1/2}$	1368.8420±0.0007 MHz	1371.1±1.2 MHz	−1.9	fast beam, rf	4.5.2(a)	176
$4D_{5/2}-4S_{1/2}$	1692.790±0.001 MHz	1693.0±0.4 MHz	−0.5	bottle, ac (2nd order)	3.3.4(b)	109
$4D_{5/2}-4P_{3/2}$	454.7459±0.0003 MHz	455.7±1.6 MHz	−0.6	fast beam, rf	4.5.2(a)	176
$4F_{5/2}-4D_{3/2}$	456.2242±0.0002 MHz	456.8±1.6 MHz	−0.4	fast beam, rf	4.5.2(a)	176
$4F_{7/2}-4D_{5/2}$	227.7061±0.0001 MHz	227.96±0.41 MHz	−0.6	fast beam, rf	4.5.2(a)	176
$5S_{1/2}-5P_{1/2}$	68.1997±0.0007 MHz	64.6±5.0 MHz	+0.7	fast beam, rf	4.5.2(a)	176
$5P_{3/2}-3S_{1/2}$	633.8178±0.0007 MHz	622.4±10.1 MHz	+1.1	fast beam, rf	4.5.2(a)	176
$5D_{3/2}-5P_{1/2}$	700.8356±0.0004 MHz	704.3±7.1 MHz	−0.5	fast beam, rf	4.5.2(a)	176
$5D_{5/2}-5P_{3/2}$	232.8193±0.0002 MHz	232.2±2.9 MHz	+0.2	fast beam, rf	4.5.2(a)	176
$5F_{7/2}-5D_{5/2}$	116.5822±0.0001 MHz	117.2±1.5 MHz	−0.4	fast beam, rf	4.5.2(a)	176
D $1S_{1/2}-1S_{1/2}^p$	8172.24±0.12 MHz	8300±300 MHz	−0.4	two-photon absorption	2.3	29
$2S_{1/2}-2S_{1/2}^p$	0.27260 cm^{-1}	0.262±0.038 cm^{-1}	+0.3	Lyα wavelength	2.3	28
	0.03488 cm^{-1}	0.0369±0.0016 cm^{-1}	−1.3	interferometric	2.1	21[g]
		0.0300±0.0050 cm^{-1}	+1.0	(interferometric)	2.1	33[e]
$2S_{1/2}-2P_{1/2}$	1059.273±0.015 MHz	1059.28±0.06 MHz	−0.1	slow beam, ac	3.2.2(a)	58, corr. 59
		1058.996±0.064 MHz	+4.2	slow beam, rf	3.2.1	51, corr. 12
		1059.6±1.1 MHz	−0.3	fast beam, anisotropy	4.2	145, corr. 148
$2P_{3/2}-2S_{1/2}$	9912.761±0.015 MHz	9912.607±0.056 MHz	+2.7	slow beam, rf (αa)	3.2.1	52, corr. 12
		9912.803±0.094 MHz	−0.4	slow beam, rf (αc)	3.2.1	52, corr. 12
$3S_{1/2}-3S_{1/2}^p$	0.01040 cm^{-1}	0.0083±0.002 cm^{-1} −0.003	+1.0	interferometric	2.1	22[g]
$3S_{1/2}-3P_{1/2}$	315.301±0.004 MHz	315.3±0.4 MHz	−0.0	bottle, rf	3.3.3(b)	81[f,d]
$3P_{3/2}-3P_{1/2}$	3250.976±0.002 MHz	3250.7±1.0 MHz	+0.3	bottle, rf (indirect)	3.3.3(b)	81[f,d]
$3P_{3/2}-3D_{3/2}$	5.3370±0.0005 MHz	5.0±5.0 MHz	+0.1	bottle, rf (indirect)	3.3.3(b)	81[f]
$4S_{1/2}-4P_{1/2}$	133.2560±0.0018 MHz	133.0±5.0 MHz	+0.1	bottle, rf	3.3.3(b)	81[f,d]

continued overleaf

Table 3 (continued)

Atom	Interval	Calculated values[a] (±1σ)	Measured value[b] (±1σ)	Difference[b]	Method[c]	Text (section number)	Ref., Notes
T	$2S_{1/2}-2S_{1/2}$[p]	0.03489 cm^{-1}	$0.037^{+0.002}_{-0.004}$ cm^{-1}	−0.5	interferometric		209[g]
^{3}He$^+$	$3S_{1/2}-3P_{1/2}$	139.594±0.007 mK	140.0±1.0 mK	−0.4	interferometric ($n=4$ →$n=3$)	2.1	18
	$4S_{1/2}-4P_{1/2}$	59.018±0.003 mK	60.3±2.0 mK	−0.6	interferometric ($n=4$ →$n=3$)	2.1	18
			57.4±4.9 mK	+0.3	interferometric ($n=5$ →$n=4$)	2.1	20
			57.4±4.1 mK	+0.4	interferometric ($n=6$ →$n=4$)	2.1	20
	$5S_{1/2}-5P_{1/2}$	30.249±0.001 mK	37.7±9.5 mK	−0.8	interferometric ($n=5$ →$n=4$)	2.1	20
	$6S_{1/2}-6P_{1/2}$	17.515±0.001 mK	31.5±8.3 mK	−1.7	interferometric ($n=6$ →$n=4$)	2.1	20
^{4}He$^+$	$2S_{1/2}-2P_{1/2}$	14045.1±0.4 MHz	14046.2±1.2 MHz	−0.9	bottle, rf	3.3.2(a)	72
			14040.2±1.8 MHz	+2.7	bottle, rf	3.3.2(a)	70
		0.46849±0.00001 cm^{-1}	0.48±0.02 cm^{-1}	−0.6	optical ($n=3$ →$n=2$)	2.1	17[g]
	$3S_{1/2}-3P_{1/2}$	4184.54±0.11 MHz	4183.17±0.54 MHz	+2.5	bottle, rf	3.3.3(d)	89
			4184.2±2.4 MHz	+0.1	bottle, ac	3.3.3(d)	91[h]
		139.581±0.004 mK	138.5±1.6 mK	+0.7	interferometric, beam ($n=4$→$n=3$)	2.1	16[i]
			139.1±2.0 mK	+0.2	interferometric ($n=4$→$n=3$)	2.1	18
	$3P_{3/2}-3S_{1/2}$	47843.29±0.10 MHz	47844.05±0.48 MHz	−1.5	bottle, rf	3.3.3(d)	89
	$3D_{5/2}-3P_{3/2}$	17255.373±0.030 MHz	17259±5 MHz	−0.7	bottle, ac	3.3.3(d)	91[h]

Transition				Method		Ref.
$4S_{1/2}$–$4P_{1/2}$	1769.15 ± 0.05 MHz	1769.0 ± 2.0 MHz	$+0.1$	bottle, rf	3.3.3(e)	98
		1768 ± 5 MHz	$+0.2$	bottle, rf	3.3.3(e)	97
		1766 ± 7.5 MHz	$+0.4$	bottle, rf (pulsed)	3.3.3(e)	95
		1751 ± 13 MHz	$+1.4$	bottle, rf	3.3.3(e)	96[f]
		1755 ± 44 MHz	$+0.3$	fast beam, rf	4.5.2(b)	177
		60.0 ± 4.0 mK	-0.2	interferometric, beam ($n = 4 \rightarrow n = 3$)	2.1	16[i]
	59.012 ± 0.002 mK	57.1 ± 2.0 mK	$+1.0$	interferometric ($n = 4 \rightarrow n = 3$)	2.1	18
		60.3 ± 2.9 mK	-0.4	interferometric ($n = 5 \rightarrow n = 4$)	2.1	20
		58.9 ± 3.3 mK	$+0.0$	interferometric ($n = 6 \rightarrow n = 4$)	2.1	20
$4P_{3/2}$–$4S_{1/2}$	20179.99 ± 0.04 MHz	20180.6 ± 0.8 MHz	-0.8	bottle, rf	3.3.3(e)	99
		20179.7 ± 1.2 MHz	$+0.2$	bottle, rf	3.3.3(e)	97
$4D_{3/2}$–$4S_{1/2}$	20143.27 ± 0.04 MHz	20145.3 ± 3.9 MHz	-0.5	bottle, ac (2nd order)	3.3.4(b)	113[f]
$4D_{5/2}$–$4S_{1/2}$	27459.02 ± 0.04 MHz	27455.3 ± 3.7 MHz	$+1.0$	bottle, ac (2nd order)	3.3.4(b)	113[f]
$4F_{5/2}$–$4S_{1/2}$	27446.06 ± 0.04 MHz	27435.7 ± 6.9 MHz	$+1.5$	bottle, ac (3rd order)	3.3.4(b)	112[f]
$4F_{5/2}$–$4D_{3/2}$	7302.786 ± 0.003 MHz	7300.2 ± 1.5 MHz	$+1.7$	bottle, rf	3.3.3(e)	100
$4F_{7/2}$–$4S_{1/2}$	31103.84 ± 0.04 MHz	31098.8 ± 5.2 MHz	$+1.0$	bottle, ac (3rd order)	3.3.4(b)	112[f]
$5S_{1/2}$–$5P_{1/2}$	906.75 ± 0.02 MHz	902 ± 15 MHz	$+0.3$	fast beam, rf	4.5.2(b)	177
	30.246 ± 0.001 mK	36.6 ± 4.9 mK	-1.3	interferometric ($n = 5 \rightarrow n = 4$)	2.1	20
$5P_{3/2}$–$5S_{1/2}$	10331.15 ± 0.02 MHz	10332.9 ± 1.4 MHz	-1.2	bottle, rf	3.3.3(f)	99
$5D_{3/2}$–$5S_{1/2}$	10312.18 ± 0.02 MHz	10309.8 ± 7.3 MHz	$+0.3$	bottle, ac (2nd order)	3.3.4(b)	114[f]
$5D_{5/2}$–$5S_{1/2}$	14057.85 ± 0.02 MHz	14055.4 ± 6.3 MHz	$+0.4$	bottle, ac (2nd order)	3.3.4(b)	114[f]
$5F_{5/2}$–$5S_{1/2}$	14051.15 ± 0.02 MHz	14058.2 ± 7.8 MHz	-1.0	bottle, ac (3rd order)	3.3.4(b)	114[f]
$5F_{7/2}$–$5S_{1/2}$	15923.94 ± 0.02 MHz	15924.5 ± 5.4 MHz	-0.1	bottle, ac (3rd order)	3.3.4(b)	114[f]
$5G_{7/2}$–$5S_{1/2}$	15920.45 ± 0.02 MHz	15924.4 ± 5.9 MHz	-0.7	bottle, ac (4th order)	3.3.4(b)	79
$5G_{9/2}$–$5S_{1/2}$	17044.11 ± 0.02 MHz	17040.5 ± 8.0 MHz	$+0.4$	bottle, ac (4th order)	3.3.4(b)	79
$6S_{1/2}$–$6P_{1/2}$	525.047 ± 0.014 MHz	530 ± 10 MHz	-0.5	fast beam, rf ($n = 6 \rightarrow n = 4$)	4.5.2(b)	177

continued overleaf

Table 3 (continued)

Atom	Interval	Calculated values[a] ($\pm\sigma$)	Measured value ($\pm1\sigma$)	Difference[b]	Method[c]	Text (section number)	Ref., Notes
			524 ± 13 MHz	$+0.1$	fast beam, rf ($n=6\to n=3$)	4.5.2(b)	177
	17.5137 ± 0.0005 mK		20.7 ± 7.7 mK	-0.4	interferometric ($n=6\to n=4$)	2.1	20
$6P_{3/2}-6S_{1/2}$	5978.342 ± 0.013 MHz		5979.1 ± 1.2 MHz	-0.6	bottle, rf	3.3.3(f)	99
$6D_{5/2}-6S_{1/2}$	8134.96 ± 0.13 MHz[i]		8133.4 ± 0.8 MHz	$+1.9$	bottle, rf (2γ)	3.3.4(b)	117[k]
$6G_{9/2}-6D_{5/2}$	1728.135 ± 0.001 MHz		1728.6 ± 3.4 MHz		fast beam, rf (2γ)	4.5.3	186[l]
$6H_{9/2}-6F_{5/2}$	1730.767 ± 0.001 MHz						
$6G_{9/2}-6F_{7/2}$	648.235 ± 0.003 MHz		648.5 ± 5.0 MHz		fast beam, rf	4.5.2(b)	177
$6H_{9/2}-6G_{7/2}$	649.009 ± 0.003 MHz						
$6H_{11/2}-6D_{5/2}$	2160.385 ± 0.001 MHz		2161.2 ± 6.0 MHz	-0.1	fast beam, rf (3γ)	4.5.3	186[m]
$6H_{11/2}-6G_{9/2}$	432.250 ± 0.002 MHz		437 ± 10 MHz	-0.5	fast beam, rf	4.5.2(b)	177
$7S_{1/2}-7P_{1/2}$	330.761 ± 0.009 MHz		333 ± 8 MHz	-0.3	fast beam, rf	4.5.2(b)	177
$7G_{9/2}-7P_{3/2}$	2446.321 ± 0.003 MHz		2452.8 ± 6.6 MHz		fast beam, rf (3γ)	4.5.3	186[m]
$7H_{9/2}-7D_{3/2}$	2452.501 ± 0.001 MHz						
$9H_{9/2}-9P_{1/2}$	3077.530 ± 0.003 MHz		3097.2 ± 20.0 MHz	-1.0	fast beam, rf (4γ)	4.5.3	186[n]
^{6}Li^{++}	$2S_{1/2}-2P_{1/2}$	62762 ± 9 MHz	62765 ± 21 MHz	-0.1	bottle, rf (pulsed)	3.3.2(c)	78
			62790 ± 70 MHz	-0.4	fast beam, rf	4.5.2(c)	179
			63031 ± 327 MHz	-0.8	fast beam, lifetime	4.1.2(a)	129
C^{5+}	$2S_{1/2}-2P_{1/2}$	783.64 ± 0.23 GHz	780.1 ± 8.0 GHz	$+0.4$	fast beam, lifetime	4.1.2(b)	136
O^{7+}	$2S_{1/2}-2P_{1/2}$	2204.6 ± 1.3 GHz	2202.7 ± 11.0 GHz	$+0.2$	fast beam, lifetime	4.1.2(c)	137
			2215.6 ± 7.5 GHz	-1.4	fast beam, lifetime	4.1.2(c)	138

F^{8+}	$2P_{3/2}-2S_{1/2}$	68833.9 ± 2.4 GHz	68853.9 ± 35 GHz	-0.6	fast beam, laser	4.7	194
Ar^{17+}	$2S_{1/2}-2P_{1/2}$	39.01 ± 0.16 THz	38.3 ± 1.1 THz	$+0.6$	fast beam, lifetime	4.1.2(e)	140[o]

[a] Calculated by G. W. Erickson, private communication, January 1977. The values shown differ slightly from those communicated in September 1976 and published in Ref. 8 as a result of using more recent values for some of the fundamental constants [especially $\alpha^{-1} = 137.035987(29)$ of Ref. 212].

[b] Difference calculated minus measured value in units of $\sigma = \sqrt{(\sigma_{calc}^2 + \sigma_{meas}^2)}$.

[c] rf: radio frequency; ac: anticrossing; 2γ: two-photon (rf) transitions.

[d] See also Ref. 12.

[e] Pasternack (Ref. 33) noted that a $2S$ shift of this magnitude would fit the discrepancies between the Dirac theory and various measurements.

[f] Uncertainty used here is one-half the published limit of error.

[g] See also Ref. 10.

[h] Uncertainty reduced from the published value. A. Eibofner, private communication (communication from Professor G. W. Erickson, January 1977).

[i] All measured line components of the $n = 4 \rightarrow n = 3$ complex were found to agree with theory.

[j] Uncertainty estimated, only statistical uncertainty given (± 0.04 MHz).

[k] Only statistical error given.

[l] ± 2 MHz systematic uncertainty included.

[m] ± 3 MHz systematic uncertainty included.

[n] ± 4 MHz systematic uncertainty included.

[o] Quoted experimental value from private communication H. Gould (communication by G. W. Erickson, January 1977).

[p] QED value minus Dirac value.

[q] corr. used before a reference number means a corrected experimental value is given in that reference.

[r] Reference 8.

Table 4. Lamb Shifts with $j > \frac{1}{2}$ in MHz

Atom	Interval	Theory[a]	Experiment[b]	Composition[c]
H	$3P_{3/2}-3D_{3/2}$	5.3313 ± 0.0006	3.2 ± 1.4	$(P_{3/2}-S_{1/2})^{176}_{exp}-(D_{3/2}-S_{1/2})^{115}_{exp}$
			3.0 ± 2.9	$-(D_{5/2}-P_{3/2})^{176}_{exp}+(D_{5/2}-D_{3/2})_{th}$
			5.3 ± 1.1	$-(D_{5/2}-P_{3/2})^{109}_{exp}+(D_{5/2}-D_{3/2})_{th}$
			5.0 ± 0.8	$-(D_{3/2}-P_{1/2})^{115}_{exp}-(S_{1/2}-P_{1/2})^{172}_{exp}+(P_{3/2}-P_{1/2})_{th}$
			4.7 ± 1.2	$-(D_{3/2}-S_{1/2})^{115}_{exp}-(S_{1/2}-P_{1/2})^{176}_{exp}+(P_{3/2}-P_{1/2})_{th}$
			4.9 ± 1.2	$-(D_{3/2}-S_{1/2})^{115}_{exp}-(S_{1/2}-P_{1/2})^{109}_{exp}+(P_{3/2}-P_{1/2})_{th}$
			3.0 ± 1.4	$-(D_{5/2}-S_{1/2})^{115}_{exp}+(P_{3/2}-S_{1/2})^{176}_{exp}+(D_{5/2}-D_{3/2})_{th}$
	$4P_{3/2}-4D_{3/2}$	2.28 ± 0.02	2.8 ± 2.1	$(P_{3/2}-S_{1/2})^{87}_{exp}-(D_{3/2}-S_{1/2})^{109}_{exp}$
			1.3 ± 1.6	$-(D_{5/2}-P_{3/2})^{176}_{exp}+(D_{5/2}-D_{3/2})_{th}$
			0 ± 1.2	$-(D_{3/2}-P_{1/2})^{176}_{exp}+(P_{3/2}-P_{1/2})_{th}$
	$4D_{5/2}-4F_{5/2}$	0.82 ± 0.02	1.8 ± 0.5	$-(D_{5/2}-S_{1/2})^{109}_{exp}+(P_{3/2}-S_{1/2})^{87}_{exp}+(D_{5/2}-D_{3/2})_{th}$
			0.56 ± 0.41	$-(F_{7/2}-D_{5/2})^{176}_{exp}+(F_{7/2}-F_{5/2})_{th}$
D	$3P_{3/2}-3D_{3/2}$	5.3369 ± 0.0006	5.0 ± 5.0	Not known, Ref. 52[d]
$^4He^+$	$3P_{3/2}-3D_{3/2}$	85.60 ± 0.03	81.9 ± 5.0	$-(D_{5/2}-P_{3/2})^{91,e}_{exp}+(D_{5/2}-D_{3/2})_{th}$
	$4P_{3/2}-4D_{3/2}$	36.5 ± 0.3	35.3 ± 4.0	$(P_{3/2}-S_{1/2})^{99}_{exp}-(D_{3/2}-S_{1/2})^{113,d}_{exp}$
			34.4 ± 4.1	$(P_{3/2}-S_{1/2})^{97}_{exp}-(D_{3/2}-S_{1/2})^{113,d}_{exp}$
			40.9 ± 3.8	$-(D_{5/2}-S_{1/2})^{113,d}_{exp}+(P_{3/2}-S_{1/2})^{99}_{exp}+(D_{5/2}-D_{3/2})_{th}$
			40.0 ± 3.9	$-(D_{5/2}-S_{1/2})^{113,d}_{exp}+(P_{3/2}-S_{1/2})^{97}_{exp}+(D_{5/2}-D_{3/2})_{th}$
			34.8 ± 4.4	$-(D_{3/2}-S_{1/2})^{113,d}_{exp}-(S_{1/2}-P_{1/2})^{98}_{exp}+(P_{3/2}-P_{1/2})_{th}$

$4D_{5/2}-4F_{5/2}$	13.2 ± 0.3	19.7 ± 7.8	$(D_{5/2}-S_{1/2})^{113,d}_{exp} - (F_{5/2}-S_{1/2})^{112,d}_{exp}$ $-(F_{5/2}-D_{3/2})^{100}_{exp} + (D_{5/2}-D_{3/2})_{th}$
		15.5 ± 1.5	
		9.9 ± 5.6	$(D_{5/2}-S_{1/2})^{113,d}_{exp} - (D_{3/2}-S_{1/2})^{113,d}_{exp} - (F_{5/2}-D_{3/2})^{100}_{exp}$
		14.4 ± 6.4	$(D_{5/2}-S_{1/2})^{113,d}_{exp} - (F_{7/2}-S_{1/2})^{112,d}_{exp} + (F_{7/2}-F_{5/2})_{th}$
$5P_{3/2}-5D_{3/2}$	18.8 ± 0.2	23.1 ± 7.4	$(P_{3/2}-S_{1/2})^{99}_{exp} - (D_{3/2}-S_{1/2})^{114,d}_{exp}$
		23.2 ± 6.4	$-(D_{5/2}-S_{1/2})^{114,d}_{exp} + (P_{3/2}-S_{1/2})^{99}_{exp} + (D_{5/2}-D_{3/2})_{th}$
$6P_{3/2}-6D_{3/2}$	11.0 ± 0.1	13.3 ± 1.4^{f}	$(P_{3/2}-S_{1/2})^{99}_{exp} - (D_{5/2}-S_{1/2})^{117}_{exp} + (D_{5/2}-D_{3/2})_{th}$

a G. W. Erickson.[8]

b Only results are shown that carry an uncertainty of less than the full theoretical interval (uncertainties combined in quadrature).

c Combinations of two experimental results are shown first, followed by combinations of one experimental with one theoretical interval [only theoretical intervals of the type $P_{3/2}-P_{1/2}$ ($\Delta l = 0$) are used, which depend little on QED effects]. Combinations of three intervals are also shown as far as they provide better or comparable accuracy. The numbers attached to the intervals refer to the reference list.

d Uncertainty used is one-half the published limit of error.

e See footnote h of Table 3.

f Only the statistical error is entered for the interval $(D_{5/2}-S_{1/2})_{exp}$.

to represent approximately 1 standard deviation (68% confidence level). The differences between theoretical and experimental results are also given in multiples of a "standard deviation" represented by the sum of the uncertainties taken in quadrature. In cases where several related measurements were reported by the same group of authors, only the most recent published result is entered. In general, only quantities are shown that were measured directly and were not obtained by combining several measured or calculated results. The last three columns give an indication of the method used for the measurement and refer to the appropriate section of this article and to the list of references.

5.2. Lamb-Shift Intervals with $j > \frac{1}{2}$

QED shifts of hydrogenic states with $j > \frac{1}{2}$ result mainly from the anomalous magnetic moment of the electron affecting the energies through spin–orbit coupling. The corresponding Lamb-shift intervals like $P_{3/2}-D_{3/2}$ or $D_{5/2}-F_{5/2}$ are very small, smaller in fact than the combined natural widths of the states. They can thus be investigated with rf or similar methods only in a magnetic field of sufficient strength to fully separate the states. However, the fairly small population differences, the fairly large natural widths, and the small electric dipole matrix elements between states with $\Delta j = 0$ all combine to make such measurements not very feasible. To my knowledge, Richardson and Hughes[84] reported the only rf signal of this type between $3^2P_{3/2}$, $m_j = \frac{3}{2}$ (a) and $3^2D_{3/2}$ $m_j = \frac{1}{2}(H)$ in hydrogen centered at 1165 G with a frequency of 2966 MHz. No result was given for the interval, and it should be noted that the substate "H" in a magnetic field is a superposition of $D_{3/2}(H)$ and $D_{5/2}(C)$ contributions. The normal way to "measure" Lamb-shift intervals with $j > \frac{1}{2}$ is therefore by combining values of suitable measured (and often calculated) intervals. A list of such results is given in Table 4. In view of the smallness of the intervals, the accuracies are not overwhelming and only results with errors less than the intervals themselves are shown.

5.3. The Fine-Structure Constant α

Precision fine-structure and Lamb-shift measurements of H and D have been of great importance for the determination of the fine-structure constant $\alpha \simeq 1/137$. α can be derived fairly directly[13] and with little influence from QED effects from experimental values of ΔE which were obtained most accurately for $n = 2$ of H and D (e.g., by combining the measured values of $\mathscr{S}$ and $\Delta E - \mathscr{S}$). The procedure was simplified when

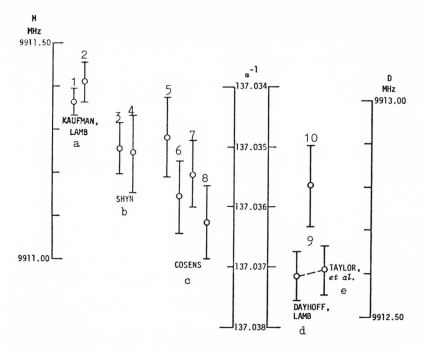

Figure 33. (From Ref. 64.) Measured fine-structure intervals $2^2S_{1/2}-2^2P_{3/2}$ $(\Delta E - \mathscr{S})$ of H and D (all transitions shown separately). Comparing these with the theoretical expression for $\Delta E - \mathscr{S}$,[12,13] the values shown are obtained for the fine-structure constant α.[64] (a) Reference 75 (cf. Section 3.3.2(b)]; (b) Reference 55 [cf. Section 3.2.2(b)]; (c) Reference 61 [cf. Section 3.2.2(b)]; (d) Reference 52 (cf. Section 3.2.1); (e) value of Reference 52 revised in Reference 12. (1) αa; (2) αb; (3) β^+b$^+$; (4) β^+d$^+$; (5) β^-b$^-$; (6) β^+b$^+$; (7) β^-d$^-$; (8) β^+d$^+$; (9) αa; (10) αc.

Cohen and Taylor,[13] in their 1973 adjustment of fundamental constants, decided that the theoretical values of $\mathscr{S}^{(6,8)}$ were well enough confirmed and much more accurate than the experimental values, so that it would be appropriate to extract α directly from the measured values of $\Delta E - \mathscr{S}$ using the theoretical expression for $\Delta E - \mathscr{S}$. This procedure may have to be revised again in the light of the improved recent Lamb-shift measurements on H, $n = 2$[180,181] and the slight differences between the leading two theoretical values.[7,8] However, the changes would not be too drastic since there are no major discrepancies between the theoretical and experimental Lamb-shift results (cf. Figure 29). The experimental results of $\Delta E - \mathscr{S}$ of H and D, $n = 2$ are thus shown in Figure 33 together with the corresponding values of α as compiled by Cohen.[64] All transitions investigated are

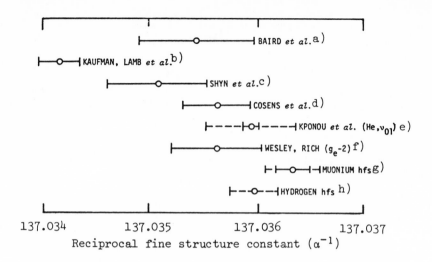

137.034   137.035   137.036   137.037
Reciprocal fine structure constant (α^{-1})

Figure 34. (From Ref. 64.) Measured values of the fine-structure constant α as listed in the 1973 adjustment of fundamental constants (Table 23.1 of Ref. 13). The values a–d are derived from measurements of the hydrogenic fine structure, the values e–h from other experiments. The experimental uncertainty is shown by the solid error bars while the total uncertainty, including the contribution from the shortcomings of the present theoretical calculations, is shown by the broken bars. (a) Reference 124 (cf. Section 3.3.5); (b) Reference 75 [cf. Section 3.3.2(b)]; (c) Reference 55 [cf. Section 3.2.2(b)]; (d) Reference 61 [cf. Section 3.2.2(b)]; (e) Reference 202; (f) References 203, 204; (g) References 205, 206, and see Reference 13; (h) References 207, 208.

shown separately. Figure 34 puts the values of the fine-structure constant α based on these measurements alongside the result based on the direct measurement of ΔE of $H^{(124)}$ (cf. Section 3.3.5) and results obtained in different ways. These values were used to determine the current value of $\alpha = 1/137.03604(11).^{(13)}$

References and Notes

1. A. Sommerfeld, *Ann. Phys. (Leipzig)* **51**, 1 (1916).
2. H. A. Bethe and E. E. Salpeter, *Quantum Mechanics of One- and Two-Electron Atoms*, Springer-Verlag, Berlin (1957).
3. W. E. Lamb, Jr., and R. C. Retherford, *Phys. Rev.* **72**, 241 (1947).
4. G. W. Erickson and D. R. Yennie, *Ann. Phys. (N.Y.)* **35**, 271, 447 (1965).
5. T. Appelquist and S. J. Brodsky, *Phys. Rev. A* **2**, 2293 (1970).
6. G. W. Erickson, *Phys. Rev. Lett.* **27**, 780 (1971).
7. P. J. Mohr, *Phys. Rev. Lett.* **34**, 1050 (1975).
8. G. W. Erickson, *J. Phys. Chem. Ref. Data* **6**, 831 (1977).

9. W. E. Lamb, Jr., *Rep. Prog. Phys.* **14**, 19 (1951).

10. G. W. Series, *The Spectrum of Atomic Hydrogen*, Oxford University Press, London (1957).

11. R. T. Robiscoe, *Cargèse Lectures in Physics*, Vol. 2, Ed. M. Levy, Gordon and Breach, New York, pp. 3–53 (1968).

12. B. N. Taylor, W. H. Parker, and D. N. Langenberg, *Rev. Mod. Phys.* **41**, 375 (1969).

13. E. R. Cohen and B. N. Taylor, *J. Phys. Chem. Ref. Data* **2**, 663 (1973).

14. S. J. Brodsky and R. G. Parsons, *Phys. Rev.* **163**, 134 (1967).

15. S. L. Kaufman, pp. 401–402 of Ref. 15a (1971).

15a. *Precision Measurement and Fundamental Constants*, Eds. D. N. Langenberg and B. N. Taylor, National Bureau of Standards (U.S.), special publication No. 343 (1971).

16. H. P. Larson and A. W. Stanley, *J. Opt. Soc. Am.* **57**, 1439 (1967).

17. G. Herzberg, *Z. Phys.* **146**, 269 (1956).

18. H. G. Berry and F. L. Roesler, *Phys. Rev. A* **1**, 1504 (1970).

19. H. G. Berry, *J. Opt. Soc. Am* **61**, 123 (1971).

20. E. G. Kessler, Jr., and F. L. Roesler, *J. Opt. Soc. Am.* **62**, 440 (1972).

21. H. Kuhn and G. W. Series, *Proc. R. Soc. London* A **202**, 127 (1950).

22. G. W. Series, *Proc. R. Soc. London* A **208**, 277 (1951).

23. E. G. Kessler, Jr., *Phys. Rev. A* **7**, 408 (1973).

24. B. P. Kibble, W. R. C. Rowley, R. E. Shawyer, and G. W. Series, *J. Phys. B* **6**, 1079 (1973).

25. T. W. Hänsch, I. S. Shahin, and A. L. Schawlow, *Nature Phys. Sci.* **235**, 63 (1972).

26. T. W. Hänsch, M. H. Nayfeh, S. A. Lee, S. M. Curry, and I. S. Shahin, *Phys. Rev. Lett.* **32**, 1336 (1974).

27. L. J. Boya, An. Fis. (*Spain*) **66**, 421 (1970).

28. G. Herzberg, *Proc. R. Soc. London A* **234**, 516 (1956).

29. T. W. Hänsch, S. A. Lee, R. Wallenstein, and C. Wieman, *Phys. Rev. Lett.* **34**, 307 (1975).

30. C. Wieman and T. W. Hänsch, Book of Abstracts, 5th International Conference on Atomic Physics, Berkeley, 1976, pp. 194–195 (1976).*

31. B. E. Lautrup, A. Petermann, and E. de Rafael, *Phys. Rep.* **3**, 193 (1972).

32. K. R. Lea, in *Atomic Masses and Fundamental Constants* 4, Eds. J. H. Sanders and A. H. Wapstra, pp, 355–372, Plenum Press, London (1972).

33. S. Pasternack, *Phys. Rev.* **54**, 1113 (1938).

34. O. Betz, *Ann. Phys.* (*Leipzig*) **15**, 321 (1932).

35. E. Schneider, Thesis, Freiburg (1933).

36. H. Klumb, *Phys. Z.* **33**, 445 (1932).

37. T. Haase, *Ann. Phys.* (*Leipzig*) **23**, 657 (1935).

38. A more recent estimate is given in Appendix I of Ref. 47.

39. R. Reinecke, Thesis, Berlin (1932).

40. W. E. Lamb, Jr., and T. M. Sanders, Jr., *Phys. Rev.* **119**, 1901 (1960).

41. H. Wieder and T. G. Eck, *Phys. Rev.* **153**, 103 (1967).

42. G. W. Series, *Phys. Rev.* **136**, A684 (1964).

43. Appendix II of Ref. 47.

44. Section 67γ of Ref. 2.

45. Appendix IV of Ref. 49.

46. R. T. Robiscoe, *Phys. Rev.* **168**, 4 (1968).

47. W. E. Lamb, Jr., and R. C. Retherford, *Phys. Rev.* **79**, 549 (1950).

* The proceedings of the International Conferences on Atomic Physics cited here and below have been published under the title *Atomic Physics 1, 2, . . .* (Plenum Publishing, New York, 1969–1977).

48. W. E. Lamb, Jr., and R. C. Retherford, *Phys. Rev.* **81**, 222 (1951).
49. W. E. Lamb, Jr., *Phys. Rev.* **85**, 259 (1952).
50. W. E. Lamb, Jr., and R. C. Retherford, *Phys. Rev.* **86**, 1014 (1952).
51. S. Triebwasser, E. S. Dayhoff, and W. E. Lamb, Jr., *Phys. Rev.* **89**, 98 (1953).
52. E. S. Dayhoff, S. Triebwasser, and W. E. Lamb, Jr., *Phys. Rev.* **89**, 106 (1953).
53. R. T. Robiscoe, *Phys. Rev.* **138**, A22 (1965).
54. R. T. Robiscoe and B. L. Cosens, *Phys. Rev. Lett.* **17**, 69 (1966).
55. T. W. Shyn, T. Rebane, R. T. Robiscoe, and W. L. Williams, *Phys. Rev. A* **3**, 116 (1971).
56. R. T. Robiscoe and T. W. Shyn, *Phys. Rev. Lett.* **24**, 559 (1970).
57. R. T. Robiscoe, pp. 373–376 of Ref. 15a (1971).
58. B. L. Cosens, *Phys. Rev.* **173**, 49 (1968).
59. T. V. Vorburger and B. L. Cosens, pp. 361–365 of Ref. 15a (1971).
60. B. L. Cosens and T. V. Vorburger, *Phys. Rev. Lett.* **23**, 1273 (1969).
61. B. L. Cosens and T. V. Vorburger, *Phys. Rev. A* **2**, 16 (1970).
62. T. W. Shyn, W. L. Williams, R. T. Robiscoe, and T. Rebane, *Phys. Rev. Lett.* **22**, 1273 (1969).
63. T. W. Shyn, R. T. Robiscoe, and W. L. Williams, pp. 355–360 of Ref. 15a (1971).
64. E. R. Cohen, CODATA *Newsl.* **13**, 10 (Sept. 1974).
65. E. S. Gillespie, *J. Phys. B* **5**, 1916 (1972), and references therein.
66. R. J. Anderson, E. T. P. Lee, and C. C. Lin, *Phys. Rev.* **160**, 20 (1967).
67. M. Skinner and W. E. Lamb, Jr., *Phys. Rev.* **75**, 1325A (1949).
68. W. E. Lamb, Jr., and M. Skinner, *Phys. Rev.* **78**, 539 (1950).
69. R. Novick, E. Lipworth, and P. F. Yergin, *Phys. Rev.* **100**, 1153 (1955).
70. E. Lipworth and R. Novick, *Phys. Rev.* **108**, 1434 (1957).
71. M. A. Narasimham and R. L. Strombotne, pp. 393–402 of Ref. 15a (1971).
72. M. A. Narasimham and R. L. Strombotne, *Phys. Rev. A* **4**, 14 (1971).
73. S. L. Kaufman, W. E. Lamb, Jr., K. R. Lea, and M. Leventhal, *Phys. Rev. Lett.* **22**. 507 (1969).
74. S. L. Kaufman, W. E. Lamb, Jr., K. R. Lea, and M. Leventhal, pp. 367–371 of Ref. 15a (1971).
75. S. L. Kaufman, W. E. Lamb, Jr., K. R. Lea, and M. Leventhal, *Phys. Rev. A* **4**, 2128 (1971).
76. M. Leventhal, R. T. Robiscoe, and K. R. Lea, *Phys. Rev.* **158**, 49 (1967).
77. M. Leventhal and P. E. Havey, *Phys. Rev. Lett.* **32**, 808 (1974).
78. M. Leventhal, *Phys. Rev. A* **11**, 427 (1975).
79. H.-J. Beyer, H. Kleinpoppen, and J. M. Woolsey, *Phys. Rev. Lett.* **28**, 263 (1972).
80. W. E. Lamb, Jr., and T. M. Sanders, Jr., *Phys. Rev.* **103**, 313 (1956).
81. L. R. Wilcox and W. E. Lamb, Jr., *Phys. Rev.* **119**, 1915 (1960).
82. H. Kleinpoppen, *Z. Phys.* **164**, 174 (1961).
83. M. Glass-Maujean and J. P. Descoubes, *C.R. Acad. Sci.* **273**B, 721 (1971).
84. C. B. Richardson and R. H. Hughes, *Bull. Am. Phys. Soc. II* **15**, 45 (1970).
85. C. B. Richardson, D. A. Mitchell, and R. H. Hughes, *Bull. Am. Phys. Soc.* **15**, 1509 (1970).
86. R. A. Brown and F. M. Pipkin, pp. 383–388 of Ref. 15a (1971).
87. R. A. Brown and F. M. Pipkin, *Ann. Phys. (N.Y.)* **80**, 479 (1973).
88. M. Leventhal, K. R. Lea, and W. E. Lamb, Jr., *Phys. Rev. Lett.* **15**, 1013 (1965).
89. D. L. Mader, M. Leventhal, and W. E. Lamb, Jr., *Phys. Rev. A* **3**, 1832 (1971).
90. M. Baumann and A. Eibofner, *Phys. Lett.* **34A**, 421 (1971).
91. A. Eibofner, *Z. Phys.* **249**, 58 (1971).
92. M. Baumann and A. Eibofner, *Phys. Lett.* **33A**, 409 (1970).
93. G. W. Series and W. N. Fox, *J. Phys. Rad.* **19**, 850 (1958).

94. K. R. Lea, M. Leventhal, and W. E. Lamb, Jr., *Phys. Rev. Lett.* **16**, 163 (1966).
95. L. L. Hatfield and R. H. Hughes, *Phys. Rev.* **156**, 102 (1967).
96. H.-J. Beyer and H. Kleinpoppen, *Z. Phys.* **206**, 177 (1967).
97. R. R. Jacobs, K. R. Lea, and W. E. Lamb, Jr., *Phys. Rev. A* **3**, 884 (1971).
98. A. Eibofner, *Phys. Lett.* **58A**, 219 (1976).
99. A. Eibofner, *Z. Phys.* **A277**, 225 (1976).
100. A. Eibofner, *Phys. Lett.* **49A**, 335 (1974).
101. T. Hadeishi, *Phys. Rev. Lett.* **21**, 957 (1968).
102. T. Hadeishi, *Phys. Rev. Lett.* **22**, 815 (1969).
103. G. W. Series, *Phys. Rev.* **178**, 429 (1969).
104. T. G. Eck and R. J. Huff, *Phys. Rev. Lett.* **22**, 319 (1969).
105. M. Baumann and A. Eibofner, *Phys. Lett.* **43A**, 105 (1973).
106. A. Eibofner, *Phys. Lett.* **47A**, 399 (1974).
107. H.-J. Beyer, Thesis, Stirling University (1973).
108. M. Glass-Maujean and J. P. Descoubes, *Opt. Comm.* **4**, 345 (1972).
109. M. Glass-Maujean, Thesis, Université de Paris (1974).
110. T. G. Eck and R. J. Huff, in *Beam-Foil Spectroscopy*, Ed. S. Bashkin, pp. 193–202, Gordon and Breach, New York (1968).
111. H.-J. Beyer, H. Kleinpoppen, and J. M. Woolsey, *J. Phys. B* **6**, 1849 (1973).
112. H.-J. Beyer and H. Kleinpoppen, *J. Phys. B* **4**, L129 (1971).
113. H.-J. Beyer and H. Kleinpoppen, *J. Phys. B* **5**, L12 (1972).
114. H.-J. Beyer and H. Kleinpoppen, *J. Phys. B* **8**, 2449 (1975).
115. M. Glass-Maujean, *Opt. Comm.* **8**, 260 (1973).
116. M. Glass-Maujean, T. Dohnalik, and J. P. Descoubes, Book of Abstracts, 3rd International Conference on Atomic Physics, Boulder, 1972, pp. 93–95 (1972).
117. A. Eibofner, private communication (1976) and *Phys. Lett.* **61A**, 159 (1977).
118. F. D. Colegrove, P. A. Franken, R. R. Lewis, and R. H. Sands, *Phys. Rev. Lett.* **3**, 420 (1959).
119. P. A. Franken, *Phys. Rev.* **121**, 508 (1961).
120. M. E. Rose and R. L. Carovillano, *Phys. Rev.* **122**, 1185 (1961).
121. L. C. Himmell and P. R. Fontana, *Phys. Rev.* **162**, 23 (1967).
122. H. Metcalf, J. R. Brandenberger, and J. C. Baird, *Phys. Rev. Lett.*, **21**, 165 (1968).
123. J. C. Baird, J. Brandenberger, K.-I. Gondaira, and H. Metcalf, pp. 345–353 of Ref. 15a (1971).
124. J. C. Baird, J. Brandenberger, K.-I. Gondaira, and H. Metcalf, *Phys. Rev. A* **5**, 564 (1972).
125. W. H. Wing, Thesis, University of Michigan (1968), quoted in Ref. 12.
126. K.-J. Kollath and H. Kleinpoppen, *Phys. Rev. A* **10**, 1519 (1974).
127. G. Lüders, *Z. Naturforsch.* **5a**, 608 (1950).
128. I. A. Sellin, *Phys. Rev.* **136**, A1245 (1964).
129. C. Y. Fan, M. Garcia-Munoz, and I. A. Sellin, *Phys. Rev.* **161**, 6 (1967).
130. H. K. Holt and I. A. Sellin, *Phys. Rev. A* **6**, 508 (1972).
131. G. W. F. Drake and R. B. Grimley, *Phys. Rev. A* **8**, 157 (1973).
132. C. Y. Fan, M. Garcia-Munoz, and I. A. Sellin, *Phys. Rev. Lett.* **15**, 15 (1965).
133. M. Leventhal, *Nucl. Instrum. Methods* **110**, 343 (1973).
134. M. Leventhal and D. E. Murnick, *Phys. Rev. Lett.* **25**, 1237 (1970).
135. D. E. Murnick, M. Leventhal, and H. W. Kugel, *Phys. Rev. Lett.* **27**, 1625 (1971).
136. H. W. Kugel, M. Leventhal, and D. E. Murnick, *Phys. Rev. A* **6**, 1306 (1972).
137. M. Leventhal, D. E. Murnick, and H. W. Kugel, *Phys. Rev. Lett.* **28**, 1609 (1972).
138. G. P. Lawrence, C. Y. Fan, and S. Bashkin, *Phys. Rev. Lett.* **28**, 1612 (1972).
139. D. E. Murnick, M. Leventhal, and H. W. Kugel, Abstracts, 3rd International Conference on Atomic Physics, Boulder, 1972, pp. 182–184 (1972).

140. H. A. Gould and R. Marrus, Book of Abstracts, 5th International Conference on Atomic Physics, Berkeley 1976, p. 207 (1976).
141. W. L. Fite, W. E. Kauppila, and W. R. Ott, *Phys. Rev. Lett.* **20**, 409 (1968).
142. W. R. Ott, W. E. Kauppila, and W. L. Fite, *Phys. Rev. A* **1**, 1089 (1970).
143. G. W. F. Drake and R. B. Grimley, *Phys. Rev. A* **11**, 1614 (1975).
144. A. van Wijngaarden, G. W. F. Drake, and P. S. Farago, *Phys. Rev. Lett.* **33**, 4 (1974).
145. G. W. F. Drake, P. S. Farago, and A. van Wijngaarden, *Phys. Rev. A* **11**, 1621 (1975).
146. G. W. F. Drake, A. van Wijngaarden, and P. S. Farago, in *Electron and Photon Interactions with Atoms*, Eds. M. R. C. McDowell and H. Kleinpoppen, pp. 339–347, Plenum Press, New York (1976).
147. E. Bentz, T. Liedtke, and E. Salzborn, *Verhandl DPG(VI)* **11**, 89 (1976).
148. G. W. F. Drake and Chien-ping Lin, *Phys. Rev. A* **14**, 1296 (1976).
149. H. J. Andrä, *Phys. Scr.* **9**, 257 (1974).
150. S. Bashkin, W. S. Bickel, D. Fink, and R. K. Wangsness, *Phys. Rev. Lett.* **15**, 284 (1965).
151. W. S. Bickel and S. Bashkin, *Phys. Rev.* **162**, 12 (1967).
152. W. S. Bickel, *J. Opt. Soc. Am.* **58**, 213 (1968).
153. I. A. Sellin, C. D. Moak, P. M. Griffin, and J. A. Biggerstaff, *Phys. Rev.* **184**, 56 (1969).
154. I. A. Sellin, C. D. Moak, P. M. Griffin, and J. A. Biggerstaff, *Phys. Rev.* **188**, 217 (1969).
155. H. J. Andrä, *Phys. Rev. Lett.* **25**, 325 (1970).
156. I. A. Sellin, J. R. Mowat, R. S. Peterson, P. M. Griffin, P. Laubert, and H. H. Haselton, *Phys. Rev. Lett.* **31**, 1335 (1973).
157. A. Gaupp, H. J. Andrä, and J. Macek, *Phys. Rev. Lett.* **32**, 268 (1974).
158. H. J. Andrä, *Phys. Rev. A* **2**, 2200 (1970).
159. E. H. Pinnington, H. G. Berry, J. Desesquelles, and J. L. Subtil, *Nucl. Instrum. Methods* **110**, 315 (1973).
160. A. van Wijngaarden, E. Goh, G. W. F. Drake, and P. S. Farago, *J. Phys. B* **9**, 2017 (1976).
161. J. Bourgey, A. Denis, and J. Desesquelles, *J. Phys. (Paris)* **38**, 1229 (1977).
162. M. J. Alguard and C. W. Drake, *Nucl. Instrum. Methods* **110**, 311 (1973).
163. M. J. Alguard and C. W. Drake, *Phys. Rev. A* **8**, 27 (1973).
164. H. J. Andrä, P. Dobberstein, A. Gaupp, and W. Wittman, *Nucl. Instrum. Methods* **110**, 301 (1973).
165. Yu. L. Sokolov, *Sov. Phys. JETP Lett.* **11**, 359 (1970).
166. Yu. L. Sokolov, *Sov. Phys. JETP* **36**, 243 (1973).
167. T. Hadeishi, W. S. Bickel, J. D. Garcia, and H. G. Berry, *Phys. Rev. Lett.* **23**, 65 (1969).
168. T. Hadeishi, *Nucl. Instrum. Methods* **90**, 337 (1970).
169. H. J. Andrä, *Phys. Lett.* **32A**, 345 (1970).
170. G. Newton, P. J. Unsworth, and D. A. Andrews, *J. Phys. B* **8**, 2928 (1975).
171. C. W. Fabjan and F. M. Pipkin, pp. 377–381 of Ref. 15a (1971).
172. C. W. Fabjan and F. M. Pipkin, *Phys. Rev. A* **6**, 556 (1972).
173. M. P. Silverman and F. M. Pipkin, *J. Phys. B* **5**, 1844 (1972).
174. M. P. Silverman and F. M. Pipkin, *J. Phys. B* **7**, 704, 730 (1974).
175. C. W. Fabjan and F. M. Pipkin, *Phys. Rev. Lett.* **25**, 421 (1970).
176. C. W. Fabjan, F. M. Pipkin, and M. Silverman, *Phys. Rev. Lett.* **26**, 347 (1971).
177. S. Churassy, M. L. Gaillard, and J. D. Silver, *Phys. Rev. Lett.* **33**, 185 (1974).
178. D. D. Dietrich, P. Lebow, B. Da Costa, R. de Zafra, and H. Metcalf, *Bull. Am. Phys. Soc.* **19**, 572 (1974).
179. D. Dietrich, P. Lebow, R. de Zafra, and H. Metcalf, *Bull. Am. Phys. Soc.* **21**, 625 (1976).
180. D. A. Andrews and G. Newton, *Phys. Rev. Lett.* **37**, 1254 (1976).
181. S. R. Lundeen and F. M. Pipkin, *Phys. Rev. Lett.* **34**, 1368 (1975).
182. D. A. Andrews and G. Newton, *J. Phys. B* **8**, 1415 (1975).

183. D. A. Andrews and G. Newton, *J. Phys. B* **9**, 1453 (1976).

184. P. B. Kramer, S. R. Lundeen, B. O. Clark, and F. M. Pipkin, *Phys. Rev. Lett.* **32**, 635 (1974).

185. D. A. Andrews and G. Newton, *Phys. Lett.* **57A**, 417 (1976).

186. J. R. Brandenberger, S. R. Lundeen, and F. M. Pipkin, *Phys. Rev. A* **14**, 341 (1976).

187. C. W. Fabjan and F. M. Pipkin, *Phys. Lett.* **36A**, 69 (1971).

188. N. F. Ramsey, *Phys. Rev.* **78**, 695 (1950); and in *Molecular Beams*, Clarendon Press, Oxford (1956).

189. S. R. Lundeen, Y. L. Yung, and F. M. Pipkin, *Nucl. Instrum. Methods*, **110**, 355 (1973).

190. F. M. Pipkin, *Comm. Atom. Mol. Phys.* **5**, 45 (1975).

191. F. Borkowski, G. G. Simon, V. H. Walther, and R. D. Wendling, *Z. Phys.* **A275**, 29 (1975).

192. B. O. Clark, F. M. Pipkin, and D. A. van Baak, Book of Abstracts, 5th International Conference on Atomic Physics, Berkeley, 1976, pp. 196–197.

193. A. Bertin *et al.*, *Phys. Lett.* **55B**, 411 (1975).

194. H. W. Kugel, M. Leventhal, D. E. Murnick, C. K. N. Patel, and O. R. Wood II, *Phys. Rev. Lett.* **35**, 647 (1975).

195. H. W. Kugel, M. Leventhal, D. E. Murnick, C. K. N. Patel, and O. R. Wood II, Book of Abstracts, 5th International Conference on Atomic Physics, Berkeley, 1976, pp. 208–209 (1976).

196. G. W. F. Drake, *J. Phys. B* **10**, 775 (1977).

197. M. Leventhal, *Phys. Lett.* **20**, 625 (1966).

198. G. W. Erickson, private communication. Computer printout dated 8 February (1974); see also Ref. 8.

199. G. W. Series, *Contemp. Phys.* **14**, 49 (1974).

200. P. J. Mohr, Beam Foil Conference, Sept. 15–19, 1975 (preprint No. LBL-3876) (communication from Professor G. W. Erickson, January 1977); similar value in Ref. 7.

201. P. J. Mohr, 5th International Conference on Atomic Physics, Berkeley, 1976 (communication from Professor G. W. Erickson, January 1977).

202. A. Kponou, V. W. Hughes, C. E. Johnson, S. A. Lewis, and F. M. J. Pichanick, *Phys. Rev. Lett.* **26**, 1613 (1971).

203. J. C. Wesley and A. Rich, *Phys. Rev. A* **4**, 1341 (1971); and *Rev. Mod. Phys.* **44**, 250 (1972).

204. S. Granger and G. W. Ford, *Phys. Rev. Lett.* **28**, 1479 (1972).

205. P. A. Thompson, P. Crane, T. Crane, J. J. Amato, V. W. Hughes, G. zu Putlitz, and J. E. Rothbert, *Phys. Rev. A* **8**, 86 (1973).

206. D. Favart, P. M. McIntyre, D. Y. Stowell, V. L. Telegdi, and R. DeVoe, *Phys. Rev. Lett.* **27**, 1336 (1971).

207. H. Hellwig, R. F. C. Vessot, M. W. Levine, P. W. Zitzewitz, D. W. Allan, and D. J. Glaze, *IEEE Trans. Instrum. Meas.* **IM-19**, 200 (1970).

208. L. Essen, R. W. Donaldson, M. J. Bangham, and E. G. Hope, *Nature* **229**, 110 (1971); M. J. Bangham and R. W. Donaldson, National Physical Laboratory report No. Qu 17 (March, 1971).

209. P. S. Kireyev, *Sov. Phys. Dokl.* **2**, 8 (1957).

210. H. J. Andrä, *Nucl. Instrum. Methods* **90**, 343 (1970).

211. G. W. Erickson, Table D of Ref. 8 (preprint, October, 1976) and private communication, 6 January (1977). See also footnote *a* of Table 3.

212. P. T. Olsen and E. R. Williams, Fifth International Conference on Atomic Masses and Fundamental Constants, Paris (1975).

213. D. A. Andrews, R. Golub, and G. Newton, *J. Phys. G* **3**, L91 (1977), and G. Newton, private communication (1977).

13
Anticrossing Spectroscopy

H.-J. Beyer and H. Kleinpoppen

1. Introduction

About thirty years ago, new ways of investigating the level structure of excited states were opened by the radio-frequency experiment of Lamb and Retherford[1] and by the optical double resonance (ODR) technique[2] suggested by Brossel and Kastler.[3] Together with the level-crossing[4,5,7] and the anticrossing techniques,[6,9] these methods enabled spectroscopists to study the structure of excited states with unprecedented accuracy. Further, they allowed the extraction of information about the strength of internal couplings in atomic states, lifetimes, the influence of external perturbations, and also provided much insight into the interaction between atoms and photons. Of course, all these methods are interrelated, and Figure 1 summarizes some of their characteristics. The ODR and rf techniques make use of changes in the population differences between the two states when a resonant radio-frequency field is applied. Changes in the population are detected as intensity changes in the resonance fluorescence radiation. In a similar way, the populations of the two states tend to equalize in anticrossing experiments when the two states approach each other. In level-crossing experiments, on the other hand, interference takes place between the amplitudes describing the rates for the absorption and the emission of the resonance line at the intersection of the Zeeman levels.

As a method of investigating energy differences in excited states, the level-crossing technique was first employed by Colegrove et al.,[4] yet this kind of interference effect was actually observed much earlier as depolarization of resonance fluorescence radiation in a magnetic field

H.-J. Beyer and H. Kleinpoppen • Institute of Atomic Physics, University of Stirling, Stirling, Scotland.

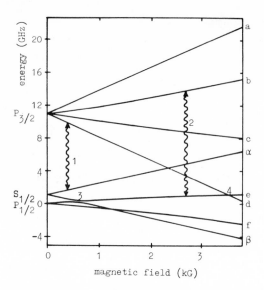

Figure 1. Spectroscopic techniques illustrated on hydrogen, $n = 2$ by radio-frequency transitions ("1" electric dipole transition induced by an electric radio-frequency field, "2" magnetic dipole transition induced by a magnetic radio-frequency field) and by crossings ("3" anticrossing, "4" level crossing). The resonance signal can be detected in all cases in the Ly_α radiation; the anticrossing shown is induced by a static electric field ($E \perp H$).

("Hanle effect")[49] which is now known as zero-field level crossing. The finite-field level crossings[4] represent not just another method but provide, in many cases, the same information with accuracy comparable to that obtainable from radio-frequency methods while requiring less sophisticated techniques. The same holds for the anticrossing method, which, in addition, extends the range of applicability of crossing techniques to more states. The first anticrossing signal was in fact observed accidentally by Eck et al.[6] during a level-crossing investigation of the fine and hyperfine structure of the 2^2P term of lithium, when, in spite of a "misoriented" polarizer, a crossing signal was observed that did not result from the Hanle-type interference effect. The levels involved were prevented from crossing by an internal perturbation (magnetic dipole interaction) which couples the two magnetic substates. The substates repel each other and their wave functions interchange their identities as the magnetic field is varied through the region of closest approach. At the field strength of closest approach, the "crossing position" (at which the "level crossing" would occur in the absence of the perturbing interaction), the wave function of each state is a mixture of the wave functions of the uncoupled states. Even though the crossing of the substates does not actually occur, intensity changes in the line emitted from the corresponding substates are observable as a result of the state mixing at and near the crossing position. Thus, the anticrossing technique has some similarity with the optical double resonance techniques and the Lamb–Retherford experiment in which population differences between the two substates are reduced by a radio-frequency field (Series[8]). Some characteristics of anticrossing phenomena were already revealed in the experiments of Lamb and Retherford,[1] where hydrogen $2S_{1/2}$ substates were heavily quenched by motional electric fields in the vicinity of S–P crossings.

2. Theory of Anticrossing Signals

A general theory has been given by Wieder and Eck[9] describing anticrossing and level-crossing signals in a combined equation. In their paper, they discussed the problem under the assumption of a resonance fluorescence process; the results of their calculation, however, are also valid for electron impact excitation.

The problem is reduced to a calculation of the steady-state resonance fluorescence signal arising from the decay of two coupled states $|a\rangle$ and $|b\rangle$ which are excited at uniform excitation rates. A time-dependent perturbation theory is required for this problem, which leads to the following time-dependent Schrödinger equation:

$$i\hbar \frac{\partial |t\rangle}{\partial t} = (H_0 + H_D + H')\,|t\rangle \tag{1}$$

H_0 is the Hamiltonian of the unperturbed atom with eigenstates $|a\rangle$, $|b\rangle$ and the eigenvalues $E_a = \hbar\omega_a$ and $E_b = \hbar\omega_b$, respectively. The radiative decay of the two eigenstates is introduced by a damping Hamiltonian H_D with the elements $-\frac{1}{2}i\hbar\Gamma_a$ for $|a\rangle$ and $-\frac{1}{2}i\hbar\Gamma_b$ for $|b\rangle$ (with $\Gamma_a = 1/\tau_a$ and $\Gamma_b = 1/\tau_b$, where τ_a and τ_b are the lifetimes of the states). The states are coupled by the time-independent perturbation H'. Following the discussion of Wieder and Eck,[9] the wave function of the above Schrödinger equation may then be solved by the following Ansatz satisfying the required conditions for $t = 0$:

$$|t\rangle = a(t)\,e^{-i\omega_a t}|a\rangle + b(t)\,e^{-i\omega_b t}|b\rangle \tag{2}$$

with $a(t)$ and $b(t)$ as time-dependent amplitudes. As we shall discuss in the following sections, the perturbation operator H' may represent an external perturbation (e.g., induced by an electrostatic field), an internal perturbation (e.g., induced by spin–orbit interaction), or even a mixture of both types of perturbations. For the moment we do not specify the perturbation, which is a general potential of the type $V = \langle a|H'|b\rangle/h$ coupling the two states. Setting $t = 0$ in Eq. (2) and taking $a(0) = f_{am}$ and $b(0) = f_{bm}$ to define the dipole matrix elements for the excitation of the two excited states from the ground state $|m\rangle$ with photons of polarization $\mathbf{f}$, we have

$$\begin{aligned} a(0) &= \langle a|\mathbf{f}\cdot\mathbf{r}|m\rangle = f_{am} \\ b(0) &= \langle b|\mathbf{f}\cdot\mathbf{r}|m\rangle = f_{bm} \end{aligned} \tag{3}$$

The intensity of the detected resonance fluorescence light is proportional to the square of the electric dipole matrix element $\langle m'|\mathbf{g}\cdot\mathbf{r}|t\rangle$ for the spontaneous decay to the final state $|m'\rangle$ via photons of polarization $\mathbf{g}$. Taking into account Eq. (3) we obtain a quantity

$$|a(t)g_{m'a}e^{-i\omega_a t}+b(t)g_{m'b}e^{-i\omega_b t}|^2 \tag{4}$$

which is proportional to the detected signal and which assumes that the atom is excited at $t=0$. The steady-state signal of the resonance fluorescence intensity from a collection of atoms can be found by integrating the above expression from $-\infty$ to t. This yields[9]

$$S = (1/\gamma_a)\sum\underbrace{[|f_a|^2|g_a|^2]}_{1}+(1/\gamma_b)\sum\underbrace{[|f_b|^2|g_b|^2]}_{2}$$

$$+(\gamma_a\gamma_b/\bar{\gamma}D)\sum\underbrace{[f_af_b^*g_ag_b^*+f_a^*f_bg_a^*g_b]}_{3}$$

$$-(i\gamma_a\gamma_b\Delta\nu/\bar{\gamma}^2D)\sum\underbrace{[f_af_b^*g_ag_b^*-f_a^*f_bg_a^*g_b]}_{4}$$

$$-(2|V|^2\gamma_a\gamma_b/\bar{\gamma}D)\sum\underbrace{[fg]}_{5}$$

$$+(2/\bar{\gamma}D)\sum\underbrace{[(V^*f_af_b^*+Vf_a^*f_b)(Vg_ag_b^*+V^*g_a^*g_b)]}_{6}$$

$$+(\Delta\nu\gamma_a\gamma_b/\bar{\gamma}^2D)\sum\underbrace{[f(Vg_ag_b^*+V^*g_a^*g_b)+g(V^*f_af_b^*+Vf_a^*f_b)]}_{7}$$

$$+(i\gamma_a\gamma_b/\bar{\gamma}D)\sum\underbrace{[f(Vg_ag_b^*-V^*g_a^*g_b)+g(V^*f_af_b^*-Vf_a^*f_b)]}_{8} \tag{5}$$

with

$$D = \gamma_a\gamma_b+|2V|^2+(\gamma_a\gamma_b/\bar{\gamma}^2)\Delta\nu^2, \qquad \omega=\omega_a-\omega_b=2\pi\,\Delta\nu$$

$$\Gamma_a=2\pi\gamma_a,\;\Gamma_b=2\pi\gamma_b, \qquad \bar{\gamma}=\tfrac{1}{2}(\gamma_a+\gamma_b)$$

$$f=(|f_a|^2/\gamma_a)-(|f_b|^2/\gamma_b), \qquad g=(|g_a|^2/\gamma_a)-(|g_b|^2/\gamma_b)$$

The symbol for summation means that the quantities within the bracket are to be summed over all the relevant initial and final levels m and m', respectively. All other factors like the "instrumental" parameters for intensity and spectral distribution of the resonance lamp, acceptance aperture of the light detector, etc. are included in the matrix elements.

Obviously, the first two terms of Eq. (5) represent the nonresonant background intensity of light scattering from the states $|a\rangle$ and $|b\rangle$. For $V=0$, the third and fourth terms describe the normal level-crossing signals. In order to be observable, a level-crossing signal requires "coherence" in both the excitation and detection parts of the experiment (interference terms for f_a, f_b, g_a and g_b). In the presence of the perturbation ($V\neq0$) in the resonance denominator D, the crossing signal is modified. Term 5 represents the "pure" anticrossing signal, which does not require coherence in either the excitation or detection parts of the experiment. In contrast to the level-crossing signal the anticrossing signal vanishes for $V = 0$ and also if either f or g equals zero (for $\Sigma_m |f_a|^2/\gamma_a = \Sigma_m |f_b|^2/\gamma_b$ the states $|a\rangle$ and $|b\rangle$ have equal steady-state populations and for $\Sigma_{m'} |g_a|^2/\gamma_a = \Sigma_{m'} |g_b|^2/\gamma_b$ equal emission rates). The terms 6–8 represent mixed level-crossing and anticrossing signals, requiring the perturbation $V \neq 0$ and simultaneously coherent excitation or deexcitation. Term 6 requires coherence in both excitation and deexcitation and terms 7 and 8 require coherence at least in one step. No level crossing or anticrossing signals occur when γ_a or $\gamma_b = 0$. If coherence is absent in both the excitation and detection step, only the background (terms 1 and 2) and the pure anticrossing signal (term 5) are left.

As a function of the separation $\Delta\nu$ of the undisturbed sublevels $|a\rangle$ and $|b\rangle$ the most common pure anticrossing signal [term 5 of Eq. (5)] is represented by an absorption Lorentzian curve centered at $\Delta\nu = 0$:

$$S(\Delta\nu) = d\frac{A}{1 + \Delta\nu^2/B^2} \tag{6}$$

where the amplitude A and the full width at half-maximum $2B$ are given by

$$A = -\frac{(\Sigma_{m'} |g_a|^2/\gamma_a - \Sigma_{m'} |g_b|^2/\gamma_b)(\Sigma_m |f_a|^2/\gamma_a - \Sigma_m |f_b|^2/\gamma_b)|2V_{ab}|^2}{(\gamma_a + \gamma_b)(1 + |2V_{ab}|^2/\gamma_a\gamma_b)} \tag{7}$$

$$2B = (\gamma_a + \gamma_b)(1 + |2V_{ab}|^2/\gamma_a\gamma_b)^{1/2} \tag{8}$$

d takes account of the instrumental parameters already mentioned; $(\Sigma_{m'} |g_a|^2/\gamma_a - \Sigma_{m'} |g_b|^2/\gamma_b)$ is the difference of the branching ratios (including polarization) describing the degree of discrimination between $|a\rangle$ and $|b\rangle$ in the decay to the selected states m'; $(\Sigma_m |f_a|^2/\gamma_a - \Sigma_m |f_b|^2/\gamma_b)$ is the steady-state population difference of levels $|a\rangle$ and $|b\rangle$. Particle (e.g., electron) excitation is used in most anticrossing experiments, and in this case, $\Sigma|f_a|^2$ and $\Sigma|f_b|^2$ may be replaced by the excitation rates r_a and r_b, respectively. V_{ab} is the perturbation element between $|a\rangle$ and $|b\rangle$.

3. Experimental Studies of Anticrossing Signals

Anticrossing experiments have been carried out either by photon or electron impact excitation. In the first studies,[6,9] the experimental apparatus was similar to that used in level-crossing investigations. A photon beam excited a resonance transition in alkali atoms, and on variation of an external magnetic field a change of the resonance fluorescence intensity

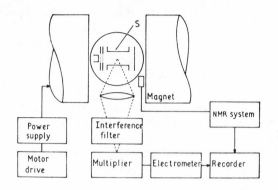

Figure 2. Typical experimental arrangement used in anti-crossing experiments based on electron excitation.

was observed as a result of an anticrossing induced by internal perturbations.

Since coherent excitation of anticrossing substates is not required [it may actually distort the anticrossing signal by contributions from terms (6–8) of Eq. (5)], most subsequent investigations used excitation by an electron beam parallel to the magnetic field. This also allows the population of states that cannot easily be excited with light. A typical anticrossing apparatus[10] with electron impact excitation is illustrated in Figure 2. A vacuum system contains an electron gun to excite atoms of the gas with which the system is filled (gas pressure $\sim 10^{-3}$ Torr). The gun consists of a cathode, accelerating grids, an anode, and, if necessary, "Stark" plates S through which a static electric field can be applied. An optical interference filter or a monochromator selects appropriate spectral lines, the relative line intensity of which is measured by the photomultiplier and recorded as a function of the magnetic field produced by an external magnet.

3.1. Anticrossings Induced by External Perturbations

3.1.1. First-Order Signals

A simple external perturbation is provided by a static homogeneous electric field $\mathbf{F}$. This results in electric dipole coupling between states differing in l by 1, so that the perturbation in the anticrossing equations (6)–(8) is in this case

$$V_{ab} = (1/h)\mathbf{F} \cdot \langle a|e\mathbf{r}|b \rangle \tag{9}$$

Such electric-field-induced anticrossing signals were employed in precision measurements of the $2^2S_{1/2}-2^2P_{1/2}$ Lamb-shift intervals of H and D[11–13] and in investigations of some other fine-structure intervals of H[14] and He$^+$.[15,16] Details are given in another chapter of this book.[18]

3.1.2. Higher-Order Signals

In fine-structure systems consisting of many nearly degenerate states with different l values, such as hydrogenic systems or the helium atom at high n, the coupling is restricted to states with $\Delta l = \pm 1$ only for small electric fields. Stronger electric fields (of the order of 100 V/cm) induce noticeably higher-order Stark mixing, so that higher-order anticrossing signals with $\Delta l = 2, 3, \ldots$ become observable. The higher-order coupling is achieved through one or more intermediate states, each step being governed by the appropriate dipole selection rules (a homogeneous electric field can only produce electric dipole coupling). Usually there are several possible channels to connect $|a\rangle$ and $|b\rangle$, and the sum over all possible intermediate states has to be taken to evaluate V_{ab}. For example, for $|a\rangle$ and $|b\rangle$ belonging to $S(l=0)$ and $F(l=3)$, respectively, the lowest-order interaction element at the crossing is of the form

$$V_{ab} \propto \sum_i \sum_j \langle a|e\mathbf{F}\cdot\mathbf{r}|i\rangle\langle i|e\mathbf{F}\cdot\mathbf{r}|j\rangle\langle j|e\mathbf{F}\cdot\mathbf{r}|b\rangle \propto |\mathbf{F}|^3 \qquad (10)$$

where i and j are substates with $l=1$ and $l=2$, respectively. An *ab initio* calculation of a second-order anticrossing signal ($S–D$ in H, $n=3$) has been carried out by Glass-Maujean and Descoubes[19] in a three-level approximation.

Higher-order electric-field-induced anticrossing signals were first observed by Eck and Huff[20] on the $n=4$ state of He^+. Detailed investigations of such signals were carried out by Beyer and Kleinpoppen on the $n=4$[21] and $n=5$[22,23] states of He^+ [cf. Section 3.1.2(a) below] by Glass-Maujean on the $n=3$[24] and $n=4$[25] states of H [cf. Section 3.1.2(b)] by Beyer and Kollath[26] on states with $n \geq 6$ and high L of the He atom [cf. Section 3.1.2(c)], and by Adler and Malka[47] on the 4^2D state of Li [cf. Section 3.1.2(d)].

(a) He^+. Figure 3 shows the Zeeman splitting of the $n=4$ fine-structure system of He^+ in a magnetic field alone. Some of the crossovers of substates are converted into anticrossings if a static electric field is applied perpendicular to the magnetic field. The corresponding crossovers between substates of the S state and those of the P, D, and F states are marked in Figure 3 by squares, circles, and arrows, respectively. In addition the electric field affects the crossing position by Stark interaction with other substates so that anticrossing signals can also be used to investigate the low-field Stark effect, which is hardly accessible with ordinary spectroscopic methods.

The behavior of anticrossing substates near their crossing position can be calculated by diagonalizing the energy matrix of the full fine-structure system (a 32×32 matrix for $n=4$) in combined electric and magnetic fields. The calculation was carried out using the time-indepen-

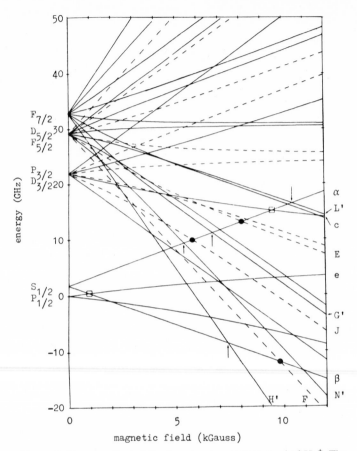

Figure 3. Zeeman effect of the fine-structure system $n = 4$ of He^+. The
D substates are shown as dashed lines, the others as solid lines.
Anticrossings occurring between substates from S and substates
from P, D, and F (on application of an electric field perpendicular to
the magnetic field) are marked by squares, circles, and arrows,
respectively.

dent approach, and as example the results of the eigenvalues for the
substates α and G' (see notation in Figure 3) near their crossing position
are shown in Figure 4 for various electric field strengths (electric field F
perpendicular to the magnetic field H). In this approach, the separation of
the perturbed substates at the crossing position is $2V_{ab}$ which, according to
Eqs. (7) and (8), governs the amplitude and width of the anticrossing
signals.

Anticrossing signals in $n = 4$ of He^+ were observed as intensity
changes in the transition line complex from $n = 4 \rightarrow n = 3$ at 4686 Å. The

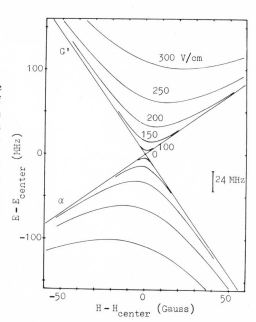

Figure 4. Eigenvalues of the substates α and G' of $n = 4$ of He$^+$ near their crossing position for various electric field strengths ($\mathbf{F} \perp \mathbf{H}$), calculated from time-independent matrix diagonalization. The crossing center is normalized to a common position (E_0, H_0) for all electric fields. The separation of the eigenvalues at H_0 represents $2|V_{\alpha G'}| \cdot \gamma_\alpha \gamma_{G'} = 24$ MHz is marked by the bar. Near the crossing the time-dependent results[42] would depart strongly (producing a real crossing) for $2|V_{\alpha G'}| \lesssim 12$ MHz and show little change for $2|V_{\alpha G'}| > 24$ MHz.

discussed electric field dependence of the signal amplitude and the signal width as well as the Stark shift are demonstrated in Figure 5 by the set of recorded signals $\beta H'$ taken with the mixing electric field as parameter. The signal amplitude (normalized to an extrapolated saturation value) and the signal width (absolute), measured in dependence of the electric field, are compared with the calculated data in Figure 6 for the anticrossing signal $\alpha G'$. The comparison indicates that the assumed process of stepwise dipole coupling is reasonable.

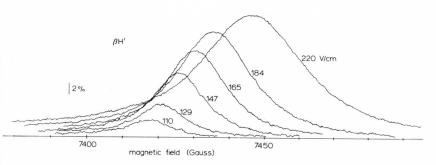

Figure 5. Superposition of several recordings of the anticrossing $\beta H'$ in $n = 4$ of He$^+$ showing clearly the influence of the electric field on amplitude, width, and position of the anticrossing signal. The bar indicates a change of approximately 2% of the total light intensity.

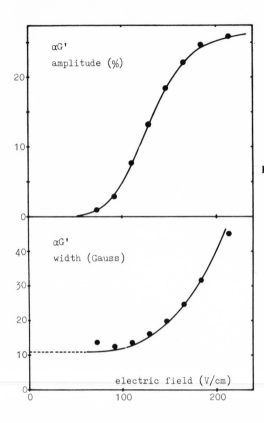

Figure 6. Signal amplitude and signal width (FWHM) for the anticrossing $\alpha G'$ in $n = 4$ of He^+ as a function of the electric field. The points are measured. The theoretical lines are obtained using the interaction elements $V_{\alpha G'}$ (Figure 4) calculated by matrix diagonalization. The theoretical amplitudes are normalized to the experimental saturation value of 27.5%. The point size approximately represents the estimated experimental uncertainty, except for the lowest and highest field points, where the error could be larger.

Mixed level-crossing-anticrossing signals following terms 6–8 of Eq. (5) can also be observed in He^+ signals. As indicated in Figure 7, interference effects are possible in the transition lines from $4S$ and $4D$ since they may decay to common substates of $3P$. The electron impact excitation with an electron beam parallel to the magnetic field did not result in a noticeable coherent excitation of appropriate substates of $4S$ and $4D$, and no level-crossing signal could be observed without application of an external electric field. Coherence between the anticrossing substates of $4S$ and $4D$ may, however, be introduced by the static electric field and an example of the resulting mixed signals is shown in Figure 8. Under the chosen experimental condition, the interference (level-crossing) part has a dis-

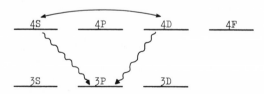

Figure 7. The basis for mixed level-crossing and anticrossing signals between $4S$ and $4D$ states (see text).

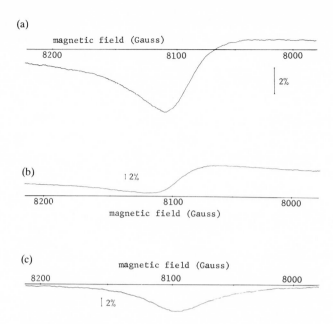

Figure 8. Recorder traces of the anticrossing αE $(4S{-}4D)$ in He$^+$. (a) unpolarized light, (b) σ light, (c) π light. This illustrates how the dispersion admixture from the interference term modifies the absorption-shaped pure anticrossing signal at different states of polarization.

Table 1. Fine-Structure Intervals of He$^+$ (in MHz)

He$^+$, interval	Experiment	Theory[e]
$4^2F_{5/2}{-}4^2S_{1/2}$	27435.7 ± 13.8^a	27446.06 ± 0.13
$4^2F_{7/2}{-}4^2S_{1/2}$	$31098.2 \pm 10.5^{a,b}$	31103.84 ± 0.13
$4^2D_{3/2}{-}4^2S_{1/2}$	20145.3 ± 7.8^a	20143.51 ± 0.28
$4^2D_{5/2}{-}4^2S_{1/2}$	27455.4 ± 7.4^a	27459.25 ± 0.28
$5^2F_{5/2}{-}5^2S_{1/2}$	14058.2 ± 13.6^c	14051.13 ± 0.14
$5^2F_{7/2}{-}5^2S_{1/2}$	15924.5 ± 10.7^c	15923.92 ± 0.14
$5^2D_{3/2}{-}5^2S_{1/2}$	10309.8 ± 14.6^c	10312.27 ± 0.20
$5^2D_{5/2}{-}5^2S_{1/2}$	14055.4 ± 12.6^c	14057.93 ± 0.20
$5^2G_{7/2}{-}5^2S_{1/2}$	15924.4 ± 5.9^d	15920.38 ± 0.09
$5^2G_{9/2}{-}5^2S_{1/2}$	17040.5 ± 8.0^d	17044.04 ± 0.09

[a] Beyer and Kleinpoppen, Ref. 21.
[b] Slightly different value from Ref. 21.
[c] Beyer and Kleinpoppen, Ref. 23.
[d] Beyer *et al.*, Ref. 22.
[e] G. W. Erickson, private communication (1974 and 1976, computer printout dated February 8, 1974).

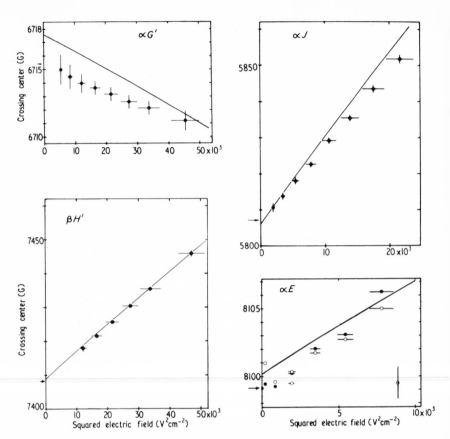

Figure 9. Position of the observed anticrossings in $n = 4$ of He^+ as a function of the squared electric field. The line in each graph represents the theoretical shift, starting at the theoretical crossing point in zero electric field. Extrapolation of the measured points leads to the value marked by an arrow, which determines the measured fine-structure interval.

person Lorentzian shape. It can be suppressed if the observation conditions are chosen such that no decay exists to common substates of $3P$. In the example of Figure 8 this is achieved by selecting π-light only for the detection. The resulting pure anticrossing signal (Figure 8c) is easier to analyze than the total signal (Figure 8a), but shows a less favorable signal-to-noise ratio.

The results for the fine-structure intervals of $n = 4$ of He^+ obtained from these anticrossing measurements are summarized in Table 1. The low-field Stark shifts of the anticrossing positions are illustrated in Figures 9 and 10 and compared with the shift calculated by the matrix diagonaliza-

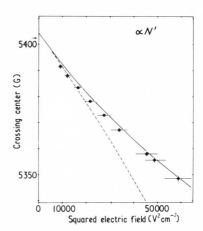

Figure 10. As for Figure 9. This figure shows the start of a change from the quadratic to the linear Stark effect. The pure quadratic Stark shift would follow the dashed line.

tion discussed before. As expected, the observed low-field Stark shift is predominantly quadratic in the electric field strength. Only the anticrossing $\alpha N'$ of Figure 10 departs from the quadratic dependence as a result of two of the interacting substates being so close together that the Stark shift becomes comparable with their separation even at the low electric field used, so that the beginning transition to the linear Stark effect is observed.

In the same way, but with a less favorable signal-to-noise ratio, S-D and S-F signals in $n = 5$ of $He^{+(23)}$ were investigated by observing the 3203-Å line ($n = 5 \rightarrow n = 3$). The results are also shown in Table 1.

Furthermore, anticrossing signals between substates of $5S$ and $5G$ were investigated from the effect on the 4686-Å line ($n = 4 \rightarrow n = 3$). The basis for detecting such "cascading anticrossings" is illustrated in Figure

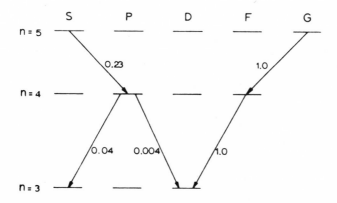

Figure 11. Simplified level diagram of He^+ with the branching ratios of interest for the cascading $5S$-$5G$ anticrossings.

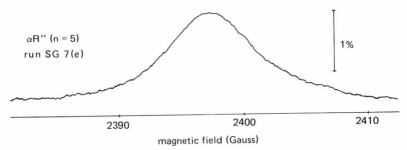

Figure 12. Recorder trace of the cascading anticrossing between the $5S_{1/2}$ ($m_j = 1/2$) and $5G_{7/2}$ ($m_j = -7/2$) states of He^+ observed in the line 4686 Å ($n = 4 \rightarrow n = 3$).

11. The fraction of He^+ ions in the $5S$ state that decays into $4P$ with subsequent emission of the 4684-Å line is very small ($\approx 1\%$), whereas all ions in the $5G$ state decay into $4F$ and then into $3D$. In fact, the difference of the branching ratios which determines the ultimate signal amplitude [Eq. (7)] (together with the population difference which is unaffected by the detection process), is more favorable for cascade observation in the line 4686-Å, where it is $+0.99$, than for direct observation in the line 3202 Å, where it is -0.31. The constant background signal is larger, though, for the cascading signal, but this is compensated by the better overall detection efficiency which can be achieved for the line 4686 Å. An example for such a cascading anticrossing signal is given in Figure 12, which shows a recorder trace of the signal between the substates $5^2S_{1/2}(m_j = 1/2)$ and $5^2G_{7/2}$ ($m_j = -7/2$) observed in the 4686-Å line ($n = 4 \rightarrow n = 3$). The S–G results are also given in Table 1.

 (b) *Hydrogen.* Anticrossing signals S–D were studied on $H, n = 3$, and $n = 4$ by Glass-Maujean.[24,25] As in the measurements described above, a sealed glass system was used containing H_2 gas at a pressure of about 10^{-3} Torr and an electron gun to produce excited H atoms via a

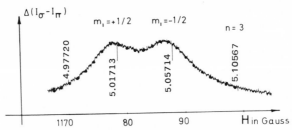

Figure 13. (From Ref. 25.) Recorder trace and computed points of the αE anticrossing signal $3^2S_{1/2}(m_j = 1/2)$–$3^2D_{5/2}(m_j = -3/2)$ of hydrogen taken as the difference between the intensities of σ and π light in the H_α line. The vertical bars are magnetic field calibration marks.

Table 2. Fine-Structure Intervals of Hydrogen (in MHz)

Interval	Experiment	Theory
$3^2D_{3/2}$–$3^2S_{1/2}$	2930.3 ± 0.8^a	2929.859 ± 0.005^c
$3^2D_{5/2}$–$3^2S_{1/2}$	4013.8 ± 0.8^a	4013.195 ± 0.007^c
$4^2D_{3/2}$–$4^2S_{1/2}$	1235.0 ± 2.1^b	1235.727 ± 0.085^d
$4^2D_{5/2}$–$4^2S_{1/2}$	1693.0 ± 0.4^b	1692.760 ± 0.085^d

[a] M. Glass-Maujean, Ref. 24.
[b] M. Glass-Maujean, Ref. 25.
[c] G. W. Erickson, private communication (1976).
[d] G. W. Erickson, private communication (1971).

dissociative state H_2^*. The system was placed in a set of Helmholtz coils and the variation of intensity or polarization of the H_α (H_β) line was recorded as a function of the magnetic field, to locate the anticrossings. No external electric field was applied to couple the states since the velocity of the excited H atoms after the dissociation is sufficient to create the necessary state mixing through motional electric fields. The recorded signals, an example of which is shown in Figure 13 for $n = 3$, are double peaked because of the hyperfine structure of the S state. It will be noted that no mixed level crossing–anticrossing contributions are observed here, in contrast to the S–D signals of He^+ described above (cf. Fig. 8). This is a result of the isotropic distribution of the motional field around the magnetic field axis (compared with the fixed electric field direction in the He^+ experiments). When averaged over all atoms the coherence contributions vanish in such an isotropic field distribution. In the $n = 4$ states of H, the much higher sensitivity to static electric fields causes the S–D signals to be nearly saturated and considerably broadened at the motional fields experienced. The results for $n = 3$ and 4 are collected in Table 2.

(c) *Helium.* The fine structure of excited states of the He atom is much larger than that of corresponding states of hydrogenic systems. It is therefore necessary to go up in n and L before suitable substates can be tuned to degeneracy with fields available from conventional electromagnets. Beyond about $n = 6$, however, higher-order anticrossing signals can be studied in a way similar to that discussed in Sections 3.1.2(a) and 3.1.2(b). The D states have proved to form the most suitable basis for such investigations since they are conveniently separated from (higher L) fine-structure states and at the same time decay to $2P$ with the emission of light in the visible or near-ultraviolet region while higher L states emit only in the difficult infrared region. The corresponding section of the fine structure of $n = 8$ in He is shown in Figure 14. The 8^3P, 8^1S, 8^3S states are far below the ^3D state [the nearest (^3P) is at ~ -850 GHz], and the 8^1P state is ~ 150 GHz above the $8^{1,3}K$ state. Up to $L = 2$ the exchange energy is much larger than the relativistic fine structure shown in the ^3D state insert, but for $L \geqslant 3$ the exchange energy is smaller than the relativistic fine structure resulting in singlet–triplet mixing as shown in the insert for the $^{1,3}G$ states. Thus, for $L \geqslant 3$, the ^1L and ^3L states are considered as one structure.

In a magnetic field there are many crossovers between substates of ^1D (or ^3D) and substates of $^{1,3}F, G, H, \ldots$ and some of these crossings are converted into anticrossings by an electric field perpendicular to the magnetic field (it was noted during the experiment that, owing to singlet–triplet mixing and probably small electric field components parallel to the magnetic field, the usual selection rules were not strictly fulfilled, in particular for higher n). The position of each anticrossing is determined by the corresponding interval $n^{1(3)}D-nL'$ and the relative slope $\Delta k = k - k' \simeq (m_L + 2m_S) - (m'_L + 2m_S) = \Delta m_L$ (assuming $\Delta m_S = 0$). Since the same value of Δk (Δm_L) can usually be obtained by connecting any of the five m_L values of the D states with appropriate m'_L values of L', each anticrossing is made up of five components. In the linear Zeeman effect these components would coincide, but the degeneracy is slightly lifted if the relativistic fine structure, the quadratic Zeeman effect, and differential Stark shifts are taken into account. The structure of the signals is further complicated by the fact that each anticrossing component is again made up of up to four contributions if singlet–triplet mixing of the higher L state and transitions for all m_S values ($\Delta m_S = 0$) are considered. All these contributions and components could not be resolved in the experiment and formed only single and broadened anticrossing signals. The signals $n^{1(3)}D-$

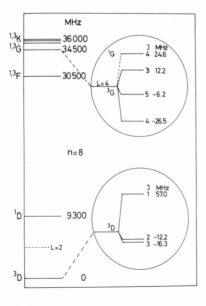

Figure 14. Energy spectrum of some of the $n = 8$ states of the He atom showing the electrostatic fine structure of states with different L and of the ^1D and ^3D states which are separated by the exchange energy, $2K$ (K becomes very small for $L > 2$). The bottom insert shows the relativistic fine structure of the ^3D state and the upper insert shows the structure of the $^{1,3}G$ state including the relativistic fine structure of ^3G, the exchange energy (negligible), and the off-diagonal spin-orbit interaction mixing of $^1G(J = 4)$ and $^3G(J = 4)$.

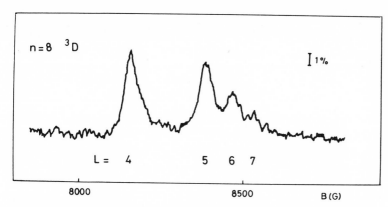

Figure 15. Recorder trace of the group of electric-field-induced anticrossings $8^3D-8^{1,3}G, H, I, K$ with relative slope $\Delta k = 3$ between the Zeeman branches, observed as an intensity change of the transition line $8^3D \rightarrow 2^3P$. The bar indicates 1% of the total light intensity. A static electric field of 40 V/cm is applied. The electron beam current was 200 μA.

$n^{1,3}G, H, I, K, \ldots$ with the same relative slope Δk form groups as shown in Figure 15 for the example $8^3D-8^{1,3}G, H, I, K$. A considerable number of such groups, both for ^1D and ^3D and for various relative slopes, have been investigated by Beyer and Kollath[26] for $n = 7–10$. The experimental setup was as shown in Figure 2 and the signals were recorded as intensity changes of the $n^1D \rightarrow 2^1P$ or $n^3D \rightarrow 2^3P$ lines isolated by a monochromator. In carrying out the analysis the relative positions of all the various contributions and components were calculated (considering the effects mentioned above) and averaged to obtain the fine-structure intervals shown in Table 3. Only a few of these intervals could be measured before.

(d) *Lithium.* Electric-field-induced anticrossing signals of second order were also employed by Adler and Malka[47] to measure the fine structure $4^2D_{5/2}–4^2D_{3/2}$ of ^6Li and ^7Li. A beam of Li atoms was crossed with an electron beam (directed parallel to the external magnetic field, H) and the decay light $4^2D_{3/2,5/2} \rightarrow 2^2P_{1/2,3/2}$ at 4603 Å was detected in a direction perpendicular to the plane formed by the two particle beams. In a magnetic field of up to 400 G, five crossings occur between fine-structure substates of $D_{3/2}$ and $D_{5/2}$, three of which are converted into anticrossings by an electric field applied parallel to the direction of light detection ($\perp H$). The coupling is provided in second order through the nearby (6.8 cm^{-1}) F tates. Including the hyperfine structure (I and J decoupled), the selection ules for these anticrossings are $\Delta L = 0$, $\Delta M_I = 0$, $\Delta M_J = \pm 2$. The coresponding three electric-field-induced anticrossing signals were observed rom the change of the polarization of the light when the magnetic field was aried through the crossing positions. The remaining two fine-structure

Table 3. Fine-Structure Intervals $n^1D-nL'_{av}$ and $n^3D-nL'_{av}$ (Relative to the Center of Gravity of the ^3D States) (in MHz)

			Experiment		Theory	
n	L	L'	Beyer and Kollath[a]	MacAdam and Wing	Chang and Poe[e]	Deutsch[g]
7	^1D	G	37163 ± 60	37224.3 ± 0.1[b]	36610	32800
		H	38599 ± 30			34100
		I	39100 ± 50			34500
	^3D	G	50812 ± 20	50858.1 ± 0.3[c]	50154[f]	
		H	52250 ± 30			
		I	52728 ± 50			
8	^1D	G	25084 ± 30		24699	22000
		H	26085 ± 45			22900
		I	26423 ± 65			23200
		K	26712 ± 90			23300
	^3D	G	34397 ± 40		33968[f]	
		H	35418 ± 55			
		I	35743 ± 55			
		K	36018 ± 105			
9	^1D	F	14967 ± 160			12800
		G	17691 ± 65	17679.8 ± 0.1[b]		15500
		H	18422 ± 60			16100
		I	18617 ± 80			16300
		K	18703 ± 100			16400
9	^3D	G	24368 ± 55			
		H	25062 ± 55			
		I	25329 ± 85			
		K	25447 ± 110			
10	^1D	F	10945 ± 120	10893.7 ± 0.8[d]		
		G	12930 ± 60			
		H	13467 ± 60			
		I	13661 ± 110			
	^3D	G	17819 ± 67[a]	17821.6 ± 0.4[b]	17595[f]	

[a] Reference 26. The value 10^3D-10G is derived from the interval 10^1D-10G and the 10^1D-10^3D interval measured previously (Ref. 30).
[b] Ref. 43 adjusted to L'_{av} using theoretical $^3L'$ splitting and $^1L'-^3L'$ interaction.
[c] As (b), combined with the measured $^1D-^3D$ interval (Ref. 38, see Table 4).
[d] As (b), in addition the theoretical exchange energy of $K = 12.15$ MHz (Ref. 33) has been taken into account. The error has been increased to account for a slight difference between the experimental and theoretical $^1F_3-^3F_3$ intervals.
[e] Reference 33.
[f] Reference 33 gives the values for $n^3D_2-nL'_{av}$ and these have been transformed into $n^3D_{mean}-nL'_{av}$ (mean referring to the center of gravity of ^3D) using the experimental separation $7^3D_{mean}-7^3D_2$ (Ref. 38) and $1/n^3$ scaling.
[g] Reference 44 and private communication, calculated from the values T_{nl} without R^{-7} correction.

crossings are in fact also converted into anticrossings by internal coupling (in the same way as discussed in Section 3.2), and one of these overlaps with two of the electric-field-induced signals. No resolution of the hyperfine structure was possible, but the fine structure $4^2D_{3/2}-4^2D_{5/2}$ was determined to be 400 ± 10 MHz for both ^{6}Li and ^{7}Li. A discussion of the various measurements of this fine-structure interval and a new experimental value was recently given by Fredriksson *et al.*[48]

3.2. Anticrossings Induced by Internal Perturbations

Internal perturbations of atomic states occur, for example, through interaction between the electron and the nuclear magnetic moment, resulting in the hyperfine structure, or through electron spin–orbit interaction, resulting in the relativistic fine structure. Off-diagonal matrix elements of these interactions, although of little influence on the zero magnetic field energies, may have a very noticeable effect if appropriate substates are tuned to near degeneracy by an internal magnetic field. In this case crossings of substates may be converted into anticrossings so that anticrossing signals can be detected if the conditions regarding the population of the states and the detection (cf. Section 2) are fulfilled.

3.2.1. Lithium

The internal magnetic hyperfine interaction formed the basis for the first anticrossing measurement carried out by Eck *et al.*[6] and Wieder and Eck[9] on the $2^2P_{1/2,3/2}$ terms of ^{7}Li and ^{6}Li. The second paper[9] also contains a discussion of many aspects of the anticrossing signals. The crossing $2^2P_{3/2}(m_J = -3/2)$ with $2^2P_{1/2}(m_J = -1/2)$ at approximately 4800 G was studied and, under appropriate conditions, would permit an observation of a level-crossing signal ($\Delta m_J = 1$). When the hyperfine structure is included ($I = 3/2$ for ^{7}Li, $I = 1$ for ^{6}Li) the crossing is divided into several components of which the candidates for level-crossing signals with $\Delta m_J = 1$, $\Delta m_I = 0$ are indicated by circles in Figure 16 for ^{7}Li. However, the picture is changed by the strong coupling near the crossings between states with $\Delta m = 0 (m = m_J + m_I)$ which is caused by off-diagonal matrix elements of the magnetic hyperfine interaction. The states including this coupling are drawn as continuous lines in Figure 16. Two of the previous crossing components have disappeared and the other two are shifted to the positions indicated by arrows. In addition, three anticrossing positions have appeared, which are marked by squares. Thus, both level-crossing and anticrossing signals can be observed (under appropriate conditions) when the magnetic field is varied through the crossing complex. Note, however, that no mixed level-crossing–anticrossing signals of terms 6–8 of Eq. (5) can be detected since the states coupled by the interaction have $\Delta m_I = 1$ and cannot be excited in electric dipole transitions (with $\Delta m_I = 0$) from a common state, nor do they decay to a common state.

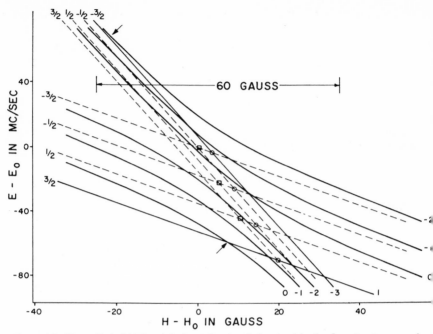

Figure 16. (From Ref. 6.) The eight hyperfine levels involved in the fine-structure crossing $2^2P_{3/2}(m_J = -3/2)-2^2P_{1/2}(m_J = -1/2)$ of ^{7}Li. To show more clearly the details of the crossing, the spacing of the four m_I levels with $m_J = -3/2$ has been increased by a factor of approximately 2.5. The spacing of the m_I levels with $m_J = -1/2$ is correct as shown.

The experimental setup is that of a classical resonance fluorescence experiment with the detection of the fluorescence light in a direction perpendicular to that of the exciting light and the magnetic field at right angles to the plane formed by the two light beams. Linear polarizers are inserted in both beams and, fortunately, these can be set such that only the

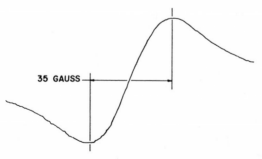

Figure 17. (From Ref. 6.) Recorder trace of the derivative of the anticrossing signal in the $2P$ state of ^{7}Li.

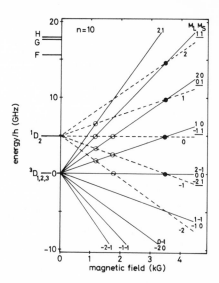

Figure 18. Zeeman splitting of the $10D$ states of helium (the fine-structure splitting of the ^3D states is too small to be seen on this scale). The selection rules for singlet–triplet anticrossings induced by internal spin–orbit coupling provide four nearly degenerate crossings marked by full circles. The crossings marked by open circles may also be converted into anticrossings if, in addition, an external electric field is applied perpendicular to the magnetic field (cf. Section 3.3.2).

level crossing or the anticrossing signals are detected. The recorder curve of a pure anticrossing signal, obtained in this way for ^{7}Li, is shown in Figure 17. The three components are not resolved. Apart from measuring the crossing position, it is also possible to derive the strength of the interaction causing the anticrossing, and this was found to be in very good agreement with the calculated interaction strength.

3.2.2. Helium

(a) Spin–Orbit Coupling. Anticrossing signals between ^1D and ^3D states of the helium atom, which are based on the spin–orbit interaction, have been investigated by Miller et al.[27,28] and Derouard et al.[29] for low n, and by Beyer and Kollath[30,31] for up to $n = 20$. The coupling between appropriate substates of ^1D and ^3D is provided by the off-diagonal spin–orbit operator[27,32] $H_3' = \mathscr{A} \mathbf{L} \cdot \mathscr{S}$, $\mathscr{S} = \mathbf{s}_1 - \mathbf{s}_2$. The matrix elements in the $|n, Lm_L, Sm_S\rangle$ representation appropriate to the magnetic field applied correspond to hV_{ab} in the anticrossing equations (6)–(8) and are

$$\langle n, Lm_L, Sm_S | H_3' | n, L'm_L', S'm_S' \rangle$$

$$= \mathscr{A} \sum_{\nu=-1}^{1} (-1)^{L-m_L+S-m_S} \delta_{LL'} \, \delta_{SS'\pm1} \, [3(2L+1)(L+1)L]^{1/2}$$

$$\times \begin{pmatrix} L & 1 & L' \\ -m_L & \nu & m_L' \end{pmatrix} \begin{pmatrix} S & 1 & S' \\ -m_S & -\nu & m_S' \end{pmatrix} \quad (11)$$

The selection rules for singlet–triplet coupling are therefore $\Delta L = 0$, $\Delta S =$

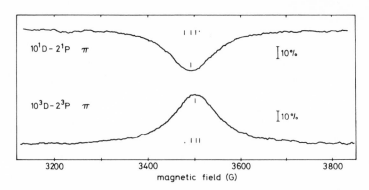

Figure 19. Recorder traces of the singlet–triplet anticrossing signal in $n = 10$ of He observing singlet decay (upper trace) and triplet decay (lower trace), both in π polarization. The theoretical positions of the four anticrossings are indicated below and above the recorded signals using lines for components contributing to the observed signal and dots for the components not detected in π light. The bars indicate 10% of the baseline intensity of the observed line.

± 1, $\Delta M = 0 (M = m_L + m_S)$. Only some of the corresponding substates [those with $\Delta m_L = m_L(^3D) - m_L(^1D) = +1$, $\Delta m_S = -1$] actually cross in the magnetic field and are converted into anticrossings. They are marked by full circles in Figure 18, which shows part of the $n = 10$ fine structure of helium and the Zeeman splitting of the D states. To first approximation the four allowed anticrossings are centered at the same magnetic field. However, the degeneracy is lifted by the relativistic fine structure of the ^3D states, the quadratic Zeeman effect, and the Stark shifts.

The experimental arrangements used for these measurements are essentially as shown in Figure 2. They differ somewhat in the magnetic field facilities (for low n values Bitter coil magnets were employed), and the group at Grenoble (Derouard *et al.*)[29] used a radio-frequency gas discharge instead of electron impact excitation. The anticrossing signals can be detected both in $n^1D \to 2^1P$ and $n^3D \to 2^3P$ transition lines, naturally with opposite sign of the intensity change. Recorder traces of the 10^1D–10^3D signals are shown in Figure 19. The four anticrossings whose approximate positions are indicated near the respective baselines are not resolved since the signals are strongly broadened by the (given) strength of the perturbation.* The four anticrossings carry different weights depending on whether the signal is observed in singlet or triplet lines and in π or σ light. This gave hope that the individual components could be extracted by a least squares fit, and considerable effort was put into this problem by Beyer and

---

* The coefficient $\mathscr{A}$ which determines the perturbation has been derived from the width of such signals via Eq. (8) by Miller *et al.*[27,28] and Derouard *et al.*[29]

Table 4. n^1D–n^3D Intervals of He Relative to the Center of Gravity of the ^3D levels (in MHz)

n	Experiment		Optical data[f] (Martin)	Parish and Mires[g]	Theory	
	Beyer and Kollath	Other			Chang and Poe[h]	Quantum defect[i] (Seaton)
3		102130 ± 200^d	102300	318200	101800	
4		59050 ± 80^d	59100	184700	58760	
5	34066.3 ± 7.2^a	34078 ± 45^d	34400	106800	33870	
6	20918.0 ± 9.2^a	20920 ± 30^d	20800	65600	20790	
7	13632.8 ± 5.3^a	$\begin{cases}13629\pm28^d\\13633.8\pm0.3^e\end{cases}$	14800	42800	13540	13620
8	9326 ± 20^b	9334 ± 35^d	10800	29300	9271	9322
9	6644 ± 20^b		7200	20900	6608	6641
10	4889 ± 15^b		5700	15400	4865	4891
11	3696 ± 15^b		4500		3684	3701
12	2862 ± 15^c		3900			2867
13	2258 ± 15^c		300			2265
14	1812 ± 15^c		1800			1820
15	1475 ± 20^c		1500			1484
16	1215 ± 20^c		300			1225
17	1013 ± 20^c		1200			1023
18	852 ± 30^c		2100			863
19	707 ± 40^c					735
20	596 ± 50^c					631

[a] Electric-field-induced singlet–triplet anticrossings (EFISTAC), Ref. 36 (cf. Section 3.3.2).
[b] Singlet–triplet anticrossings (STAC), Ref. 30 (He pressure contribution to the uncertainty subtracted).
[c] STAC, Ref. 31.
[d] STAC, Derouard et al., Ref. 29 (Table VI).
[e] Radio-frequency transitions, MacAdam and Wing, Ref. 38.
[f] Reference 39.
[g] Reference 40. Includes only first-order exchange contributions.
[h] Reference 33. Includes second-order exchange contributions.
[i] Calculated using the quantum defect $\mu = C_2 + C_3 e$ (e term energy with respect to the first ionization energy). The coefficients C_2 and C_3 for ^1D and ^3D were derived by Seaton (Ref. 41) from a fit to low-n optical data.

Kollath (unpublished) and by Derouard *et al.*,[29] who used a different experimental procedure but basically the same idea. However, no satisfactory fit was obtained. In Stirling this may have been the result of the unknown relative populations of the various substates excited by electron bombardment and of pressure effects. Results for the n^1D-n^3D intervals (with appropriate error allowance) were thus obtained from the overall signals under the assumption of equal population differences for all four anticrossings and with theoretical corrections to account for the relativistic fine structure and the quadratic Zeeman effect. At Stirling (Beyer and Kollath[30,31]) Stark shifts were corrected experimentally. The results are collected in Table 4 and are all much more accurate than previous optical data. They agree quite well with the theoretical results of Chang and Poe,[33] which appear to underestimate the measured results consistently by approximately 0.5%.

(*b*) *Spin–Spin Coupling.* Jost and Lombardi[17] have recently considered the spin–spin interaction between the two electrons of the helium atom, and, on this basis, predicted anticrossing signals with the selection rules $\Delta S = 0 (S \geq 1)$, $\Delta L = 2$ and $\Delta M_L = \pm 1$, $\Delta M_S = \mp 1$ (relative slope 1) or $\Delta M_L = \pm 2$, $\Delta M_S = \mp 2$ (relative slope 2). Owing to the $\Delta L = 2$ selection rule, spin–spin anticrossing signals, if observed, would provide a convenient way of investigating intervals between states with different L without requiring an external electric field to couple the states. Following the discussion of the corresponding electric-field-induced anticrossings in Section 3.1.2(c), intervals $nD-nG$ are best suited for anticrossing investigations in He, and Jost and Lombardi observed a number of $5D-5G$ and $6D-6G$ signals which they interpreted as spin–spin anticrossings. However, it cannot yet be ruled out with certainty that the signals were (at least partly) induced by an electric field (an rf discharge was used in the experiment, and there were also considerable motional electric fields present since the low relative slopes of the substates require high magnetic fields for the anticrossings). In a similar way, the electric-field-induced anticrossing signals ^3D-G with relative slopes 1 or 2, discussed in Section 3.1.2(c) could partly be spin–spin signals. Both types of signals are made up of several different components, and taking into account the "second-order" effects (relativistic fine structure, quadratic Zeeman effect, Stark effect), the observed mean signals would be expected at different magnetic field values. Unfortunately, this difference appears to be insufficient for the discrimination. Even the observation of signals from ^1D states may not be decisive because of possible $^1D-^3D$ mixing by spin–orbit interaction.

Attempts have also been made[45] to observe spin–spin anticrossing signals $F-H$ in He through cascade effects in the $4^3D \rightarrow 2^3P$ line, but no signals could be found at the expected positions, while various other known higher n signals were observed in cascades.

3.2.3. Ytterbium

A type of mixed level-crossing–anticrossing signal, to which the name "coherent anticrossing" was given, was detected by Budick and Snir[34] when recording the resonance fluorescence radiation $(6s)^2\,{}^1S_0 \leftrightarrow 6s6p\,{}^1P_1$ of ^{173}Yb as a function of the magnetic field. The relevant section of the Zeeman diagram of the $6\,{}^1P_1$ state is shown in Figure 20. In addition to the two $\Delta m = 2$ level crossings, an anticrossing is visible between the $F = \frac{7}{2}$, $m_F = -\frac{3}{2}$ and $F = \frac{3}{2}$, $m_F = -\frac{3}{2}$ substates. The coupling between these states is provided in second order via the $F = \frac{3}{2}$, $m_F = -\frac{3}{2}$ substate (not shown). In spite of the apparent separation of the two substates, induced by the perturbation (in fact the substates are mixed), a level-crossing signal following term 3 of Eq. (5) can still be observed if the experimental arrangement provides coherence in both the excitation and decay processes. The pure anticrossing term 5 of Eq. (5) contributes very little under the conditions chosen, but there are some contributions from the mixed level-crossing–anticrossing terms 6 and 7. The resulting mixed level-crossing–anticrossing signal is, therefore, predominantly a level-crossing signal of two anticrossing substates with additional level-crossing–anticrossing contributions, while the mixed signals between S and D states of He$^+$, described in Section 3.1.2(a) were in the first place pure anticrossing signals with additional mixed level-crossing–anticrossing contributions.

The single $\Delta m = 0$ level-crossing–anticrossing signal of ^{173}Yb can conveniently be separated from the stronger, but overlapping, $\Delta m = 2$ level-crossing signals by appropriate setting of the polarizers, thus allowing

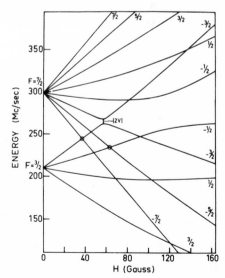

Figure 20. (From Ref. 34.) Zeeman levels of the $^1P_1$ state of ^{173}Yb. The hyperfine state $F = 5/2$ has an energy of -540 MHz and is not shown. The mixed anticrossing–level-crossing is indicated by the perturbation. The two $\Delta m = 2$ level crossings are marked by circles.

the precise determination of the signal position. A more detailed investigation of these level-crossing and level-crossing–anticrossing signals has been carried out recently by Bauer *et al.*[46] and has confirmed the hyperfine term order on which Figure 20 is based.

3.3. Anticrossings Induced by Combined Internal and External Perturbations

Anticrossings that combine both internal and external perturbations (sometimes called "forbidden anticrossings") require at least two steps of coupling and are, therefore, at least of second order. They rely on the admixture of a small component of another state function into one of the crossing states by the internal perturbation, which is then coupled to the other crossing state by the external perturbation. The combined coupling allows the observation of anticrossings which cannot be induced by internal or external perturbations alone, and there is the further advantage over the mere internal coupling, which is fixed, that the degree of coupling can be controlled by the external field, so that, within limits, the amplitude and the width of the anticrossing signals can be adjusted.

Second-order anticrossings with combined internal and external perturbations were first investigated by Leventhal[35] on $n = 2$ states of hydrogen. Even more complex third-order anticrossings with combined coupling have been reported recently by Beyer and Kollath[36] between ^1D and ^3D substates of helium.

3.3.1. Hydrogen

In his study, Leventhal[35] (cf. Chapter 12, p. 565) observed anticrossing signals between Zeeman levels $\alpha(2^2S_{1/2}, m_J = 1/2)$ and $d(2^2P_{3/2}, m_J = -3/2)$ of atomic hydrogen (see Figure 21). In the electric dipole approximation of first order the coupling between these two levels is zero because the selection rule $\Delta m_J = 0, \pm 1$ forbids coupling of states with $\Delta m_J = \pm 2$. However, Leventhal showed that in second order an anticrossing occurs through a hyperfine interaction combined with a Stark interaction from an external electric field. The hyperfine interaction causes the doubling of all Zeeman levels and also results in the mixing of a small amount of the hyperfine states $c_-(2^2P_{3/2}, m_J = -1/2, m_I = -1/2)$ and $f_-(2^2P_{1/2}, m_J = -1/2, m_I = -1/2)$ into the $d_+(2^2P_{3/2}, m_J = -3/2, m_I = +1/2)$ state and also a small amount of the hyperfine state $\beta_+(2^2S_{1/2}, m_J = -1/2, m_I = +1/2)$ mixed into the $\alpha_-(2^2S_{1/2}, m_J = 1/2, m_I = -1/2)$ state, while the states d_- and α_+ remain pure. Thus, the selection rules $\Delta m_J = 0, \pm 1$ and $\Delta m_I = 0$ for electric dipole coupling now lead to a conversion of the crossing $\alpha_- d_+$ into an anticrossing in a transverse electric field. This

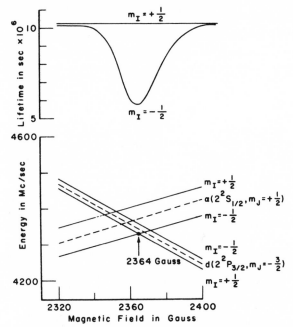

Figure 21. (From Ref. 35.) Upper part, lifetimes of the $\alpha_-(2^2S_{1/2}, m_J = 1/2, m_I = -1/2)$, and $\alpha_+(2^2S_{1/2}, m_J = 1/2, m_I = +1/2)$ states of atomic hydrogen in the vicinity of the αd anticrossing. Lower part, Zeeman splitting of $\alpha_\pm$ and $d_\pm(2^2P_{3/2}, m_J = -3/2, m_I = \pm 1/2)$ states near the $\alpha_- d_+$ anticrossing at 2364 G.

causes the metastable $\alpha_-(2^2S_{1/2})$ state to be selectively quenched, yielding Lyman-α radiation ($\lambda = 1216$ Å) in the decay to the ground state of atomic hydrogen. The change of the $\alpha_\pm(2^2S_{1/2})$ lifetimes has been calculated by Leventhal (see the upper part of Figure 21) and detected by observing the Lymann-α radiation as a function of the magnetic field near 2360 G.

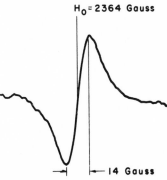

Figure 22. (From Ref. 35.) Anticrossing signal (dispersion-type signal due to magnetic field modulation) between the α_- and d_+ states of atomic hydrogen, $n = 2$. 3 Gauss modulation; 2-sec time constant; 2 mA in electron beam; pressure 3 mTorr.

Figure 22 demonstrates a typical $\alpha_- d_+$ anticrossing as dispersion signal due to magnetic field modulation. A bottle-type arrangement was used in the experiment, and metastable hydrogen atoms were produced by simultaneous excitation and dissociation of H_2 by electron bombardment.

3.3.2. Helium

Anticrossing signals in He between certain substates of ^1D and ^3D not coupled directly by the internal spin–orbit interaction have been investigated by Beyer and Kollath[36] making use of a higher-order process in which the states are both coupled by the external (electric field) perturbation to intermediate states which are mixed by the internal (spin–orbit) perturbation. As will be seen, these signals provide a better resolution than the purely internal ^1D–^3D anticrossings discussed in Section 3.2.2.

In the high field $|n, Lm_L, Sm_S\rangle$ representation of excited helium states appropriate to the magnetic fields applied in the experiments, the linear Zeeman effect for the quantization axis in the direction of the magnetic field B is described by

$$W_z = (g_L m_L + g_S m_S)\mu_B B \approx k\mu_B B \tag{12}$$

where $k = m_L + 2m_S$ gives the slope of a particular Zeeman branch. As shown in Figure 18 for $n = 10$, the Zeeman branches of the ^1D and ^3D states with the same relative slope $\Delta k = k(^3D) - k(^1D)$ cross at about the same magnetic field, and there are groups of such crossings with $\Delta k = 1, 2, \ldots, 6$. The group with relative slope $\Delta k = 1$ forms the singlet–triplet anticrossings described in Section 3.2.2 (allowed anticrossings are marked by full circles), while the groups with $\Delta k = 2$–6 remain crossings, where no signals can be detected since there is no perturbation to couple the states.

If, however, an external electric field is applied perpendicular to the magnetic field, some of these crossings can be turned into anticrossings. The coupling is provided by the combination of the standard electric field perturbation $H_F = e\mathbf{F} \cdot \mathbf{r}$ discussed in Section 3.1 and the spin–orbit perturbation $H_3' = \mathscr{A}\mathbf{L} \cdot \mathscr{S}$ discussed in Section 3.2.2. Taking as example the states $|Lm_L, Sm_S\rangle = |2\,0, 0\,0\rangle$ and $|2\,2, 1\,0\rangle$, which have relative slope $\Delta k = 2$, a coupling between the states is possible in three steps through the matrix elements $\langle 2\,0, 0\,0|H_F|3\,1, 0\,0\rangle, \langle 3\,1, 0\,0|H_3'|3\,1, 1\,0\rangle$, and $\langle 3\,1, 1\,0|H_F|2\,2, 1\,0\rangle$ using the singlet–triplet mixing of the intermediate F states. Similarly, a coupling can be found for two more crossings with relative slope 2 between ^1D and ^3D states with $\Delta m_L = 2, \Delta m_S = 0$ and for four crossings with relative slope 3 between states with $\Delta m_L = 1$ and $\Delta m_S = 1$. These are indicated in Figure 18 by open circles.

Figure 23 shows a recorder trace of the electric-field-induced singlet–triplet anticrossing signals with relative slope $\Delta k = 3$ for $n = 7$ using, as

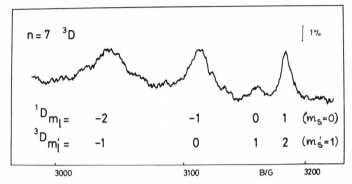

Figure 23. Recorder trace of the group of electric-field-induced singlet–triplet anticrossings (EFISTAC) with relative slope $\Delta k = 3$ between the Zeeman branches of 7^1D and 7^3D. The $7^3D \rightarrow 2^3P$ line is observed, and a static electric field of 150 V/cm is applied. The bar indicates 1% of the total light intensity. The electron beam current was 500 μA.

before, electron bombardment to excite the helium atoms. In contrast to the ordinary singlet–triplet anticrossings (cf. Figure 19) where the signals of the group with relative slope $\Delta k = 1$ were not resolved, Figure 23 shows the four electric-field-induced singlet–triplet anticrossings fully separated. The better resolution is partly due to the smaller width of the new signals and partly due to the additional separation caused by the Stark effect. The two signals involving $|m_L| = 2$ levels could be positively identified using the property that these levels cannot decay by π light in D–P transitions and that the two signals vanish when the respective lines are observed in π polarization.

The electric field applied to induce the signals causes a specific Stark shift of each anticrossing from which the differential Stark shift ΔW_F of the corresponding pair of ^1D and ^3D substates can be determined. The quadratic Stark constants C_{St}, defined by $\Delta W_F = C_{St}F^2$, were derived from the experimental data and are given in Table 5 in comparison with theoretical values calculated from second-order perturbation theory using hydrogenic wave functions and taking into account contributions from nD–nP and nD–nF interactions. The results are in satisfactory agreement.

The anticrossing positions B_c without electric field perturbation are obtained from the experimental data by extrapolation to zero electric field. Apart from the known linear Zeeman effect [Eq. (12)], each anticrossing position depends on four parameters. These are the ^1D–^3D interval (taken to be relative to the ^3D center of gravity), the anisotropic magnetic susceptibility χ_A describing the relevant part of the quadratic Zeeman effect[37] and the two parameters A and b describing the relativistic fine structure of

Table 5. Stark Constants $-C_{St}$ of the Electric-Field-Induced $^1D\text{-}^3D$ Anticrossings in He [in $kHz/(V/cm)^2$][a]

	Relative slope 2			Relative slope 3		
n	$m_L(^1D)$	Exp.	Theory	$m_L(^1D)$	Exp.	Theory
5	0	0.95(0.12)	0.85	1	0.95(0.23)	0.80
	−1	2.05(0.23)	1.86	0	—	1.05
	−2	—	3.23	−1	1.80(0.30)	1.66
				−2	2.87(0.41)	2.64
6	0	4.03(0.61)	3.64	1	3.56(0.69)	3.30
	−1	8.6(1.1)	7.85	0	4.6(1.1)	4.4
	−2	14.7(2.7)	13.5	−1	7.7(1.1)	6.9
				−2	11.7(1.5)	10.9
7	0	12.7(1.7)	11.5	1	11.3(1.4)	10.7
	−1	28.2(3.2)	24.8	0	16.8(2.1)	14.3
	−2	—	42.6	−1	25.0(2.8)	22.4
				−2	37.6(4.1)	35.1

[a] Beyer and Kollath, Ref. 36.

the ^3D states. With the measured anticrossing positions, an overdetermined system of linear equations is obtained for these parameters which thus have been extracted by a least-squares fit. The results for the $n\,^1D\text{-}n\,^3D$ intervals, $n = 5\text{-}7$, are listed in Table 4. They agree within the errors with previous measurements and show a considerably improved accuracy when compared with the pure singlet–triplet anticrossings also shown in Table 4. The difference between the experimental values and the theoretical results of Chang and Poe[33] is of the usual order of 0.5% probably owing to the neglect of higher-order terms in the calculation.

References

1. W. E. Lamb, Jr., and R. C. Retherford, *Phys. Rev.* **72**, 241 (1947); **79**, 549 (1950); **81**, 222 (1951).
2. J. Brossel and F. Bitter, *Phys. Rev.* **86**, 308 (1952).
3. J. Brossel and A. Kastler, *C.R. Acad. Sci.* **229**, 1213 (1949); A. Kastler, *J. Phys. Radium*, **11**, 255 (1950).
4. F. D. Colegrove, P. A. Franken, R. R. Lewis, and R. H. Sands, *Phys. Rev. Lett.* **3**, 420 (1959).
5. P. A. Franken, *Phys. Rev.* **121**, 508 (1961).
6. T. G. Eck, L. L. Foldy, and H. Wieder, *Phys. Rev. Lett.* **10**, 239 (1963).
7. K. C. Brog, T. G. Eck, and H. Wieder, *Phys. Rev.* **153**, 91 (1967).
8. G. W. Series, *Phys. Rev. Lett.* **11**, 13 (1963).
9. H. Wieder and T. G. Eck, *Phys. Rev.* **153**, 103 (1967). See also the discussion given by K. E. Lassila, *Phys. Rev.* **135**, A1218 (1964).

10. H.-J. Beyer and H. Kleinpoppen, *J. Phys. B* **4**, L129 (1971).
11. R. T. Robiscoe, *Phys. Rev.* **138**, A22 (1965); **168**, 4 (1968).
12. R. T. Robiscoe and T. W. Shyn, *Phys. Rev. Lett.* **24**, 559 (1970).
13. B. L. Cosens, *Phys. Rev.* **173**, 49 (1968).
14. M. Glass-Maujean and J. P. Descoubes, *C. R. Acad. Sci.* **273**, B721 (1971).
15. M. Baumann and A. Eibofner, *Phys. Lett.* **33A**, 409 (1970).
16. A. Eibofner, *Z. Phys.* **249**, 58 (1971).
17. R. Jost and M. Lombardi, Abstracts of the Fifth International Conference on Atomic Physics, Berkeley 1976, pp. 230–231 (1976).
18. H.-J. Beyer, Chapter 12 of this work.
19. M. Glass-Maujean and J. P. Descoubes, *Opt. Commun.* **4**, 345 (1972).
20. T. G. Eck and R. J. Huff, in *Beam Foil Spectroscopy*, Ed. S. Bashkin, pp. 193–202, Gordon and Breach, New York (1968); *Phys. Rev. Lett.* **22**, 319 (1969).
21. H.-J. Beyer and H. Kleinpoppen, *J. Phys. B***4**, L129 (1971); *J. Phys. B* **5**, L12 (1972).
22. H.-J. Beyer, H. Kleinpoppen, and J. M. Woolsey, *Phys. Rev. Lett.* **28**, 263 (1972).
23. H.-J. Beyer and H. Kleinpoppen, *J. Phys. B* **8**, 2449 (1975).
24. M. Glass-Maujean, *Opt. Commun.* **8**, 260 (1973).
25. M. Glass-Maujean, Thesis, Université de Paris (1974).
26. H.-J. Beyer and K.-J. Kollath, *J. Phys. B* **11**, 979 (1978).
27. T. A. Miller, R. S. Freund, F. Tsai, T. J. Cook, and B. R. Zegarski, *Phys. Rev. A* **9**, 2474 (1974).
28. T. A. Miller, R. S. Freund, and B. R. Zegarski, *Phys. Rev. A* **11**, 753 (1975).
29. J. Derouard, R. Jost, M. Lombardi, T. A. Miller, and R. S. Freund, *Phys. Rev. A* **14**, 1025 (1976).
30. H.-J. Beyer and K.-J. Kollath, *J. Phys. B* **8**, L326 (1975).
31. H.-J. Beyer and K.-J. Kollath, *J. Phys. B***9**, L185 (1976).
32. H. A. Bethe and E. E. Salpeter, *Quantum Mechanics of One- and Two-Electron Atoms*, Berlin, Springer-Verlag (1957).
33. T. N. Chang and R. T. Poe, *Phys. Rev. A* **10**, 1981 (1974); *Phys. Rev. A* **14**, 11 (1976).
34. B. Budick and J. Snir, *Phys. Rev. Lett.* **20**, 177 (1968).
35. M. Leventhal, *Phys. Lett.* **20**, 625 (1966).
36. H.-J. Beyer and K.-J. Kollath, *J. Phys. B* **10**, L5 (1977).
37. T. A. Miller and R. S. Freund, *Phys. Rev. A* **4**, 81 (1971).
38. K. B. MacAdam and W. H. Wing, *Phys. Rev. A* **12**, 1464 (1975).
39. W. C. Martin, *J. Phys. Chem. Ref. Data* **2**, 257 (1973).
40. R. M. Parish and R. W. Mires, *Phys. Rev. A* **4**, 2145 (1971).
41. M. J. Seaton, *Proc. Phys. Soc. London* **88**, 815 (1966).
42. W. E. Lamb, Jr., *Phys. Rev.* **85**, 259 (1952).
43. K. B. MacAdam and W. H. Wing, *Phys. Rev. A* **15**, 678 (1977).
44. C. Deutsch, *Phys. Rev. A* **13**, 2311 (1976).
45. H.-J. Beyer, unpublished.
46. M. Baumann, H. Liening, and H. Lindel, *Phys. Lett.* **59A**, 433 (1977).
47. A. Adler and Y. Malka, *Z. Physik* **266**, 219 (1974).
48. K. Fredriksson, H. Lundberg, and S. Svanberg, *Z. Physik A* **283**, 227 (1977).
49. W. Hanle, *Z. Physik* **30**, 93 (1924).

14

Time-Resolved
Fluorescence Spectroscopy*

J. N. Dodd and G. W. Series

1. Introduction: Coherence and Superposition States

We may learn about the force fields that drive dynamical systems by studying their evolution in time from nonequilibrium configurations. What is to be discussed in this article is the kind of information that can be obtained about atoms from measurements of time variations with micro- or nanosecond resolution of the intensity of fluorescent light from an ensemble. All the atoms whose fluorescence is to be observed must have been prepared in the same way at the same time, to within the desired time resolution of the experiment.

Equivalent information may, of course, be obtained by direct spectral analysis, in the radio- or microwave regimes, of the variations in the fluorescent intensity. Again, a steady-state version of such experiments is provided by the level-crossing technique, which exploits the spatial interferences between radiations from different states when these are tuned

* Preliminary Remark. Shortly after the authors had agreed to write this article, but before it was actually written, they were privileged to see a copy of the excellent review of the same subject written by S. Haroche for *High Resolution Laser Spectroscopy*, Ed. K. Shimoda, Springer, Berlin (1976). While the two articles share a great deal of common ground, that of Haroche is more detailed, reports more applications, and gives more technical information. Our article has more to say about the relation between "quantum beats" and other phenomena of coherence, about the geometrical features of the experiments, and about the effects of pulsed magnetic and electric fields.

J. N. Dodd • Department of Physics, University of Otago, New Zealand.
G. W. Series • J. J. Thomson Physical Laboratory, University of Reading, Reading, U.K.

through degeneracy by the application of magnetic fields (see the article by Happer and Gupta, Chapter 9 of this work). Fundamentally, the characteristics of time-resolved measurements of intensity, of the spectral analysis of intensity variations, and of level-crossing experiments, are the same.

These studies of *intensity* variations are to be carefully distinguished from conventional spectral analyses where one studies (by way of the Fourier transform) the evolution in time of the amplitude of the field, that is to say, of the electric dipole moments of atoms oscillating at optical frequencies. The variations in *intensity* recorded by a detector at some position fixed in relation to the source arise partly on account of variations in the number of excited atoms (as, for instance, by radiative decay) and partly because of variations in the spatial orientation of the radiating dipoles. It is these variations in orientation that we shall be mainly concerned with. The detector registers variations of intensity as the dipoles in different atoms of the ensemble precess and nutate in synchronism. It is not a question of coherence between the fields oscillating at optical frequencies, but of the superposition of the anisotropic radiation patterns of the individual atomic dipoles. The synchronous precession may be brought about, for example, by the sudden application of an external field to an assembly of atoms fluorescing under steady excitation, so that the resulting modulation of the light recorded by the detector is a direct demonstration of Larmor precession, a "searchlight" effect. Less obviously, the precession may result from internal coupling of the orbital motion of the electron to its spin or to the nucleus. While under steady excitation these motions are randomly phased, they can be made synchronous in different atoms if these latter are excited simultaneously by a short pulse of polarized light. The intensity of the fluorescence is then modulated at fine or hyperfine frequencies.

The precessional motions can, of course, be determined by measuring the splittings of lines in the optical spectra, but we gain a very important advantage by studying the modulation of the intensity directly, which is that, to all intents and purposes, the Doppler effect is eliminated. The Doppler broadening of spectral lines in the optical region, being proportional to the frequency of the radiation, is of the order of 1 GHz. The Doppler broadening of the intensity variations, on the other hand, is smaller by the ratio of optical to radio or microwave frequencies—a factor of 10^6 or 10^8—and is negligibly small in the present context.

1.1. Simple Theory of Quantum Beats

For a quantum-mechanical understanding of the modulation of the intensity of fluorescent light, the elementary notion of the decay of excited atoms from energy eigenstates is inadequate. An essential notion is the preparation of atoms into superpositions of energy eigenstates and their decay from such superposition states. The preparation is easy to achieve. One aims to apply a sudden perturbation which breaks the symmetry of the preexisting environment. The atoms will evolve into equilibrium with the new environment—usually an exponential change characterized by a rate constant—but during the transient period there will exist atoms that were prepared in eigenstates of the old environment and that have evolved coherently from these states. In terms of eigenstates of the new environment, they are in a coherent superposition and will remain so for the duration of the transient.

The superposition state is the formal mathematical entity from which may be deduced the synchronous precession we spoke of in the last section.

To calculate the intensity of fluorescent light from an ensemble of atoms in superposition states it is plausible to use the ordinary formula for spontaneous emission from energy eigenstates with the superposition state taking the place of the initial eigenstate.

The ordinary formula reads

$$I \propto \sum_f |\langle f|\hat{D}_\alpha|e\rangle|^2 \tag{1}$$

where $\hat{D}_\alpha$ is a component of the electric dipole operator, $|e\rangle$ is the initial (excited) state of the transition, and $\{|f\rangle\}$ is the set of final states that may connect with $|e\rangle$ (for example, hyperfine states of the ground level)—it being understood that $|e\rangle$ and the $|f\rangle$ are eigenstates of whatever Hamiltonian describes the system.

In place of (1) we now write

$$I(t, t_0) \propto \sum_f |\langle f|\hat{D}_\alpha|t, t_0\rangle|^2 \tag{2}$$

with

$$|t, t_0\rangle = \sum_e a_e(t, t_0)|e\rangle \tag{3}$$

The $|e\rangle$ and the $|f\rangle$ are eigenstates of the Hamiltonian under the *new* environment. Equation (3) shows the superposition state at time t generated by a sudden perturbation at time t_0. The coefficients $a_e(t, t_0)$ may be calculated with knowledge of the previous Hamiltonian, of the nature of the perturbation, and also of the Hamiltonian under the new environment. This latter yields time-evolution factors $\exp[(-i\omega_e - \frac{1}{2}\Gamma_e)(t - t_0)]$, where $\hbar\omega_e$ is the energy and Γ_e the damping constant of the state $|e\rangle$. In many cases the states $|e\rangle$ will belong to the same hyperfine or Zeeman structure and will have the same damping constant. Accordingly, we shall drop the subscript on Γ.

Insertion of (3) into (2) yields a sum of terms of the type

$$I_f(t, t_0) \propto \sum_{e,e'} [A + B \cos \omega_{ee'}(t - t_0) + C \sin \omega_{ee'}(t - t_0)] \exp [-\Gamma(t - t_0)] \qquad (4)$$

where (here, as later) we write $\omega_{ee'} = \omega_e - \omega_{e'}$. It is often possible to choose the method of preparation so that either B or C is equal to zero. Equation (4) shows how the Bohr frequencies between levels of the excited superposition state appear as frequencies of modulation of the fluorescent light. Notwithstanding the very simple derivation offered here, the equation is indeed sufficient to describe the fluorescent light from free atoms in an optically thin vapor. It is implicit in many detailed studies of the interaction of light and atoms, for example, in the work of Breit[1] and of Barrat and Cohen-Tannoudji.[2,3] It disregards the differences in the time taken by light from different atoms to reach the detector, but these differences are negligibly small on the nanosecond scale for samples a few millimeters in extent.

When Eq. (4) is derived from first principles by means of the quantum theory of radiation the quantity we have called $I_f(t, t_0)$ appears as the probability of occurrence of a process after which the atom is left in a state $|f\rangle$ and the field is left as a coherent superposition of energy eigenstates for each of which the photon occupation number has increased by one unit. It is legitimate, therefore, to speak of one photon in the emitted field, but it is a mistake to suppose that the quantum energy of this photon is uniquely specified. Coherence between different energy eigenstates of the field implies coherence between time-dependent wave functions oscillating at different frequencies. A consequence of this coherence is that the probability that such a field will liberate a photoelectron at a detector is modulated. The modulation may be displayed by recording a large number of photoelectrons accumulated in a sequence of repetitions of the experiment, or as a transient current in a single-shot experiment. Since the modulation arises from coherence in the wave functions that enter into single quantum events, the term "quantum beat spectroscopy" has come into use to describe this kind of experiment.

Before leaving Eq. (4) it is to be noted that the coefficients A, B, and C are not necessarily time independent. Whether they are or are not time-varying depends on the state of affairs before the sudden perturbation—the Hamiltonian might contain an oscillatory term—but in most spectroscopic applications these coefficients do not depend on time: The sole time dependence arises from the evolution of the system *after* the application of the perturbation.

1.2. Methods of Preparation into Superposition States: Sudden Perturbations

The "suddenness" of the symmetry-breaking perturbation is of the essence of the preparation of the system into a superposition state. (The preparation of the excited superposition state in a level-crossing experiment might appear to be an exception to this statement, but it is not, since the states that are coherently excited are eigenstates of the Hamiltonian exclusive of the interaction with the exciting light, that is, they are not eigenstates of the total Hamiltonian.) Among the common methods of preparation are excitation by pulses of light, of electrons, of ions, or by the sudden change of some parameter such as the electric or magnetic field. An important class of "sudden" preparations arises when an atom is formed from an ion by the capture of an electron, as in "beam-foil" experiments. Coherence effects generated in this and other ways are discussed in J. Andrä's article, Chapter 20 of this work.

Excitation by pulses is distinguishable from excitation by changes in some parameter since in the former case the environment before and after the pulse is the same, whereas in the latter this is not so. The former case takes on the characteristics of the latter if the pulse duration is long compared with the relaxation time of the atoms: the *essential* requirement is that the cutoff of the pulse be sudden, though it is *usually* desirable also that the duration of the pulse be short compared with the period of any modulation that is to be studied. Section 2 is mainly concerned with pulsed excitation, and Section 3 with sudden changes in magnetic and electric fields.

1.3. Steady-State Excitation and Modulated Excitation: Applications

It is instructive and useful to study the connection between pulsed excitation and steady-state excitation, on the one hand, and between pulsed excitation and modulated excitation on the other. Steady-state excitation, as, for example, by a resonance lamp emitting light at a constant rate, or by a uniform current of electrons, may be analyzed as a sequence of excitation pulses at random times, each pulse producing modulated fluorescence according to Eq. (4). In the case of the resonance lamp the pulses are constituted by wave packets of light whose coherence time is given roughly by the reciprocal of the spectral width. (Pulses of nanosecond duration, such as we have been considering until now, would contain many such wave packets). The interaction of the atoms with a succession of randomly phased wave packets is an incoherent set of processes so that the result of irradiation producing on average N excitations per second is given by

$$I(t) \propto \int_0^t NI(t, t_0)\, dt_0 \tag{5}$$

with

$$I(t, t_0) \propto \sum_{e,e'} [A + B \cos \omega_{ee'}(t - t_0)] \exp[-\Gamma(t - t_0)] \tag{6}$$

a simplified version of Eq. (4). We have supposed the irradiation to have started at $t_0 = 0$ and, by using the average rate of excitation, we have smoothed out the fluctuations. (A profound discussion of the phase problem and of the fluctuations of intensity in this type of calculation has been given by Durrant.[4])

Carrying out the integration in (5) we have, for $(t - t_0) \gg \Gamma^{-1}$,

$$I \propto N\left(\frac{A}{\Gamma} + \sum_{e,e'} \frac{B\Gamma}{\Gamma^2 + \omega_{ee'}^2}\right) \tag{7}$$

This expression is characteristic of level-crossing experiments. The interference terms (those with coefficient B) vanish when the $\omega_{ee'} \gg \Gamma$.

Suppose now that the sequence of pulses is not random but that the rate of excitation is modulated according to $N(t_0) = N_0(1 + m \cos \omega t_0)$. Using this expression instead of N in Eq. (5) we obtain

$$
\begin{aligned}
I(t) \propto\ & N_0\left(\frac{A}{\Gamma} + \sum_{ee'} \frac{B\Gamma}{\Gamma^2 + \omega_{ee'}^2}\right) \\
& + mN_0\left[\frac{A\Gamma}{\Gamma^2 + \omega^2} + \sum_{ee'} \frac{B\Gamma/2}{\Gamma^2 + (\omega - \omega_{ee'})^2} + \sum_{ee'} \frac{B\Gamma/2}{\Gamma^2 + (\omega + \omega_{ee'})^2}\right] \cos \omega t \\
& + mN_0\left[\frac{A\omega}{\Gamma^2 + \omega^2} + \sum_{ee'} \frac{B(\omega - \omega_{ee'})/2}{\Gamma^2 + (\omega - \omega_{ee'})^2} + \sum_{ee'} \frac{B(\omega + \omega_{ee'})/2}{\Gamma^2 + (\omega + \omega_{ee'})^2}\right] \sin \omega t
\end{aligned} \tag{8}
$$

It is seen that modulation of the rate of excitation has the consequences (a) that the intensity of fluorescent light is modulated at the same rate—this, of course, is to be expected, and (b) that resonances occur when $\omega \approx \pm\omega_{ee'}$. The resonance is to be seen in changes in the amplitude of modulation of the intensity variations. We notice that, if we are exploring a resonance with $\omega \approx \omega_{ee'} \gg \Gamma$, the expression for the intensity reduces to

$$I(t) \propto \frac{N_0 A}{\Gamma} + \tfrac{1}{2} \sum_{e,e'} \frac{mN_0 B}{[\Gamma^2 + (\omega - \omega_{ee'})^2]^{1/2}} \cos(\omega t - \phi)$$

where

$$\tan \phi = (\omega - \omega_{ee'})/\Gamma$$

The interference that is responsible for the resonance terms derives from a coherent superposition of the nondegenerate states $|e\rangle$, $|e'\rangle$. The coherence results from the periodicity of the exciting pulses, not, as formerly, from the sudden application of a single pulse or short group of pulses.

Consider now the practical application of Eq. (7) (steady-state excitation), which is elaborated in the article on the level-crossing technique by Happer and Gupta (Chapter 9 of this work). "Resonances" are found at values of an applied magnetic field such that any of the $\omega_{ee'}$ are zero. But to determine a fine or hyperfine interaction constant one needs to know a g factor, in addition to the experimentally determined magnetic field value, and this is normally done by carrying out a double-resonance experiment. However, this brings complications that are sometimes serious. It is necessary to provide a strong oscillating magnetic field in the MHz or GHz range. At best, this perturbs the atoms; at worst, it can set up a discharge in the cell that is difficult to control. However, the modulated-excitation technique can provide an elegant solution to this problem. The "resonance" may be shifted, according to Eq. (8), without the application of a strong, additional field on the sample. All that is necessary is to modulate the exciting light and to study the relationship between the frequency of the modulation and the shift of the resonance in the applied, static field. The validity of Eq. (8) has been established experimentally by Corney and Series[5,6] and by Skalinski et al.[7]

The technique of modulated excitation can be particularly useful in molecular spectroscopy, where one may be interested in measuring some small interval that responds only feebly to magnetic tuning. This situation arises, for example, where there is no electronic contribution to the magnetic moment but where, nevertheless, there may be paramagnetism on account of the rotation of the molecule. Magnetic moments arising in this way are of the order of nuclear magnetons. To explore the interval by double resonance would require an oscillating magnetic field of extremely high amplitude. The difficulty may be avoided by using modulated excitation, using an arrangement that generates a coherent superposition of the states whose separation is to be found. A recent application of this technique to I_2 has been made by Broyer, Lehmann, and Vigué.[8]

1.4. Coherences between Different Atoms

1.4.1. Forward Scattering

In all that has gone before we have calculated the fluorescent intensity from many atoms in an ensemble by adding the intensities from individual atoms. This is a correct procedure for lateral scattering since the optical paths to a point on the detector are random, but it is not correct for forward scattering in collimated light. Here, the paths from a plane-wave front in the incident beam to a parallel wave front in the scattered beam are independent of the position of the scattering atom, so that the fields scattered by all the atoms are coherent. This is the basis of the classical, atomic theory of dispersion. Because of this coherence it is possible to explore, by means of modulation techniques, classes of structures of greater generality than we have hitherto supposed.

The structures we have so far been concerned with are structures in the excited states of single atoms. In forward scattering we may expect to find evidence of structures in the lower states of atoms of the same kind, or, indeed, structures arising from differences between atoms of different kinds. As an example of this last class of structure, consider spectroscopic isotope shifts. Imagine a vapor consisting of a mixture of two isotopes excited by a plane wave front of light, pulsed, and of sufficient spectral width to excite corresponding transitions in the two kinds of atom. The atoms would respond resonantly at different frequencies, and the forward-scattered light from the two kinds of atom, being coherent, would beat at the difference frequency, that is, one would expect to find modulation at the frequency of the isotope shift. Since the individual resonances would be Doppler-broadened (and possibly pressure broadened), one would expect the modulation to damp out in a time given roughly by the reciprocal of the Doppler width.

An experiment of this kind has not, so far as we are aware, been performed, but a detailed analysis of forward scattering carried out by Corney et al.[9] showed that the level-crossing effects and the resonances in modulated light which are to be expected in double-resonance experiments should, indeed, be Doppler broadened for optically thin vapors, but that for optically thick vapors "coherence narrowing" should supervene (see Section 2.2). This was demonstrated experimentally for the zero-field level crossings in excited states of mercury and sodium. A very thorough study of level-crossing effects in forward scattering was carried out by Durrant and Landheer.[10]

That spectroscopic intervals between atoms of different kinds could be exhibited in forward scattering was shown by Hackett and Series,[11] who carried out the "level-crossing" equivalent of the modulation experiment described above. Changes in the intensity of forward-scattered light from a mixture of isotopes were observed when the Zeeman splitting of a particular transition was tuned into coincidence with the isotope shift. Since the interpretation is based on degeneracies in the transition frequencies rather than in the energy levels of single atoms, this kind of experiment was described as "line crossing" rather than "level crossing."

The possibility of observing coherence effects arising from structure in lower levels has recently been discussed by Chow *et al.*[12] The conditions that these authors derive are those that are valid in forward scattering, namely, that it should in principle be impossible to determine which atom of the assembly was responsible for the scattering, and also that any modulation effects should be damped through dephasing on account of Doppler broadening.

1.4.2. Selective Reflection

Closely related to forward scattering is the phenomenon of selective reflection: the boundary of a vapor confined in a vessel behaves like a mirror if the vapor density is sufficiently high. The reflection coefficient is appreciable only in the region of an atomic absorption line, where a significant amount of light is scattered by the atoms. The light scattered from different atoms is coherent when the mean distance between neighboring atoms is substantially smaller than one wavelength, and when the ordinary geometrical conditions governing incident and scattered wavefronts are satisfied. Of course, the back-scattered light contains an incoherent component also: the relative intensity of the coherent and incoherent components depends on the vapor density.

The phenomenon of selective reflection is usually analyzed by treating the vapor as a homogeneous medium characterized by a (complex) refractive index derived from the polarizability of individual atoms. The reflection coefficient at the boundary can then be calculated by using classical electromagnetic theory. This treatment has been adequate for many investigations, though there are complications on account of the fact that the behavior of atoms near the wall of the vessel is not properly characterized by the bulk polarization of the medium (Cojan,[13] Schuurmans[14]).

All those classes of structure that were mentioned in the last section as capable of being studied in forward scattering, namely, structures in excited states and ground states of single atoms, and between atoms of different kinds—as, for example, spectroscopic isotope shifts—should also manifest themselves in specular reflection. Very little work has been done in this field, though the detailed analysis for level crossing and double resonance in excited states was carried out by Series[15] on the basis of the conventional theory, and the corresponding level-crossing experiment was done by Hanle and Stanzel.[16] The theory predicted that the level-crossing curve in reflected light under steady-state illumination of the vapor with white light should be similar to the ordinary (Lorentzian) level-crossing curve, except that it should be Doppler and pressure broadened. Explicit equations were given for the case of Doppler broadening. In the experiment, pressure broadening predominated. Reflection from mercury vapor was studied under illumination with a mercury lamp. Zero-field level-crossing curves were obtained, several kG in width. The pressure-broadened damping constant under the conditions of the experiment (460 Torr) was 6.6 GHz, corresponding to a linewidth of 3.2 kG.

The more recent studies of Stanzel[17] and of Siegmund and Scharmann[18] show that the level-crossing curves are actually narrower than Series' theory predicts. The theory of Schuurmans is able to explain this narrowing in terms of the wall effect. (The coherence narrowing of level-crossing curves found in forward-scattered light is a different phenomenon. Its explanation lies in multiple scattering in the forward direction, not in a wall effect.)

There is little doubt that modulation phenomena could, in principle, be observed in light reflected from the boundary of a vapor under pulsed or modulated excitation. Study of such effects might at first sight seem impracticable on account of the very rapid loss of coherence arising from Doppler or pressure broadening. As to the latter, the vapor pressure need not be so high as in the example quoted (see, for example, Hansen and Webb[19]); as to the former, the Doppler effect could be eliminated by the techniques of saturated absorption and dispersion. The prospect is not so unattractive as would appear at first sight.

2. Pulsed Excitation: Lifetimes and Quantum Beats*

Modulation of the light emitted from hydrogen atoms leaving a gas discharge through a channel in the cathode (canal rays) was observed about 50 years ago by van Traubenberg and Levy,[20] by Hertel,[21] and by Walerstein.[22] The modulation depended on the application of magnetic or electric fields to the canal rays and is an example of the effects we have been discussing. The phenomenon was interpreted in general terms on the basis of the known polarization properties of the Zeeman and Stark effects and on the classical model which Hanle[23,24] and others had used to describe the depolarization of resonance fluorescence by external fields, but the actual situation in hydrogen is, in fact, quite complicated and a detailed explanation could not be given.

With the revival of interest in resonance fluorescence in the late 1950s the phenomenon of modulation following pulsed excitation was discovered and studied in systems chosen for their simplicity. These we shall briefly describe below.

Since the early nineteen seventies, with the advantages offered by fast-pulse technology and tunable, pulsed lasers, the method of quantum

* See also Chapter 20 of this work (J. Andrä).

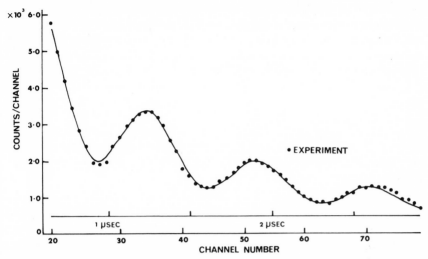

Figure 1. Modulation of fluorescence from the 5^3P_1 state of Cd. The points give the observed counts following a 200-ns pulse of optical excitation in a steady magnetic field of 34.5 μT (from Dodd *et al.*[28]).

beats has been increasingly applied to the determination of fine and hyperfine structures and Stark splittings. We refer the reader to Haroche's review[25] for an account of this work. In the main part of this section we shall discuss the theory of quantum beat signals and shall single out for attention certain geometrical factors which arise also in steady-state and in modulated-fluorescence experiments. These geometrical factors are especially important in relation to the measurement of lifetimes.

2.1. Early Observations of Quantum Beats

2.1.1. Excitation by Resonance Radiation

Observation of modulation in fluorescent light from atoms excited by a pulse of resonance radiation was first reported by Alexsandrov[26] and, independently, by Dodd, Kaul, and Warrington.[27] The technique was improved by Dodd, Sandle, and Zisserman,[28] whose experimental results are shown in Figure 1. The system studied was the vapor of Cd. The sinusoidal modulation at twice the Larmor frequency corresponds to the interval $2g\mu_B B/\hbar$ between the levels $M_J = \pm 1$ of the excited state 5^3P_1. Figure 2 shows the levels and the transitions involved in excitation and in fluorescence.

The transition $J = 0$ to $J = 1$ lends itself particularly well to a classical interpretation. To achieve coherent excitation of the levels $M_J = \pm 1$ the incident light is polarized at right angles to the magnetic field as shown in

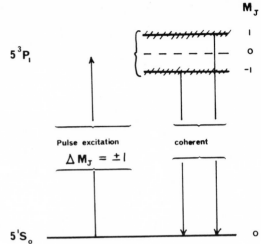

$$M_J$$

$$5\,^3P_1$$

Pulse excitation   coherent

$$\Delta M_J = \pm I$$

Figure 2. Energy levels and transitions involved for the experiment illustrated by Figure 1.

$$5\,^1S_0$$

Figure 3a. Oscillating electric dipoles excited by light of this polarization experience a torque under the field and precess round it (Figure 3b). The anisotropy of the radiation pattern from a group of dipoles excited by a pulse of light results in a modulated response of the detector at twice the frequency of precession (twice the Larmor frequency). A field strength is chosen so that several such periods occur within the mean lifetime of the excited atoms. Clearly, the depth of modulation is dependent on the ratio of the duration of the exciting pulse to the period of precession. It will be noticed from the caption of Figure 1 that the duration of the pulse, about 200 ns, was substantially smaller than the period of the precession, about 700 ns, but not

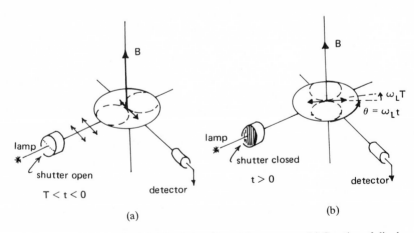

Figure 3. Classical model to illustrate modulated fluorescence. (a) Creation of dipoles by a short pulse of light. (b) Precession of bunched dipoles.

negligibly so. The depth of modulation is seen to be less than 100%. A measurement of the frequency of modulation gave the Landé factor of the excited state and the mean lifetime could be ascertained from the envelope of the modulation.

2.1.2. Excitation by Electrons

It was first demonstrated by Hadeishi and Nierenberg[29] that quantum beats could be produced by excitation of atoms with a short pulse of electrons. Again, the system studied was cadmium vapor. The states $M_J = \pm 1$ of 5^3P_1 were excited coherently and the modulation observed in the fluorescent light corresponded to the Bohr frequency between them.

A more recent exploratory study of beats excited by electron impact has been reported by Bagaev et al.[30] The Zeeman components of $n\ ^1D_2$ ($n = 4, 5, 6$) of He I were excited coherently in the gas by pulses of about 10 ns duration. The modulated fluorescence was studied for a range of magnetic fields of the order of tens of gauss, and it was concluded that, in experiments of this kind, unknown g factors could be determined to an accuracy of about 1 in 10^4. This figure indicates the kind of accuracy one might hope to attain in a set of carefully executed experiments, but it should not be taken as indicating the best that could be achieved if the method were pushed to its limit.

The method of excitation by electron impact with the object of measuring lifetimes was applied to helium in the early nineteen fifties by Heron et al.[31] Since then it has been used extensively. Recent examples are to be found in the series of elegant experiments by King, Adams, and their co-workers.[32-34] A highly monoenergetic beam of electrons impinges on a jet of atoms to be studied. The energy of an inelastically forward-scattered electron is recorded. This defines the state of excitation of the target atom and the instant of excitation. The intensity of the fluorescent light as a function of time is studied by recording delayed coincidences between exciting electrons and fluorescent photons. Impressive accuracy has been achieved in the determination of the mean lifetime of a number of excited states in Hg I and Cd II.

For further examples the reader is referred to the article by R. G. Fowler, Chapter 26 of this work.

2.2. Excitation by Light, Theory: Geometrical Characteristics

In this section we amplify the simple expressions given in Section 1 and give an equation for the intensity of fluorescent light in terms of sums over

STATES   SUBSTATES   BOHR FREQUENCIES

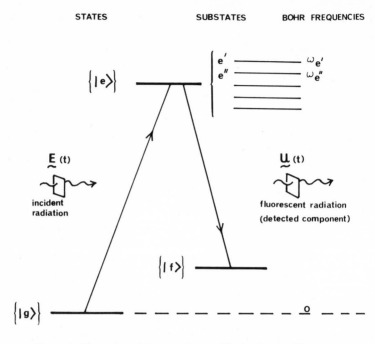

Figure 4. Illustration of the notation used in the theory of fluorescence.

eigenstates of energy and component of angular momentum, (α, J, M_J or α, F, M_F, for example). We then show how, by using expansions in terms of irreducible spherical tensors, the geometrical factors may be separated and all irrelevant quantum numbers eliminated.

Figure 4 shows the scheme we have in mind. The initial, excited, and final states (g, e, f, respectively) may each have structure, and we use primes to distinguish substates. The substates have been shown explicitly only for the excited state: the superposition-states of the analysis will be formed from these. When summations over g and f occur, these will describe incoherent processes.

We describe the light fields classically by $\mathbf{E}(t)$ for the incident light and by $\mathbf{U}(t)$ for that component of the fluorescent light that is detected. It is supposed that the spectrum of $\mathbf{E}(t)$ is sufficiently broad to extend uniformly over the complex of transitions from the $|g\rangle$ to the $|e\rangle$, making due allowance for Doppler broadening of the absorption line. $\mathbf{e}$ and $\mathbf{u}$ are unit vectors corresponding to $\mathbf{E}$ and $\mathbf{U}$.

The following expression can be obtained by first-order perturbation theory for the intensity of fluorescent light at time t from N atoms supposed initially to be distributed uniformly over the states $|g\rangle$:

$$I(t) = G \int_{-\infty}^{t} \int_{-\infty}^{t} dt_0' \, dt_0 E^*(t_0') E(t_0)$$

$$\times \sum_{\substack{e,e' \\ f,g}} \langle g|\mathbf{e}^* \cdot \hat{\mathbf{D}}|e'\rangle \exp\left[(i\omega_{e'} - \tfrac{1}{2}\Gamma)(t - t_0')\right]\langle e'|\mathbf{u}^* \cdot \hat{\mathbf{D}}|f\rangle$$

$$\times \langle f|\mathbf{u} \cdot \hat{\mathbf{D}}|e\rangle \exp\left[(-i\omega_e - \tfrac{1}{2}\Gamma)(t - t_0)\right]\langle e|\mathbf{e} \cdot \hat{\mathbf{D}}|g\rangle \qquad (9)$$

The constant G is the accumulation $(\mu_0/4\pi)(N\omega^4/8\pi c\hbar^2 r^2)$, where ω is the (average) optical frequency and r is the (average) distance from the sample. Equation (9) may easily be interpreted. Reading from right to left, the matrix element $\langle e|\mathbf{e} \cdot \hat{\mathbf{D}}|g\rangle$, together with a factor $(i/\hbar) \, dt_0 \, E(t_0)$ is the increment of probability amplitude of $|e\rangle$ which is created in time interval dt_0 by the interaction Hamiltonian $-\mathbf{E}(t_0) \cdot \hat{\mathbf{D}}$; $\hat{\mathbf{D}}$ is the operator representing electric dipole moment. Next, the exponential term represents the evolution of $|e\rangle$ from t_0 to t (Bohr frequency ω_e, amplitude damping coefficient $\tfrac{1}{2}\Gamma$). Then the matrix element $\langle f|\mathbf{u} \cdot \hat{\mathbf{D}}|e\rangle$, together with factors from G, gives the element of electric field polarized along $\mathbf{u}$ at distance r in emission from $|e\rangle$. Summation over the $|e\rangle$ and integration over t_0 gives the total field. The terms next following from right to left are the Hermitian conjugate expressions, which, correspondingly summed and integrated, give the complex conjugate field. The product gives the intensity. The prime on e' exhibits the possibility of interference between channels.

2.2.1. The Pulse Approximation

If the field $E(t_0)$ were specified the integrations could be carried out, but in fact the measurable properties of light fields are expressible in terms of the statistically averaged quantities $\Phi_{\alpha\beta}(t_0, \tau) = \langle E_\alpha^*(t_0 + \tau)E_\beta(t_0)\rangle$ (subscripts denote space components), and the observed $I(t)$ will similarly be the result of a statistically averaged equation (9). For the broadband light that we have postulated it will be a good approximation to write $\langle E^*(t_0')E(t_0)\rangle = \Phi(t_0)\delta(\tau)$, where $\tau = t_0' - t$. $\Phi(t_0)$ is the statistically averaged intensity of the field $\mathbf{E}$ at time t_0. Inserting $\Phi(t_0)\delta(\tau)$ in Eq. (9), we may carry out one of the time integrations and the equation has then been brought to the form used in Section 1, Eq. (5). What we have done here is to justify the so-called "pulse approximation."

It should be understood that, since the above treatment rests on a first-order perturbation theory, it does not incorporate stimulated emission, but this does not mean that it cannot be used for the irradiation of atoms by laser light. The first point to notice is that we have specialized to broadband light, so that there is, in this approximation, no temporal coherence. The second point to notice is that, in the context of quantum beats, the light is detected *after* the irradiation has ceased: Thus the observed fluorescence arises from spontaneous emission. However, both strong and weak light fields may, by optical pumping, change the distribution of population over the substates of $|g\rangle$. This can be taken into account, if necessary, by incorporating a weighting factor in the summation over g.

2.2.2. Excitation and Monitoring Operators

To return to Eq. (9), it may now be observed that the summed expression may be written as the trace of the product of two operators:

$$\sum_{\substack{e,e' \\ f,g}} <\cdots> = \text{Tr}\,\{\hat{\rho}(\mathbf{e}, t, t_0)\hat{L}(\mathbf{u})\} \qquad (10)$$

with

$$\hat{\rho}(\mathbf{e}, t, t_0) = \sum_g |e\rangle\langle e|\mathbf{e} \cdot \hat{\mathbf{D}}|g\rangle \exp\,[(-i\omega_{ee'} - \Gamma)(t - t_0)]\langle g|\mathbf{e}^* \cdot \hat{\mathbf{D}}|e'\rangle\langle e'| \qquad (11)$$

$$\hat{L}(\mathbf{u}) = \sum_f \mathbf{u}^* \cdot \hat{\mathbf{D}}|f\rangle\langle f|\mathbf{u} \cdot \hat{\mathbf{D}} \qquad (12)$$

$\hat{\rho}(\mathbf{e}, t)$ is the density operator of the excited states at time t. (Its matrix is not normalized to unity because of the damping coefficients.) $\hat{L}(\mathbf{u})$ is called a "monitoring operator" for the emission of electric dipole radiation which leaves the atoms in the states $|f\rangle$. Equation (10) is often used when the density operator has been obtained for more complicated situations and by more elaborate methods.

For our purposes we wish to separate $\hat{\rho}(t, t_0)$ into factors that represent the excited system at time t_0 and the subsequent evolution from t_0 to t, as follows:

$$\hat{\rho}(\mathbf{e}, t, t_0) = \hat{P}_e \hat{F}(\mathbf{e})\hat{P}_{e'} \exp\,[(-i\omega_{ee'} - \Gamma)(t - t_0)] \qquad (13)$$

with

$$\hat{F}(\mathbf{e}) = \sum_g \mathbf{e} \cdot \hat{\mathbf{D}}|g\rangle\langle g|\mathbf{e}^* \cdot \hat{\mathbf{D}} \qquad (14)$$

an expression formally identical with that for $\hat{L}(\mathbf{u})$. We call $\hat{F}(\mathbf{e})$ the excitation operator. $\hat{P}_e$, $\hat{P}_{e'}$ are the projection operators $|e\rangle\langle e|$, $|e'\rangle\langle e'|$. This factorization has reduced the expression for the intensity, Eq. (2.1), to the form

$$I(t, t_0) = G \int_{-\infty}^{t} dt_0\, \Phi(t_0)\,\text{Tr}\,\{\hat{P}_e \hat{F}(\mathbf{e})\hat{P}_{e'}\hat{L}(\mathbf{u}) \exp\,[(-i\omega_{ee'} - \Gamma)(t - t_0)]\} \qquad (15)$$

We have already discussed the time dependence. We now concentrate on the geometrical factors which are to be found in the excitation and monitoring operators $\hat{F}(\mathbf{e})$ and $\hat{L}(\mathbf{u})$.

2.2.3. Irreducible Spherical Tensor Operators

The summations over g and f that occur in $\hat{F}$ and $\hat{L}$ suggest that it may be possible to find a basis for these operators in which the summation is automatically carried out, and this is indeed the case. The basis that serves this purpose is that of the standard components of irreducible spherical tensor operators. The simplification that is obtained derives from the fact that the spin-1 light vector is a simpler object that the combination of atomic angular momenta that it couples. We can make use of this simplification in respect of those states that are effectively degenerate, as the g and f states are in our analysis, but it is of the essence of our problem that the nondegenerate excited states must be distinguished, so we must expect to find in our final equations the quantum numbers that differentiate these states from one another and factors that represent their projection on to the basis of the spherical tensors. Similar factors for the substates $|g\rangle$ would arise if weighting factors on account of optical pumping were present. In what follows we shall suppose that the $|g\rangle$ are equally weighted.

2.2.4. Quantum Beats at Hyperfine Frequencies

The construction of spherical tensor operators is explained in the article by K. Blum, Chapter 2 of this work. The operators are labeled by integers k, q. When formed from angular momentum states $|J, M_J\rangle$ the values of k run from $J + J'$ to $|J - J'|$, and the values of q from k to $-k$. Operators for the hyperfine states $|F, M_F\rangle$ are constructed similarly. Tensors that represent the polarization of the light are formed from spherical unit vectors in three-dimensional physical space. The possible values of k are 2, 1, 0. Thus, the k, q component of the tensor that represents the incident light is

$$E_q^k = (2k+1)^{1/2} \sum_{\mu,\mu'} \begin{pmatrix} 1 & 1 & k \\ -\mu & -\mu' & q \end{pmatrix} e_{-\mu} e_{-\mu'}^* \qquad (16)$$

The e_μ ($\mu = 0, \pm 1$) are components of the vector $\mathbf{e}$ along the unit vectors in the spherical basis. In terms of the Cartesian components of $\mathbf{e}$, the e_μ are

$$e_0 = e_z, \qquad e_{\pm 1} = \mp(e_x \pm i e_y)/2^{1/2} \qquad (17)$$

For the k, q tensor component of the fluorescent light we shall use the symbol U_q^k, formed from the components of $\mathbf{u}$.

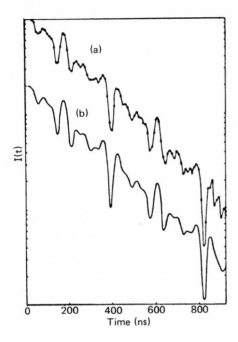

Figure 5. Quantum beats arising from hyperfine structure in $9^2D_{3/2}$ of ^{133}Cs. $I(t)$ is plotted on a logarithmic scale. (a) Experimental points. (b) Evaluation of (18) for this case (from Deech et al.[36]).

The orthogonal properties of the tensor operators result in a separation of terms having the same value of k. Hence the expression for $I(t, t_0)$ is reduced to a sum of nine terms only for each pair of states, $|e\rangle$, $|e'\rangle$, namely, for $k = 2$, five terms, for $k = 1$, three terms, and for $k = 0$, one term.

We give the result of the analysis for the case of atoms having hyperfine structure, in zero magnetic field. For this case the spherical tensor analysis is particularly appropriate, since the hyperfine structure of the states $|g\rangle$ and $|f\rangle$ is quite irrelevant to the problem of determining the hyperfine structure of the states $|e\rangle$. Equation (9) requires the summation over all these structures explicitly, whereas the corresponding expression in spherical tensor analysis contains no reference to the structures $|g\rangle$ and $|f\rangle$. Moreover, we find in the spherical tensor expression no reference to the space-quantized hyperfine states $|M_F\rangle$ for e, g, or f, since, the atom being in zero magnetic field, these are all degenerate. On the contrary, the $|e\rangle$ in Eq. (9) would be the individual states labeled $|J_e, F_e, M_{F,e}\rangle$, and similarly for the $|g\rangle$ and the $|f\rangle$.

The expression for the intensity in spherical tensor notation is

$$I(t, t_0) = G|\langle J_e\|\hat{\mathbf{D}}\|J_g\rangle|^2 \times |\langle J_e\|\hat{\mathbf{D}}\|J_f\rangle|^2$$

$$\times \sum_{\substack{k,q, \\ F_e, F_e'}} (-1)^{F_e'-F_e+q} E_q^k U_{-q}^k A^k(F_e, F_e') B^k(F_e', F_e)$$

$$\times \exp\{[-i\omega(F_e, F_e')-\Gamma](t-t_0)\} \tag{18}$$

We distinguish the states e, e' by their hyperfine quantum numbers F_e, F'_e. The initial and final states appear specified only by their electronic quantum numbers, J_g and J_f.

In the first line of (18) we find the reduced matrix elements of the electric dipole operator between the $|g\rangle$ and the $|e\rangle$, on the one hand, and between the $|e\rangle$ and the $|f\rangle$, on the other.

In the second line we find the light tensors E_q^k and U_{-q}^k, defined by Eq. (16).

The quantities A^k, B^k in the second line are, effectively, $6-j$ symbols which project the $|e\rangle$ and the transition dipoles on to the k, q basis. They are

$$A^k(F_e, F'_e) = (-1)^{3I+2J_g-J_e-F_e}$$

This may be incorrect since for integer J_g and J_f, there is no corresponding $(-1)^{J_g-J_f}$ as in Luypaert.[35]

$$\times [(2F_e + 1)(2F'_e + 1)(2J_e + 1)(2J_g + 1)]^{1/2}$$

$$\times \begin{Bmatrix} k & J_e & J_e \\ I & F'_e & F_e \end{Bmatrix} \begin{Bmatrix} k & J_e & J_e \\ J_g & 1 & 1 \end{Bmatrix} \tag{19}$$

$B^k(F_e, F'_e)$ is the same expression with J_g replaced by J_f. It is to be noticed that the quantity that occurs in Eq. (18) is $B^k(F'_e, F_e)$, not $B^k(F_e, F'_e)$.

The last line of Eq. (18) exhibits the modulation of the fluorescent light at the hyperfine frequencies $\omega(F_e F'_e) = [E(F_e) - E(F'_e)]/\hbar$, and radiative decay of the free atom at the rate Γ.

Details of the derivation of Eq. (18) have been given by Luypaert.[35] In Figure 5 we show experimental results for ^{133}Cs which are compared, in the lower curve, with an evaluation of Eq. (18) for this case. [The numerical results were, in fact, checked against an evaluation of Eq. (15) and found to be identical.]

An example of the application of spherical tensors to Zeeman structures is to be found in Carrington and Corney.[37]

In Sections 2.2.5–2.2.7 we consider some characteristics of the multipoles and show how, by paying attention to the geometrical aspects of an experiment, we can select those multipoles that convey the information we require. The discussion is not limited to hyperfine structures: indeed, it applies to level-crossing and other atomic coherence effects, not solely to quantum beats.

2.2.5. Characteristics of the Multipoles

The multipoles $k = 0$, 1 and 2 have the following interpretation:

(a) $k = 0$. A scalar quantity related to the excited state population. The $k = 0$ component of the intensity is measurable, either by integrating the fluorescence over all directions, or by biasing the polarization components of the emission in some particular direction so as to represent a space average.

(b) $k = 1$. A vector part related to the degree of orientation in the excited state and therefore reflected in the content of circular polarization of the fluorescence.

(c) $k = 2$. A tensor part related to the degree of alignment in the excited state and therefore reflected in the content of linear polarization of the fluorescence.

The properties of the 6-j symbols involving k and the quantum numbers F_e (for example) lead to the result that, for terms having $k = 0$, $F_e = F'_e$. Modulation effects are therefore not to be expected either in the integrated light from the assembly or for those geometrical configurations that allow only the $k = 0$ component to be recorded. Similarly, terms with $|F_e - F'_e| = 2$ arise only in association with $k = 2$, and terms with $|F_e - F'_e| = 1$ only with $k = 1$.

Terms with $k = 0$, 1, and 2 arise in Eq. (18) if circularly polarized light is used in the exciting and fluorescent beams, but with linearly polarized light one finds only $k = 0$ or 2. These results may be obtained as shown below.

2.2.6. Linearly Polarized Light

Recall that **e** and **u** are unit vectors along the polarization directions of the exciting and detected beams. Let these directions be chosen as principal axes of the tensors E and U, respectively. Then the only nonvanishing components of E and U are

$$E_0^k = (2k + 1)^{1/2} \begin{pmatrix} 1 & 1 & k \\ 0 & 0 & 0 \end{pmatrix} = (U')_0^k \tag{20}$$

where the prime on U reminds us that we are referring to U to a principal axis different from that of E. Referred to the same axis as that of E, we have

$$U_0^k = \sum_q R_{q0}^k (U')_q^k = R_{00}^k (U')_0^k \tag{21}$$

(since components of U' having $q \neq 0$ are zero), where the R_{q0}^k are components of the matrix that describes a transformation of axes from **e** to **u**.

Looking back now to Eq. (18) we see that U_0^k is the only component of U^k that we need, since it is to couple with E_0^k, which is the only nonvanishing component of E^k. Finally, the only terms in (18) that do not vanish are those containing the factor

$$E_0^k U_0^k = (2k + 1) \begin{pmatrix} 1 & 1 & k \\ 0 & 0 & 0 \end{pmatrix}^2 P_k(\cos \theta) \tag{22}$$

where we have written $P_k(\cos \theta)$, the Legendre polynomial, as the rotation matrix element. θ is the angle between **e** and **u**.

From the symmetry properties of the $(3\text{-}j)$ symbol we learn that $\begin{pmatrix} 1 & 1 & k \\ 0 & 0 & 0 \end{pmatrix}$ vanishes for $k = 1$. Equation (22) tells us, therefore, that the use of linear polarizers and analyzers cannot produce terms other than $k = 0$ or 2. Moreover, we read from the $P_k(\cos \theta)$ term that there is no angular dependence for $k = 0$ and that the angular dependence for $k = 2$ is as $(3 \cos^2 \theta - 1)$. If θ is chosen to make this expression vanish (54.7°) the light reaching the detector will be represented by the $k = 0$ terms only. With pulsed excitation the light will be unmodulated and will represent the decay of excited-state population. With steady-state excitation there will be no level-crossing effect.

2.2.7. Circularly Polarized Light

Let the directions of the principal axes of E and U now be chosen along the directions of propagation of the exciting and fluorescent beams. By this choice, again, the polarization vector of the light is one of the basis vectors (e_1 or e_{-1} in this case), and the only nonvanishing components of the tensor are those with $q = 0$, namely,

$$E_0^k = (2k+1)^{1/2} \begin{pmatrix} 1 & 1 & k \\ -1 & +1 & 0 \end{pmatrix} = (U')_0^k \tag{23}$$

As before we have

$$E_0^k U_0^k = (2k+1) \begin{pmatrix} 1 & 1 & k \\ -1 & +1 & 0 \end{pmatrix}^2 P_k(\cos \theta) \tag{24}$$

The conclusion is that the fluorescent light may contain components $k = 0$, 1, and 2, but that, for $\theta = 90°$ the $k = 1$ component will vanish because of the vanishing of $P_1(\cos \theta) \equiv \cos \theta$. This is an important result in relation to the study of hyperfine structures in S states. Electronic alignment ($k = 2$) cannot exist in such states, so if modulation (or coherence) effects are to be studied one must select the $k = 1$ component of fluorescence. Since this is zero for $\theta = 90°$ one must choose some geometrical configuration where, *not only* is the light circularly polarized, *but also* the exciting and fluorescent beams are at some angle other than 90°.

2.2.8. Density-Dependent Effects: Multipolar Damping Coefficients

The equations we have given so far relate to the fluorescence of an assembly of isolated atoms. As has been shown by (among others) D'yakonov and Perel',[38,39] Omont,[40] Happer and Mathur,[41] and Carrington *et al.*,[42] the method of expansion in multipoles is powerful also when the atomic density is increased to the point that other interactions occur. There are two important effects. First, when the atomic density increases or the vapor cell volume is large enough, radiation trapping may occur: The fluorescence from one excited atom may excite others before the radiation leaves the cell and is detected. This is particularly likely when the lower state of the fluorescence is the atomic ground state. It will be appreciated that this multiple scattering of radiation in all directions has the effects (i) of increasing the lifetime of the fluorescence, and (ii) of changing the polarization character of the light that escapes.

The second effect is that of collisions between atoms. In simple cases the same electronic excitation (J) will be preserved, though it may be transferred from one atom to another, but the orientation or alignment of $\mathbf{J}$ will become randomized (transfer between levels of different M_J or F or M_F). In more complicated cases the excitation may be transferred to levels of different electronic excitation and the fluorescence may be lost to the detector.

The effects of radiation trapping and of the simple cases of collisions may be described formally by admitting the existence of different decay constants for the different multipoles. Thus, in Eq. (18), Γ is to be replaced by Γ^k. Three different damping constants, that is, three different lifetimes, make their appearance in the fluorescent light. Γ^0 is not affected by collisions (in the simple cases) and, according to D'yakonov and Perel',[38] the effect of radiation trapping is expressed by the equation

$$\Gamma^0/\Gamma = \pi^{-1/2} \int_{-\infty}^{\infty} \exp{(-t^2)} \exp{[(L/l_0)\exp{(-t^2)}]}\, dt \qquad (25)$$

an integral which is tabulated by Mitchell and Zemansky.[43] L is a characteristic dimension of the scattering cell and l_0 is the mean free path of a photon whose wavelength is at the peak of the optical transition. Γ^0 can be determined in the case of the field-free situation we have analyzed by setting polarizer and analyzer at 54.7° to one another. When a damping constant is determined from Hanle-effect experiments using linear polarizers it is Γ^2 that is measured, but Γ^0 can be determined directly at finite magnetic fields, as was shown by Gunn and Sandle,[44] by orienting the field in the (111) direction relative to Cartesian axes determined by the directions of linear polarizer and analyzer.

The decay constants Γ^0 and Γ^2 for mercury atoms in zero magnetic field were determined over a range of vapor densities by Deech and Baylis[45] by direct measurements of the decay of fluorescence. With an analyzer parallel to the polarizer, two exponential components were present, characterized by Γ^0 and Γ^2. The former could be isolated by setting the analyzer at 54.7° with respect to the polarizer, and the latter evaluated by subtraction. It was confirmed that the values of Γ^2 obtained from these decay curves were in satisfactory agreement with values obtained from Hanle-effect experiments over the limited range of vapor density where reliable values could be obtained from both kinds of experiment. At low densities the subtraction procedure was impracticable. Values of Γ^2 obtained from Hanle-effect experiments at these low densities tended towards the same limit at zero density as did values of Γ^0.

The general expression for the Γ^k in terms of Γ, the free-atom decay constant, is

$$\Gamma^k = \Gamma - C^k(\Gamma - \Gamma^0) + \Gamma^k_{(\text{coll})} \qquad (26)$$

where

$$C^k = 1 \qquad \text{for } k = 0$$

$$C^k = \tfrac{3}{10}(2J_e + 1)[6 + (-1)^k] \left\{ \begin{matrix} 1 & 1 & k \\ J_e & J_e & J_g \end{matrix} \right\}^2 \qquad \text{for } k = 1, 2$$

it being understood that this applies only when the initial and final states are the same ($J_g = J_f$). Estimates of $\Gamma^k_{(\text{coll})}$ can be made from expressions given in the literature cited.

In the so-called "quenching" collisions the energy of excitation is lost by transfer out of the original state to some different electronic state of either of the collision partners. Part of the energy may be lost as kinetic energy, or as vibrational energy if the colliding particle is a molecule of a foreign gas, or in a variety of other conceivable ways. The decay constant for the population of atoms in the initially excited state is, for these "quenching" collisions, represented by

$$\Gamma^0_{\text{total}} = \Gamma^0 + \Gamma^0_{(\text{coll})}, \qquad \text{where } \Gamma^0_{(\text{coll})} \propto \nu \qquad (27)$$

Γ^0 is determined by measuring the total decay constant as a function of density of the perturbing species, ν, and extrapolating to zero density.

For some types of quenching collisions the reverse collisions contribute significantly to the population of the initially excited state. The decay of fluorescent light then ceases to follow a single exponential. It may be represented by a sum of exponential terms, each characterized by a different rate constant. An example is reported in a recent paper by Pendrill.[46] The rate constant of the dominant term tends to the Γ^0_{total} of the single-exponential decay at lower pressures. The rate constants of the other terms are related to the decay of the partner states and to the collisional rate constants. All these rate constants appear also in the equations describing sensitized fluorescence.

2.2.9. Experimental Techniques and Results

For details of the experimental realization of the foregoing analyses and for results that have been obtained we must refer the reader to Haroche's review,[25] to the original papers already cited, and to the article by R. G. Fowler, Chapter 26 of this work.

2.3. Superposition States of Mixed Parity

The modulation effects that have been described hitherto have been ascribed to the synchronous precessions of radiating atomic dipoles, motions that find quantum-mechanical expression in terms of coherent superpositions of states of different electronic orientation. It was supposed that the spontaneous decay constants for all states forming the superposition were identical. When this is not the case modulation can arise in the integrated intensity of the light received. Clearly, this can no longer be interpreted as a "searchlight effect" or as the spatial interference of components of different polarization, but it may be regarded as a temporal interference: radiation from each excited atom is periodically restrained and released by the coherent mixing of dipoles of different strengths oscillating at different frequencies. An example of such a situation is to be found in the superposition of states of opposite parity: the superposition of the $2S$ and $2P$ states of atomic hydrogen is a case in point. An early analysis of the effects to be expected by pulse excitation and by modulated excitation of these states was given by Series[47] and is represented in Figure 6. It was supposed that the atoms were in an electric field so that the eigenstates of the Hamiltonian for $n = 2$ were superpositions of the S and P states. (For the purposes of the present argument we may neglect the spin.) The symmetry-breaking perturbation was the application of Lyman-α light which was to excite atoms from the ground state $1S$ to the P components of the two nondegenerate $2S$–$2P$ superpositions. Modulation of the emitted Lyman-α light was to be expected at the frequency interval between the Stark-shifted S–P levels.

The experimental realization of this kind of superposition-state has been achieved by Bashkin et al.,[48] by Sellin et al.,[49] by Andrä[50] and co-workers, and most recently by van Wijngaarden et al.[51] These last authors allow a beam of hydrogen atoms in the metastable state $2\,{}^2S_{1/2}$ to enter a region where they suddenly experience an electric field that is very accurately known. Atoms in pure S states as they enter the electric field are no longer in eigenstates of the atom in the field but are in superpositions of the S–P eigenstates. They begin to radiate modulated Lyman-α light, which appears as a periodic spatial luminosity of the beam and which is measured by a movable detector. The modulation may be envisaged as a

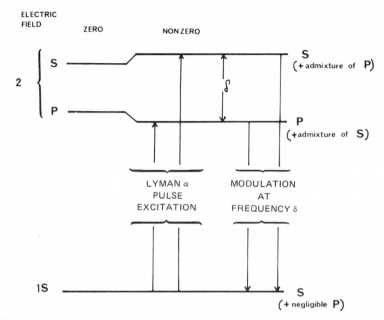

Figure 6. Representation of modulated fluorescence from superposition of S and P states in hydrogen.

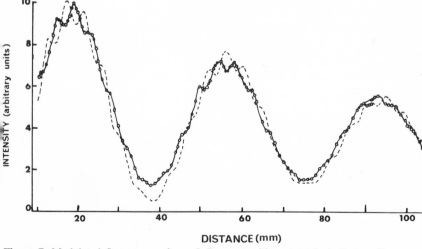

DISTANCE (mm)

Figure 7. Modulated fluorescence from S–P superposition states in hydrogen (from van Wijngaarden et al.[51]). The low-frequency modulation arises from $S_{1/2}$–$P_{1/2}$ superpositions; the high-frequency from $S_{1/2}$–$P_{3/2}$. The broken curve represents the results of a theoretical analysis in which the onset of the field was supposed to be sudden. Closer agreement with the experimental curve was obtained when allowance was made for the fringing field.

periodic oscillation of population between pure S and P states, the latter being strongly radiative and the former metastable. The modulation frequencies that have been observed in this experiment (Figure 7) correspond to the whole complex of Bohr frequencies in the $n = 2$ state, that is to say, the intervals between all the hyperfine components of $2\,^2S_{1/2}$, $2\,^2P_{1/2}$, and $2\,^2P_{3/2}$.

The fascination of these experiments is that the modulation of the fluorescent light provides a direct record of the interval between the S and P states in hydrogen, which is the Lamb-shift interval (slightly modified by the electric field). No claim is made for the experiments that have been described that they give a value for the Lamb shift better than has been obtained by other methods, but it is hoped that the method—or something related to it—may be applied to hydrogenlike ions of high Z for which accurate values of the Lamb shift are needed. If the behavior of hydrogen under a symmetry-breaking pulse can be understood in detail, it will be possible to interpret with confidence the behavior of hydrogenlike ions under similar circumstances.

3. Pulsed (Stepped) Magnetic and Electric Fields

3.1. Pulsed Magnetic Field

The transients observed in the intensity of fluorescence of atoms subjected to a pulse of some excitation process were analyzed in Section 2. Similar transients can be observed under conditions of constant excitation but following a sudden change in some physical quantity that affects the atomic structure. For example, a sudden change in the steady value of an electric or a magnetic field that determines the energies of the state of the atom gives rise to a transient in the intensity as it settles to a new steady-state value.

An early application was reported by Dehmelt.[52] A beam of circularly polarized sodium D light is passed through a cell containing sodium vapor, the dimensions and vapor density being such that about 50% absorption occurs. The intensity $I + \Delta I$ of the transmitted beam is greater than the initial intensity I because the optical pumping process removes sodium atoms from those ground-state Zeeman sublevels which more strongly absorb the circularly polarized light. A sudden reversal of the magnetic field (about 50 μT parallel to the optical beam) reverses the role of the ground-state sublevels. The intensity drops suddenly to $I - \Delta I$ and then relaxes back to $I + \Delta I$ with a lifetime dependent on the rate of the pumping process (proportional to I) and of other relaxation processes such

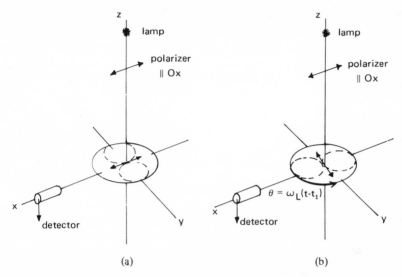

(a)   (b)

Figure 8. The stepped magnetic field experiment. (a) Magnetic field zero: Detector records zero intensity. (b) At time t the dipoles have precessed through $\theta = \omega_L(t - t_1)$. The intensity at the detector is modulated at the angular frequency $2\omega_L = -2g\mu_B B/\hbar$.

as wall collisions and collisions between atoms of sodium and the buffer gas in the cell.

An application of the principle in the observation of resonance fluorescence was made by Dodd et al.[53] Here the resonance fluorescence of the $6\,^1S_0 \leftrightarrow 6\,^3P_1$ (254 nm) line in mercury was observed. The method may readily be appreciated by appealing to a classical model. Consider the case of incident radiation, polarized parallel to the x axis, exciting a sample of atoms situated in a cell at the origin. This is illustrated in Figure 8. The dotted contour represents the polar distribution of fluorescent intensity in the x–y plane appropriate to a $J = 1 \leftrightarrow J = 0$ transition as in Figure 3a. Under conditions of zero magnetic field the axis of the equivalent dipole remains stationary in space. A detector placed along the x axis records zero intensity—the zero of the normal Hanle signal.

When a steady magnetic field B along the z axis is suddenly switched on at time $t = t_1$, all previously established equivalent classical dipoles (i.e., atoms excited with $t_0 < t_1$), begin to precess about the z axis at the Larmor frequency $\omega_L = -g\mu_B B/h$; at time t they lie at an angle to the detector given by $\theta = \omega_L(t - t_1)$. As these dipoles rotate, radiate, and decay, other dipoles are excited and rotate, their angle to the detector at time t being given by $\theta = \omega_L(t - t_0)$. Under the transient condition the intensity oscillates with frequency $2\omega_L$, damped with a mean lifetime $\tau^{(2)} = 1/\Gamma^2$, and

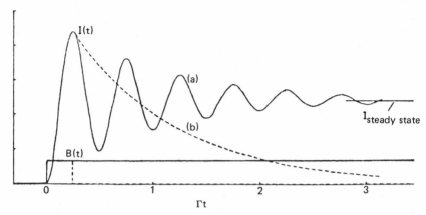

Figure 9. Predicted intensity in the stepped magnetic field experiment. (a) The modulation case: The magnetic field is such that two periods of modulation occur in one mean lifetime ($4\pi/2\omega_L = \tau$). (b) The exponential case: The magnetic field is switched off after a time interval equal to a half period of modulation $[(t_2 - t_1) = \pi/2\omega_L = \tau/4]$.

settles to a new steady-state intensity appropriate to a point on the wings of the Hanle signal in the field B. If the magnetic field is suddenly switched off the intensity decays exponentially back to zero with the same mean lifetime. Figure 9 shows the two kinds of transients. The experiment yields the values of g and $\tau^{(2)}$ independently (it is the product $g\tau^{(2)}$ that is determined from the width of the Hanle curve).

A quantum-mechanical description of the above experiment can readily be given by adapting the theory in Sections 1 and 2. The intensity $I(t, t_0)$ is calculated as before, the interaction with the field B being written into the Hamiltonian. The geometrical arrangement shown in Figure 8 generates alignment ($k = 2$), the formation of a superposition state from $J = 1$, $M_J = \pm 1$, and so leads to the expression

$$I(t, t_0) \propto \sin^2 \theta \exp\left[-\Gamma(t - t_0)\right] \tag{28}$$

where $\theta = \omega_L(t - t_1)$ for $t_0 \le t_1$ and $\theta = \omega_L(t - t_0)$ for $t_0 \ge t_1$. The intensity at time t is obtained by integrating over t_0 from $-\infty$, through the discontinuity at t_1, up to t. The result is

$$I(t) \propto \frac{4\omega_L^2}{\Gamma^2 + 4\omega_L^2} - \frac{2\omega_L}{(\Gamma^2 + 4\omega_L^2)^{1/2}} \cos\left[2\omega_L(t - t_1) - \phi\right] \exp\left[-\Gamma(t - t_1)\right] \tag{29}$$

with $\tan \phi = \Gamma/2\omega_L$. The first term on the right is just the steady-state Hanle signal; the second is the damped oscillation of the transient signal. The exponential decay following sudden switching of the field to zero is obtained by introducing a further discontinuity at t_2 ($> t_1$), and, for $t > t_2$, fixing θ at the values reached at $t = t_2$. For a more complicated atomic system, the method of expansion in multipoles would lead to a generalization of Eq. (28). Transients of other kinds may be described by a generalization of Eq. (29).

In the experiments of Dodd et al.[53] the detector recorded single photoelectric events and measured the time delay between the step of magnetic field and the arrival of the first photoelectron. Use of a time-to-pulse-height converter and a multichannel analyzer, with proper correction for "pulse pileup" and system linearity, yielded results for $\tau^{(2)}$ with better than 1% accuracy. A typical result is shown in Figure 10. The method was used to study the variation of $\tau^{(2)}$ with atomic density and thereby to check theories cited earlier for how this quantity is influenced by the phenomena of radiation trapping and atomic collisions. The technique was further developed by Piper and Sandle[54,55] to study the effects of cell geometry, atomic collisions, and isotopic constitution of the vapor in the resonance cell and in the lamp.

3.2. Pulsed Electric Field

The method is obviously applicable to the case where a step or pulse of electric field is applied to an atomic sample initially in a region free of electric and magnetic fields. In this case one obtains a measurement of the quadratic differential Stark coefficient rather than the Landé g coefficient. The principal advantage of the pulsed method is that, with an electric field applied for only a short time, the effects of discharge in the vapor are largely eliminated. An experiment of this kind has been performed by Sandle et al.[56] and is reported in the article by Kollath and Standage, Chapter 21 of this work.

3.3. Pulsed Radio-Frequency Field

In the traditional double-resonance method a radio-frequency magnetic field creates a coherent superposition of states from a pure state excited by appropriately polarized incident light. A transient experiment may obviously be developed by applying the rf field in a pulse. Pulses or steps of rf field have been extensively used in optical pumping experiments to study the relaxation of orientation in the ground state. We refer only to the early experiments of Dehmelt,[52] and of Cagnac and Brossel,[57] and a more recent paper of Gibbs and White.[58] These experiments are more concerned with relaxation processes than with atomic properties. Of more direct application to atomic properties, the method has been used by Jacobson[59] for a study of the nutational motion of aligned mercury atoms in the metastable $6\,^3P_2$ state. In this case, Hg atoms were excited in a

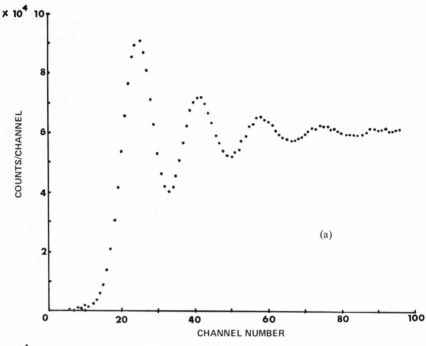

(a)

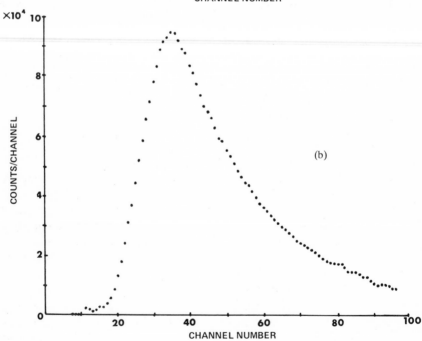

(b)

resonance vessel by electron impact parallel to the static magnetic field, thus producing an alignment in the metastable state. The pulsed rf field, perpendicular to the steady field, disturbed this alignment. The transient character of the disturbance was monitored by absorption of linearly polarized light, $\lambda = 546$ nm, corresponding to the transition $6\,^3P_2$ to $7\,^3S_1$.

4. Resolution within the Natural Width

By Fourier transformation of the time-resolved intensity variations we obtain spectra whose individual components are Lorentz curves centered on frequencies corresponding to the intervals $\omega_{ee'}$ between excited levels, and whose widths, provided the vapor density is sufficiently low, depend inversely on the radiative lifetime of the excited states. (The full width at half-height is $2\Gamma = 2/\tau$ if the lifetimes of the two levels are equal.)

It is possible to change the shape of the curves by biasing the signal in favor of a particular group of atoms: in particular, a bias in favor of the longer-lived atoms produces narrower curves. This can readily be shown analytically and has been demonstrated experimentally on a number of occasions. The technique has been applied in Mössbauer spectroscopy and in nuclear physics as well as in atomic physics. It appears to have been discovered independently by a number of different authors.

4.1. Theory

A simple form of the quantum beat signal, Eq. (4), is

$$I(T) = (A + B \cos \omega_0 T) \exp (-\Gamma T) \qquad (30)$$

where T is the time elapsed after the pulse and ω_0 is the interval between the excited states. The Fourier cosine transform of (30) is

$$I(\omega) = A\Gamma/(\Gamma^2 + \omega^2) + \tfrac{1}{2}B\Gamma/[\Gamma^2 + (\omega - \omega_0)^2] + \tfrac{1}{2}B\Gamma/[\Gamma^2 + (\omega + \omega_0)^2] \qquad (31)$$

For ω, $\omega_0 > 0$ the second term gives the expected Lorentzian centered at $\omega = \omega_0$. The peak of $I(\omega)$ is not exactly at $\omega = \omega_0$ owing to overlap from the other terms, but this can be allowed for if necessary.

Figure 10. Typical experimental results in the stepped magnetic field experiment. (a) The modulation case: ^{198}Hg in cell at $-21.6°$C (5.7×10^{17} atom m^{-3}); $B = 190\ \mu T$; 7.78 ± 0.02 ns per channel. (b) The exponential case. ^{198}Hg in cell at $-11.4°$C (1.09×10^{18} atom m^{-3}); B and time calibration as in (a) (from Dodd et al.[53]).

A biasing function $f(T)$ can be introduced at the same time as the Fourier transform is calculated:

$$G(\omega) = \int_0^\infty I(T) f(T) \cos \omega T \, dT \tag{32}$$

Various forms of $f(T)$ have been used. A simple form that corresponds to the experimental technique of delayed detection is the step function

$$f(T) = \begin{cases} 0, & T < T_1 \\ 1, & T \geq T_1 \end{cases} \tag{33}$$

This form of biasing, though easy to realize experimentally, introduces oscillatory structure (of period $1/T_1$) into the wings of the spectral lines. It has been studied by Schenk et al.[60] and by Figger and Walther.[61]

An attractive function is the increasing exponential

$$f(T) = K e^{\gamma T}, \qquad \gamma < \Gamma \tag{34}$$

which preserves the Lorentzian shape of the lines but narrows them from 2Γ to $2(\Gamma - \gamma)$. It is, however, unrealistic to use a biasing function which increases indefinitely, and the function must be cut off at some time, which again introduces spurious oscillatory structure.

The Gaussian Biasing Function

A biasing function that has greater appeal is the displaced Gaussian

$$f(T) = \exp\left[-(T - a)^2 / b^2\right] \tag{35}$$

It is easy to show that, if the width, b, of the Gaussian is properly chosen in relation to the displacement of its peak, a, the oscillatory structure is suppressed. The required relation is

$$b = (2a/\Gamma)^{1/2} \tag{36}$$

With $f(T)$ chosen according to Eqs. (35) and (36) the biased transform

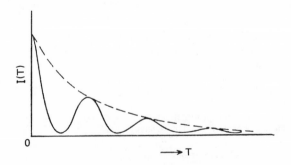

Figure 11. The function $I(T)$.

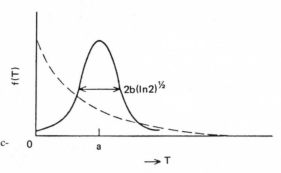

Figure 12. Gaussian biasing function, $\exp[-(T-a)^2/b^2]$.

$G(\omega)$ of $I(T)$ [Eqs. (32) and (30)] becomes

$$G(\omega) = \tfrac{1}{2}b\pi^{1/2} \exp(-\Gamma^2 b^2/4)\{A \exp(-\omega^2 b^2/4)$$
$$+\tfrac{1}{2}B \exp[-(\omega-\omega_0)^2 b^2/4] + \tfrac{1}{2}B \exp[-(\omega+\omega_0)^2 b^2/4]\} \qquad (37)$$

This is to be compared with Eq. (31) with $A = B = 1$. The Lorentzians peaked at $\omega = \pm\omega_0$ have been transformed into Gaussians whose full width at half-intensity is $4(\ln 2)^{1/2}/b$. The condition under which this is less than the width of the Lorentzian is

$$b > 2(\ln 2)^{1/2}/\Gamma = 1.67\tau \qquad (38)$$

Figures 11–13 show the functions $I(T)$, $f(T)$, $I(\omega)$, and $G(\omega)$.

When b is chosen so that the widths are equal the height of the Gaussian is about 0.7 times the height of the Lorentzian. It will be noticed that the height of the Gaussian decreases exponentially with b^2, that is, with a, the time interval between the initiating pulse and the peak of the Gaussian. This implies a degradation of signal relative to noise and is the price one pays for the line-narrowing. On the other hand, the steeper fall of

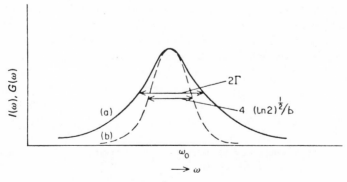

Figure 13. Narrowing achieved by biasing. (a) The unbiased transform. $I(\omega)$. (b) The biased transform, $G(\omega)$.

the wings of the Gaussian relative to the Lorentzian gives an important advantage when one is trying to improve the resolution of overlapping spectral lines.

4.2. Practicalities and Examples

The advantages gained by the technique of time-biasing have to be weighed against the increased uncertainties introduced by the degraded signal-to-noise ratio. If the problem were simply to obtain the best value for the center of a symmetrical line it could not be maintained that narrowing the width at the expense of increasing the uncertainty would lead to an improvement. On the other hand, if the problem is to find the relative positions of overlapping lines where the individual line profiles are not known with confidence, then improved resolution—even at the expense of signal-to-noise ratio—is to be preferred to an analysis based on profile synthesis, the alternative method. And although the Fourier-transformed signals of quantum beats are in principle Lorentzian curves, one cannot always rely on this on account of instrumental nonlinearities and unknown complicating factors. It is a matter of practical experience in general that the methods of profile analysis can lead to errors, and resolution of components, where possible, is to be preferred. A particular case, the hyperfine structure of $3\,^2P_{3/2}$ in sodium, was studied in detail by Deech et al.[62] It was argued that the results obtained by line narrowing were more reliable than those obtained by profile analysis of experimental results obtained under comparable conditions.

An improvement in accuracy of an important spectroscopic interval, the Lamb shift in hydrogen, has been reported by Lundeen and Pipkin[63] in an application of the time-biasing technique. They measured electric dipole resonances between hyperfine components of $2\,^2S_{1/2}$ and $2\,^2P_{1/2}$ in a fast beam of hydrogen atoms. The complete pattern of resonances is complicated, but was simplified by a technique of state selection which very largely suppressed all but one of the transitions. Resonance curves are obtained by scanning the frequency of the rf field, so this is not strictly an experiment in which time-resolved signals are obtained, but the principles of time biasing still apply. Under ordinary conditions the width of the resonance curves is determined mainly by the inverse of the very short lifetime of the P states—the S states are metastable. In Lundeen and Pipkin's experiment a spectacular narrowing of the resonance curve was achieved by employing Ramsey's well-known two-field arrangement to elicit a time-biased signal. Atoms in a beam pass successively through two interaction regions before reaching the detector. The main part of the signal corresponds to atoms that undergo the transition from S to P in the first interaction region, and this resonance is broad. It does not depend on

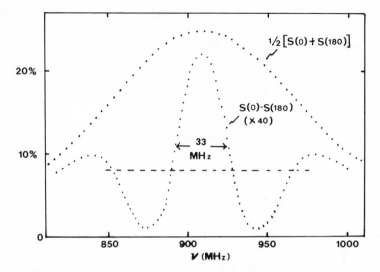

Figure 14. Lamb-shift resonance in H. The upper curve is typical of "ordinary" resonance curves, the lower shows the response of atoms that have interacted with the first field and have survived to interact with the second (from Lundeen and Pipkin[63]).

the phases of the rf fields. The total signal contains also the contribution from atoms that survive to interact with the field in the second region. The sign of this contribution depends on the relative phase of the fields, and the contribution is separated from the total signal by taking the difference between $S(0)$ and $S(180)$, signals obtained with the fields in phase and in antiphase, respectively.

The narrowing achieved is seen by comparing the difference curve, which is the lower curve of Figure 14 with the upper curve, which is the average of $S(0)$ and $S(180)$ and so does not reflect the time biasing. The width of the difference curve is about three times less than the natural width of the P states and the authors claim to have gained a factor of 3 in precision over earlier measurements of the Lamb shift. It will be noticed that the signal-to-noise ratio in the time-biased curve is very high, notwithstanding the reduction in absolute signal strength. The oscillatory structure at the sides of the main peak will also be noticed: this reflects the spatial (and therefore the temporal) envelope of the rf fields.

4.3. Conceptual Problems

It is sometimes asked whether the narrowing of resonance curves below the "natural" width does not violate the uncertainty principle. Of course it does not. Curves having the natural width are obtained by taking unweighted signals from all members of an assembly of atoms decaying by spontaneous emission of radiation. The narrowed curves are obtained by taking signals from a selected sample whose mean lifetime is longer than that of all members of the assembly. The relation between mean lifetime and spectral linewidth is simply the reciprocal relationship of Fourier transform pairs.

A more well-founded objection is to point out that all atoms, irrespective of how long they have survived in the excited state, have the same probability of decay. Does not this probability govern the spectral linewidth of the emitted radiation? How, then, can the linewidth be reduced by taking the light from a biased sample of atoms? Indeed, the spectral linewidth as determined by an ordinary spectrometer *is* independent of the time of observation after an exciting pulse. But the significant point is that the spectra obtained in an ordinary spectrometer are transforms of the amplitude of the light field correlated against itself, whereas the spectra we are concerned with here are transforms of the intensity of the light field correlated against a biasing function which we are free to choose. The lines are narrowed because we bias against the wings. The information to which we attach most weight is that which relates to the region near line center.

5. Concluding Remarks

We have attempted in this article to show how the time evolution of the intensity of fluorescent light from atoms reflects the dynamical processes in the atoms themselves. By implication, and by means of examples, we have shown how atomic parameters may be determined by measurements of fluorescent light, and we have shown how such determinations are free of Doppler broadening. We have shown also how the time-evolving intensity is related to the steady-state intensity and how the same atomic parameters enter into both, but we have not discussed the techniques in sufficient detail to assess their relative merits or to allow a critical comparison between them and alternative Doppler-free techniques such as radio-frequency optical double resonance, laser fluorescence or absorption from atomic beams, saturated absorption spectroscopy, multiple-photon laser spectroscopy, and others. Some comments on the relation between double resonance and the technique of modulated excitation were made in Section 1.4 and on certain advantages of the technique of pulsed electric fields in Section 3.2.

The impact of lasers on high-resolution spectroscopy has been explosive: several of the techniques just mentioned did not exist before lasers. The steady-state version of coherence spectroscopy (level-crossing spectroscopy), however, did exist and a great deal of valuable work was accomplished using conventional light sources. As for time-resolved coherence spectroscopy, the principles were known and the pioneer experiments had been done in the prelaser age, but vastly more territory in this domain now lies open for exploration with lasers on account of their intensity, their tunability and the pulse technology that has been developed. The spectral purity offered by lasers is not useful in this work since the spectral range has to span the whole structure that is to be explored.

Level-crossing spectroscopy has been invigorated by lasers and many new structures have been explored both by this technique and also by time-resolved spectroscopy. The results have been gratifyingly in agreement. The choice of one or the other method is often determined by the equipment available to the experimenter—he or she requires magnetic fields and steady-state detection on the one hand, and short-pulse technology and counting techniques on the other. Where time-biasing is to be applied, time-resolved detection is essential (or, if atomic beams are used, space-resolved detection). It is sometimes useful to combine the two techniques, as when the intervals in zero field are uncomfortably large (time-resolution uncomfortably small) for pulse-counting but can be made manageable by the application of a magnetic field. Then, time-resolved measurements are made for a set of values of magnetic field, and for each field point an integration is made over time, incorporating a biasing function if required.[62] In this way one obtains a set of values of intensity as a function of magnetic field, which constitutes a level-crossing curve.

Acknowledgments

The authors, in conclusion, wish to record their indebtedness to their colleagues and collaborators, many of whom have allowed their work to be quoted as examples to illustrate this text. They also wish to thank the authors specified in the figure legends and references for permission to use published material, and also the owners of copyright as follows: for Figures 1, 5, 7, 9, and 10, the Institute of Physics; for Figures 11, 12 and 13, the Rank Prize Funds and Academic Press; and for Figure 14, the American Institute of Physics.

References

1. G. Breit, *Rev. Mod. Phys.* **5**, 91 (1933).
2. J. P. Barrat and C. Cohen-Tannoudji, *J. Phys. Radium* **22**, 329, 443 (1961).
3. C. Cohen-Tannoudji, Thesis, University of Paris (1962).

4. A. V. Durrant, *J. Phys. B* **5**, 133 (1972).
5. A. Corney and G. W. Series, *Proc. Phys. Soc. London* **83**, 213 (1964).
6. A. Corney, *J. Phys. B*. **1**, 458 (1968).
7. T. Skalinski, A. Kopystynska, and K. Ernst, *Bull. Acad. Pol. Sci.* **13**, 851 (1965).
8. M. Broyer, J.-C. Lehmann, and J. Vigué, *J. Phys. (Paris)* **36**, 235 (1975).
9. A. Corney, B. P. Kibble, and G. W. Series, *Proc. R. Soc. London A* **293**, 70 (1966).
10. A. V. Durrant and B. Landheer, *J. Phys. B* **4**, 1200 (1971).
11. R. Q. Hackett and G. W. Series, *Opt. Commun.* **2**, 93 (1970).
12. W. W. Chow, M. O. Scully, and J. O. Stoner, *Phys. Rev. A* **11**, 1380 (1975).
13. J. L. Cojan, *Ann. Phys. (Paris)* **9**, 385 (1954).
14. M. F. H. Schuurmans, *J. Phys. (Paris)* **37**, 469 (1976); *Z. Phys.* **A279**, 243 (1976).
15. G. W. Series, *Proc. Phys. Soc. London* **91**, 432 (1967).
16. W. Hanle and G. Stanzel, *Z. Naturforsch.* **25a**, 309 (1970).
17. G. Stanzel, *Z. Phys.* **A270**, 361 (1974).
18. W. Siegmund and A. Scharmann, *Z. Phys.* **A276**, 19 (1976).
19. J. M. Hansen and H. W. Webb, *Phys. Rev.* **72**, 332 (1947).
20. R. van Traubenberg and S. Levy, *Z. Phys.* **44**, 549 (1927).
21. K. L. Hertel, *Phys. Rev.* **29**, 848 (1927).
22. I. Walerstein, *Phys. Rev.* **33**, 800 (1929).
23. W. Hanle, *Z. Phys.* **30**, 93 (1924).
24. W. Hanle, *Ergeb. Exakt Wiss.* **4**, 214 (1925).
25. S. Haroche, *High Resolution Laser Spectroscopy* (Ed. K. Shimoda), pp. 253–313. Springer, Berlin (1976).
26. E. B. Aleksandrov, *Opt. Spektrosk.* **17**, 957 (1964) [English transl. **18**, 522 (1964)].
27. J. N. Dodd, R. D. Kaul, and D. M. Warrington, *Proc. Phys. Soc. London* **84**, 176 (1964).
28. J. N. Dodd, W. J. Sandle, and D. Zissermann, *Proc. Phys. Soc. London* **92**, 497 (1967).
29. T. Hadeishi and W. A. Nierenberg, *Phys. Rev. Lett.* **14**, 891 (1965).
30. S. A. Bagaev, V. B. Smirnov, and M. P. Chaika, *Opt. Spektrosk.* **41**, 166 (1976) [English transl. *Opt. Spectrosc.* **41**, 98 (1976)].
31. S. Heron, R. W. P. McWhirter, and E. H. Rhoderick, *Proc. R. Soc. London A* **234**, 565 (1956).
32. G. C. King and A. Adams, *J. Phys. B* **7**, 1712 (1974).
33. G. C. King, A. Adams, and D. Cvejanovic, *J. Phys. B* **8**, 356 (1975).
34. D. A. Shaw, A. Adams, and G. C. King, *J. Phys. B* **8**, 2456 (1975).
35. R. Luypaert, Thesis, University of Reading (1976). See also R. Luypaert and J. Van Craen, *J. Phys. B* **10**, 3627 (1977).
36. J. S. Deech, R. Luypaert, and G. W. Series, *J. Phys. B* **8**, 1406 (1975).
37. G. G. Carrington and A. Corney, *J. Phys. B* **4**, 849 (1971).
38. M. I. D'yakonov and V. I. Perel', *Zh. Eksp. Teor. Fiz.* **47**, 1483 (1964). [English transl. *Sov. Phys. JETP* **20**, 997 (1965)].
39. M. I. D'yakonov and V. I. Perel', *Zh. Eksp. Teor. Fiz.* **48**, 345 (1965) [English transl. Soviet Phys. JETP **21**, 227 (1965)].
40. A. Omont, *J. Phys. (Paris)* **26**, 26 (1965).
41. W. Happer and B. S. Mathur, *Phys. Rev.* **163**, 12 (1967).
42. C. G. Carrington, D. N. Stacey, and J. Cooper, *J. Phys. B* **6**, 417 (1973).
43. A. C. G. Mitchell and N. W. Zemansky, *Resonance radiation and excited atoms*, Cambridge University Press, Cambridge (1934).
44. H. I. Gunn and W. J. Sandle, *J. Phys. B* **4**, L1 (1971).
45. J. S. Deech and W. E. Baylis, *Can. J. Phys.* **49**, 90 (1971).
46. L. R. Pendrill, *J. Phys. B* **10**, L469 (1977).
47. G. W. Series, *Phys. Rev.* **136**, A684 (1964).

48. S. Bashkin, W. S. Bickel, D. Fink, and R. K. Wangsness, *Phys. Rev. Lett.* **15**, 284 (1965).
49. I. A. Sellin, P. M. Griffin, and J. A. Biggerstaff, *Phys. Rev. A* **1**, 1553 (1970) (and earlier papers referenced there).
50. H. J. Andrä, *Phys. Scr.* **9**, 257 (1974).
51. A. van Wijngaarden, E. Goh, G. W. F. Drake, and P. S. Farago, *J. Phys. B* **9**, 2017 (1976).
52. H. G. Dehmelt, *Phys. Rev.* **105**, 1487 (1957).
53. J. N. Dodd, W. J. Sandle, and O. M. Williams, *J. Phys. B* **3**, 256 (1970).
54. J. A. Piper and W. J. Sandle, *J. Phys. B* **3**, 1357 (1970).
55. J. A. Piper and W. J. Sandle, *J. Phys. B* **5**, 377 (1972).
56. W. J. Sandle, M. C. Standage, and D. M. Warrington, *J. Phys. B* **8**, 1203 (1975).
57. B. Cagnac and J. Brossel, *C.R. Acad. Sci.* **249**, 253 (1959).
58. H. M. Gibbs and C. W. White, *Phys. Rev.* **188**, 180 (1969).
59. E. Jacobson, *J. Phys. B.* **8**, 869 (1975).
60. S. Schenk, R. C. Hilborn, and H. Metcalf, *Phys. Rev. Lett.* **31**, 189 (1973).
61. H. Figger and H. Walther, *Z. Phys.* **267**, 1 (1974).
62. J. S. Deech, P. Hannaford, and G. W. Series, *J. Phys. B* **7**, 1131 (1974).
63. S. R. Lundeen and F. M. Pipkin, *Phys. Rev. Lett.* **34**, 1368 (1975).

15
Laser High-Resolution Spectroscopy

W. DEMTRÖDER

1. Introduction

One of the essential limitations imposed on experimental information about finer details of atomic structure is determined by the attainable spectral resolution.

The classical absorption—or emission—spectroscopy, which has extensively contributed to our knowledge about atomic spectra, is mainly limited in its spectral resolution by the resolving power of the wavelength-dispersing instruments. The Doppler width may only be approached by interferometric methods. The introduction of Doppler-free techniques, such as level-crossing spectroscopy or optical double resonance (see Chapter 9), has greatly enhanced the experimental capability and resulted in resolution of hyperfine structure splittings or Zeeman sublevels. Because of intensity problems, however, these methods have been restricted mainly to atomic resonance transitions, where intense resonance lines from hollow-cathode lamps or rf-excited gas discharges can be used to obtain a sufficiently large population, in selected upper states.

The introduction of lasers to atomic spectroscopy has considerably improved the situation concerning intensity and spectral resolution. Since the spectral brightness of existing lasers may be several orders of magnitude higher than that of spectral lamps, experiments that had failed with conventional techniques because of lack of intensity, can now be done with good signal-to-noise ratios.

W. DEMTRÖDER • Fachbereich Physik, Universität Kaiserslautern, Germany.

With fixed-frequency lasers, atomic laser spectroscopy was restricted to fortuitous coincidences between atomic transitions and laser lines and has been performed mainly on the laser medium itself. The first experiments in atomic laser spectroscopy were concerned with laser transitions between the excited states of neon which are observed in the He–Ne laser.[1] Compared to conventional gas discharge spectroscopy, where the fluorescence spectrum is measured, these laser spectroscopic techniques can utilize the induced emission as a monitor presenting the advantages of much higher intensity, sensitivity, and accuracy.

The great breakthrough in atomic laser spectroscopy came with the development of tunable lasers, which can be tuned within a certain range to the center of any atomic transition.[2a] The most important type of laser used so far in atomic spectroscopy is the dye laser,[2b] either in its pulsed operation (with nitrogen lasers or flashlamps as pumping sources) or in its argon-laser-pumped cw version.[3] Using different dyes the accessible wavelength range extends from about 340 to 1000 nm and may be even further enlarged by frequency-doubling[4] or frequency-mixing techniques.[5] Some experiments have been reported in the near-infrared using tunable semiconductor lasers[6] or optical parametric oscillators.[7]

The most useful property of the tunable lasers for high-resolution spectroscopy is the extremely small linewidth, which they can be made to produce by using appropriate techniques. In combination with the attainable spectral brightness, spectroscopists can now use a new class of Doppler-free techniques, based on saturation phenomena, multiphoton transitions, or on the linear interactions of monochromatic light with atoms in highly collimated atomic beams. These laser-specific methods supplement the Doppler-free techniques using Doppler-limited excitation, such as level crossing or double-resonance spectroscopy.

In order to give some insight into the experimental demands imposed on the laser as an outstanding spectroscopic light source, we will first briefly discuss some basic properties of the spectral characteristics of lasers in multimode and single-mode operation. When used in Doppler-free spectroscopy, the laser linewidth should be so small that the resolution is not limited by the laser. This requires wavelength stabilization. Continuous wavelength tuning of this stabilized laser is not trivial and will be explained in Section 2.3.

In Section 3, methods using multimode lasers as pumping sources to perform high-resolution excited-state spectroscopy are discussed. The application of lasers to absorption spectroscopy allows the use of some very sensitive detection methods, thereby enabling the detection sensitivity to be increased by several orders of magnitude; these methods are discussed in Section 3.4.

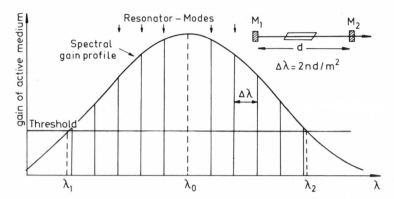

Figure 1. Gain profile and fundamental cavity modes of a laser.

Finally, in Section 4, some techniques of Doppler-free laser spectroscopy are explained. These techniques have led to a revolution of experimental methods in high-resolution atomic spectroscopy and will certainly bring us to a more thorough understanding of atomic structure. Several excellent reviews on the applications of lasers to atomic spectroscopy[8-11,72a] have been published recently and partly overlap with the contents of this chapter. The reader is referred to these articles and to the references cited for further details.

2. Laser Linewidth, Stability, and Tuning

2.1. Multimode and Single-Mode Operations

The different wavelengths at which laser oscillation is possible are determined by the spectral-gain profile of the amplifying medium and by the eigenresonances of the laser cavity. Without any further restriction the laser will oscillate on all cavity modes within that spectral part of the gain profile where the amplification exceeds the total losses. Because of their larger diffraction losses the higher transversal modes can be suppressed by a proper choice of the cavity geometry, so that the laser will only oscillate on longitudinal resonator modes. This is illustrated in Figure 1. The wavelengths, λ, of these longitudinal modes in a resonator with mirror separation d and mean refractive index n are given by

$$m\lambda = 2nd \tag{1}$$

where m is a large integral number. The frequency separation $\Delta\nu$ between adjacent cavity modes is found from (1) with $\nu = c/\lambda$ to be

$$\Delta\nu = c/2nd \tag{2}$$

If there is no coupling between different laser modes, simultaneous stable oscillation on all of these modes is achieved, as can be observed, for example, in the He–Ne laser. Many of the known laser media, however, do exhibit such coupling phenomena. They are caused, for example, by gain competition between different modes. Since in homogeneously broadened transitions all molecules can contribute to the amplification of a light wave with arbitrary wavelength within the threshold points λ_1 and λ_2 (Figure 1), a strong mode may "eat up" the inversion of other modes. This decreases their gain and prevents them from oscillation. Since this gain saturation is most pronounced at the maximum of the electric field of the standing wave inside the resonator, the inversion of the active medium will show a local variation with a modulation period of $\lambda/2$, where λ stands for the wavelength of the oscillating mode. This effect, which is called *spatial hole burning*, allows other modes, which have their maximum field around the nodes of the first mode, to find enough gain for oscillation. Because of fluctuations in the optical path length, nd, the local position of maxima and nodes also fluctuates, resulting in a randomly changing mode competition. If the homogenous width of the gain profile is larger than the separation of these competing modes, the output of the laser consists of a superposition of all possible modes, where the amplitudes and relative phases of the different modes are randomly fluctuating.

The spatial hole burning effect is especially pronounced if the active medium only fills a small fraction of the resonator length, as, for instance, in the cw-dye laser. The time-averaged spectral output can be described by the envelope of all oscillating modes, together with an averaged spectral half-width that is determined by several parameters, for example, the width of the gain profile, the flow time of dye molecules through the pumping region, and the wavelength-selecting elements of the resonator.

For many experiments in atomic physics in which a laser is used as an optical pumping source, the fluctuations of the frequency spectrum within a stable time-averaged envelope are of no importance provided that the integration time of the detection system is long compared to the fluctuation period. The fluctuations may, however, cause detectable effects in cases where the coherence properties of excited states are used, such as in level-crossing experiments.[12]

The spectral width of a dye laser that does not have any wavelength-selecting elements is about a few nanometers which is too broad for most applications in atomic spectroscopy. The spectral width of the envelope can be narrowed down to about 1 Å by using either diffraction gratings, interference filters, prisms, or Lyot filters. Further linewidth reduction can be achieved with slightly tilted Fabry–Perot interferometers inside the resonator, which are often used in the form of solid etalons with two plane parallel reflecting surfaces.

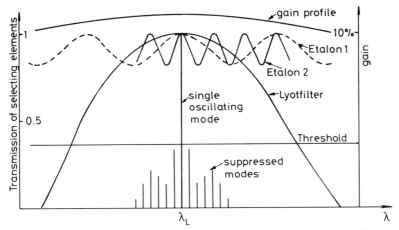

Figure 2. Schematic illustration of mode suppression with etalons and prefilters to achieve single-mode operation of a cw-dye laser.

For investigations in which the spectral resolution is limited by the Doppler width of the atomic transitions, a reduction of the laser linewidth down to 0.01 Å is generally sufficient. With flashlamp-pumped dye lasers this can be achieved, for example, by a grating and one tilted etalon inside the laser resonator. The lower limit for the linewidth of pulsed lasers is mainly determined by the Fourier transformation of the time-dependent laser intensity, which yields, for laser pulses with a duration Δt, a spectral linewidth given by $\Delta \nu = 1/(2\pi \Delta t)$. A nitrogen-laser-pumped dye laser, for example, with 2-nsec pulses has a minimum spectral width of about 100 MHz.

In order to obtain stable single-mode operation two etalons with different optical thickness nt are generally required inside the cavity. The transmission peaks, λ_p, of such an etalon, tilted by an angle α against the resonator axis, are given by

$$q\lambda_p = 2nt \cos \alpha \qquad (3)$$

When the angles of both etalons have been adjusted to equalize λ_p and the wavelength, λ_c, of one of the cavity modes, the total losses have a minimum of λ_p and single-mode operation can be achieved at this wavelength. The conditions for single-mode operation are illustrated in Figure 2, which shows the transmission curves for both etalons, the gain profile, and the resonator modes.

2.2. Wavelength Stabilization

Unless further precautions are taken, the wavelength of a single-mode laser is not sufficiently stable for applications in Doppler-free spectroscopy.

Any changes $\Delta(nd)$ of the optical length of the laser resonator will cause a corresponding wavelength change $\Delta\lambda$; it follows from Eq. (1) that

$$\Delta\lambda/\lambda = -\Delta\nu/\nu = \Delta d/d + \Delta n/n \qquad (4)$$

If, for example, the frequency fluctuations must be kept below 5 MHz then at a laser frequency $\nu_L = 5 \times 10^{14}\,5^{-\lambda}$ (corresponding to $\lambda_L = 600\,nm$), a relative stability $\Delta\nu/\nu \leq 10^{-8}$ must be obtained, which implies a stabilization of the cavity length $d = 1$ m to better than 100 Å!

Acoustic vibrations of the cavity mirrors and fluctuations of the refractive index inside the cavity are the main reasons for laser-frequency fluctuations. In cw-jetstream dye lasers, for example, density fluctuations in the liquid jet and thickness modulations due to surface waves cause changes in nd. Rigid mounting of all optical components on a vibrationally isolated table will partly reduce the wavelength fluctuations but cannot completely remove them. A more efficient method uses an electronic stabilization system, which tries to eliminate the changes $\Delta(nd)$ by applying a compensating change in the geometrical mirror separation, d. This can be achieved by mounting one of the resonator mirrors onto a piezoceramic element, which changes its length when a voltage is applied to it. Typical values of the expansion coefficient are a few nanometers per volt.

A commonly used stabilization scheme (see Figure 3) locks the laser wavelength to the slope of a transmission peak of a Fabry–Perot interferometer FPI 1, which has to be temperature–stabilized and pressure-tight. The photodiode PhD 1 then produces a signal proportional to the transmitted intensity, I_t, which is compared to a reference signal I_R; the difference $I_T - I_R$ is then amplified and fed to the piezo element. If the laser frequency drifts, I_T changes and produces a voltage at the piezo element which casuses the cavity length to change until the laser wavelength is brought back to its proper value.

By using this or similar stabilization techniques a short-term frequency stability of about 1 MHz can be achieved without too much effort. Special designs have been reported which have obtained stabilities of better than 50 KHz.[13] The reference interferometer, which determines the laser

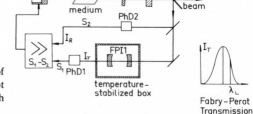

Figure 3. Wavelength stabilization of lasers using a stable Fabry–Perot interferometer as wavelength reference.

wavelength, may in spite of its temperature stabilization, suffer long-term drifts of several MHz, which may not be tolerable in experiments where the laser frequency must be kept on the center of a Doppler-free atomic line for several hours. In these cases the atomic line itself can serve as the reference. When the laser is drifting away from the line center, the resulting decrease of laser-excited fluorescence can be used to activate the stabilization feedback system. For more detailed discussion see Ref. 14.

2.3. Continuous Wavelength Tuning

In both Section 2.1 and Figure 3 it was shown that the wavelength of a single-mode dye laser is determined by the cavity length nd and the transmission peaks of several optical components in the resonator. If the laser has only to be tuned over a small range (up to about 100 MHz) then, as long as the transmission of the etalons is nearly constant over this range, it is often sufficient to merely change the cavity length d using piezoelectric tuning. For larger tuning ranges the etalons must be tilted while changing the cavity length in synchronism. The shift of the transmission peaks against the value λ_0 for $\alpha = 0$ is obtained for small tilting angles α directly from Eq. (3):

$$\Delta\lambda = \lambda_0 - \lambda = 2nt(1 - \cos \alpha)/q \approx 2nt\alpha^2/2q = \lambda_0\alpha^2/2 \tag{5}$$

The shift is independent of the thickness, t, of the etalon,[15] so both etalons can be mounted on the same tilting device.

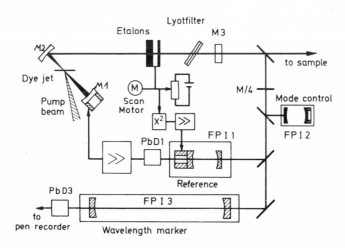

Figure 4. Experimental arrangement for high-resolution spectroscopy with a single-mode wavelength-controlled dye laser.

High-resolution spectroscopy with tunable lasers requires that the laser wavelength λ_L follow the desired wavelength tuning sequel smoothly and without fluctuations. This demands that the laser should be stabilized while being tuned, which can be performed by piezoelectric tuning of the reference interferometer to which the laser wavelength is locked. To maintain synchronization between the etalons the following technique can be used: The etalons are tilted by a motor, which turns a potentiometer, producing a voltage proportional to the tilting angle α; this voltage is electronically squared, amplified, and fed to the piezo element of the reference interferometer. The amplification is adjusted to obtain a synchronous wavelength shift of the etalon and the reference interferometer (see Figure 4). Since the laser wavelength is locked to the reference by the feedback control, it can be smoothly scanned and kept stabilized, over a frequency range of more than 10 GHz.

Another method of synchronous scanning utilizes the dependence of the transmission maxima of the selecting elements on the refractive index by changing the pressure in a pressure-tight tank which contains the laser resonator.[13a]

2.4. Wavelength Calibration

An important problem in high-resolution spectroscopy is the accurate determination of spectral line spacings and level splittings. We will discuss three different methods that have been used to solve this problem.

If a small part of the laser output is split from the main beam and is sent through a long Fabry–Perot interferometer with fixed mirror separation (FPI 3 in Figure 4) the photodiode PhD 3 will produce a signal every time the laser wavelength is tuned through a transmission peak of FPI 3. While the laser wavelength is tuned through the spectral region of interest, wavelength marks are obtained which are separated by the free spectral range of FPI 3. With a confocal FPI with $d = 75$ cm, for instance, these marks have a frequency separation of $\Delta \nu = c/4d = 100$ MHz corresponding to $\Delta \lambda = 10^{-3}$ Å at $\lambda = 600$ nm. The separation of atomic lines between these marks is obtained by linear extrapolation.

When the wavelength scan of the laser is not linear with respect to the voltage applied to the piezoelement, this linear extrapolation introduces slight errors, which can be avoided by another technique. Here part of the laser intensity is modulated at a frequency, f, which generates two side-bands, $\nu_L + f$. The laser is now stabilized on one of these sidebands using the stabilization techniques discussed in Section 2.2. Variation of f tunes the laser frequency, ν_L, since $\nu_L + f$ is fixed. The modulation frequency f can be measured with high accuracy, so this method allows a very accurate determination of line spacings. The tuning range is limited by the upper frequency limit of the modulation device and can be as high as a few gigahertz.

The third method, which is comparable in accuracy to the modulation technique, uses two lasers, which are separately stabilized on two atomic transitions which share a common level. The outputs of both lasers are mixed in a nonlinear optical crystal, thereby generating the difference frequency, which again can be measured directly. A special frequency offset locking technique, developed by Hall et al.,[16] uses a stabilized laser as reference and a second tunable laser, which is locked to the reference with a variable frequency offset, controlled by a sophisticated electronic circuit. The techniques mentioned above allow the determination of line *spacings*. The *absolute* values of laser-wavelengths can be obtained either with interferometric methods or by measuring the absolute *laser frequency* $\nu = c/\lambda$. Such measurements have been performed with heterodyne techniques as follows: The output of the laser with unknown frequency ν_x is superimposed on that of another laser with known frequency ν_R and both waves are mixed in a nonlinear detector. The difference frequency, $\nu_x - m\nu_R$, between ν_x and the mth harmonics of the reference frequency, ν_R, is measured, using microwave technology.[16a]

3. Atomic Spectroscopy with Multimode Lasers

3.1. Optical Pumping with Lasers

Optical pumping is the process of preparation of atoms in selectively excited states (see Chapter 9 of this work). Multimode and single-mode lasers are very effective pumping sources with respect to different aspects of optical pumping:

The first aspect is concerned with the increase of the population density n_k of a selected level k with energy E_k. Because of their high intensity, lasers are clearly superior to spectral lines from gas discharge lamps. When the laser wavelength is tuned to an atomic transition $E_i \rightarrow E_k$, population densities of the upper state E_k can be reached, which are comparable to that of the lower state E_i. This should be compared with typical figures of 10^{-3}–10^{-4} for the fraction of excited atoms, as obtained by optical pumping with incoherent lamps.

Large population densities considerably facilitate all experiments that rely on increased populations of selected levels, such as double-resonance experiments, lifetime measurements, stepwise excitation or spectroscopic investigations of collision processes (see Chapter 29 of this work). The spectral characteristics of the laser radiation are of minor importance for these kinds of experiments as long as the laser linewidth is sufficiently small to avoid overlapping with more than one atomic transition. This means that multimode lasers can be used as pumping sources.

The second aspect of optical pumping is related to the generation of alignment or orientation of excited atoms. This implies that only certain magnetic sublevels are populated by absorption of polarized laser light. These sublevels may be, for instance, Zeeman levels or hyperfine structure levels. The selection may be performed by choosing the appropriate polarization of the pumping light, or, in case of nondegenerate levels, by selective excitation with single-mode lasers using one of the methods developed in Doppler-free spectroscopy (see Section 4).

In contrast to optical pumping with conventional light sources, which are only able to induce orientation and alignment in atomic levels, the strong pumping rates involved in laser optical pumping experiments are responsible for the appearance of higher-order nonlinear effects, such as the production of hexadecapole moments or other tensorial quantities of order higher than 2.[17a]

The third aspect of optical pumping is related to the coherent preparation of the excited state, which means that several sublevels are simultaneously excited and phase relations are established between the wave functions of the different sublevels. Experiments that rely on this coherent excitation are observations of level crossing, quantum beats, and photon echoes. These kinds of experiments generally require "broadband excitation" because the linewidth of the optical pumping source has to be larger than the splitting of the sublevels. Since the spectral features of multimode lasers and their time dependence (see Section 2.1) may influence the experimental results, they have to be carefully inspected.[12,17]

For most experiments in time-resolved spectroscopy (see Chapter 14 of this work) such as lifetime measurements or quantum-beat spectroscopy, pulsed or mode-locked lasers[18] offer so many advantages that they are rapidly replacing conventional pulsed light sources.

The usefulness of lasers as optical pumping sources will be illustrated by some examples.

3.2. Stepwise Excitation and Excited-State Spectroscopy

Because of the large population density that can be achieved in an atomic state excited by optical pumping with lasers, this state can serve as initial state for further excitation (see Figure 5).

In one group of experiments the absorption spectrum of atoms in selectively excited states is measured. For the second excitation either a flashlamp or a second laser can be employed. McIlrath and Carlsten[19] used a dye laser at $\lambda = 657, 278 \; nm$, which was pumped with a frequency-doubled Nd glass laser, to populate the $^3P_1$ state of calcium. In spite of its low transition probability, the spin-forbidden transition $4s^2 \, 2 \, 1 \, ^1S_0 \rightarrow$

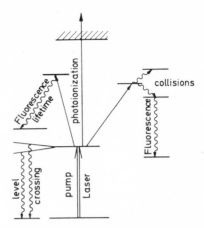

Figure 5. Stepwise excitation and excited-state spectroscopy.

$4s4p\,^3P_1$ could be completely saturated. A fast flashlamp, which was fired immediately after the laser pulse, served as light source for absorption spectroscopy. Transitions of the type $4s4p\,^3P \to 4sns\,^3S$ and $4s4p\,^3P \to 4sn'd\,^3D$ could be measured with n up to 14 and n' even up to 22. Investigations of the absorption spectrum of the ^3D metastable state of barium allowed the fundamental triplet series $6s5d\,^3D \to 6snf\,^3F$ to be measured up to $n = 32$. In addition the absolute photoionization cross section can be determined between 303 and 250 nm; it is found to have a constant value of about 18×10^{-18} cm^2 in this range.[21]

Bradley et al.[20] populated the $5\,^2P_{3/2}$ state of rubidium with a narrow-band dye laser pumped by a giant pulse ruby laser. The ruby-laser simultaneously pumped a second broadband dye laser, which could be used to detect the absorption spectrum of high transitions. Even the brilliant spontaneous emission from dye cells excited by the fourth harmonics of a Nd laser can serve as light source with a spectral continuum for excited-state absorption spectroscopy.[22]

Several experiments have been performed in which the first excitation step into metastable states was achieved by other pumping sources, e.g., electron impact, but the tunable laser was used for further excitation. The studies of xenon atoms in high Rydberg states by Stebbings et al.[23] will be given as an example. In these experiments xenon atoms in a thermal beam are photoexcited by a pulsed dye laser from the metastable $^3P_0$ state into $np(1/2)$ states with $11 \le n \le 16$ and $nf(3/2)$ states with $8 \le m \le 40$. The radiative lifetimes of these states and the field ionization characteristics can then be determined. The photoionization of argon and krypton metastable atoms was also studied with this apparatus.[24] The results clarified the important role of autoionizing $p^5(^2P_{1/2})nf'$ and $p^5(^2P_{1/2})np'$ levels, which were selectively excited by the dye laser and which autoionize into the underlying $P_{3/2}$ continuum.

Selective photoionization by two-step-laser excitation can also be used for isotope separation, if the energy of the level populated in the first step, is sufficiently different for the different isotopes to allow isotope-specific excitation.[25] An interesting method of measuring absolute photoionization cross sections of excited atomic states has been demonstrated by Ambartzumian *et al.*[25a] The authors excited rubidium atoms into the $6P_{1/2}$ and $6P_{3/2}$ states by a dye laser which was pumped by the second harmonic of a ruby laser. Part of the frequency-doubled ruby laser output, I_2 was used to photoionize the excited rubidium atoms. The ion yield was measured as a function of the ionizing intensity, I_2, and was found to saturate at sufficiently high intensities. The photoionization cross section, σ_i can be evaluated from the saturation condition $\sigma_1 \cdot I_2 \cdot \tau \approx 2$, where τ is the spontaneous lifetime of the excited $6P$ state.

A large group of experiments is aimed towards measurements of transition probabilities for transitions between excited states by using stepwise excitation. Either both steps are induced by absorption of light from two different lasers tuned to the corresponding transitions, or the lower one of the two excited levels is populated by other excitation methods. Since pulsed lasers offer the opportunity of generating very intense short pulses, they are ideally suited for lifetime measurements, thereby allowing the transition probabilities to be deduced.

The lifetime measurements of the highly excited states n^2S and $n^{12}D$ in Na and Ca, which have been reported by Gornik *et al.*[26,27] will be given as an example of stepwise excitation using two lasers. A flashlamp-pumped dye laser is tuned to the D line of Na to populate the 3^2P state and a second, nitrogen-laser-pumped dye laser, with a pulsewidth of 5 nsec, is used to further excite one of the higher-lying n^2S or n^2D states.

One experimental technique, which is especially suitable for spectroscopy of highly excited states of ions or atoms, uses a combination of collisional and laser excitation in fast ion beams. The ions, which have been accelerated to about 100 keV, are excited or further ionized by collisions

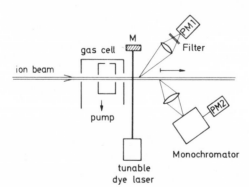

Figure 6. Experimental arrangement for lifetime measurements in fast ion beams using selective laser excitation.

with atoms in thin foils or in gas chambers; alternatively, they can be transferred into highly excited atoms by charge-transfer collisions. A few centimeters behind the excitation region the ion beam crosses a dye laser beam which has its wavelength tuned to the desired transitions between excited states. The fluorescence intensity from the laser-excited upper level is measured as a function of distance downstream from the crossing point. Together with the use of several precautions, this method yields very accurate cascade-free lifetimes[28-30] (Figure 6).

3.3. Laser Photodetachment

Accurate values of electron affinities of atoms are important for the understanding of negative ion structure, atomic polarizabilities, resonance effects in atom–electron scattering, and in many other fields of atomic physics. Of the many different methods developed to measure these quantities, the technique of laser photodetachment has proved to be the most accurate one so far. It may be realized experimentally in several different ways: A beam of negative ions is illuminated by the intracavity field of a fixed-frequency laser. Those electrons that are photodetached into the acceptance angle of a hemispherical electrostatic condenser are energy-analyzed and detected.[31]

Using tunable lasers, the threshold law for photodetachment can be very accurately verified by measuring the intensity of the photoionized negative ions (which are neutral atoms) as a function of laser wavelength. The fine-structure splitting of S^- has been determined by this method.[32] Experiments in which the laser flux was high enough to saturate the photodetachment have been used (see previous section) to determine absolute cross sections for the photodetachment of Pt^-.[33]

Many atoms and ions have been studied, and a compilation of them is given in Ref. 8, p. 31; see also the review of W. C. Lineberger.[34]

3.4. Spectroscopy of Atomic Laser Media

By profiting from the high intensity of induced emission, wavelength measurements of numerous atomic or ionic laser lines[35] have yielded very precise data on atomic energy levels, hfs splittings, and Zeeman-shifts. Even in the spectral ranges where sensitive detectors are not available, the available intensity makes signal-to-noise problems negligible. This has been displayed in many infrared laser lines originating from transitions between high-lying, atomic Rydberg states. These states have been the object of renewed attention in the last few years since they are very sensitive to various kinds of collision processes.[36] The accuracy can be increased if the laser oscillation is restricted to a single mode which is stabilized on the center of the atomic transition. The attainable precision has been demonstrated by Vetter,[37] who measured isotope shifts of 15 infrared Xe-laser transitions for the isotopes ^{128}Xe–^{136}Xe with a precision of 10^{-5} cm^{-1}.

Schäfer[38] obtained an accuracy of $\Delta\nu/\nu \leqslant 2 \times 10^{-9}$ using optical heterodyne techniques (see Section 4.4). He used two different Xe lasers and stabilized each laser on the center of a different Xe transition. The output from both lasers was superimposed and mixed in a photodetector mixer. The frequency difference was measured using an rf-spectrum analyzer.

Any shift of the atomic line center due to Zeeman or Stark effect, or caused by collisions with neutral atoms can be detected with high sensitivity and precision by the corresponding shift of the laser line, which is stabilized on line center. This opens the possibility of determining dipole moments and the polarizabilities of states, connected with a laser transition.[39]

A wide class of experiments on optical pumping, level crossings, and other linear and nonlinear phenomena observable on laser transitions between excited atomic states in laser media have also been performed. They are partly reviewed (with emphasis on neon transitions) by Decomps et al.[17a]

3.5. High-Sensitivity Absorption Spectroscopy

When a monochromatic light wave $E = E_0 \exp(i\omega t - kx)$ travels through an absorbing sample with absorption coefficient $\alpha(\omega)$ and length L, the intensity I_0 of the incident wave decreases to $I(\omega) = I_0(\omega) \exp[-\alpha(\omega)L]$. If $\alpha L \ll 1$ the relative attenuation can be expressed by $\exp(-\alpha L) \approx 1 - \alpha L$ since $(I_0 - I)/I_0 = \alpha L$. In order to measure α, one has to measure a small difference of two large quantities, which becomes less and less accurate with decreasing αL.

Placing the absorbing sample inside the laser resonator may increase the sensitivity by several orders of magnitude.[40,41] This is due to several factors. The first is simply the fact that the laser photons are being reflected back and forth between the resonator mirrors, thus traversing the sample m times. If one mirror has the reflectivity $R_1 = 1$, the transmitting mirror $R_2 < 1$, the enhancement factor is readily seen to be $m = 1/(1 - R_2)$, which yields $m = 100$ for $R_2 = 0.99$.

If the laser is operated close above threshold, the unsaturated gain of the active medium is only slightly higher than the losses. Any additional losses introduced by the absorbing sample will result in a drastic decrease of the laser output power. This brings about a further enhancement of sensitivity. An additional increase of sensitivity can be achieved when the laser is oscillating simultaneously on many modes which are coupled by gain competition (see Section 2.1). Increasing the losses of one of the modes decreases its output power; this increases the gain for the other modes, which increases their intensity. The gain for the "absorption" mode is then further decreased because the other modes "eat up" its inversion. Even a diminutive additional absorption $L\alpha(\omega)$ may therefore completely suppress the mode, oscillating at the frequency ω.[42,43]

This intracavity absorption technique has found wide applications in the detection of absorption spectra of spurious elements in flames[44] or excited reaction products from chemical reactions. It can be used to measure the oscillator strengths of weak atomic transitions.[45] Maeda *et al.*[44] found many unknown absorption lines of rare-earth atoms with this technique.

4. Doppler-Free Laser Spectroscopy

The spectroscopic investigations discussed in Section 3 are mostly suitable for use with multimode lasers because, in most cases, the attainable spectral resolution is not limited by the laser linewidth. In this chapter we will consider some new techniques of Doppler-free absorption spectroscopy in which the extremely small linewidths of stabilized single-mode lasers and the tunability of their wavelengths are essential factors.

The elimination of the Doppler width is achieved either by selection of a subgroup of those atoms that have velocity components in the direction of light propagation within a narrow interval around $V_z = 0$ (laser spectroscopy in collimated atomic beams and saturation spectroscopy), by simultaneous absorption of several photons, $\hbar\omega_i$, with momenta $\mathbf{k}_i$ and $\Sigma_i\mathbf{k}_i = 0$ by the same atom (Doppler-free multiphoton spectroscopy), or by application of optical–optical double resonance techniques. We will briefly discuss the basic ideas and some recent applications of these methods. For more details the reader is referred to the extensive literature on this subject.[8–11,46–50,72a]

4.1. Laser Spectroscopy in Collimated Atomic Beams

When a parallel laser beam is crossed perpendicularly with a collimated atomic beam and the laser wavelength is tuned across an atomic absorption line, the Doppler width of the resulting absorption profile is reduced by a factor that depends on the collimation angle of the atomic beam (see Figure 7). This is due to the fact that the thermal distribution of the velocity components v_z is reduced by the collimating apertures to $N(v_z) \propto \exp\left(-mv_z^2/\theta \cdot 2KT\right)$ with $\theta = d/L$. Two apertures with a slit width

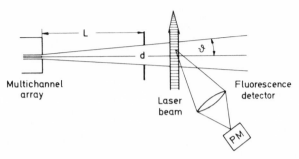

Figure 7. Reduction of Doppler width by single-mode laser
absorption in collimated atomic beams.

$d = 1$ mm at a distance $L = 100$ mm allow for instance a reduction of the
Doppler broadening by a factor $\theta = 0.01$. For many atomic transitions this
reduced Doppler width is already below the natural linewidth. In highly
collimated beams the atomic density is generally low and the absorbed
fraction of the incident laser intensity may be too small to be measured
directly. The fluorescence from atoms, however, which have been excited
by absorption of laser radiation, can, however, serve as a very sensitive
monitor of the absorption. The fluorescence intensity $I_F(\lambda_L)$, which is
monitored as a function of laser wavelength, λ_L, while the laser is tuned
across the atomic absorption spectrum, reflects this absorption spectrum
and is often called the *"excitation spectrum."* Figure 8 shows as an example
the excitation spectrum of the Na $^3P_{1/2} \leftarrow {}^3S_{1/2}$ transition, where the
hyperfine structure of the D line is clearly resolved.[51]

Similar high-resolution measurements in atomic beams have been
performed for different Na isotopes[52] yielding values for isotope shifts
and quadrupole moments. Even short-lived radioactive isotopes can be
measured this way (Figure 9). Other examples are studies of hyperfine-
structure and isotope shifts for ytterbium[53] and barium isotopes.[54] The
accuracy of these measurements is about one order of magnitude better
than that obtained by conventional spectroscopy.

Instead of using laser-induced fluorescence one may also employ the
recoil which the atom suffers when absorbing a laser photon with momentum $\hbar\mathbf{K}$ as a monitor. The resultant slight deflection of the atomic beam can
be detected as a change of beam intensity behind a narrow slit in front of
the detector. This method has been used by several authors with sodium[55]
or barium atoms.[56]

Since all atoms in the absorbing state can contribute to the measured
fluorescence signal, Doppler-free spectroscopy in atomic beams is a very
sensitive technique which allows spectra to be measured with reasonable
signal-to-noise ratios down to atom densities of $10^9/\text{cm}^3$. Already, at
modest laser intensities (a few milliwatts), saturation of the atomic tran-

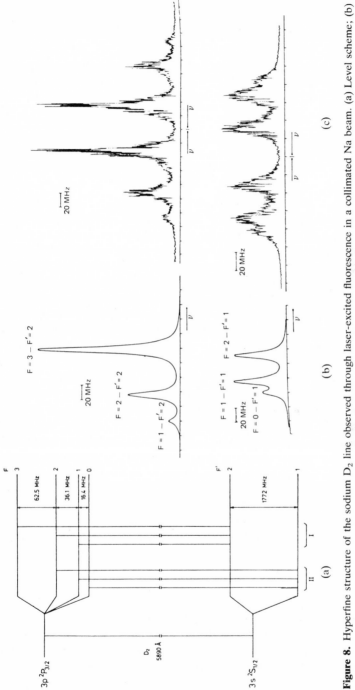

Figure 8. Hyperfine structure of the sodium D_2 line observed through laser-excited fluorescence in a collimated Na beam. (a) Level scheme; (b) calculated line profiles; (c) experimental results (from Ref. 51).

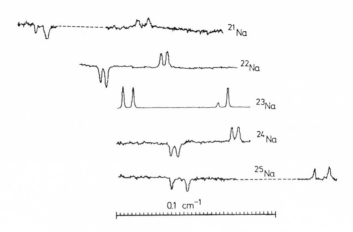

Figure 9. Isotope shifts of different Na isotopes (from Ref. 52).

sition can be reached. This means that most atoms in the atomic beam passing the laser beam do absorb a laser photon. If the spontaneous lifetime of the upper level is short enough, each atom may even be excited several times during its interaction time with the laser beam, if the laser intensity is sufficiently high. This multiple excitation can enlarge the fluorescence signal considerably. For sodium atoms, for instance, the spontaneous lifetime of the 3^2P level is 16 nsec; the time of flight through a laser beam of diameter 2 mm, however, is about 4 μsec. Each sodium atom may therefore be excited more than hundred times. One must, of course, take care that the lower state is not permanently depleted by optical pumping effects; this can be avoided either by choosing the right hyperfine transitions $^3S_{1/2} F = 2 \to {}^3P_{3/2} F' = 3$ for Na atoms or by simultaneous pumping of all hfs levels of the ground state.[57,57a]

Strong pumping influences the line shape of the atomic absorption profile because the atom is subjected to a strong radiation field with a linewidth much narrower than the natural linewidth. Weisskopf showed many years ago[58] that under these conditions the fluorescence should have the same spectral profile as the exciting radiation. This has been shown experimentally by Walther and his group,[59] who used a sodium beam with a collimation ratio of 1 : 500 which was crossed perpendicular with the beam from a single-mode stabilized cw-dye laser. The crossing point was in the center of a confocal Fabry–Perot, which allowed the spectral distribution of the atomic fluorescence to be measured for different values of $\omega_L - \omega_0$ when the laser frequency ω_L had been detuned from the line center ω_0. Similar investigations on the 2852-Å transition in Mg atoms, excited by the frequency-doubled output from a dye laser have been performed by Gibbs and Venkatesan.[60]

An interesting variation of high-resolution laser spectroscopy in fast beams of ions or atoms which promises a resolution of a few megahertz has been proposed by Kaufman.[61] Contrary to the previous examples in beams with thermal velocities where the perpendicular velocity components, v_z, had been restricted by collimating slits, this proposal is going to utilize the bunching effect obtained for velocity components in the *beam direction*, v_x, by acceleration of the ions.

Consider, for example, two ions in the ion source with thermal velocity components $v_{x1} = 0$ and $v_{x2} = (2KT/m)^{1/2}$. After being accelerated in the x direction through a voltage U their final velocity components are $v_{a1} = (2cU/m)^{1/2}$ and $V_{a2} = (v_{x2} + 2cU/m)^{1/2} \approx v_{a1} + v_{x2}^2/(2v_{a1})$. The difference in velocities $v_{a2} - v_{a1} = v_{x2}R$ has been reduced by a factor $R = \frac{1}{2}(KT/eU)^{1/2}$. For $T = 2000°K$ and $U = 10\,kV$, $R = 2.1 \times 10^{-3}$. Atoms can be produced by charge transfer. The additional velocity spread due to charge transfer collision can be shown to be of minor importance.

This method exhibits some unique features which make it very attractive. The laser beam has to be directed parallel to the ion beam. The hyperfine structure of a spectral line can then be scanned by tuning either the laser frequency or the accelerating voltage. One may therefore use fixed-frequency lasers, which generally have more power, and tune the atomic spectrum by utilizing the large Doppler shift. The method should be especially suited to measuring quadrupole moments and isotope shifts of short-lived unstable isotopes.

For further examples of high-resolution laser spectroscopy in atomic beams and its various applications see the comprehensive review of Jacquinot.[46]

4.2. Saturation Spectroscopy

The Doppler-free method discussed in Section 4.1 achieved the selection of a narrow interval $|v_z| \le \Delta v_z$ of the atomic velocity components by collimating slits, forming an atomic beam with low divergence. We will now present a technique in which this selection is accomplished for atoms with randomly oriented velocities in a thermal gas cell by velocity-selective saturation of an atomic transition by a monochromatic light wave.

If such a wave $E = E_0 \exp(i\omega_L t - Kz)$ is traveling through a gaseous sample with a Doppler-broadened absorption profile centered at ω_0, the laser frequency, ω, as seen in the reference frame of an atom moving with a velocity component v_z is Doppler-shifted to $\omega = \omega_L - kv_z$. Only those atoms that are just shifted into resonance with ω can absorb the monochromatic light. This means that

$$\omega_L - kv_z = \omega_0 \pm \Delta\omega_n \tag{6}$$

where $\Delta\omega_n$ stands for the homogeneous linewidth of the transition.

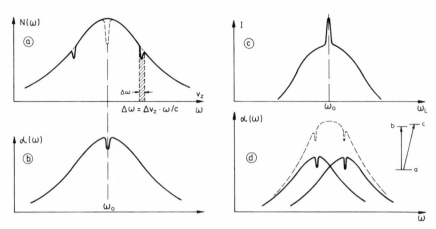

Figure 10. Hole burning in Doppler-broadened transitions by a monochromatic light wave.

At sufficiently high laser intensities the transition starts to saturate and the population density of atoms with velocity components in the interval $v_z \pm \Delta v_z = (\omega_L - \omega_0 \pm \Delta \omega_m)/k$ in the absorbing state decreases. This results in a selective "hole," which is burned into the thermal velocity distribution (see Figure 10a). The width $\Delta v_z = \Delta \omega_n/k$ of this hole resembles the homogeneous width of the atomic transition and approaches the natural linewidth if pressure broadening and power broadening can be neglected. In the visible range this width is about 2–3 orders of magnitude smaller than the Doppler width. Several techniques have been developed to detect this small hole (often called a Bennet hole),[62] and to utilize it in Doppler-free spectroscopy.[63]

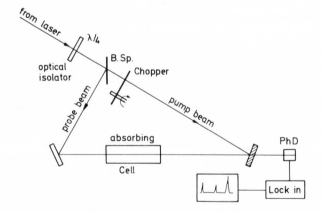

Figure 11. Experimental setup used for saturation spectroscopy.

One of several possible experimental realizations uses the same laser for saturating and probing the Lamb-dip. The laser beam is split into two beams, which are sent antiparallel through the probe (Figure 11). If the laser frequency ω_L is detuned from the line center ω_0 the frequencies of both beams, as seen in the reference frame of the moving atoms, are Doppler-shifted into opposite directions to $\omega_+ = \omega_L + k v_z$ and $\omega_- = \omega_L - k v_z$. They are therefore absorbed by two different groups of atoms moving with $-v_z + v_z$ respectively (see Figure 10a). If, however, the laser is tuned to the center, ω_0, both saturation holes coincide because both waves are now being absorbed by the same group of atoms with velocity components around $v_z = 0$ [Eq. (6)]. In this case the probe wave finds less atoms in the absorbing ground state, which means that its absorption decreases at the center of the Doppler broadened absorption line (Lamb-dip, Figure 10b). The absorption coefficient $\alpha(\omega)$ can be described by[64]

$$\alpha(\omega) = \alpha_0 \left[1 - \frac{6}{2} \frac{\gamma_L^2}{(\omega_0 - \omega)^2 + \gamma_L^2} \right] e^{(\omega_0 - \omega)/(\Delta\omega_0)} \qquad (7)$$

The halfwidth γ_L of the Lamb-dip depends on the intensity of the saturating wave (power broadening) and its depth on the saturation parameter $G = I/I_{sat}$, where I_{sat} is that intensity which saturates $\alpha(\omega_0)$ to $\frac{1}{2}$ of its unsaturated value. For sufficiently low intensities, I, and at low pressures γ approaches the natural linewidth of the absorbing transition. A more detailed treatment of saturated absorption line profiles can be found in Ref. 65a.

The sensitivity of Lamb-dip detection can be considerably increased when the absorbing probe is placed inside the laser resonator[66] (see Section 3.4). In this case the saturating and probing waves have the same intensity and are superimposed to form a standing wave. When the laser frequency is tuned across the absorption dip, the laser intensity will exhibit a sharp peak because of the reduced losses in the Lamb-dip (Figure 10c).

The importance of these narrow Lamb-dips for high-resolution spectroscopy is illustrated in Figure 10d. If two closely spaced atomic transitions (for instance hfs components of an atomic line) overlap within their Doppler width, the corresponding Lamb-dips are still clearly resolved. The resolution capability of the saturation method has been demonstrated by Hänsch et al.,[67] who measured the hfs splittings of the Na D lines using a narrow-band tunable dye laser which was pumped by a nitrogen laser.

One advantage of saturation spectroscopy compared to atomic-beam spectroscopy stems from its capability to study the effects of collisions on the line profile. When using pulsed lasers relaxation phenomena can be investigated by introducing a variable time delay between saturating pulse and probe pulse.[68,69]

Utilizing the high spectral resolution achieved with Doppler-free saturation spectroscopy, Hänsch *et al.*[70] determined the Rydberg constant with an order-of-magnitude improvement in precision. In this experiment the absolute wavelengths of resolved fine-structure components of the Balmer α line in hydrogen and deuterium, excited in a gas discharge, were measured with a Fabry–Perot interferometer. A He–Ne laser, stabilized on the line center of an iodine hfs transition, served as the wavelength standard for the calibration of the Fabry–Perot distance. Because of the small linewidth the centers of the saturation dips, induced by a pulsed dye laser tuned to the Balmer α line, could be determined with high accuracy. In addition the experiments yielded the isotopic shift between H and D.

A modified saturated amplification method for high-resolution spectroscopy of hyperfine structure in Xe atoms has been reported by Cahuzus and Vetter,[71] who used two Xe lasers. The outputs from the two lasers are sent from opposite sides through a Xe discharge tube, which amplifies the xenon laser lines. The strong, saturating laser decreased the population of the upper level and therefore the amplification for the probe laser; it was stabilized on the Lamb dip of its own active medium. The weak probe laser was tunable within its gain profile. By measuring the amplification change for the probe laser due to the saturation dip produced by the pump laser, the different hfs components could be resolved with linewidths down to 10 MHz.

Further examples and more detailed information about saturation spectroscopy can be found in a very recent article by Letokhov[72,65a] and in an excellent work on nonlinear spectroscopy by Th. Hänsch.[72a]

4.3. Doppler-Free Two-Photon Spectroscopy

If the laser intensity is sufficiently high, atomic transitions induced by the simultaneous absorption of two photons by the same atom may be observed. The probability, A_{if}, that an atom will absorb two photons, $\hbar\omega_1$ and $\hbar\omega_2$, from two incident light waves, $\mathbf{E}_1 = A_1\hat{\mathbf{e}}_1 \exp{(\omega_1 t - \mathbf{k}_1 \cdot \mathbf{r})}$ and $\mathbf{E}_2 = A_2\hat{\mathbf{e}}_2 \exp{(\omega_2 t - \mathbf{k}_2 \cdot \mathbf{r})}$, resulting in a transition from an initial state i to a final state f, can be calculated from second-order perturbation theory.[73] For free atoms, moving with velocity $\mathbf{v}$, which have discrete initial and final states, A_{if} is given by the expression

$$A_{if} \propto \frac{\gamma_{if}}{\left[\omega_{if} - \omega_1 - \omega_2 + (\mathbf{k}_1 + \mathbf{k}_2) \cdot \mathbf{v}\right]^2 + (\gamma_{if}/2)^2}$$
$$\times \left| \sum_K \frac{\mathbf{M}_{ik} \cdot \hat{\mathbf{e}}_1 \mathbf{M}_{kf} \cdot \hat{\mathbf{e}}_2}{\omega_{ki} - \omega_1 + \mathbf{k} \cdot \mathbf{v}} + \frac{\mathbf{M}_{ik} \cdot \hat{\mathbf{e}}_2 \mathbf{M}_{kf} \cdot \hat{\mathbf{e}}_1}{\omega_{ki} - \omega_2 + \mathbf{k}_2 \cdot \mathbf{v}} \right|^2 I_1 I_2 \qquad (8)$$

The first factor describes the line profile of the two-photon transition and has a maximum at $\omega_1 + \omega_2 - (\mathbf{k}_1 + \mathbf{k}_2) \cdot \mathbf{v} = \omega_{if}$. If both photons are absorbed out of the same light wave, both wave vectors are equal $(\mathbf{k}_1 = \mathbf{k}_2)$ and one obtains a normal Doppler-broadened line profile by integrating over the velocity distribution of the atoms in a gaseous sample. The Doppler width is the same as that of linear absorption at a frequency $\omega_{if} = \omega_1 + \omega_2$.

If, however, the two photons are absorbed from two different light waves traveling in opposite directions the term $(\mathbf{k}_1 + \mathbf{k}_2) \cdot \mathbf{v}$, which contributes to the Doppler broadening, is reduced. It is exactly zero if $\mathbf{k}_1 = -\mathbf{k}_2$, which implies $\omega_1 = \omega_2$. In this case the first factor in Eq. (8) describes a Lorentzian line profile with a half-width γ_{if} equal to the homogeneous width of the transition $i \to f$. At sufficiently low pressures this is equal to the natural linewidth. The elimination of the Doppler width can be immediately understood as follows: Assume the two light waves with equal frequencies $\omega_1 = \omega_2 = \omega$ are traveling in the $\pm z$ direction. In the reference frame of an atom, moving with a velocity component v_z, the frequency of the wave $\hat{\mathbf{e}}_1 \exp(\omega, t - kz)$ is Doppler-shifted to $\omega_- = \omega - kv_z$; the frequency of the opposite wave, $\hat{\mathbf{e}}_2 \exp(\omega t + kz)$, is however, shifted to $\omega_+ = \omega + kv_z$. If two photons, one out of each wave, are simultaneously absorbed by the atom, the sum frequency is $\omega_+ + \omega_- = 2\omega$ *and does not depend on the velocity components, v_z*. This implies that *all atoms* with arbitrary velocities can absorb the two photons if $2\omega = \omega_{if}$ equals the transition frequency $\omega_{if} = (E_f - E_i)/\hbar$ between two levels, with energies E_i and E_f.

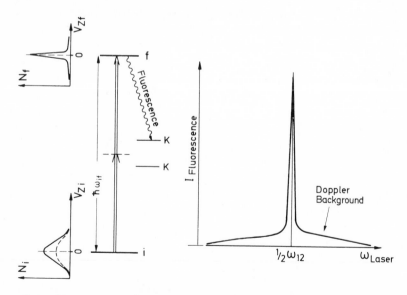

Figure 12. Schematic term diagram of levels involved in two-photon absorption and line profile with (exaggerated) Doppler-broadened background.

The actual transition probability for the two-photon transition $i \to f$ is governed by the second factor in Eq. (8), which is derived by assuming the dipole approximation. $\mathbf{M}_{ik}$ and $\mathbf{M}_{kf}$ are the respective dipole transition moments and the sum extends over all possible atomic states. If both photons have the same frequency and polarization, both terms in the sum can be combined. The terms may become large if ω equals an atomic eigenfrequency, ω_{ik}. The situation is often described by the introduction of "virtual states" (see Figure 12), which are reached by the absorption of the first photon. The probability of a two-photon transition strongly depends on the energy difference between this virtual state and real atomic states. The resonance case is a real stepwise excitation process.

The second factor in Eq. (8) does not vanish only if neither of the two transition dipole moments $\mathbf{M}_{ik}$ and $\mathbf{M}_{kf}$ are zero. Both transitions $i \to k$ and $k \to f$ therefore have to be allowed dipole transitions, which implies that the two-photon-induced transition $i \to f$ causes a parity change by 0 or 2, and induces transitions with $\Delta l = 0$ or $\Delta l = \pm 2$. The polarization of the two light waves determines which of both types can be observed (for instance, $s \to s$ transitions with left and right circular polarization and $s-d$ transition with linear polarization). The important point is that only those excited states which cannot be populated by linear absorption from the groundstate can be reached by two-photon absorption.

The third factor, $I_1 \cdot I_2$, in Eq. (8) asserts that the total probability A_{if} is proportional to the product of the intensities of both light waves. In spite of the second factor, which is generally very small, this may bring about a total probability of two-photon absorption, comparable to that of linear absorption at sufficiently large intensities.

Many experimental realizations of Doppler-free two-photon spectroscopy have used a cw single-mode tunable dye laser. The output from the laser is split into two beams traveling in opposite direction through the gaseous sample. In order to increase the intensities, both beams are focused. The two-photon absorption in the common focus is monitored by observing the fluorescence intensity from the upper level f as a function of laser wavelength (see Figure 13). Since the probability that both photons

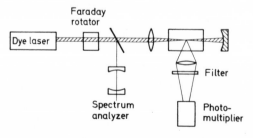

Figure 13. Experimental arrangement for observation of Doppler-free two-photon spectroscopy.

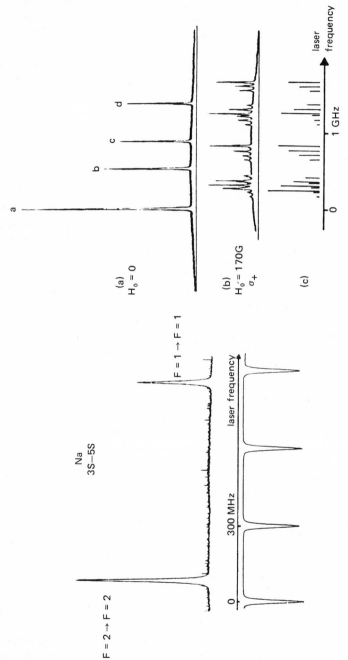

Figure 14. Two-photon transition $3S$–$5S$ in sodium (left) and Zeeman effect on the two-photon transition $3S \rightarrow 4D$ in sodium (right) at two different magnetic fields (a and b) and calculated pattern (c) (from Ref. 78).

come from the same beam is one half of the probability that just one photon comes from each beam, a Doppler-free two-photon transition shows a broad, Doppler-broadened background and a sharp Doppler-free peak at $2\omega = \omega_{if}$ (see Figure 12). Since the area under the Doppler-free line is twice as large as that under the Doppler-broadened curve (its width, however, is about two orders of magnitude smaller), the peak of the Doppler-free line is far above the background, which can be generally neglected.

Most experiments on Doppler-free two-photon absorption of atoms have been performed on alkali atoms where the hfs of higher-lying S levels,[74] the fine structure of D levels[75,76] or Zeeman and Stark splittings of $S \rightarrow D$ transitions[77,101a] were studied. Several of these results have been reviewed by Cagnac.[78] Figure 14 demonstrates the attainable resolution and the excellent signal-to-noise ratio of two-photon sodium transitions. Some recent papers report the precise measurements of isotope shifts in neon,[79] and rubidium[80] and of hfs splittings in the 7^2P states of thallium.[81]

A remarkable experiment on atomic hydrogen, using Doppler-free two-photon spectroscopy of the $1S \rightarrow 2S$ transition to determine the Lamb shift of the $1S$ ground state, has been reported by Hänsch et al.[82] The authors used the frequency-doubled output from a pulsed narrow-band dye laser tuned to 4860 Å to simultaneously record the $1S$–$2S$ transition by two-photon absorption $(2 \times 2\omega_L)$ and the Balmer β transition $2S, P \rightarrow 4S, P$ by linear absorption of the undoubled laser frequency. From the difference between both absorption frequencies the Lamb shift of the $1S$ ground state was calculated to be 8.3 ± 0.3 GHz for deuterium and 8.6 ± 0.8 GHz for hydrogen.

This technique of Doppler-free spectroscopy can be extended to more than two-photon processes (see Chapter 2, Section 13, of this work) and will certainly give a great impetus to the quest for a thorough understanding of nonlinear interactions between atoms and light. For a detailed discussion see the review of Bloembergen and Levenson.[73]

4.4. Laser Optical Double-Resonance Spectroscopy

The spectroscopic applications of the saturation phenomena discussed in Section 4.2 are not confined to the interaction of a monochromatic wave with an atomic two-level system; their extension to the study of three or more level systems interacting simultaneously with several light waves offers definite advantages for a detailed study of the nonlinear response of atoms to electromagnetic fields. An example of this kind of experiment is the investigation of coupled atomic transitions in the presence of two laser waves with different frequencies ω_1 and ω_2.[83]

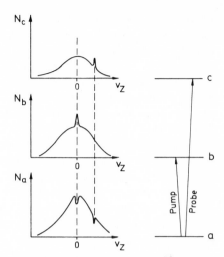

Figure 15. Selective saturation of two coupled Doppler-broadened transitions.

Consider, for example, two Doppler-broadened transitions sharing a common lower level (a) (Figure 15). An incident monochromatic laser wave tuned to a frequency ω_1 within the Doppler width of transition $a \to b$ interacts with atoms in a velocity interval Δv_z around $\omega_0 - k_1 v_z$ and produces a saturation hole in the velocity distribution of atoms in the absorbing state (a) (see Section 4.2). When a second laser wave is tuned across the Doppler profile of the transition $a \to c$ it will probe this hole causing an absorption dip or a corresponding dip in the fluorescence intensity $I_{Fc}(\omega_2)$.

The monochromatic light waves may be represented by two different oscillating modes of a multimode laser. Assume that the two upper levels are Zeeman sublevels of a degenerate excited state. If these levels are split by a magnetic field to such an extent that the level splitting equals the frequency separation between two laser modes, i.e., $(E_b - E_c)/\hbar = \omega_1 - \omega_2$, then both laser modes can be simultaneously absorbed by the same group of atoms and cause a decrease of total absorption due to selective saturation.

These coupling phenomena can occur also in the active medium of gas lasers. Assume that inversion has been achieved between an upper level (b) with $J = 0$ and a lower level (a) with $J = 1$. By using an external magnetic field the lower level can be split into three Zeeman sublevels. If the Zeeman splitting equals the separation of two resonator modes, the two corresponding laser waves are amplified by the same group of atoms in level (b) within the velocity interval $|v_z| \lesssim \Delta v_z$. This results in a saturation of the amplification at a lower power level of each mode and causes a decrease of the total laser output. Tuning the magnetic field therefore causes resonance dips in the laser intensity. The width of these dips reflects the homogeneous width Γ_{ab} of the atomic transition $a \rightarrow b$ and gives information on the natural linewidth and pressure and power-broadening effects. This method of magnetic mode crossings has been used[84] to measure the Landé factor of atomic levels. One example is the precise determination of g (= 1.3 005 ± 0.1%) for the $^2P_4$ level in neon by Herman et al.[85]

Monitoring atomic level crossings through their influence on the saturation of stimulated emission is a special case of level-crossing spectroscopy and has been therefore called "stimulated level-crossing."[86] It can be applied either to atomic transitions in active laser media or to external passive samples. One of its early applications was in the observation of hfs level crossings in Xe^{129} levels, using the 3.37 μm Xe laser transition.[87] The experimental result gave an accurate determination of the hfs interaction, which was measured to be 829.5 ± 0.8 MHz.

The investigation of coupling phenomena between different atomic transitions may be also performed with two widely spaced laser lines which share a common level. Examples are the 1.15-μm and 0.63-μm He–Ne laser lines, which share the $^2P_4$ neon level as common lower level. Toschek and his group[63] used a special experimental technique, called "tuned differential spectroscopy," to study extensively the different coupling phenomena between these lines. This technique may be summarized as follows:

A neon absorber cell inside the resonator of a single-mode He–Ne laser tunable around 0.63 μm is probed by a second single–mode laser oscillating at 1.15 μm. The population difference $N_a - N_b$ in the absorption cell is controlled by selecting the proper discharge and pressure conditions. The velocity-selective population change of the lower $^2P_4$ neon level, produced by the 0.63-μm laser line is then probed by the corresponding change of absorption for the 1.15-μm laser line.

It turns out that, besides the saturation effect discussed above, other coupling phenomena, such as a resonant Raman effect, are present. Because of its nonlinear interaction with the neon atoms, the 0.63-μm laser radiation generates Stokes radiation at 1.15μm. This couples both transitions and enables the 0.63-μm line to influence the phase relations of the $^2S_2$ wave functions. Another effect, which had been known already as the "Autler–Townes effect" in microwave spectroscopy is a slight splitting of the atomic levels induced by the light wave.[88]

Besides these studies of nonlinear interaction phenomena between atoms and monochromatic light waves, the optical double-resonance experiments allow the detailed study of various kinds of collision processes, such as depolarizing collisions, velocity-changing collision, or inelastic collisions. Because of the high spectral resolution, long-distance collisions, which only slightly affect the line shape, are no longer masked by the Doppler width but can be detected with high sensitivity.[89]

4.5. Polarization Spectroscopy

A new sensitive technique of Doppler-free spectroscopy, based on light-induced birefringence and dichroism of an absorbing gas, has been demonstrated by Hänsch et al.[90] If monochromatic circularly polarized light of sufficient intensity is absorbed by the gas atoms, an optical anisotropy is induced in the sample because the m-sublevel equilibrium distribution is altered due to selective saturation. This causes a difference, $\alpha^+ - \alpha^-$, of absorption coefficients for right and left circularly polarized light, respectively, and a corresponding difference, $n^+ - n^-$, of the refractive index. The basic setup for polarization spectroscopy may be summarized as follows:[90]

A single-mode tunable laser sends a linearly polarized probe beam through a gas sample. After passing through a nearly crossed linear polarizer only a small fraction of this beam reaches the photodetector. If an optical anisotropy is induced in the sample by absorption of a second circularly polarized pump beam, sent in the opposite direction through the cell, the probe polarization will alter. This will change the light flux through the polarizer, which can be detected with high sensitivity. If both beams have the same frequency, ω_L, then they interact with the same velocity group of atoms around $v_z = 0$, if the laser frequency, ω_L equals the center frequency, ω_0, of the atomic transition (see Section 4.2). As in conventional saturation spectroscopy a resonant probe signal is therefore expected only near the center of the Doppler-broadened transition.

The sensitivity of this method was demonstrated by studying the fine structure of the hydrogen Balmer β line. Polarization spectra of this line, recorded with a cw dye laser at 4860 Å, revealed the Stark splitting of single fine-structure components in the weak axial electric field of a Wood-type gas discharge. The observed Stark pattern changes drastically if the laser beams are displaced from the discharge-tube axis, indicating the presence of additional radial electric fields. At low electric fields, the natural linewidth of the quasi forbidden $2S$–$4S$ transition is only about 1 MHz. With a frequency-stabilized dye laser than the H–D isotope shift should be measurable to better than 0.1 MHz. This would improve the determination of the ratio m_e/m_p of electron mass to proton mass. An absolute wavelength measurement to better than 6 MHz would also yield an improved value for the Rydberg constant.[101b]

A polarization–rotation effect, caused by the dispersion associated with a $3S$–$5S$ two-photon transition in sodium vapor, has been demonstrated by Liao and Bjorklund.[91] A rotation of the polarization of a linearly polarized dye-laser beam of wavelength λ_1 can be achieved with a second circularly polarized laser beam of wavelength λ_2, if the wavelengths are chosen such that the sum of the frequencies of the two lasers is near but not equal to the $3S$–$5S$ two-photon transition. A combination of this two-photon polarization spectroscopy applied to the $1S$–$2S$ transition in hydrogen with the one-photon polarization spectroscopy of the Balmer β line should allow a substantial improvement in the accuracy of the value of the $1S$ Lamb shift.

Recent theoretical considerations of parity violation in atoms have predicted that these effects should show up as polarization effects in atomic transitions:[92] A linearly polarized light beam should rotate its plane of polarization when transversing a sample of free atoms, because of a slight difference in the refractive indices n^+ and n^- for right-hand, or, respectively, left-hand circularly polarized light, induced by the parity violation of the weak interaction. In heavy atoms, such as thallium,[93] bismuth,[93a] or cesium,[93b] this effect should be observable if polarization differences of about 10^{-6} can be safely detected. Preliminary results which just reach the demanded accuracy indicate, however,[93b] that the effect must be smaller than predicted. For the latest results, see References 101c, d.

4.6. The Ultimate Resolution Limit

As a result of the impressive progress in spectral resolution and accuracy achieved through Doppler-free laser spectroscopy the question arises as to which basic physical principles limit the utmost attainable resolution. With sufficient care and expenditure for electronic equipment the stabilization of the laser frequency can be perfected to such an extent that the residual laser linewidth does not impose a serious limitation to the attainable resolution.

For most atomic transitions the natural linewidth determines a natural resolution limit, but it may be partly outwitted by some experimental tricks such as the Ramsay method,[94,94a] where only those excited atoms are detected that have survived for times long compared to the mean lifetime.[95,95a,101e]

Regarding ultrahigh resolution, those atomic transitions that have small transition probabilities are especially interesting, such as parity-forbidden $1S$–$2S$ transitions which are accessible by two-photon spectroscopy. In these cases the natural linewidth may be extremely small and other broadening effects will predominate.[96] One of these contributions to line broadening, which is essential in spectroscopy with collimated laser beams, is the finite interaction time of atoms with the laser field, which is limited by the transit time of atoms with thermal velocities through the laser beam.

One method of reducing this difficulty is the enlargement of the laser beam, but restrictions are imposed by the decreasing power density and the curvature of the wavefronts of a "plane wave," caused by diffraction.[97,101f]

Recently some interesting proposals to enlarge the interaction time by cooling and trapping atoms in standing light waves have been made. These

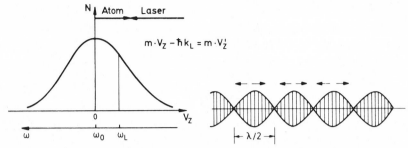

Figure 16. Cooling and trapping of atoms by photon recoil.

proposed working schemes utilize the recoil that each atom suffers when it absorbs a photon. Assume that the frequency, ω_L, of a monochromatic laser wave is tuned to the lower-frequency side of a Doppler-broadened transition with center frequency ω_0. Only those atoms with velocity components $V_z = (\omega_0 - \omega_L)/k_L$ opposite to the propagation of the laser wave are Doppler-shifted into resonance with ω_L and can therefore absorb a laser photon. Their velocity component decreases because of the recoil. If the laser frequency of a standing wave is continuously increased from $\omega_L < \omega_0$ towards ω_0, all velocity components v_z are compressed into a small interval Δv_z around $v_z = 0$. The width of $\Delta v_z = \Delta \omega_n / K$ is determined by the natural linewidth, $\Delta \omega_n$. With $\Delta \omega_n \approx 1/50$ of the Doppler width this corresponds to a residual Doppler width at a temperature of 0.24°K.[98] All three components of a three-dimensional standing wave may be cooled down this way. The calculated cooling time for a magnesium transition with a transition probability $A = 5 \times 10^8$ is of the order of 10^{-5} sec.

Letokhov[99] has pointed out that these cooled atoms can be trapped in the nodes of a three-dimensional standing wave of sufficient intensity. (Figure 16). In these nodes the absorption is zero and no momentum is transferred. As soon as an atom diffuses out of the nodes, it is pushed back by the recoil of absorption.

At a constant detuning, $\omega_L - \omega_0 \approx \Delta \omega_n$, the atoms will be further cooled down to a velocity interval corresponding to the momentum transfer for single-photon absorption or emission. The residual velocity spread would correspond to a temperature of 5×10^{-6} °K! The verification of these proposals may bring resolution to a point where other broadening effects such as Stark or Zeeman shifts from spurious electric or magnetic fields, become essential.[100] It will, however, still take a long time until the principal linewidth limit of ideally stabilized lasers, which is below 10^{-3} Hz, can be reached in practical experiments.

Note Added in Proof. Since the field of laser high-resolution spectroscopy is rapidly expanding, many papers have been published since the completion of this chapter that could not be mentioned here. For a survey on recent work, see the proceedings of the Third International Conference on Laser Spectroscopy (published under the title *Laser Spectroscopy III* [101]).

References and Notes

1. R. L. Fork, L. E. Hargrove, and M. A. Pollack, *Phys. Rev. Lett.* **12**, 705 (1964).
2a. For example see M. J. Colles and C. R. Pidgeon, *Rep. Prog. Phys.* **38**, 329 (1975).
2b. For example see F. P. Schäfer, Ed., *Dye Lasers*, Springer-Verlag, Berlin (1973).
3. F. Y. Wu, R. E. Grove and S. E. Ezekiel, *Appl. Phys. Lett.* **25**, 73 (1974).
4. D. Fröhlich, L. Stein, H. W. Schröder, and H. Welling, *Appl. Phys.* **11**, 97 (1976).
5. A. S. Pine, *J. Mol. Spectrosc.* **54**, 132 (1975).
6. J. L. Picqué and H. H. Stroke, *Comm. Atom. Mol. Phys.* **V**, 141, (1976).
7. R. L. Byer, R. L. Herbsts, and R. N. Fleming, in *Laser Spectroscopy*, Eds. S. Haroche *et al.*, Springer-Verlag, Berlin (1975).
8. *Laser Spectroscopy of Atoms and Molecules*, Ed., H. Walther Springer-Verlag, Berlin, (1976).
8a. H. Walther, *Phys. Scr.* **9**, 197 (1974).
9. *High Resolution Laser Spectroscopy*, Ed. K. Shimoda, Springer-Verlag, Berlin (1976).
10. P. Jacquinot, in *Very High Resolution Spectroscopy*, Ed. R. A. Smith, Academic Press, London (1976).
11. W. Lange, J. Luther, and A. Steudel, in *Advances in Atomic and Molecular Physics*, Ed. D. R. Bates, Vol. 10, p. 173, Academic Press, New York (1974).
12. C. Cohen-Tannoudji, in *Atomic Physics 4*, Eds. G. zu Putlitz E. W. Weber, and A. Winnacker, Plenum Press, New York, (1975).
13. M. Steiner, H. Walther, and C. Zygan, *Opt. Commun.* **18**, 2 (1976).
13a. R. Wallenstein and T. W. Hänsch, *Opt. Comm.* **14**, 355 (1975); M. M. Saloun, *Opt. Comm.* **22**, 202 (1977).
14. K. M. Baird and G. R. Hanes, *Rep. Prog. Phys.* **37**, 927 (1974).
15. M. Hercher, *Appl. Opt.* **8**, 1103 (1969).
16. J. Hall, in *Fundamental and Applied Laser Physics*, Esfahan Symposium 1971, Wiley Interscience, New York (1973).
16a. K. M. Evenson and F. R. Petersen, in *Laser Spectroscopy of Atoms and Molecules*, Ed. H. Walther, Springer-Verlag, Berlin (1976).
17. I. Hertel, in *Advances in Atomic and Molecular Physics*, Eds. D. R. Bates and B. Bederson, Academic Press, New York (1977).
17a. B. Decomps, M. Dumont, and M. Ducloy, *Laser Spectroscopy of Atoms and Molecules*, Ed. H. Walther, Springer-Verlag, Berlin (1976).
18. For a review on mode-locked lasers see for instance P. W. Smith, M. A. Duguay, and E. P. Ippen, *Progress in Quantum Electronics* Vol. 3, Part 2, Pergamon Press, New York (1974).
19. T. J. McIlrath and J. L. Carlsten, *Proc. Phys. Soc. London* **6**, 697 (1973).
20. D. J. Bradley, G. M. Gale, and P. D. Smith, *Proc. Phys. Soc. London* **3**, L11 (1970).
21. J. L. Carlsten, T. J. McIlrath, and W. H. Parkinson, *J. Phys. B. Atom. Mol. Phys.* **8**, 38, (1975).
22. D. J. Bradley *et al.*, *Phys. Rev. Lett.* **31**, 263 (1973).
23. R. F. Stebbings, C. J. Latimer, W. P. West, F. B. Dunning, and T. B. Cook, *Phys. Rev.* **A12**, 1453 (1975).
24. F. B. Dunning and R. F. Stebbings, *Phys. Rev.* **A9**, 2318 (1974).
25. U. Brinkmann, W. Hartig, H. Telle, and H. Walther, *Appl. Phys.* **5**, 109 (1974).
25a. R. V. Ambartzumian, N. P. Furzikov, V. S. Letokhov, and A. A. Puretsky, *Appl. Phys.* **9**, 335 (1976).
26. W. Gornik, D. Kaiser, W. Lange, J. Luther, H. Radloff, and H. Scholz, *Appl. Phys.* **1**, 285 (1973); *Phys. Lett.* **45A**, 219 (1973).
27. W. Gornik *et al.*, in Summaries of the Vth EGAS-conference, Lund (1973).
28. H. Harde and G. Guthöhrlein, *Phys. Rev.* **A10**, 1488 (1974).
29. A. Arnesan, *Phys. Lett.* **56A**, 355 (1976).

30. D. Schulze-Hagenest, H. Harde, W. Brand and W. Demtröder, *Z. Phys. A* **282**, 149 (1977).
31. H. Hotop, R. A. Bennet, and W. C. Lineberger, *J. Chem. Phys.* **58**, 2373 (1973).
32. W. C. Lineberger and B. W. Woodward, *Phys. Rev. Lett.* **25**, 424 (1970).
33. H. Hotop and W. C. Lineberger, *J. Chem. Phys.* **58**, 2379 (1973).
34. W. C. Lineberger, in *Laser Spectroscopy*, Eds. R. G. Brewer and A. Mooradian, Plenum Press, New York (1974).
35. R. Beck, W. Englisch, and K. Gürs, *Table of Laser Lines in Gases and Vapours*, Springer, Berlin (1976).
36. W. H. Smith, K. R. Lea, and W. Lamb Jr, in *Atomic Physics* 3, Eds, St. Smith and G. K. Walthers, p. 119, Plenum Press, New York (1973).
37. R. Vetter, *Phys. Lett.* **31**A, 559 (1970); *C. R. Acad. Sci.* **267**, 1007 (1968).
38. J. H. Schäfer, *Phys. Rev.* A3, 752 (1971).
39. F. J. Mayer, *IEEE J. Quant. Electr.* **QE3**, 690 (1967).
40. R. A. Keller, E. F. Zalewski, and M. C. Peterson; *J. Opt. Soc. Am.* **62**, 319 (1972).
41. T. W. Hänsch, A. L. Schawlow, and P. E. Toschek, *IEEE J. Quant. Electr.* **QE8**, 802 (1973).
42. W. Brunner and H. Paul, *Opt. Commun.* **12**, 252 (1974).
43. K. Tohama, *Opt. Commun.* **15**, 17 (1975).
44. M. Maeda *et al.*, *Opt. Commun.* **17**, 302 (1976).
45. E. N. Antonov *et al.*, *Opt. Commun.* **15**, 99 (1975).
46. P. Jacquinot. "Atomic Beam Spectroscopy," in *High Resolution Laser Spectroscopy*, Ed. K. Shimoda, Springer-Verlag, Berlin (1976).
47. Sh. Ezekiel, *Proc. Phot. Opt. Soc.* **49**, 23 (1974).
48. *Laser Spectroscopy*, Eds. J. Haroche *et al.*, Springer-Verlag, Berlin (1975).
49. *Laser Spectroscopy*, Eds. R. G. Brewer and A. Mooradian, Plenum Press, New York (1974).
50. K. M. Evenson and F. R. Petersen; in *Laser Applications to Optics and Spectroscopy*, Eds. St. Jacobs *et al.*, Addison-Wesley, New York (1975).
51. W. Lange, J. Luther, B. Nottbeck and H. W. Schröder, *Opt. Commun.* **8**, 157 (1973).
52. G. Hüber *et al.*, *Phys. Rev. Lett.* **34**, 1209 (1975).
53. J. H. Broadhurst *et al.*, *J. Phys. B* (*At. Mol. Phys.*) **7**, L513 (1974).
54. W. Rasmussen, R. Schieder, and H. Walther, *Opt. Commun.* **12**, 315, (1974).
55. P. Jacquinot, S. Liberman, J. L. Picque, and J. Pinard; *Opt. Commun.* **8**, 163 (1973).
56. A. F. Bernhardt, D. E. Duerre, J. R. Simpson, and L. L. Wood, *Opt. Commun.* **16**, 166 (1976).
57. G. M. Carter, D. E. Pritchard, and T. W. Ducas, *Appl. Phys. Lett.* **27**, 498 (1975).
57a. Y. M. Deventer, B. Mijinders, G. Nienhuis, and F. van der Walk, *Opt. Commun.* **18**, 60, (1976).
58. V. Weisskopf, *Ann. Phys.* (Leipzig) **9**, 23 (1931).
59. H. Walther, in *Laser Spectroscopy*, Eds. S. Haroche *et al.*, Springer-Verlag, Berlin, (1975). pp. 358 ff.
60. H. M. Gibbs and T. N. C. Venkatesan, *Opt. Commun.* **17**, 87 (1976).
61. S. L. Kaufman, *Opt. Commun.* **17**, 309 (1976).
62. W. I. Lamb, Jr., *Phys. Rev.* **134A**, 1429 (1964).
63. P. Toschek, *International Conference on Doppler-Free Spectroscopy*, Aussois, France 1973, p. 13, C.N.R.S., (1973).
64. A. Yariv, *Quantum Electronics*, Wiley, New York (1974).
65. S. Haroche, *Phys. Rev.* A6, 1280 (1972).
65a. V. S. Letokhov and V. P. Chebotayev, *Nonlinear Laser Spectroscopy*, Springer-Verlag, Berlin–Heidelberg–New York (1977).
66. H. Maeda and K. Shimoda, *J. Appl. Phys.* **47**, 1069 (1976).

67. T. W. Hänsch, I. S. Shahin, and A. L. Schawlow, *Phys. Rev. Lett.* **27**, 707 (1971).
68. P. W. Smith and T. W. Hänsch, *Phys. Rev. Lett.* **26**, 740 (1971).
69. I. S. Shahin and T. W. Hänsch, *Opt. Commun.* **8**, 312 (1973).
70. T. W. Hänsch *et al.*, *Phys. Rev. Lett.* **32**, 1336 (1974).
71. Ph. Cahuzac and R. Vetter, *Phys. Rev. Lett.* **34** 1070 (1975).
72. V. S. Letokhov, in *High Resolution Laser Spectroscopy*, Ed. K. Shimoda, Springer Verlag, Berlin (1976).
72a. Th. W. Hänsch, *Nonlinear High Resolution Spectroscopy of Atoms and Molecules*, Enrico Fermi Summer School LXIV (1977).
73. N. Bloembergen and M. D. Levinson, in *High Resolution Laser Spectroscopy*, Ed. K. Shimoda, Springer-Verlag, Berlin (1976).
74. F. Biraben, B. Cagnac, and G. Grynberg, *Phys. Lett.* **49***A*, 71 (1974).
75. C. D. Harpner and M. Levinson, *Phys. Lett.* **56***A*, 361 (1976).
76. T. W. Hänsch *et al.*, *Opt. Commun.* **11**, 50 (1974).
77. K. C. Harvey, R. T. Hawkins, G. Meisel, and A. L. Schawlow *Phys. Rev. Lett.* **34**, 1073 (1975).
78. B. Cagnac, in *Laser Spectroscopy*, Eds. S. Haroche *et al.*, p. 165, Springer-Verlag, (1975).
79. F. Biraben, G. Grynberg, and E. Giacohino, *Phys. Lett.* **56***A*, 441 (1976).
80. D. E. Roberts *et al.*, *Opt. Commun.* **14**, 332 (1975).
81. A. Flussberg, T. Mossberg, and S. R. Hartmann, *Phys. Lett.* **55***A*, 403 (1976).
82. T. W. Hänsch, S. A. Lee, R. Wallenstein, and C. Wieman, *Phys. Rev. Lett.* **34**, 307 (1975).
83. V. P. Chebotayev, in *High Resolution Spectroscopy*, Ed. K. Shimoda, p. 201ff, Springer-Verlag, Berlin (1976).
84. M. Dumont, J. *Phys.* (*Paris*) **33**, 971 (1972).
85. G. Hermann and A. Scharmann, *Z. Phys.* **254**, 46 (1972).
86. M. S. Feld, A. Sanchez, and A. Javan, in *International Colloquium on Doppler-free Spectroscopy*, Ausois 1973, pp. 87 ff., C.N.R.S., Paris (1973).
87. J. S. Levine, P. A. Bonczyk, and A. Javan, *Phys. Rev. Lett.* **22**, 267 (1969).
88. T. W. Hänsch, R. Keil, A. Schabert, Ch. Schmelzer, and P. Toschek, *Z. Phys.* **226**, 293, (1969).
89. T. W. Hänsch, and P. Toschek, *IEEE. Quant. Electr.* **QE5**, 61 (1969).
90. C. Wieman and T. W. Hänsch, *Phys. Rev. Lett.* **36**, 1170 (1976).
91. P. F. Liao and G. C. Bjorklund, *Phys. Rev. Lett.* **36**, 584 (1976).
92. J. C. Brodsky and C. Karl, *Comm. Atom. Mol. Phys.* **5**, 63 (1976).
93. C. C. Bouchiat and M. A. Bouchiat, *Phys. Lett.* **48B**, 111 (1974).
93a. D. S. Soreide *et al.*, *Phys. Rev. Lett.* **36**, 352 (1976).
93b. M. A. Bouchiat and L. Pottier, *Phys. Lett.* **62B**, 327, (1976).
94. Kleppner, H. M. Goldenberg, and N. F. Ramsey, *Phys. Rev.* **126**, 603 (1962).
94a. J. C. Bergquist, S. A. Lee, and Y. L. Hall, *Phys. Rev. Lett.* **38**, 159 (1977).
95. H. Figger, and H. Waltzer. *Z. Phys.* **267**, 1 (1974).
95a. R. E. Grove, F. Y. Wu, and S. Ezekiel, *Opt. Commun.* **18**, 61 (1976).
96. J. L. Hall, in *Fundamental and Applied Laser Physics*, Wiley & Sons, New York (1973).
97. J. L. Hall, The Ultimate Resolution Limit, in *International Colloquium on Doppler-Free Spectroscopy*, Aussois 1973, C.N.R.S., Paris (1973).
98. T. W. Hänsch and A. L. Schawlow, *Opt. Commun.* **13**, 68 (1975).
99. V. S. Letokhov, V. G. Minogin, and B. D. Paulik, *Opt. Commun.* **19**, 92 (1976).
100. V. P. Chebotayev, in *Laser Spectroscopy*, Eds. S. Haroche *et al.*, pp. 150ff., Springer-Verlag, Berlin (1975).
101. a–f. *Laser Spectroscopy III*, Eds. J. L. Hall and J. L. Carlsten, Springer-Verlag, Berlin–Heidelberg–New York (1977); (*a*) S. Svanberg, p. 183; (*b*) C. Wieman and T. W. Hänsch, p. 39; (*c*) M. A. Bouchiat and L. Pottier, p. 9; (*d*) P. G. H. Sandars, p. 21; (*e*) see, e.g., Chap. 3, p. 121ff; (*f*) C. J. Bordé, p. 121.

Index